Holocene Marine Sedimentation
in the
North Sea Basin

Holocene Marine Sedimentation in the North Sea Basin

EDITED BY S.-D. NIO,
R. T. E. SHÜTTENHELM AND
Tj. C. E. VAN WEERING

SPECIAL PUBLICATION NUMBER 5 OF THE
INTERNATIONAL ASSOCIATION OF SEDIMENTOLOGISTS
PUBLISHED BY BLACKWELL SCIENTIFIC PUBLICATIONS
OXFORD LONDON EDINBURGH
BOSTON MELBOURNE

First published 1981

British Library
Cataloguing in Publication Data

Holocene marine sedimentation in the North Sea basin.
  – (Special publication/International Association of
Sedimentologists, ISSN 0141–3600; 5)
1. Marine sediments – North Sea
2. Sedimentation and deposition
I. Nio, Swie-Djin
551.46 083'336   GC 591
ISBN 0–632–00858–X

Printed and bound in Great Britain at the
University Press, Cambridge

Contents

Spec. Publs int. Ass. Sediment. (1981) **5**, 1-3

Holocene marine sedimentation in the North Sea Basin: an introduction

S.-D. NIO*, R. T. E. SCHÜTTENHELM† *and* Tj. C. E. VAN WEERING‡

**Comparative Sedimentology Division, State University of Utrecht* †*Department of Marine Geology, Geological Survey of the Netherlands, Haarlem and* ‡*Netherlands Institute for Sea Research, Texel*

Research on Holocene marine sedimentation in the North Sea basin started early this century. Research activities have increased tremendously during the last decade and we can now assume that the North Sea basin is probably the most measured and surveyed marine basin in the world. Most of these research activities were and still are related to the economic and social importance of the North Sea to the surrounding countries.

Among the tremendous amount of gathered data some can be regarded as 'pacemakers in the study of marine tidal basins'. The basis of modern sedimentological research into the marine Holocene of the North Sea are Van Veen's publications (1936, 1938), in which he described the occurrence and distribution of megabedforms on the sea-floor. He related the occurrence and formation of these bedforms to certain hydrodynamic conditions, developing relationships which are still useful today.

A new upsurge in research into megabedforms on the North Sea floor was brought about by the publication of Stride (1963). The principles of research defined in his publication became a guideline for future research carried out by many marine research institutions of countries around the North Sea. These activities resulted in a series of important publications on the distribution, geometry and genesis of megaripples, sand waves and sand ridges on the floor of the southern North Sea (e.g. Stride & Cartwright, 1958; Jones, Kain & Stride, 1965; Houbolt, 1968; McCave, 1971; Terwindt, 1971 a; Caston & Stride, 1973). In the northern North Sea where conditions are quite different, important oceanographic and sedimentological studies have been carried out by Helland-Hansen (1907), Helland-Hansen & Nansen (1909), O. Holtedahl (1929, 1940) and Pratje (1949).

Very extensive research was also carried out in inshore tidal areas. Van Veen (1950), Van Straaten (1950a, b, 1952, 1953) and Schafer (1962) did pioneering work on tidal deposits of the inshore areas of the southern North Sea. One of the important aspects dealing with the settling and behaviour of fines in the estuarine environment has been discussed by Postma (1960, 1961, 1967). Somewhat later, Reineck's school provided the sedimentologists with a wealth of information on clastic tidal deposits along the North Sea coast of Germany (e.g. Reineck & Singh, 1967; Reineck *et al.*, 1967; Reineck & Wunderlich, 1969). An important contribution to the understanding of intertidal flat sedimentation was made by the studies of the Wash, U.K. by Evans (1958, 1965).

In the Netherlands, however, research into inshore tidal sedimentation has emphasized the relation between processes and also between the bedform and internal structural organization (Terwindt, de Jong & Van der Wilk, 1963; Boersma, 1969; Terwindt & Breusers, 1972; Terwindt, 1971 b, 1973; Boersma & Terwindt, 1981).

Most of the studies mentioned above had the purpose of obtaining a better understanding of the different sedimentological features produced in tidally influenced areas and estuaries. Another important aspect of these studies was the collection of qualitative data, which can be used to interprete analogous ancient deposits. In practice, however, it is obvious that, despite the wealth of information we have on recent and relict sediments of the North Sea, it is very difficult to apply such information to ancient deposits. Studies on the preservation potential of recent deposits are necessary before we can use data collected from them. A start in such comparative studies has shown a promising development in the research into marine tidal deposits (e.g. Van Straaten, 1954; Wunderlich, 1970; De Raaf & Boersma, 1971; Nio, 1976).

0141-3600/81/1105-0001 $02.00

We have mentioned only a few aspects of the sedimentological research activities in the North Sea basin. Studies on sea-level fluctuations during the Holocene, chemical and biological processes and their products and also the faunal distribution within the North Sea are of course equally important for a better understanding of recent marine processes. Recent reviews on aspects of studies of the marine Holocene of the North Sea were presented in Banner, Collins & Massie (1979) and in Oele, Shüttenhelm & Wiggers (1979). It was because of this wide variety of research topics that the Texel meeting on Holocene marine sedimentation in the North Sea basin was organized in September 1979.

The purpose of the meeting, which was sponsored by the International Association of Sedimentologists, Mobil Producing Netherlands Inc., The Royal Geological and Mining Society of the Netherlands, the Ministry of Education and Sciences, the Netherlands Institute for Sea Research, the Geological Survey of the Netherlands and the State University of Utrecht was to bring together different research topics. We came to realize that many problems still remain unsolved, for example the knowledge of near-bottom current velocities and directions and the impact on the sediment distribution is still very poor. This again emphasizes the fact that a multidisciplinary approach is necessary to solve the problems in the marine research of the North Sea.

Comparative studies of similar marine basins are also helpful in understanding processes. It is for this reason that some chapters in this volume are devoted to studies of marine basins other than the North Sea. Papers in this volume present the 'state of the art' in research into the Holocene North Sea basin.

We wish to thank the Bureau of the International Association of Sedimentologists for allowing publication of this volume. Thanks are also due to the authors themselves who made publication of this volume possible. We are indebted to G. P. Allen, A. C. Bloom, J. C. Duinker, D. Eisma, R. W. Feyling-Hansen, U. Forstner, P. F. Friend, B. M. Funnel, I. N. McCave, G. V. Middleton, H. Postma, H. G. Reading, G. V. Sharma, C. Siegenthaler, D. J. P. Swift, J. H. J. Terwindt, M. J. Tooley, A. M. Winkelmolen, W. H. Tagwijn and J. F. Zimmerman for refereeing the papers.

REFERENCES

BANNER, F.T., COLLINS, M. B. & MASSIE, K. S. (1979) *The North-West-European Shelf Seas: the Sea Bed and the Sea in Motion. Part I: Geology and Sedimentology.* Elsevier Oceanography Series 24A. Elsevier, Amsterdam, 300 pp.

BOERSMA, J.R. (1969) Internal structure of some mega-ripples on a shoal in the Westerschelde estuary, the Netherlands. *Geologie Mijnb.* **48**, 409–414.

BOERSMA. J.R. & TERWINDT, J.H.J. (1981) Neap–spring tide sequences of intertidal shoal deposits in a mesotidal estuary. *Sedimentology,* **28**, 151–170.

CASTON, V.N.D. & STRIDE, A.H. (1973) Influence of older relief on the location of sand waves in a part of the southern North Sea. *Est. coastal Mar. Sci.* **1**, 379–386.

EVANS, G. (1958) Some aspects of recent sedimentation in the Wash. *Eclog. geol. Helv.* **51**, 508–515.

EVANS, G. (1965) Intertidal flat sediments and their environments of deposition in The Wash. *Q. Jl geol. Soc. Lond.* **121**, 209–241.

HELLAND-HANSEN, H. (1907) Current measurements in Norwegian fjords, the Norwegian Sea and the North Sea in 1906. *Bergens Mus. Årb.* **15**, 1–61.

HELLAND-HANSEN, H. & NANSEN, F. (1909) The Norwegian Sea. *Report Norw. Fishery mar. Invest.* **II–2**, 360 pp.

HOLTEDAHL, O. (1929) On the geology and physiography of some Antarctic and sub-Antarctic islands. With notes on the character of fjords and strandflats of some northern lands. *Sci. Results Norweg. Antarctic Expedition,* **3**, 1–172.

HOLTEDAHL, O. (1940) The submarine relief off the Norwegian coast. *Det Norske. Videnskaps Academi i Oslo, Spec. Publ.* 1–43.

HOUBOLT, J.J.H.C. (1968) Recent sediments in the southern bight of the North Sea. *Geologie Mijnb.* **47**, 245–273.

JONES, N.S., KAIN, J.M. & STRIDE, A.H. (1965) The movement of sand waves on Warts Bank, Isle of Man. *Mar. Geol.* **3**, 324–336.

McCAVE, I.N. (1971) Sand waves in the North Sea off the coast of Holland. *Mar. Geol.* **10**, 199–225.

NIO, SWIE-DJIN (1976) Marine transgressions as a factor in the formation of sand wave complexes. *Geologie Mijnb.* **55**, 18–40.

OELE, E., SCHÜTTENHELM, R.T.E. & WIGGERS, A.J. (1979) The Quaternary of the North Sea. *Acta Universitatis Upsaliensis.* Uppsala, 248 pp.

PRATJE, O. (1949) Die Bodenbedeckung der nordeuropaischen Meere. *Handbuch der Seefischerei Nordeuropas,* pp. 1–3. Schweizerbart, Stuttgart.

POSTMA, H. (1960) Einige Bemerkungen uber den Sinkstofftransport im Ems-Dollart Gebiet. *Verh. K. ned. Geol.-mijnb. Genoot., Geol. Ser.* **19**, 103–110.

POSTMA, H. (1961) Transport and accumulation of suspended matter in the Dutch Wadden Sea. *Neth. J. Sea Res.* **1**, 149–190.

POSTMA, H. (1967) Sediment transport and sedimentation in the estuarine environment. In: *Estuaries* (Ed. by G.H. Lauff), pp. 158–179. American Association for the Advancement of Science, Washington, D.C.

RAAF, J.F.M. DE & BOERSMA, J.R. (1971) Tidal deposits and their sedimentary structures (seven examples from Western Europe). *Geologie Mijnb.* **50**, 479–504.

REINECK, H.-E., DORJES, J., GADOW, S. & HERTWECK, G. (1967) Sedimentologie, Faunenzonierung und Faziesabfolge vor der Ostkuste der inneren Deutschen Bucht. *Senckenb. Leth.* **49**, 261–309.

REINECK, H.-E. & SINGH, I.B. (1967) Primary sedimentary structures in the Recent sediments of the Jade, North Sea. *Mar. Geol.* **5**, 227–235.

REINECK, H.-E. & WUNDERLICH, F. (1969) Die Entstehung von Schichten und Schichtbanken im Watt. *Senckenb. Mar.* **1**, 85–106.

SCHAFER, W. (1962) *Aktuopalaontologie nach Studien in der Nordsee.* Kramer, Frankfurt a.M. 666 pp.

STRAATEN, L.M.J.U. VAN (1950a) Environment of formation and facies of the Wadden Sea sediments. *Koninkl. Aardrijk. Genootsch.* **67**, 94–108.

STRAATEN, L.M.J.U. VAN (1950b) Giant ripples in tidal channels. *Koninkl. Aardrijk. Genootsch.* **67**, 336–341.

STRAATEN, L.M.J.U. VAN (1952) Biogene textures and the formation of shell beds in the Dutch Wadden Sea. *Proc. K. ned. Akad. Wet.* B **55**, 500–516.

STRAATEN, L.M.J.U. VAN (1953) Megaripples in the Dutch Wadden Sea and in the Basin of Arcachon (France). *Geologie Mijnb.* **15**, 1–11.

STRAATEN, L.M.J.U. VAN (1954) Sedimentology of Recent tidal flat deposits and the psammites du Condroz (Devonian). *Geologie Mijnb.* **16**, 25–47.

STRIDE, A.H. & CARTWRIGHT, D.E. (1958) Sand transport at the southern end of the North Sea. *Dock Harb. Auth.* **38**, 323–324.

STRIDE, A.H. (1963) Current-swept sea floors near the southern half of Britain. *Q. Jl geol. Soc. Lond.* **119**, 175–199.

TERWINDT, J.H.J. (1971a) Sand waves in the southern bight of the North Sea. *Mar. Geol.* **10**, 51–67.

TERWINDT, J.H.J. (1971b) Lithofacies of inshore estuarine and tidal inlet deposits. *Geologie Mijnb.* **50**, 515–526.

TERWINDT, J.H.J. (1973) Sand movement in the in- and offshore tidal area of the SW part of the Netherlands. *Geologie Mijnb.* **52**, 69–77.

TERWINDT, J.H.J. & BREUSERS, H.N.C. (1972) Experiments on the origin of flaser, lenticular and sand-clay alternating bedding. *Sedimentology,* **19**, 85–98.

TERWINDT, J.H.J., JONG, J.D. DE & WILK, E. VAN DER (1963) Sediment movement and sediment properties in the tidal area of the Lower Rhine (Rotterdam waterway). *Geol. Serie,* **21**, 244–258.

VEEN, J. VAN (1936) Onderzoekingen in de hoofden, in verband met de gesteldheid der Nederlandse kust. *Algemeene Landsdrukkerij.* 'sGravenhage, 252 pp.

VEEN, J. VAN (1938) Die unterseeische Sandwuste in der Nordsee. *Geologie Meere Binnengewäss.* **2**, 62–86.

VEEN, J. VAN (1950) Eb- en vloedschaarsystemen in de Nederlandse getijdewateren. *Tijdschr. K. ned. aardrijksk. Genoot.* **66**, 303–325.

WUNDERLICH, F. (1970) Genesis and environment of the 'Nellenkopfchenschichten' (lower Emsian, Rheinian Devon) at locus typicus in comparison with modern coastal environment of the German Bay. *J. sedim. Petrol.* **40**, 102–130.

Spec. Publs int. Ass. Sediment. (1981) **5**, 4–26

Origin and sequences of sedimentary structures in inshore mesotidal deposits of the North Sea

J. H. J. TERWINDT

Department of Geography, Heidelberglaan 2, 3508 TC Utrecht, the Netherlands

ABSTRACT

The inshore subtidal and intertidal areas in mesotidal basins of the North Sea coasts are analysed for bedding types and sedimentary structures. Seven lithofacies can be distinguished, based on the intensity of the tidal currents and the wave action. Variation of the tidal currents produces distinctive responses in the lateral and vertical sequences of sedimentary structures within and between the lithofacies produced by megaripples. The diagnostic features of inshore mesotidal deposits appear to be bi-directional foresetting and in places the occurrence of reactivation and slackening structures; the consistent occurrence of tidal bundles and lateral neap-spring tide sequences in megaripple deposits and finally the high frequency of erosional contacts and abrupt facies changes coupled with the absence of a general vertical macro-sequence.

INTRODUCTION

Studies of processes of sedimentation and sedimentary features in tidal areas along the North Sea coast started as early as the 1930s. Numerous publications have been produced highlighting the provenance of the sediments, the bedforms, the textural properties, the faunal characteristics and the sedimentary structures. Important contributions and summaries may be found in Elliott (1978), Evans (1965, 1975), Ginsburg (1975), Postma (1954, 1961, 1967), De Raaf & Boersma (1971), Reineck (1967, 1975), Reineck & Singh (1973) and Van Straaten (1954).

The tidal basins of the North Sea are usually classified as mesotidal. In this paper we will focus on the inshore subtidal and intertidal zones only. Morphographically these zones are characterized by ebb-, flood- or indifferent channels surrounded by sandy shoals, tidal flats and tidal marshes.

The bedding types and sedimentary structures can be categorized into seven lithofacies. The distinction

between each category is a function of the intensity of tidal currents and wave action. Typically the change in tidal currents has a distinct effect upon the variation within and between lateral and vertical sequences of the lithofacies. The different characteristics of these sequences are described below.

THE LITHOFACIES

As defined by Reading (1978) 'a lithofacies should refer to an objectively described rock unit...that forms under certain conditions of sedimentation, reflecting a particular process or environment'. In accordance with this definition seven lithofacies can be distinguished in the subtidal and intertidal areas along the North Sea:

Lithofacies STRO CUR	(strong currents)
Lithofacies MED CUR	(medium currents)
Lithofacies WEAK CUR	(weak currents)
Lithofacies STRO WAV	(strong waves)
Lithofacies MED WAV	(medium waves)
Lithofacies WEAK WAV	(weak waves)
Lithofacies STORM DEP	(storm deposits).

0141-3600/81/1205–0003 $02.00

Some of these lithofacies can be subdivided into subfacies.

Lithofacies STRO CUR

Characteristics:

coarse- to medium-grained sands;
mud pebbles, imbricated or not;
shell (debris) layers, imbricated or not;
peat lumps and detritus;
x-bedding*, set thickness: decimetre-metre scale, some with mud drapes on the foresets;
structures:

(semi-)tabular sets	STRO CUR 1,
trough sets (festoon bedding)	STRO CUR 2,
fill structures	STRO CUR 3.

This lithofacies typically reflects a period of intensive transport in the form of megaripples. The megaripples are either type I (almost straight crestlines, no scour pits in the troughs, large length to height ratios) or type II (winding crestlines, bowl-shaped troughs with scour pits and smaller length-to-height ratios) according to a classification proposed by Dalrymple, Knight & Lambiase (1978). Measurements taken in intertidal as well as subtidal areas indicate that a threshold velocity of approximately 0.5–0.6 m sec^{-1} is needed to initiate migration of this type of bedforms (Allen & Friend, 1976; Boersma & Terwindt, 1981; Boothroyd & Hubbard, 1975; Terwindt, 1970). Mean grain sizes which exist in the mesotidal areas under study range between 200 and 450 μm. Numerous current observations over such megaripple areas reveal the presence of these bedforms when the maximum current velocity in a tidal cycle measured at 0.5 m above the bed ($U_{0.5max}$) exceeds 0.65–0.70 m sec^{-1} (Terwindt, 1970).

Megaripples type I show only (semi-)tabular sets (STRO CUR 1). Megaripples type II, however, possess all three types of structures (STRO CUR 1, 2 and 3), with (semi-)tabular and trough sets occurring more frequently.

The fill structures (STRO CUR 3) are defined here as infillings of decimetre–metre scaled scour pits, not being megaripple troughs. Such scour pits have on occasion been observed on the stoss sides of megaripples. The bedding in some of these pits is entirely concordant with the pit form, especially

* x-bedding in this paper refers to cross-bedding produced by migrating megaripples, while x-lamination is used for cross-bedding produced by ripples.

when looking perpendicular to the current direction. In sections parallel to the current direction some infilling shows foreset bedding discordant with the lower pit profile (Terwindt, 1971, fig. 5).

Lithofacies MED CUR

Characteristics:

medium- to fine-grained sands and clay layers;
x-lamination, set thickness: centimetre-decimetre.
structures:

tabular sets	MED CUR 1,
trough sets	MED CUR 2,
herringbone x-lamination	MED CUR 3,
simple flaser bedding	MED CUR 4,
wavy flaser bedding	MED CUR 5.

This lithofacies is generated by bi- or unidirectional moving ripples (MED CUR 1, 2 and 3). The ripples either travel in the troughs of larger megaripples or over rather extensive horizontal bedding planes. The characteristic differences between the internal structures of both types of ripples are not yet clear. Therefore even though the possibility exists that the ripples are connected with two different microenvironments, there is no further differentiation made here. The mud drape in the ripple trough forming flaser bedding (MED CUR 4) is particularly interesting. Wherever the term 'mud' is used in this paper it describes a mixture of clay, silt, fine sand and organic material assembled in, and deposited as, flocs. This mud drape marks a slack water period (Reineck & Wunderlich, 1969). The thickness of the mud layer in flaser bedding is generally a few millimetres. This is in accordance with calculations based on the observed near-bed suspended mud concentrations, which are usually some 100 mg l^{-1} and the measured mean fall velocity of the mud flocs which is 1–1.5 m h^{-1}. It is also in accordance with the time available for mud deposition during slack water which is usually between 0.5–3 h, when the current velocities are below 0.2 m sec^{-1}. It should be noted that the critical current velocity for flocculated mud deposition used here, namely 0.2 m sec^{-1}, differs from the 0.04 m sec^{-1} mentioned by Little & Reineck (1974). However, the value of 0.2 m sec^{-1} is supported by numerous flume and field measurements which in the author's view justifies its use in this paper.

The maximum thickness of an unconsolidated, water-rich (80 %) mud deposit, calculated on the basis of an extremely high near-bed suspended sediment concentration (1000 mg l^{-1}) and exceptionally long slack water period (3 h), is about 1 cm.

Flume experiments further revealed that this freshly deposited mud exhibits a rather rapid initial consolidation and with time an appreciable increase in the critical shear velocity (Migniot, 1968; Terwindt & Breusers, 1972). Application of these data to field conditions yields the result that mud drapes, deposited during slack water are eroded when the current velocity during the next tide, measured at 0·5 m above the bed ($U_{0·5}$) exceeds approximately 0·45 m sec^{-1}. Terwindt & Breusers (1972) suggested a threshold value of $U_{0·5}$ of 0·6 m sec^{-1}. However, in light of newer data now available the threshold value should be somewhat lower. Thus simple flaser (MED CUR 4) bedding may be formed when $U_{0·5\ max}$ during one tide just exceeds 0·45 m sec^{-1}. The mud drape may be removed from the ripple crest, but is preserved in the ripple trough due to a slight reduction in current velocity here (Reineck & Wunderlich, 1968). Wavy flaser bedding (MED CUR 5) could be formed when $U_{0·5\ max}$ is just below the critical value of 0·45 m sec^{-1}. In that case the entire mud drape may be preserved, although the thickness of the drape will usually be less on top of the ripple crest than in the trough, as a result of a slight erosion. It should be noted that the threshold velocity which would cause such erosion of the mud drape exceeds the velocity needed for the formation and migration of ripples. Hence, even if no erosion of the mud drapes takes place, migrating ripples, originating from elsewhere, can still cover them. The sand layers between the mud drapes show a delicate, unidirectional and herringbone x-lamination (De Raaf & Boersma, 1971; Reineck, 1960; van Straaten, 1954; Terwindt, 1971). Another interesting feature is the sharp lower and upper boundaries of the mud layers (van Straaten, 1954). The sharp lower boundary could be attributed to the fact that mud deposition starts only after migration of the ripples has already stopped. The sharp upper boundary may have been formed when insufficiently consolidated mud was washed away after the turn of the tide. Current velocities in this instance are below the threshold for ripple formation and movement. They may, however, be slightly higher than the threshold velocity during the time it takes for ripples to penetrate the mud draped area from elsewhere. A remaining question is how this water-rich mud layer, which has been only briefly consolidated, can support a sharply outlined layer of sand. Laboratory experiments by Migniot (1968) and Terwindt & Breusers

(1972) have demonstrated that initial consolidation is such that after 3–4 h of consolidation time a mud layer of a few millimetres thickness has gained enough strength to support a sand layer 0·5 cm thick. Increasing overburden by gradually thickening the sand layer will facilitate further consolidation of the mud layer. It should be remembered that insufficiently consolidated mud has been washed away prior to the deposition of the sand layers.

This rather rapid initial consolidation is more likely to occur in thin mud layers, surrounded by sand layers which can take up the excess water as it is expelled from the mud.

A conclusion from the foregoing may be that lithofacies MED CUR incorporates deposits of migrating ripples which are periodically halted during slack water. Ultimately a mud drape may then be deposited and subsequently be either preserved in some form or completely eroded. Thus this lithofacies is characterized by current velocities of sufficient strength to generate ripple migration during every single tide.

Lithofacies WEAK CUR

Characteristics:

fine grained sands and mud layers;
x-lamination in millimetre–centimetre scale;
structures:
lenticular bedding   WEAK CUR 1,
sand–mud alternations   WEAK CUR 2.

In this lithofacies mud and sand deposition alternate. The thickness of the mud layers, normally ranging between 0·5 and 3 cm, militates against deposition during one slack-water period only, even taking account of the fact that in areas of weak currents the slack-water period will last longer and more mud might have been deposited. A characteristic of this lithofacies is therefore that several tides are needed for the sedimentation of the mud layers. During this period $U_{0·5\ max}$ remains below the critical value for the erosion of the total mud layer. Such is also true for both the ebb and flood stages, where insufficiently consolidated top layers of the mud will be washed away at higher current velocities. As a result the total mud layer is made up of several laminae, some of which are separated by a sand layer of almost one grain's thickness (Reineck, 1958; Reineck & Singh, 1973). During these weak current conditions the supply of sand is minimal and sometimes absent as evidenced by uniform mud layers of several centimetres thickness. At other times sand is introduced into the mud deposition area either by invasion of ripples moving in from elsewhere or by outfall from suspension. In the latter case, the sand is possibly swept together forming ripples. This phenomenon has been observed in many flume studies where thin sand layers on a rigid boundary were sheared by currents, generating patterns of isolated long-crested and moving ripples. The ripples produced in either of these ways, are halted during slack water and are subsequently covered by mud. As a result, they have a relatively short life. In this manner sand lenses which show a ripple x-lamination and which are incorporated in the single lenticular bedding are formed (De Raaf & Boersma, 1971; Reineck, 1967; Reineck & Wunderlich, 1968, 1969). Where more sand is available trains of ripples are formed which give rise to the connected lenticular bedding (Reineck & Wunderlich, 1968). Only unidirectional foresetting is, to my knowledge, found in the sand lenses produced by current ripples. Herringbone structures have not been reported as yet, such also being an indication of the relatively short life in one phase of a single tide.

The sand–mud alternating bedding (WEAK CUR 2) consists of horizontally bedded sand and mud layers, 1–3 cm thick (Reineck & Wunderlich, 1968). The sand layers are very elongated lenses with lengths of decimetre–metre scale and heights of some millimetres. They have a rather irregular shape. The internal structure, as far as it can be distinguished, is mostly made up of parallel bedding but individual laminae wedge out over rather short distances. In some cases low-angle x-lamination can be seen. It should be noted, however, that the internal structures are barely visible on the available lacquer peels, and therefore the structural properties remain uncertain.

It could be that the weak currents, being below the threshold for ripple formation, cause the difference between the lenticular bedding and this type of deposit where the sand supplied by suspension fall-out is hardly rippled (Reineck & Singh, 1973). However, during deposition of the sand, the current velocities exceed the threshold value for mud deposition ($U_{0 \cdot 5} > 0 \cdot 2$ m sec^{-1}) as the sand layers may have surprisingly low mud contents.

In summary the mud layers of the WEAK CUR lithofacies are not formed during one slack-water period but encompass a number of tides. The sand lenses and layers, on the contrary, can be deposited during one phase of a single tide.

Lithofacies STRO WAV

Characteristics:

fine- to coarse-grained sands;
structures: horizontal or low-angle parallel lamination.

This lithofacies represents strong wave-action producing parallel lamination due to wash and backwash of shoaling and breaking waves. This parallel lamination is formed under sheet flow conditions with high values of the orbital velocity near the bed (Davidson–Arnott & Greenwood, 1976; Clifton, Hunter & Phillips, 1971; Reineck, 1963). Applying the findings of Dingler (1974), cited in Clifton (1976), the threshold value of the maximum orbital velocity needed to generate sheet flow depends upon the grain diameter. This threshold value is approximately $0 \cdot 7$–$1 \cdot 2$ m sec^{-1} for the mean grain diameters encountered in the inshore tidal areas of the North Sea. It therefore appears that lithofacies STRO WAV is generated when the maximum orbital velocities near the bed (U_{m}) exceed approximately 1 m sec^{-1}.

Lithofacies MED WAV

Characteristics:

medium- to fine-grained sands and mud drapes;
structures:
wave x-lamination MED WAV 1,
flaser bedding in wave ripple troughs MED WAV 2.

This lithofacies comprises wave ripples and their structures (Boersma, 1970; Newton, 1968; De Raaf, Boersma & Van Gelder, 1977; Reineck, 1961).

According to Komar & Miller (1973) the threshold value of the maximum orbital velocity for the formation of wave ripples is directly related to the grain size and the wave period. In the inshore tidal areas of the North Sea this value is $0 \cdot 15$–$0 \cdot 25$ m sec^{-1}. Above this value wave ripples form only when not destroyed by stronger tidal or drift currents. Flaser bedding (MED WAV 2) can be formed when the maximum orbital velocity is temporarily below $0 \cdot 2$ m sec^{-1}, the upper limit for mud deposition (Reineck & Singh, 1973). The tides may be a factor when considering the large range in orbital velocities. During low tide there may be a different wave pattern. Wave heights and periods can be lower than during high tide due to filtering of the waves over shoals.

Fig. 1. Lithofacies STORM DEP. Vertical microsequence, consisting of wavy flaser bedding, followed by an even laminated set (STORM DEP), followed by x-lamination and flaser bedding. Note: infilling of lower mud draped trough by almost even lamination. Haringvliet estuary, 14 m below M.S.L.

Lithofacies WEAK WAV

Characteristics:

fine-grained sands and muds;
structure: lenticular bedding with wave ripple lamination in the sand lenses.

This lithofacies encompasses occasional low wave activity during periods where mud deposition is normally taking place. Here, as in the case of lenticular bedding produced by currents, sand that had been deposited over the mud layer by suspension fall out is swept together and piled up by wave action. These lenses are made up of one single ripple only, without connections between them (Reineck & Singh, 1973). Therefore it may be that the period during which these ripples were active was short. This period was both preceded as well as followed by periods of mud deposition in the course of several tides.

Lithofacies STORM DEP

Characteristics:

medium- to fine-grained sands;
structure: horizontal parallel lamination.

This lithofacies, often encountered far below the wave base, is considered to have been generated by settling out of heavily loaded suspensions produced elsewhere by severe wave attack on submerged shoals. Such suspension clouds may be transferred by tidal currents, probably agitated by wave-induced turbulence and may deposit material outside the area of provenance (Gadow & Reineck, 1969; Reineck; 1963; Terwindt, 1971; Wunderlich, 1969).

The almost total absence of mud in this lithofacies indicates that deposition of sand was either rather rapid with hardly any admixture of mud or took place under rather intensive turbulence. Individual laminae can be traced over a distance of up to 2 m. This then may be evidence that sedimentation occurred under conditions where shearing by currents or ripple building, were non-existent (Gadow & Reineck, 1969). In some sections it was observed that parallel-bedded sand had been deposited immediately on top of a mud-layer of wavy flaser bedding. The ripple trough had also been filled in by parallel lamination (see Fig. 1), a possible indication that the suspension cloud arrived shortly after slack water being just the timing that one would expect. During absence of tidal currents, there may be less dispersion of the suspension and the suspension clouds may be more coherent. Weak currents just after slack water may then cause the sediment clouds to drift away to deeper parts of the channels.

Table 1. Summary of lithofacies

Lithofacies	Texture	Structures	Hydrodynamic conditions
STRO CUR	Coarse–medium sand	x-bedding, tabular, festoon, fills	$U_{0.5\,max} > 0.65$ at 0.70 m sec^{-1}
MED CUR	Medium–fine sand, mud drapes	x-lamination, tabular, herringbone, flaser	$U_{0.5\,max} > 0.45$ m sec^{-1}
WEAK CUR	Fine sand, mud	Lenticular + sand/clay alternation	$U_{0.5\,max} \leqslant 0.45$ msec^{-1} periodically
STRO WAV	Coarse–fine sand	Even lamination	$U_m > 0.7$ at 1.2 m sec^{-1}
MED WAV	Medium–fine sand, mud drapes	Wave x-lamination, flaser bedding	$0.15 < U_m < 0.7$ m sec^{-1}
WEAK WAV	Fine sand, mud layers	Wave x-lamination, lenticular bedding	Periodically $U_m < 0.2$ and $U_{0.5\,max} < 0.45$ m sec^{-1}
STORM DEP	Medium–fine sand	Even lamination	Heavy turbulence

A summary of the characteristics of the above described lithofacies is presented in Table 1.

THE SEQUENCES

A sequence is defined here as a gradual change in:

(1) structures;
(2) grain size;
(3) thickness of the mud layers;
(4) number of mud layers within or between the lithofacies.

A sequence thus reflects a progressive change in environmental conditions. Such sequences may be lateral or vertical.

In a tidal area the environmental conditions are rather variable with time. The question is how this variability shows itself in the sequences, and furthermore whether there are sequences characteristic of the tidal environment.

The lateral sequences

The magnitude of the tidal currents in a particular location may change in time due to (Fig. 2):

the tidal effect;

the diurnal inequality of the tide: a succession of higher and lower tidal amplitudes;

the neap/spring-tide cycle with high and low spring-tides alternating;

extra inflow or outflow due to wind setup or setdown;

shifts in the streamline pattern either related or unrelated to the shifting of the channel. The streamline of the maximum current velocity may change its position in the channel due to slight variations in the gradient of the water levels. It may traverse systematically due to an overall migration tendency of the channel system.

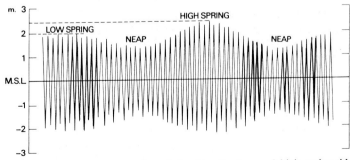

Fig. 2. Tidal curves, showing the diurnal inequality of the tide, the low and high spring-tides and the almost equal neap-tides. Westerschelde estuary.

10   *J. H. J. Terwindt*

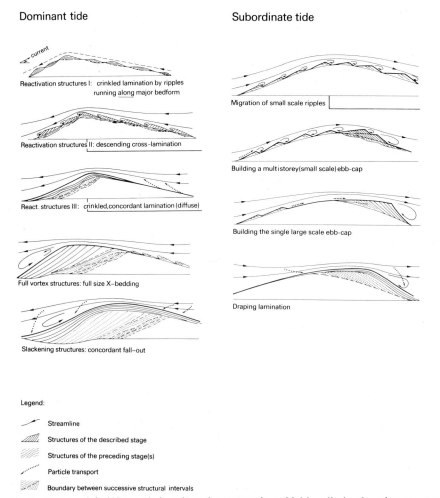

Fig. 3. Structural (sub)intervals in a lateral sequence in a tidal bundle (explanation, see text).

The effects of these variations in current strength are clearly discernible in the sedimentary structures. Boersma & Terwindt (1981) described some basic lateral sequences for the succession of structures during a single tide and the neap/spring-tide cycle for intertidal megaripples. The same basic sequences may also be found in subtidal deposits. It should be noted, however, that the link between the intertidal and subtidal sequences is based on analogies. On intertidal shoals it is possible to measure accurately both the flow conditions and the structural responses associated with them. The area is accessible during low tide, thus allowing the determination of changes in bedforms, migration, structures, etc. In subtidal deposits accurate measurement is much more difficult.

Lateral sequences develop in deposits of migrating megaripples and thus refer to lithofacies STRO CUR in the lee of megaripples and MED CUR in the trough of megaripples. It is interesting to note that megaripple deposits in tidal areas show a large predominance of one tide over the other as evidenced by a prevailing unidirectional foresetting and subordinate or non-existent counter-current effects.

The tidal bundle

During the dominant tide a lateral succession of cross-strata, called a tidal bundle, is generated (Boersma, 1969). This tidal bundle is laterally enclosed by pause planes, i.e. erosional or non-erosional sur-

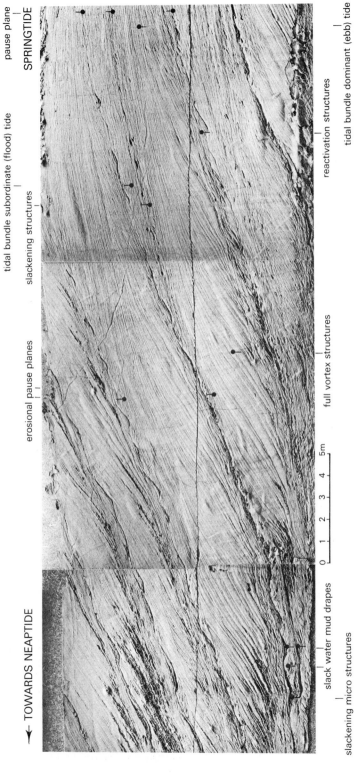

Fig. 4. Tidal bundles in megaripple deposits, showing reactivation, full vortex and slackening structures. Note: alternation of dominant tide (x-bedding) and subordinate counter directional tide (x-lamination); decrease in length of tidal bundles towards neap-tide; topsets occurring only during spring-tide; erosional pause planes and flood bundles running over the megaripple crests during spring-tide and non-erosional pause planes towards neap-tide. Lacquer peel, Oosterschelde 12 m—M.S.L.

2-2

faces representing the stand-still phase of the mega-ripple migration during the subordinate tide.

Within the tidal bundle three structural intervals arranged in a lateral sequence can be distinguished (Fig. 3):

(1) The first interval comprises the so-called *reactivation structures*. It represents an acceleration of the flow after the turn of the tide. Within the first interval a further distinction of three subintervals can be made.

(a) This subinterval represents the feeble current shortly after slack water. Ripples are moving along as well as across the resident megaripple. This results in an indistinct crinkled lamination, occasionally with x-lamination.

(b) This subinterval reflects an acceleration of the dominant tide. Small ripples start moving in the megaripple's trough. The lee face of the mega-ripple, which may have been more or less de-graded during the previous subordinate tide, possibly has a lower angle of slope than the angle of repose for the material. As a consequence ripples reaching the upstream edge of the dune's lee face will not feed their material to an ava-lanche. They retain their identity while moving down the face. This may produce x-lamination.

(c) In the third subinterval the angle of repose is restored at the lee of the megaripple. Ripples reaching the edge of the large slip face lose their material to avalanches moving down the front. These avalanches are still imperfect as they tend to be suppressed by the relatively large amount of material falling out from suspension on the slip face. This suppression is due to the imperfect flow separation which only gradually develops into a vortex. A poor grading of grain sizes in the foreset laminae also confirms this.

The three subintervals which together build the reactivation structures represent the pre-vortex ac-celeration stage of the dominant tide (Fig. 4).

(2) The middle interval is made up of robust, sharply delineated cross-bedding which is sloping at the angle of repose. The material shows a good fore-set grading. It is generated while intermittent ava-lanching occurs over the entire lee front of the megaripple under full vortex activity. The structures in this interval are therefore called *full vortex structures*. A further subdivision of three successive subintervals can be made (Kohsiek & Terwindt, 1981):

(a) A subinterval that shows angular x-bedding, consisting of straight laminae with dip angles between 30° and 35°.

(b) A subinterval with tangential x-bedding, and straight foreset laminae with dip angles between 25° and 30° merging into concave toesets.

(c) A subinterval with concave x-bedding, consisting of laminae showing a gradual downward de-crease in dip angle.

(3) The terminal interval in the tidal bundle has cross-strata, showing a poorer foreset grading and less distinct stratification than the preceding interval. Two subintervals can be distinguished.

(a) A subinterval with sigmoidal x-bedding which in places can be traced into topsets. The angle of foresetting is lower than the angle of repose.

(b) A subinterval which usually shows crinkled lamination with diffuse, thin, imperfectly defined cross-strata.

These structures in this interval are called the *slackening structures*. They represent the deceleration stage of the tidal current, when the extensive suspen-sion load, introduced into the stream during pre-ceding higher energy levels, gradually settles out. The reduced vortex activity, which finally drops to zero, is no longer capable of whirling up the suspension clouds from the lee side trough of the megaripple. This results in a progressive deposition in the toe area at a gradually lower angle. Finally only ripples move over the lee front and trough, more or less in the direction of the main flow.

The flow reversal

The reversal of the flow during the next subordin-ate tide is reflected in one or more of the following stages:

Stage 1. Accentuation of the pause plane by erosion of the top of the megaripple as its lee side becomes the stoss side. This process causes erosional pause planes, which dip at lower angle than the underlying foresets.

Stage 2. Generation of upslope and obliquely climbing ripples on top of the pause planes giving rise to opposite ripple foresetting and crinkly bedding.

Stage 3. Formation of an ebb or flood cap on top of the megaripple, formed when small ripples override each other along horizontal or descending planes, over the crest.

This basic tidal sequence is not always developed completely, depending on bedform characteristics

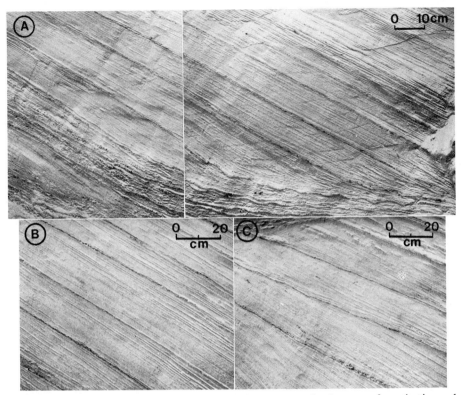

Fig. 5. (A) Tidal bundles, bounded by slack water mud drape, showing poor development of reactivation and slackening structure. Note: rising megaripple trough by deposits of sigmoïdal toesets and ripple x-laminated sets. Note: reducing thickness of tidal bundles. (B) Detail of same section near the toe of the megaripple. Hardly or non-erosional pause planes. Reactivation and slackening structures almost absent. Oosterschelde 12 m below M.S.L. (C) Detail of same section near the crest of the megaripple. Erosional pause planes with crinkled bedding (reactivation) and angular foresets of full vortex stage.

(particularly height), current strength, available material, etc.

The structural organization in tidal bundles is illustrated in Fig. 4. In the case of high megaripples (Fig. 5A) with set thicknesses of over 0·5 m such as those which may be present in subtidal deposits, the reactivation and the slackening structures are rather thin. Such megaripples are apparently formed under rather high current velocities with quick acceleration and deceleration. The reactivation is mostly encountered in the forms of some crinkled bedding followed by angular sets over the pause planes. This is especially the case in the upper parts of the megaripple (Fig. 5C). In the lower parts reactivation structures may be absent (Fig. 5B). Slackening structures are characterized by form-concordant toe sets and lesser foreset grading. The infilling of the trough is limited and the megaripple retains its steep lee side front. In other locations the

middle subinterval of the full vortex structures, i.e. concave bedding, is missing and there is a gradual transition from angular to low-angle tangential and finely to sigmoidal bedding in the toe sets. This is attributable to maximum current velocities, that are reduced, although still capable of producing a full vortex action (see Kohsiek & Terwindt, 1981).

The slack-water period after the dominant tide leaves hardly any trace in the intertidal area, but in subtidal regions a few mud drapes have been deposited concordant with the last slackening structure. Most of these drapes are a few millimetres thick, which is in accordance with the above-mentioned calculated maximum possible thickness during one or two slack-water periods. The mud drape may be traced upwards to the topsets when a non-erosional pause plane is present. Sometimes the subordinate tide has no effect at all. Evidence of reversed currents is completely absent in the structures. This is a frequent occurrence in higher megaripples as they develop mainly where one direction of the tide dominates. However, in smaller megaripples the effects of the reversed current are also frequently absent. In other cases, the subordinate tide has had an effect where it has produced erosional pause planes.

Although ebb and flood caps appear to have a low preservation potential they do occur in both intertidal and subtidal deposits. Tidal bundles may be arranged in many different ways. The diurnal inequality of the tides, producing a succession of higher and lower maximum current velocities of the dominant tide, is reflected in a series of thicker and thinner tidal bundles (Figs 4, 12).

These may most likely form in areas with long-crested, two-dimensional, type I, megaripples. In locations with the more lingoid, type II, megaripples a succession is more difficult to establish. The effects of the diurnal tidal inequality may be obscured by additional wind-caused in- or outflow, which may also influence the dimensions of the tidal bundles. There are, however, many examples that show without any doubt a variation in thickness of the tidal bundles that is attributable to diurnal inequality.

Neap/spring-tide cycle

Another important factor in the arrangement of the tidal bundles is the neap/spring-tide cycle. From neap to spring there is a gradual increase in the maximum current velocities of the dominant as well as the subordinate tide. The reverse is true from spring to neap. This increase and subsequent decrease of the current intensity is reflected in the tidal bundles:

Prevalent structural signs of increasing energy are:

(1) more pronounced full vortex stage, reflected in an increased length and height of the x-bedded interval; the appearance and increasing importance of the middle (tangential) subinterval in the full vortex structures;

(2) lowering of the lower set boundary, which is the result of more intense and prolonged vortex action;

(3) more significant suspension fall-out in the deceleration stage and subsequent thickening of the slackening structures;

(4) more easily recognizable reactivation and slackening structures:

(5) increasing magnitude of ebb or flood caps, and of erosive pause-planes.

The above tendencies are reversed from spring to neap-tide, when a decrease in energy of the flow gives rise to:

(6) diminishing of the full vortex structures and increasing relative importance of the reactivation and slackening structures;

(7) rise of the lower set boundary as a result of the decreasing vortex action;

(8) gradual shift from erosive pause planes to non-erosive pause planes.

This basic neap–spring sequence may vary with local conditions, i.e. current strength and bedform dimensions.

Some illustrative examples are given below.

In larger megaripples with set heights of 0·5–2·5 m, produced by high current velocities, the neap–spring–neap sequence mainly causes the lower set boundary to move downwards and upwards and increases and decreases the length of the tidal bundles (Fig. 12).

Furthermore, towards neap-tide the slackening structures exhibit an increasing amount of sigmoidal toesets, due to suspension fall-out. During neap-tide the megaripple is still present. There is still some avalanching continuing with mega-foresetting although the trough level has been gradually rising in the trough, current ripples are present showing x-lamination in the direction of the dominant current. The section of Fig. 12 further does not show any traces evidencing the reversal of the current. Mud drapes mark the pause planes. The diminishing strength of the currents of the subordinate tide from spring to neap-tide is demonstrated in the decreasing erosion as evidenced by the pause planes. Especially near the upper part of the megaripple the intensity of the erosion, witnessed by the erosional pause planes, gradually decreases, ultimately leading to non-erosive pause planes around neap-tide.

Another example (Figs 6, 7) shows a very rapid filling of the megaripple trough accompanied by a decrease of foreset angles and sigmoidal toe sets. At the same time a growing preponderance of slakening structures would indicate a sharp reduction in the vortex action. Such filling of the megaripple trough towards neap is commonly observed in both intertidal and subtidal deposits.

Another example (Fig. 8) shows a slightly different version of the same rapid infilling of the trough towards neap-tide. This section, taken in the Haringvliet estuary, presents many megaripple troughs some of those completely filled up. Here a great

Fig. 6. Lateral sequence in megaripple deposit showing spring-neap-tide sequence. Arrows indicate pause planes (mud drapes or reactivation structures). Note: rise of lower set boundary and rather rapid infilling of the trough. Length of scale rule: 50 cm. Haringvliet 7·5 m below M.S.L.

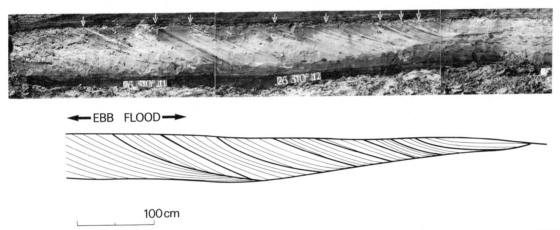

Fig. 7. Termination of megaripple towards neap-tide. Arrows mark the pause planes. Note: rising trough level, decreasing foreset angles, increasing occurrence of sigmoïdal toesets towards neap-tide. Length of scale rule: 50 cm. Haringvliet 13 m below M.S.L.

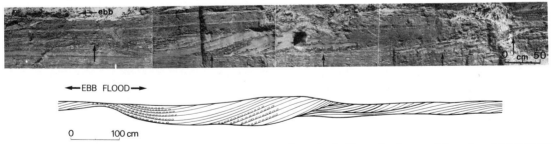

Fig. 8. Neap-spring-tide sequence (arrows indicate lower set boundary). Note: flat lower set boundary (right) at the onset of the spring-tide and the steeper lower set boundary towards full spring-tide. Further note the choking up of the megaripple trough by mud pebbles. Haringvliet 6–7 m below M.S.L.

abundance of mud pebbles represents a complicating factor as these pebbles have been known on occasion to completely choke up the trough. The pebbles, after deposition, form a layer resistant to subsequent erosion and prevent a lowering of the set boundary at the upstream end becoming steeper further downstream. This may be attributed to an increase in vortex action and greater erosive power to dredge out the trough. This section of Fig. 8 also shows the normal tidal characteristics: neap/spring-tide bundles, slack water mud drapes on the toe and bottomsets, and in the megaripple trough a lowering

Fig. 9. Section parallel to dune crest showing festoon bedding and slack-water mud drapes. Note indications of crinkled bedding above mud drape and further the asymmetrical infilling of the trough. Haringvliet 5 m below M.S.L. Length of scale rule: 50 cm.

and rising of the lower set boundaries. Noteworthy is the astonishing fact that mud pebbles accumulate in the trough during decreasing current intensity. There is no explanation so far as to how the clay pebbles are transported while relatively low current velocities exist. It should be noted, that lateral and neap–spring sequences are well exposed in sections parallel to the direction of migration of the megaripple, but these sequences are very hard to distinguish in sections perpendicular to the migration. When dealing with two-dimensional megaripples, transversal sections show a repetition of parallel or inclined bedding with occasional slack water mud drapes and possibly a delicate, small-scale, ripple lamination above and below the drape, especially near the trough of the dune.

Three-dimensional megaripples are more complicated. These bedforms have spurs and bowl-shaped depressions in the trough of the megaripple, the latter filled in by the migrating megaripple creating festoon bedding (Fig. 9). This filling is mostly asymmetrical, i.e. more material is delivered to one side of the trough than to the opposite. It is obvious that tidal and neap–spring sequences are difficult to establish in such sections. Sometimes such depressions are filled in during both tides causing the formation of bidirectional x-bedding (Fig. 10).

Megaripple troughs

In the preceding section we specifically emphasized the features of the lee front of the megaripple (lithofacies STRO CUR), although some attention was given to the trough of the megaripple.

We will now focus our attention especially on the trough: i.e. the domain of the MED CUR lithofacies. Obviously the sedimentation in the trough is influenced by the events occurring on the lee of the megaripple. We may therefore anticipate gradual transitions between the megaripple front and the trough and between lithofacies STRO CUR and MED CUR.

There is a difference in the facies associations in an erosive megaripple trough as compared with a depositional trough.

In the erosive trough, megaripple as well as the trough are migrating. Vortex action and turbulence excavate the trough of the megaripple. Ripples which may have been present in the trough and on the stoss-side of the next megaripple are eroded, thus barely leaving any traces except on the toe of the migrating megaripple front in the spring-tide bundles and higher up the lee slope in the neap-tide bundles. Ripples in the neap-tide bundles are very difficult to analyse because of the variation of the direction of migration giving rise to crinkled lamination. In the spring-tide bundles, however, ripple x-lamination may be present on the lower parts of the foresets and the toesets. These ripple structures show a specific lateral and vertical sequence which is a continuation of the tidal megaripple sequence. The latter ends with the slackening structures, which in the first subinterval are made up of form-concordant low angle x-bedded suspension fall-out structures. These structures are overlain by the second subinterval consisting of co-flow ripples migrating downward over the megaripple in the direction of the dominant tide. Most commonly only the lower parts of the foresets are preserved as small

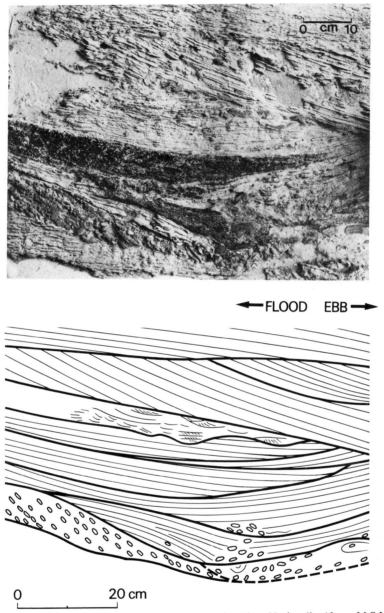

FLOOD ← EBB →

Fig. 10. Bidirectional trough filling. Laquer peel section. Haringvliet 13 m—M.S.L.

sets which are erosional at the top. However, towards slack water these ripples come to a standstill showing exquisite '*slackening-micro-structures*' made up of gradually upfanning foresets and infilling of their troughs (see Fig. 11). In places the complete morphological form can be preserved by a slack-water mud drape. Afterwards the subordinate counter-directional tide moves in, producing ripples and x-lamination. Here too, towards the next slack water, the slackening microstructures are present, either capped or not by a mud drape. These slackening microstructures then are found just below the slack-water marks, i.e. either in the form of a mud drape or a discontinuity plane. The absence of a mud drape can be due to a slight erosion afterwards or to a slack-water period too short in duration to

ebb

0   10 cm

Fig. 11. Depositional megaripple troughs towards neap-tide, showing toesets made up of an alternation of x-laminated sets produced by the dominant (a) and the subordinate tide (b) separated by (mud draped) pause planes (c). Note the slackening micro-structures (d) and absence of x-lamination during neap-tide (e). Laquer peel. Oosterschelde 12 m — M.S.L.

create sufficient initial consolidation of the deposited mud. Short slack-water periods are a normal feature because of asymmetrical tidal velocity curves.

Where the megaripple troughs are depositional, a different lateral and vertical sequence may form. To the author's knowledge no observations of such sequences exist in intertidal deposits, but there are several examples of subtidal sediments from both the Haringvliet estuary and the Oosterschelde tidal inlet. Depositional troughs are present if there is great net sedimentation and reduced action of the vortex and/or turbulence (Fig. 12). Under these circumstances, the trough of the megaripple, or even the whole megaripple, may be preserved. As the deposits in the trough are left behind, the trough level will gradually rise. Typically, the sediments in the troughs originate from the reactivation and slackening of the megaripple migration. During the full vortex stage the deposition in the trough is minimal. The full vortex structures mostly wedge out and are angular or subangular to the toe of the megaripple. During the acceleration of the mega-ripple movement the sediments in the trough are thought to be primarily supplied by fall-out from suspensions. This may explain why we often observe

form-concordant bedding with less demixing of the grain sizes in the side of the trough just near the megaripple toe. Further into the trough (Fig. 13) we notice formation of ripples producing small, some-times climbing, x-laminated sets, indicating that several ripples have moved one on top of the other. Further away from the megaripple front we observe fewer ripple trains, ultimately leading to one ripple level only. It would therefore seem that sediments supplied by suspension fall-out are reworked by ripple formation further down in the trough. To-wards slack water, here too, slackening micro-structures are present below the slack-water mud drape. The subordinate counter-directional tide essentially produces the same kind of structures. Some ripple trains moving towards the former lee, now stoss, side of the megaripple also gradually wedge out into one ripple level. Here too, slackening microstructures are found below the slack-water mud drape.

General remarks

The sequence of events described above may vary from place to place but the general trend is rather consistent.

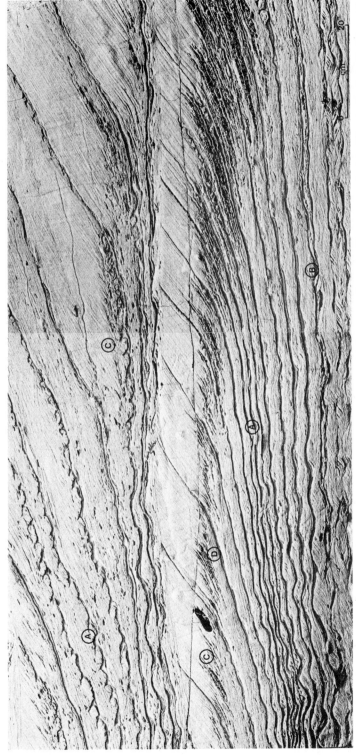

Fig. 12. Depositional megaripple trough. (A) Neap-tide bundles, showing sigmoidal foresets and toesets. Form concordant bedding in the toesets and small-scale ripple x-lamination in the trough. (B) Slack-water mud drapes separating the dominant and the subordinate tide. Note: the reversed direction of x-lamination, sometimes climbing ripples. (C) Angular full vortex foresets merging into sigmoidal slackening structures. Note: toesets are mainly made up of reactivation and slackening structures. (D) Non-erosional pause planes, single slack water mud drape. Laquer peel. Oosterschelde 12 m—M.S.L.

Fig. 13. Depositional megaripple trough showing deposition in the trough made up of superimposed (overriding) x-laminated sets gradually merging into one x-laminated set (sometimes with stoss-side deposition). Laquer peel. Oosterschelde 12 m—M.S.L.

In some locations it was observed that during neap-tide the current conditions in the trough are clearly below the threshold for ripple building. In these instances form-concordant, low-angle, parallel bedding was also observed in the trough (Figs 11, 12). However, in other sections evidence was found that ripples were still on the move during neap-tide.

Another interesting feature of depositional troughs during neap-tide is the slump structures which occur in some of the mud drapes covering the neap-tide bundles (Figs 11, 12). Due to a longer slack-water period the mud drapes are thicker and apparently some material with low initial consolidation remain. The mud layer is incapable of supporting the sand layer and load-casting or even slumping is the result. Slumping may also occur in thinner mud layers on steeper foresets.

Some other unusual aspects should be mentioned. Almost all the ripples in the megaripple trough are

co-flow ripples either generated by the dominant or by the subordinate tidal currents. During the acceleration and deceleration of the dominant tide therefore co-flow conditions in the trough necessarily point to an absence, or at the most a very weak presence, of a vortex. A vortex in the trough is apparently only active during the full vortex stage of the megaripple migration. In that case true foreset x-beds are present, with foreset laminae approaching normal angles of repose, with foreset-grading and angular or tangential toesets. It seems that, in most cases, back-flow velocities in the vortex do not have much strength and are incapable of generating back-flow ripples. Such back-flow ripples (Boersma, Van der Meene & Tjalsma, 1968) in trough deposits may have been theoretically possible, but are to the author's knowledge rather exceptional. When comparing ripple beds of this lithofacies in intertidal with subtidal deposits we note that intertidal sediments show far more crinkled bedding. This occurs because ripples move not only in the ebb or flood direction but also in a direction perpendicular to the crestline of the megaripple during drainage and flooding of the emerging and submerging shoal.

As mentioned before, lateral sequences are most likely to develop in areas where the bedforms consist of megaripples. In the outcrops studied, lateral sequences could not be established in ripple and plane bed environments. This absence may be due to the limited size of the exposures as lateral sequences of MED CUR 1, 2, 3 to 4 and 5 merge into WEAK CUR 1 and 2, or the opposite may be anticipated in these environments as a result of a gradual change in current conditions perpendicular to the axis of the tidal channels. These variations work out over several tens of metres while in megaripples the sequences occupy only a few metres.

The vertical sequences

Two types of vertical sequences may be distinguished:

> micro-sequences in the decimetre-metre scale;
> macro-sequences in the tens-of-metres scale.

We will consider the micro-sequences first.

(1) *The coarsening upward megaripple micro-sequence* (*MED CUR–STRO CUR*)

It is obvious that in depositional megaripple troughs the vertical upbuilding of megaripple trough deposits (MED CUR) overlain by megaripple front deposits (STRO CUR) forms a vertical coarsening upward sequence. The lower boundary of the micro-sequence is made up of the lowest trough line of the megaripples. The vertical succession is illustrated in Fig. 12. Here we have two MED CUR–STRO CUR micro-sequences above each other, separated by an erosional surface.

(2) *The fining upward megaripple micro-sequence* (*STRO CUR–MED CUR*)

This micro-sequence is encountered in areas where megaripple troughs are almost completely filled in. In the final filling stage the megaripple fades away and the vortex, if still present at all, is too weak to leave any traces. In the megaripple trough sigmoidal lamination produced by suspension fall-out merges into ripple beds (MED CUR 1, 2 or 3), some of it even into flaser beds (MED CUR 4) (see Figs 7 and 8, left side).

Such infilling of troughs of the megaripples is normally observed on intertidal shoals going from spring- to neap-tide. However, preservation would be the exception because of subsequent erosion during the next sequence although it does remain a rare phenomenon.

Analysis of numerous megaripple sections leads to the conclusion that in most cases the top of the STRO CUR lithofacies is erosional showing a sharp boundary with the next lithofacies. Even where lithofacies STRO CUR is followed by lithofacies MED CUR, there is generally no gradual transition but a sharp contact precluding labelling the sequence as a micro-sequence.

(3) *The STRO WAV–MED WAV–MED CUR micro-sequence*

Wunderlich (1969) indicated that the intertidal and shallow subtidal banks of tidal channels might occasionally be subjected to strong wave action during heavy gales. This results in inclined parallel lamination, erosive at the base and intercalated into other lithofacies. If the top of this parallel lamination is erosive too, or is followed upwards by an abrupt change in lithofacies, then we have no micro-sequence but a random intercalation of a STRO WAV lithofacies into another lithofacies. When there exists, however, a gradual transition between the parallel laminated set and the lithofacies on top of it, then it is reasonable to describe it as a micro-sequence. Wunderlich (1969) demonstrated such a STRO WAV–MED WAV–MED CUR micro-sequence. The parallel lamination is overlain by wave-ripple lamination which gradually merges upwards into current-ripple lamination, indicating a reduced wave action after cessation of the gale and finally restoration of the normal tidal current action.

(4) *The fining upward STORM DEP–MED CUR micro-sequence*

A similar transition from wave to current action is observed in water depths below the wave base. Here, we may find some STORM DEP–MED CUR micro-sequence comprised of horizontal parallel lamination, non-erosive at the base, followed by trough and tabular x-laminated sets, which may be overlain by flaser bedding (see Fig. 1). This micro-sequence is made up of material falling out from suspensions. These suspensions are produced in shallower water by breaking waves as evidenced by the slightly or non-erosional lower set boundary, followed by x-lamination produced by current

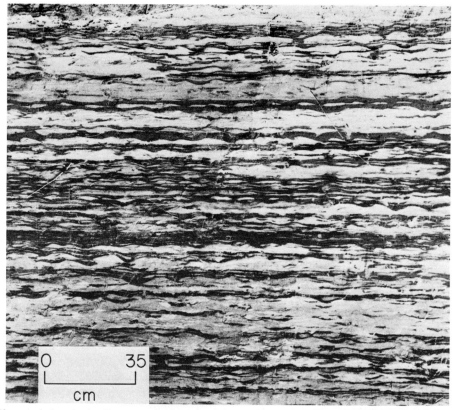

Fig. 14. Minor variations in the distance and thickness of the mud layers in a flaser bedding. Haringvliet 15 m—M.S.L.

ripples which rework the top layers of the parallel lamination. Eventually, the intensity of the currents decreases in time and this is reflected in the flaser bedding.

(5) *The fining upward MED CUR–WEAK CUR micro-sequence*

This sequence starts with x-laminated sets, many with herringbone structures (MED CUR 1, 2 and 3), which gradually merge into simple flaser bedding (MED CUR 4) and wavy bedding (MED CUR 5) and finally into lenticular (WEAK CUR 1) and sand-clay alternating bedding (WEAK CUR 2) (see, for example, De Raaf & Boersma, 1971, fig. 2A, and Terwindt, 1971, fig. 11).

Superimposed upon the general trend of this micro-sequence we see some minor variations. In the MED CUR 4 and 5 the number and thickness of the mud layers of the flaser beds is not uniform. In a particular horizon this total number and the thickness may be greater than in the neighbouring higher and lower horizons. Thus, horizons of sand-clay

alternating bedding are preceded or followed by horizons of lenticular bedding. Wavy bedding horizons are intercalated between lenticular or flaser bedding (see Fig. 14). In this case we have a sequence in which superimposed on the general trend of a gradual decreasing current intensity normally associated with this micro-sequence, there are fluctuations in current strength which may be attributed to neap–spring tide variations. During neap-tide more mud deposition takes place, and during spring-tide there is more deposition and movement of sand. It may be remembered, however, that clay layers in lithofacies WEAK CUR are deposited during several consecutive slack water periods. This factor makes counting the clay layers for the purpose of detecting neap–spring-tide cycles in this micro-sequence difficult. In a semi-diurnal tide there are 27 slack waters from one spring-tide to the next. Everywhere in this micro-sequence we do, however, see layers indicating higher current strength alternating with layers of implying lower current strength in some decreasing or increasing trend. It may be noted that we also

Fig. 15. Coarsening–fining upward micro-sequence made up of MED CUR 5–MED CUR 4–MED CUR 3–MED CUR 4–MED CUR 5–WEAK CUR 1. Haringvliet 7·5 m—M.S.L.

meet many of these fining upward sequences in intertidal and supratidal deposits (Evans, 1975; Reineck, 1975). However, there a more or less strong bioturbation is present, caused by specific organisms in particular horizons (Dörjes *et al.*, 1970a; Reineck, 1958; Van Straaten, 1954).

(6) *The fining upward MED WAV–WEAK WAV micro-sequence*

Such micro-sequences might be expected in areas with very low current action and periodically limited and then diminishing wave action. However, the author has never found evidence in the field for this micro-sequence nor has he found publications describing these.

(7) *The coarsening WEAK CUR–MED CUR micro-sequence*

This micro-sequence is the reverse of that described under (5). It is made up of a gradual transition of sand-clay alternating bedding and/or lenticular bedding and/or wavy and flaser bedding into x-lamination (see Terwindt, 1971, fig. 11; Sellwood, 1975, figs 11–3). Here too we have the same minor variations upon the general trend as described above.

(8) *The fining coarsening MED CUR–WEAK CUR– MED CUR micro-sequence and the coarsening- fining WEAK CUR–MED CUR–WEAK CUR micro-sequence*

Both sequences are combinations of various micro-sequences mentioned before. These may be attributed to a gradual change in the overall current conditions during deposition (Fig. 15, see also Reineck, 1967, fig. 13, lower part).

(9) *The fining–coarsening–fining (MED CUR–WEAK CUR–MED CUR–WEAK CUR) and the coarsening–fining–coarsening (WEAK CUR–MED CUR– WEAK CUR–MED CUR) micro-sequences (Fig. 16)*

Both types of micro-sequences have been observed in subtidal deposits (Terwindt, 1975).

(10) *The frequency of occurrence of the micro-sequences*

A general impression of how frequently the micro-sequences occur may be obtained from sections and borings (Terwindt, 1975). It should be kept in mind that the analysis of lithofacies in core samples is in some instances very difficult and not accurate. Particularly the distinction between current and wave x-lamination is very difficult. Furthermore, the data gathered pertain primarily to conditions of the area studied. However, such analysis may produce an order of magnitude of the frequencies concerned. In total some 5000 m of undisturbed continuous cores were analysed for micro-sequences. The cores were taken in the seaward part of the tidal inlets in the south-west part of the Netherlands.

Many of the described micro-sequences are rare (frequency < 5 %). Particularly the MED CUR– STRO CUR, the STRO CUR–MED CUR, the STRO WAV–MED WAV–MED CUR, the STORM DEP–MED CUR and the WEAK CUR–MED CUR –WEAK CUR–MED CUR micro-sequences occur infrequently.

Most of the observed micro-sequences (approximately 40 %) are fining upward MED CUR–WEAK CUR. The corsening upward WEAK CUR–MED CUR and the fining coarsening MED CUR–WEAK CUR–MED CUR micro-sequences both have a frequency of about 20 % while the fining coarsening fining (MED CUR–WEAK CUR–MED CUR– WEAK CUR) accounts for approximately 7 %.

It can be generally concluded that the micro-sequences are normally made up of gradual transi-

Fig. 16. Coarsening–fining–coarsening–fining–coarsening upward micro-sequence made up of WEAK CUR 2–WEAK CUR 1–MED CUR 5–MED CUR 4–MED CUR 5–WEAK CUR 1–MED CUR 5–MED CUR 4–MED CUR 5–WEAK CUR 1–WEAK CUR 2–WEAK CUR 1–MED CUR 5–MED CUR 4–MED CUR 2. Length of scale rule: 50 cm. Haringvliet 14·5 m—M.S.L.

tions between lithofacies MED CUR and WEAK CUR in several combinations.

The macro-sequence

In tidal areas there are no channels which are continuously silting up from their deepest part up to the intertidal and supratidal zone. If this were the case we might expect a fining upward macro-sequence, implying a gradually decreasing current intensity near the bottom with decreasing water depth (Dörjes *et al.*, 1970b); Oomkens & Terwindt, 1960) and an increase in wave effects and bioturbation (Dörjes *et al.*, 1970a, b; Van Straaten, 1954). However, tidal channels are usually migrating, or they may be temporarily more or less abandoned and afterwards reactivated. It also happens that the maximum stream line shifts in the channel, or in some cases the tidal volume changes due to variations in the overall channel patterns.

These large-scale variations mean that in tidal channels deposition is followed by erosion and non-deposition. Furthermore, even in shallow channels of small dimensions, current velocities may be as high as in large channels, producing similar sedimentary structures.

The same is true for the shoals. Here too, erosion and sedimentation alternate.

All these processes mean that overall fining upward macro-sequences should not be expected in inshore tidal deposits and have not been observed so far. Of course, changing conditions in the channels

and on the shoals may produce micro-sequences, but these are all cut off by erosional surfaces and abrupt changes in lithofacies. These erosional contacts and abrupt facies changes are the most striking features in observing exposures of inshore tidal sediments. Such rapid changes in lithofacies not only occur vertically but also laterally, indicating important lateral variations in micro-environments (Terwindt, 1971).

We agree therefore with De Raaf & Boersma (1971), who stated that: 'tidal deposits differ sequentially from fluviatile (fining upward) as well as from deltaic sediments (coarsening upward)'. We should further add that tidal deposits are most probably characterized by the absence of a macro-sequence.

DIAGNOSTIC STRUCTURAL FEATURES IN INSHORE TIDAL DEPOSITS

Diagnostic features are discussed by Klein (1970) and De Raaf & Boersma (1971). Three phenomena are characteristic of the tidal environment:

> acceleration and deceleration and a reversal of flow during a full tidal cycle, on a diurnal or semi-diurnal time-scale;
>
> increase and decrease of the maximum current velocities in the neap–spring–neap cycle on a fortnightly time-scale;
>
> rather rapid and important change in the channel and shoal systems, on a time-scale covering several months to years.

The structural features indicative of the tidal environment should reflect these hydraulic conditions.

The reversal of tidal flows is reflected by the most characteristic sedimentary feature in inshore tidal deposits, namely the bidirectional foresetting. The foresetting may be herringbone structures in megaripple or ripple cross-bedding, or bidirectional trough filling or current ripple x-lamination of a subordinate tide covering the x-bedding produced by the opposite dominant tide. Finally the remains of ebb or flood caps on megaripples may appear, but it is unlikely that these would be preserved.

Bidirectional foresetting is, however, not always present in tidal deposits. There are many outcrops in sediments of undoubtedly tidal origin which exhibit only unidirectional foresetting. The absence of bidirectional foresetting does not therefore exclude a tidal origin of the sediments.

The acceleration and deceleration of the tidal flow is reflected by the x-bedded (megaripple) sets in the tidal bundles comprising the reactivation, full vortex and slackening structures. This bundle is rather short (decimetre–metre scale) and bounded by erosional or non-erosional pause planes, some capped by a slack water mud drape. Indications of deceleration of the flow may be found in slackening-microstructures.

These tidal bundles show lateral sequences or trends in their length, height and structural organization. Such trends are the short–long–short succession of the width of the bundles reflecting in the daily inequality of the tides but particularly the neap/spring-tide lateral sequence. Boersma & Terwindt (1981) considered the consistent occurrences of these tidal and neap/spring-tide lateral sequences a good diagnostic feature of intertidal deposits. It may be added that this is true for subtidal deposits too.

The rapid and important changes in the channel and shoal system are reflected in the absence of a general fining upward macro-sequence. Furthermore, although there are many types of microsequences, tidal outcrops exhibit many erosional surfaces and abrupt changes in lithofacies.

This feature is rather vaguely defined and is probably not a unique feature of the tidal environment. However, in combination with both the other diagnostic features, it characterizes tidal deposits.

Summarizing, it seems that all the data gathered by numerous studies of tidal sediments around the North Sea confirm that the diagnostic features for inshore mesotidal deposits are:

bidirectional foresetting; consistent occurrence of tidal bundles with reactivation and slackening structures and lateral neap–spring-tide sequences in megaripple deposits;

high frequency of erosional contacts and abrupt facies changes coupled with the absence of a general vertical macro-sequence (in combination with both the other items).

ACKNOWLEDGMENTS

The author is greatly indebted to all people who and institutions which provided great assistance during many years of work on tidal sediments in the south-west part of the Netherlands. The cooperation of Rijkswaterstaat–Deltadienst–Bureau Hellevoetsluis is gratefully acknowledged. I am grateful to Professor J. F. M. De Raaf, Drs J. R. Boersma, S.-D. Nio and J. H. van den Berg and many other colleagues for stimulating discussions and demonstrations. During the preparation of the manuscript technical assistance for drawing, photographing and typing was kindly provided by R. L. Rieff, G. H. Huijgen, T. Lekkerkerker and L. M. Butteling. Mrs A. Philips is thanked for editing the text.

REFERENCES

ALLEN, J.R.L. & FRIEND, P.F. (1976) Changes in intertidal dunes during two spring–neap cycles, Lifeboat Station Bank, Wells-next-the-Sea, Norfolk (England). *Sedimentology*, **23**, 329–346.

BOERSMA, J.R. (1969) Internal structure of some tidal megaripples on a shoal in the Westerschelde estuary, The Netherlands. *Geologie Mijnb.* **48**, 409–414.

BOERSMA, J.R. (1970) *Distinguishing features of wave-ripple cross-stratification and morphology.* Ph.D. Thesis, University of Utrecht, 65 pp.

BOERSMA, J.R., VAN DER MEENE, E.A. & TJALSMA, R.C. (1968) Intricated cross-stratification due to interaction of a megaripple with its leeside system of backflow ripples. *Sedimentology*, **11**, 147–162.

BOERSMA, J.R. & TERWINDT, J.H.J. (1981) Neap-spring tide sequences of intertidal shoal deposits in a mesotidal estuary. *Sedimentology*, **28**, 151–170.

BOOTHROYD, J.C. & HUBBARD, D.K. (1975) Genesis of bedforms in mesotidal estuaries. In: *Estuaries Research*, vol. II (Ed. by E. Cronin), pp. 217–234. Academic Press, New York.

CLIFTON, H.E. (1976) Wave-formed sedimentary structures: a conceptual model. In: *Beach and Near-shore Sedimentation* (Ed. by R. A. Davis and R. L. Ethington). *Spec. Publ. Soc. econ. Paleont. Miner., Tulsa*, **24**, 126–148.

CLIFTON, H.E., HUNTER, R.E. & PHILLIPS, R.L. (1971) Depositional structures and processes in the non-barred high energy near-shore. *J. sedim. Petrol.* **41**, 651–670.

DALRYMPLE, R.W., KNIGHT, R.J. & LAMBIASE, J.J. (1978) Bedforms and their hydraulic stability relationships in a tidal environment, Bay of Fundy, Canada. *Nature*, **275**, 100–104.

DAVIDSON-ARNOTT, R.G.D. & GREENWOOD, B. (1976) Facies relationships on a barred coast, Kouchibouguac Bay, New Brunswick, Canada. In: *Beach and Nearshore Sedimentation* (Ed. by R.A. Davis and R.L. Ethington). *Spec. Publ. Soc. econ. Paleont. Miner., Tulsa*, **24**, 149–168.

DINGLER, J.R. (1974) *Wave-formed ripples in nearshore sands*. Ph. D. Thesis, University of South California, 148 pp.

DÖRJES, J., GADOW, S., HERTWECK, G., REINECK, H.E. & WUNDERLICH, F. (1970a) *Das Watt Ablagerungs- und Lebensraum.* Kramer Verlag, Frankfurt. 142 pp.

DÖRJES, J., GADOW, S., REINECK, H.E. & SINGH, I.B. (1970b) Sedimentologie und Makrobenthos der Nordergründe und der Aussenjade (Nordsee). *Senckenberg Mar.* **2**, 31–59.

ELLIOTT, T. (1978) Clastic shorelines. In: *Sedimentary Environments and Facies* (Ed. by H. G. Reading), pp. 143–177. Blackwell Scientific Publications, Oxford.

EVANS, G. (1965) Intertidal flat sedimentation and their environments of deposition in the Wash. *Q. Jl geol. Soc. Lond.* **121**, 209–245.

EVANS, G. (1975) Intertidal deposits of the Wash, Western margin of the North Sea. In: *Tidal Deposits* (Ed. by R.N. Ginsburg), pp. 13–20. Springer-Verlag, Berlin.

GADOW, S. & REINECK, H.E. (1969) Ablandiger Sandtransport bei Stormfluten. *Senckenberg Mar.* **1**, 63–78.

GINSBURG, R.N. (Ed.) (1975) *Tidal Deposits.* Springer-Verlag, Berlin, 428 pp.

KLEIN, G. DE VRIES (1970) Depositional and dispersal dynamics of intertidal sand bars. *J. sedim. Petrol.* **40**, 1095–1127.

KOHSIEK, L.H.M. & TERWINDT, J.H.J. (1981) Characteristics of foreset and topset bedding in megaripples related to hydrodynamic conditions on an intertidal shoal. In: *Holocene Marine Sedimentation in the North Sea Basin* (Ed. by S.-D. Nio *et al.*). *Spec. Publs int. Ass. Sediment.* **5**, 27–37. Blackwell Scientific Publications, Oxford, 524 pp.

KOMAR, P.D. & MILLER, M.C. (1973) The threshold of sediment movement under occillatory water waves. *J. sedim. Petrol.* **43**, 1101–1110.

LITTLE-GADOW, S. & REINECK, H.E. (1974) Diskontinuierliche Sedimentation von Sand und Schlick in Wattensedimenten. *Senckenberg. Mar.* **6**, 149–159.

MIGNIOT, C. (1968) Etude des propriétés physiques de differents sediments très fins et de leur comportement sous des actions hydrodynamiques. *Houille blanche*, **23**, 591–620.

NEWTON, R. S. (1968) Internal structure of wave ripple marks in the nearshore zone. *Sedimentology*, **11**, 275–292.

OOMKENS, E. & TERWINDT, J.H.J. (1960) Inshore estuarine sediments in the Harinvbliet (Netherlands). *Geologie Mijnb.* **39**, 701–710.

POSTMA, H. (1954) Hydrography of the Dutch Wadden Sea. *Inst. Neerl. Zoöl.* **10**, 405–511.

POSTMA, H. (1961) Transport and accumulation of suspended matter in the Dutch Wadden Sea. *Neth. J. Sea Res.* **1**, 148–190.

POSTMA, H. (1967) Sediment transport and sedimentation in the estuarine environment. In: *Estuaries* (Ed. by G.H. Lauff). *Publs Am. Ass. Advant Sci.* **83**, 158–179.

DE RAAF, J.F.M. & BOERSMA, J.R. (1971) Tidal deposits and their sedimentary structures. *Geologie Mijnb.* **50**, 479–504.

DE RAAF, J.F.M., BOERSMA, J.R. & VAN GELDER, A. (1977) Wave generated structures and sequences from a shallow marine succession, Lower Carboniferous, County Cork, Ireland. *Sedimentology*, **24**, 451–483.

READING, H.G. (1978) Facies. In: *Sedimentary Environments and Facies* (Ed. by H.G. Reading), pp. 4–14. Blackwell Scientific Publications, Oxford.

REINECK, H.E. (1958) Wühlbau-Gefüge in Abhängigkeit von Sedimentumlagerungen. *Senckenberg. Leth.* 1–24.

REINECK, H.E. (1960) Über die Entstehung von Linsen und Flaserschichten. *Abh. dt. Akad. Wiss. Berl.* **3**, 370–374.

REINECK, H.E. (1961) Sedimentbewegungen an Kleinrippeln in Watt. *Senckenberg. Leth.* **39, 42** 51–61.

REINECK, H.E, (1963) Sedimentgefüge im Bereich der südlichen Nordsee. *Abh. senckenb. naturforsch. Ges.* **505**, 138 pp.

REINECK, H.E. (1967) Layered sediments of tidal flats, beaches and shelf bottoms of the North Sea. In: *Estuaries* (Ed. by G. H. Lauff). *Publs Am. Ass. Advent Sci.* **83**, 141–206.

REINECK, H.E. (1975) German North Sea tidal flats. In: *Tidal Deposits* (Ed. by R.N. Ginsburg), pp. 5–12. Springer-Verlag, Berlin.

REINECK, H.E. & SINGH, I.B. (1972) Genesis of laminated sand and graded rhymites in storm-sand layers of shelf mud. *Sedimentology*, **18**, 123–128.

REINECK, H.A. & SINGH, I.B. (1973) *Depositional Sedimentary Environments.* Springer-Verlag, Berlin. 439 pp.

REINECK, H.E. & WUNDERLICH, F. (1968) Classification and origin of flaser and lenticular bedding. *Sedimentology*, **11**, 99–104.

REINECK, H. E. & WUNDERLICH, F. (1969) Die Entstehung von Schichten und Schichtbanken im Watt. *Senckenberg. Mar.* **1**, 85–106.

SELLWOOD, B.W. (1975) Lower Jurassic tidal-flat deposits, Bornholm, Denmark. In: *Tidal Deposits* (ed. by R.N. Ginsburg), pp. 93–102. Springer-Verlag, Berlin.

STRAATEN, L.M.J.U. VAN (1954) Composition and structure of recent marine sediments in the Netherlands. *Leidse geol. Med.* **19**, 1–110.

TERWINDT, J.H.J. (1970) Observations on submerged sandripples with heights ranging from 30 to 200 cm occurring in tidal channels of SW Netherlands. *Geologie Mijnb.* **49**, 484–501.

TERWINDT, J.H.J. (1971) Lithofacies of inshore estuarine and tidal inlet deposits. *Geologie Mijnb.* **50**, 515–526.

TERWINDT, J.H.J. (1975) Sequences in inshore subtidal deposits. In: *Tidal Deposits* (Ed. by R.N. Ginsburg), pp. 85–89. Springer-Verlag. Berlin.

TERWINDT, J.H.J. & BREUSERS, H.N.C. (1972) Experiments on the origin of flaser, lenticular and sand-clay alternating bedding. *Sedimentology*, **19**, 85–98.

WUNDERLICH, F. (1969) Studien zur Sedimentbewegung: 1. Transportformen und Schichtbildung im Gebiet der Jade. *Senckenberg. Mar.* **1**, 107–146.

Spec. Publs int. Ass. Sediment. (1981) **5**, 27–37

Characteristics of foreset and topset bedding in megaripples related to hydrodynamic conditions on an intertidal shoal

L. H. M. KOHSIEK* *and* J. H. J. TERWINDT

Department of Physical Geography, Heidelberglaan 2, 3508 TC, Utrecht, the Netherlands

ABSTRACT

The succession in the shape of the foreset laminae in cross-bedded sets with increasing current velocity so well documented from flume experiments is also observed in tidal deposits. During the full vortex stage of megaripple migration this succession consists ideally of angular, followed by tangential and concave cross-bedding. In the deceleration stage of the megaripple the concave cross-bedding is followed by sigmoidal cross-bedding produced mainly by suspension fall-out under conditions of decreasing flow velocities. Depending on the magnitude of the maximum velocity of the current in a tidal cycle this succession may be more or less complete.

Topsets may be generated during deceleration and full vortex action under conditions of high maximum flow velocities.

Thresholds could be established of the current velocity for the generation of the different types of foresets and topsets.

INTRODUCTION

Bedforms of different size, form and orientation are encountered on the Ossenisse intertidal shoal located in the Westerschelde estuary, the Netherlands. The major bedforms consist of megaripples which may be classified according to Dalrymple, Knight & Lambiase (1978) as type I (= linear magaripples) and type II (sinuous megaripples). The internal structure of such megaripples have been described by several authors, namely Klein (1970), Boersma (1969), Gellathly (1969) and De Raaf & Boersma (1971).

A summary of the findings is given in Boersma & Terwindt (1981). This paper attempts to analyse in detail the characteristics of the cross-bedding of these megaripples and to correlate the specific phenomena with the hydrodynamic conditions. For that aim current velocity measurements were per-

* Present address: Rijkswaterstaat–Deltastienst, van Alkemadelaan 400 The Hague, the Netherlands.

0141-3600/81/1205–0027 $02.00

formed over megaripples fields, lacquer peels were made and the distance of migration was determined by spreading luminophore tracers on the megaripple front during the dry period of the shoal.

THE HYDRAULIC CONDITIONS

Many observations from areas of different tidal ranges and flow patterns reveal that the morphology, net migration and the organization of the internal structures of megaripples depend on the intensity and direction of the tidal currents (Boothroyd & Hubbard, 1975; Klein, 1970; Allen & Friend, 1976; Dalrymple *et al.*, 1978; McCave & Geiser, 1978). It appears that the net migration is in the direction of the dominant tidal current. This dominance not only reflects the magnitude of the maximum current velocity (mean over the vertical) of one tide compared with the opposite tide but more specifically the dominance of the sand transport. The magnitude of the maximum current velocity as well as the dominance may change in time due to the diurnal

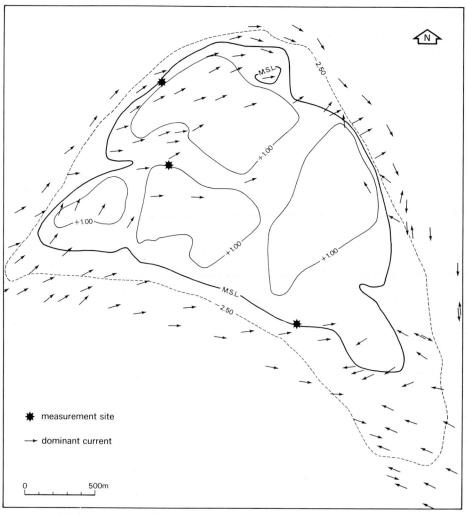

Fig. 1. Dominance of the tidal currents over the Ossenisse shoal during spring-tide.

inequality of the tides and, more important, to the neap/spring-tide variation.

The tidal dominance over the Ossenisse shoal around spring-tide is indicated on Fig. 1. On the shoal two measurement platforms were installed one over a linear megaripple field and one over a sinuous field (Fig. 2). The mean grain size of the bed material in both fields ranges between 150 and 200 μm. Both measurement sites were located in flood-dominated areas. On the sites the vertical current velocity distribution was measured at 10 min intervals throughout the tidal cycle for two complete neap/spring-tide cycles. Fig. 3 gives the maximum values of current velocity $\overline{U}_{max}$ (mean over the vertical) for the successive tides during the neap/spring-tide

cycles. It appears that the linear megaripple field shows lower values of $\overline{U}_{max}$ especially during spring-tide than those of the sinuous megaripple field.

THE STRUCTURES

The net accretion of the megaripples occurs in the form of tidal bundles, bounded by pause planes. The latter are formed during the subordinate tide. They may be erosional or non-erosional, depending on the ability of the subordinate tidal currents to erode the crestal part of the resident bedform which has remained after the dominant tide. These pause planes represent the stand still periods of

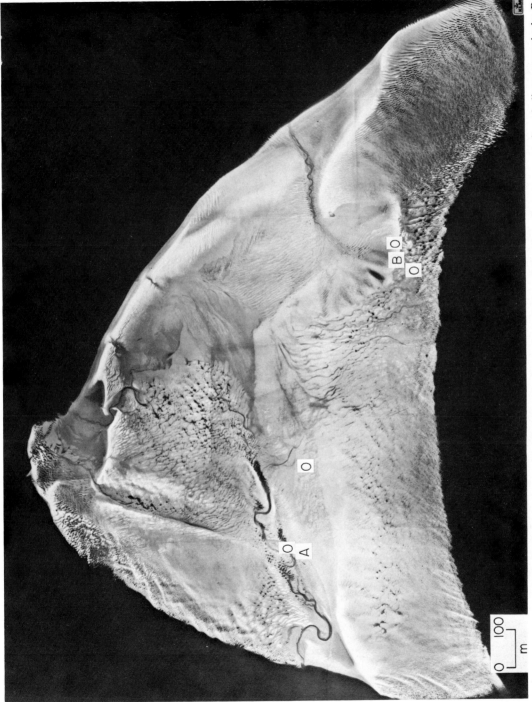

Fig. 2. Aerial photograph of the Ossenisse shoal in 1976 showing megaripple fields and location of measurement sites. A, linear megaripple; B, sinuous megaripple; O, laquer peel sections. (Photograph by **KLM Airocarto**, courtesy of Rijkswaterstaat.)

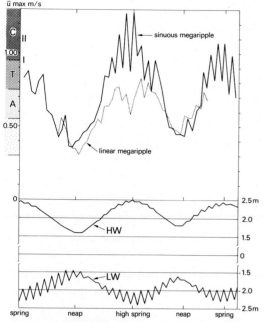

Fig. 3. The values of $\overline{U}_{max}$ in successive tides over a linear and sinuous megaripple field. Threshold values for angular, A, tangential, T, and concave, C, cross-bedding (explanation, see text).

the bedform. Boersma & Terwindt (1981) analysed in detail the structural properties of the tidal bundles and the pause planes. They distinguished three families of structures in the tidal bundle, namely the reactivation, the full-vortex and the slackening structures reflecting respectively the acceleration, the condition of a fully developed vortex during high flow velocities and the deceleration of the tidal flow (see also Terwindt, 1981). These structural associations make up a lateral sequence. In this paper we will restrict ourselves to the properties of the large-scale cross-bedding, produced during the migration of megaripples. The cross-laminated sets that are

formed by small-scale ripples in the first part of the acceleration and final part of the deceleration of the flow will not be considered further.

The cross-bedded sets

The cross-bedding in tidal bundles show a systematic change in foreset angle and form as illustrated in Fig. 4. The succession starts with angular cross-bedding, followed by tangential and concave cross-bedding and finally ends with sigmoidal cross-bedding.

Angular cross-bedding is defined here as consisting of straight foreset laminae with dip angles between 30° and 35°, making a sharp angle with the underlying sets. Tangential cross-bedding is made up of foreset laminae with dip angles between 25° and 30° merging into concave toesets. Concave cross-bedding consists of foreset laminae showing a gradual downward decrease in dip angle but having still a small angle between the toesets and the underlying sets. The sigmoidal cross-bedding shows no longer straight dipping foresets but instead a very gradual reduction of dip angle towards the toesets which are parallel with the lower set boundary.

The succession angular–tangential–concave foresets is essentially the same as described by Jopling (1963), based on his flume studies on a micro-delta. This succession was attributed (Jopling, 1963, see also the discussion in Reineck & Singh, 1973) to an increasing proportion of suspended load versus bed load in the sedimentation on the leeside of the megaripple and furthermore to a growing importance of the vortex and the associated strength of the back-flow with increasing current velocities.

Sigmoidal cross-bedding was attributed by Boersma & Terwindt (1981) to suspension fall-out occurring in a decelerating flow under conditions of high sediment load. Material derived from this settling suspension is draped as a blanket over the tangential or concave foresets. Sigmoidal sets on

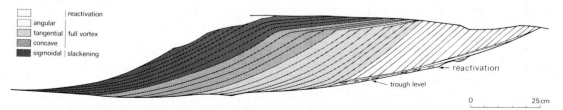

Fig. 4. Tidal bundle, made up of reactivation, full-vortex and slackening structures and showing the succession of angular, tangential and concave cross-bedding belonging to the full vortex structures and the sigmoidal cross-bedding belonging to the slackening structures (drawn after lacquer peel).

concave sets are the normal feature and sigmoidal sets on tangential ones are very rare.

The topsets

Topsets can be observed on some lacquer peels especially from sinuous megaripples (Fig. 4).

We may distinguish two different types of topsets:

topsets connected with the deceleration of the flow;
topsets associated with the full-vortex stage of the flow.

Boersma & Terwindt (1981) pointed out that during deceleration of the tidal flow material may fall out from highly loaded suspensions. This suspension is generated by the previous higher current velocities. There is some time lag between the reduction of the tidal flow and the adjustment of a new equilibrium in the sediment load. During this time sedimentation takes place producing nearly horizontal even laminated or convex topsets merging into sigmoidal foresets and toesets (Fig. 4). These deceleration topsets lie over the full-vortex structures and are further characterized by a very gradual transition in dip angle without any sharp knickpoint.

This type of deceleration topset is only observed when the maximum current velocity in the tidal cycle ($\overline{U}_{max}$) has exceeded 0·9 m sec^{-1}. This indicates that only under conditions of high velocities and high sediment load is the suspension fall-out great enough to produce topsets to such an extent that these may be preserved.

The second type of topset is associated with the full-vortex stage of the tidal flow because of their position in the tidal bundle. These topsets pass into foresets, belonging to the full-vortex structures and are confined between the reactivation and slackening structures. The full-vortex topsets are made up of very-low-angle laminae (6–10°) which pass with a sharp knick into steeper foresets (Figs 5, 6). Many topset laminae have a wedge-shape, starting somewhere in the middle of the topset and increasing in thickness towards the brinkpoint. Discontinuity of erosional surfaces can be observed in some places in the sets. Low-angle parallel lamination is present again on these erosional surfaces. The brinkpoint remains almost on the same level. The foreset laminae show a better demixing and a gradual coarsening towards the tangential toeset.

Some subtypes may be distinguished. Subtype 1 is illustrated in Fig. 5. It shows rather regular extending topsets in the direction of the migration. The brinkpoint lies relatively low and there are only a few discontinuity surfaces. This contrasts with subtype 2 which is characterized by a pattern which is discontinuous showing an alternation of extending topsets and erosional or non-depositional surfaces (Fig. 6). The brinkpoint is relatively high.

Subtype I, although of a larger scale, very much resembles Jopling & Walker's (1968, fig. 12) type B ripple drift cross-lamination. Their fig. 12 also shows fine-grained, wedge-shaped, low-angle topset laminae which pass over a more or less well-defined brinkpoint into coarser-grained steeper foresets. Similar properties are described by Stanley (1974) for small-scale silt draped ripples.

CURRENT VELOCITY THRESHOLDS (ASSOCIATED WITH THE DIFFERENT TYPES OF FORESETS AND TOPSETS)

We were able to connect the measured current velocities over the megaripple fields to particular parts of the lacquer peels by using different coloured tracers to mark the pause planes and to date the accretion of the bedforms. Thus threshold values for the mean current velocity in the vertical ($\overline{U}$) could be established for the foreset and topset types.

We found the thresholds shown in Table 1.

These threshold values are mean values with 5% scatter above and below. They apply to conditions of gradually varying tidal flow under relative small water depths (1–3 m) and mean grain sizes of the bed material ranging between 150–200 μm.

Table 1 indicates that full-vortex topsets occur in association with concave foresets. Furthermore deceleration topsets can be, but not necessarily should be, present over concave foresets and full-vortex topsets.

Table 1. Thresholds for $\overline{U}$ (m sec^{-1}) for foreset and topset types:

Set type	$\overline{U}$
Angular foreset	0·55
Tangential foresets	0·75
Concave foresets	0·95
Decelerating topsets	0·90
Full-vortex topsets	
Subtype 1	0·95
Subtype 2	1·10

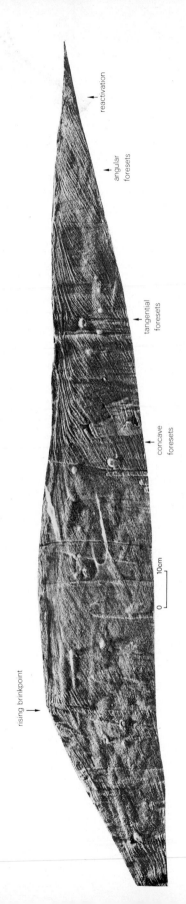

Fig. 5. Topsets subtype I passing into concave foresets. Note wedge-shape of topset laminae, increasing in thickness towards the brinkpoint and the erosional and non-depositional surfaces. Low-lying brinkpoint. Lacquer peel section.

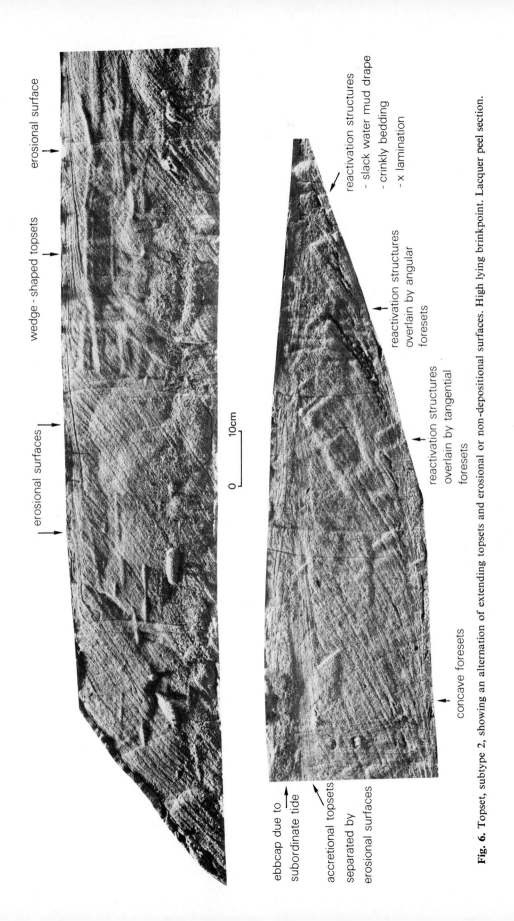

Fig. 6. Topset, subtype 2, showing an alternation of extending topsets and erosional or non-depositional surfaces. High lying brinkpoint. Lacquer peel section.

34   *L. H. M. Kohsiek and J. H. J. Terwindt*

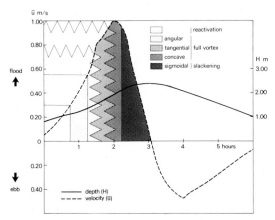

Fig. 7. Observed foreset and topset types in relation to the measured mean current velocity in the vertical ($\overline{U}$) and water depth (H) during a single tide in a sinuous mega-ripple field (loc B).

In a tidal cycle $\overline{U}$ may cross these thresholds producing the observed alterations in foreset and topset types (Fig. 7).

In the neap/spring-tide cycle there is a connection between the length of the tidal bundle and the set types within it. It turns out that there is a good correlation between the length of the tidal bundle (measured halfway between the upper and lower set boundaries) and the value of $\overline{U}_{max}$: the maximum current velocity during a tidal cycle (Fig. 8). Thus, going from neap- to spring-tide, $\overline{U}_{max}$ increases (Fig. 3) and the thresholds may be crossed successively. So we observed first relatively small tidal bundles with no full-vortex but only reactivation and slackening structures. Further in the neap/spring-cycle we observed small tidal bundles having angular cross-bedding. Then with increasing length of the tidal bundles we see angular foresets followed by tangential ones in the same bundle; the tangential foresets come more and more to the fore. Finally, only in some places where the spring-tide values of $\overline{U}_{max}$ are high enough we see concave foresets and topsets in the largest tidal bundles. In this situation the angular sets are usually rather thin, most probably because the threshold for the tangential foresets is reached soon after the turn of the tide.

In the neap- to spring-tide half cycle the $\overline{U}_{max}$ increases and also the effects of the deceleration of the tidal flow become more and more apparent in the tidal bundles. This is reflected in an increase in length of well-developed sigmoidal foresets and in some cases deceleration topsets.

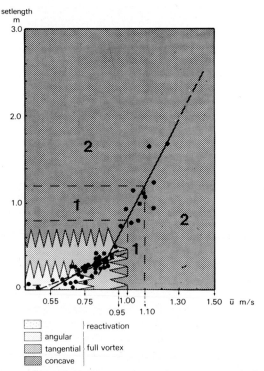

Fig. 8. The relation between the maximum current velocity during a tidal cycle ($\overline{U}_{max}$) and the length of the tidal bundle and the observed foreset types and topset subtypes (1 and 2).

A reverse succession is present going from spring- to neap-tide.

The succession described above is more or less ideal. In fact deviations do occur depending on local conditions, so sometimes these set types are surprisingly thin and even missing. Nevertheless the basic succession was established in almost all lacquer peels, so we have confidence in the applicability of the findings in tidal deposits generally.

POSSIBLE ORIGIN OF THE TOPSETS

We can only speculate on the origin of these topsets. We were unable to observe and measure the flow system and sand transport character over the stoss side, the crest and the leeside of these megaripples during varying flow.

Nevertheless we think that the type of megaripple in which topsets form differs from the well-known one. The latter has a rather straight stoss side, a well-defined brinkpoint and flow separation below the

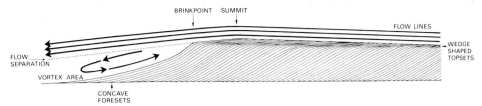

Fig. 9. Schematic representation of flow lines over a bedform during topset formation.

brinkpoint resulting in a vortex in the trough of the megaripple. This is the picture so well described by Jopling (1963, 1967), Allen (1968) and many others.

It is also a well-established fact that under high current velocities the megaripples show a more streamlined form (see, for example, Middleton & Southard, 1978, p. 7.21). The stoss side merges into a rather flat and rounded crest which more or less gradually passes into the inclined lee side. The brinkpoint does not coincide with the summit point but is located somewhat downstream.

It is further well known that flow separation occurs if the angle between the flow lines and the bed exceeds 6–8° (Rouse, 1961). The topsets in our lacquer peels make an angle with the horizontal of about 6–10°.

Combining these data we think that if the downward slope of the lee side of the crest between the summit and the brinkpoint is less than about 8° there will be no flow separation here.

In this case the flow lines are inclined to the crestline and follow the slope from summit to brinkpoint (Fig. 9).

So we think that during migration of this kind of megaripple topsets are generated under conditions in which the flow is still attached to the crest. Because the brinkpoint lies lower than the summit point, the water depth increases downstream of the summit and the current velocity will decrease. As a consequence material passing over the summit may be deposited on the lee side. This deposition will be greater towards the brinkpoint and this may explain the wedge-shape of the lamination (Fig. 5).

Flow separation occurs below the brinkpoint. Here we have the normal avalanching, foreset grading, etc.

In order to substantiate the above idea, a potential flow sheet analogue was used (see Rouse, 1961). The potential flow theory takes into consideration the influence of the gravity forces only and neglects the frictional forces. Although the analogue represents only a rough approximation of reality it may be applied to get a general idea of the order of magnitude of the reduction in velocity and sand transport near the bed in the area between the summit and the brinkpoint. This may indicate whether the described

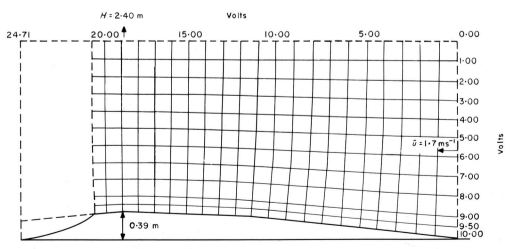

Fig. 10. Potential flow sheet analogue showing stream net over a megaripple indicating a reduction in near-bed current velocity from summit to brinkpoint.

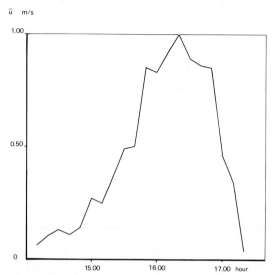

Fig. 11. Measured mean current velocites ($\overline{U}$) over the vertical during a tidal cycle showing fluctuations in the velocity.

bedforms and the sedimentation. The acceleration results in an increase in the transport capacity of the current. It takes time to fill up this transport capacity or in other words to reach an equilibrium in the vertical sediment concentration gradient. So there is always a time lag between the increase in velocity and the accompanying sand transport. This time lag is the reason why *over* the crest, although there is a slight deceleration due to an increase in water depth the sediment concentration is still not in equilibrium with the flow and is thus able to keep the sediment in the flow or even to erode some sediment.

During deceleration of the main flow the opposite tendency is present leading to conditions in which topset deposition can be expected. We think that the properties of topsets deserve more attention because they may be useful in palaeoflow reconstruction.

ACKNOWLEDGMENTS

The authors are very grateful to the hydraulic Research Section of Rijkswaterstaat, Vlissingen, for their great field assistance. The paper benefitted much from discussions with H. J. Geldof and J. R. Boersma.

The fieldwork was a part of a larger investigation on intertidal megaripples in which M. Brouwer, J. K. Lemkes, F. Berben and F. Steyaert participated. We thank them for their help in the field and during the data analysis.

The technical staff of the Geographical Institute, G. Ouwerkerk, Ph.L. Rieff and G. H. Huygen did a wonderful job in taking the lacquer peels and performing the drawings and photographs. L. M. Butteling and E. Gelderman are acknowledged for the typewriting.

process of topset deposition may be considered as a realistic possibility.

The boundary conditions for the analogue such as megaripple dimensions and form, water depth and mean current velocity in the vertical profile were derived from measurements in the field. The flow net was obtained by using the electric method. It is illustrated in Fig. 10. It turns out that the reduction of near bed flow velocity between summit and brink-point is in the order of 10 %.

This holds for maximum current velocity condition. The rate of sand transport is a power function of the velocity, so considerable reduction in the rate of sand movement may be anticipated in this area.

A further confirmation of the process described is needed by carefully conducted flume experiments.

As mentioned before in many topset units an alternation of topset deposition and a slight erosion or non-deposition can be observed in the lacquer peels (Fig. 6). This alternation may be attributed to acceleration (discontinuity or erosion) and deceleration (topset formation) of the main flow. Records of current velocities in a tidal cycle show many times that the velocity curve is not smooth but that minor fluctuations are superimposed on the general trend (Fig. 11). The time-scale of these fluctuations is of the order of quarters of hours and is therefore long enough to have an effect on the

REFERENCES

ALLEN, J.R.L. (1968) *Current Ripples*. North Holland, Amsterdam.
ALLEN, J.R.L. & FRIEND P.F. (1976) Changes in intertidal dunes during two spring-neap cycles, Lifeboat Station Bank, Wells-next-the-sea, Norfolk, England. *Sedimentology*, **23**, 329–346.
BOERSMA, J.R. (1969) Internal structure of some tidal megaripples on a shoal in the Westerschelde estuary, The Netherlands. *Geologie Mijnb.* **48**, 409–414.
BOERSMA, J.R. & TERWINDT, J.H.J. (1981) Neap-spring-tide sequences of intertidal shoal deposits in a mesotidal estuary. *Sedimentology*, **28**, 151–170.

BOOTHROYD, J.C. & HUBBARD, D.K. (1975) Genesis of bedforms in mesotidal estuaries. In: *Estuaries Research*, vol. II (Ed. by E. Cronin), pp. 217–234. London.

DALRYMPLE, R.W., KNIGHT, R.J. & LAMBIASE, J.J. (1978) Bedforms and their hydraulic stability relationships in a tidal environment, Bay of Fundy, Canada. *Nature*, **275**, 100–104.

GELLATHLY, D.C. (1970) Cross-bedded tidal megaripples from King Sound (NW. Australia). *Sedim. Geol.* **4**, 185–191.

JOPLING, A.V. (1963) Hydraulic studies on the origin of bedding. *Sedimentology*, **2**, 115–121.

JOPLING, A.V. (1967) Origin of laminae deposited by the movement of ripples along a streambed, a laboratory study. *J. Geol.* **75**, 287–305.

JOPLING, A.V. & WALKER, R.G. (1968) Morphology and origin of ripple-drift cross-lamination with examples from the Pleistocene of Massachusetts. *J. sedim. Petrol.* **38**, 971–984.

KLEIN, G.DeV. (1970) Depositional and dispersal dynamics of intertidal sand bars. *J. sedim. Petrol.* **40**, 1095–1127.

McCAVE, I.N. & GEISER, A.C. (1978) Megaripples, ridges and runnels on intertidal flats of the Wash, England. *Sedimentology*, **26**, 353–369.

MIDDLETON, G.V. & SOUTHARD, J. B. (1978) Mechanics of sediment movement. *Lecture Notes SEPM Short Course*, no. 3.

RAAF, J.F.M. De & BOERSMA, J.R. (1971) Tidal deposits and their sedimentary structures. *Geologie Mijnb.* **3**, 479–504.

REINECK, H. E. & SINGH, I.B. (1973) *Depositional Sedimentary Environment*. Springer-Verlag, Berlin.

ROUSE, H. (1961) *Fluid Mechanics for Hydraulic Engineers*. Dover, New York.

STANLEY, K.O. (1974) Morphology and hydraulic significance of climbing ripples with super-imposed micro-ripple-drift cross-lamination in lower Quaternary lake silts, Nebraska. *J. sedim. Petrol.* **44**, 472–483.

TERWINDT, J.H.J. (1981) Origin and sequences of sedimentary structures in inshore mesotidal deposits of the North Sea. In: *Holocene Marine Sedimentation in the North Sea Basin* (Ed. by S.-D. Nio *et al.*). *Spec. Publs int. Ass. Sediment.* **5**, 3–26. Blackwell Scientific Publications, Oxford, 524 pp.

Spec. Publs int. Ass. Sediment. (1981) **5**, 39–49

Berms on an intertidal shoal: shape and internal structure

J. R. BOERSMA* *and* J. H. J. TERWINDT†

*Institute of Earth Sciences, Budapestlaan 4, 3508 TA Utrecht, the Netherlands and
†Department of Physical Geography, Heidelberglaan 2, 3508 TC Utrecht, the Netherlands

ABSTRACT

On the intertidal part of the Ossenisse shoal, Westerschelde estuary, the Netherlands, elevated ridges or strips ('berms') appear which separate areas with different bedform configuration and hydrographic character. The ebb is the dominant tide on one side of the berm; the flood on the other. Ebb and flood currents make an appreciable angle with each other and supply material to the berm which gives rise to its elevated position. The different morphological appearances of berms, varying between a high and sharp negative step and a narrow dome-like elevation, are associated with specific internal structures, in which cross-bedding and cross-lamination participate in a way slightly different from ordinary sandwaves or dunes. Spring- and neap-tides can be distinguished in the typical berm structures by juxta- or superposition of cross-bedded or cross-laminated bundles sets and cosets. Often the cross-bedding rests on sloping faces.

Berms are expected to have a relatively high preservation potential.

INTRODUCTION

Sandy shoals occurring in inshore tidal environments such as estuaries and tidal flats often show more or less sharply delineated ridges which separate areas of different bedform configuration. These ridges, here called berms, appear on aerial photographs (see Fig. 1) as light-coloured strings, due to their elevated position, which permits quick drying-out as the tide falls. The height of these ridges found on the Ossenisse shoal, Westerschelde Estuary, the Netherlands, varies between a few decimetres and a metre.

In this paper we will describe berms from several locations on the shoal. We will focus on their external shape and configuration, the bedforms found at both sides of them and their internal sedimentary structures. We will attempt to correlate these features with the strength and direction of the tidal currents involved.

The Ossenisse shoal is located in the middle of the Westerschelde Estuary, about halfway inland

from the mouth (Fig. 2). The shoal measures about 60,000 m² at low water. It is surrounded to the south and to the north-west by a flood channel and to the north-east by an ebb channel. The tide in the Westerschelde is semi-diurnal. In the vicinity of the shoal the mean tidal range during spring-tide is 4·90 m and during neap-tide about 3·50 m. The average depth over the berm areas is about 2·5 m at high-water spring and about 0·5–1·0 m at high-water neap. Wave action, due to the limited fetch, is small and of a very subordinate effect as compared to the tides. Its effect tends to be restricted to a slight planing off of the large bedforms shortly before their ebb-emergence.

The mean grain size of the bed material is about 200 μm. Some mud may be encountered under neap-tide conditions on isolated spots, usually the central more elevated part of the shoal.

DESCRIPTION OF BEDFORMS

Apart from berms the Ossenisse shoal is populated by dunes, sandwaves and ripples. Dunes and

0141-3600/81/1205–0039 $02.00

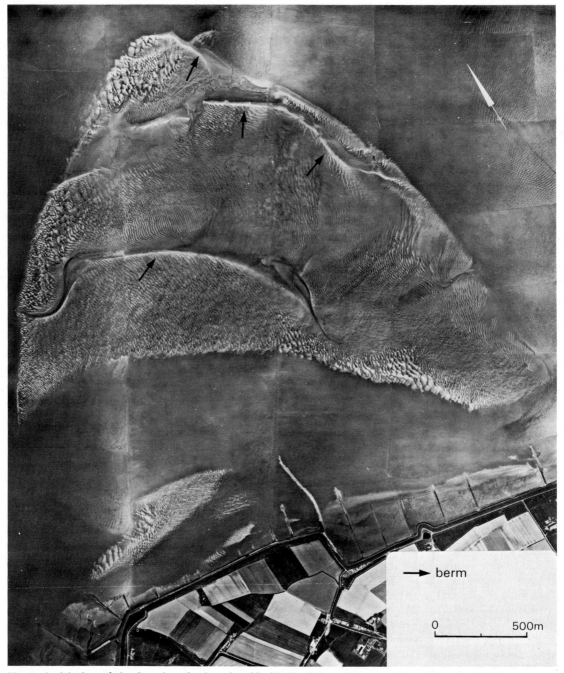

Fig. 1. Aerial view of the Ossenisse shoal, spring-tide (1970). Oblong dimension 3 km. Photo by **KLM** Aerocarto, courtesy of Rijkswaterstaat.

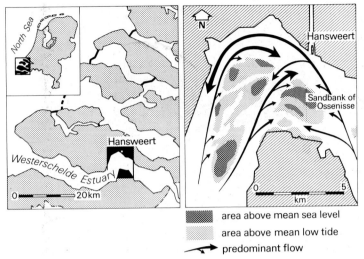

Fig. 2. Setting of the study area.

sandwaves are large-scale bedforms of essentially different size and shape and associated with different hydraulic conditions. According to Harms *et al.* (1975) sandwaves are characterized by a high length–height ratio and a continuous straight or weakly undulating crestline. They occur at current strengths below that of the dunes. Dunes tend to have a low length–height ratio and strongly winding, frequently interrupted crestlines. Their troughs often are inhabited by scour pits (e.g. Collinson, 1978). Ripples are found in the most elevated areas where low current strengths are unable to produce large-scale bedforms.

Sandwaves tend to produce cross-bedded sets extending many metres in which the effect of one tide is represented by a tidal bundle (Boersma, 1969). These tidal bundles are enclosed by pause planes marking the turn of the tide. The tidal bundles are thickest and longest around spring-tide; toward neap-tide they become thinner and shorter. Within each tidal bundle the effect of acceleration, full (vortex) stage and deceleration of the tidal flow is witnessed by specific structures (Fig. 3).

A neap–spring–neap cycle of bedform development and migration together with the internal structure left behind constitutes one of the major sedimentological characteristics on the shoal (Boersma & Terwindt, 1981). It will be demonstrated below that the changing hydrographic conditions over the neap/spring-tide period make themselves felt in the internal structure of the berm

in a way slightly different from that of the regular bed-forms.

DESCRIPTION OF THE BERMS

Location 1

This berm (Fig. 4) attains a height of about 1 m. Its major steep face, sloping less than the angle of repose, points north (flood direction), while a smaller superimposed face is directed southward. The area lying north to the berm is covered by ebb-directed sandwaves. The ebb current flowing parallel to the berm-foot tends to cause some erosion. A component of the ebb flow deflected south across the berm can be held responsible for a kind of wash-over at the berm crest which creates the superimposed steep flank mentioned above.

The higher elevated area to the south of the berm is governed by the flood. Current strength is relatively low as is demonstrated by the smaller size of the bedforms: small flood-orientated sandwaves and ripples.

In summary it may be stated that the present berm separates two areas of different current strength and direction. The ebb and flood currents run obliquely into each other, both piling up material on the berm. From this berm no record of sedimentary structures is available. The major steep flank of the present berm seems to be the effect of berm-front erosion by the ebb current flowing alongside and supply of material across the berm by the flood.

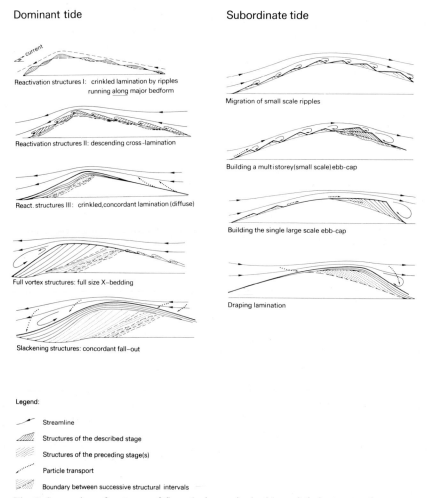

Fig. 3. Succession of patterns of flow during a single tide and their structural response.

Locations 2–4

Here an almost continuous berm is encountered with variable morphological features and flanked by areas of changing hydrographical character.

At location 2 the berm is sharp and steep. The height is 0·5–1 m. Sandwaves of flood origin coming from the south-west move at an angle of about 45° toward the berm crest where they contribute their material to an avalanche face. The ebb currents being weaker than the flood are largely active to the north of the berm where they move ripples in a direction essentially parallel to the berm. These ebb currents (as in location 1) are deflected by the berm which by effect is subjected to erosion along its foot.

At location 3 the berm has a less steeply sloping lee face (about 10°) while ebb and flood currents trend more opposite to each other than at location 2. The flood area to the south is inhabited by sand-waves only during a few days around spring-tide. In the remaining period this area has a mixed ebb-flood character at low flow strengths. Ripples are the resident bedform type here. To the north of the berm the ebb currents predominate which, however, in the immediate vicinity of the berm, do not produce bedforms, leaving a flat sloping flank probably subject to current erosion and wave action. Further away sandwaves are found largely below low tide level. During peak flooding the sandwaves coming from the south cross the berm and move downward over the steep ebb flank of the

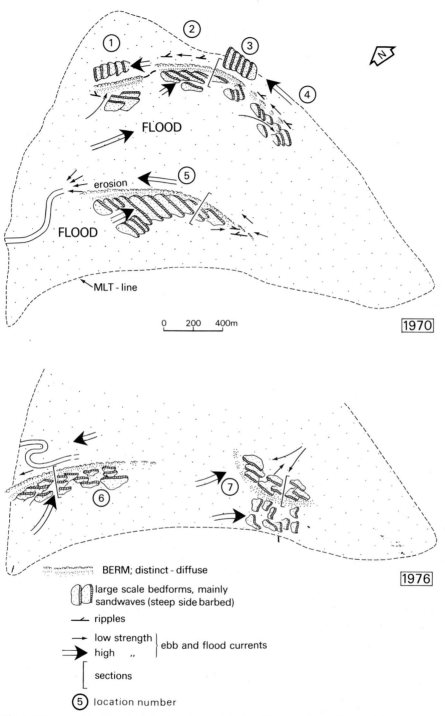

Fig. 4. Berms on the Ossenisse shoal and associated bedform configurations and flow pattern.

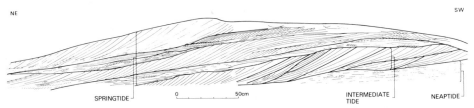

Fig. 5. Internal structure of the berm at location 2, characterized by x-bedded sets and x-laminated intervals resting on sloping bedding planes. Note the occurrence of two types of ∥-lamination; one being the topsets of x-bedded sets, the other separate intercalations between the sets. The latter is probably formed by wave uprush. (Drawing after lacquer peel.)

berm. In doing so they appear to retain their identity.

The internal structure of this berm as demonstrated in a lacquer peel (Fig. 5) is in accordance with the above picture of current systems and bedform morphology. It is characterized by flood-directed x-bedded* sets 'moving down' over sloping bedding planes. Characteristically, the sets are

* Throughout the paper, x-bedding is short for large-scale cross-stratification, x-lamination is short for small-scale cross-stratification, and ∥-lamination is short for even, parallel lamination.

separated by erosion surfaces associated with x-lamination and ∥-lamination. The x-lamination may be both ebb- and flood-directed. The ∥-lamination is possibly formed by wave uprush. Another type of ∥-lamination, constituting the topsets of x-bedded sets does occur. This type is identical to that encountered in ordinary dunes on the shoal. Its generation was attributed to deceleration of the flow toward the end of the dominant tide (Boersma & Terwindt, 1981).

Furthermore, on the lacquer peel of this berm three structural intervals resting on top of each

Fig. 6. Berm at location 5. Downstepping train of sandwaves.

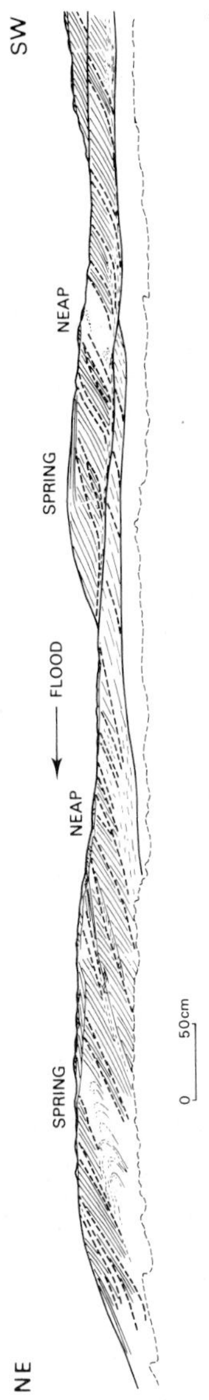

other can be seen. The lower interval is largely composed of x-lamination, made up by ripples moving in various directions in front of the berm. This interval represents neap-tide conditions. The second, middle interval shows the accretion of large-scale tidal bundles one in front of the other, thus building a kind of lateral sequence of accretionary stages by the dominant tide interrupted by pause planes marking the subordinate tide.

The upper interval is made up of a few superimposed sets produced by successive dominant tides. The subordinate tide builds thin bands with x- and ‖-lamination, which is largely found in the lower levels. Apparently during the subordinate tide berm crest is subject to erosion by the combined effect of slow currents and wave uprush which remove the material downslope. This upper interval represents spring-tide conditions.

At location 4 the above-mentioned berm gradually loses height and peters out. The current directions as deduced from the orientation of bedforms at both sides of the berm, appear to be more or less opposite. Current strengths are relatively low. Sandwaves existing to the south-west of the berm are relatively small and active only a few days around spring-tide. At the other side of the berm the smooth strip which separates the berm from the ebb-directed sandwave field is wider than at locations 2 and 3. Small ripples and an even surface, the latter produced by wave uprush, are the major features of this strip.

The gradual petering out of the berm is attributed primarily to the fact that the currents, instead of performing a flanking movement running obliquely into each other as is the case at the previous location, now come to run almost parallel to each other. Because of this the piling-up of material from both sides of the berm is greatly reduced.

Location 5

The configuration at location 5 is comparable with locations 3 and 4. The berm attains a height of approximately 2 m in the west, decreasing eastward. This considerable height, more than at locations 2 and 3, seems to be the combined effect of the higher current velocities at both sides of the berm and the considerable angle they make. The flood-directed sandwaves to the south are more pronounced and retain their shape and size throughout the whole neap–spring period.

The sandwaves are active 6–8 days around spring-

Fig. 7. Internal structure of the berm at location 5 made of a downstepping train of sandwaves. Each sandwave contains a lateral sequence of tidal bundles separated by pause planes. (Drawing after lacquer peel.)

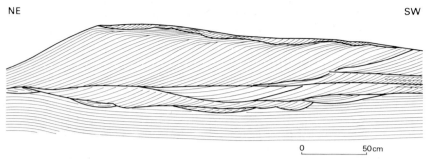

Fig. 8. Internal structure of the berm at location 6. Large-scale foresetting over entire berm height. Considerable accretion during a single (spring-) tide is witnessed by the long set along the top of the drawing. (Drawing after lacquer peel.)

tide. The ebb, being active to the north of the berm concentrates in a drainage channel which runs along the foot of the berm, where it performs some erosion. Deflection of the ebb current into a direction obliquely across the berm is witnessed by a recurving of the sandwaves near the berm crest. The relative importance of the ebb here is also demonstrated by the occurrence of ebb caps on top of the flood sandwaves.

The flood sandwaves seem to be the principal contributors to the berm. Their material is fed to the steep lee of the berm, where an avalanche face is built. This condition of large-scale avalanching goes for the western part of the berm of location 5. Eastward along the berm where it recurves southward, the mode of material supply changes, together with a reduction of the lee-side angle. Here a train of sandwaves arranged in a downstepping manner constitutes the berm (Fig. 6). This change in berm configuration from a single avalanche face to a

composite train of sandwaves is attributed to two factors. First, the less-confined action of the ebb, covering an area rather than a channel, thus causing less berm-foot erosion. Secondly, the reduced current strength going eastward entails smaller bedforms. It cannot be judged whether the change in angle between the ebb and flood currents plays an additional role.

The internal structure of the berm (see Fig. 7) made by the downstepping train of sandwaves appears to consist of a pile of long sets each being identically structured by tidal bundles. Individual bundles differ in size according to the diurnal inequality of the tides, and more important, the course of the neap–spring cycle. The pause planes between successive bundles are usually not associated with x-lamination which indicates that either the ripples were not active or that erosion prevailed during or after their activity. In view of the slightly erosive character of the pause planes we surmise

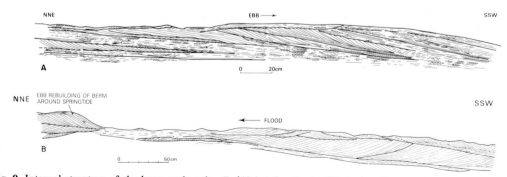

Fig. 9. Internal structure of the berm at location 7. (A) is taken in the 'fan' extending southward over a dune field. It shows an alternation of ebb-directed x-bedded sets and x-lamination sets resting on sloping faces. The x-bedding belongs to spring-tide conditions and the x-lamination to the neap-tide. (B) is taken at the place where the lower-lying dune field climbs and peters out against the fan. Note the considerable set length produced by a single flood during spring-tide. Sections (A) and (B) lie approximately in one line, (A) ± 20 m to the north-east from (B).

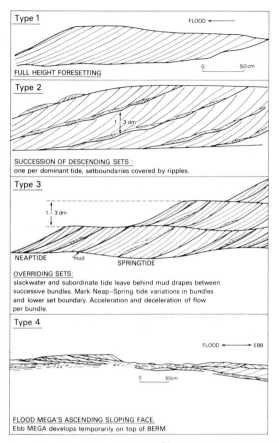

Fig. 10. Schematic representation of the four different types of berm structures encountered on the Ossenisse shoal.

that the latter condition was the case. From the fact that superimposed sets do not erode each other, or hardly so, it can be concluded that the sets 'climb' along horizontal paths away from the sloping face of the berm.

The main distinguishing feature of this type of berm deposits as compared to that of ordinary sandwaves is the arrangement in several, almost identical superimposed sets. Another feature may constitute the local occurrence of slumped bundles. For further details, see Fig. 7.

Location 6

This berm is flanked to the south by a field of dunes which are propelled by strong flood currents. These dunes feed their material into an avalanche face. Leeward of the steep face to the north of the berm an ebb-governed drainage channel exerts some

erosion at the foot, and removes or throws back part of the material supplied by the flood. The ebb currents are relatively weak.

Here, unlike most other locations, the two competing current systems are more or less at right angles to each other. The internal structure of this berm (Fig. 8) clearly shows the large-scale foresetting over the entire berm height and the considerable setlength produced during one flood tide.

Location 7

This berm is of an unusual type. It expresses itself as a broadly shouldered elevation which coincides with the front of a 'fan' extending southward over an upsloping surface (see Fig. 1). The 'fan' appears to be ebb-governed. It bears bedforms of the sandwave type which are best visible around spring-tide.

Around neap they are almost obscured, as the depressions between the sandwaves are filled in by ripples. The area to the south of the fan is flood-dominated and covered with irregularly crested dunes, which retain their identity over the neap-spring period. The flood largely parallels the fan's front; a side-branch of it is deflected up the fan. This takes place during the late stages of the rising tide when the shoal is completely submerged and current velocities are decreasing again.

The internal structure of this berm (Fig. 9) is unique in the way that cross-bedding is directed toward the berm front from both sides. On the ebb-dominated 'fan' (Fig. 9A) gently sloping x-bedded sets formed during a few days around spring-tide are found to be enclosed between x-laminated layers, which are related to the longer neap-tide period. During these days around spring-tide the berm re-erects itself while it is subject to degrading in the remaining period.

On the flood area to the south of the fan flood-directed x-bedding resting on ascending surfaces is the main structure encountered in a lacquer peel (Fig. 9B). The sets tend to peter out upslope and be replaced by x-lamination of the berm foot area.

As with the other berms two important conditions are fulfilled here: the ebb and flood currents make a considerable angle and take material to the berm from both sides.

This description has revealed two major differences with the berms already dealt with above, namely:

(1) both tides instead of only one produce large-scale structures;

(2) the foot of the berm is not a site of erosion as in most other cases but of deposition.

CONCLUSIONS

Flow pattern, bedform shape, size and orientation as well as internal structure point to the fact that berms develop at the boundary between ebb- and flood-dominated areas.

Conditions favouring their occurrence are:

(*a*) the ebb and flood currents should make a considerable angle (90–180°) with each other. The more pronounced berms develop in areas where the ebb and flood perform a kind of obliquely flanking movement (locations 1, 2, 5 and 6). Strictly opposing currents on the other hand usually produce only a slightly elevated strip (location 4);

(*b*) current strength should exceed the threshold for the formation of dunes or sandwaves at least on one side of the berm.

Berms become more pronounced where erosion is taking place at their foot.

The internal structures of berms may be of four types (Fig. 10):

(1) foresetting over the full height of the berm (at places where supply as well as foot erosion is high);

(2) x-bedded sets descending gently sloping faces (at places of intermediate strength of both current systems);

(3) x-bedded sets overriding each other along more or less horizontal planes (at places where the supply system is powerful but foot erosion is weak);

(4) x-bedded sets approaching the berm from two sides, along ascending faces.

The internal structure of berms differ from that of the ordinary bedforms in the following aspects:

(*a*) they usually show sloping sets of x-bedding and cosets of x-lamination, while ordinary bedforms give rise to essentially horizontal ones. Where the berm accretional sets are horizontal they occur in multistorey repetitions of almost identical sets in which the neap-spring-tide cycle is represented in the lateral sequence. In the latter aspect no difference exists with the ordinary sandwaves;

(*b*) the erosion surfaces separating successive descending sets are usually straight and low angled which contrast with the spoon-shaped lower set boundaries of the ordinary tidal bedforms.

Spring- and neap-tide effects are equally well discernible in the berms as well as in the bedforms. They reflect themselves in a high proportion of x-bedding and x-lamination respectively.

In those instances that berms give rise to foresetting over the full height of their steep lee face it may be difficult to distinguish them from ordinary dunes.

The situation in our study area with berms prograding into depressions suggest that they have a relatively high preservation potential.

ACKNOWLEDGMENTS

The authors wish to express their gratitude to all who contributed to this study. The technical assistance of Rijkswaterstaat Vlissingen is kindly acknowledged. During the field survey many students helped us much in making observations and lacquer peels. Field and laboratory assistance was kindly provided by G. H. Ouwerkerk and drawing, photography and typewriting by Ph. L. Rieff, G. H. Huygen, T. Lekkerkerker and E. Gelderman.

REFERENCES

BOERSMA, J.R. (1969) Internal structure of some tidal megaripples on a shoal in the Westerschelde Estuary, The Netherlands. *Geologie Mijnb.* **48**, 409–414.

BOERSMA, J.R. & TERWINDT, J.H.J. (1981) Neap–spring tide sequences of intertidal shoal deposits in a mesotidal estuary. *Sedimentology*, **28**, 151–170.

COLLINSON, J.D. (1978) Alluvial sediments. In: *Sedimentary Environments and Facies* (Ed. by H.G. Reading), pp. 15–60. Blackwell Scientific Publications, Oxford. 576 pp.

HARMS, J.C., SOUTHARD, J.B., SPEARING, D.R. & WALKER, R.G. (1975) Depositional environments as interpreted from primary sedimentary structures and stratification sequences. *Lecture Notes, S.E.P.M. Short Course 2.*

Spec. Publs int. Ass. Sediment. (1981) **5**, 51–64

Ripple, megaripple and sandwave bedforms in the macrotidal Loughor Estuary, South Wales, U.K.

T. ELLIOTT *and* A. R. GARDINER

Department of Geology, University College of Swansea, Singleton Park, Swansea SA2 8PP, Wales, U.K.

ABSTRACT

The macrotidal Loughor Estuary in South Wales exhibits a wide range of ripple, megaripple and sandwave bedforms which, with one exception, are produced by tidal currents. The bedforms occur in ebb- and flood-oriented fields and are extensively exposed on channel floors, channel flanks and the intervening sand-bars following an ebb tide. *Ripples* are small scale bedforms which are either asymmetrical linguoid forms produced by the tidal currents, or straight crested symmetrical or asymmetrical forms produced by local wind-generated waves. *Megaripples* are intermediate scale bedforms which occur in two classes distinguished on a range of criteria: low amplitude, long span, two-dimensional forms devoid of scour pits (Type 1); and higher amplitude, shorter span, three-dimensional forms with scour pits (Type 2). *Sandwaves* are substantially larger, two-dimensional bedforms divisible into rippled and megarippled types. This suite of bedforms closely resembles that described from the Bay of Fundy, though there are several significant differences.

In addition, the megaripples and sandwaves display a range of modification features which result from several scales of unsteadiness in the tidal flow. Recognition of these features is an essential prerequisite to considering the bedforms as reflectors of the principal flow conditions.

INTRODUCTION

Flume studies have made significant contributions to bedform theory, particularly with respect to lower flow regime transverse bedforms. However, in recent years numerous researchers have stressed certain limitations of flume studies, such as their use of steady flow conditions and the small scale of experimental flow systems, which limits flow depth and the number of bedforms under observation (Allen, 1973; Jackson, 1975; Middleton & Southard, 1978). Increased awareness of these limitations has renewed the stimulus for studying bedforms produced by larger-scale, unsteady flows in natural settings. Sand-dominated estuaries have proved popular in this respect since in intertidal areas bedforms are exposed, albeit briefly, after each ebb tide. Most studies of this type have been based in microtidal (< 2 m tidal range) or mesotidal (2–4 m) estuaries and inlets, with

the Bay of Fundy providing the only macrotidal example (> 4 m). Ripples and megaripples (or dunes) have been described from numerous estuaries (van Straaten, 1950; Klein, 1970; Allen & Friend, 1976b; Boothroyd, 1978), and larger-scale transverse bedforms termed sandwaves which have not been recognized in flume studies have been described from the Bay of Fundy (Klein, 1970; Dalrymple, Knight & Lambiase, 1978). Observations in these settings have contributed towards the understanding of bedform production by tidal currents, and attempts have been made to define flow regime schemes for tidal current bedforms (Boothroyd & Hubbard, 1975). The extent to which problems of unsteady flow have been considered in previous studies varies. Klein (1970) and Knight (1972) have described a number of features attributed either to waning ebb flow conditions or flow reversal, which slightly modified the appearance of the bedforms. Allen & Friend

0141-3600/81/1205-0051 $02.00

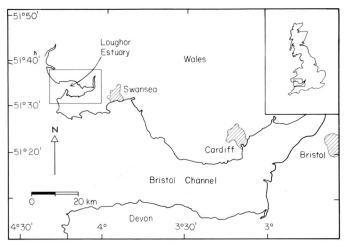

Fig. 1. Location map for the Loughor Estuary with the area of Fig. 2 outlined.

(1976a,b) addressed the problem in terms of the 'relaxation time' of dune or megaripple bedforms and the bedform population dynamics under a given tidal regime. Tidal currents are characterized by several scales of unsteadiness and complete understanding of the bedforms requires that the full range of effects is recognized.

This paper presents the initial results of research into tidal current bedforms of the macrotidal Loughor Estuary, South Wales, concentrating on descriptions of the bedform types and the range of features due to unsteady flow.

THE LOUGHOR ESTUARY

The Loughor Estuary (or Burry Inlet) is located on the northern side of the Bristol Channel in a drowned valley partially filled by unconsolidated Pleistocene and Recent deposits (Fig. 1). The estuary is broadly funnel-shaped, but is constricted at the mouth by a beach-spit which extends from the southern side of the estuary. Behind the beach-spit on the landward side is an extensive area of supratidal and intertidal flats traversed by tidal channels and creeks which drain northwards into the main part of the estuary (Fig. 2). The estuary *sensu stricto* comprises a series of east–west trending ebb- and flood-dominated channels separated by extensive sand-flats or sand-bars. The main channel occupies a central position in this complex except towards the estuary mouth where it is deflected northwards by the beach-spit. At high water the entire area (45 km²)

is inundated, but at low water the bar top areas, channel flanks and the floors of relatively shallow channels are exposed, leaving only the deeper parts of channels covered by water. Channels in the lower part of the estuary have an average depth of 6 m below local mean sea-level though towards the mouth of the estuary the main channel deepens to 15 m. The intervening bar top areas range from 1 to 4 m below mean sea-level. Seaward of the estuary mouth is a well-developed ebb-tidal delta, and the area immediately landward of the mouth bears some resemblance to a flood-tidal delta. Hayes (1976) considers that estuaries with conspicuous ebb- and flood-tidal deltas occur exclusively in mesotidal settings, but the tidal range in the Loughor Estuary falls well within the macrotidal range (see below). Hubbard, Oertal & Nummedal (1979) argue that well-developed ebb-tidal deltas form where the ebb flow is countered by incoming refracted waves. This interpretation seems more applicable to the study area, though it seems probable that the fundamental, overriding control on the development of ebb-tidal deltas is the constriction and subsequent expansion of tidal flow which occurs where estuary mouths are confined by beaches or barrier islands.

Tidal processes predominate in the estuary and on the tidal flats. The volume of the tidal prism consistently exceeds discharge from the rivers entering the estuary by several orders of magnitude, thus denying a significant role for fluvial processes and causing the estuarine waters to be permanently well mixed and laterally homogeneous (Moore, 1976). Waves are of limited importance as they approach

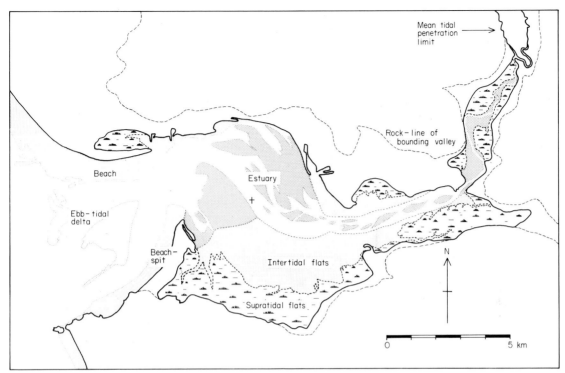

Fig. 2. The Loughor Estuary, including the wave-dominated beach-spit, the supratidal and intertidal flats, and the estuary *sensu stricto* with its ebb-tidal delta. Based on 1:50,000 Ordnance Survey sheet no. 159 (1974 edition) and with reference to aerial photograph coverage. + denotes the station for the hydrographic data in Fig. 3.

primarily from the south-west and are therefore dissipated on the seaward faces of the beach-spit and ebb-tidal delta (Fig. 2). The few long-fetch waves which enter the estuary break against the front of the most seaward bars. Wave action is otherwise limited to short-period, locally generated waves which periodically affect the shallow waters covering the bar top areas and intertidal sand-flats.

The tides are semi-diurnal with slight diurnal inequality. Mean tidal range values of 4·0 m during neap periods and 8·1 m during spring periods place the estuary permanently in the macrotidal range. Hydrographic work by Moore (1976) and Collins, Moore & Al-Ramadhan (1981) demonstrates that the tidal curve is generally symmetrical near the estuary mouth, but becomes progressively asymmetrical towards the head and has an accentuated asymmetrical form throughout the estuary at spring periods. Near the mouth salinity maxima and minima coincide with high and low water respectively, whereas ebb and flood current velocity maxima occur approximately mid-way between low and high water (Fig. 3). This is characteristic of a standing tidal

wave (Dyer, 1973) which is presumably inherited from the standing tidal wave which occurs on the northern side of the Bristol Channel (Heaps, 1968; Moore, 1976). Towards the head of the estuary the tidal wave is modified slightly and acquires certain characteristics of a progressive wave. The distance from the estuary mouth to the tidal penetration limit of mean spring tides is 16 km. Current velocity data

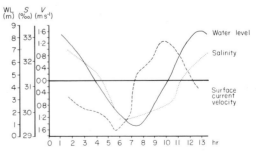

Fig. 3. Tidal wave characteristics of the Loughor Estuary surveyed during a spring tide at a site slightly landward of the estuary mouth; modified after Collins *et al.* (1981).

are limited at present, but surface current velocities of 0·7 m s⁻¹ (ebb) and 0·9 m s⁻¹ (flood) during neap periods, and 1·7 m s⁻¹ (ebb) and 1·2 m s⁻¹ (flood) during spring periods have been measured (Collins *et al.*, 1981). Due to the high tidal range, rates of water level fluctuation attain values of 1·5 m h⁻¹ for water level fall and 2·0 m h⁻¹ for water level rise during spring periods.

The sediment in the estuary is considered to be thoroughly reworked fluvioglacial outwash material derived from offshore and driven landwards during the Holocene transgression and later (Bridges, 1975). In the main area of bedform development immediately landward of the estuary mouth the sediment is a very well-sorted, fine-grained quartz sand with local accumulations of whole and fragmented shells providing the only coarser material.

BEDFORM TYPES

Following ebb tides, exposed areas of the estuary can be seen to be ornamented by an extremely diverse assemblage of bedforms present as distinct bedform fields. Each field is composed of bedforms of a certain type and orientation, and is separated from neighbouring fields by narrow zones (10–20 m wide) of hybrid bedforms. Repeated visits to selected areas near the estuary mouth reveal that these bedform fields are permanent, but not static, features of the estuary. In response to contrasts between neap and spring tidal flow, fields progressively change their internal composition and shift their boundaries. Bedforms in the vicinity of field boundaries are therefore frequently in long-term disequilibrium and are considered separately.

At present there is only a partial consensus on the definition and terminology of lower flow regime transverse bedforms in tidal settings. General agreement exists that the main types are ripples, megaripples (or dunes) and sandwaves, but the definition of each class, in particular sandwaves, often varies between different sets of researchers. In this paper the scheme proposed by Dalrymple *et al.* (1978) for intertidal bedforms in the Bay of Fundy is adopted since it uses a diverse array of factors to discriminate between bedform types and is therefore preferable to the main alternative schemes which only use wavelength (e.g. Boothroyd & Hubbard, 1975). The terminology applied to individual bedforms is generally that of Allen (1968, pp. 59–66), though the term wavelength is preferred to chord as a description of the horizontal distance between successive troughlines or crestlines.

Direct observation of bedforms is naturally limited to periods following ebb tides, but it appears from observations made during these periods that only ripple bedforms reverse with each successive tide whilst megaripples and sandwaves retain their orientation. Critical evidence in support of this assertion is provided by exposed flood-oriented megaripples and sandwaves which are virtually unmodified by the ebb flow, suggesting that these bedforms at least do not reverse. With regard to ebb-oriented megaripples and sandwaves, their internal structures contain no features suggestive of repeated bedform reversal, and distinctive features of the bedforms persist between ebb tides enabling individual bedforms to be identified easily despite the intervening flood tides. This contrasts with evidence from several broadly comparable settings where it appears that megaripples reverse with successive tides and only sandwaves retain their orientation (Boothroyd & Hubbard, 1975; Dalrymple, Knight & Middleton, 1975; Boothroyd, 1978). The probable explanation for this contrast lies in the degree of difference between dominant and subordinate tidal flows at given sites which may be greater in the study area than in other areas, causing bedform reversal to be limited to the smallest class of bedforms.

Ripples

Ripples are the smallest bedforms observed. Generally they are ebb-oriented, asymmetrical, linguoid forms which undoubtedly reverse with the tide. They exist in association with larger bedforms (see below), but also form extensive fields in their own right on topographically high sand-flat areas of the intertidal bars. Larger-scale bedforms are probably absent because these areas do not experience maximum tidal currents due to the mid-tide timing of these events (Fig. 3). During maximum ebb current velocities the sand-flats have already been exposed, whilst during maximum flood current velocities they have not yet been inundated.

During periods of moderate to strong winds the sand-flats are covered by straight-crested, short-wavelength, symmetrical and asymmetrical ripples. These ripples are produced by local, short-period, wind-driven waves generated in the shallow waters which cover the sand-flats around high-water period.

On sand-flat areas the ripples are often partially

Fig. 4. Typical appearance of Type 1 megaripples (ebb-oriented) with superimposed linguoid ripples. Note the straight crests, constant crest height and lack of scour pits. Trowel 0·39 m long. In the background is an aeolian dune system capping the beach-spit at the mouth of the estuary.

Fig. 5. Typical appearance of Type 2 megaripples (ebb-oriented). Note the sinuous crests, fluctuating crest height, lee-side spurs and well-developed scour pits.

Table 1. Dimensions of megaripple and sandwave bedforms together with values for vertical form index (wavelength divided by height) and horizontal form index (span divided by wavelength). Single figures refer to mean values, bracketed values state the range of values. Based on detailed measurements of 65 representative bedforms to date. The values for megarippled sandwaves are stated as ranges only due to the small sample number at the time of writing. Discrimination between bedform types does not rely solely on these data, but also includes a range of morphological features—see text for full discussion

Megaripples	Height (m)	Wavelength (m)	Span (m)	Vertical form index	Horizontal form index
Type 1	0·19 (0·08–0·36)	5·32 (2·23–10·65)	52·02 (7·20–140·0)	31·52 (8·7–76·7)	10·0 (2·3–23·9)
Type 2	0·38 (0·15–0·66)	6·86 (2·75–10·65)	21·49 (3–56)	20·10 (8·57–59·16)	3·61 (0 6–13·68)
Sandwaves					
Rippled	0·46 (0·3–0·63)	9·68 (6·6–12·15)	72·09 (34–96)	22·61 (14·6–35·7)	7·65 (3·74–11·97)
Megarippled	0·82–1·25	25–30	80–140(+)	—	—

overprinted by the burrowing activities of *Arenicola marina*.

Megaripples

Megaripples are intermediate in scale between ripples and sandwaves and are developed on channel floors, channel flanks and some bar top areas. They vary considerably in scale and exhibit an extremely wide range of forms. By using the same criteria as Dalrymple *et al.* (1978) it is currently felt that the variety of megaripples can be grouped into classes corresponding to their Type 1 and Type 2. However, lateral transitions between these types are commonly observed in the study area and a more rigorous study of their form may therefore provide evidence of a continuum rather than discrete classes.

Type 1 megaripples

Type 1 megaripples are relatively simple, two-dimensional forms with low amplitude, moderate wavelength and long span (Fig. 4, Table 1). They are generally straight-crested, but occasionally include minor components of sinuosity superimposed on this general trend. Crest height remains constant along the span and there are no well-developed scour pits in the trough. Slipfaces dip at angles of 26–30° and have a simple form devoid of spurs except where the bedforms have been modified by a change in flow direction as the water level falls (see below). Superimposed bedforms are limited to ripples arranged in ripplefan patterns on the stoss sides (see below).

Many of the sandwaves of other workers closely resemble Type 1 megaripples (e.g. Boothroyd & Hubbard, 1975; Boothroyd, 1978).

Type 2 megaripples

Type 2 megaripples are more three-dimensional, lunate forms with higher amplitude and shorter span (Fig. 5, Table 1). Crest height varies substantially along the span and the slipfaces generally exhibit pronounced spurs which project down-current. Between the spurs are well-developed scour pits which cause the troughline to fluctuate considerably in height. Slipfaces dip at 26–30° and attain heights of 0·7–0·8 m in scour pits. Superimposed on the stoss surfaces are well-developed ripplefans localized around the scour pits (see below). At times the upper parts of the stoss surfaces are also littered with smaller-scale, isolated megaripples. Often these superimposed megaripples have limited positive relief above the general level of the stoss surface, but instead are characterized by deep, localized scour pits with angle-of-repose slipfaces and discrete ripplefans (Fig. 6). They appear abruptly during the approach to spring periods and persist for several tidal cycles. They migrate faster than the bedform on which they are superimposed, and eventually intersect and locally deflate the main slipface. At present it is not known whether the two scales of megaripples are active at the same time, or whether the smaller-scale forms appear after the main slipface has ceased to migrate (Jackson, 1975; Dalrymple *et al.*, 1978; Allen & Collinson, 1974). These features closely

Fig. 6. Small-scale ebb-oriented megaripples superimposed on the upper stoss surface of an ebb-oriented Type 2 megaripple which has its slipface in the foreground. The superimposed megaripples are characterized by deep localized scour pits, but limited positive relief.

Fig. 7. Typical appearance of sandwaves (flood-oriented). Note the scale of the bedforms (compare with Figs 4 and 5), the simple two-dimensional form and the lack of scour pits. Type 2 megaripples superimposed on the upper stoss surface have been partially planed off by the preceding ebb flow. In the foreground is an ebb apron formed from slight reworking of the bedform crest.

resemble the 'minor dunes' of Allen & Friend (1976b) which developed on the upper stoss surfaces of tidal current dunes (megaripples) as the rate of sediment transport increased during spring periods. The appearance of these minor bedforms was felt to herald changes in the bedform population.

Sandwaves

Sandwaves appear to be confined to the flanks and floors of flood-dominated channels and to a lesser extent ebb-dominated channels, and are the largest-scale bedforms observed in the study area (Table 1). They are high-amplitude, long-wavelength, long-span bedforms with a relatively simple, two-dimensional form. Their crests are straight or of very low sinuosity and maintain a constant height along the span. Slipfaces are devoid of spurs, and scour pits are not developed in the troughs (Fig. 7). In keeping with observations in the Bay of Fundy by Dalrymple et al. (1978), sandwaves in the study area appear to fall into two classes: a relatively small-scale set with superimposed ripples (termed rippled sandwaves); and a larger-scale set with Type 2 megaripples superimposed on the upper stoss surfaces of the sandwaves (termed megarippled sandwaves; Table 1). However, despite close similarities in the sandwaves from these areas there are also several important differences. In the Bay of Fundy megarippled sandwaves only occur in sediments coarser than medium sand, whereas in the Loughor Estuary they are developed in fine sand (0·2–0·25 mm). The lee sides of sandwaves in the Bay of Fundy dip at angles of 10–20° which are substantially below angle-of-repose values. This was attributed to the effects of reversing tidal flows, though the orientation of the sandwaves was not stated (Dalrymple et al., 1978). In the Loughor Estuary the majority of the sandwaves are flood-oriented yet their slipfaces have steep angle-of-repose values (26–30°) comparable to those of ebb-oriented megaripples. This may reflect a more pronounced difference between dominant and subordinate tides in this sandwave field than is the case in the Bay of Fundy example. A final difference is that rippled sandwaves in the Bay of Fundy do not form stable bedform fields in their own right and are considered to be a disequilibrium, decaying form of megarippled sandwaves (Dalrymple et al., 1978). In the Loughor Estuary rippled sandwaves constitute a stable bedform field in a shallow, flood-dominated channel and do not appear to prograde or retrograde into another bedform type.

OBSERVED EFFECTS OF UNSTEADY FLOW

Tidal processes contain several scales of unsteadiness of which the most important are: (1) fluctuating depth and velocity within ebb and flood periods; (2) flow reversal associated with successive ebb and flood periods; (3) variations between neap and spring tidal flow; and (4) seasonal variations in tides between equinox and solstice (the equinoctial cycle). At a point in time individual bedforms and the overall bedform population may display features due to one or more of these scales of unsteadiness. If the bedforms are to be considered as quasi-equilibrium forms reflecting principal flow and sediment transport conditions it is imperative that the effects of unsteadiness are subtracted as far as possible. The following section describes the range of features associated with (1–3) of the above scales of unsteadiness in the study area. At present no information is available on bedform response to the equinoctial cycle.

Modification features due to fluctuating velocity and depth within ebb and flood periods

For megaripples and sandwaves modification processes commence when flow velocity falls below the value required for active slipface migration as ebb or flood currents wane from peak flow conditions. Waning flood periods involve a progressive reduction in flow velocities accompanied by an increase in flow depth. In contrast, during waning ebb periods the decrease in flow velocities is accompanied by decreasing flow depth. In intertidal areas this culminates in emergence and often includes a brief period of high flow velocities as the flow shallows immediately prior to emergence. For convenience the term 'falling stage' will be used when referring to these flow changes. Since direct observation of the bed is limited to low-water periods after ebb flows the following descriptions relate to modification features produced during the falling stage of ebb periods.

(1) Skewed spurs

Some Type 1 megaripples are distinctive by virtue of possessing sharp-crested spurs which project downcurrent from the slipface at an angle of up to

Fig. 8. Ebb-oriented Type 1 megaripple with sharp-crested spurs attached to the slipface at an angle of 45° and mantled by ebb-oriented falling-stage ripplefans. The spurs result from a change in flow direction during the falling stage which causes helical flow cells to develop in front of the slipface—see text for full explanation. Trowel 0·28 m long.

Fig. 9. Ripplefan associated with an ebb-oriented Type 2 megaripple which has its slipface in the bottom left corner. Note the continuous, arcuate-crested ripples near the slipface (locally overprinted by an interference ripple pattern) which pass downcurrent into linguoid ripples. Trowel 0·28 m long.

45° (Fig. 8). The spacing between the spurs ranges from 1 to 2 m and they persist for 2–3 m downcurrent before fading out. Very slight scour pits occur between spurs and contain ripplefans which either surmount the crests of the spurs or die out near the crests. Megaripples exhibiting these skewed spurs occur on bar top areas and channel flanks. Where they are developed on channel flanks there is a consistent relationship whereby the slipfaces are directed across the channel flank at a fine angle whilst the spurs are parallel to the channel axis.

Previous descriptions of these distinctive megaripples include Allen (1968, pp. 78–80), Knight (1972) and Boothroyd (1978, fig. 18c), though the latter does not discuss the spurs. Allen (1968, pp. 271–279) performed a series of experiments in which a rigid, skewed slipface was imposed across the floor of a flume. This revealed that the scours and spurs were created by pairs of contra-rotating vortices with helical-spiral motion which were aligned parallel to the flow and therefore at an angle to the slipface. However, the origin of skewed slipfaces in natural settings was not discussed and the bedforms were considered to be unmodified, equilibrium forms. In contrast, Knight (1972) argued that the skewed vortices result from a change in flow direction during the falling stage of the ebb flow. This interpretation seems more appropriate to the examples under discussion and the following explanation is proposed. Early, deeper flows propagate the slipfaces across the channel flank, then as the water level falls the flow direction changes (? abruptly) as the influence of channels and local slopes increases. Vortices develop in front of the skewed, moribund slipface and produce evenly spaced swept areas and spurs. Finally, the ripplefans are superimposed as the water level falls further (see below), and the bedforms emerge.

(2) Falling stage ripplefans

Exposed megaripples in the study area exhibit ordered patterns of asymmetrical ripples termed ripplefans (Allen, 1968, pp. 84–87) superimposed on their stoss surfaces (Fig. 9). In Type 2 megaripples the ripplefans are localized in spur-defined scour pits, whereas in Type 1 they are more extensive along the trough, though still with an overall fan shape. In both cases ripples near the trough-line have continuous, arcuate crestlines which fan away from the slipface of the main bedform. As the stoss side of the downcurrent bedform is ascended these ripples break down to form a field of linguoid ripples which normally extends to the brinkpoint. Actively migrat-

ing megaripples may have superimposed ripplefans radiating from the point or line along which the flow reattaches to the bed in the lee of the bedform. However, two observations suggest that the exposed ripplefans reflect falling stage conditions rather than active bedform migration. First, the point from which the ripples emanate is always located close to the base of the slipface, suggesting that the separation eddy had contracted substantially. Secondly, where the major bedforms occur on a sloping channel flank the ripplefans are skewed downslope, reflecting increasing influence of the slope as the water level fell.

(3) Planed-off crests

The crest regions of megaripples and sandwaves sometimes appear as plane-bed surfaces devoid of ripple bedforms, or with scattered patches of ripples. These surfaces can be seen to form during a brief period of plane-bed sediment transport in the shallow waters above the bedform crest immediately prior to bedform emergence. Pre-existing ripple bedforms are largely obliterated as rapid deflation of the bedform crest takes place. Planed-off megaripples have also been recognized by Boothroyd & Hubbard (1975) and Dalrymple et al. (1978), though the former allocate them a specific position in a flow regime scheme for intertidal bedforms and therefore presumably consider them to be a separate class of bedform rather than a modified version of an existing bedform type.

(4) Dissection channels and micro-deltas

The slipface spurs of Type 2 megaripples are sometimes cut by narrow, steep-sided channels trending across the spurs. Sediment eroded by these channels is redeposited at the mouth of the channel as a steep-fronted micro-delta which progrades into the adjacent scour pit. These couplets form preferentially on sloping channel flanks during late-stage run-off as water-filled scour pits drain downslope. Type 1 megaripples and sandwaves on sloping surfaces exhibit extremely shallow, micro-braided channel systems which extend along the troughline for considerable distances, reworking the upper few centimetres of sediment in the trough (see Fig. 10).

(5) Strandlines

Slipfaces of megaripples and sandwaves are sometimes etched by closely spaced parallel lines which reflect slight reworking of the slipface by successively lower water levels (Klein, 1970). These features are

Fig. 10. Small-scale, ebb-oriented apron on the crest of a flood-oriented rippled sandwave—main bedform orientation to the right, ebb apron orientation to the left. Note also the shallow micro-braided channel system extending along the troughline of the bedform. Spade 0·97 m long.

Fig. 11. Decaying form of ebb-oriented Type 1 megaripples covered by linguoid ripples with the slipface of the mega-ripple just discernible in places. Trowel 0·28 m long.

particularly common on the lower slipfaces of Type 2 megaripples as pools of slightly agitated water are trapped in the scour pits as the water level falls.

Modification features due to flow reversal

As noted earlier, bedform reversal in the study area is limited, and fields of flood-oriented megaripples and sandwaves are exposed in several areas following the ebb tide. These bedforms generally exhibit slight rounding of the crests accompanied by small, asymmetrical, ebb-oriented aprons (Fig. 10), as described by Klein (1970). These features reflect the limited reworking ability of the subordinate ebb tide. The aprons occur as localized, low-angle mounds a few centimetres high where the ebb flow is extremely weak, or as more continuous, 0·05– 0·15 m slipfaces where the ebb flow is less subordinate.

Bedform changes due to contrasts between neap and spring conditions

Megaripple fields in the study area change their composition and shift their boundaries in response to contrasts between neap and spring tidal flow conditions. These changes generate a suite of bedforms which differ from those described above and are frequently ignored in studies of this type. Two types are worthy of mention: decaying megaripples produced within bedform fields, and partly reversed and symmetrical megaripples produced in boundary zones between fields of opposed direction.

(1) Decaying megaripples

During the neap–spring cycle megaripples within fields retain their orientation, but wax or wane according to whether conditions become more or less favourable for the bedform type. During unfavourable periods even the dominant tide may become incapable of sustaining formerly active megaripples. Under these conditions Type 1 megaripples increase in wavelength, diminish in height and therefore become less distinct. In some cases continued sediment transport via ripple bedforms assists in the decay of these bedforms and they degenerate into low-amplitude 'megaripples' which are entirely covered by linguoid ripples (Fig. 11). In these examples the remnant of a slipface can just be discerned, but in places it is absent due to ripples having broken through the slipface position. Type 2 megaripples decrease dramatically in height and become

simple, weakly developed asymmetrical forms with lee-side angles of 9–11°.

(2) Partly reversed and symmetrical 'megaripples'

In response to contrasts in flow over the neap–spring cycle, megaripples in the vicinity of field boundaries gradually change over several semidiurnal tidal cycles as the fields shift. Megaripples near boundary zones separating fields of opposed direction often lose their asymmetry and gradually reverse. Partly reversed megaripple-scale bedforms are common in these zones. The degree of reversal ranges from forms with ebb aprons (see earlier) to composite forms in which bedforms of opposed direction are superimposed. In the latter case the troughline of the superimposed bedform is located some distance in front of its toeline and the low-angle surface between them is the remnant stoss surface of the earlier, oppositely oriented bedform. Low-amplitude forms devoid of slipfaces and with symmetrical to quasi-symmetrical profiles also occur in these zones. These forms are produced during periods when ebb and flood flows are equal or nearly equal, and the ebb flow at least is incapable of sustaining active megaripples.

In other study areas characterized by a greater equivalence between dominant and subordinate tidal flows these bedform types can also result from flow reversal during successive ebb and flood tides. Partly reversed megaripples produced in this way have been described previously (Klein, 1970; Knight, 1972), but the occurrence of these and symmetrical bedforms in discrete boundary zones between fields of opposed direction has not been documented before. Although the boundary zones are narrow at a point in time (10–20 m), their composite extent over a neap–spring cycle is much greater and their contribution to the total bedform population can therefore be substantial.

DISCUSSION

The macrotidal Loughor Estuary exhibits a wide range of ripple, megaripple and sandwave bedforms which with one exception (wave-produced ripples) are the product of tidal currents operating in the estuary. The bedforms bear a close resemblance to those of the Bay of Fundy recently described by Dalrymple *et al.* (1978), but differ in several important respects. This is particularly true of the sandwaves, which in the Loughor Estuary form in finer sediments and have steeper slipfaces. In the Bay of Fundy, Dalrymple *et al.* (1978) demonstrated a substantial overlap between Type 2 megaripples and megarippled sandwaves in terms of mean tidal current speed and water depth. Mean sediment grain size was the only factor which was felt to be unique for each of these bedforms, with sandwaves being restricted to sediments coarser than 0·308 mm medium sand. No physical reasoning was offered in explanation of this point, but the sandwaves from the Loughor Estuary clearly contradict its generality. Local factors in the Bay of Fundy may be required to explain this observation, rather than there being a fundamental grain-size control underlying the distinction between Type 2 megaripples and sandwaves.

Comparisons of information on tidal current bedforms from different study areas will ultimately provide further insights into bedform genesis by tidal currents. However, such insights are only valid if the modifying influences of flow instability or unsteadiness can be identified, and their effects subtracted from observations of bedform characteristics. Megaripples and sandwaves exposed at low water following an ebb tide present evidence for the peak, productive flow conditions which is masked by a range of features produced by unsteadiness of the flow system. In intertidal settings the effects of the falling stage of the ebb flow and flow reversal can be assessed with comparative ease. However, the effects of longer-term neap–spring variations in tidal flow are more difficult to discern, particularly when the study area is visited and sampled intermittently. Tidal current bedforms in subtidal areas will also be modified to some degree by the effects of unsteady flow. The range of features will not be as great because the bedforms do not emerge, but the effects may still be substantial. Increased awareness gained from studies of exposed intertidal bedforms may therefore also assist studies of sub-tidal bedforms.

ACKNOWLEDGMENTS

Thanks are due to the Department of Geology, University College of Swansea for the provision of equipment, and Llanelli Borough Council and the harbour-master at Burry Port for mooring facilities. Advice and information provided by Dr M. B. Collins (Department of Oceanography, University College of Swansea) assisted the study, and discussions with other researchers at the I.A.S. Conference at Texel during September 1979 proved extremely beneficial.

REFERENCES

ALLEN, J.R.L. (1968) *Current Ripples: their Relation to Patterns of Water and Sediment Motion.* North-Holland, Amsterdam. 433 pp.

ALLEN, J.R.L. (1973) Phase differences between bed configuration and flow in natural environments, and their geological relevance. *Sedimentology,* **20,** 323–329.

ALLEN, J.R.L. & COLLINSON, J.D. (1974) The superimposition and classification of dunes formed by unidirectional aqueous flows. *Sediment. Geol.* **12,** 169–178.

ALLEN, J.R.L. & FRIEND, P.F. (1976a) Relaxation time of dunes in decelerating aqueous flows. *J. geol. Soc. London,* **132,** 17–26.

ALLEN, J.R.L. & FRIEND, P.F. (1976b) Changes in intertidal dunes during two spring–neap cycles, Lifeboat Station Bank, Wells-next-the-Sea, Norfolk (England). *Sedimentology,* **23,** 329–346.

BOOTHROYD, J.C. (1978) Mesotidal inlets and estuaries. In: *Coastal Sedimentary Environments* (Ed. by R.A. Davies Jr), pp. 287–360. Springer-Verlag, New York. 420 pp.

BOOTHROYD, J.C. & HUBBARD, D.K. (1975) Genesis of bedforms in mesotidal estuaries. In: *Estuarine Research, Vol. II, Geology and Engineering* (Ed. by L. E. Cronin), pp. 217–234. Academic Press, New York.

BRIDGES, E.M. (1975) Geomorphology of the Burry Inlet. In: *Problems of a Small Estuary* (Ed. by A. Nelson-Smith and E. M. Bridges), pp. 1:2/1–1:2/14. Quadrant Press, Swansea.

COLLINS, M.B., MOORE, N.H. & AL-RAMADHAN, B.L. (1981) Some aspects of the physical oceanography of the Loughor Estuary. In preparation.

DALRYMPLE, R.W., KNIGHT, R.J. & LAMBIASE, J.J. (1978) Bedforms and their hydraulic stability relationships in a tidal environment, Bay of Fundy, Canada. *Nature,* **275,** 100–104.

DALRYMPLE, R.W., KNIGHT, R.J. & MIDDLETON, G.V. (1975) Intertidal sand bars in Cobequid Bay (Bay of Fundy). In: *Estuarine Research, Vol. II, Geology and Engineering* (Ed. by L.E. Cronin), pp. 292–307. Academic Press, New York.

DYER, K.R. (1973) *Estuaries: a physical introduction.* Wiley, London. 140 pp.

HAYES, M.O. (1976) Morphology of sand accumulation in estuaries: an introduction to the symposium. In: *Estuarine Research, Vol. II, Geology and Engineering* (Ed. by L.E. Cronin), pp. 3–22. Academic Press, New York.

HEAPS, N.S. (1968) Estimated effects of a barrage on tides in the Bristol Channel. *Proc. Instn civ. Engrs,* **40,** 495–509.

HUBBARD, D.K., OERTAL, G. & NUMMEDAL, D. (1979) The role of waves and tidal currents in the development of tidal-inlet sedimentary structures and sand body geometry: examples from North Carolina, South Carolina and Georgia. *J. sedim. Petrol.* **49,** 1073–1092.

JACKSON, R.G. II (1975) Hierarchical attributes and a unifying model of bedforms composed of cohesionless material and produced by shearing flow. *Bull. geol. Soc. Am.* **86,** 1523–1533.

KLEIN, G. DE V. (1970) Depositional and dispersal dynamics of intertidal sand bars. *J. sedim. Petrol.* **40**, 1095–1127.

KNIGHT, R.J. (1972) Cobequid Bay sedimentology project: a progress report. *Mar. Sedim.* **9**, 45–60.

MIDDLETON, G.V. & SOUTHARD, J.B. (1978) *Mechanics of Sediment Movement*. Short Course No. 3. Society of Economic Paleontologists and Mineralogists, Tulsa.

MOORE, N.H. (1976) Physical oceanographic and hydrological observations in the Loughor Estuary (Burry Inlet). In: *Problems of a Small Estuary* (Ed. by A. Nelson-Smith and E. M. Bridges), pp. 1:3/1–1:3/15. Quadrant Press, Swansea.

VAN STRAATEN, L.M.J.U. (1950) Giant ripples in tidal channels. *Tijdschr. K. ned. aardrijksk. Genoot.* **67**, 76–81.

Spec. Publs int. Ass. Sediment. (1981) 5, 65–80

Sediment transport by tidal currents and waves: observations from a sandy intertidal zone (Burry Inlet, South Wales)

P. A. CARLING*

Department of Oceanography, University College of Swansea, Singleton Park, Swansea, U.K.

ABSTRACT

Measurements of tidal current velocities and suspended sediment concentrations have been made at fifteen stations on a sandy intertidal zone in South Wales, U.K. These data have been interpreted in terms of tidal circulation and sediment transport.

Self-recording current meters yielded continuous data on tidal current velocities at 0·50 m above the bed, for periods of up to one month. In particular, measurements of tidal current speed made simultaneously at a series of points in a vertical plane, provided information on the structure of the boundary layer.

In general, velocity data fit a logarithmic distribution of current speed with height above the bed. Coefficients of determination better than 0·80 are achieved for 81 % of the time during flooding tides and 75 % of the time during ebbing tides. Deviations from the model occur primarily, close to the beginning and the end of a tidal half-cycle.

Throughout most of a tidal cycle, variations in suspended sediment concentration and grain-size distributions within the water column are related to the first-order fluctuations in tidal current velocity and the shear stress exerted at the sediment–water interface. Generally, average sediment concentrations are reduced exponentially as a function of height above the sediment–water interface. However, an exponential relationship is inadequate for limited periods. Close to low water, wind–wave-induced resuspension, producing reversed concentration gradients, is a more influential factor when compared with tidal current scour. Conversely, gravitational settling dominates close to high water, producing 'bulges' in the instantaneous vertical concentration gradient.

For those periods dominated by tidal current resuspension, close agreement exists between observed sediment transport rates and transport estimated from time-integrated stream-power values. At stations where no suspended sediment concentration data are available, sediment transport may be estimated from stream-power relationships.

Conclusions drawn from this investigation should be capable of extrapolation to other similar intertidal zones.

INTRODUCTION

Investigations of contemporary intertidal water circulation and sediment transport assist in the understanding of processes of sedimentation and aid in the construction of type-models of stratigraphy. As an example, basic models of *on-shore to off-shore* motion of intertidal water masses and the associated

* Present address: Freshwater Biological Association, The Ferry House, Far Sawrey, Ambleside, Cumbria, U.K.

0141–3600/81/1205–0065 $02.00

sediment transport have been developed (Postma 1961, 1967; Groen, 1967); these in turn may be related to simplified stratigraphic models (e.g. Straaten, 1950; Bouma, 1963; Evans, 1965). Few investigators have been concerned with the complex spatial variations in water circulation (Boon, 1975; Pethick, 1971; Bayliss-Smith et al., 1979).

In the Wash, Eastern England, Evans (1965) proposed that a rectilinear tidal circulation system normal to the coast over the higher intertidal flats was replaced by a confused circulatory system over the seaward intertidal flats (Evans, 1965, fig. 3). This complexity was related to the proximity of a rectilinear system parallel to the coast in the subtidal channel off-shore. Amos (1974) extended this investigation and showed that intertidal vectors on both the flood and the ebb are progressively normal to the coast along an onshore transect. In contrast, Kestner (1975) produced a hypothetical model with the flow lines normal to the coast in intertidal creeks but progressively parallel to the coast on the adjacent flats.

Inglis & Kestner (1958), Kestner (1963, 1976), Evans & Collins (1975) and Collins, Amos & Evans (1981) have examined suspended sediment transport and intertidal circulation in the Wash. Recently, McCave & Geisler (1979) have described the dynamics of intertidal bedforms and have derived estimates of bedload flux. There are, however, few data concerning the hydraulics of the intertidal boundary layer (Parker, 1971; Hamilton, 1979). Similarly, sediment transport equations derived from laboratory investigations and fluvial systems have not been tested in the intertidal environment. This is in contrast to a substantial body of literature related to the shallow marine environment; notably continental shelf seas (e.g. McCave, 1974; Sternberg, 1968; Dyer, 1980).

In this paper, the results of an integrated investigation of tidal circulation, hydraulics and sediment transport are presented. Comparisons are drawn with existing models and information on intertidal sediment dynamics where appropriate.

The area chosen, in many hydrodynamic and sedimentological characteristics, is similar to intertidal zones elsewhere on the European continental shelf. The results presented should therefore be pertinent to the interpretation of sediment dynamics in other similar areas.

Physical setting

The Burry Inlet, the estuary of the River Loughor (South Wales, U.K.) has a total area of some 42 km² (Fig. 1). Salinity characteristics (Moore, 1976) indicate that the estuary is vertically and sectionally homogeneous, with only minor freshwater inputs in relation to the tidal excursion. The estuary is a macrotidal environment (cf. Hayes, 1975) with a mean spring tidal range of 8 m and a mean high water mark at 7 m above Admiralty Chart Datum

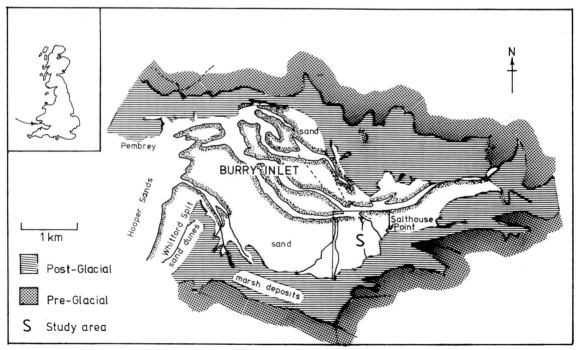

Fig. 1. Location of the study area—Burry Inlet, S. Wales, U.K.—showing intertidal sandflats and marshes which are Flandrian post-glacial deposits.

(ELW). Depths at mean high water are only 4–5 m over the intertidal banks and marginal flats. At extreme low water springs the estuary effectively dries out. During other tidal states, depths in subtidal channels may reach 15 m.

Intertidal flats and salt marshes, with an overall gradient of 0·06, reach a maximum breadth of 4 km on the southern side of the estuary. Deposits, including those of the marshes, consist predominantly of well-sorted fine sands (modal grain size 125 μm, $\sigma = 0·66$). Variations in sediment facies are indistinct in comparison with the Wash embayment, on the east coast of England, and only broad physiographic differences may be deduced (Table 1).

The presence of sponge spicules, shell fragments and echinoid spines in the sediment record indicates a marine origin for much of the intertidal deposits. Freshwater inputs of sediments to the estuary, at present of the order of 4–5 tonnes day^{-1}, are insignificant in the overall sedimentary regime. ^{14}C dates and pollen analysis of preserved brackish-water deposits in the sedimentary record indicate a dominance of marine influence since 5500 yr BP (Carling, 1978).

The Whitford Point sand-dune complex and the subtidal Hooper Sands to the west inhibit the penetration of swell waves from Carmarthen Bay into the estuary. Consequently, the local wave climate is controlled by local wind patterns and, with the progression of the semi-diurnal tide, variable fetch.

METHODS

Fifteen stations were occupied on the intertidal zone within a 2 km^2 study area (Fig. 2). For comparative purposes data were collected from one station in the subtidal system.

At stations 5–15 Plessey self-recording current meters were deployed mounted on scaffolding rigs at 0·50 m above the bed.

Data on current speed and direction recorded on magnetic tape at 10 min intervals were filtered to remove wind-run and erroneous data. Useful data were collected for 196 spring tides and 150 neap tides (springs > 6·4 m at Milford Haven, Admiralty Tide Tables, neaps < 6·4 m at Milford Haven), during 1975 and 1976.

Direct-reading Braystoke meters were used at stations 1–4 and 16. Velocity records were made at surface, mid-depth and 0·5 m above the bed throughout 12 tidal cycles.

Boundary-layer flow measurements were made at stations 1 and 11 over six spring and neap tidal cycles in August 1976 and February 1977. A velocity-gradient rig consisting of four self-aligning Braystoke meters logarithmically spaced 0·08, 0·24, 0·44 and 0·77 m above the bed was used. Data were recorded from digital read-out every 40 sec.

Suspended sediment samples were collected every 15 min throughout tidal cycles at a level of 0·06 m above the bed. Automatic water samplers and a tidal depth recorder were mounted on light-alloy towers (Collins, 1975). During calm weather, data on tidal-current-induced suspensions were collected for 28 tidal cycles at various current metering stations. These data were supplemented by sampling throughout the water column at additional reference levels using a horizontal 'Nansen-type' sampler. Further samples were collected during unsettled meteorological periods when wave activity resuspended the sediment.

Concentration data were determined by filtration through 0·45 μm pore-size filter pads (Eaton, Likens & Bormann, 1969) and grain-size distributions determined using a Coulter Counter Model TA II (Swift, Schubel & Sheldon, 1972).

Movement of sand as a traction load was estimated using fluorescent sand as a tracer (Ingle, 1966).

Table 1. Comparison of the physiographic sub-environments of east and west coast (U.K.) intertidal zones

East coast—Wash embayment	West coast—Burry Inlet (this study)
Butterwick Low (Evans, 1965). Facies normal to the coast	Facies normal to the coast
Equivalent to	
(a) Salt marsh	(1) Primary marsh
	(2) Secondary marsh
(b) Higher mudflat	Absent
(c) Inner sandflat	(3) High sandflat
(d) *Arenicola* sandflat	*Arenicola* zone indistinct
(e) Lower mudflat	Absent
(f) Lower sandflat	(4) Lower sandflat Equivalent in position seaward to (e) and is characterized by increased gradient to seaward
(g) Subtidal channel	(5) Subtidal channel

N.B. (e) is commonly absent in areas other than Butterwick Low (McCave & Geisler, 1979).

Fig. 2. Schematic representation of the intertidal zone with a summary circulation pattern. Vector axes are scaled 1 mm equivalent to 10 cm sec⁻¹. A low current speed zone exists in the vicinity of station 9. Notably the marsh sward (pecked line) reaches its most seaward (northerly) extent in this area and in similar low current speed areas to the east and west. Scale: 6 cm equals 1 km.

Wave height, period and length were recorded using a wave staff. Azimuths of wave trains were determined by compass sightings. Because of the limitations of the method, data were averaged for successive wave trains over 5 min periods. Salinity, water temperature and wind-speed data were collected as required.

Full details of data collection and reduction are described elsewhere (Carling, 1978).

RESULTS

Tidal circulation pattern

Individual results are not discussed for each station. Vector plots, based on all the data for individual stations, summarize the data concisely (Fig. 2). In Fig. 2 the mean vector flow directions are not differentiated for neap and spring tides, but the major axes of individual vector plots may be taken as representative of average tidal streamlines.

During the flood tide, flow directions are rotatory in character at stations to seaward (stations 11 and 12). This pattern reflects the strong lateral shear between easterly up-estuary flow and the flow on to the intertidal zone to the south. As the flood tidal current speed initially strengthens in the subtidal system to the north, current vectors at stations on the seaward flats are aligned east to south-east. Subsequent slackening of the up-estuary currents, during the latter part of the flood tide, results in flow being diverted on to the intertidal zone, and vectors swing to the south. This process is less evident in the intertidal creeks where sedimentary banks (stations 13 and 14) direct flow to the south throughout the flood tide.

To landward (station 9), flow divergence occurs; current vectors in the creek (station 8) become rectilinear (southerly) as tidal waters are channelled and accelerated into the narrowing marsh creeks to landward. On the high sandflats flow directions (south-easterly) are less variable than to seaward as the effects of the subtidal current vectors are reduced (stations 9 and 10). To the landward of the divergence zone current speeds remain low, average current speeds (at 0·5 m above the bed) being less than 10 cm sec^{-1} and reaching a maximum of 30 cm sec^{-1}; this is in contrast to 25 cm sec^{-1} and a maximum of 45 cm sec^{-1} at station 10 and values of 15 and 50 cm sec^{-1} at station 8.

During the ebb tide the gross circulation pattern of the flood tide is reversed and a zone of tidal convergence occurs to seaward of station 9. However, tidal current vectors have poor directional stability on the ebb tide compared with the flood. The broad spread of ebb current directions at stations 9 and 10 is partially related to a progressive swing of current direction from north-west to north as the offshore energy gradient becomes steeper than the gradient toward the creek. Nevertheless, erratic direction changes are common and may be related to the presence of an adverse pressure gradient

during ebbing tides, this is in contrast to flooding tides which propagate against the slope of the intertidal zone.

In terms of maximum current speeds, flood currents dominate over ebb currents in the seaward portions of tidal creeks and across the sandflat. Flood dominance is reduced to landward in the creeks as water that flowed on to the marsh surface directly from the sandflat during the flood is evacuated during the ebb via the creek system. This is demonstrated in Fig. 2, where the ebb vector for station 5 is directed to the north-west into the creek system. The situation arises whereby, during spring tides, the ebb discharge in the creek system is commonly up to 26% greater than the flood discharge.

Within the overall framework, the (easterly) longshore residuals of velocity components increase offshore; by 20% between stations 10 and 11 on the sandflat. However, within the creek system, e.g. stations 13 and 14, onshore to offshore motion dominates over longshore residuals.

Boundary layer

The bottom shear stress (τ_0) and friction velocity (U_*), related in the form

$$U_* = (\tau_0/\rho)^{0.5} \qquad (1)$$

are basic to many practical evaluations of sediment transport (Graf, 1971). Time-averaged values of U_* or τ_0 may be derived where a logarithmic distribution of velocity with depth exists. In a turbulent flow where the depth Z is much greater than z_0, a suitable relationship may be written as

$$\ln Z = (\chi/U_*)\bar{U}_z + \ln z_0, \qquad (2)$$

where $\bar{U}_z$ is the mean velocity at a height Z above the bed, χ is the Karman coefficient (0·40 in sediment-free water), z_0 is a function of both the roughness length describing the bed topography and the roughness Reynolds number ($R_r = U_*\kappa_s/\nu$).

U_z values (time integrated over 40 sec periods) for each of the four reference levels in each profile were fitted with the respective heights above the bed (Z) to equation (2), using a least-squares procedure and the iterative method of Inman (1963). Coefficients of determination (r^2) indicate the quality of fit achieved by the regression. Shear velocities evaluated from the gradients of logarithmic profiles were adjusted for small changes in the Karman coefficient in sediment-laden water (McCave, 1974). A mean boundary

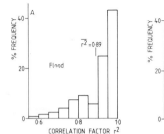

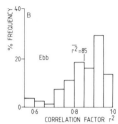

Fig. 3. (A) Histogram of distribution of r^2 values. Flood tide 4 February 1977. (B) Histogram of distribution of r^2 values. Ebb tide 4 February 1977.

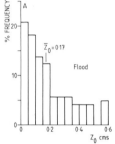

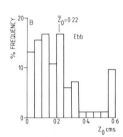

Fig. 4. (A) Histogram of distribution of z_0 values. Flood tide 4 February 1977. (B) Histogram of distribution of z_0 values. Ebb tide 4 February 1977.

roughness was calculated for flood and ebb tides using the relationship $\kappa_a' = 30\bar{z}_0$ (Nikuradse, 1933).

Velocity profiles with r^2 values greater than 0·80 were regarded as within acceptable velocity deviation limits to derive U_* values. In the example given (Fig. 3), typical of spring tides at station 1, the occurrence of logarithmic profiles with r^2 values better than 0·80 corresponds to a frequency of 81% of the flood tide duration (Fig. 3A). Sixty-six per cent of occurrences were within the range 0·90 to 1·00. During ebb tides the frequency distribution is more widely spread, with r^2 values better than 0·80 for 75% of the time, and only 42% of these are within the range 0·90–1·00 (Fig. 3B). This reflects an increase of non-logarithmic profiles during ebbing tides (cf. Collins *et al.*, 1981).

z_0 values obtained from the y intercept in the regression analysis are a measure of the notional height above the bed where the velocity is zero. A mean z_0 value for the flood tide (Fig. 4A) is 0·17 cm, therefore steep velocity profiles and consequently low shear velocity values might be expected. During ebbing tides (Fig. 4B) z_0 values are distributed over a greater range than is observed in the case of the flood. However, an average value is 0·22 cm. Very high (erroneous) values of z_0 occur close to slack

water when profiles are non-logarithmic; these are not represented in Fig. 4. A mean z_0 value of 0·19 cm for the tidal cycle detailed above yields a κ_a' value of 5·7 cm, which describes the order of magnitude of roughness of the bed. This value subsumes the influence of grain roughness, ripple and bar geometry on the characteristics of the tidal flow.

Shear velocities calculated from the slope of the velocity profiles change throughout a tidal cycle in sympathy with tidal current velocity near the bed, $\bar{U}_{50}$ (Fig. 5). During the flood tide maximum shear velocities reach 3·5 cm sec⁻¹ approximately 1½ hr before high water, and are maintained at other times at values in the range 2·0–3·5 cm sec⁻¹. During ebb tides maximum shear velocities reach 5·0 cm sec⁻¹.

A detailed analysis of the relationship between bed roughness, drag coefficients and suspended sediments will be presented in the near future (Carling & Handyside, in preparation).

Suspended sediments

The modal grain size of deposits on the sandy portion of the intertidal zone (125 μm) was found to be in suspension when the shear velocity on the bed exceeded ~2·0 cm sec⁻¹ (Carling & Handyside, in preparation). This is in broad agreement with published relationships between U_* values and the initiation of suspension (e.g. Inman, 1949). A lower flow regimen occurred throughout all tidal cycles monitored. Material appeared to move as an intense 'suspension' load in the lower flow depths, the mode of transport being possibly similar in nature to an intermittent suspension mechanism described from laboratory experiments (Francis, 1973).

Tracer experiments, using fluorescent sand of a similar grain-size distribution to the natural sediment, indicated that traction transport was very limited (Carling, 1978). The well-sorted bed material easily went into full or temporary suspension under tidal action. A threshold of traction transport was of the order of 1·5 cm sec⁻¹; those grains moving wholly or mainly in traction travelled at speeds in the range 0·01–0·07 cm sec⁻¹ and constituted 1–2% of the total sediment load.

Mid-depth and surface concentrations of suspended material frequently were less than 50 mg l⁻¹, whilst near-bed concentrations 6 cm above the sediment–water interface varied with tidal state and ranged from less than 50 mg l⁻¹ to greater than 300 mg l⁻¹ (Fig. 6). Within the marsh creeks, con-

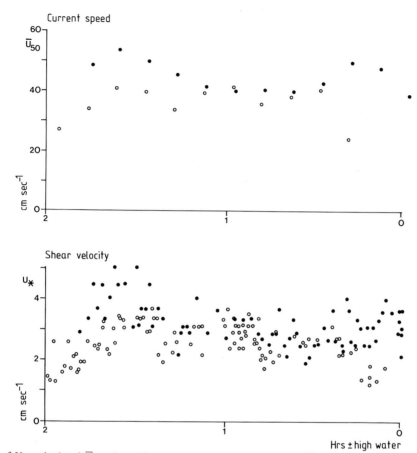

Fig. 5. Plot of U_* and related $\overline{U}_{50}$ values. The sympathetic response of U_* to $\overline{U}_{50}$ should be noted as should the slight time lag involved. $\bigcirc$ = flood tide values, $\bullet$ = ebb tide values.

centrations occasionally rose to greater than 1000 mg l^{-1}. These concentrations, typical of settled meteorological conditions, were of similar orders of magnitude to values reported from other sandy temperate intertidal zones (Evans & Collins, 1975; Anderson, 1973).

An exponential reduction in the mean suspended sediment concentration with distance from the sediment–water interface was distinct during those phases of the tide dominated by tidal currents. This corresponds to Rouse's (1937) approximation,

$$C_z = C_a \left(\frac{D-z}{D-a} \cdot \frac{z}{a} \right)^{z_1}, \qquad (3)$$

where C_z is an unknown concentration at point z in the flow depth D and C_a is a known concentration at depth a. A value for the exponent z_1 is commonly obtained from the surrogate value $z_2 = v_s/\chi U_*$,

where v_s is the mean settling velocity of the sampled grains. However, measured values of z_1 are commonly less than z_2 for a variety of physical reasons discussed elsewhere (Task Committee, 1963). Exponents for measured distributions in this investigation ranged from -0.74 to -1.15 and were commonly less than calculated z_2 values. An exponential model is not appropriate at the beginning of the flood tide or at the end of the ebb tide, nor close to slack water. Examples of concentration gradients which are exponentially distributed with depth may be found in Fig. 7.

Distinct peaking in concentrations of suspended solids at the beginning of the flood tide or close to low water on the ebb tide (Fig. 6B) may be related to wave resuspension of deposited material. At these times tidal currents are commonly too weak to resuspend the bed material. However, low amplitude

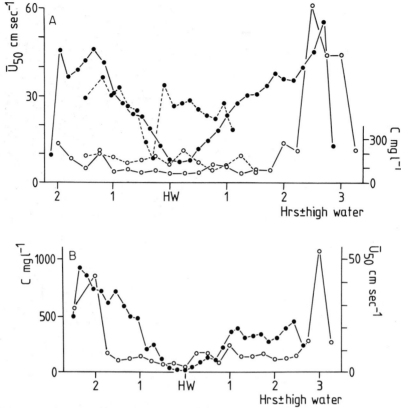

Fig. 6. (A) Suspended sediment concentrations and related current speed. ●—●, current speed in creek; ○—○. suspended sediment in creek; ●--------● current speed on high sandflat; ○--------○, suspended sediment on high sandflat. (B) Suspended sediment concentrations and related current speed. Note the distinct peak in concentration at the end of the ebb tide. Peaking of this nature commonly occurs after tidal currents have waned and may be related to wave-induced suspension (see text). ●—●, current speed; ○—○, suspended sediment concentration.

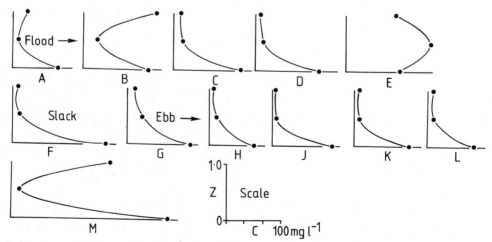

Fig. 7. Series of concentration profiles with depth typifying suspended sediment distribution throughout a tidal cycle. A and B are wave induced profiles, C and D are flood-current-induced profiles, E represents a 'gravitational bulge' near slack water, F is a slack water profile, G–L are ebb tidal-current-induced profiles, M is a wave-induced profile near to low water.

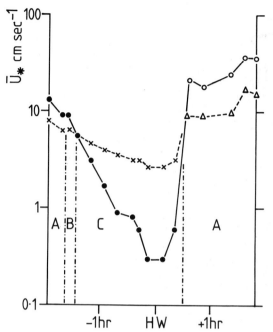

Fig. 8. Wave-induced orbital velocities and the threshold of motion (for material of 250 μm grain size) under waves of 1 and 2 sec period throughout a spring tidal cycle. ●—●, orbital velocity 1 sec period; ○—○, orbital velocity 2 sec period; ×-------×, threshold orbital velocity 1 sec period; △-------△, threshold orbital velocity 2 sec period. A = period of wave-induced suspension, B = period of wave-induced traction, C = period of no motion. Thresholds calculated from equations of Komar & Miller (1973).

waves shoaling on the intertidal zone produce near-bed orbital velocities in excess of the threshold of motion (Komar & Miller, 1973) (Fig. 8). At other tidal states, waves in the Inlet are deep-water Airy waves, which generate orbital velocities near the bed below threshold values.

Close to slack water on the flood tide, gravitational 'bulges' in the 'concentration curve with depth' occur as material begins to settle. Conversely, a 'bulge' travels back up through the water column on the first phase of the ebb (Fig. 7). At slack water, when horizontal currents are virtually zero, changes in concentrations at reference levels throughout the water column indicate vertical fluxes toward the sediment–water interface of some 100–190 mg m^{-2} sec^{-1} and a mean settling velocity of 0·68 mm sec^{-1}. For a concentrated sediment load in a still water body the deposition rate may be written:

$$\Delta = v_s \overline{C}_{z=a}. \qquad (4)$$

Following Maude & Whitmore (1958) near-bed contrations (125 μm mean grain size) of the order of 185 mg l^{-1} should settle at 13·89 mm sec^{-1}, yielding concentration fluxes in the range 2–3 × 10^3 mg m^{-2} sec^{-1}. Clearly, residual turbulence hinders rapid settling. An appropriate relationship yielding the correct order of magnitude of observed deposition rates at slack water is (Parker, 1978):

$$\Delta = (v_s^2/\epsilon)\zeta, \qquad (5)$$

where v_s is the settling rate, ϵ is the vertical eddy diffusivity equal to $0·077 U_* D$ over flat sand beds (Engelund, 1970) and ζ is the depth-integrated suspended sediment concentration (Parker, 1978). An approximate value for ϵ was found to be of the order of $2·01 \times 10^{-3}$ m^2 sec^{-1} at slack water.

Sediment frequently does not settle out completely, but forms a concentrated suspended sediment layer (Fig. 7F) close to the bed (typically 0·06 m thick). This layer may represent up to 4 % of the tidal depth. The fluid nature of this layer is clearly demonstrated by rapid dispersion of the near-bed concentration back through the water column on the ebb tide (Fig. 7). This may be achieved before U_* values have reached threshold values to resuspend deposited material.

Sediment transport model

Sediment transport caused by tidal currents was calculated for tidal cycles where data on depth-integrated suspended sediment concentrations and time-integrated sediment transport rates were available.

$$\zeta = \int_{z=a}^{D} C(z)\, dz; \quad I_i = \int_0^T I_{\mathrm{m}}(t)\, dt$$

are the vertically integrated concentration and time-integrated transport rate where $I_m = \zeta \overline{U}_z$. Extrapolation of concentration data toward the bed ($z = 0$) was terminated a few grain diameters above the bed ($z = a$) as, under lower flow regime conditions, this is the usual maximum extent of fine sand movement as bedload (Einstein, 1950). Traction transport was neglected owing to its small contribution to the total and the absence of direct measures of transport rates.

Measured, immersed weight, sediment transport across a metre width of bed was typically 6·71 kg sec^{-1}, ranging between 0·017 and 21·42 kg sec^{-1}. Values of these orders of magnitude typified tidal

6

transport at all the monitored stations on the inter-tidal zone.

Sediment transport may be proportional to time-integrated stream-power values (Swift & Ludwick, 1976), where stream power per unit bed width may be written

$$\omega = \bar{U}_z \gamma RS = \bar{U}_z r_0. \qquad (6)$$

In this investigation suspended sediment transport was related to stream power (Fig. 9) by an empirical relationship:

$$I_m \propto (\omega - \omega_0)^{1\cdot49}. \qquad (7)$$

This relationship is close to the power relationship proposed by Bagnold (1977) whereby *bedload* transport is proportional to the 1·5 power of the available stream power. According to Bagnold, transport of sand-size material is proportional to both the available stream power above a threshold and the relative depth of flow.

$$I_e = 1\cdot6\frac{(\omega - \omega_0)^{1\cdot5}}{\omega_0^{0\cdot5}}(y/d)^{-0\cdot66}. \qquad (8)$$

In equations (6, 7 and 8), I_e is the estimated immersed weight transport, R is the hydraulic radius, S is the energy slope and $\bar{U}_z$ is a flow speed representative of the flow depth, ω is a stream-power value and ω_0 is a threshold value for suspension of a given grain size. y is the flow depth and d is equal to the median grain size of the bed deposits (d_{50}) in flat

sandbed conditions. The use of d_{50} to characterize the bed roughness is arbitrary and frequently does not apply for larger bed material (Francis & Ackers, 1977), or where bed roughness may be associated with current rippling as in this investigation. Here κ'_s (the bed roughness) was substituted for d, with a threshold of appreciable sediment flux, 0·013 kg m^{-1} sec^{-1}, defined from field observations. Plotting y/κ'_s or y/d against $\omega - \omega_0$ for a constant transport rate, the term $(y/d)^{-0\cdot66}$ could not be substantiated for the values of y/d observed (Fig. 10). Bagnold (1977) and Williams (1970) noted that this might be the case in deep flows when y/d is greater than 2400. However, only a narrow range of y/d values were available for consideration (Fig. 10). Additional data would be required to verify or modify equation (8) for use in the marine environment. In spite of the lack of theoretical justification, equation (8) continues to yield excellent results when compared to field data (Fig. 11).

General transport model

For tidal cycles where boundary layer data were unavailable, τ_0 was estimated using equation (1) and a general model for the logarithmic boundary layer (Sternberg, 1968):

$$\log Z = \frac{\bar{U}_z + \log \bar{z}_0}{\bar{U}_* 5\cdot75}. \qquad (9)$$

In equation (9), values of $\bar{U}_{50}$ were substituted as representative flow speeds; similarly, z_0 values of

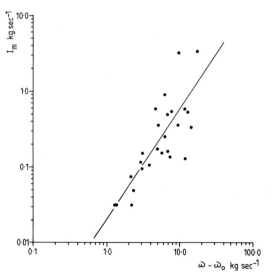

Fig. 9. Plot of measured sediment transport I_m (kg m^{-1} sec^{-1}, immersed weight) and the associated available stream power $\omega - \omega_0$ (kg m^{-1} sec^{-1}), $I_m \propto (\omega - \omega_0)^{1\cdot49}$.

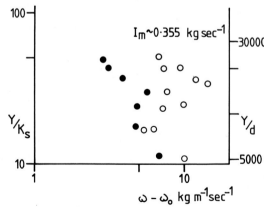

Fig. 10. Plot of values of relative depth, y/K_s and y/d against the stream power necessary to maintain a constant transport rate, approximately 0·35 kg m^{-1} sec^{-1}. $\bigcirc$, flood values; $\bullet$, ebb values.

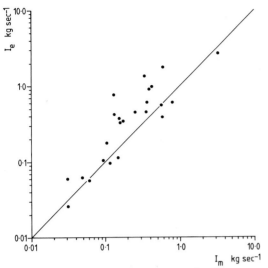

Fig. 11. Relationship between measured sediment transport I_m and values I_e estimated from equation (8). The solid line is the line of potential perfect agreement.

Table 2. Potential sediment transport across the intertidal zone—Burry Inlet, U.K.

Location	Flood transport	Ebb transport	Residual transport
Station 5 (S)	38·40	0·52	+39·05
(N)	1·51	0·35	
Station 6 (S)	33·70	23·00	+7·43
(N)	7·83	11·10	
Station 7 (S)	11·10	26·60	−12·78
(N)	10·90	8·18	
Station 8 (S)	48·60	55·60	−14·11
(N)	0·79	7·90	
Station 9 (S)	6·56	1·61	+5·69
(N)	1·15	0·41	
Station 10 (S)	15·40	10·90	+12·56
(N)	9·92	1·86	
Station 11 (S)	30·60	1·19	+32·95
(N)	3·90	0·36	
Station 12 (S)	37·50	0·47	+49·44
(N)	12·50	0·10	
Station 13 (S)	40·50	3·51	+45·53
(N)	8·56	0·02	
Station 14 (S)	48·00	15·00	+?
(N)	(no data)	(no data)	
Station 15 (S)	34·80	26·80	+12·31
(N)	12·40	8·09	

(S) = spring tide values, (N) = neap tide values. + = onshore residual, − = offshore residual. Values are immersed weight transport, kg ($\times 10^1$) flood^{-1} or ebb^{-1} throughout the total depth and for 1 m width of bed.

0·17 cm for flood tides and 0·22 cm for ebb tides were introduced. Stream-power values ($\overline{U}_{50} \tau_0$, representative of a 10 min integrated period) were integrated over 196 spring and 150 neap tidal cycles. Average total tidal transport for flood or ebb was evaluated from equation (8) for each station. Residual values of sediment transport are presented in Table 2. The intertidal zone is inundated for periods of 2–6 h depending on tidal height. The calculated transport rates for flood or ebb tide therefore compare favourably with the observed transport rates (p. 74).

Estuarine suspended-sediment transport residuals are usually in the direction of residual tidal currents (Halliwell & O'Connor, 1966); therefore the scalar values in Table 2 have been given directional components equal to the current velocity vectors (Fig. 2), yielding sediment transport vectors (Fig. 12A, B). Depositional gradients may be inferred for flood or ebb, spring or neap tidal conditions, based on the premise that a reduction in the potential sediment transport in the same vectorial direction is indicative of deposition along a gradient between stations.

DISCUSSION

Circulation pattern

In general terms the pattern outlined above is similar to the circulation pattern deduced for the intertidal flats of the Wash (Evans, 1965; Amos, 1974), a confused flow pattern over the lower intertidal zone and a progressive realignment of current vectors shorewards. Vectors lie subparallel to the coast at offshore stations and become increasingly rectilinear, normal to the shore at stations on-shore. This process is most evident in the creeks. Further, with the progression of the tides from neaps to springs, the major axes of current vectors are increasingly directed shore-normal.

As the flood flow crosses the intertidal zone it divides into largely *separate* flow bodies, propagating *independently* over the sandflat and through the creek system. Distinct low current speed areas define the 'boundary' between these water masses. This pattern is dissimilar to the hypothetical model presented for the Wash (Kestner, 1975), where creeks 'feed' the flow over the sandflats. During the ebb tide the lateral exchange of tidal waters from the marsh surface to the creek system is notable and an area of

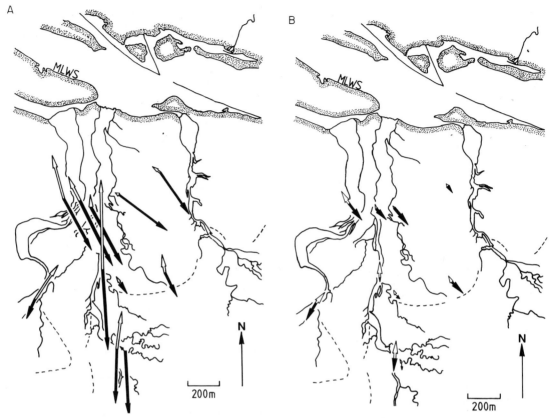

Fig. 12. (A) Summary sediment transport pattern for spring tides. Black vectors are flood tide transport residuals, white vectors are ebb tide residuals. Vector lengths represent (approx.) transport values in Table 2. See Fig. 2 for station numbers. (B) Summary sediment transport pattern for neap tides, details as for spring tides.

low current speed is re-established in the convergence zone to seaward. General observations suggest that a similar pattern occurs to seaward of each of the creeks along this coast. As a result, the flow pattern across the intertidal zone may be envisaged as an alongshore series of contiguous cells bounded by creeks.

The low current speed areas, where tidal scour is reduced, are readily colonized by marsh phanerogams. Consequently, the circulation pattern may be a primary cause of the cuspate marsh-edge (Fig. 2).

Boundary layer

Smith (1969), Nece & Smith (1970) andd Luwick (1974) have commented on the theoretical limitations of the velocity profile model. However, the model presented here proved adequate within the limits of this investigation. Similarly, Sternberg (1968) con-

cluded that relationships of the form of equation (2) fit the observed distribution of velocity within the boundary layer 61–100 % of the time. The excellent fit of the model to the observed boundary-layer data is reflected in the values of r^2 and the percentage of time during flood or ebb tides when the model is valid.

Sternberg made two other pertinent observations; roughness lengths usually fall in the range 0·001–3·00 cm; and an empirical value for the Karman coefficient of 0·40 is applicable to the marine environment. However, other investigators have concluded that the latter assumption is invalid. In laboratory studies, Einstein & Chien (1952, 1965) and later Vanoni & Nomicos (1959) and McCave (1974) from investigations in the shallow marine environment, have demonstrated that suspended sediment loads may decrease the value of the Karman coefficient.

In this investigation z_0 and κ_s values were calculated using corrected Karman coefficients (McCave, 1974), a method that was found to be beneficial in reducing scatter in the U_* data. The calculation of a corrected κ'_s value (5·7 cm) from a mean tidal z_0 value (0·19 cm), in order to derive an appropriate Karman coefficient, indicates the inadequacy of the use of $\kappa_s = d$ for the intertidal environment. Grain-size analyses of the natural bed material at station 1 indicated that 20% of the bed material was less than 100 μm, with a modal peak at 125 μm. The use of 125 μm as an equivalent grain roughness would therefore be between one and two orders of magnitude in error. Any subsequent correction of χ, utilizing κ'_s/z_0, would result in further errors. Consequently, Karman coefficients would be underestimated by approximately 50%. Assuming $\chi = 0·40$ (Sternberg, 1968) leads to greater inaccuracies than the process adopted here of continually re-evaluating the Karman coefficient. z_0 values within the range 0·001–0·05 cm on the flood phase of the tide agree with observations over an intertidal flat in the Severn Estuary, U.K. (0·04 cm—Hamilton, 1979). However, average z_0 values (0·17 cm for the flood and 0·22 cm for the ebb) and the range of z_0 values are similar to those reported for the shallow marine environment (Sternberg, 1968; Dyer, 1980) and for an open coast intertidal zone (Parker, 1971).

Suspended sediments

Although equations of the type proposed by Rouse (1937) are acceptable for estimates of suspended sediment distribution throughout the water column, they should not be relied upon to calculate total transport. Whenever possible the exponent z_1 should be evaluated from field data and not from the relationship $z_2 = V_s/\chi U_*$. In particular the presence of reversed concentration gradients and non-exponential concentration profiles should be noted. Sediment transport during periods dominated by wave-induced sediment transport must be treated independently. Failure to consider the influence of low-amplitude wave action in resuspending and transporting sediments may lead to errors in the residual transport of the order of 100% (Carling, unpublished data). Sharp peaking in concentrations of suspended sediments, related here to wave action, is similar to peaking in suspended sediment concentrations described on another intertidal zone (Anderson, 1972).

The limited flux of sediment as a traction load is paralleled by Kestner's (1975) observations in the Wash, where less than 10% of the total sediment transport was attributed to bedform migration. McCave (1971), in an investigation of subtidal sandwaves, observed that, as here, well-sorted bed material finer than 200 μm readily went into suspension and thus limited bedform migration.

Presently, although data on concentration levels of suspended sediments in intertidal areas are not limited, information on the sediment distribution in the vertical and variations in grain size during transport is very limited.

Sediment transport model

Sediment transport equations developed in controlled laboratory conditions are frequently poor predictors of sediment transport monitored in the marine environment. Bagnold's (1977) equation derived from laboratory and field data to predict *bedload* transport correctly predicts *total* sediment transport rates within the intertidal environment with a mean discrepancy ratio of $\pm 0·37$. Commonly, summing transport over a flood or ebb tide yielded results less than 16% in error when compared to monitored sediment transport data. The applicability of a bedload equation to total transport predictions may possibly be explained by the ease with which deposited sediments (modal grain size, 125 μm) on the intertidal zone are put into suspension. Bagnold (1973) showed that sand with a mean grain size of 100 μm went directly into suspension when the threshold of bedload transport was exceeded. It is possible that Bagnold's (1977) relationship is a total load transport equation when fine sand beds are under consideration.

The application of the stream-power concept to sediment transport in the dynamic intertidal environment has considerable potential, as stream-power is highly sensitive to rapid changes in energy slope and discharge (Bull, 1979). Modest changes in stream power should reflect change in sediment transport rates, as suspended sediment concentrations are highly sensitive to changes in discharge (Leopold, Wolman & Miller, 1964) and U_* (Dyer, 1980).

Bagnold's approach, in particular, is attractive in the marine environment, requiring no measure of energy gradient as is required by other stream-power approaches (Yang, 1972). Although measures of energy gradient have been obtained at sea (Bowden & Fairbairn, 1952), these are not usually practical.

Defining a threshold stream-power for any given grain size is believed to be especially important at very low suspended-sediment concentrations (Yang, 1979). This is problematical, as the critical power varies as a function of sediment load, grain-size and hydraulic roughness of the flow. However, in the intertidal environment, characterized by well-sorted fine sand, the use of a mean static threshold value is not inappropriate. The resulting discrepancy in calculating I_e is usually negligible for fine sandbeds at low threshold values and the high values of ω at which appreciable transport occurs.

General transport model

The most striking observation to be made from the estimated sediment transport at a station is the considerable variation in values when comparing the flood or ebb of a spring or neap tide (Table 2). The range of values far exceeds any differences that may be attributed to averaging errors. A simple onshore decrease in potential energy (cf. Postma, 1961) is an incomplete model. For example, flood or ebb values of I_e (Table 2) range through four orders of magnitude. Station 13 in the seaward portion of the creek has a very low potential transport capacity on ebb neap tides (0.02×10^4 kg) compared to station 8 on ebb spring tides (55.60×10^4 kg) in the creek close to the marsh sward. The complexity of sediment transport potential is perhaps greater than might have been expected. The high transport rate within the creek close to the edge of the marsh highlights an important location in terms of sediment transport on the intertidal zone: the marsh/sandflat interface (Fig. 2).

The differences in sediment transport residuals comparing adjacent creeks (stations 7 and 8, Table 2), draining adjacent marsh catchments, reflect the complexity of intertidal marsh sediment budgets. In this respect, the evidence suggesting that individual marsh creeks are either net sediment importers or net sediment exporters (Pestrong, 1973; Schultz, 1973; Boon, 1973; Settlemeyre & Gardner, 1977) may only present partially complete sediment budgets for marsh complexes drained by a series of creek systems.

Despite the importance of salt marshes and sandflats in nutrient cycling in coastal waters (Odum, 1961, 1973) and for coastal defence, very little is known of the sediment dynamics at the salt marsh/sandflat interface and the relationship that tidal creeks have to play in the overall sedimentary budget of the intertidal zone.

CONCLUSIONS

(1) Circulation of tidal waters and sediments over a channelled intertidal zone is more complex than might be supposed from a consideration of existing simplified intertidal models. Notably, flow vectors and potential sediment transport display distinct temporal and spatial variability.

(2) Cuspate marsh forelands are related to the gross tidal circulation pattern, in particular the existence of distinctive contiguous cells of tidal circulation.

(3) Simple boundary-layer models are adequate for the intertidal environment, when wave action is negligible. However, the choice of values of the roughness lengths and bed roughness needs to be carefully assessed.

(4) Distribution of suspended-sediment concentration with depth is frequently non-exponential due to wave action or settling effects at slack water.

(5) The action of low-amplitude wind waves on the resuspension process may be invoked to explain sharp peaking in suspended sediment concentration data. Any complete intertidal sediment transport model should carefully consider wave-induced sediment transport.

(6) There is a dearth of information concerned with the detailed dynamics of resuspension and deposition on intertidal zones.

(7) Bagnold's (1977) relationship relating stream power to bedload sediment transport rate yields good results when compared to monitored suspended sediment transport of fine sands in the intertidal zone.

ACKNOWLEDGMENTS

This investigation was carried out whilst the author was in receipt of a Natural Environment Research Council research studentship, supervised by Dr M. B. Collins. Dr M. B. Collins commented on draft scripts, whilst Mr T. Handyside made valuable contributions, in discussion, to the interpretation of the boundary-layer data. Dr A. D. Heathershaw (Institute of Oceanographic Sciences) afforded computer analysis facilities for data reduction. The F.B.A. kindly defrayed expenses to present this paper at Texel.

Mr N. A. Reader drafted the figures, Mrs D. Jones and Mrs J. Hawksford typed the manuscript.

REFERENCES

Amos, C.L. (1974) *Intertidal flat sedimentation in the Wash, Eastern England.* Unpublished Ph.D. Thesis, University of London. 404 pp.

Anderson, F.E. (1972) Resuspension of estuarine sediments by small amplitude waves. *J. sedim. Petrol.* **42**, 602–607.

Anderson, F.E. (1973) Observations of some sedimentary processes acting on a tidal flat. *Mar. Geol.* **14**, 101–116.

Bagnold, R.A. (1973) The nature of saltation and of bedload transport in water. *Proc. R. Soc.* A, **332**, 473–504.

Bagnold, R.A. (1977) Bedload transport by natural rivers. *Wat. Resour. Res.* **13**, 302–312.

Bayliss-Smith, T.P., Healey, R., Lailey, R., Spencer, T. & Stoddart, D.R. (1979) Tidal flows in salt marsh creeks. *Est. coastal Mar. Sci.* **9**, 235–256.

Boon, J.D. (1973) Optimized measurements of discharge and suspended sediment transport in a salt marsh drainage system. *Int. Symp. Interrelationships of Estuarine and Continental Shelf Sedimentation*, pp. 67–74. Bordeaux, France, 9–14 July.

Boon, J.D. (1975) Tidal discharge asymmetry in a salt marsh drainage system. *Limnol. Oceanogr.* **20**, 71–80.

Bouma, A.H. (1963) A graphical presentation of the facies model of saltmarsh deposits. *Sedimentology*, **2**, 122–129.

Bowden, K.F. & Fairbairn, L.A. (1952) A determination of the friction forces in a tidal current. *Proc. R. Soc.* A, **214**, 371–392.

Bull, W.B. (1979) Threshold of critical power in streams. *Bull. geol. Soc. Am.* **90**, 453–464.

Carling, P.A. (1978) *The influence of creek systems on intertidal flat sedimentation.* Unpublished Ph.D. Thesis, University of Wales. 258 pp.

Collins, M.B. (1975) Suspended sediment sampling towers as used in the intertidal flats of the Wash, Eastern England. *Est. coastal Mar. Sci.* **4**, 45–47.

Collins, M.B., Amos, C.L. & Evans, G. (1981) Observations of some sediment-transport processes over intertidal flats, the Wash, U.K. In: *Holocene Marine Sedimentation in the North Sea Basin* (Ed. by S.-D. Nio *et al.*). *Spec. Publs int. Ass. Sediment.* **5**, 81–98. Blackwell Scientific Publications, Oxford. 524 pp.

Dyer, K.R. (1980) Velocity profiles over a rippled bed and the threshold of movement of sand. *Est. coastal Mar. Sci.* **10**, 181–199.

Eaton, J.S., Likens, G.E. & Bormann, F.M. (1969) Use of membrane filters in gravimetric analysis of particulate matter in natural waters. *Wat. Resour. Res.* **5**, 1151–1156.

Einstein, H.A. (1950) The bedload function for sediment transport in open channel flows. *Tech. Bull. U.S. Dep. Agric. 1026*, 70 pp. (Reprinted as Appendix B, in: *Sedimentation* (Ed. by H.W. Shen). 1972, Fort Collins, Colorado.)

Einstein, H.A. & Chien, N. (1952) Second approximation to the solution of the suspended load theory. *Univ. California Inst. Engng Res. U.S. Army Corp. Engrs, Missouri River Div. M.R.D. Sediment Ser. No. 3*, 30 pp.

Einstein, H.A. & Chien, N. (1955) Effects of heavy sediment concentration near the bed on the velocity and sediment distribution. *Univ. California Inst. Engng Res. U.S. Army Corp. Engrs, Missouri River Div. M.R.D. Sediment Ser. No. 8*, 76 pp.

Engelund, F. (1970) Instability of erodible beds. *J. Fluid Mech.* **42**, 225–244.

Evans, G. (1965) Intertidal flat sediments and their environments of deposition in the Wash. *Q. Jl geol. Soc. Lond.* **121**, 209–245.

Evans, G. & Collins, M.B. (1975) The transport and deposition of suspended sediment over the intertidal flats of the Wash. In: *Nearshore Sediment Dynamics and Sedimentation* (Ed. by J.R. Hails and A. Carr). Wiley, Chichester.

Francis, J.R.D. (1973) Experiments on the motion of solitary grains along the bed of a water stream. *Proc. R. Soc.* A, **332**, 443–471.

Francis, J.R.D. & Ackers, P. (1977) Developments in sediment transport theories. *Proc. Inst. civ. Engrs* **62**, 343–346.

Graf, W.H. (1971) *Hydraulics of Sediment Transport.* McGraw-Hill, New York.

Groen, P. (1967) On the residual transport of suspended matter by an alternating tidal current. *Neth. J. Sea Res.* **3**, 564–574.

Halliwell, A.R. & O'Connor, B.A. (1966) Suspended sediment in a tidal estuary. *10th Conf. Coast. Engineering*, pp. 687–706. American Society of Civil Engineers.

Hamilton, D. (1979) The high energy sand and mud regime of the Severn Estuary, S.W. Britain. In: *Tidal Power and Estuary Management* (Ed. by R.T. Severn, D. Dinely and L.E. Hawker). Scientechnica, Bristol.

Hayes, M.C. (1975) Morphology of sand accumulations in estuaries. In: *Estuarine Research* (Ed. by L.E. Cronin), pp. 3–22. Academic Press, New York.

Ingle, J.C. (1966) *The Movement of Beach Sand.* Elsevier, Amsterdam.

Inglis, C.C. & Kestner, F.J.T. (1958) The long-term effects of training walls, reclamation and dredging in estuaries. *Proc. Inst. civ. Engrs* **9**, 193–216.

Inman, D.L. (1949) Sorting of sediments in the light of fluid mechanics. *J. sedim. Petrol.* **19**, 51–70.

Inman, D.L. (1963) Sediments; physical properties and mechanics of sedimentation. In: *Submarine Geology* (Ed. by F.P. Shepard), 2nd edn, pp. 101–147. Harper & Row, New York.

Kestner, F.J.T. (1963) The supply and circulation of silt in the Wash. *Proc. int. Ass. Hydraul. Res. Congr., London*, pp. 231–238.

Kestner, F.J.T. (1975) The loose boundary regimen of the Wash. *Geogr. J.* **141**, 388–414.

Kestner, F.J.T. (1976) The effect of training works on the loose boundary regimen of the Wash. *Geogr. J.* **143**, 490–504.

Komar, P.D. & Miller, M.C. (1973) The threshold of sediment movement under oscillatory water waves. *J. sedim. Petrol.* **43**, 1101–1110.

Leopold, L.B., Wolman, M.G. & Miller, J.P. (1964) *Fluvial Processes in Geomorphology.* San Francisco, Freeman. 522 pp.

Ludwick, J.C. (1974) Tidal currents and zig-zag shoals in a wide estuary entrance. *Bull. geol. Soc. Am.* **85**, 717–726.

McCave, I.N. (1971) Sandwaves in the North Sea off the coast of Holland. *Mar. Geol.* **10**, 109–225.

McCave, I.N. (1974) Some boundary layer characteristics of tidal currents bearing sand in suspension. *Mém. Soc. r. Sci. Liège*, **VI**, 187–206.

McCave, I.N. & Geisler, A.C. (1979) Megaripples, ridges and runnels on intertidal flats of the Wash, England. *Sedimentology*, **26**, 353–371.

Maude, A.D. & Whitmore, R.L. (1958) A generalised theory of sedimentation. *Br. J. appl. Phys.* **9**, 101–112.

Moore, N.H. (1976) *Physical oceanographic and hydrological observations in the Loughor Estuary (Burry Inlet).* Unpublished M.Sc. Thesis, University of Wales.

Nece, R.E. & Smith, J.D. (1970) Boundary shear stress in rivers and estuaries. *J. WatWays Harb. Div. Am. Soc. civ. Engrs* **96**, 335–358.

Nikuradse, J. (1933) Stromungsgesetze in rauhen Rohren. *Forach. Geb. IngWes.* V, 361.

Odum, E.P. (1961) The role of tidal marshes. *Conservationist*, June/July, 12–15, 35.

Odum, E.P. (1973) The pricing system. *G. Cons. Mag.* Fourth Quarter, 8–10.

Parker, G. (1978) Self-formed straight rivers with equilibrium banks and mobile bed. Part 1. The sand–silt river. *J. Fluid Mech.* **89**, 109–125.

Parker, W.A. (1971) *Formby Point, Lancashire.* Unpublished Ph.D. Thesis, University of Liverpool.

Pestrong, R. (1973) Ebb–flood sedimentation within a tidal marsh system. *Int. Symp. Interrelationships of Estuarine and Continental Shelf Sedimentation*, pp. 61–66. Bordeaux, France, 9–14 July.

Pethick, J.S. (1971) *Salt marsh morphology.* Unpublished Ph.D. Thesis, University of Cambridge.

Postma, H. (1961) Transport and accumulation of suspended matter in the Dutch Wadden Sea. *Neth. J. Sea Res.* **1**, 149–190.

Postma, H. (1967) Sediment transport and sedimentation in the estuarine environment. In: *Estuaries* (Ed. by G.H. Lauff). *Publs Am. Ass. Adv. Sci.* **83**, 158–179.

Rouse, H. (1937) Modern conceptions of the mechanics of fluid turbulence. *Trans. Am. Soc. civ. Engrs* **102**, 436–505.

Schultz, D.M. (1973) Fatty acid composition of organic detritus from *Spartina alterniflora. Est. coastal Mar. Sci.* **1**, 177–190.

Settlemeyre, J.L. & Gardner, L.R. (1977) Suspended sediment flux through a salt marsh drainage basin. *Est. coastal Mar. Sci.* **5**, 653–664.

Smith, J.D. (1969) Studies of non-uniform boundary layer flows. In: *Investigations of Turbulent Boundary Layer and Sediment Transport Phenomena as Related to the Shallow Marine Environment, part 2. Rep. Dep. Oceanog. Univ. Washington RLO—1752–13*, 13 pp.

Sternberg, R.W. (1968) Friction factors in tidal channels with differing bed roughness. *Mar. Geol.* **6**, 243–260.

Straaten, L.M.J.U. van (1950) Environment of formation and facies of the Wadden Sea sediments. *Tijdschr. K. ned. aardrijksk. Genoot.* **67**, 354–368.

Swift, D.J.P. & Ludwick, J.C. (1976) Substrate response to hydraulic process: grain-size frequency distributions and bedforms. In: *Marine Sediment Transport and Environment Management* (Ed. by D.J. Stanley & D.J.P. Swift), chapter 10, pp. 159–196.

Swift, D.J.P., Schubel, J.R. & Sheldon, R.W. (1972) Size analysis of fine grained suspended sediments: a review. *J. sedim. Petrol.* **42**, 122–134.

Task Committee (1963) Sediment transportation mechanics: suspension of sediment. *Proc. Am. Soc. civ. Engrs, J. Hydr. Div.* **89**, 45–76.

Vanoni, V.A. & Nomicos, C.N. (1959) Resistance properties of sediment laden streams. *Proc. Am. Soc. civ. Engrs, J. Hydr. Div.* **85**, 77–107.

Williams, G.P. (1970) Flume width and water depth effects in sediment transport experiments. *Prof. Pap. U.S. geol. Surv. 562-H.*

Yang, C.T. (1972) Unit stream power and sediment transport. *Proc. Am. Soc. civ. Engrs, J. Hydr. Div.* **98**, 1805–1826.

Yang, C.T. (1979) Unit stream power equations for total load. *J. Hydrol.* **40**, 123–138.

Spec. Publs int. Ass. Sediment. (1981) **5**, 81–98

Observations of some sediment-transport processes over intertidal flats, the Wash, U.K.

M. B. COLLINS*, C. L. AMOS† *and* G. EVANS‡

**Department of Oceanography, University College of Swansea, Wales, U.K.,
†Geological Survey of Canada, Bedford Institute of Oceanography, Canada,
and ‡Department of Geology, Imperial College, London, U.K.*

ABSTRACT

The intertidal flats of the Wash are areas of active present day sedimentation. The net flux of sediment is from seaward to landward. The decrease in grain size and the decrease in magnitude of the sedimentary structures faithfully record the decrease in tidal current speeds in the same direction. However, the directional properties preserved in the sediments do not clearly reflect the net movement of sediment and caution needs to be used in deducing the direction of net sediment transport from such features in ancient marine sequences.

INTRODUCTION

The Wash is a relatively shallow macrotidal embayment (the average tidal range is 6·5 m on springs and 3·6 m on neaps) bordering the Lincolnshire and Norfolk coastlines, on the eastern shore of England. Its maximum depth is 35·5 m (below extreme low water level) in Lynn Well at the mouth of the embayment. It is approximately 20 km wide at the mouth and 30 km in length and has an area of 615 km² of which 325 km² are exposed during extreme low tide. During the low-tide periods, a broad intertidal zone, which is 2·6 km locally, though at its widest it exceeds 10 km, emerges around the margin of the bay; this grades from salt-marsh into mudflats, sandflats, sometimes further mudflats, and channel-margin sandflats from landwards to seawards (see Fig. 1). The slope of this intertidal zone is extremely gentle, with a gradient of 1 in 800 on the more landward parts and 1 in 100 further seawards. As pointed out by Evans (1965), the lower mudflat is not always present, as is true of the area under discussion. In the remainder of this discussion the terms inner, mid and outer sandflats have been used to describe the areas

seaward of the omnipresent salt-marsh and higher mudflats.

Rapid sedimentation occurs in the intertidal zone and this has been substantially aided by man. Because of the engineering significance of this sedimentation as well as the biological interest in the area, it has been the site of substantial amounts of research (Kestner, 1963). However, in spite of this, it is still a poorly understood region.

The shape of the Wash has resulted from successive reclamations of the wide salt-marshes which accumulate to a level of 0·71 m below highest high-water (Kestner, 1975). Reclamation began with the so-called 'Roman Bank' constructed during the Middle Ages. This activity has continued until the present day. A total of 1245 km² have been reclaimed and this area is now used for agricultural purposes.

The rivers draining into the Wash have a catchment area of $1·3 \times 10^6$ hectares. These rivers (the Witham, the Welland, the Nene, and the Great Ouse) all have sluice-controlled outlets with entrained estuarine channels. Siltation within the mouths of the Fenland ports of Boston, Spalding, Wisbech and Kings Lynn has been a problem throughout historical times (Darby, 1940). During periods of low fresh-

0041–3600/81/1205–0081 $02.00

water flow, dredging on the seaward side of the sluices has proved to be necessary (Clark, 1959). Collins (1972) determined that 0.15×10^6 tonnes annum^{-1} of sediment are derived from fluvial sources (see Wilmot & Collins, 1981, for further details). This material is predominantly of silt and clay sizes and constitutes 20% of the total volume of sediment deposited annually on the intertidal flats of the Wash. There is much evidence to indicate that the source of the major part of sand-size material found on these intertidal flats of the Wash appears to have been derived from a marine source. There is some dispute on the exact nature of this source (see Evans, 1965; Evans & Collins, 1975), although Kestner (1975) believes that this material is derived by reworking and scour within the Wash.

The intertidal flat sedimentation is significant from both geological and engineering viewpoints; consequently, studies were made on part of this zone. The purpose of this investigation was: to measure the dominant processes in a representative area, to measure the responding transport and distribution of suspended solids and bed material, and finally to relate the properties of the accumulating sediment to such processes.

The relationship between active processes and the response to these are obviously of significance to engineering schemes such as the construction of engineering works on intertidal flats, land reclama-tion schemes, or channel entrainment (Ruxton, 1979). In addition, the relationship between surface processes and sedimentary characteristics should aid the geologist in the understanding of the depositional environment of ancient rocks.

THE NATURE OF THE STUDY

The investigation was carried out over a two-year period along three transects of the intertidal zone in the south-west corner of the Wash, east England (see Evans & Collins, 1975, for details of available data). The transects are assumed to be representative of the intertidal zone in general and an attempt has been made to avoid the influences of features such as creeks or mussel-banks, which occur locally. The location of the reference lines, which were surveyed from highest high-water to the low-water mark to seawards, and the reference stations quoted throughout the text are shown in Fig. 1. The majority of the data on tidal elevation, current speed and direction, and wave characteristics were gathered along line 1, while accretional measurement and the internal structures of the flats were examined along all three lines. The outer portion of line 1 is covered by every tidal inundation (i.e. 706 yr^{-1}) whereas further landward the number of inundations per year decreases progressively (line 1 station 5 and line 1 station 4—

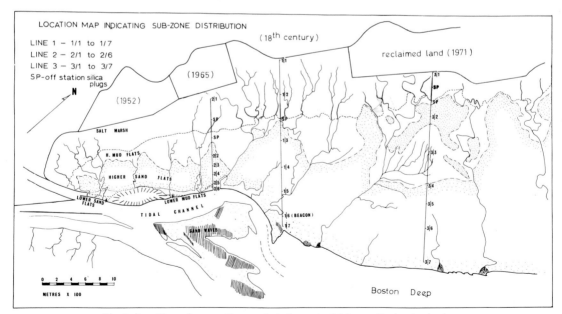

Fig. 1. Location of survey lines and stations at which monitoring took place.

706 inundations; line 1 station 3—669 inundations; line 1 station 2—381 inundations; line 1 station 1—179 inundations). The sedimentological and physical oceanographical data collected were: suspended sediment concentration (mg l⁻¹), tidal elevation (m), tidal current speed and direction, wave height and period, the accretion or erosion of the intertidal flat surface, the grain-size distribution (spatially and temporally), the surface bedform morphology and direction of orientation of surface ripples and internally preserved structures.

Evans & Collins (1975) have already presented some data on the transport of suspended sediment and Amos & Collins (1978) have discussed the relationship of surface bedforms to the combined influence of waves and tidal currents in this area.

Surface sediments were collected over a period of $2\frac{1}{2}$ years. These were taken by skimming the sediment surface over an area of 1 m² at each sampling site. Grain-size analyses were made using a comparable technique to Evans (1965) and are presented using Inman's (1952) statistics. The maximum variations in repeated sampling at any one station were: mean $\pm 0.05\ \phi$; sorting $\pm 0.005\ \phi$; skewness $\pm 0.07\ \phi$, and median $\pm 0.05\ \phi$. The general trend of the mean grain size of the sediments is shown in Fig. 2.

Intertidal water samples and their suspended sediment were collected solely from line 1 using a series of mechanical sampling towers (Collins, 1976). The design was such that sampling could be carried out at a pre-set time and depth. Mid-depth sampling was carried out during the flood and ebb, while spectrum sampling was carried out mainly during high-water stage. Samples were collected at various stations along the line at approximately 2-month intervals. These were filtered for suspended particulate matter using Oxoid membranes of 0·45 μm pore size, after initial filtration through a 63 μm nylon mesh sieve.

Current velocity measurements were made using a Toho-Dentan and Ott current meter; salinity and temperature measurements were made using an Electronic Instruments Ltd (MC5) T/S bridge and wave measurements were made using a graduated staff. These latter data were collected periodically from boats moored across the intertidal zone.

The morphology of intertidal bedforms was measured at 25 m intervals along each of the reference lines. Measurements were made repeatedly throughout the year using a similar device to Newton (1968) and were classified according to the schemes put forward by Tanner (1967) and Reineck & Singh (1973).

Internal bedforms were measured from box cores collected along the intertidal profiles. Cores were cut and impregnated with resin to bring out the various structures and the relative frequency of particular structures was determined.

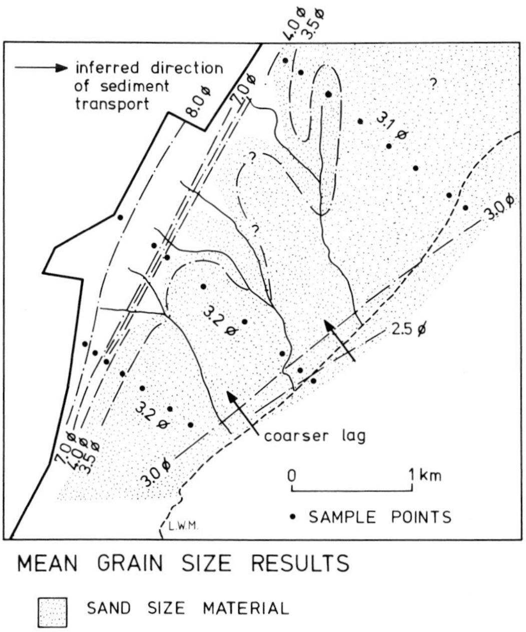

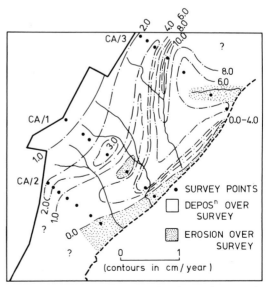

Fig. 2. The mean grain size of the surface sediment.

Fig. 3. The pattern of erosion and sedimentation over the intertidal zone.

Finally, erosion or accretion of the intertidal zone was measured using silica plugs on the salt-marsh and mudflats (Inglis & Kestner, 1958) and over the sandflats by reference to the changing depth of aluminium plates (30 cm × 30 cm) buried within the sediment. The locations of these measurements are shown in Fig. 1. Generally, the greater part of the area was one of accretion (Fig. 3); however, due to movements in the adjacent low-water channel, since the original description of the area by Evans (1965), erosion had set in on the lower part of the intertidal zone (this has now stopped and a reversal to the earlier condition has occurred).

RESULTS: PHYSICAL CHARACTERISTICS OF THE WATER MASS

The water mass flowing over the intertidal zone is generally well mixed except for the first-phase flood waters which are of low salinity and high temperature (Fig. 4). Generally, salinities ranged from 26‰ to 35‰, with the highest salinities being found at the outer stations. Similar results were obtained by Anderson (1972) working in a small tidal cove in New Hampshire, and Amos & Long (1980) working in the Bay of Fundy.

The measurements of the tidal current that flows over the intertidal flats show that (Evans & Collins, 1975): (1) the tidal waters do not simply advance and retreat over the intertidal zone, but there is a marked clockwise rotation in the waters, throughout a tidal cycle; (2) vectorial resolution of the velocity data has demonstrated that the alongshore component is relatively consistent in strength throughout a tidal cycle and is of equal magnitude to the onshore–offshore component: and (3) on an individual tide, there is a velocity gradient across the intertidal flats (Fig. 5). This is true of the onshore–offshore and alongshore components, progressing from maxima at the outer station (on the sandflats) to minima at the inner station (on the salt-marsh (see Evans & Collins, 1975)).

The tides of the Wash are standing tidal waves with their velocity maxima at mid-flood and mid-ebb (Amos, 1974). However, across the intertidal zone,

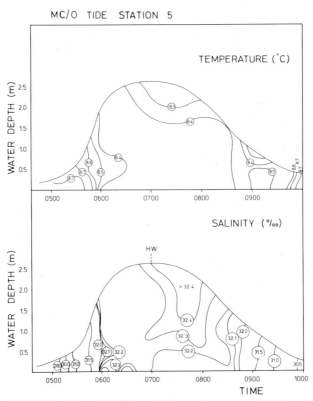

Fig. 4. Variations of salinity and temperature over a tidal inundation.

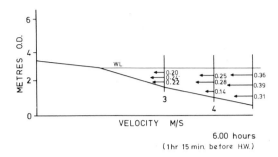

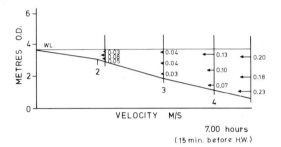

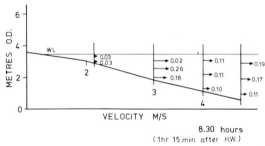

Fig. 5. The distribution of tidal current velocities over the intertidal flats. (25.8.72, tidal height 3.8 m O.D.)

current velocity maxima occur during the first phase of the flood and the final stage of the ebb (see Fig. 6). During the flood, the velocity diminishes towards the bed. At high tide, the velocity profile is abrupt or irregular, and during the ebbing tide, no distinct velocity profile was apparent. Generally, these velocity observations fit the tidal logarithmic–velocity profile calculated using the Karman–Prandtl equation:

$$\frac{U_z}{U_*} = \frac{1}{K}\ln(Z-Z_0)$$

U_z = mean velocity at height Z above the bed,

K = Von Karman's constant, taken as 0·4,

Z_0 = roughness length,

U_* = shear velocity.

It was assumed that the boundary layer extended from the bed to the sea surface. The correlation coefficients (r) for three different tidal cycles are plotted in Table 1 and were calculated for each half-hourly interval throughout the tidal inundation. The results show a consistently good correlation during the rising tide, but it is poor during the remaining part of the tidal inundation. The ebb variability is probably related to cross-flows during lateral draining of the intertidal flats. Carling (1978) and Handyside (1977) working in the Loughor Estuary, South Wales, demonstrated that logarithmic velocity profiles (with $r^2 > 0·80$) were present during 81% of the time occupied by the flood-tide, compared to less than 75% of the time occupied by the ebb-tide (see also Carling, 1981).

There is a definite change in the symmetry of the tidal curve from the low-water to the high-water mark. The curve is nearly symmetrical over the outer flats (i.e. the ratio of rising time taken by the rising tide as against that taken by the falling tide is 0·92) and becomes progressively more skewed to landward, e.g. the ratio of time taken by the rising as against the falling tide for Station 1 is 0·36 (see Fig. 7). This asymmetry is associated with a progressive increase in the peak flood velocity (relative to the peak ebb velocity). Such modifications to the tidal wave are similar to those experienced as a tidal wave progresses up an estuary (McDowell & O'Connor, 1977).

SUSPENDED SEDIMENT DISTRIBUTIONS

A general interpretation of the levels of concentration of suspended sediment over the intertidal flat has already been given by Evans & Collins (1975). Levels were found to be usually less than 200 mg l^{-1}, although extreme maxima ranged up to 2190 mg l^{-1}. All the results obtained are shown in Fig. 8. In this figure, the samples collected during the first-phase flood and last-phase ebb, due to their very high values (see below), have been separated from those collected during the remainder of each of the tidal cycles. Generally, it was found that the concentration increased from surface to bottom throughout the water column and invariably from high- to low-water stage (see Fig. 6). The concentration bore no relationship to tidal range, phase of the tide, or season, but there were some very high values related to individual storm events (cf. Kamps, 1963). The first-phase flooding and final-stage ebbing waters

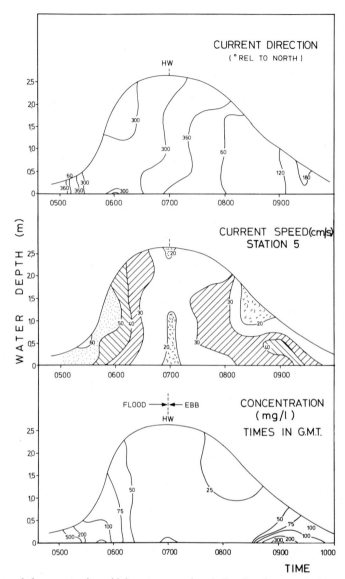

Fig. 6. Suspended concentration, tidal current speed and direction for a typical intertidal flat station.

Table 1. Correlation coefficient for the relationship between observed velocities and a logarithmic–velocity profile

	HW$-1\frac{3}{4}$	HW$-1\frac{1}{2}$	HW-1	HW$-\frac{1}{2}$	HW	HW$+\frac{1}{2}$	HW$+1$	HW$+1\frac{1}{2}$
Tide N	0·62	0·72	0·99	0·95	0·17	0·86	0·78	0·81
Tide O	—	0·94	0·82	0·86	0·45	0·57	−0·22 (0·99)	−0·21
Tide P	0·98	0·97	0·88	0·99	0·10	0·78	0·16 (0·85)	0·17

Note: bracketed values show the correlation-coefficient (r) when surface readings are excluded.

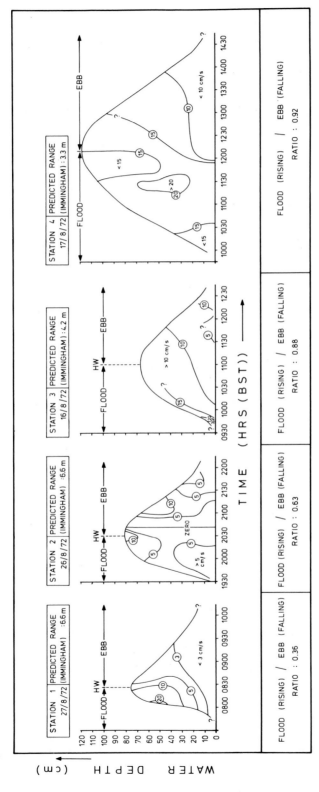

Fig. 7. Variations in asymmetry of the tidal curve from seaward to landward (stations 4–1). This figure also shows the distribution of current velocities at each station over a typical tidal cycle.

have high concentrations varying by an order of magnitude from those of the remainder of the tide (cf. Kamps, 1963; Postma, 1961; Carling, 1978). The lowest concentrations of suspended sediment occur at high tide (Fig. 6). During the flooding tide, sedi-

ments are mixed throughout the water column, while layering of the water column occurs during the ebb (Fig. 6). There is a high temporal variance in concentration, and plumes of turbid water were observed which were associated with boils or 'bursts' of turbulence. Anderson (1972) related these short-term phenomena to local small-amplitude wave action and Kamps (1963) to the effects of local winds. Indeed, the effects of wave motion at the bed is often significant (Amos & Collins, 1978).

The high and instantaneous levels of suspended sediment concentration measured over the intertidal flats corresponded with estuarine and nearshore levels in the Wash, presumed to be related to reworking by local waves and tidal currents. Also, the general levels (200 mg l^{-1}) correspond with those from offshore and North Sea coastal observations.

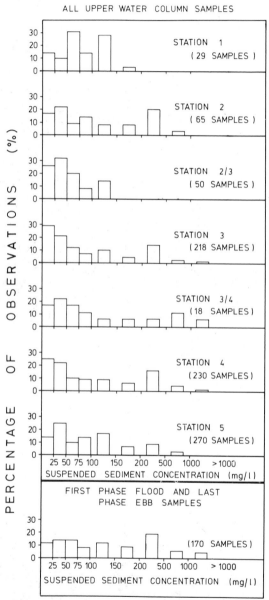

Fig. 8. Overall variation of suspended sediment concentration. Upper water column concentrations and those at the base of the water column, representing first-phase flood and last-phase ebb, are shown in the diagram (see text for discussion).

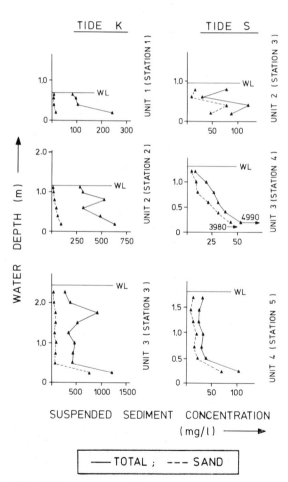

Fig. 9. Instantaneous variation of suspended sand and total suspended sediment with depth in the water column, for two tides, K and S.

Measurements in 1962 (by the Hydraulics Research Station) in the approach channels to the River Nene, recorded levels which were generally less than 100 mg l⁻¹. This survey followed the collection of samples from inshore, mid-Wash and at the entrance to the Wash in 1957; these demonstrated that under spring tides and 'storm' conditions, concentrations of up to 1000 mg l⁻¹ could be recorded inshore (in the Old Lynn Channel) whilst the offshore levels were low (< 250 mg l⁻¹). Similarly, the results of surveys based on surface sampling over complete tidal cycles at Lynn Point, the Great Ouse, appear to show that the concentrations are dependent upon the weather, but that the higher values occur on the ebb phase of the tide, with maximum values approaching 540 mg l⁻¹. In subsequent offshore surveys, carried out as part of the Wash Feasibility Study, offshore levels were generally less than 100 mg l⁻¹ (Evans & Collins, 1975), but with localized values of up to 1200 mg l⁻¹.

The waters crossing the intertidal flats show an increase in concentration towards the bed of both total load and percentage of sand. Such a distribution is shown in Fig. 9, which is an example of the distribution of suspended sediment concentration with depth for two tidal elevations of 4·30 m (tide K)

and 3·06 m (tide S), respectively. These data mean, of course, that the concentration of silt and clay remains constant throughout the water column. Generally, it was found that the mid-depth concentration was close to the depth averaged concentration at that time (see Collins, 1976).

The general decrease in the concentration of sand above the bed and the uniformity of concentration of silt and clay is in general agreement with the theoretical equation given by Rouse (1950):

$$\frac{C}{C_a} = \left\{\frac{z-a}{y-a}\right\}^{w/[(k\sqrt{\tau/\rho})]}$$

where C is the concentration at a height Z above the bed, C_a is the concentration at reference level a above the bed, w is the settling velocity, K is Von Karman's constant (0·4) and $\sqrt{(\tau/\rho)}$ is the shear velocity (U_*).

In view of the high magnitude and hence the potential significance on total suspended sediment transport of the first-phase flood and on the last-phase ebb, an attempt has been made (whenever comparative data were available) to compare concentrations during these periods (Fig. 10). The results (plotted to a log–log scale) show the total sediment concentrations represented as a bivariate

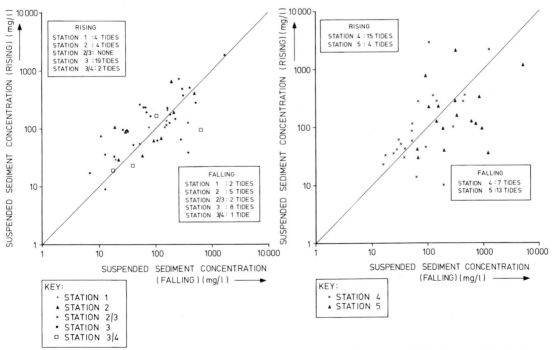

Fig. 10. Relationship between suspended sediment concentration on the first phase of the rising and the last phase of the falling tides. The flood- or ebb-dominance in concentration is listed.

7

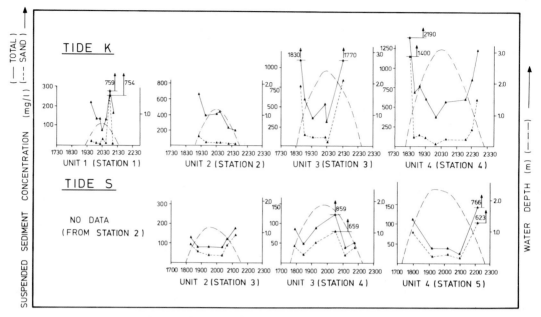

Fig. 11. Temporal variations of suspended sand and total suspended sediment, over tides K and S.

plot; the line on the diagrams represents the function of $x = y$ or equality between the samples. The data show that the largest variations between early flood and late ebb concentrations occur at the outer two stations (stations 4 and 5). The results from each of the stations, over a total of 76 tidal inundations, are compared. Nevertheless, in view of the comparatively short period of the observations, this may not represent the net transport at each of the stations. It appears that the main areas of flood-dominance, in terms of suspended sediment concentration, are at stations 3 and 4, whilst ebb-dominance is maintained at station 5. Kamps (1963) has attributed such ebb- and flood-dominance in the Dutch Wadden Sea to the action of wind.

The percentages of sand to silt and clay show striking variations both during the tidal inundation and across the intertidal zone. The mid-depth concentrations of sand and silt and clay are plotted for two tidal inundations at stations 1 to 4 in Fig. 11. Station 1 is located on the salt-marsh: station 2 on exposed mudflat and stations 3 and 4 on sandflats. The most dramatic variations in concentration occur at the stations on the sandflats. These increases are purely a result of an increase in the volume of sand in suspension. For example, at station 4 the first phase flood concentration of suspended sand is 1400 mg l⁻¹ compared with a concentration of 100 mg l⁻¹ during the remainder of the tide. Clearly, inter-

mittent suspension of the sand at the bed is taking place when there is only a shallow cover of water. A similar pattern can be seen at station 3. At station 2, located on the mudflat, the amount of suspended sand is below 100 mg l⁻¹ and shows a much lower variation in the concentration of sand. In fact, at this site, the majority of the variation in concentration is due to changes in concentration of silt and clay. At station 1, on the salt-marsh, the variation in concentration throughout the inundation is erratic, with no obvious trends. Assuming that it is the same water mass which passes over each of the stations, then it is apparent that the sand, suspended during the first phase of the inundation of the sandflat, drops out of suspension before passing on to the mudflat at station 2. However, this is probably not true during storm conditions when sand is moved into the adjacent mudflats, as can be seen from the inter-layered sand/mud deposits which characterize this area. The lack of sand is presumably a consequence of decrease in current speed landward and the lack of local supply as well as the cohesive nature of the bed at the latter station.

Throughout all the tides excepting one in February 1972 (which is thought to reflect storm conditions), for which particle size data are available, the content of silt and clay in suspension remains consistently between 10 and 100 mg l⁻¹, irrespective of sample location. Thus, no distinctive gradient in the dis-

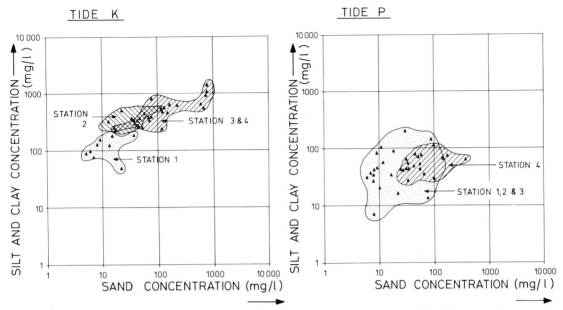

Fig. 12. Overall variations of sand and silt plus clay for two tides, one a storm period tide (K) and one during quieter conditions (P).

tribution of fine-grained material exists as would be expected under the tidal current transport models of Postma (1954) and Van Straaten & Kuenen (1957). Evidently the content of silt and clay is independent of variations in flow conditions. Kamps (1963) reached a similar conclusion from his studies of the eastern Dutch Wadden Sea.

Suspended sand, on the other hand, increases in concentration seaward from 10 to 100 mg l^{-1}, to approximately 500 mg l^{-1} at station 4. The concentration of suspended sand varies from tide to tide, no doubt, in response to changes in tidal and wave conditions. Any seaward increase in the concentration of total suspended matter creates a gradient which contrasts, and is the reverse of, the gradual increase in a shoreward direction found in the Dutch Wadden Sea (Postma, 1954).

The importance of storm conditions on the concentration of material in suspension is shown by comparing the results of a storm tide (K) to those of a tide of similar range under quieter conditions (tide P) (Fig. 12). Under quiet conditions (tide P) the sand contents at stations on the marsh, mudflat and sandflats are similar but with only a slight increase on the outer parts, and the content of silt and clay ranges from 10 to 100 mg l^{-1}. In contrast, under a following storm condition (K), with winds predominantly from the north-east, a tide of similar magni-

tude (and therefore with similar tidal current speeds) shows an order of magnitude increase in the concentration of silt and clay (100–1000 mg $^{-1}$) and only a slight increase in concentration of sand. Furthermore, in the latter case, a distinct gradient in distribution of sand and silt and clay in suspension has developed across the flats. The results obtained from this single occasion, when storms were active, seem to indicate the great importance of such events, in spite of their infrequency, on intertidal-flat sedimentation. Also, it shows the importance of obtaining continuous day-to-day measurements, as such events are easily missed if there is a lower frequency of sampling.

The combination of sediment concentrations with current speeds and water depth allows the calculation of sediment flux. The importance of the first-phase flood and last-phase ebb is perhaps even more apparent when such calculations are made (Fig. 13). Here it is clearly seen that the greatest amount of sediment moves over a particular site (base 5) during these earlier and late stages of the tidal cycle.

Calculations have been made of the flux of suspended material over the various positions on the intertidal flats. These show that the resultant sediment flux, of between 456 and 912 tonnes moves, per tidal cycle (comparable, in magnitude, to those found by Carling, 1981, for the Burry Inlet), obliquely

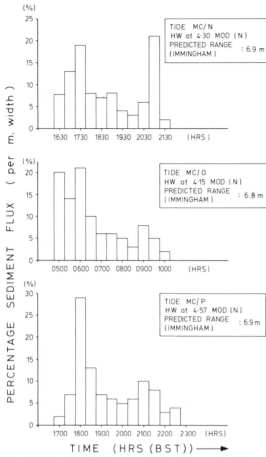

Fig. 13. Variations of suspended sediment flux during various tidal inundations. Notice the high percentage of the total flux transport during the first-phase flood and late-stage ebb periods. (Key MOD(N)—metres above datum (Newlyn).)

shorewards across the sampling line 1 (see Fig. 1), towards N 22° E.

BED LOAD TRANSPORT ACROSS THE INTERTIDAL ZONE

The presence of bedforms across the intertidal sandflats of the Wash reflects the movement of material as traction load. Two-dimensional, asymmetrical ripple marks occur across the upper intertidal flats; larger lunate ripples occur in tidal creeks and at the edges of the low-water tidal channel; and flood-oriented megaripples are exposed during extreme low water on banks in the centre of the low-water tidal channel (Fig. 1). The distribution of these

structures of various magnitudes, based on the bedform phase diagrams of Guy, Simmons & Richardson (1966) and Vanoni (1974), indicates that bed shear stresses increase seawards with a proportionate increase in sediment movement as traction load; this is indeed true, as is shown by the distribution of current velocities. Tidal creeks are exceptional in sometimes developing bedforms (anti-dunes) indicative of local high shear stresses.

Calculations of sediment moved as tractive load are described below; these are based on velocity profiles measured at half-hourly intervals and grain-size data and have been made for four stations along the intertidal transect (line 1 station 3, line 1 station 4, line 1 station 5, line 1 station 6). These sites are the same as those used to monitor the suspended sediment as described earlier.

In addition, direct measurements of the movement of bed material were made using a sediment trap designed for the purpose (Amos, 1974). The latter results showed, albeit in a semi-quantitative manner, that material was moving as bed-load across the entire intertidal sandflats and that the greater part of this material was moving alongshore.

The nature and distribution of sand-flat sediments

The majority of the sandflats are composed of well- to very well-sorted, fine to very fine sand. The mean size ($d_{50} = 0.125$ mm) is constant across the sandflats but increases to $d_{50} = 0.180$ mm in the low-water tidal channel. The mean grain size and the character of the size distribution generally remain constant with time. The grain-size changes only occur when the absolute magnitude of tidal flow increases beyond the critical stress for suspension. For example, the grain-size characteristics of a site on the outer sandflats were measured repeatedly over a three-year period from 1971 to 1973. During mid-1972 a significant increase in grain size occurred in response to the scouring effects of a nearby, migrating tidal creek. The mean grain size used in the calculation of sediment discharge was 0.125 mm. According to the phase diagram published by Allen (1970, p. 51 (after A. Shields and R. A. Bagnold)) this material should go immediately into suspension under flow conditions which exceed the critical threshold of movement, and indeed, local intermittent suspension does take place. Bedform phases for varying grain sizes and shear stresses, summarized by Middleton & Southard (1977), tend to be ill defined below a $d_{50} = 0.150$ mm. Yet bedforms in

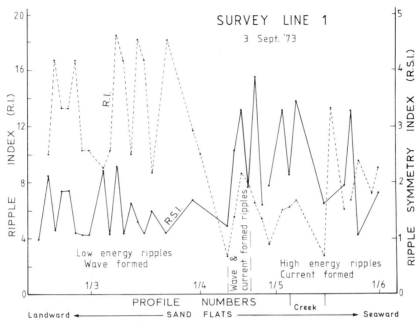

Fig. 14. Ripple mark characteristics across the intertidal zone (profile 1—see Fig. 1 for location of stations).

such material exist on the intertidal flats of the Wash.

Middleton & Southard (1977) suggested that sands with $d_{50} = 0.125$ mm accumulate in a zone of active transport and not in regions of sub-critical flow. The inner boundary of such an active transport zone is defined by the critical shear stress for this particular grain size (d_{50}) and the outer boundary by the shear velocity (U_*), which is equivalent to the Stokes settling velocity (w) of the material.

According to ASCE (1975), the settling velocity (w) of material of $d_{50} = 0.125$ mm through water at a temperature of 10°C is 1.5 cm sec^{-1}. By extrapolation, this is equivalent to a mean current 1 m above the bed ($\bar{U}_{100}$) = 0.4 cm sec^{-1}, which was exceeded by over 100% for considerable portions of the tidal cycle over the sandflats of the Wash. Indeed, at the outer boundary of the intertidal sandflat, the maximum $\bar{U}_{100} = 0.8$–1.1 m sec^{-1} (i.e. U_*(max) = 4.5–6.1 cm sec^{-1}); this is three times greater than the equivalent value of w. The outer boundary of active transport on the intertidal flat of the Wash, defined by the low-water tidal exposure, corresponds with a region where the maximum shear velocity (U_*(max)) is approximately $3w$.

In contrast to the maximum, the mean shear velocity $\bar{U}_*$ (averaged over the entire tidal inundation) is equal to 2.0 cm sec^{-1}; this is, indeed, close to the settling velocity (w) of the very fine sand deposits.

The relative grain size (d_{50}/h; where h is the mean water depth) of the intertidal sandflats varies from 6.05×10^{-5}, on the outer sandflats, to 2.5×10^{-4}, on the inner parts. The corresponding Froude number ($F = \bar{u}\sqrt{gh}$) maxima are 0.16 and 0.17, respectively. The observed phases of dune and ripple bedform development (Vanoni, 1974) do, in fact, correspond with those which might be predicted for such sediments and flow conditions (ASCE, 1975).

Surface sedimentary structures

The primary sedimentary structures were measured at 50 m intervals along an intertidal transect (line 1). The Ripple Indices (RI = length/height) and Ripple Symmetry Indices (RSI = wavelength/lee slope length) were determined and the bedforms classified after Tanner (1967) and Reineck & Singh (1973) (Fig. 14). The most common sedimentary structures measured were asymmetrical (RSI = 10–12), sinuous, low-energy ripples between 1 and 5 cm in height. According to Tanner (1967) these features are created by the combined influence of waves and currents. Landwards, on the inner parts of the sandflats, less asymmetric ripples occur. The crest lines anastomose and, in local depressions, are sharp crested and symmetrical. These features are wave-formed but have been reactivated by the ebb current.

On the lower parts of the sandflats and within the creeks linguoid and lunate high-energy current ripples occur. These bedforms increase in size seawards in response to the increase in grain size. On the bed of the low-water tidal channel megaripples occur.

The relative magnitude of wave oscillatory motion (U_{wave}) and unidirectional flow (expressed in terms of the friction velocity, U_*) have been related dimensionlessly in the form U_{wave}/U_* (Amos & Collins, 1978). For values greater than approximately 10, wave-formed bedforms are created: for values less than 1, current-formed bedforms are created. Based on the most common conditions of wave activity in the Wash (Driver & Pitt, 1974) intermediate ratios and bedforms would be expected to occur over the majority of the intertidal sandflats.

Over any one tidal cycle, the ratio U_{wave}/U_* varies. For example, on 23 August 1972, flows were wave-dominant on the inner parts of the sandflats during the first and last stages of tidal inundation (cf. Carling, 1981, fig. 8). During the remainder of the tidal inundation, intermediate flows occurred. For the same tide over the middle and outer part of the sandflats, flows were intermediate during the initial and final stages of tidal inundation and current-dominant during the remainder of the inundation. The critical velocity for entrainment of sediment as bed load (see later section) is exceeded during all stages of the tide, except for a period around the high-water stage. Thus it appears that a variety of types of bedform may be generated over a tidal inundation and, of these, only those produced during the last-stage ebb flow may be visible during low-water stage (see Fig. 15).

Bed-load movement

Two methods were used to calculate the net transport of bed material across the intertidal sandflats; the Sternberg (1972) method and the Engelund & Hansen (1972) method. Both methods were based on current measurements made synoptically at four sites along line 1 between 23 and 30 August 1972.

Sternberg's method was applied using a coefficient of drag (C_{100}) of 3.0×10^{-3} and using Sundborg's (1967) critical entrainment velocity ($\overline{U}_{100}$) which was 0.35 m sec^{-1} for the size of sediment covering the intertidal flats. The computation is based on the DuBoy's excess shear stress method and assumes a Von Karman's constant (K) of 0.4 for the logarithmic current velocity profile. The velocity profiles

(though the ebb does show variations) converge at a height above the bed, i.e. giving a roughness coefficient (Z_0) of 0.09 cm (this is comparable in magnitude to a Z_0 of 0.04 cm found for an intertidal sandflat in the Severn Estuary (Hamilton, 1979)). The mean Reynolds number at maximum tidal flow was 5×10^5 and flows were consequently hydrodynamically rough; hence the values of C_D and Z_0 were assumed to be constant throughout the tidal inundation.

The Engelund & Hansen method involves the solution of the following computation derived by ASCE (1975):

$$g_s = 0.05\delta_s U^2 \sqrt{\frac{d_{50}}{g(\delta_s/\delta - 1)}} \left[\frac{\tau_0}{(\delta_s - \delta)d_{50}}\right]^{\frac{3}{2}} \text{ (kg m}^{-1}\text{)},$$

where $\delta_s =$ sediment specific weight $= 2.65 \times 9.81$ (m sec^{-2}), $\delta =$ water specific weight $= 1.02 \times 9.81$ (m sec^{-2}), $d_{50} =$ mean grain size $= 0.000125$ m, $\tau_0 = C_D . \rho . |U|^2$ (N m^{-2}) and $U =$ depth-integrated velocity (m sec^{-1}).

The equation was derived for dune phase discharges and was originally calibrated for material coarser than 0.18 mm. However, both ASCE (1975) and Graf (1971) recommend its general use. The critical velocity for entrainment used in the formula is based on Shield's dimensionless shear stress for fully turbulent flows:

$$\frac{\tau_{crit}}{(\delta_s - \delta)d_{50}} = 0.06: \text{ from which } \tau_{crit} = 0.172 \text{ N m}^{-2};$$

this yielded a lower value for the critical entrainment velocity ($U_{100} = 0.23$ m sec^{-1}) than that derived from Sundborg's (1967) graph.

The hydraulic data used in the computations and the net bed-load transport for each of the four stations have been computed for a tidal inundation on 23 August 1972. The results, calculated in half-hourly intervals, are presented in Table 2. The Sternberg method predicts bed-load transport only across the lower parts of the sandflats (i.e. stations 3, 4/5 and 5 (see Fig. 1)). In this region transport occurs for a period of 0.5–1.5 h of the first phase of the flood and over 0.5 h of the final stages of the ebb. No transport occurs over the high-tide stage. The resultant bed-load transport of sediment is onshore and alongshore at discharge rates of 31 g cm^{-1} tide^{-1} towards N $14°$ E (station 4/5) and of 72 g cm^{-1} tide^{-1} at N $10°$ W (station 5), respectively, for the outer stations.

The Engelund & Hansen method predicts transport across the entire sandflat at all stations over a

Table 2. Bed-load computations

	U_{10}(m sec^{-1})	U_{100}(m sec^{-1})	U_*(cm sec^{-1})	τ(N m^{-2})	(°T)	Sediment discharge (g cm^{-1} sec^{-1})	
						Sternberg	Engelund & Hansen
				Station 2			
1	0.20	0.30	1.65	0.277	335		1.34×10^{-3}
2	0.20	0.30	1.65	0.277	280		1.34×10^{-3}
3	0.16	0.25	1.31	0.199	30		7.28×10^{-4}
4	0.11	0.17	0.93	0.089	65		
5	0.08	0.13	0.67	0.052	80		
6	0.12	0.18	0.98	0.100	120		
7	0.19	0.28	1.55	0.241	120		1.05×10^{-3}
				Station 3			
1	0.21	0.31	1.75	0.305	30	4.3×10^{-5}	1.54×10^{-3}
2	0.16	0.25	1.31	0.199	340		7.28×10^{-4}
3	0.16	0.25	1.31	0.199	360		7.28×10^{-4}
4	0.11	0.17	0.93	0.092	15		
5	0.12	0.18	0.98	0.100	40		
6	0.11	0.17	0.93	0.089	90		
7	0.13	0.20	1.06	0.123	120		
8	0.19	0.28	1.57	0.241	135		1.05×10^{-3}
				Station 4/5			
1	0.53	0.80	4.40	1.970	335	3.0×10^{-2}	4.16×10^{-2}
2	0.38	0.57	3.15	0.999	20	1.4×10^{-3}	1.27×10^{-2}
3	0.20	0.30	1.65	0.277	12		1.34×10^{-3}
4	0.16	0.23	1.31	0.163	25		
5	0.16	0.23	1.31	0.163	40		
6	0.15	0.23	1.23	0.163	73		
7	0.10	0.15	0.84	0 069	91		
8	0.18	0.26	1.47	0.208	95		8.14×10^{-4}
9	0.20	0.30	1.65	0.277	130		1.34×10^{-3}
10	0.52	0.76	4.30	1.776	130	2.0×10^{-2}	3.47×10^{-2}
				Station 5			
1	0.55	0.80	4.55	1.97	360	4.0×10^{-2}	4.16×10^{-2}
2	0.30	0.45	2.48	0.623	345	2.5×10^{-4}	5.55×10^{-3}
3	0.23	0.36	1.90	0.399	305	6.2×10^{-5}	2.54×10^{-3}
4	0.18	0.26	1.47	0.208	320		8.14×10^{-4}
5	0.16	0.25	1.31	0.192	5		
6	0.10	0.15	0.84	0.069	50		7.09×10^{-4}
7	0.17	0.26	1.40	0.208	80		8.14×10^{-4}
8	0.20	0.30	1.65	0.277	120		1.34×10^{-3}
9	0.20	0.30	1.65	0.277	120		1.34×10^{-3}
10	0.35	0.54	2.88	0.897	122	7.0×10^{-4}	1.05×10^{-2}

greater part of the tidal inundation (see Table 2). Transport rates are both higher and longer in duration during the flood than during the ebb. Consequently, the resultant bed-load transport of sediment is onshore and alongshore. The rates of bed-load transport for the four stations from the inner to the outer intertidal zone are 1·48 g cm^{-1} tide^{-1} at N 70° W (station 2), 4·25 g cm^{-1} tide^{-1} at N 66° E (station 3), 57·1 g cm^{-1} tide^{-1} at N 29° (station 4/5) and 75·8 g cm^{-1} tide^{-1} at N 10° W (station 5).

The two methods used to compute the bed-load transport vary according to which of the appropriate critical velocities are used; however, results are comparable (average difference $-0\cdot001$ g cm^{-1} sec^{-1}) at flows which exceed critical threshold conditions. When the computed bed-load transport is plotted against the prevailing friction velocity (U_*), both plots are linear and converge at a U_* value of 4·5. The resultant bed-load transport determined for the complete tidal inundation is remarkably similar

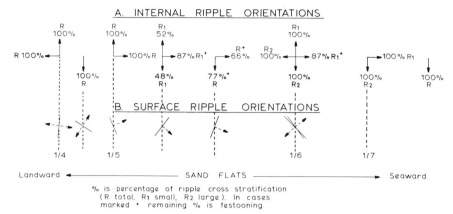

Fig. 15. The orientation of surface ripple marks and internal ripple cross-stratification.

using either method at stations 4 and 5. This is greatest on the lower intertidal sandflat and decreases shorewards. The resultant transport direction is onshore on the inner and outer parts of the flats and alongshore on the central portion of the sand-flats.

Internal sedimentary structures

Internal sedimentary structures were measured at 11 locations along line 1 (Fig. 1) as well as on other adjacent lines, using a Senckenburg box sampler ($30 \times 40 \times 5$ cm). Samples were collected in pairs, perpendicular and parallel to the shoreline.

The orientation of the preserved cross-laminae is shown in Fig. 15. In this figure the percentages show the cross-laminae as a percentage of the total surface area of the box core dipping in the direction of the arrow. In some cores the cross-stratification has been differentiated into various sizes: R, R_1 and R_2 which represent the apparent dip directions of the total, small-scale and large-scale cross-stratification respectively.

When the internal cross-stratification (Fig. 15A) is compared with the surface ripple orientations (Fig. 15B), it appears that the structures preserved appear dominantly to reflect this late-phase ebb run-off stage of the flow.

Only in a few box cores is any bimodal cross-stratification seen; and only in a few is the evidence of flood-tide transport obvious. If the preserved internal sedimentary structures were observed in ancient deposits this would be taken to indicate a dominant seaward movement of sand with only minor reversals: neither of these would appear to be correct.

CONCLUSIONS

The intertidal flats of the Wash are today areas of active sedimentation. Sediment is transported shorewards on to the intertidal flats from the neighbouring submarine areas. This phenomenon is taken advantage of and aided by man, and many thousands of hectares have been reclaimed and turned into productive agricultural land.

The decrease in grain size, from low- to high-water mark, of the intertidal flat sediments faithfully records the decrease in tidal current speed in that direction. Surface ripple marks, when observed at low water, naturally reflect the water movements of the last phase of the ebbing tide; however, their magnitude and general form clearly record the varying magnitude of tidal and/or wave-induced currents over the intertidal zone.

The landward or oblique landward net transport of sediment does not, however, appear to be reflected in the preserved sedimentary structures in the sand-flat sediments. These structures do not clearly record the circulation pattern of bed material and its seaward source. Instead, they record mainly the particular direction of bed-load sediment at the time of its inclusion and preservation in the geological record; this is selective and occurs largely during the ebbing phase of the tide, and thus the structures only record a small part of the tidal event. Such results suggest that caution should be exercised in conclusions, such as direction of net sediment transport, drawn from such features in ancient marine sediments.

ACKNOWLEDGMENTS

The writers would like to thank N.E.R.C. for supporting the study which led to this paper and also for supporting C. L. Amos during his Ph.D. studies at Imperial College. Also, they wish to record their gratitude to Mr Keith Naylor and Miss Mary Pugh for drawing the figures and Miss Amanda Hartland and Ms J. Greengo for typing the manuscript.

REFERENCES

ALLEN, J.R.L. (1970) *Physical Processes of Sedimentation.* Unwin University Books, London. 248 pp.

AMERICAN SOCIETY OF CIVIL ENGINEERS (1975) *Sedimentation Engineering* (Ed. by V.A. Vanoni). New York. 745 pp.

AMOS, C.L. (1974) *Intertidal flat sedimentation of the Wash—E. England.* Unpublished Ph.D. Thesis, University of London. 404 pp.

AMOS, C.L. & COLLINS, M.B. (1978) The combined effects of wave motion and tidal currents on the morphology of intertidal ripple marks: the Wash, U.K. *J. sedim. Petrol.* **48**, 849–856.

AMOS, C.L. & LONG, B.F.N. (1980) The sedimentary character of the Minas Basin, Bay of Fundy. In: *The Coastline of Canada: Littoral Processes and Shore Morphology* (Ed. by S.B. McCann). Geological Survey of Canada. Paper 80-10, 153–180.

ANDERSON, F.E. (1972) Resuspension of estuarine sediments by small amplitude waves. *J. sedim. Petrol.* **42**, 602–607.

CARLING, P.A. (1978) *The influence of creek systems on intertidal sedimentation.* Unpublished Ph.D. Thesis, University of Wales. 258 pp.

CARLING, P.A. (1981). Sediment transport by tidal currents and waves: observations from a sandy intertidal zone (Burry Inlet, South Wales). In: *Holocene Marine Sedimentation in the North Sea Basin* (Ed. by S.-D. Nio *et al.*). *Spec. Publs int. Ass. Sediment.* **5**, 65–80. Blackwell Scientific Publications, Oxford. 524 pp.

CLARK, H.W. (1959) In discussion of Inglis, C.C. and Kestner, F.J.T. (1958). *Proc. Instn civ. Engrs* **13**, 393–407.

COLLINS, M.B. (1972) *The Wash-sediment contribution from freshwater sources.* Interim Report to Natural Environment Research Council, U.K.

COLLINS, M.B. (1976) Suspended sediment sampling towers as used on the intertidal flats of the Wash, Eastern England. *Est. Coastal Mar. Sci.* **4**, 46–57.

DARBY, H.C. (1940) *The Drainage of the Fens.* Cambridge University Press, London.

DRIVER, J.S. & PITT, J.D. (1974) Wind and wave relationship in a shallow water area. *14th Conf. Coastal Engr.* **1**, 146–163.

ENGELUND, F. & HANSEN, E. (1972). *A Monograph on Sediment Transport in Alluvial Streams.* Technical University of Denmark, 62 pp.

EVANS, G. (1965) Intertidal flat sediments and their environments of deposition in the Wash. *Q. Jl geol. Soc. Lond.* **121**, 209–245.

EVANS, G. & COLLINS, M.B. (1975) The transportation and deposition of suspended sediment over the intertidal flats of the Wash. In: *Nearshore Sediment Dynamics and Sedimentation* (Ed. by J. Hails and A. Carr), pp. 273–304. Wiley, London.

GRAF, W.H. (1971) *Hydraulics of Sediment Transport.* McGraw-Hill, New York. 513 pp.

GUY, H.P., SIMMONS, D.B. & RICHARDSON, E.V. (1966) Summary of alluvial channel data from flume experiments, 1956–1961. *Prof. Pap. U.S. geol. Surv.* **462**, 96 pp.

HAMILTON, D. (1979) The high energy, sand and mud regime of the Severn Estuary, S.W. Britain. In: *Tidal Power and Estuary Management* (Ed. by R. T. Severn, D. Dinely and L.E. Hawker). Scientechnica, Bristol. 296 pp.

HANDYSIDE, T. (1977) *An investigation of hydrodynamical parameters of an intertidal sand flat in the Loughor Estuary using velocity gradient profiles.* Unpublished B.Sc. Dissertation, Department of Oceanography, University College of Swansea. 55 pp.

INGLIS, C.C. & KESTNER, F.J.T. (1958) Changes in the Wash as affected by training walls and reclamation works. *Proc. Instn civ. Engrs* **11**, 435–466.

INMAN, D.L. (1952) Measures for describing the size distribution of sediments. *J. sedim. Petrol.* **22**, 51–70.

KAMPS, L.F. (1963) Mud distribution and land reclamation in the Eastern Wadden Shallows. *Int. Inst. Land Reclamation Improvement*, **9**, 91 pp. Wageningen, the Netherlands.

KESTNER, F.J.T. (1963) The supply and circulation of silt in the Wash. *Proc. Int. Ass. Hydraul. Res. Congr. London*, pp. 231–238.

KESTNER, F.J.T. (1975) The loose-boundary regime of the Wash. *Geogrl J.* **141**, 388–414.

McDOWELL, D.M. & O'CONNOR, B.A. (1977) *Hydraulic Behaviour of Estuaries.* Macmillan Press, London. 292 pp.

MIDDLETON, G.V. & SOUTHARD, J.B. (1977) *Mechanics of Sediment Movement.* Short Course no. 3, Society of Economic Palaeontologists and Mineralogists, Tulsa.

NEWTON, R.S. (1968) A new device for measuring ripple mark profiles underwater. *Mar. Geol.* **6**, 73–75.

POSTMA, H. (1954) Hydrography of the Dutch Wadden Sea. *Archs. néerl. Zool.* **10**, 405–511.

POSTMA, H. (1961) Transport and accumulation of suspended matter in the Dutch Wadden Sea. *Neth. J. Sea Res.* **1**, 148–190.

REINECK, H.-E. & SINGH, I.B. (1973) *Depositional Sedimentary Environments.* Springer-Verlag, Berlin. 439 pp.

ROUSE, H. (Ed.) (1950) *Engineering Hydraulics.* Wiley, New York. 1039 pp.

RUXTON, D. (1979) Physical aspects of reclamation for freshwater reservoirs. In: *Estuarine and Coastal Land Reclamation and Water Storage* (Ed. by B. Knights and A.I. Phillips), pp. 152–176. Saxon House, Farnborough, England.

STERNBERG, R.W. (1972) Predicting initial motion and bed load transport of sediment particles in the shallow marine environment. In: *Shelf Sediment Transport* (Ed. by D.J.P. Swift, D.B. Duane and O.H. Pilkey), pp. 61–79. Dowden, Hutchinson & Ross, Stroudsburg.

SUNDBORG, A. (1967). Some aspects on fluvial sediments and fluvial morphology. General views and graphic methods. *Geogr. Annlr.* **49**(A), 333–343.

TANNER, W.F. (1967) Ripple mark indices and their uses. *Sedimentology,* **9**, 89–104.

VANONI, V.A. (1974) Factors determining bed forms in alluvial channels. *Am. Soc. civ. Engrs* **100**, 363–377.

VAN STRAATEN, L.M.J.U. & KUENEN, Ph. H. (1957) Accumulation of fine grained sediments in the Dutch Wadden Sea. *Geologie Mijnb.* **19**, 329–354.

WILMOT, R.D. & COLLINS, M.B. 1981. Contemporary fluvial sediment supply to the Wash. In: *Holocene Marine Sedimentation in the North Sea Basin* (Ed. by S.-D. Nio *et al.*). *Spec. Publs int. Ass. Sediment.* **5**, 99–110. Blackwell Scientific Publications, Oxford. 524 pp.

Spec. Publs int. Ass. Sediment. (1981) **5**, 99–110

Contemporary fluvial sediment supply to the Wash

R. D. WILMOT* *and* M. B. COLLINS†

Department of Geology, Imperial College, London, U.K.

ABSTRACT

The area draining into the Wash is one of generally low relief covering an area of 12,500 km² and includes the flat, artificially drained region of the Fenland. Attempts have been made to determine the amount of material supplied from this drainage basin to the Wash, and to compare this with the amount of suspended sediment within the Wash. Fluvial contributions have been assessed by the use of prediction equations established for similar areas, and by the use of a variety of data collected within the drainage basin. The latter includes the results of a three-year depth-integrated suspended sediment sampling programme, as well as longer-term surface bucket sampling carried out by the Anglian Water Authority and its predecessors. The surface bucket data covering the complete discharge range have been used to construct rating curves relating sediment concentration to water discharge. Such rating curves, combined with the appropriate flow duration curves, are used to calculate annual sediment yields at several sites, these being extrapolated by load/area or load/discharge relationships to non-sampled sites. These various methods give annual sediment loads ranging from 43,000 to 173,000 tonnes, but are within the same order of magnitude. These loads are several orders of magnitude lower than the amount of sediment carried in suspension over the intertidal flats of the Wash each year, confirming quantitatively the results of other studies, both in the Wash and the Wadden Sea, which have suggested that fluvial contributions are of minor importance in intertidal deposits of the Southern North Sea.

INTRODUCTION

The Wash embayment in east England (Fig. 1) is the remnant of a basin now partially infilled by the Flandrian peats and muds of the Fenland. The sediments of the Wash have been the object of several studies (e.g. Evans, 1965; Shaw, 1973; Kestner, 1975) and various erosional sources for this material have been suggested (Evans, 1965); the east coast of Lincolnshire; Pleistocene boulder clay within the deeper parts of the Wash; the north Norfolk coastline; and the bed of the North Sea. Material from each of these sources would subsequently be trans-ported to the intertidal zone of the Wash by tidal currents and longshore drift. Interpretations of the available data suggest that the rivers contribute little sediment to the Wash; however this has never been proven quantitatively. The present paper describes the results of a variety of approaches to quantify the contemporary fluvial contribution and is the result both of investigations into the potential siltation in bunded reservoirs within the Wash (Collins, 1975) and as part of a study on the evolution of the Wash drainage basin (Wilmot, 1981).

Four methods have been used to calculate the annual sediment supply of the four principal rivers (the Witham, Welland, Nene and Ouse) which drain into the Wash:

(a) by reference to pre-existing prediction equations, derived for rivers in temperate climate zones;

Present addresses: *Applied Mineralogy Unit, Institute of Geological Sciences, 64 Gray's Inn Road, London WC1, U.K. and †Department of Oceanography, University College, Swansea, Wales, U.K.

0141-3600/81/1205-0099 $02.00

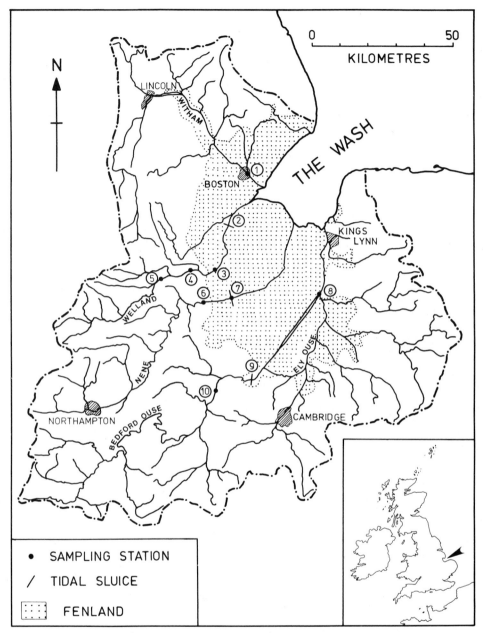

Fig. 1. The drainage basin of the Wash. Inset shows general location map. (1) Grand Sluice, Boston. (2) Marsh Road Sluice, Spalding. (3) Peakirk. (4) Tallington. (5) Tinwell. (6) Orton. (7) Dog-in-a-Doublet Sluice. (8) Denver Sluice. (9) St Ives. (10) Offord.

(b) combining mean suspended sediment concentration data, based on the results of a depth-integrating programme at the outfalls, with average annual mean daily water discharges;

(c) extrapolating and interpolating a sediment yield/catchment area relationship for the Wash rivers, based on the analysis of short-term flow and suspended sediment concentration data from two sampling sites; and, finally,

(d) analysis of longer-term suspended sediment and discharge data from six sampling sites.

These methods of approach to the problem, and the results obtained are described in this paper.

PHYSIOGRAPHY OF THE DRAINAGE BASIN

The region draining into the Wash covers an area of 12,500 km² and is divisible into two major regions (Fig. 1). The smaller and more seaward of these, with an area of 3500 km², is known as the Fenland, and is underlain by freshwater peats and freshwater, brackish and intertidal silts and clays. The Fenland reaches a maximum height of 15 m above sea-level and has been subjected to extensive drainage; consequently the rivers crossing it all flow in artificial channels. River gradients here are very low (approximately 1 in 10,000) and, as a result of peat wastage after drainage, many of the rivers run above the level of the surrounding land, being supplied by pumping stations. The outfalls of these rivers, into the tidal reaches, are controlled by engineering structures (e.g. sluices) which only allow freshwater discharge at low tide. These tidal sluices are at varying distances from the Wash itself; being very close (10 km) at Boston on the River Witham, but almost at the edge of the Fenland at St Ives on the Bedford Ouse and Dog-in-a-Doublet on the River Nene (Fig. 1).

The remainder of the drainage basin, which reaches a maximum height of 247 m above sea-level, has an area of 9000 km². This region is underlain by a variety of Jurassic and Cretaceous clays and limestones, while much of it is covered by Quaternary gravels, sands and tills. River gradients are steeper here than in the Fenland (typically, 1 in 2500) and although locks and sluices are used to control flooding and improve navigation, there is a naturally developed drainage network.

The average annual rainfall over the drainage basin is approximately 570 mm. Monthly rainfall maxima occur during summer (June and August) and winter (November and January), with minima in spring (March and April) and autumn (October). Varying soil moisture contents and rates of evapotranspiration modify this temporal distribution of rainfall, to produce a monthly runoff or discharge distribution with a single maximum during February and a single minimum during September. The average runoff, based on the sum of the freshwater discharges at the artificial controls, is 185 mm (Wilmot, 1981).

METHODS

(a) Use of pre-existing prediction equations

A variety of equations for predicting sediment loads have been established (for examples see Gregory & Walling, 1973) The majority of such equations are based on multiple regression analysis of annual load against a number of geomorphological and climatological variables; however, a number based solely on the relationship between load and catchment area have been developed by Fleming (1969). The latter utilized world-wide data from some 253 catchments with a variety of vegetational types. The equation applicable to the British Isles (converted into metric units, with load (y) in tonnes annum^{-1} and area (x) in km²) is:

$$y = 156 \cdot 0 \, x^{0 \cdot 96}.$$

The catchments considered in the original analysis (Fleming, 1969) were much larger than those which commonly occur in the British Isles, and a modified equation using previously unpublished data from British catchment areas in Sussex and Yorkshire, together with a limited amount of American data relating to small catchments, was produced by Collins (1970). This modified equation (in metric units) is:

$$y = 60 \cdot 0 \, x^{0 \cdot 78}.$$

In this study, the latter equation has been used to predict annual loads for the rivers of the Wash drainage basin.

(b) Depth-integrated suspended sediment sampling programme

Between January 1972 and September 1974 a programme of depth-integrated suspended sediment sampling was undertaken in association with the then Lincolnshire, Welland and Nene, and Great

Ouse River Authorities. Sampling stations were established at control structures draining into the tidal reaches, at Fenland pumping stations and at Offord on the Bedford Ouse (being the most seaward convenient location on this river). Outfall sluices and pumping stations are effectively constricted sections, so that samples collected at these points should measure the 'total load', including bed material brought into suspension. Full details of the sampling programme and the equipment used during the survey are presented elsewhere (Collins, 1972, 1975).

The period covered by this sampling programme was one when rainfall was below average (e.g. 456 mm in 1972 and 475 mm in 1973) and consequently discharges were low. Because of this restricted range of discharge, an 'average' suspended sediment concentration (60 ppm) derived from this sampling programme was adopted and combined with monthly and annual mean daily water discharges in order to provide an estimate of average annual sediment loads. The water discharge information was provided by the River Authorities referred to above, based on measurements at gauging structures on the rivers (or indeed estimates where such data were unavailable), or pump settings and water levels in the case of drainage channels.

(c) Analysis of short-term suspended sediment concentration and water discharge data

Because, generally, only low discharges were sampled by the programme described in (b), other suspended sediment concentration data, representing higher discharges and collected by the Welland and Nene River Authority over the period 1968–70, were also analysed. These data are limited to surface bucket samples from two sites: Tinwell on the River Welland and Orton on the River Nene. Such samples could be less accurate than those collected using the depth-integrating technique and might underestimate sediment concentrations, particularly of larger grain sizes whose concentration varies appreciably with depth (Leopold, Wolman & Miller, 1964).

The suspended sediment concentration data and associated mean daily discharges were used to construct rating curves (Campbell & Bauder, 1940), between concentration and water discharge; this overcomes any spurious correlations which might develop if sediment load is related to water discharge (e.g. Gregory & Walling, 1973). The general form of the rating curve is:

$$\log C = A + B \log Q$$

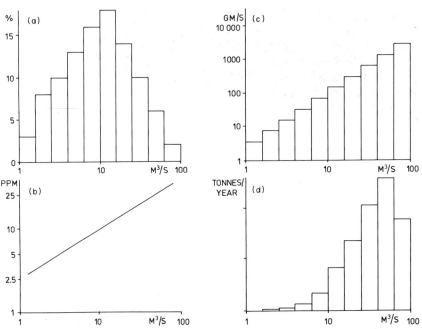

Fig. 2. Derivation of annual loads. (a) Flow-duration histogram. (b) Suspended sediment rating curve. (c) Suspended sediment discharge within each discharge increment. (d) Suspended sediment load within each discharge increment. (Total load is the sum of the heights of the histogram bars.)

where C is sediment concentration (ppm), Q is water discharge (cumecs) and A and B are constants.

Curves were fitted by linear least squares regression analysis to the logarithmically transformed data.

Annual (suspended) sediment loads were derived by combining the sediment concentration rating curves with the flow duration curve covering the same period. In this study, the two curves were combined by: (1) dividing the discharge range into 41 equally spaced logarithmic divisions; (2) using the rating curve to predict the sediment concentration corresponding to each discharge division; and (3) multiplying this concentration by the average discharge within the division and the corresponding frequency of occurrence of flows. The annual load is the sum of these subtotals within each division. The principles of this technique are illustrated in Fig. 2 and the computer program used will be described elsewhere (Collins & Wilmot, in preparation).

The sediment loads for Tinwell and Orton were calculated using this technique; based on these an approximate annual load/catchment area relationship was established and applied subsequently to the remaining catchments of the Wash drainage basin.

(d) Analysis of longer-term suspended sediment and water discharge data

In an attempt to improve the accuracy of the predicted loads as derived from the technique outlined above, use is made of longer-term data from a larger number of sites. Sediment concentrations were based on surface bucket samples and discharges on flow through gauging structures. Table 1 indicates the availability of this sediment and discharge data for sampling/gauging stations on the Welland, Nene and Ouse rivers, as supplied by the Anglian Water Authority.

Rating curves were, once again, established by least squares linear regression analysis on the logarithmically transformed data. Overall, annual and seasonal (April–September and October–March) rating curves were constructed for each site.

Rating curves were combined with the flow-duration curves in the same manner as described for method (c); however, different combinations of the two types of curve were analysed:

(1) an 'overall' rating curve combined with a long-term (overall) flow-duration curve;

(2) an 'overall' rating curve combined with yearly flow-duration curves;

Table 1. Periods of availability of suspended sediment and water discharge information, used for method (d)

River	Station (see Fig. 1)	Catchment area (km²)	Period of availability Sediment concentration	Water discharge
Welland	Tinwell	530	1968–78	1968–78*
	Peakirk	750	1972–78	1968–78†
Nene	Orton	1630	1967–78	1968–78
	Dog-in-a-Doublet	1680	1974–78	1968–78‡
Ely Ouse	Denver	3570	1969–78	1926–75
Bedford Ouse	Offord	2570	1969–78	1935–76

* Based on flows at Tallington (see Fig. 1) multiplied by 0·6.
† Assumed equal to flows at Tallington.
‡ Assumed equal to flows at Orton.

(3) annual rating curves combined with the corresponding yearly flow-duration curves;

(4) seasonal rating curves combined with seasonal flow-duration curves.

Note: methods (a), (b) and (c), above, all used total suspended sediment data; however, for method (d) both total sediment concentrations and those after ashing at 550 °C were available. Although some of the difference between these two values can be attributed to dehydration of clay minerals or breakdown of carbonates, the main weight loss is due to oxidation of the organic matter present. Using both concentrations, an approximate ratio of organic/inorganic material reaching the Wash has been determined.

RESULTS

Methods (a), (b) and (c)

Table 2 presents the annual loads derived using methods (a), (b) and (c) for all the freshwater outfalls into the tidal reaches of the Wash.

The results of the depth-integrated suspended sediment sampling programme (method (b)) are presented, as histograms, in Fig. 3. With the proviso that discharges were below average for much of the time, suspended sediment concentrations at both constricted and unconstricted sections rarely exceeded 100 ppm. In the absence of concurrent discharge data at many of these localities, rating curves could not be established; consequently, a *mean* sediment concentration was used in the calculations.

Table 2. Predicted annual loads (tonnes), using various methods

Site	River	Method of computation		
		(a)*	(b)†	(c)‡
Denver	Ely Ouse	35,000	30,000	26,000
Offord	Bedford Ouse	27,000	27,000	22,000
Dog-in-a-Doublet	Nene	19,500	24,000	16,400
Marsh road	Welland	11,000	10,000	10,000
Grand Sluice	Witham	22,000	23,000	18,600
Sluices on drainage channels§		25,000	14,000	24,000
Totals		139,500	128,000	117,000

* Suspended sediment, based on prediction equation.
† Total load, based on depth-integrated sampling at constricted sections.
‡ Total load, based on prediction of suspended sediment load (Fig. 4) and including an estimate of bed-load as 20 % of the total.
§ Including St German's Pumping Station, Black Sluice (South Forty Foot Drain) and Hobhole Sluice (see Collins, 1975, fig. 1).

This was estimated to be 60 ppm and was applied to discharge data from all stations.

The equations of the rating curves and the predicted annual loads for the upstream sampling stations at Tinwell and Orton, using method (c), are presented in Table 3. Further, Fig. 4 represents the relationship between these loads and catchment area (see Table 1), from which loads at the outfalls were derived.

Method (d)

A summary of the water discharge suspended sediment concentration data used in method (d) is presented in Table 4; this is based on averages of all the available data. In general, an increase in catchment area corresponds with an increase in the annual average mean daily discharge and a decrease in the variability of discharge values; the latter being due to an increase in the storage capacity of the basin. An increase in catchment area also corresponds with a decrease in average sediment concentration (the only exception to this generalization being Dog-in-a-Doublet Sluice, which is probably affected by the intrusion of estuarine water).

Table 4 also presents the percentage loss in weight upon ashing of the suspended sediment sample; this represents the organic content of the sediments and remains fairly constant in the larger rivers. More detailed analyses have shown (Wilmot, 1981) that

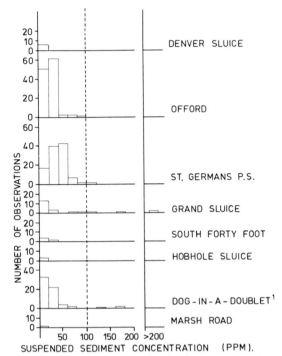

Fig. 3. Histograms relating number of observations from depth-integrating programme to suspended sediment concentration (method (b)).

Table 3. Regression equations for rating curves and annual suspended sediment loads (1968–70) at Orton and Tinwell, on the Rivers Nene and Welland respectively

Station	Year	Intercept (a)	Gradient (b)	Annual load (tonnes)
Orton	1968	0·34	0·94	21,000
	1969	0·54	0·66	16,000
	1970	0·76	0·59	14,000
Tinwell	1968	1·02	0·76	6000
	1969	1·14	0·81	11,000
	1970	1·27	0·72	6000

organic matter concentrations are greatest during low discharges in summer, and lowest during the winter months.

The coefficients of regression of the overall rating curve (using method (d)) at each site are given in Table 5, together with annual loads predicted from these using the overall flow-duration curve (method d(1) above). A typical example of the use of an 'overall' rating curve is illustrated in Fig. 5, together with the observations on which it was based. In all

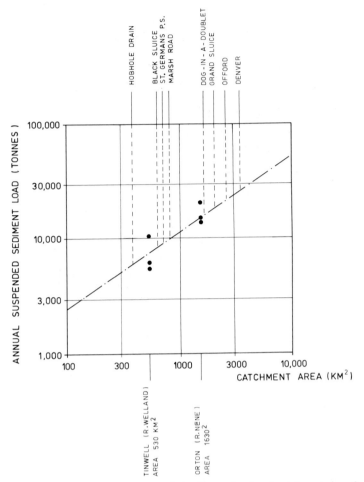

Fig. 4. Approximate relationship between annual suspended sediment load and catchment area, for upstream sampling locations. Interpolation and extrapolation of the relationship is applied to the outfalls.

Table 4. Summary of water discharge and suspended sediment concentration data

| Site | Discharge (cumecs) | | Suspended sediment (ppm) | | | Ashed mean | Loss on ashing at 550 °C (%) |
	Mean (1)	SD (1)	Max.	Min.	Mean		Mean
				Total			
Denver	11·9	1·41	149·0	3·2	17·9	11·5	36
Offord	6·6	1·50	191·0	4·0	20·1	13·1	36
Orton	6·9	1·51	193·0	2·0	23·4	15·1	35
Dog-in-a-Doublet	—	—	206·0	7·5	27·0	17·9	34
Peakirk	—	—	773·0	0·5	25·1	18·4	27
Tallington	1·5	1·78	—	—	—	—	—
Tinwell	—	—	692·0	0·5	29·2	25·3	15

(1) Derived from the logarithmically transformed data.

cases (Table 5) the gradient of the rating curve is less than unity, which differs from most other studies where it generally lies between 1·0 and 2·0 (Gregory & Walling, 1973). In other systems with a readily available supply of sediment, sediment concentration would be expected to increase with water velocity and discharge (see Leopold *et al.*, 1964); similarly, the rating curve gradient would be expected to be greater than unity. In contrast, sediment concentration in the drainage basin of the Wash appears to be controlled by sediment availability rather than the transporting capacity of the river waters.

Annual suspended sediment loads calculated using method (d(2)), although not presented here, differ only slightly ($\pm 5\%$) from those using method (d(1)). These differences arise from approximations and the groupings of discharge data into various increments during the calculations.

An uneven sampling distribution during the period of investigation limits the number of annual rating curves which can be derived. Some years have too few sampling occasions or unevenly distributed data: further, years of low discharges, such as 1973 and 1976, are associated with high levels of organic matter in the water, which are not directly related to the corresponding water discharge. An example of

Table 5. Intercepts (a) and gradients (b) of rating curves relating suspended sediment concentration (y, in ppm) to water discharge (x, in cumecs) of the form $\log y = a + b \log x$. Annual loads are predicted by combining these with long-term flow duration curves (method d(1))

Station	Intercept (a)	Gradient (b)	Annual load (tonnes)
Tinwell	1·00	0·69	2800
Peakirk	1·06	0·16	1800
Orton	0·90	0·37	8300
Dog-in-a-Doublet	1·20	0·15	8400
Denver	1·10	0·05	8100
Offord	0·94	0·38	7300

the variation between annual rating curves at one sampling site (Orton on the River Nene) is shown in Fig. 6: overall and winter rating curves are included for comparison (the influence of organic matter at low discharges means that a summer curve cannot be derived). Because low discharge/high organic matter values are largely excluded from the rating curve analyses drawn in Fig. 6, the significant (at the 5% level, using the F test) annual rating curves are generally steeper than the overall rating curve (no. 1

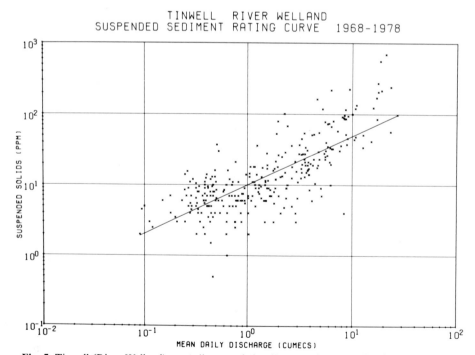

Fig. 5. Tinwell (River Welland): overall suspended sediment rating curve for the period 1968–78.

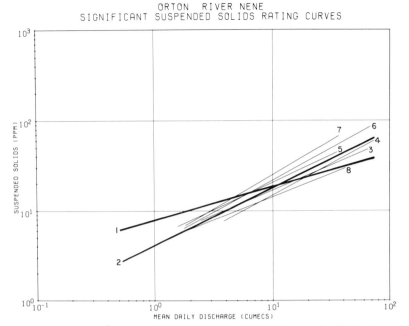

Fig. 6. Orton (River Nene): annual suspended sediment rating curves significant at the 5 % level, together with the overall and winter curves for comparison. (1) Overall, (2) winter, (3) 1968, (4) 1969, (5) 1970, (6) 1971, (7) 1972, (8) 1977.

in Fig. 6); this, in turn, leads to an increased sediment load prediction. For example at Orton, the mean annual sediment load predicted using method (d(3)) is 11,100 tonnes, while at Offord it is 8800 tonnes (compared with values of 7700 and 8300 tonnes, respectively, using method (d(1)) (Table 5). At Tinwell, there is less difference between predictions using the various methods, probably because organic material is less abundant at this site and its influence on the gradient of the rating curve (by increasing the low discharge concentrations) is less important.

The small amount of variation between the annual rating curves (Fig. 6) would appear to indicate that the most important factor in controlling annual load variability is water discharge. Although not presented here, there is a general correlation between sediment load and the annual (arithmetic) average mean daily discharge; however, because the flow-duration curves approximate to log-normality and a power function is used to relate discharge to concentration, a better predictor of sediment load is the annual logarithmic average mean daily discharge. This relationship is shown in Fig. 7.

Once again, because of the influence of organic matter at low flows, significant (at the 5 % level) seasonal rating curves for both summer and winter

could only be established at Offord (Bedford Ouse) and at Tinwell (River Welland); in contrast, winter curves were significant at all the sites. The combination of the seasonal curves with the appropriate seasonal water discharge data demonstrates that, at Tinwell, the winter proportion of the load is some 10 times greater than the summer proportion; similarly the total load using this method is some 13 % greater than a prediction based on methods (d(1)) to (d(3)) above. At Offord, differences are far less, with an increase of 3 times in the winter contribution over that of the summer, and of only 3 % over the load predicted from the overall curve. Such differences, both at stations and between sites, are considered to arise because of differences in the magnitude and frequency of floods. For example at Tinwell floods are fairly frequent during the winter and corresponding water levels may be up to 3 m higher than those during summer. Conversely, at Offord the much larger catchment area (Table 1) dampens out any short-period floods; corresponding differences between summer and winter water levels may only be a matter of 1 m.

The relationships shown in Fig. 7 were used to predict the suspended sediment loads of non-sampled rivers, and the total (suspended load and bedload (see below)) annual fluvial sediment con-

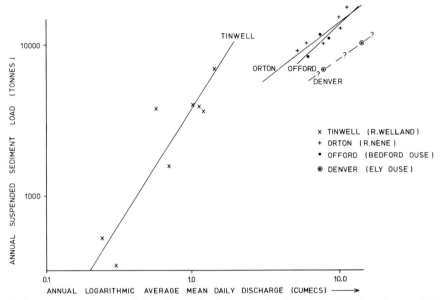

Fig. 7. Relationship between annual suspended sediment load (predicted using method d(3)) and annual logarithmic average mean daily discharge, showing the variation between catchment areas.

tribution to the Fenland is estimated to be 54,000 tonnes; of this, 43,000 tonnes is thought to reach the tidal reaches (Wilmot, 1981). This figure, based on method (d), is somewhat less than the predicted totals of 173,000, 129,000 and 146,000 tonnes estimated by methods (a), (b) and (c), respectively (see Collins, 1975, table 6); however, because of the aims of the study, these latter predictions were all specifically intended to predict maximum loads.

The use of prediction equations (method (a)) related to suspended sediment; however, the total load has been derived based on the assumption that the suspended load contribution is 80% of the total (Collins, 1972). Method (b), based on sampling at constructed sections, was considered to have included bed load in the original concentration measurements. Methods (c) and (d), on the other hand, were based entirely upon measurements of suspended sediment. Although no direct measurements of bed load transport have been made as part of the study, examination of the bed material distribution along the whole of the river system (Wilmot, 1981) demonstrates that transport of sand-grade material is occurring up to the edge of the Fenland. In order to account for this in the results of method (d), the load reaching the Fenland, as presented above, includes an estimate of 20% for bed load as a proportion of the total. This same figure was used

in conjunction with method (c) (Collins, 1975) for the bed load component at the outfalls; examination of the bed material however suggests that little, if any, sand-grade material crosses the Fenland, so that this value is probably an overestimate. Taking into account that silt-sized material may be transported as bed load and that the surface sampling used in method (d) probably underestimates suspended load, a revised estimate of 10% bed load was used in method (d) for the load reaching the estuaries.

DISCUSSION AND CONCLUSIONS

A variety of methods have been used for predicting annual sediment loads of the Wash rivers, representing the contemporary fluvial contribution to the estuarine system and the tidal reaches. These predictions of total load vary only by an order of magnitude and range from 43,000 to 173,000 tonnes. Such estimates of annual transport through the artificial sluices can be compared with 200,000 tonnes presented in the Crown Estates Commissioners Report (1969).

Within the context of the significance of the freshwater sediment supply to the overall pattern of fine-grained sediment transport and deposition within the Wash embayment, these annual loads can be compared with (Evans & Collins, 1975): (a) minima of 30,000 and 120,000 tonnes of sediment carried over the intertidal flats of the Wash in suspension during each neap and spring tide respectively; and (b) a total of approximately 5×10^7 tonnes annum^{-1} of sediment transported over the intertidal flats of the Wash (based on a general analysis of suspended sediment data collected over the intertidal zone).

Some of the material in suspension over the intertidal flats may be the result of *in situ* resuspension and, for this and other reasons, the net rate of deposition in the Wash is difficult to determine. Nonetheless, the comparative orders of magnitude of the transport processes over the intertidal zone and the freshwater sediment contribution appear to provide quantitative evidence for the previously held assumption that the material being transported, and hence that eventually deposited, within the Wash is *not* derived primarily from the river systems.

Additional evidence for the offshore source of at least much of the sediment of the intertidal zone is provided by the composition of the sediments. These contain displaced marine organisms (MacFadyen, 1938), and have a heavy mineralogy related to that of the adjacent seafloor (Baak's, 1936, E-Group), and the coastal exposures (Chang, 1971). The clay mineralogy (Shaw, 1973) is not indicative of a particular source, but shows that material from a variety of sources is well-mixed by repeated reworking of deposited material.

Comparison can be made between this conclusion for the origin of material in the Wash, located within the western coastline of the North Sea, and that reached by Dutch workers for the material of the eastern coastline. Referring to the Eastern (Dutch) Wadden Sea shallows, Kamps (1963) summarizes the investigations as follows.

(1) 'Concerning the question of the origin of the shoal mud of a larger grain size than 10 microns, it must be concluded that it consists mainly of marine material. The possibility of a direct supply of mud from the Rivers Ems, Weber and Elbe to the shallows must be denied, although perhaps a part of the mud which is discharged into the N. Sea is mixed with purely marine material and then brought onto the shoals through the sea gaps. But when they are thinned out so extensively that it is not shown mineralogically' (Dr Cromueliu's conclusion).

(2) 'From textural and mineralogical research (photographic X-rays) of a number of samples of mud and water from the shoals and the Ems, Weber and Elbe, it has appeared that there are differences between the samples from the two shoals and those

of the rivers, on the basis of which the sedimentation of recent Ems, Weber and Elbe mud on the Groningen shoals must be considered unlikely' (Dr Favejee's conclusion).

The freshwater contributions around the North Sea, therefore, would not appear to be major source areas for the supply of material to the various intertidal zones. Rather, an offshore source is inferred.

An examination of the 40 to 50-year water discharge data from Offord and Denver shows that there have been no significant changes in annual average mean daily discharges over this period. There have however been slight changes in the annual maximum mean daily discharges. At Offord a decrease in maximum discharges since 1960 is interpreted as being due to improved river management, while an increase in annual maxima at Denver since 1960 probably results from the opening of the Cut-off Channel; a catchwater drain which diverts high flows around the Fenland direct to Denver. In general, however, suspended sediment loads calculated using water discharge data from the past 10 years are within $\pm 3\%$ of those using data from the past 50 years. From the point of view of sediment transport, therefore, the last 10 years can be regarded as representative of, at least, the past 50 years.

Deposition of sand at the edge of the Fenland accounts for some of the difference between the load entering the Fenland and that reaching the estuaries. The remainder results from flood-plain and channel storage of silt and clay. Most of the Fenland reaches of the rivers retain, between the channels and the embankments, areas of flood-plain which, by increasing the channel width at high flows, accommodate floods safely and are sites of sediment deposition. The sluices at the outfalls impound river water both for periods of a tidal cycle and over several months when flows are very low and water is abstracted for irrigation; consequently, although the absence of silting behind the sluices has often been cited as evidence against appreciable sediment transport by the rivers, some channel deposition will occur during such periods of slack water. Much of this material will be retransported when the sluices are reopened. This sequence of movement has been examined by Collins (1975), who sampled at half-hourly intervals after the opening of the sluice gates, and found that there was a steady increase in sediment concentration until a maximum was reached at about $2\frac{1}{2}$ h after the sluice (Grand Sluice, Boston, on the River Witham) opening.

Sediment is also stored in the flood-plains and channels of the upper reaches of any river system; in the case of the Wash rivers this is increased by river management. The latter, attempting to limit flooding and improve navigation, has resulted in over-deepening of the channels, and this in turn not only means the direct removal of material from the channels, but also, by lowering water velocities, results in the deposition of fine-grained material which would otherwise leave the system as suspended load (Wilmot, 1981).

Holeman (1968) gives an average denudation rate of 35 tonnes km^{-2} yr^{-1} for Europe, which would be increased by the agricultural practices found in the present area. This rate is much greater than that predicted from the sediment load reaching the edge of the Fenland (method (d)), which gives a denudation rate of 6 tonnes km^{-2} yr^{-1}. The difference between these two rates reflects the amount of material being stored within the drainage basin. The Fenland itself is undergoing greater rates of denudation, but this is almost entirely the result of wind erosion of the heavily used soils.

In conclusion, the annual fluvial sediment contribution to the Wash over the past 50 years at least is of the order of 100,000 tonnes. Even without the influence of man, which has reduced sediment loads in this area, this annual fluvial load would be several orders of magnitude less than the amount of sediment in suspension over the intertidal flats of the Wash in a year.

ACKNOWLEDGMENTS

We would like to thank the Anglian Water Authority and its predecessors, the Lincolnshire, Welland and Nene and Great Ouse River Authorities, for their help and cooperation, and for supplying much of the data on which this study is based. The work was undertaken while R.D.W. was in receipt of an NERC Research Studentship, and M.B.C. was supported by the NERC and the Water Resources Board (through the Hydraulics Research Station, Wallingford). Keith Naylor (Department of Oceanography, Swansea) kindly redrew the original figures.

REFERENCES

BAAK, J.A. (1936) *Regional petrology of the Southern North Sea.* Thesis, Wageningen, the Netherlands.
CAMPBELL, F.B. & BAUDER, H.A. (1940) A rating curve method for determining the silt discharge of streams. *Trans. Am. Geophys. Un.* 21, 603–607.
CHANG, S.C. (1971) *A study of the heavy minerals of the coastal sediments of the Wash and the adjacent areas.* M. Phil. Thesis, University of London.
COLLINS, M.B. (1970) Discussion of design curves for suspended load estimation. *Proc. Instn civ. Engrs* 46, 85–88.
COLLINS, M.B. (1972) *The Wash—sediment contribution from freshwater sources.* Interim Report, Imperial College, London. 45 pp.
COLLINS, M.B. (1975) *The Wash—sediment contribution from freshwater sources.* Final Report, Imperial College, London. 44 pp.
CROWN ESTATE COMMISSIONERS REPORT (1969) *Wash Land Reclamation, Technical and Economic Aspects of Reclamation at Wingland.* Grontmij n.v., the Netherlands.
EVANS, G. (1965) Intertidal flat sediments and their environments of deposition in the Wash. *Q. Jl geol. Soc. Lond.* 121, 209–245.
EVANS, G. & COLLINS, M.B. (1975) The transportation and deposition of suspended sediment over the intertidal flats of the Wash. In: *Nearshore Sediment Dynamics and Sedimentation; an Interdisciplinary Review* (Ed. by J. Hails and A. Carr). Wiley, London. 316 pp.
FLEMING, G. (1969) Design curves for suspended load estimations. *Proc. Instn civ. Engrs* 43, 1–9.
GREGORY, K.J. & WALLING, D.E. (1973) *Drainage Basin Form and Process: a Geomorphological Approach.* Edward Arnold, London. 457 pp.
HOLEMAN, J.N. (1968) The sediment yield of major rivers of the world. *Wat. Resour. Res.* 4, 737–747.
KAMPS, L.F. (1963) *Mud Distribution and Land Reclamation in the Eastern Wadden Shallows.* Publication No. 9. International Institute for Land Reclamation and Improvement, Wageningen, the Netherlands. 91 pp.
KESTNER, F.J.T. (1975) The loose-boundary regime of the Wash. *Geogrl J.* 141, 388–414.
LEOPOLD, L.B., WOLMAN, M.G. & MILLER, J.P. (1964) *Fluvial Processes in Geomorphology.* Freeman, San Francisco. 522 pp.
MACFADYEN, W.A. (1938) Post glacial foraminifera from the English Fenland. *Geol. Mag.* 75, 409–417.
SHAW, H.F. (1973) Clay mineralogy of Quaternary sediments in the Wash Embayment, Eastern England. *Mar. Geol.* 14, 29–45.
WILMOT, R.D. (1981) *The nature of the fluvial sediment contribution to the Fenland coastal plain.* Ph.D. Thesis, University of London, in preparation.

Spec. Publs int. Ass. Sediment. (1981) **5**, 111–121

More skewed against than skewing

MALCOLM W. CLARK

Imperial College Computer Centre, Exhibition Road,
London SW7 2BH, U.K.

ABSTRACT

Many beach sands are found to have negative skewness when analysed in terms of the weight frequency of their phi values. Tanner suggested a number of mechanisms which could lead to the modification of a size-frequency distribution: truncation, filtering and mixing. Truncation can be regarded as an extreme form of filtering. The concept of a filter may be broadened to identify two types, which differ in the range of application of the filter.

The 'mixing' hypothesis has received some popular support, notably in the form proposed by Visher, where a number of individually truncated distributions may be mixed. However, it is possible to derive similar results from filtering mechanisms; it is also possible to support these mechanisms by physical processes related to deposition in the nearshore. Much of the material in the nearshore passes through a suspension phase, where the key filters which would modify the distributions are (1) concentration, (2) turbulence and (3) shape. The most effective filter is probably associated with the increased viscosity due to wave breaking.

These several filters may act in concert, but the measurement base used in the analysis also contributes to the asymmetry. If the size-frequency distribution is converted to a settling velocity distribution, its shape becomes more symmetrical, although some skewness may still be present. While there is an implicit assumption behind most of these transformations that the 'fundamental' or 'intrinsic' distribution of sand sizes is lognormal, or at least symmetrical in its logarithms, this need not be the case. Analyses show that negative skewness could be derived from positively skewed intrinsic distributions.

Filtering can provide a more parsimonious and physically justifiable model for the modification of grain size curves than mixing.

INTRODUCTION

Naturally deposited beach sediments in the sand size range are seldom found to have size frequencies which are symmetrical in their logarithms. Commonly, the phi distribution (where phi is defined as $-\log_2$ (grain diameter in mm/standard particle size of 1 mm)) is slightly negatively skewed. This may be interpreted as a lack of fine material. Tanner (1964) suggested a number of mechanisms which could lead to the modification of a size frequency distribution, including modification from a symmetric to a negatively skewed distribution. These modifications may be summarized as:

0141–3600/81/1205–0111 $02.00

(1) Truncation: if truncation occurs, all material below a certain size is absent.

(2) Filtering: a filter removes a proportion of the material present in a systematic way. Two major types of filter may be recognized (Fig. 1). Type 1 filters are those which change in efficiency monotonically with size. They may have no effect on the distribution over some size range, but increase in effect as the size increases (or decreases). Such filters may inhibit the deposition of material completely – in effect truncating the distribution. Type 2 filters are more selective and their efficiency is restricted to a particular range. While type 1 filters will always modify a distribution in a similar and

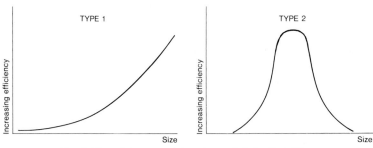

Fig. 1. Type 1 (monotonic) and type 2 (selective) filters.

consistent way, the effect of type 2 filters will depend on the characteristics of the size distribution and the filter. Some type 1 filters may actually be type 2 if the full range of effect is examined. In both cases it may be more useful to discuss filters with respect to a range of operation. Compound filters can be formed by combination.

(3) Mixing: if two or more symmetric distributions are mixed the resulting mixture may be skewed. The sign and extent of the skewness will be determined by the parameters (i.e. means, standard deviations and proportions) of the component distributions. Some type 2 filters may be seen as 'negative' mixing, where one distribution is removed from a full distribution.

The mixing hypothesis conveniently combines the notion of lognormal components with the observed asymmetry of many deposits. Visher (1969) has developed the notion of mixing further, by introducing individually truncated segments. A major element of such analysis derives from the apparent ease with which a sediment distribution, plotted on probability paper, may be reduced to a number of straight line segments. There are several inherent problems with such analyses; it is difficult to place a straight line through only a few points with any certainty and reliability; there is an apparent need to justify every single segment with a different transport mechanism; and each consecutive segment joins at a mutual truncation point, with no overlap. Despite such functional drawbacks and difficulties of interpretation, Visher's model has found wide acceptance. Within Tanner's system, it may be seen as a combination of mixing and truncation.

While the negative skewness of many beach sands could be explained by some variant of Visher's model, physically more pleasing and parsimonious explanations can be suggested, involving settling and filtering.

For a strict mixing model, with overlapping components, it is difficult to see how two (or more) components can retain their integrity, especially in the nearshore zone, unless the components are simply artifacts of sampling procedures which sample several horizons. If a mixing model is to be used to explain the presence of skewness, it must be accompanied by a physical explanation for the presence of the components. A truncation or filtering model requires only one component which is modified according to some physical process. Such models are pleasingly parsimonious.

It is possible to approach the problem of nearshore sedimentation by restricting examination to the zone shorewards of the breakers: and by considering only material deposited from suspension. In doing this, any processes are ignored which operate on the beach face when it is not covered by water. The deflation of the surface by wind action is not considered. Similarly, the activity of marine organisms is ignored. The organisms of the nearshore have been nominated by Howard & Dorjes (1972) as major contributors of the fine material found on tidal flats.

Within the surf zone, that is, landward of the breakers, there are processes operating which return the water transmitted through radiation stress, back beyond the breakers. Rip currents are well known, but even when they are not present, any movement of water mass onshore must be balanced by a movement offshore. Any such offshore movement will tend to carry any undeposited (suspended) material. Eddy (or turbulent) diffusion may also contribute to the removal of material from the nearshore. The velocity asymmetry observed in the nearshore, where the period of less intense flow offshore exceeds the period of the flow shorewards may also tend to contribute to the seawards movement of suspended material.

Miller & Ziegler (1958) hinted at a model which

was more fully developed by Dean (1973). When material is in suspension, for any time increment, a portion of sediment will settle out. The characteristics of a fraction settling out are a function of the distribution of sizes available in suspension, the concentration profile, the time interval, and the characteristics of the fluid. Consider the situation where a breaking wave resuspends material: if the water depth is 1 m, and the finest material in suspension is about 0·15 mm, the suspended material would take at least 80 sec to be deposited. If another 'suspension event' takes place before all deposition takes place, the finer material will tend to stay in suspension. Note that some of the finest material will have been deposited, since it will have been near the bottom of the water/sediment column.

The figure of 80 sec is evaluated with respect to the settling velocity of single smooth spheres in still water. This is a convenient, readily calculated reference, which represents a kind of ideal situation.

The filters are closely linked to wave characteristics; plunging breakers are likely to be those which take the greatest proportion of sediment into suspension, while the wave frequency is one of the key variables in determining whether sediment will be deposited before the next breaker.

There has been a long debate on the relative roles of suspension and bed load as transporting processes in the nearshore. On the one hand, Komar (1976, p. 127) dismisses suspension as being of only marginal importance. In a series of papers Thornton (cf. Thornton & Morris, 1977) maintains a more prominent role for suspension.

It is a common observation on the North Sea and English Channel coasts of the United Kingdom that there is suspended material in the nearshore. It may be significant that these are areas classed as high-energy coasts, with an appreciable tidal range. Laboratory experiments show very clearly that the rippled nature of the bed in the nearshore leads to suspension (cf. Nielsen, 1979), and in the field, where the water is clear enough, it is often possible to see the bursts of sediment from ripple crests as the waves pass.

Brenninkmeyer (1976) showed that even the coarsest sand may be taken into suspension, noting that the presence of a hydraulic jump caused by interference between the swash and backwash can help to carry material into suspension.

However, arguments about the relative importance of suspension and other transport mechanisms are misleading. We are concerned only that material does pass into suspension at some point, and that as a result of this suspension 'event', the normal sedimentation processes occur. It is not even the total amount of material in suspension which is important, but the ability of the 'suspension' processes to modify the distribution. Sediment in the nearshore moving as bed load gives little indication of selectivity; everything moves. Segregation may indeed occur within the bed load, but it is more likely to be in a vertical sense than a horizontal one (cf. Clifton, 1969).

By restricting discussion to deposition from suspension, it is possible to identify a number of factors as relevant; (1) concentration, (2) turbulence, (3) grain shape. A further consideration is: (4) the measurement base.

It is assumed, for the sake of a convenient reference, that the underlying (or 'intrinsic') sediment size distribution is lognormal (i.e. phi-normal). This is not a fundamental part of the argument, but merely provides a convenient reference. If the intrinsic distribution is symmetric in its logarithms it will be possible to identify the filters that lead to removal of fines (i.e. increase the negative skewness) fairly readily. But even when the distribution is already skewed the filters will tend to reduce positive skewness, and increase negative skewness.

The modifications which lead to a reduction in the fine part of the distribution will produce this effect. Truncation of the fines fits this requirement, but so also may any kind of filter which leads to a reduction in settling velocity, since any process which leads to the settling velocity of the sediment being reduced gives greater opportunity for the finer portion of the sediment to be removed from the nearshore area. If all the material is allowed to settle out, no modification can take place.

A filter may be described in terms of the change in settling velocity which it promotes, over a given size range. The ratio, w_f/w_0, where w_f is the settling velocity with the filter in operation, and w_0 is the still water single sphere settling velocity, allows comparisons to be made between filters from different processes.

CONCENTRATION

Concentration effects are best known from their presence in settling tube analysis. But these effects must also be present in 'natural' systems. Kranenburg & Geldof (1974) provide a succinct review and analysis, identifying two major processes:

(1) hindered settling, where the settling velocities of particles in suspension are less than that of solitary particles;

(2) settling convection, where groups of particles may fall at a relatively high rate.

Hindered settling is reasonably well known. McNown & Lin (1952) showed that concentrations of as little as 0·01 by volume were sufficient to reduce settling velocities by as much as 20 %.

Maude & Whitmore (1958) developed a theoretical model, and the following relationship

$$w_c = w_0(1-c)^n, \qquad (1)$$

where w_c is the settling velocity of the particle in a suspension of concentration c, and n is an exponent which depends on the Reynolds number, Re. Although they note that the size distribution of the sediment will be important, they do not indicate how this may be readily introduced into the equation.

In order to examine the relationship between settling velocity and concentration more closely, it is possible to use an approximate expression for Re and n:

$$\left.\begin{array}{ll} n = 4\cdot65 & Re \leqslant 1, \\ n = 4\cdot65 - 0\cdot75 \log_{10} Re & 1 < Re < 1000, \\ n = 2\cdot40 & Re \geqslant 1000. \end{array}\right\} \quad (2)$$

The Reynolds number may be defined as

$$Re = w_0 d/v, \qquad (3)$$

where w_0 is the terminal settling velocity of a single particle of diameter d, and v is the kinematic viscosity of the fluid.

John & Goyal (1975) have also developed an equation for the estimation of settling velocity which need not be restricted to particles of uniform size. But their solution refers to the mean size, and is developed in the Stokes range. The relationship may be reformulated and extended further (cf. Appendix).

Fig. 2 gives an indication of the reduction in settling velocity that may be expected for moderate concentrations. Broadly, it indicates that the finer sizes are retarded somewhat more than the coarser ones. The correspondence between the two alternative approaches is encouraging, but a reservation must be noted: the applicability of either expression to individual fractions of the size distribution remains unproved.

Settling convection has been inferred in natural systems by Bradley (1965) and Kuenen (1968), who

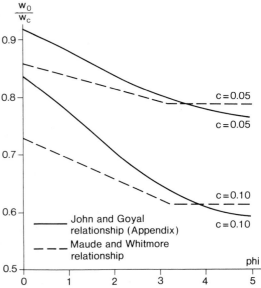

Fig. 2. w_c/w_0; reduction in settling velocity as a function of concentration.

both comment on the geological significance. No fully developed theoretical work is available, although there are a number of tantalizing observations, and some theory for simpler cases (cf. Happel & Brenner, 1965).

Kranenburg & Geldof (1974) have also analysed settling convection (together with hindered settling), but their major objective lay in improving settling tube analysis, and although they develop a model for the settling of mixtures of grain sizes, it is not possible to reformulate this for the present requirement.

It is difficult to incorporate settling convection into the discussion in a quantitative way. This is unfortunate, since Happel & Brenner (1965, p. 413) note that 'no single equation relating settling rate to concentration is adequate for dilute suspensions', attributing at least some of the inadequacy to failure to account for settling convection. A further complication in examining concentration effects for natural sediments is the change in effective concentration which must take place as the material settles out. The value of c may be best defined locally.

Settling convection does provide a likely mechanism for the incorporation of fine material into many deposits. Until more quantitative details are known of the significance of settling convection, this must remain speculative.

TURBULENCE

Turbulence may be considered in two main ways:

(1) Bagnold (1963) formulated the notion of auto-suspension. In a sense this is not strictly a turbulence effect, but since it is confined to conditions of turbulent fluid flow, it is appropriate to consider it here. Briefly, Bagnold deduces that the suspended load would be expected to transport itself to an unlimited degree when $\tan \beta$, the gravity gradient, or slope, exceeds the internal friction gradient, $\tan \theta$. He develops this concept to the inequality

$$w_0 < u . \tan \beta, \qquad (4)$$

where w_0 is the particle settling velocity, u is the mean horizontal fluid velocity, and β is the beach slope. This indicates that material below the diameter which has a settling velocity w_0 will not be deposited, but will remain in 'auto'-suspension. A necessary consequence of this notion is that the material in autosuspension will be uniformly distributed throughout the water/sediment column. Such a process would lead, at first glance, to the truncation of the grain size distribution. Examination of actual size distributions does not indicate that material is suddenly truncated at a certain point. This is not really surprising, since the horizontal fluid velocity in the nearshore is not a constant value; it is a stochastic variable, with a distribution. A possible modification of the expression would be

$$w_0 < u' . \tan \beta, \qquad (5)$$

where u' is the instantaneous horizontal fluid velocity. Therefore if u' has a distribution, so too must w_0. This may then be viewed as a filter process. What is the distribution of u'? Note that it is the absolute value of u' which is important. Some measurements of velocities in the surf and swash zones are available. Schiffman (1965) suggests a reasonably symmetric distribution for u' in both zones, while Wright's (1976) data provide evidence for a more skewed distribution in the swash zone. A feature of the lognormal distribution is that two such distributions with equal variances, but different means, will have rather different shapes, when plotted on an untransformed axis. The larger the mean (relative to the variance), the more symmetric the distribution will appear. The distribution with the lower mean will be more positively skewed. Since Schiffman's velocities are about twice the

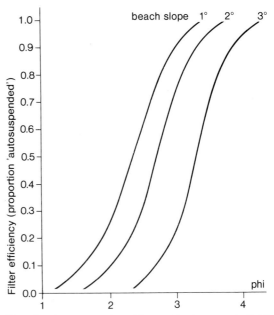

Fig. 3. Autosuspension filter with lognormal fluid velocities (mean $0 \cdot 5$ m sec^{-1}, standard deviation $0 \cdot 3$ m sec^{-1}).

magnitude of Wright's, both distributions may be approximated as lognormal, with Schiffman's having a higher mean but about the same variance. Huntley & Bowen (1974) give graphs of velocities in the surf zone which appear to fall within the structure of this kind of model.

Adopting a lognormal distribution of velocities, with a mean of $0 \cdot 5$ m sec^{-1} and a standard deviation of $0 \cdot 3$ m sec^{-1}, together with a beach slope of $3°$ (Fig. 3) it is evident that autosuspension is of very limited importance for grains coarser than about $1 \cdot 6$ phi, and that below $2 \cdot 8$ phi there is almost complete truncation. Between these values, autosuspension is a type 1 filter, permitting a proportion of the suspended material to settle.

The filters in Fig. 3 are a guide only. The auto-suspended material would be liable to diffusion out of the breaker zone, and therefore removal would tend to continue, until ultimately all the fine material may be removed. This does not occur—for example, it is possible to find material of the silt/clay grade in the runnels of a ridge and runnel beach. The substitution of u' for u may not be valid. In any event, the material taken into suspension by the high instantaneous velocities, may be permitted to settle by the subsequent lower velocities. It is not

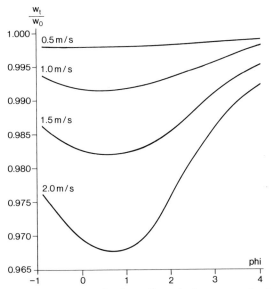

Fig. 4. w_t/w_0; reduction in settling velocity as a result of turbulence (after Murray).

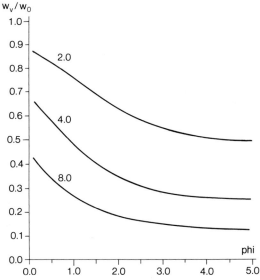

Fig. 5. w_v/w_0; reduction in settling velocity as a function of eddy viscosities 2, 4 and 8 times kinematic.

just the distribution of u' which is required, but also its autocorrelation structure.

(2) A number of models have been formulated which take a simplified model of turbulence. Settling in an oscillating fluid can be regarded as an adequate simulation of certain scales and intensities of turbulence. Through this model, it is possible to relate settling velocity to the frequency and period of the oscillation (e.g. Ho, 1964; Field, 1968; Murray, 1970). In all these cases, the conclusion, based on theoretical and laboratory work, is substantially the same. The time-averaged fall velocity is smaller than the terminal fall velocity.

Murray noted that under certain conditions turbulence could account for a marked reduction in settling velocity. Murray estimates a value of 3 Hz for the turbulence frequency. Evaluating the relationship for various horizontal velocities up to 2 m sec^{-1} (Fig. 4) shows that the greatest decrease in settling velocities is at the highest velocities. The turbulence filter is a type 2 filter, and has a similar 'shape' for different velocities. The modification it will produce in the size distribution will depend on the size distribution, and in particular the position of the mean. In the case shown here, a deposit with a mean size coarser than 0·5 phi would tend to be modified towards negative skewness, while finer deposits would tend to positive skewness. The position of the filter minimum is relatively invariant to the turbu-

lence frequency. However, the decrease in settling velocity provided by this model is so small for the velocities likely to be encountered in the nearshore that it is difficult to assign any significance to it. In these models, the conversion from spheres in an oscillating fluid to sand grains in the nearshore may not be straightforward.

Nielsen (1979) also considered the motion of a sand grain in a vortex, again showing that the settling velocity would be reduced, as grains become 'trapped' in vortices. This effect is closely related to the nature of the bed and the flow characteristics. In passing, Wood & Jenkins (1973) suggested that, under some circumstances, a particle might be suspended 'indefinitely' in an individual eddy.

Guza & Bowen (1976) noted that the turbulence associated with wave-breaking results in an eddy viscosity much larger than molecular viscosity. One of the controls on settling velocity is the viscosity. Assume that this increase in viscosity should be taken into account with the kinematic viscosity in the calculation of settling velocity; this would have the effect of making fall velocities slower, providing more time for any processes to operate. Fig. 5 demonstrates that the retardation from an increase in viscosity can be quite marked, and that it provides a type 1 filter.

It is convenient here to take into account the likely effect of the presence of bubbles in the surf zone, although not strictly a feature of turbulence. When a wave breaks, a certain amount of air may become entrapped, and will make its way back to the surface as air bubbles (cf. Miller, 1976). These bubbles may hinder the free settling of particles in suspension, especially the fines, since the bubbles and the fines are most likely to be in the upper layers of the fluid.

GRAIN SHAPE

There are two main questions which should be answered in examining grain shape: what is the relationship between grain shape and settling velocity; and what is the variation of shape in naturally occurring sands?

There are various ways in which grain shape, size and settling velocity may be examined together, and various relationships have been suggested between them. The most successful of these use some shape measurement such as Corey's shape factor, which is defined by the three orthogonal dimensions of the particle, such that c is the minimum axis, a the maximum, and b the 'medium'; thus,

$$SF = c(ab)^{-0.5}. \qquad (6)$$

Recently, Komar & Reimers (1978) and Brezina (1979) have used this shape measurement to relate the three variables. Other formulations are possible (cf. McCulloch, Moser & Briggs, 1960), but the use of Corey's shape factor is fairly well established, and correlates reasonably well with other measures (Briggs, McCulloch & Moser, 1962).

Brezina's formulation is convenient since it expresses the relationship in a single equation, covering the whole range relevant here. However, in order to achieve this, Brezina had to redefine the concept of the shape factor from its original form, of purely physical grain dimensions, to a 'shape factor' defined in terms of the hydraulic characteristics of the grain ('hydraulic-shape factor' or SF'). Brezina leaves implicit the idea that the SF and SF' values of naturally worn particles will be the same. The hydraulic-shape factor helps to account for the observed anomaly that smooth spheres settle more quickly than naturally worn particles which are spherical. This retardation is due to the surface roughness of natural particles. According to this scheme, a smooth sphere has an SF' value of 1·2 ('an impossible value of an actually measured SF').

Over the range of interest, the reduction in settling

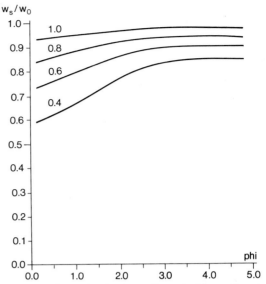

Fig. 6. w_s/w_0; reduction in settling velocity as a function of SF' (0·4–1·0).

velocity is not large, and in contrast to the other type 1 filters presented here, it is more pronounced with coarser particles.

Extensive quantitative measurements of the shape factor of naturally occurring beach sands are limited, but Poche & Goodell (1975) have analysed beach sands in this way. Although they observed no variation of shape with size, they were able to show that the SF had an approximately normal distribution with mean 0·65 and standard deviation 0·13. Similarly, the data of Harrison & Morales-Alamo (1964) analysed by the same shape factor, provide no evidence for shape differentiation by size. Moss (1962) has published a large amount of shape and size data, but not in a form which is readily recast for the purposes of this analysis. Accepting for the moment his argument that plotting the long axis of the grains against the ratio of the long to intermediate axis (a versus a/b) gives a characteristic 'elongation function', we may note that beach sands show a straight line graph, indicating that as particles become larger they also become more elongated.

Clearly there is a discrepancy between Moss's conclusion and the other results. Part of the lack of agreement may be attributed to the different ways in which 'shape' is defined.

In the particular case of beaches consisting largely of quartz grains, it is difficult to entertain this

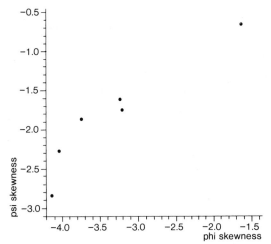

Fig. 7. Phi and psi skewness from sieving (0·25 phi intervals).

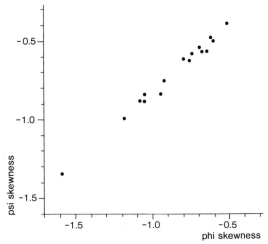

Fig. 8. Psi and phi skewness from sedimentation column analysis (0·02 psi resolution).

effect as very serious, given the limited effect of grain shape on settling velocity, and the apparently limited range of those shapes.

THE MEASUREMENT BASE

The arguments above have been presented as modifications which may occur in terms of the settling velocity. In many respects, analysis in the domain of settling velocity seems more natural and apposite to this problem, rather than analysing sediment 'size'. In fact these two measurement systems can be shown to be closely related (Kennedy & Koh, 1961), especially when both are presented by their logarithms. Middleton (1967) introduced the 'psi' notation for settling velocities, by analogy with the phi notation. Psi is defined as

$$-\log_2 \frac{\text{settling velocity in cm sec}^{-1}}{\text{standard settling velocity of 1 cm sec}^{-1}}.$$

It is especially interesting that a lognormal settling velocity distribution, when transformed to a size distribution through one of the empirical relationships (Gibbs, Matthews & Link, 1971; Reed, LeFevre & Moir, 1975; Brezina, 1979) gives a negatively skewed distribution (cf. Brezina, 1963). Pursuing this idea further, most settling velocity analysis is carried out at room temperatures and in fresh water, while the sediment was probably laid down in rather different conditions of temperature and density. This anomaly would actually introduce

some positive skewness, if the original phi distribution were normal.

The extent of the changes in the measurement base can be estimated with reference to some samples taken from the western shore of Dungeness. Six swash-zone samples were sieved at quarter-phi intervals, and a further 16 were analysed by settling tube, with a resolution of 0·02 psi. The phi and psi values were transformed using the relationship suggested by Gibbs *et al.* (1971). Fig. 7 shows the marked reduction in skewness which takes place with the sieved samples when the measurement base is transformed. The very high values of phi skewness may be partly due to the calculation of moments when the data are collected at a rather wide interval. The mean phi value of the samples is about 2·5–2·6 while the standard deviations vary from 0·3 to 0·6. Thus the bulk of each distribution is concentrated into about 6 quarter-phi intervals. This does not lend precision to the calculated values.

The psi analyses (Fig. 8) were also transformed to the phi scale and show much lower values of skewness for both bases. Again the psi-based skewness is numerically closer to zero than the phi-based value.

When the transformations between phi and psi are calculated, it is possible to compare 'laboratory' and 'natural' conditions, to see how much difference the small changes in density and viscosity will have on the skewness values. If it is assumed that 'laboratory' conditions are freshwater at 20° C (giving a density of 0·988 g ml⁻¹ and viscosity 0·009136 stokes), while 'natural' conditions are saltwater at 4° (density 1·0209 g ml^{-1} and viscosity 0·0151 stokes), the change in product moment skewness is not appreciable. The decrease is in the order of 0·01, and may be ignored for practical purposes, being well within the bounds of experimental error.

CONCLUSION

It is tempting to attribute a large proportion of the observed negative skewness to the use of the phi rather than the psi notation, but this is clearly not the full explanation. The need to invoke the filtering mechanisms was the result of the assumption of intrinsic lognormality (or log-symmetry). But it is clear that even if this does not exist, the filter mechanisms must still operate if significant suspension occurs. When deposition of any fine material is inhibited by the filter processes, the deposit will tend to become negatively skewed, whatever the 'intrinsic' distribution. Of the processes which have been outlined, only autosuspension is a true filter process, the others operating as filters only because the rate of settling of the finer material is retarded, affording greater opportunity of removal outside the nearshore system. It is difficult to erect some kind of quantitative model, since most of the discussion is, at best, semi-quantitative. These effects must operate together, and their relative significance is also obscure. It would be expecting a great deal to anticipate that the effects outlined would operate in a simple additive way.

ACKNOWLEDGMENTS

I am grateful to the Central Research Fund of the University of London, which provided some funding for this research. Dr Jiri Brezina and Dr Isobel Clark provided encouragement at crucial times. I should like to thank the referee who pointed out the likely significance of air bubbles in reducing settling velocities. The title is a paraphrase of the quotation 'more sinned against than sinning' from Shakespeare's *King Lear*. It appears in Hoban (1974).

REFERENCES

BAGNOLD, R.A. (1963) Mechanics of marine sediment-ation. In: *The Sea*, vol. 3 (Ed. M. N. Hill), pp. 507–528. Wiley (Interscience), New York.

BRADLEY, W.H. (1965) Vertical density currents. *Science*, **150**, 1423–1428.

BRENNINKMEYER, B. (1976). Sand fountains in the surf zone. In: *Beach and Nearshore Sedimentation* (Ed. by R. A. Davis and R. L. Ethington). *Spec. Publs Soc. econ. Paleont. Miner.*, Tulsa, **24**, 69–91.

BREZINA, J. (1963) Kepteyn's transformation of grain size distribution. *J. sedim. Petrol.* **33**, 931–937.

BREZINA, J. (1979). Particle size and settling rate distributions of sand sized materials (preprint). *2nd European Symp. Particle Characterization*, Nurnberg. 44 pp.

BRIGGS, L.I., McCULLOCH, D.S. & MOSER, F. (1962) The hydraulic shape of sand particles. *J. sedim. Petrol.* **32**, 645–656.

CLIFTON, H.E. (1969) Beach lamination: nature and origin. *Mar. Geol.* **7**, 553–559.

COLBY, B.C. & CHRISTENSEN, R.P. (1957) Some funda-mentals of particle size analysis. *St Anthony Falls Hydraul. Lab.*, *Rep.* **12**, 55 pp.

DEAN, R.G. (1973) Hueristic models of sand transport in the surf zone. *Proc. Conf. Eng. Dynamics of the Coastal Zone*, pp. 208–214. Sydney, Australia.

FIELD, W.G. (1968) Effects of density ratio on sedi-mentary similitude. *J. Hydraul. Div. Am. Soc. civ. Engrs* **94**, 705–719.

GIBBS, R.J., MATTHEWS, M.D. & LINK, D.A. (1971) The relationship between sphere size and settling velocity. *J. sedim. Petrol.* **41**, 7–18.

GUZA, R.T. & BOWEN, A.J. (1976) Resonant interactions for waves breaking on a beach. *Proc. 15th Conf. Coastal Engng*, pp. 560–579.

HAPPEL, J. & BRENNER, H. (1965) *Low Reynolds Number Hydrodynamics*. Noordhoff, Leiden. 553 pp.

HARRISON, W. & MORALES-ALAMO, R. (1964) Dynamic properties of immersed sand at Virginia Beach, Virginia. *US Coastal Eng. Res. Station, Tech. Mem.* **9**, 52 pp.

HO, H-W. (1964) *Fall velocity of a sphere in an oscillating fluid.* Ph.D. Dissertation, University of Iowa. 78 pp.

HOBAN, R. (1974) *Kleinzeit*. Picador, London. 191 pp.

HOWARD, J.D. & DORJES, J. (1972) Animal–sediment relationships in two beach-related tidal flats: Sapelo Island, Georgia. *J. sedim. Petrol.* **42**, 608–623.

HUNTLEY, D.A. & BOWEN, A.J. (1974) Field measure-ments of nearshore velocities. *Proc. 14th Conf. Coastal Engng*, pp. 538–557.

JOHN, P.T. & GOYAL, V.K. (1975) An empirically modified Stokes equation for mean particle size from hindered settling experiment at any one concentration. *Ind. J. T.* **13**, 69–71.

KENNEDY, J.F. & KOH, R.C.Y. (1961) The relationship between the frequency distributions of sieve diameters and fall velocities of sediment particles. *J. geophys. Res.* **66**, 4233–4246.

KOMAR, P.D. (1976) *Beach Processes and Sedimentation*. Prentice-Hall, Englewood Cliffs. 429 pp.

KOMAR, P.D. & REIMERS, C.E. (1978) Grain shape effects on settling rates. *J. Geol.* **86**, 193–209.

Kranenburg, C. & Geldof, H.J. (1974) Concentration effects on settling-tube analysis. *J. Hydraul. Res.* **12**, 337–355.

Kuenen, P.H. (1968) Settling convection and grain-size analysis. *J. sedim. Petrol.* **38**, 817–831.

McCulloch, D.S., Moser, F. & Briggs, L.I. (1960) Hydraulic shape of mineral grains (abs.). *Bull. geol. Soc. Am.* **71**, 1925.

McNown, J.S. & Lin, P.-N. (1952) Sediment concentration and fall velocity. *Proc. 2nd Conf. Fluid Mech.*, pp. 401–411. Ohio State University.

Maude, A.D. & Whitmore, R.L. (1958) A generalized theory of sedimentation. *J. appl. Phys.* **9**, 477–482.

Middleton, G.V. (1967) Experiments on density and turbidity currents. III. Deposition of sediments. *Can. J. Earth Sci.* **4**, 475–505.

Miller, R.L. (1976) Role of vortices in surf zone prediction: sedimentation and wave forces. In: *Beach and Nearshore Sedimentation* (Ed. by R. A. Davis and R. L. Ethington). *Spec. Publs Soc. econ. Paleont. Miner.*, Tulsa, **24**, 92–114.

Miller, R.L. & Zeigler, J.M. (1958) A model relating dynamics and sediment pattern in equilibrium in the region of shoaling waves, breaker zone, and foreshore. *J. Geol.* **66**, 417–441.

Moss, A.J. (1962) The physical nature of common sand and pebbly deposits. Part 1. *Am. J. Sci.* **260**, 337–373.

Murray, S.P. (1970) Settling velocities and vertical diffusion of particles in turbulent water. *J. geophys. Res.* **75**, 1647–1654.

Nielsen, P. (1979) *Some Basic Concepts of Wave Sediment Transport.* Series paper no. 20, Institute of Hydrodynamics and Hydraulic Engineering (ISVA), Technical University of Denmark. 160 pp.

Poche, D. & Goodell, H.G. (1975) Influence of grain shape on the selective sorting of sand by waves. *Mem. geol. Soc. Am.* **142**, 137–148.

Reed, W.E., LeFevre, R. & Moir, G.J. (1975) Depositional environment interpretation from settling velocity (psi) distributions. *Bull. geol. Soc. Am.* **86**, 1321–1328.

Schiffman, A. (1965). Energy measurements in the swash-surf zone. *Limnol. Oceanogr.* **10**, 255–260.

Tanner, W.F. (1964) Modification of sediment size distributions. *J. sedim. Petrol.* **34**, 156–164.

Thornton, E.B. & Morris, W.D. (1977) Suspended sediments measured within the surf zone. *Proc. 5th Symp. Waterway Port Coastal and Ocean Div. Am. Soc. civ. Engrs* (Coastal Sediments '77), pp. 655–664.

Visher, G. (1969) Grain size distributions and depositional processes. *J. sedim. Petrol.* **39**, 1074–1106.

Wood, I.R. & Jenkins, B.S. (1973) A numerical study of the suspension of a non-buoyant particle within a turbulent flow. *Proc. Int. Symp. River Mech.*, Bangkok, A38, 1–12.

Wright, P. (1976) A cine-camera technique for process measurement on a ridge and runnel beach. *Sedimentology*, **11**, 83–98.

NOTATION

a	maximum grain axis.
b	intermediate (medium) grain axis.
c	minimum grain axis; volume concentration.
c'	$(1-c)$.
C_d	drag coefficient.
d	grain diameter.
e	base of natural logarithms (2·7183).
g	acceleration due to gravity.
p_t	density of fluid.
p_s	density of solid.
p'	$p_s - p_t$.
p''	$(p_s - p_f)/p_f$.
Re	Reynolds number.
SF	Corey's shape factor.
SF'	hydraulic-shape factor.
u	horizontal fluid velocity.
u'	instantaneous fluid velocity.
v	kinematic viscosity.
w_0	settling velocity (single sphere, still water).
β	slope.
π	constant (3·1416).

APPENDIX

It is possible to generalize the empirical relationship between concentration and settling velocity developed by John & Goyal (1975) in order to extend it beyond the Stokes range.

The drag coefficient, C_d can be defined (Colby & Christensen, 1957) as

$$4dp''g/3w^2 \qquad (A\ 1)$$

This is achieved by defining the gravitational force as

$$F = \pi . d^3 p' g/6 \qquad (A\ 2)$$

and the resisting force as

$$F' = C_d \pi . d^2 p_f w^2/8. \qquad (A\ 3)$$

Since F and F' are equal, (A 1) can be derived by substitution. John & Goyal (1975) modify (A 2) slightly to take account of concentration:

$$F = \pi . d^3 p' g c'/6. \qquad (A\ 4)$$

This gives

$$C_d = 4dp''gc'/3w^2. \qquad (A\ 5)$$

Clearly, for values of concentration of zero, (A 1) and (A 5) are identical.

John & Goyal introduce.

$$v_b = v e^{5c}$$

for bulk viscosity, and for the fluid return velocity the term p'' is replaced by

$$1 + (cp'')^{1\cdot5}. \qquad (A\ 6)$$

However, in an unenclosed environment we might assume that the return velocities do not affect the descending grains.

An empirical relationship may be determined between C_d and the Reynolds number, Re:

$$C_d = f(Re).$$

This may be written (Brezina, 1979)

$$C_d = a_1 Re^{-1} + a_2 Re^{-0.5} + a_3. \qquad (A 7)$$

Therefore, ignoring return velocity, (A 5) becomes

$$a_1 Re^{-1} + a_2 Re^{-0.5} + a_3 = 4dp''c'g/3w^2. \qquad (A 8)$$

Let $b_0 = -4dc'p''g/3$, $b_1 = a_1 v_b/d$, $b_2 = a_2(v_b/d)^{0.5}$ and $b_3 = a_3$. By substituting dw/v_b for Re, (A 8) may be rewritten

$$b_1 w^{-1} + b_2 w^{-0.5} + b_3 = -b_0 w^{-2}. \qquad (A 9)$$

Multiplying through by w^2, and letting $x = w^{0.5}$, this becomes

$$b_0 + b_1 x^2 + b_2 x^3 + b_3 x^4 = 0.0, \qquad (A 10)$$

which may be solved readily by a numerical method. The fall velocity may then be calculated as $w = x^2$. Note that the equation may be modified readily to take account of return velocity by resetting

$$b_0 = -4dc'(1 + (cp'')^{1.5})g/3.$$

Similarly, for single spheres, the expressions for b_1, b_2 and b_3 must be altered to substitute v for v_b, and

$$b_0 = -4dp''g/3.$$

The relationship can be further extended to provide a unified expression linking concentration, grain shape and settling velocity: Brezina (1979) shows that (A 7) may be expanded to account for the 'hydraulic shape factor', SF', by rewriting

$$a_1 = p_2 \text{SF}'^{p_1}, \quad a_3 = p_4 \text{SF}'^{p_3}, \quad a_3 = p_6 \text{SF}'^{p_5},$$

where $p_1 \ldots p_6$ are determined empirically. The remaining equations, (A 9) and (A 10), may be found readily.

Spec. Publs int. Ass. Sediment. (1981) **5**, 123–131

Holocene palaeoenvironments of Broadland, England

BRIAN P. L. COLES *and* BRIAN M. FUNNELL*

*Olchfa Secondary School, Swansea, Wales and *School of Environmental Sciences,
University of East Anglia, Norwich NR4 7TJ, U.K.*

ABSTRACT

The sediment of the Broadland valleys of East Norfolk preserves evidence of two periods of marine transgression in the Holocene epoch. They are represented by estuarine lower and upper clay horizons, sandwiched between freshwater lower, middle and upper peat.

The lower clay is 12–16 m thick at the seaward limit, extends 20 km inland, and its upper surface, corresponding to a date of *c*.4500 BP, is at present −5·5 to −6·5 m below O.D.

The upper clay is up to 6 m thick at the coast and extends 23 km inland. Its lower surface, dated at *c*.2000 BP, is at −5·0 to −6·0 m O.D. at the coast, and the upper surface, dating from about 1500 BP, is at −0·05 m O.D.

Comparisons with the modern distribution of thecamoebans and foraminiferids in the River Yare and Breydon Water have allowed the definition of saline penetration and tidal levels during deposition of the estuarine clays.

A model is constructed combining progressive subsidence at 1·5 m ka^{-1} and cyclical sea-level changes (as inferred by Tooley, 1978), which is thought to account for the observed facies and relative sea-level changes and to have potential for projection into the future.

INTRODUCTION

The English province of East Anglia lies directly opposite the Netherlands and forms most of the western coastline of the southern North Sea. It is mainly drained by a series of eastward-flowing rivers which terminate seawards under estuarine conditions. The valleys of these rivers are filled by alluvial and estuarine deposits accumulated during the Holocene epoch. These deposits record two episodes of marine transgression separated and succeeded by regressive conditions. During the latest (Mediaeval) regressive episode previously accumulated brushwood peat was extensively excavated for fuel. The abandoned, freshwater-filled workings now form shallow lakes known as the Broads.

Only limited systematic investigations of the Holocene deposits of Broadland have previously been carried out, and those reported here are the first to be designed to elucidate more general problems than the specific one of the origin of the Broads. The earlier results of investigations into the origin and making of the Broads are extensively reported in Jennings (1952) and Lambert & Jennings (1960; see also Ellis, 1965). The present studies were carried out by B. P. L. Coles during the tenure of a Natural Environment Research Council Studentship 1972–75 (Coles, 1977). They consisted of: (*a*) a study of the distribution of thecamoebans and foraminiferids in the Yare river system between Norwich and Great Yarmouth and in intertidal sub-environments in Breydon Water (for locations see Fig. 1), and (*b*) a study of the alluvial and estuarine stratigraphy, by means of 300 hand- and power-driven boreholes, between Surlingham and Halvergate Marshes in the Yare valley.

0141–3600/81/1205–0123 $02.00

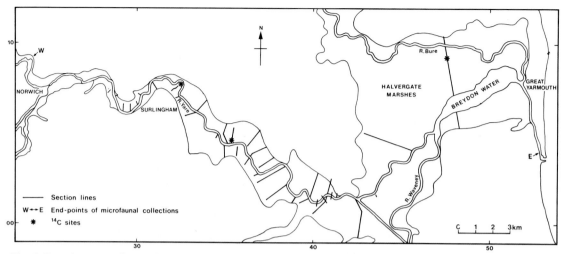

Fig. 1. Location map of central Broadland. (Margins in this figure, and in Figs 4, 5, 6 and 8, correspond to TG 23 east and TG 54 east, and TM 99 north and TG 12 north National Grid lines).

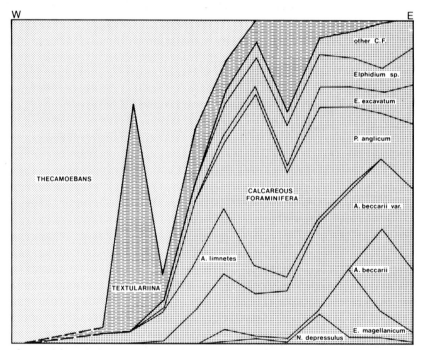

Fig. 2. Thecamoeban and foraminiferal assemblages along the course of the River Yare (vertical scale: relative percentages of taxa).

DISTRIBUTION OF THECAMOEBANS AND FORAMINIFERIDS

River Yare

The quantitative distribution of thecamoebans and foraminiferids in bottom sediments along the course of the River Yare, from fresh water near Norwich to saline, estuarine waters near Great Yarmouth is shown in Fig. 2. Detailed taxonomies and systematic descriptions of species are contained in Coles (1977) and will be published elsewhere. There is a progressive replacement of thecamoebans by agglutinating foraminiferids and then by calcareous foraminiferids passing from fresh, through brackish into increasingly saline estuarine waters. Knowledge of these distributions enables salinity conditions accompanying the estuarine transgressive episodes of clay deposition to be worked out in detail.

Breydon Water

Sampling of foraminiferid populations from modern intertidal sub-environments in the estuarine conditions of Breydon Water enabled four distinctive assemblages and sub-environments to be recognized.

(*a*) High salt-marsh, characterized by *Trochammina inflata* and *Jadammina macrescens*.

(*b*) Low salt-marsh characterized by *Jadammina macrescens*, *Trochammina inflata*, *Miliammina fusca*, *Haplophragmoides* sp., *Elphidium williamsoni*, and small, thin-walled Milioline foraminifers.

(*c*) High intertidal flats characterized by *Protelphidium anglicum*, *Elphidium magellanicum*, *Elphidium excavatum*, *Ammonia beccarii* and small percentages of *Elphidium williamsoni*, *E. gerthi* and *E. earlandi*.

(*d*) Lower intertidal flats characterized by *Protelphidium anglicum*, *Elphidium magellanicum*, *Elphidium excavatum*, *Ammonia beccarii* and 4–10% *Elphidium waddensis*.

These sub-environment assemblages can also be recognized in the clays of the Holocene transgressive estuarine episodes, enabling the progress of sedimentation in relation to contemporary tidal levels to be interpreted.

SEDIMENTARY SEQUENCE

The sequence of deposits in the Yare valley is basically very simple. There are lower, middle and upper peat deposits separated by lower and upper clay episodes (Fig. 3). The clay deposits thin landwards, and the peat deposits thin seawards (see Fig. 9). Details are given in the following account of the individual deposits.

Lower peat

The earliest Holocene deposits are a lower peat. This peat is generally a brushwood peat except near its contact with the base of the overlying lower clay where it passes into a *Phragmites* reed peat. Landward, in the inland part of the valley, this peat is only 0·5–1·0 m thick and occurs below −8·0 to −13·0 m O.D. It has been attributed by pollen analysis (Jennings, 1955, p. 200) to pollen zones V, VI and earliest VII (i.e. roughly 9000–7500 BP). Seaward, near the coast this peat generally occurs below −20·0 m O.D. and is up to 2 m thick. Salt-marsh peat sampled between −19·3 and −19·5 m O.D. near Great Yarmouth gave a ^{14}C date of 7580 ± 90 y BP (HAR 2535) (P. Murphy, personal communica-

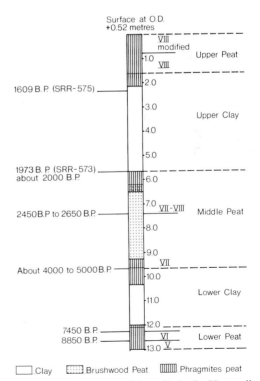

Fig. 3. General sequence of deposits in the Yare valley. (Stratigraphy and pollen zones after Jennings, 1955, p. 200 with ^{14}C dates added.)

tion). The salt-marsh peat at this point is immediately overlain by lower clay.

Basically therefore the lower peat can be seen to represent a freshwater environment, existing in the Yare valley prior to the main Flandrian transgression which occurred here at about 7500 BP.

Lower clay (Fig. 4)

The transition to brackish water conditions indicated by the establishment of *Phragmites* reed beds, and the occurrence of salt-marsh at about 7500 BP is followed by widespread estuarine clay deposition in the Lower Yare valley. The base of the clay ranges from −20·0 m O.D. near the coast to as little as −6·0 m O.D. inland. Its top occurs between −5·5 and −8·0 m O.D., although generally it occurs no lower than −6·5 m O.D. It seems likely that the lower contact of the clay is diachronous, although insufficient dating is available on materials from this level to establish this with certainty.

The lower clay is actually a very fine sand at the seaward end (where it is 12–15 m thick). The upper 4–5 m here shows a succession of foraminiferal assemblages passing from low intertidal flat through high intertidal flat, low salt-marsh and high salt-marsh conditions, culminating in *Phragmites* reed peat at the transition to the overlying freshwater middle peat. The change from low intertidal flat to high salt-marsh conditions occupies a vertical interval of up to 5 m. This compares with the present-day range between mean low water springs (MLWS), the approximate boundary between low and high tidal flats, and mean high water springs (MHWS), the approximate boundary between low and high salt-marsh (at this point on the coast) of 1·90 m. Accumulation during a period of relative sea-level rise is therefore implied.

The deposition of the lower clay is known from pollen analysis (Jennings, 1955) to have occurred in the early part of pollen zone VII (the Atlantic period). The age of the end of deposition has not been very firmly established but probably falls between 5000 and 4000 BP, say 4500 BP. If it was completed, as the evidence indicates, by the achievement of high salt-marsh conditions (at least at the seaward end), the terminal sedimentation surface, now at approximately −6·5 m O.D., must have stood originally at about +1·0 m relative to its contemporary sea-level.

Towards its inland limit the lower clay becomes a genuine clay and thins to a feather-edge at about −6·0 m O.D. The contained foraminifers indicate brackish (mesohaline) water conditions.

Middle peat (Fig. 5)

The accumulation of middle peat seems to have commenced shortly after the beginning of the Sub-Boreal period, during the latter part of pollen zone VII, i.e. early VIIb of current terminology (Jennings, 1955). Near the sea (i.e. under the present Halvergate marshes) dark brown structureless peat appears to indicate open freshwater conditions; landward fibrous peat with some wood fragments indicates open wooded conditions, whilst mainly at the inland

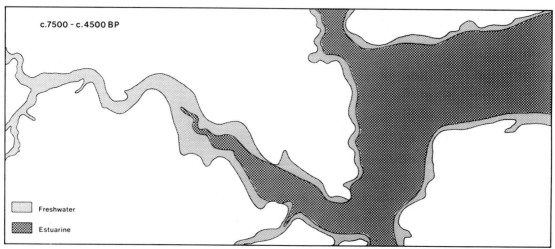

c.7500 – c.4500 BP

▨ Freshwater

▨ Estuarine

Fig. 4. Lower clay palaeogeography—*c.* 7500–4500 BP.

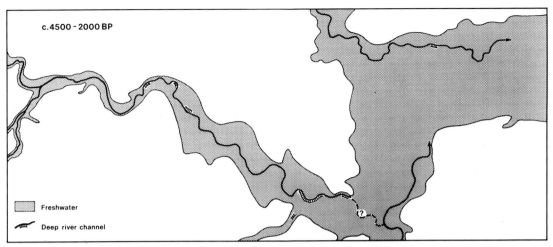

Fig. 5. Middle peat palaeogeography—*c.* 4500–2000 BP.

end brushwood peat represents a fen woodland environment. Six to eight metres of peat accumulated at the inland end of the valley during the middle peat episode, but nearer the coast it was only 3–4 m. Peat accumulation seems to have come to an end during the VII–VIII (Sub-Boreal/Sub-Atlantic) transition in the pollen zonation (Jennings, 1955) and is ^{14}C-dated at its upper limit underneath the upper clay towards the inland end of the valley at 1973 ± 50 y BP (SRR 573).

At or near the end of the accumulation of the middle peat the main river channel was enlarged and incised into the peat. It was subsequently preserved by being infilled with upper clay. Its larger cross-sectional area and steeper longitudinal profile (Fig. 9) suggest a larger freshwater discharge at that time than at the present-day. Other indications of wetter conditions include a change from brushwood peat to soft fibrous peat at the top of the middle peat towards the inland end of the valley. It is interesting to note that the present river channel west of Halvergate marshes follows very closely, but not precisely, the course of the 'fossil' channel.

Upper clay (Fig. 6)

The second estuarine episode apparently began at the seaward end in the area of the present Halvergate marshes with the establishment of sheltered estuarine conditions in what had previously been a freshwater lagoon. Coarse sandy sediment, containing open estuarine and estuarine channel foraminifers soon succeeded however, indicating both the destruction

of a pre-existent sand barrier and the strong incursion of marine water into the lower valley.

The base of the upper clay, as recorded above, is dated by ^{14}C to 1973 ± 50 y BP, towards its inland limit, and its feather-edge representing the maximum extension of estuarine influence is dated by ^{14}C to 1609 ± 50 y BP (SRR 575). At its maximum extension the upper clay reached 23 km inland to within 7 km of the site of Norwich. It consisted of medium to very fine sand nearest the coast, ranging through silt further inland to true clay furthest inland. Open estuarine intertidal mud flats and salt-marsh environments characterized the seaward limits, whereas brackish-water (mesohaline) microfaunas predominated at the inland limit.

The episode of upper clay sedimentation apparently terminated rather suddenly. Although there is a partial succession from intertidal through salt-marsh conditions in the upper clay near the coast, this process did not reach its natural end points as it did at the top of the lower clay. Over wide areas, and especially on Halvergate marshes, the present-day land surface provides clear evidence of both rapidly draining (linear) intertidal, as well as of mature (meandrine) salt-marsh relict drainage patterns (Fig. 7). The desertion and preservation of both kinds of creek system indicates that the exclusion of estuarine waters was not achieved by the completion of sedimentary accretion, as it apparently was at the termination of the lower clay episode, but that other factors were responsible for the withdrawal of marine waters from the intertidal zone. One possible explanation is that southward extension of the coastal spit (on which the town of Great Yarmouth was subsequently built) across the Bure–Yare–Waveney estuary mouth caused a reduction in tidal range in the enclosed estuary. The present-day range on the open coast is 2·8 m (MHWS to MLWS = 1·90 m). It is quite possible that a reduction to 1·9 m (MHWS to MLWS = 1·0 m) in the vicinity of Halvergate marshes could have resulted in a reduction of MHWS levels by about 0·45 m. Another possibility is that absolute sea-level was falling at that time faster than the local subsidence rate. Whatever the precise combination of factors was, estuarine waters were largely excluded from all except the area of Breydon Water by about 1500 BP.

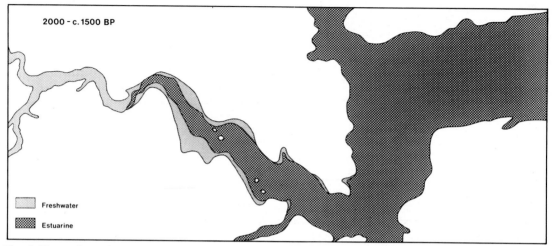

Fig. 6. Upper clay palaeogeography—*c.* 2000–1500 BP. (The small clear areas within the estuarine tract represent islands of peat.)

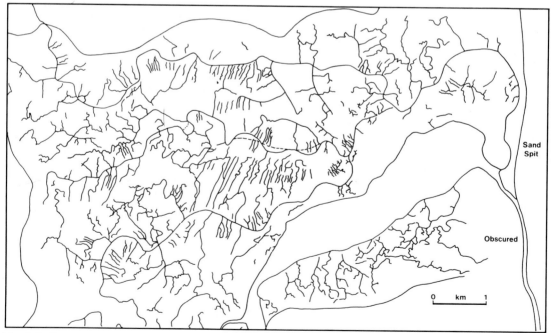

Fig. 7. Relict intertidal and salt-marsh drainage patterns on Halvergate marshes dating from the end of the upper clay transgression (drawn from aerial photographic coverage).

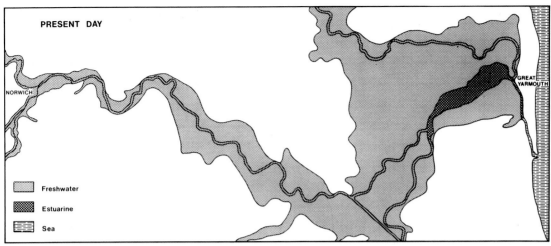

Fig. 8. Present-day distribution of Broadland environments.

Upper peat (Fig. 8)

Potentially an upper peat could have accumulated almost everywhere in the valley after the withdrawal of the upper clay estuarine transgression. In practice much of it has been lost as a result of embankment, drainage and agricultural practices leading to surface wastage. The whole record of the Broadland region, subsequent to the upper clay, is in fact largely affected, if not controlled by human activity. At some time probably shortly after 1000 BP the middle peat was extensively excavated for fuel. It is likely that initially the water table in the valley was low, either because of decreased freshwater run-off from upstream, or a lower contemporary relative sea-level or both. By about 700 BP, however, occasional difficulties were being encountered because of flooding, and by about 600 BP (i.e. the end of the fourteenth century) the peat diggings had been abandoned to become the shallow Broads of today. In places the sites of these Broads have been filled or partially filled with biogenic sediment which has then been encroached upon by accumulations of upper peat. Elsewhere they still remain as open water.

SUBSIDENCE AND SEA-LEVEL CHANGE

Apart from the detailed evidence they give of continuous environmental change during the Holocene, the sediments of Broadland also record a history of subsidence and sea-level change (Fig. 9).

This record is difficult to decipher as so many factors other than subsidence and sea-level change may also have caused or contributed to the observed environmental changes.

Evidence of subsidence

In the Broadland deposits it is possible to identify one or two levels whose position relative to contemporary sea-level can be precisely inferred.

The salt-marsh peat ^{14}C-dated to 7500 ± 90 BP and occurring at -19.3 to -19.5, O.D. near Great Yarmouth probably accumulated above MHWS, say $+1.0$ m relative to contemporary sea-level. Therefore there has subsequently been 20.4 m of apparent net subsidence, i.e. 2.69 m ka^{-1}.

The top of the lower clay, which culminates in high salt-marsh facies, also clearly accumulated up to about MHWS. Its present-day level is -6.0 m O.D. and pollen analysis indicates that it should be dated to about 4500 BP, i.e. an apparent net subsidence of 7.0 m, or 1.56 m ka^{-1}.

Finally the top of the upper clay towards the inland end of the valley also accumulated up to MHWS. (At the present-day MHWS level inland is $+0.5$ m O.D.) The top of the upper clay now occurs at -0.5 m O.D. and its maximum development is ^{14}C-dated to 1609 ± 50 BP. Apparent net subsidence since approximately 1600 BP is 1.0 m, and the rate of subsidence 0.63 m ka^{-1}.

These apparent subsidence amounts and rates are however arrived at on the assumption that global and local absolute sea-level was the same as at the present

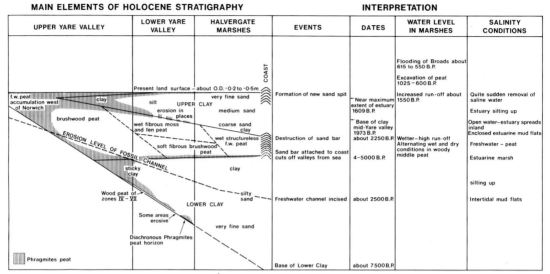

Fig. 9. Schematic section, and summary of events in the Yare valley.

day throughout the last 8000 years or so. This is not so, and an attempt to apply a correction must be made.

Relation to sea-level changes

The most detailed attempt at evaluation of sea-level changes in the United Kingdom is that of Tooley (1978). Inspection of his curve of sea-level fluctuations gives values of absolute sea-level of about −7·0 m at 7580 BP, of at least −1·0 m O.D. at 4500 BP, and of +1·75 m O.D. at 1600 BP. On this basis the actual values for subsidence would be 13·4 m since 7580 BP, i.e. 1·77 m ka⁻¹; 6·0 m since 4500 BP, i.e. 1·33 m ka⁻¹; and 2·75 m since 1600 BP, i.e. 1·72 m ka⁻¹. The precision of these rate estimates is very much dependent on the accuracy of the dates and past sea-level estimates from which they are calculated. For instance the sea-level rise immediately prior to 7580 BP is particularly steep and small miscalculations of either age or past sea-level, or their relationship to one another, could significantly affect the subsidence rate calculated from it. Nevertheless a rate of about 1·5 m ka⁻¹ subsidence is indicated by all three levels. If we make the simplifying assumption that this has been the average rate for the last 8000 or so years, we can modify the Tooley sea-level curve to give a relative sea-level curve for the Broadland region which allows for continuous subsidence at a 1·5 m ka⁻¹ rate (Fig. 10).

In Fig. 10 the freshwater/estuarine and estuarine/freshwater changes have been entered according to their determined ages. The combination of the Tooley absolute sea-level curve with an assumed local subsidence rate of 1·5 m ka⁻¹ then correctly indicates the depths at which occurrence of these facies changes may be expected at the present day. Two cycles of the Tooley (1978) curve have been speculatively projected, assuming continuing subsidence at 1·5 m ka⁻¹, to give a possible indication of potential relative sea-level changes in the Broadland region to the year 2000 AP (*c.* AD 4000).

Local physiographical influences

It is not intended to leave the impression that all environmental changes in the Broadland region are simply related to subsidence and absolute sea-level changes. Clearly local physiographical changes are very important.

Thus, although the slowing down of the general rate of absolute sea-level rise relative to sedimentation may correspond to the transition from (lower) estuarine clay to peat at 4500 BP, the indication of freshwater lagoonal conditions near the coast suggests that the estuary mouth may also have become largely sealed by a coastal barrier at that time. This interpretation seems even more likely when we consider the large quantities of sand incorporated in the base of the seaward part of the

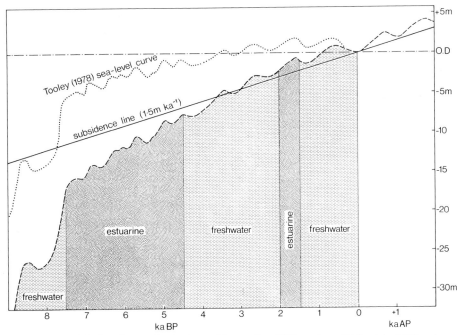

Fig. 10. Land–sea-level and facies changes in the lower Yare valley 8000 BP (= *c*. 6000 BC) to 2000 BP (= *c*. AD 4000). (For fuller explanation see main text.)

subsequent upper clay estuarine transgression, because this could well have been provided from the destruction of a coastal barrier system.

Likewise the enlargement of the river channels towards the end of the middle peat period strongly indicates a higher freshwater throughput at that time. This may have served to maintain freshwater conditions throughout the valley until the combination of subsidence and rising sea-level at around 2000 BP led to the reassertion of estuarine conditions.

At the present day, after a dry summer surge conditions (unaccompanied by rainfall) in the early autumn may lead to massive saline penetration of the Broadland rivers. This occurs in spite of the existence of the coastal spit at Great Yarmouth, the construction of which may itself have contributed to the abandonment of the intertidal flats of the upper clay, before the sedimentary sequence reached its culmination.

ACKNOWLEDGMENTS

We are grateful to the Anglian Water Authority for access to tidal records, to Peter Murphy, Centre of East Anglian Studies, and the School of Environmental Sciences Consolidated Research Fund for ^{14}C dates. We also acknowledge the loan of a specialized Hiller auger from the Sub-department of Quaternary Studies, Cambridge University for taking samples for ^{14}C dating.

REFERENCES

COLES, B.P.L. (1977) *The Holocene foraminifera and palaeogeography of Central Broadland*. Ph.D. thesis, University of East Anglia.
ELLIS, E.A. (1965) *The Broads*. Collins, London.
JENNINGS, J.N. (1952) The origin of the Broads. *Mem. R. geog. Soc.* **2**, 1–66.
JENNINGS, J.N. (1955) Further pollen data from the Norfolk Broads. Data for the study of post-glacial history. XIV. *New Phytol.* **54**, 199–207.
LAMBERT, J.M. & JENNINGS, J.N. (1960) Stratigraphical and associated evidence. In: *The Making of the Broads. Mem. R. geog. Soc.* **3**, 1–61.
TOOLEY, M.J. (1978) Interpretation of Holocene sea-level changes. *Geol. För. Stockh. Forh.* **100**, 203–212.

Spec. Publs int. Ass. Sediment. (1981) **5**, 133–145

Sedimentary sequences of the sub-recent North Sea coast of the Western Netherlands near Alkmaar

D. J. BEETS*, Th. B. ROEP* *and* J. DE JONG†

*Geological Institute, University of Amsterdam, Nieuwe Prinsengracht 130, *Amsterdam, the Netherlands, and* †*Geological Survey of the Netherlands, Spaarne* 17, *Haarlem, the Netherlands*

ABSTRACT

On the basis of sedimentary structures, the upper 4–5 m of a sub-recent beach plain sequence, exposed in pits around the town of Alkmaar in the western part of Holland, could be subdivided into three units. The upper unit (usually less than 1 m thick) consists largely of aeolian structures and is supratidal. The middle unit (*c.* 2·25 m thick) is characterized by sets of coast-directed, low-angle cross-bedding, alternating with wave-rippled strata. Assuming that the tidal range during deposition of the sequence was not much different from the present range off Alkmaar, it is concluded that this unit is largely intertidal. The lower unit consists mainly of coast-directed sets of high-angle cross-bedding. Both the low- and high-angle cross-bedded sets are interpreted as lee deposits of ridges, with the wave-rippled sand representing the runnel floors. The foresets of the ridges, in particular those of the inferred intertidal zone, show regular, lateral variations in the type of lamination, which is interpreted as due to the daily shift in wave base because of the tides. The beach plain sequence is compared to one underlying a dune ridge and to recent beaches at the Dutch coast. It is concluded that beach plain sequences are characteristic of periods of rapid progradation, whereas dune ridges represent more stagnant phases in the beach development.

INTRODUCTION

Since a few years ago the upper part of the sub-recent beach barrier complex of Holland has been studied in pits for civil construction work. Most of the work was done in pits around the town of Alk-maar (Fig. 1), and the data presented here were largely derived from that area. Some additional data come from pits near IJmuiden and The Hague.

The Holocene beach barrier complex of Holland forms a zone 5–10 km wide that roughly parallels the present coastline. Its eastern part is the 'Older Dune' landscape, which consists of a number of low dune ridges separated by depressions—swales or beach plains—covered by peat. In the western part, this landscape is covered by the much higher 'Younger Dunes' (Fig. 1), which date from the

twelfth century onwards (Van Straaten, 1965; Jelgersma *et al.*, 1970).

The older part of the barrier system in the vicinity of The Hague dates from about 4700 BP, as indicated by a C¹⁴ date of peat found leeward of the barriers (Zagwijn, 1965, p. 86), whereas an age of 4230 ± 30 yr BP (GrN-8169) comes from peat east of the barrier system near Alkmaar. Both dates show that initiation of the system coincides with the slackening of the Holocene sea-level rise (Jelgersma, 1961, 1979). From this time onward the barrier system between Alkmaar and The Hague prograded westward at a mean rate of several kilometres per 1000 yr (Van Straaten, 1965). Progradation ceased in Roman times, and from then on steepening of the beach face and local erosion set in. The origin of the 'Younger Dunes' in the twelfth century AD is related to this erosional phase (Van Straaten, 1965;

0141–3600/81/1205–0133 $02.00

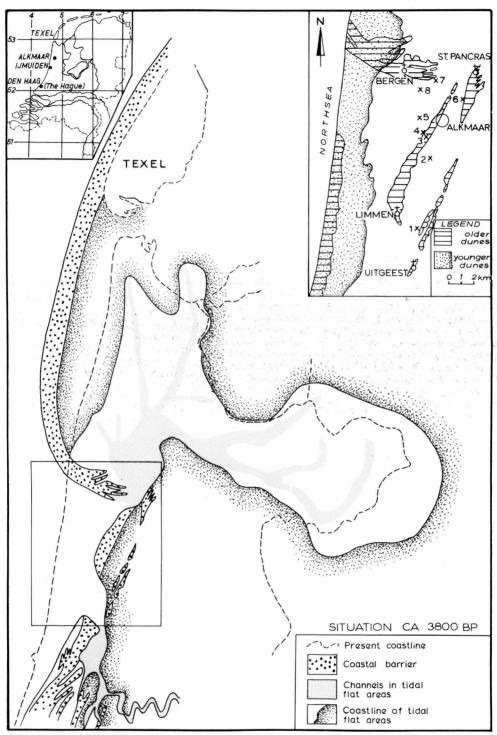

Fig. 1. Reconstruction of barrier system and tidal inlets in North Holland around 3800 BP. The inset gives localities 1–8 of studied pits. Geological information after Ente *et al.* (1975); Jelgersma *et al.* (1970); De Jong & Van Regteren Altena (1972); Pons & Wiggers (1959–60); Zagwijn (1971, 1975).

Jelgersma *et al.*, 1970). Erosion is most obvious just north of Alkmaar, where an older hooked spit complex is truncated by the present coastline (Fig. 1).

THE ALKMAAR AREA

This hooked spit complex formed the northern coast of a tidal inlet between the present towns of Bergen and Alkmaar (Fig. 1), which connected a large tidal bay that occupied the present northern part of North Holland with the North Sea (Pons & Wiggers, 1959–60; Jelgersma *et al.*, 1970; Zagwijn, 1975; Ente, Zagwijn & Mook, 1975; Roep, Beets & De Jong, 1979). The tidal inlet was active from at least 3880 ± 30 yr BP (GrN-7775). Rapid shoaling occurred around 3600–3500 BP, and the inlet closed before 3360 ± 50 yr BP (GrN-6763). The salt marshes inland were inhabited by 3275 yr BP (Van Regteren Altena, Van Mensch & Ijzereef, 1977). The southern coast of this inlet is a WNW-ward prograding barrier complex with roughly two NNE–SSW-trending dune ridges (Fig. 1) separated by a beach plain. The villages of St Pancras and Uitgeest are situated on the eastern ridge. The town of Alkmaar was founded on the western ridge. West of this ridge a flat and broad beach plain occurs, which is covered by the 'Younger Dunes' near the coast (Fig. 1).

Most pits studied were situated in the beach plains (Fig. 1, pits 1, 2, 5–8) near the mouth of the tidal bay. One pit (3) was excavated in one of the dune ridges, and one (4) at the boundary of beach plain and dune ridge.

THE BEACH PLAIN SEQUENCE

In a previous paper (Roep *et al.*, 1975), a description of this beach plain sequence was given based on an outcrop in a 6 m deep pit in the southern barrier ridge of the inlet (Fig. 1, pit 4). Since then, many new pits have been opened in this area, and they confirm the general validity of this sequence for the swales or beach plains.

In the deeper pits three units could be distinguished which show the following main characteristics.

Unit 1

The uppermost unit, which comprises the supratidal zone, varies in thickness from 0·5 to 1 m, and consists of aeolian sand covered by a soil or peat layer (Figs 2I, 3 and 4). The sand is fine-grained and shows broad, trough-shaped sets with smoothly tapering cross-bedding. Shells are scarce and restricted to the lower part of this unit, where they occur with driftwood and peat detritus on concave, seaward-dipping laminae, representing swash marks of springtide or storm surges (Fig. 2Ig). The thickness of the aeolian cover varies greatly. In pit 4 (Fig. 2I) it measures about 1 m, whereas in pit 1 (Figs 3 and 4) it is only a thin veneer of a few decimetres thickness. In the latter pit the transition to the underlying intertidal zone (unit 2) is characterized by a zone up to 50 cm thick with clay seams of a few millimetres thickness alternating with structureless or irregularly crinkled fine sand layers. The clay may have been deposited after flooding of the beach plain during spring tides or coastal set-up by storm. The crinkled layers are thought to be formed by various processes, such as: scour and deposition around stranded debris (Hill & Hunter, 1976); dry sand blown over a wet surface; hail, rain and foot imprints; disturbance of the stratification because of the escape of entrapped air (Van den Berg, 1977; De Boer, 1979).

Unit 2

This unit consists of about 2 m of predominantly fine-grained sand, characterized by up to 75 cm high, usually concave-upward, low-angle, cross-bedded sets. The sets alternate with up to 30 cm thick, wave-rippled strata, or are separated by distinct truncation planes, which are often superficially burrowed by small animals (the tubes resembled those of *Pygospio elegans*, Dr Swennen, personal communication), or are covered by thin clay seams (Figs 2, 3 and 4). Burrowing and the occurrence of clay seams are not restricted to these truncation planes, but are also occasionally found on the laminae of the cross-bedded sets. Cross-bedding is always coast-directed. This plus the low angle and the regular occurrence of wave ripples on the cross-laminae (see below) indicate that the sets represent low-angle longshore bars or ridges. The intervening wave-rippled beds are interpreted as runnel fills. Rip-current channel deposits comparable to those described for many recent barred coasts (Reineck, 1963; Davis *et al.*, 1972; Davidson-Arnott & Greenwood, 1976; Hunter, Clifton & Phillips, 1979) have not been found. However, deposits of shallow and ephemeral rip channels may

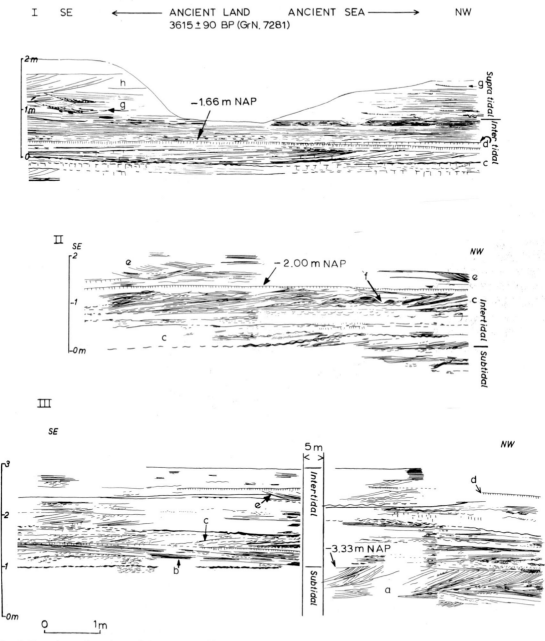

Fig. 2. Drawings of portions of the SE–NW (shore perpendicular) walls of pit 4 (for location see Fig. 1). See text for interpretation. Supratidal in these drawings is equivalent to unit 1 of text, intertidal to unit 2 and subtidal to unit 3. (a) Lee-side lamination of high-angle ridge; (b) stoss-side lamination of ridge; (c) lee-side lamination of low-angle ridge; (d) burrow levels; (e) storm-generated troughs and/or channels; (f) storm generated erosion of ridges; (g) swash marks with shells and fragments of wood and peat at dune base; (h) aeolian sand. NAP means Dutch Ordnance Level, which is roughly mean sea-level.

Fig. 3. Beach plain sequence exposed in pit 1 (Fig. 1) with characteristic low-angle lee-side lamination of the intertidal ridges. Shore is towards the left. For a schematic drawing of the sequence see Fig. 4, which also gives the scale. Subtidal unit 3, containing one set of high-angle cross-bedding, is just exposed at the bottom (A). For a better view of the same set see Fig. 6. The intertidal unit 2 (B) shows an upward decrease in lee-side slope of the ridges. Peaty soil (D) covers a thin veneer of aeolian sand, which overlies a sandy mudflat sequence. The clay layer at the top (E) was deposited during a younger transgressive phase. The lee-side lamination of the lower set of unit 2 (in between the letters A and B) shows the typical alternation of structureless and wave-rippled laminae. The latter stand out because of their finer grain size.

be concealed within the wave-rippled runnel deposits. In the lower half of the unit, seaward-dipping plane beds, probably the stoss sides of ridges, occur occasionally intercalated between the landward-dipping, low-angle lee sides (Fig. 2IIIb).

Low-angle ridges occur predominantly in unit 2 which, as will be discussed below, is interpreted as the intertidal zone of the beach plain sequence. The sets are 15–75 cm high and the dip of the foresets varies from about 15° to almost zero. In most pits, the sets near the base of unit 2 have the highest dips (Figs 2, 3 and 4). Upwards the dip gradually decreases to zero. Flattening of the dip of the foresets is often accompanied by a decrease in the height of the sets. Both flattening of the foreset dip and thinning upward of the sets could be due to a decreasing eroding capacity of the runnels in the upper part of the intertidal zone, as described by Van den Berg (1977) of a recent ridge and runnel beach at

the island of Schouwen in the south-western part of the Netherlands. In that case the truncation planes between the sets would represent the base of the runnels. Furthermore it is important that when unit 2 is overlain by a relatively thick aeolian unit 1 (as for instance in pit 4 (Fig. 2I), the foreset lamination of the ridges merges laterally and vertically into the horizontal plane bed of the foreshore, whereas in the pits with only a thin veneer of aeolian cover (as for instance pit 1), the highest ridge is overlain by a few decimetres of a sandy mudflat facies with burrow levels and occasionally *Scrobicularia plana* in living position (Figs 3 and 4). The foresets of the low-angle ridges are up to several metres long (Figs 2, 3, 4 and 6); this plus the low dip of the foresets suggests that migration of the ridges is not due to simple flow separation and avalanching. Cross-bedding of the ridges, in particular those in the lower half of unit 2, shows a regular alternation of a

SED 5

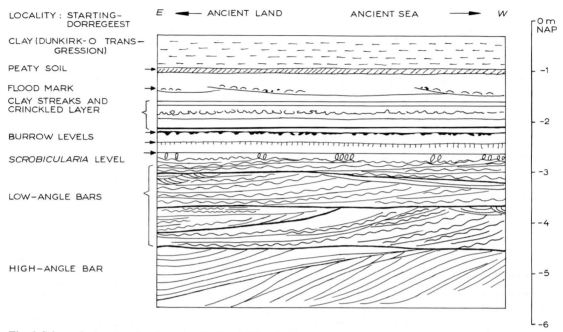

Fig. 4. Schematic drawing of east–west wall of pit 1 (Fig. 1). Photographs of the same wall are given in Figs 3 and 6.

small number of structureless, or occasionally slightly graded laminae of up to a few centimetres thickness with finer grained, wave-rippled strata (Figs 3 and 5). The former are interpreted as deposited from suspension-rich currents formed by swash surges breaking on and eroding the stoss sides of the ridges. The intercalated wave-rippled laminae represent a lower energy level, when the waves can only rework the sediment without much net transport. The thickness of these 'couples' (structureless plus wave-rippled laminae) varies between 1 and 10 cm, measured perpendicular to the surface of the lamination. The repetition of these low-angle lee sides is so characteristic that we attribute it to a highly regular phenomenon: the daily change in sea-level because of the tides. The wave ripples are believed to have formed during high tide, whereas the structureless laminae represent suspension-rich swash surges plunging into the runnels shortly before and after the emergence of the crest of the ridges. This interpretation will be discussed more fully below.

Clay drapes and burrow levels on the leeward slopes of the ridges (Fig. 2IIIc) represent periods of quiet weather without or with hardly any wave energy. Irregular erosion of the bedforms (Fig. 2IIf) and isolated troughs or channels in the upper

part of unit 2 (Fig. 2IIe, IIIe) are probably the result of severe gales.

Unit 3

The lowermost unit in the pits consists of about 1·5 m of medium- to fine-grained sand with predominantly large-scale, high-angle cross-bedding and wave-rippled beds (Figs 2III and 6). Set boundaries are roughly horizontal or dip gently seaward. The cross-bedding is interpreted as the lee-side deposits of wave-built ridges based on the observations that: (1) the cross-bedding is invariably oriented towards the ancient coast; (2) the cross-bedding is associated with small-scale wave-ripple lamination.

Although unit 2 is characterized by low-angle ridges and unit 3 by high-angle ones, this does not imply that the two ridge types are restricted to these units. High-angle cross-bedding may occur in unit 2, although always subordinately, and the same is true for low-angle cross-bedding in unit 3. In addition to these two types, ridges with convex-upward laminae occur in minor amounts in both units, preferentially near the boundary of units 2 and 3 (Fig. 7).

High-angle cross-bedding occurs in sets varying

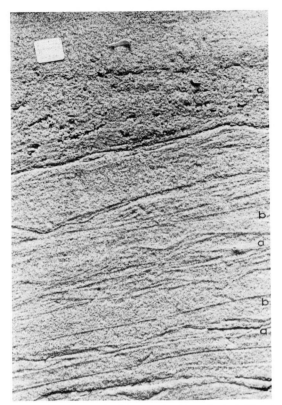

Fig. 5. Lacquer peel of low-angle ridge of pit 4 (Figs 1 and 2). The top is heavily burrowed (c) and lies at about −2·00 m NAP. Note the alternation of structureless laminae (b), inferred to be deposited at low tide, and the wave rippled laminae, inferred to be formed at high tide. The label at the top left is 2 cm wide.

in thickness from 0·1 to 1 m. The cross-bedding consists primarily of slip face lamination (angle of inclination 20–25°), which is regularly interrupted by either more gently inclined erosion planes (Figs 2IIIa and 8C), or thin wave-rippled cross-strata (Fig. 8B). Erosional unconformities are produced during periods of temporary flattening of the steep slip faces of the ridges, and are the result of a change in wave energy. Wunderlich (1972) suggested that such unconformities may be attributed to tidal fluctuations. On the other hand, Van den Berg (1977), who observed several times that the slip face of ridges was flattened after storms, connects these unconformities to changing conditions of incoming waves and/or wind. In analogy to the more obvious 'couples' in the low-angle lee-side laminations of unit 2, and because of the regular and common occurrence of these erosional surfaces

(in pit 4 an average of two per metre of slip face), we prefer the tidal fluctuations as the main cause. The erosional surfaces represent the higher energy level at low tide, whereas slip face lamination forms when the wave base rises. In the one case where we found slip face lamination alternating with wave-rippled laminae (Fig. 8B), it is obvious that the latter represents the lower wave energy at high tide. If our interpretation is correct, these two types of alternations would represent different wind conditions.

Ridges with convex-upward lamination (Fig. 7) occur subordinately in the pits of the Alkmaar area. The sets vary in height between 10 and 70 cm. They consist of fan-shaped bundles of sharply laminated, convex cross-strata, that alternate with higher-angle slip face lamination (Figs 7 and 8D). Shells on the convex laminae occur convex-side up, whereas in the slip face lamination most are concave-up. The convex shape of the laminae, the orientation of the shells, and the sharp lamination resemble a plane bed under shooting conditions. These ridges most closely resemble the first type of high-angle ridges, in which the erosional unconformities produced at low tide have been replaced by convex outbuilding (Fig. 8C, D).

Depth of deposition

The burrow levels at the top of unit 2 (Fig. 2I), in combination with the deepest aeolian excavation in unit 1, give the best criteria to establish palaeo-mean-high-tide (palaeo-MHT). The burrowing animal must have been flooded daily, whereas erosion by wind can only occur in dry sand above the groundwater table. In pit 4 (Figs 1 and 2) the deepest level of aeolian reworking was found at −1·25 m NAP (NAP = Dutch Ordnance Level, which is about equal to Mean Sea Level; −1·25 m NAP means 1·25 m below the NAP level) and the highest burrow level at −1·66 m NAP (Fig. 2Id). As wave-rippled sand was found up to a level of −1·30 NAP and air bubble sand (typical for the upper part of the intertidal zone and backshore) between −1·20 and −1·50 m NAP, we chose as palaeo-MHT the level of −1·25 m NAP (Roep, Beets & Ruegg, 1975). In pits characterized by a sandy mudflat zone upon the ridge lee side lamination (pit 1, Figs 3 and 4) the transition zone with clay seams and crinkled layers was chosen as palaeo-MHT. The total thickness of unit 2 is about 2·25 m, with a mean value for the eight pits around Alkmaar of $\bar{x}_{(8)} = 224$ cm, and a standard deviation of

Fig. 6. Detail of the lower portion of the wall of pit 1 shown in Figs 3 and 4. The hole in the wall (left below), is the same as that in Fig. 3 (centre below). Picture shows high-angle slip face lamination of a ridge of the top of unit 3 (A) overlain by low-angle lee-side lamination of a ridge of the basal part of unit 2 (B). The boundary between the two sets is the inferred mean low tide level. Note the difference in dip, structure and grain size of the two lee-side laminations.

$s = 47$ cm. The thickness of that part of unit 2 below the highest burrow level is about 1·80 m (for seven pits $\bar{x}_{(7)} = 179$ cm; $s = 36$ cm).

Using these criteria to establish palaeo-MHT, with the assumption that the tidal range during deposition of the Alkmaar sequence was roughly similar to the present range of 1·75 m along the uninterrupted part of the present coastline of the western Netherlands, one comes to the conclusion that unit 1 is supratidal, unit 2 largely intertidal, and unit 3 subtidal. The conclusion that the structural change from predominantly high-angle to predominantly low-angle lee-side lamination at the boundary of units 2 and 3 corresponds approximately to palaeo-MLT is corroborated by the observation that this boundary rises from east to west, in accordance with the known post-glacial rise in sea-level. This boundary lies at a depth of $-4·47$ m NAP in pit 1 (Figs 1 and 4) and at $-3·33$ m NAP in pit 4 (Figs 1 and 2 III). The boundary between units 1 and 2 rises similarly.

Tidal imprints on the lee-side lamination of the ridges

It has been suggested above that the internal lamination of the ridges gives a good record of the the daily shift in wave-base because of the tides. The main argument for this suggestion is that the lamination shows a rapid and regular repetition of structures indicative of different wave energy levels. This recurrence is most distinct in the low-angle ridges of the intertidal zone, where migration by suspension-rich swash surges (the higher energy level) is followed by reworking of this sand into small-scale wave ripples (the lower energy level) (Figs 3, 5 and 8A). Although the 'couples' in the foreset lamination of the high-angle ridges of the subtidal zone are more diverse (Fig. 8B, C, D), they show the same regular alternation of energy levels. As discussed, this alternation must either be due to variations in wind conditions or to changes in sea-level. We chose the latter alternative, because of the short-term character of the alternations. This is particularly obvious for

Fig. 7. Lee-side lamination of a ridge exposed in pit 8 (Fig. 1) with an internal structure consisting of bundles of slip face laminations (a) and convex upward lamination (b). A matchbox on the peat pebble (upper left of b) gives scale. Near (c) a wave-rippled layer with clay drapes and burrows can be seen.

the low-angle ridges. The thickness of the 'couples' (structureless + wave-rippled laminae) has been measured on lacquer peels of a low-angle set taken from the basal part of unit 2 in pit 5. The lacquer peels cover about 6 m continuous ridge foreset lamination, taken perpendicular to the shore. Angle of dip of the set is about 15°. In total 22 of these 'couples' have been measured perpendicular to the plane of lamination. They vary in thickness between 1 and 10 cm, averaging 4·45 cm ($s = 2·7$ cm). Assuming that each of these couples represents a tidal cycle, the mean migration rate of the bar is about 34 cm a day ($2 \times 4·45/\sin 15$). This is probably a maximum rate, as the lamination may conceal tidal cycles with no or only little migration. In comparison to migration rates of intertidal bars of most recent beaches this amount is low. For instance, Davis *et al.* (1972) give a mean migration rate of 60 cm a day for the intertidal ridge at Crane Beach, Massachusetts, and Van den Berg (1977) gives 20 m a week for intertidal ridges at the recent beach of Zandvoort in the western part of Holland; both examples are high-angle ridges of exposed coasts. On the other hand, the migration of the low-angle ridges of the Alkmaar

beach plain sequence falls in the same range as that of intertidal ridges of a sheltered beach near Schouwen (south-west Holland) for which Van den Berg (1977) gives a migration rate of less than 3 m a week. As will be discussed below, the latter beach is a good contemporary example for the Alkmaar beach plain sequence, apart from the absence of low-angle ridges in the intertidal zone.

Our data on the width of the 'couples' in the lee side lamination of the high-angle ridges (slip face lamination plus erosional unconformity) are limited. Taking those of pit 4 (Fig. 2IIIa), one comes to a mean horizontal width of each 'couple' of about 50 cm, which would give a migration rate of 1 m a day.

THE DUNE RIDGE SEQUENCE

The only pit excavated in one of the dune ridges in the Alkmaar area (pit 3, Figs 1 and 9) shows a sequence which differs considerably from that of the pits in the beach plain. The lower half of the walls of the pit consists of parallel-bedded sand with, near

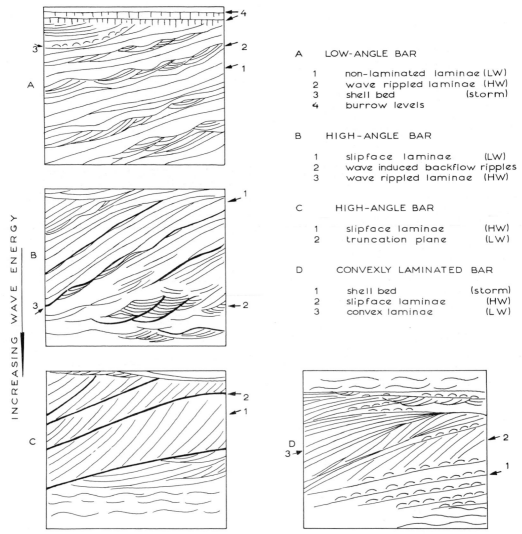

Fig. 8. Schematic summary of the different types of lee-side laminations of wave-built ridges of the Alkmaar area.

the base of the pit, one burrow level (Fig. 9a). The upper part shows broad, trough-shaped sets overlain by a peaty soil. About half-way up the walls of the pit 20 cm of irregularly bedded, crinkled layers occur. Clay seams and lee-side deposits of low-angle ridges are absent. The burrow level occurs at −2·15 m NAP and the deepest aeolian excavation at −1·39 m NAP, which gives a wide range for the inferred MHT level in this pit. However by extrapolating the more precisely defined palaeo-MHT level from nearby pit 4, it appears that the parallel-

bedded sand below the crinkled layers correlates with the upper part of intertidal unit 2 of the beach plain sequence. A similar predominance of parallel lamination in the intertidal zone was found in a slightly younger (2310 ± 35 yr BP) dune ridge sequence near the town of IJmuiden (for location see Fig. 1), and was attributed by Roep *et al.* (1975) to a more exposed position of this beach in comparison to that of the beach plain sequence of Alkmaar. As will be discussed below, recent work on the Dutch shores by Van den Berg (1977) confirms this view.

Fig. 9. Wall of pit 3 (Fig. 1) excavated in one of the dune ridges of the Alkmaar area. Aeolian sand (B) with large-scale trough-shaped sets (unit 1) resting on parallel-laminated sand (A) with burrow level (a). The rod measures 1 m. About 20 cm of irregularly bedded and crinkled layers near the top of the rod roughly represent MHT level in this pit.

COMPARISON WITH RECENT BEACHES

In the last decade, a number of important coastal studies have been published (Clifton, Hunter & Phillips, 1971; Davis *et al.*, 1972; Reineck & Singh, 1973; Davidson-Arnot & Greenwood, 1976; Van den Berg, 1977; Hunter *et al.*, 1979). Most of these studies give a vertical sequence of sediment types and internal structures that would be formed by progradation of the beach profile, mainly based on the bedforms present on the beach and foreshore during the study. Although all of these proposed vertical sequences have many structures in common with those of the Alkmaar pits, only the study of Van den Berg (1977) gives good actualistic examples for the sub-recent Alkmaar beaches. This is due to a number of factors. In the first place, the two beaches described by Van den Berg are situated on the Dutch coast, so that the wind and wave climate will not deviate much from that once existing near Alkmaar. Secondly, one of those two beaches is bordered by ebb deltas of major tidal inlets, a situation similar to that of the Alkmaar area during formation of the

beach sequences. Thirdly, his comparative study of the morphodynamics of the two beaches gives a well-documented picture of the preservation potential of the various bedforms, as it is based on a weekly or fortnightly measuring programme consisting of detailed observations of beach height and bedforms over a period of more than 10 years. The two beaches compared by Van den Berg are an exposed and 'semi-permanent ridge and runnel beach' at the village of Zandvoort, about 10 km south of IJmuiden, and a 'permanent ridge and runnel beach' at the island of Schouwen between the Oosterschelde and Grevelingen estuaries in the south-western part of the Netherlands. The latter beach is protected from severe wave attack by a series of extensive shoals situated on the tidal ebb deltas of the inlets. The morphodynamics of the beach at Schouwen are, in particular, of great relevance to the Alkmaar sequence, both because of its setting and because of the resemblance of the sedimentary sequences.

Annual measurements of the beach and nearshore variations around Schouwen from 1880 onwards by Rijkswaterstaat (the Government Board for Ways,

Waterways and Harbours) show that accretion and erosion of the beach are determined by the position of shoals and tidal channels of the ebb delta system. According to a report by De Smit & Van Malde referred to by Van den Berg (1977): 'the extensive shoals, which sometimes emerge during low tide, provide shelter against destructive storm wave attack and promote accretion; whereas active tidal channels, when situated close to the shoreline, have an erosive influence on the beach'. The weekly and fortnightly measurements, on which Van den Berg's data are based, started in 1962, following a period of beach retreat. Slow progradation from 1962 until 1967 was followed by a period of fast progradation from 1968 until 1975. In the latter period the MHT level shifted 90 m seaward and the MLT level 120 m. Fast progradation was accompanied by an overall decrease of the beach gradient, and a retarded accretion on the backshore and dunes, which caused a considerable increase in beach width. In the period of slow progradation, ridges were practically restricted to the foreshore zone. Landward movement generally ended in a welding of the ridges on a relatively steep seaward slope. The innermost ridges could be eliminated during onshore storms because of the relatively high beach gradient, and the deposits dating from this period show the predominance of a plane bed of (storm) swash and backwash in the upper half of the intertidal zone. The resulting sedimentary sequence is not very different from that predicted by Van den Berg (1977) for the exposed beach at Zandvoort and, concerning the structures in the intertidal zone, is similar to the dune ridge sequence of the Alkmaar area. A steep beachface obviously promotes dune building, as it implies relatively high wave energy on the upper part of the foreshore so that the waves can transport sand high up the backshore. The sedimentary sequence of the intertidal zone, preserved during the period of rapid progradation of the Schouwen beach (1968–75) consists predominantly of runnel deposits with small-scale cross-bedding and occasional clay drapes, and of the lee-side deposits of the ridges. The sequence is very similar to that of the Alkmaar beach plain, but differs from it in one important aspect: the ridges in the intertidal zone of the Schouwen beach show predominantly high-angle slip face lamination. According to Van den Berg (1977) lee-side lamination of the ridges changes from high-angle to low-angle, when in the course of landward migration, the ridge crest becomes so high that it is only overtopped during high tide by the swash of the larger waves.

As this occurs in the upper part of the intertidal zone at the Schouwen beach, either the tidal interval at the Alkmaar beaches was larger than the inferred 1·75 m, or conditions at the two beaches differed in a few aspects. The latter is suggested by the finer grain size of the sand in the Alkmaar beach plain sequence, by the characteristic internal structure of the low-angle ridges of the Alkmaar area (which has no present-day counterpart at the Schouwen beach), and by the occasional presence of a sandy mudflat facies in the upper part of the intertidal zone of the Alkmaar sequences. Nevertheless, despite this difference, the resemblance between the two sequences is so striking that it may be safely assumed that the Alkmaar beach plain sequence formed during periods of rapid progradation of the beach over the tidal ebb delta of the inlet. This conclusion is supported by the present data on the progradation of the Alkmaar beaches. A rough estimate of the mean progradation rate for the entire Alkmaar area (beach plains as well as dune ridges) gives 238 m/100 yr (oldest date 4230 ± 30 yr BP). Using radiocarbon dates from pit 4 (shell pairs at -2.76 m NAP: 3615 ± 90 yr BP – GrN 7281) and pit 5 (shell pairs at -2.00 m NAP: 3560 ± 40 yr BP – GrN 6309), and assuming progradation perpendicular to the Limmen–Alkmaar ridge (Fig. 1), this yields a minimum value of 500 m/100 yr (distance between the pits is 1 km; age difference about 200 yr) for the beach plain.

CONCLUSION

Beach plain sequences form in periods of rapid progradation when the beach gradient is low, and the beach is sheltered from onshore storms by shoals. Progradation is probably most rapid when the waves are dissipated so efficiently that fore- and backshore are replaced by a thin sandy mudflat facies in the upper part of the intertidal zone. Dune ridges, on the other hand, are promoted by a higher beach gradient and no or only little progradation of the beach. In these respects, the Dutch Holocene beach barrier deposits of the North Sea coast are very similar to the cheniers and plains of the northern and western coasts of the Gulf of Mexico (Gould & McFarlan, 1959; Todd, 1968). The Alkmaar beach plain and dune ridges are situated on a former ebb tidal delta, and their lateral alternation can easily be explained by changes in the position of the tidal channels. However, when dune ridges and beach plains occur far from tidal inlets, as in some stretches of the beach barrier complex between Alkmaar and The Hague, the alternation of the two must be due to changes in the supply of sediment (Van Straaten, 1965).

ACKNOWLEDGMENTS

The authors gratefully acknowledge the pleasant cooperation and help of colleagues of the Geological Survey, in particular Dr W. H. Zagwijn, Dr S. Jelgersma and G. H. J. Ruegg. Thanks are due to Professor W. G. Mook of the Laboratory of Physics of the State University at Groningen for providing C¹⁴ dates, to Miss L. Gonggrijp for typing the manuscript, and to F. Kievits and J. Wiersma for help with the illustrations. Dr R. W. Dalrymple's constructive criticism of an earlier version of this paper is highly appreciated.

REFERENCES

BERG, J.H. VAN DEN (1977) Morphodynamic development and preservation of physical sedimentary structures in two prograding recent ridge and runnel beaches along the Dutch coast. *Geologie Mijnb.* **56**, 185–202.

BOER, P.L. DE (1979) Convolute lamination in modern sands of the estuary of the Oosterschelde, the Netherlands, formed as a result of entrapped air. *Sedimentology*, **26**, 283–294.

CLIFTON, H.E., HUNTER, R.E. & PHILLIPS, R.L. (1971) Depositional structures and processes in the non-barred high-energy nearshore. *J. sedim. Petrol.* **41**, 651–670.

DAVIDSON-ARNOTT, R.G.D. & GREENWOOD, B. (1976) Facies relationships on a barred coast, Kouchibouguac Bay, New Brunswick, Canada. In: *Beach and Nearshore Sedimentation* (Ed. by R. A. Davis and R. L. Ethington). *Spec. Publ. Soc. econ. Paleont. Miner. Tulsa*, **24**, 149–169.

DAVIS, R.A., FOX, W.T., HAYES, M.O. & BOOTHROYD, J.C. (1972) Comparison of ridge and runnel systems in tidal and non-tidal environments. *J. sedim. Petrol.* **42**, 413–421.

ENTE, P.J., ZAGWIJN, W.H. & MOOK, W.G. (1975). The Calais deposits in the vicinity of Wieringen and the geogenesis of northern North Holland. *Geologie Mijnb.* **54**, 1–14.

GOULD, H.R. & MCFARLAN, E. (1959) Geologic history of the chenier plain, southwestern Louisiana. *Trans. Gulf-Cst Ass. geol. Socs* **9**, 261–270.

HILL, G.W. & HUNTER, R.E. (1976) Interaction of biological and geological processes in the beach nearshore, Padre Island, Texas. In: *Beach and nearshore Sedimentation* (Ed. by R. A. Davis and R. L. Ethington). *Spec. Publ. Soc. econ. Paleont. Miner. Tulsa*, **24**, 169–187.

HUNTER, R.E., CLIFTON, H.E. & PHILLIPS, R.L. (1979) Depositional processes, sedimentary structures and predicted vertical sequences in barred nearshore systems, southern Oregon Coast. *J. sedim. Petrol.* **49**, 711–726.

JELGERSMA, S. (1961) Holocene sea level changes in the Netherlands. *Meded. geol. Sticht. C, VI°*, 7.

JELGERSMA, S. (1979) Sea-level changes in the North Sea basin. In: *The Quaternary History of the North Sea* (Ed. by E. Oele, R. T. E. Schüttenhelm and A. G. Wiggers). *Acta Univ. Ups. Symp. Ups. Annum Q. Celebr.* **2**, 233–248.

JELGERSMA, S., JONG, J. DE, ZAGWIJN, W.H. & REGTEREN ALTENA, J.F. VAN (1970) The coastal dunes of the western Netherlands; geology, vegetational history and archeology. *Meded. Rijks geol. Dienst.* **21**, 93–167.

JONG, J. DE & REGTEREN ALTENA, J.F. VAN (1972) Enkele geologische en archeologische waarnemingen in Alkmaars oude stad. In: *Alkmaar van Boerderij tot Middeleeuwse Stad*, 25–64. Alkmaar.

PONS, L.J. & WIGGERS, A.J. (1959–60) De holocene wordingsgeschiedenis van Noord-Holland en het Zuiderzeegebied. I en II. *Tijdschr. K. ned. aardrijksk. Genoot.* **76** (1959), 104–152; **77** (1960) 3–57.

REGTEREN ALTENA, J.F. VAN, MENSCH, P.J.A. VAN & IJZEREEF, G.F. (1977) Bronze Age clay animals from Grootebroek. In: *Ex Horreo*, pp. 226–241. IPP 1951–1976. Amsterdam.

REINECK, H.E. (1963) Sedimentgefüge im Bereich der südlichen Nordsee. *Abh. senckenb. naturforsch. Ges.* **505**, 138 pp.

REINECK, H.E. & SINGH, I.B. (1973) *Depositional Sedimentary Environments.* Springer-Verlag, Berlin. 439 pp.

ROEP, TH.B., BEETS, D.J. & RUEGG, G.H.J. (1975) Wave-built structures in subrecent beach barriers of the Netherlands. *Proc. 9th Int. Sed. Congr., Nice*, pp. 141–146.

ROEP, TH.B., BEETS, D.J. & JONG, J. DE (1979) Het zeegat tussen Alkmaar en Bergen van ca. 1900 tot 1300 jaar voor Chr. *Alkmaarse Historische Reeks*, III, 8–36, De Walburg Pers, Zwolle.

STRAATEN, L.M.J.U. VAN (1965) Coastal barrier deposits in South- and North-Holland, in particular in the areas around Scheveningen and IJmuiden. *Meded. geol. Sticht.* **17**, 41–75.

TODD, T.W. (1968) Dynamic diversion: influence of longshore current – tidal flow interaction on chenier and barrier island plains. *J. sedim. Petrol.* **38**, 734–746.

WUNDERLICH, F. (1972) Georgia coastal region, Sapelo Island, U.S.A. Sedimentology and biology. III. Beach dynamics and beach development. *Senckenberg. Mar.* **4**, 47–79.

ZAGWIJN, W.H. (1965) Pollen-analytic correlations in the coastal-barrier deposits near The Hague (The Netherlands). *Meded. geol. Sticht.* **17**, 83–88.

ZAGWIJN, W.H. (1971) De ontwikkeling van het "oer-IJ" estuarium en zijn omgeving. *Westerheem*, **XX**, 11–18.

ZAGWIJN, W.H. (1975) De palaeogeografische ontwikkeling van Nederland in de laatste drie miljoen jaar. *Tijdschr. K. ned. aardrijksk. Genoot.* **3**, 181–201.

Spec. Publs int. Ass. Sediment. (1981) **5**, 147–159

Rhythmic seasonal layering in a mesotidal channel fill sequence, Oosterschelde Mouth, the Netherlands*

JAN H. VAN DEN BERG†

Comparative Sedimentology Division, Institute of Earth Sciences, Budapestlaan 4, 3508 TA Utrecht, the Netherlands

ABSTRACT

Sedimentary features of a subrecent channel fill sequence produced during abandonment of a 12 m deep tidal channel are described. The history of channel abandonment and subsequent infill is documented by detailed hydrographic charts. The sequence contains a 5 m thick layer composed of a rhythmic alternation of two facies. Evidence for a seasonal origin of this rhythmic bedding is provided by: the relationship between the thickness of individual couplets of winter and summer layers, and the annual rate of accumulation as interpreted from hydrographic charts; the growth and mortality pattern of some frequently occurring bivalve species (*Macoma balthica, Cerastoderma edule, Abra alba, Spisula subtruncata*); the different degrees of bioturbation between the two members of each couplet; the occasionally perceptible increase in diameter of internal traces (press structures) of the heart urchin (*Echinocardium cordatum*) in upward direction. Inorganic primary sedimentary structures point to remarkable low-energy conditions during deposition of the rhythmic beds. The summer facies consist of interlaminated mud, silt and fine sands. The winter layers are composed of flaser bedding. The transition from summer to following winter layer is slightly erosional, whereas a more gradual transition occurs from winter to summer layers. The thickness of individual winter–summer couplets ranges between 20 and 60 cm. Individual couplets may continue for hundreds of metres, especially in sections parallel to the axis of the channel.

INTRODUCTION

Thinly interlayered bedding caused by rhythmic seasonal changes is known from many lake deposits, both recent and ancient. Seasonal rhythmites have also been described for several recent and ancient marine stagnant and evaporite basins. In general we are dealing with thin annual couplets consisting of two members of different lithological composition. This paper documents a rather different type of thick seasonal bedding that originated in a dynamic mesotidal environment (mean spring and neap tidal ranges are 3·5 and 2·3 m respectively).

The sequence is exposed in an excavation made in the mouth of the Oosterschelde tidal inlet (for loca-

tion see Fig. 1). A general impression of the studied outcrop may be obtained from Fig. 2. Seven lithostratigraphic units have been distinguished, based largely on sedimentary structures. The sequence of rhythmic beds (unit D in Fig. 2), located at a depth of 5 to 10 m below mean sea-level, was formed in an abandoned tidal channel during the second and third decade of this century (Fig. 3).

Data in this paper were collected from two exposures, described as outcrops A and B respectively (see Fig. 1).

GENERAL CHARACTERISTICS OF THE RHYTHMIC BEDS

The rhythmic beds consist of fine sand (median grain size 110 μm) with an admixture of 5–20% mud. Following Terwindt (1967, 1977) 'mud' includes all material having a fall velocity in water less than that of a quartz grain 50 μm in diameter. The

---

* Comparative Sedimentology Division, Report No. 52.

† Present address: Rijkswaterstaat, Deltadienst, Van Veenlaan 1, Zierikzee, the Netherlands.

0141-3600/81/1205-0147 $02.00

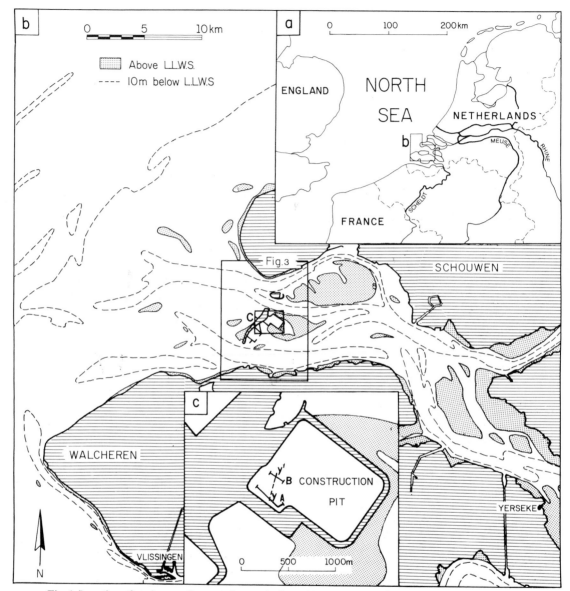

Fig. 1. Location of study area. Outcrop sites are indicated by A and B. Y–Y′: correlation line in Fig. 5.

primary sedimentary structures consist of an alternation of flaser and parallel laminated layers. The laminated layers generally have a higher mud content. A gradual transition from the flaser layers into the parallel laminated layers can be observed. The transition from parallel laminated layers into flaser layers, however, is sharp because it is slightly erosional.

A lithology of the rhythmic beds, observed in outcrop A, is given in Fig. 4. The sequence of rhythmic layers is laterally persistent. It can be found throughout the exposed wall of the pit, a length of more than 200 m. The strike of the exposed wall is more or less parallel with the axis of the abandoned channel. All individual couplets can be traced laterally without any structural and lithological change.

The thickness of the individual couplets may show slight variation, however; this variation is probably related to differences of the depositional depth. The lower part of the sequence shows more bioturbation,

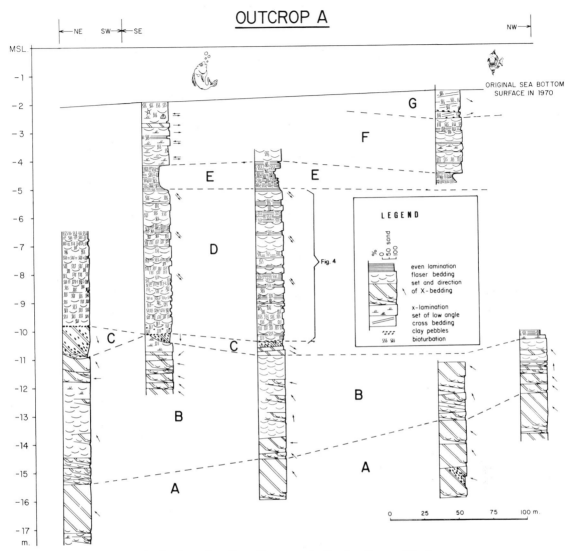

Fig. 2. General outline of outcrop A. For location see Fig. 1.

caused by 'press' structures of *Echinocardium cordatum* (Schäfer, 1962; Reineck & Singh, 1973) and the dwelling traces of various polychaete worms. The transition from laminated into flaser layers within this lower part is often marked by a population of bivalve shells which are almost completely composed of doublets. These populations are mostly dominated by specimens of one species which died at the same age (one year class). Apart from bivalve shells body fossils of the echinoderm *Echinocardium cordatum* and *Ophiura tecturata* are common. Also present are the dwelling tubes of the polychaete *Pectinaria koreni*. They can be found in large quantities within

the lower part of the sequence preserved in 'life position' or washed together in small heaps.

EVIDENCE FOR A SEASONAL ORIGIN OF THE RHYTHMIC BEDS

The agreement between the thickness of individual couplets and the annual rate of sedimentation as interpreted from hydrographic maps

The depositional history of the rhythmic beds is documented by a series of hydrographic maps (see Fig. 3). According to these maps, the water depth at

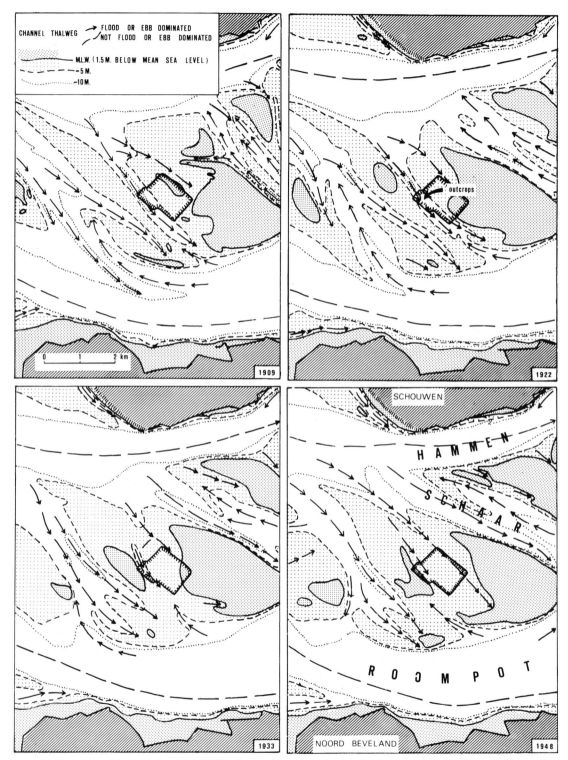

Fig. 3. Hydrography of the Oosterschelde mouth between 1909 and 1948.

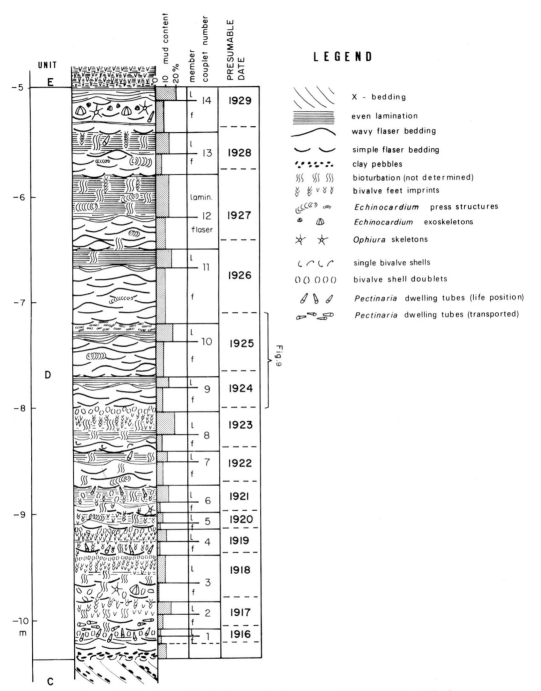

Fig. 4. Detailed log of the rhythmic beds in outcrop A. For location see Fig. 2.

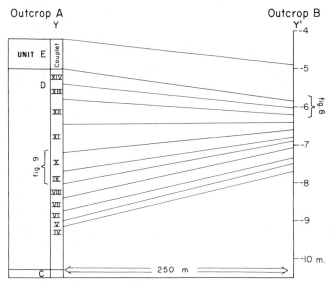

Fig. 5. Correlation of the rhythmic beds between the two outcrops. For location of Y–Y′ see Fig. 1.

the site of outcrop A was reduced from more than 9 m to about 5 m within a period of 12 years, between 1922 and 1933. A similar number of couplets can be observed between 9·8 and 5·0 m below sea-level, and this already suggests an annual origin of these couplets. In an upward direction, an increase in thickness of the rhythmic beds is noticeable. There are several indications that this increase resulted from a higher rate of annual accumulation. Although the channel was largely abandoned by the tidal flow during the deposition of the rhythmic beds, its lateral migration to the north-east continued (Fig. 3).

According to the hydrographic maps, the channel axis must have crossed the site of outcrop A in the shallowing period between 1922 and 1933, or within the period when the rhythmic bedding was formed. Thus the upward increase in thickness of the individual layers could reflect the passage of the thalweg zone. Evidence for this suggestion was found by a detailed correlation of outcrops A and B, which was possible by several key beds of echinoderms and bivalves. The correlation on the basis of the number of couplets and these key beds allows a lateral connection of the individual seasonal layers. This correlation is given in Fig. 5, clearly demonstrating the gradual inversion of the slope between both outcrop locations, caused by the passage of the deepest part of the channel.

Growth and mortality pattern of some bivalve species through successive layers

The rhythmic beds contain many bivalve remains, especially in the lower part of the sequence. The preserved shells are clearly autochthonous; they are almost always preserved as double-valved specimens, often still in life position. In shells of *Spisula subtruncata* the tender periostracum is still retained completely.

In four of the five common species, year classes could be distinguished by the fact that specimens show distinct size groups, e.g. the *Abra alba* population of couplet number 3 consists of three size classes of about 4, 10 and 18 mm shell length, interpreted as specimens in their first, second and third years of life respectively (Fig. 6).

The winter growth ring patterns locally developed in adult specimens of *Abra alba* and *Spisula subtruncata* support this conclusion. Separate year classes could be followed easily throughout the succession. They show a gradual growth through successive couplets. The growth of year classes from any couplet to the overlying one corresponds exactly with the annual growth increment. This proves again that the rhythmic cycles are annual couplets. The regular growth pattern of separate year classes through successive rhythmic beds confirms the above-mentioned continuity of deposition, without any breaks or conditions of intermittent erosion. The

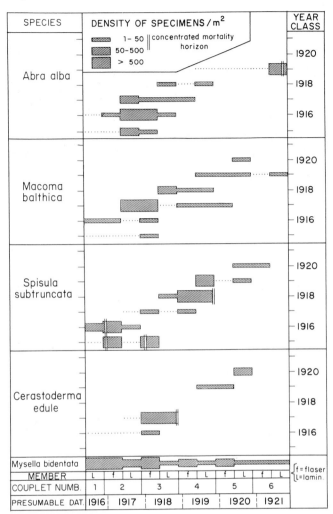

Fig. 6. Year classes and densities of specimens of some bivalve species in the lower part of the rhythmic beds (no year classes are distinguished in specimens of *Mysella bidentata*). Two valves are considered as one specimen.

transition from laminated to flaser beds is repeatedly marked by a mortality horizon, a concentration of double-valved specimens of *Spisula subtruncata* or *Cerastoderma edule* (see also Fig. 4). This horizon indicates a nearly total extermination of the year classes involved. Since couplets are annual layers it is likely that each laminated and each flaser layer represent particular parts of the year. Several considerations lead us to the conclusion that the above-mentioned mortality horizons mark the change to low water temperatures during the winter season. Water temperature measurements by Rijkswaterstaat (the Governmental Board for Ways, Waterways and Harbours) near the location of the pit in the period 1972–9 give minimum winter temperatures ranging between 0·9 and 4·9 °C. Unfortunately little is known about the effects of low water temperatures on the muscular activity of bivalves, in particular their ability to escape burial during deposition.

According to Kristensen (1957) *Cerastoderma edule* is not capable of digging itself into the sand at temperatures below 2·5 °C. It will probably also fail to dig itself out when covered by a sediment layer. Since the habitat of the bivalve populations in the abandoned channel was subjected to continuous sedimentation, such failure is precisely what might be expected during relatively cold winter periods.

During a cold spell the inactive bivalves would be buried and normal siphonal respiration processes would become impossible. Although no data are available about the survival time under anaerobic and low-water-temperature conditions, it seems likely that a long cold spell could be lethal.

In addition to the dense horizons of *Spisula* and *Cerastoderma* a less concentrated mortality layer of *Abra alba* is found in the laminated (summer) layer of couplet 6 (see Fig. 4). The different nature of this mortality layer can be explained as follows: *Spisula subtruncata* and *Cerastoderma edule* have short siphons and live near the surface, so that mass mortality in winter gives rise to a concentrated shell layer at the transition to or at the base of the flaser (winter) layer.

On the other hand *Abra alba* possess long siphons and live more dispersed in a relatively thick layer below the sediment surface. Thus, when the animals become sluggish and eventually die of asphyxia during winter, they may still be found dispersed through the summer layer. *Mysella bidentata* grows very slowly and therefore does not form distinct size classes. Since winter growth rings could not be discerned, no year classes could be distinguished; only mortality densities are given in Fig. 6. This small bivalve lives near the surface and, as could be expected, a high mortality is found in the flaser (winter) layers.

Thanks to a remarkable and continuous mortality horizon of heart urchins (*Echinocardium cordatum*) together with brittlestars (*Ophiura texturata*) present in the winter member of couplet 14, of which the probable date of origin can be established, it is possible to label all cycles with their successive years of formation.

In outcrop A mainly adult sea urchins are found in couplet 14 (30 mm test length), whereas in outcrop B together with this size class a dense population of rather juvenile specimens is present (16 mm test length). Generally many spines are still attached to the *Echinocardium* exoskeletons. The fragile skeletons of the brittlestars are also complete; this indicates that after succumbing the specimens were hardly transported. The autochthonous position of the echinoderm fauna is further confirmed by the fact that the size of the internal traces of *Echinocardium cordatum* preserved directly below the mortality horizon corresponds with the size of the body fossils. Since body and trace fossils of the heart urchin are generally found scattered at random throughout the sequence of seasonal layers, the

occurrence of only one horizon of heavy mortality points to exceptional conditions. Severe mortality of heart urchins in subtidal habitats has been reported only for the extremely cold winter of 1962–3. During this winter a long period of low water temperatures resulted in a mass mortality of many marine organisms including the heart urchin in the southern North Sea (Woodhead, 1964; Ziegelmeier, 1964). Heaviest mortalities are recorded from the coldest parts of the North Sea, such as along the south-east coast of Britain (Crisp *et al.*, 1964) and the German Bight (Ziegelmeier, 1964; Reineck, Gutman & Hertweck, 1967). According to the series of hydrographic maps, the rhythmic beds must have been deposited between about 1922 and 1933. This period includes the severe winter of 1928–9. Temperature conditions and effects on marine life (fish mortality) of this winter in the southern North Sea were comparable to those of the cold spell of 1962–3 (Lumby & Atkinson, 1929; Woodhead, 1964; Rijkswaterstaat, 1966; Anonymous, 1968). Floating ice was reported for 31 days in the eastern part of the Oosterschelde (Korringa, 1940) and was even reported during a couple of days at the lightvessel *Schouwenbank*, 20 km offshore (Dorrestein, 1964).

Thus conditions in the entrance must have been lethal to most, if not all heart urchins. Unfortunately no temperature measurements of the Oosterschelde entrance exist from this period. Approximate water temperatures can be calculated by taking into account the delayed transmission of air temperature to water masses. Results with such converted air temperatures of Vlissingen according to the equation below show a reasonable correlation with water temperatures measured in the Oosterschelde entrance (P. T. C. van der Hoeven, Royal Dutch Meteorological Institute, personal communication):

$$W_0 = (8A_0 + 4A_1 + 2A_2 + A_3)/15.$$

In this equation W_0 and A_0 are mean 'delayed' air temperature and measured air temperature of the same month respectively; A_1, A_2 and A_3 are mean air temperatures of the first, second and third months before the month of A_0. Since a long period of low water temperatures together with continuous sedimentation is postulated as the main cause of heavy bivalve mortalities, the minimum values of monthly means of winter water temperatures in the Oosterschelde mouth are compared with the position of mortality layers (Fig. 7). Also, water temperatures measured 20 km off shore at the lightvessel *Schouwenbank* are mentioned. It must be stressed here that

Couplet number	Member 1 = laminated f = flaser	Mortality horizon Species	Density m²	Lowest monthly mean of winter temperature (°C) Vlissingen*	Schouwenbank†	Date of winter
14	l					
	f	2 and 4 summer heart urchins	—	−0·8	1·3	1928/29
13	l					
	f			3·8	3·8	27/28
12	l					
	f			4·2	4·4	26/27
11	l					
	f			3·9	4·9	25/26
10	l					
	f			4·7	5·8	24/25
9	l					
	f			1·9	3·2	23/24
8	l	1 summer *Cerastoderma edule*	10,200			
	f			5·4	5·4	22/23
7	l					
	f			3·2	3·1	21/22
6	l	3 summer *Abra alba*	600			
	f			5·4	5·6	20/21
5	l					
	f			5·3	4·9	19/20
4	l	2 summer *Spisula subtruncata*	20,600			
	f			2·8	3·6	18/19
3	l	1 summer *Cerastoderma edule*	7200			
	f	3 summer *Spisula subtruncata*	1100	3·9	2·9	17/18
2	l					
	f	1 and 2 summer *Spisula subtruncata*	3200	0·1	1·2	16/17
1	l					
	f			5·1	5·2	15/16

Fig. 7. The relationship between horizons of heavy mortality and presumably lowest winter temperatures. *'Delayed' air temperatures; †water temperatures.

no definite conclusions can be made on the influence of small fluctuations between lowest winter tempera· tures on the mortality amongst bivalve species, since the number of mortality horizons is too small. Furthermore the calculations can give only an approximation of the real water temperatures. Nevertheless some interesting remarks can be made.

As can be seen in Fig. 7 the mortality horizons of *Cerastoderma edule* are connected with relatively cold winter periods. Calculated temperatures are below 3 °C, which is in fair accordance with the value of 2·5 °C at which the digging activity stops according to Kristensen (1957). The same is valid for the *Abra alba* population, which survived the rela- tively warm winters of the couplets 5 and 6 and finally perished in the somewhat colder winter spell of 1921–2. The pattern of *Spisula* mortality seems not to be influenced by differences in winter cold. Some

survived the cold winter of 1916–17 and died in relatively warm winter seasons. Obviously some un- known factor influenced their ability to survive the winter period.

The upward growth of a dense population of heart urchins through successive rhythmic beds

Traces and body remains of the heart urchin (*Echinocardium cordatum*) are found distributed throughout the whole sequence of rhythmic beds. Spines are found everywhere in fairly large quantities. Apart from their frequent occurrence in the mortality layer of couplet 14 exoskeletons are uniformly dis- tributed, although slightly more are found in the flaser layers, indicating a somewhat higher mortality in winter. On the other hand, the internal traces are more concentrated in the laminated layers, which is

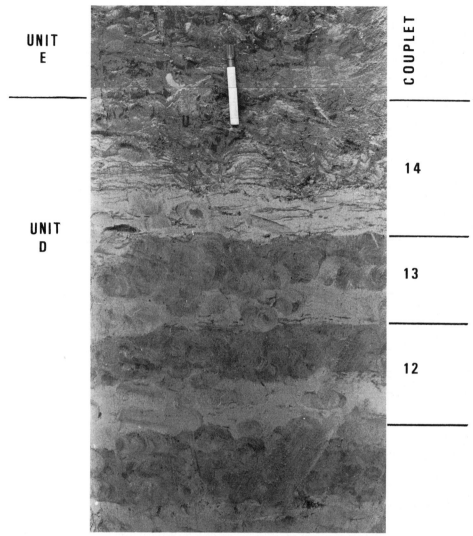

Fig. 8. Rhythmic bedding in outcrop B. Original physical sedimentary structures have almost completely disappeared due to the burrowing activity of heart urchins; only a repetition of light and dark layers remains, denoting small differences in mud content. Length of the pen is 14 cm.

in accordance with the higher activity of the animals during summer. In some places laminations have been completely destroyed by the burrowing activity of the heart urchins, particularly in outcrop B (Fig. 8). In general it was hard to distinguish size or year classes in the population of trace and body fossils. An exception is formed by a successful settlement in the upper member of couplet 10 (S in Fig. 9). Internal traces of press structures of numerous juvenile specimens impart a mottled structure to the upper part of this summer layer (Fig. 9). In a large part of outcrop A the traces of a dense population are no longer present in the overlying

winter layer. In the south-western part of the outcrop, however, a dense population continues to be present until its extermination in the mortality horizon of couplet 14. Figure 10 shows the average test length through successive couplets, as estimated from the curvature of the press structures. Comparison with data on the annual growth increment given by Buchanan (1966) for two heart urchin populations along the Northumberland coast strongly supports the annual origin of the deposits. The difference in growth rate between the two populations studied by Buchanan was considered to be mainly the result of differences in mud content between the two habitats;

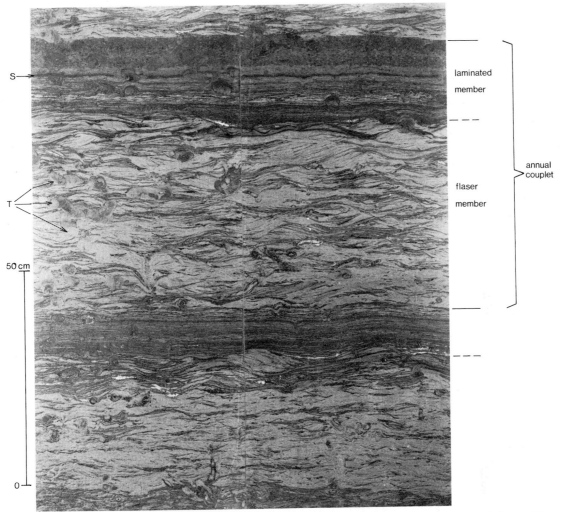

Fig. 9. Characteristic sequence of structures within annual couplets: photograph of a wet lacquer peel (wet, the mud has a dark colour by which it can be distinguished more easily from the sand). T = *Echinocardium* press structures; S = settlement of juvenile heart urchins. For location of the lacquer peel see Fig. 4.

a silty substratum would inhibit the animal's passage through the sediment and thus its food uptake, resulting in a slow growth rate. Therefore it is suggested that an increase in mud content results in a slowing down of the growth rate. The latter reasoning is in agreement with the annual growth increment of the heart urchins as measured in the rhythmic beds of the Oosterschelde, for both the mud content as well as the growth rate of the studied population in this environment is intermediate between the values in the habitats studied by Buchanan.

THE ORIGIN OF THE RHYTHMIC CHANGE IN PHYSICAL STRUCTURES

The flaser and laminated members within the annual rhythmic cycles represent a winter and a summer layer respectively. The interlaminated mud and fine sand of the summer member represent conditions of accumulation from suspended sediment on flat bottoms, periods during which wave and current activity are too low to produce ripples. Most of the flaser bedding belongs to the wavy flaser type (Reineck & Wunderlich, 1968). Their formation can be explained by the occurrence of currents which

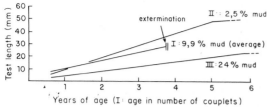

Fig. 10. Comparison of test length between the discussed population of *Echinocardium cordatum* in the rhythmic beds of the Oosterschelde (I) and at two Northumberland localities (II = intertidal, III = offshore, 40 m deep).

only partially erode the crest of previously formed ripples (Reineck & Singh, 1973). Cross-laminations contain partings of faecal pellets and fine organic matter; these features indicate that cross-laminations are produced under relatively low energy conditions. Cross-laminations show a bipolar orientation parallel to the axis of the abandoned channel in which they are formed, indicating that the ripples migrated under the influence of the reversing currents. Cross-laminated sets may be covered by multiple mud flasers infilling ripple troughs and draping ripple crests. The latter features reflect prolonged periods in which wave agitation or tidal current velocity was too low to initiate appreciable movement of grains. It is plausible to suggest that the seasonal variation in primary structures is due to temperature, for this affects the density and viscosity of the water which, in turn, involves some change in flow pattern and flow regime. For example, Harms & Fahnestock (1965) found a transition to a higher flow regime and related bedforms (from dunes to upper flat bed) after a drop in temperature of 17·5 °C in the Rio Grande river, U.S.A. However, flume experiments suggest that the influence of temperature on the threshold for the initiation of grain movement and the formation of ripples is rather small (Middleton, 1978). It is difficult to extrapolate the value of data from the extremely narrow range of flow depths in flume experiments to natural deep flows, the more so as they are all restricted to steady currents and do not take into account any superimposed wave oscillation, which in our case may be important.

It does not seem likely that a change in temperature can be held responsible for the rather drastic seasonal change in physical structures. It may be stated that in general the presence of inorganic mud particles between sand grains imparts some cohesion to the sediment. The fact that the laminated layers on the average contain more mud could therefore mean that in summer somewhat higher current velocities

would be required for the formation of ripples. Although it cannot be denied that this factor might be of some importance it cannot be the main cause, since some of the flaser and laminated layers contain about the same amounts of mud (see Fig. 4). Presumably the initial organic content of the deposits is of more importance to the seasonal change in the sedimentary structures. Most of the organic matter in settling mud is of planktonic origin. In spring, during the major bloom of phytoplankton, 50–75 % of the mud is of organic origin (Terwindt, 1977). A secondary maximum may be present in summer during the development of a second but less important plankton bloom (Drinkwaard, 1958, 1959). In the remaining part of the year amounts gradually drop to a minimum of about 10 % in winter (J. H. B. W. Elgershuizen, personal communication). It can be imagined that the high organic content of the mud in summer influences the physical properties of the sediment, in such a way that it could support the cohesive forces between the grains. Also, grains might glue together during bacterial mineralization of the organic matter. If so, this could mean that flow velocities in summer did not cross the threshold necessary for the initiation of grain movement and the formation of ripples, so that a laminated facies was produced. At this moment these assumptions are only speculation. Future research must prove their validity.

When returning to winter conditions the flat bottom changed into a rippled morphology, attended by some slight erosion of the laminated summer layer by the ripple troughs. This resulted in a sharp boundary, which is in contrast with the gradual non-erosional transition from flaser to a laminated layer.

CONCLUSIONS

The development of seasonal layering is not restricted to low energy environments, but may also occur in a highly dynamic area of tidal channels and shoals. The described seasonal layering will probably not represent a large part of the sequential complex of deposits produced in the Oosterchelde mouth. Accretional patterns caused by channel migration will play a more important role when compared with the occurrence of channel abandonment. The development of seasonal beds within these abandoned channels will be favoured by weak tidal currents and wave action together with a relatively high rate of sedimentation. Future studies must decide whether the present example should be considered as a rare exception, or that seasonal beds are produced more frequently in other areas with comparable conditions of sedimentation, for example in some delta-front environments.

ACKNOWLEDGMENTS

I wish to express my gratitude to Mr J. G. M. Raven for placing his internal report on the distribution of bivalves at my disposal. I am greatly indebted to Dr S.-D. Nio, C. Siegenthaler and A. van Gelder for critically reading the manuscript and Dr J. H. J. Terwindt for some fruitful discussions at the exposure.

REFERENCES

ANONYMOUS (1968) *Monthly Means of Surface Temperatures and Salinity for Areas of the North Sea and the North-eastern North Atlantic in 1963.* International council for the exploration of the sea, Service Hydrographique, Charlottenlund, Copenhagen, Denmark.

BUCHANAN, J.B. (1966) The biology of *Echinocardium cordatum (Echinodermata: Spatangoidea)* from different habitats. *J. mar. biol. Ass. U.K.* **46**, 97–114.

CRISP, D.J., MOISE, J. & NELSON-SMITH, A. (1964) General conclusions. In: The effects of the severe winter of 1962–63 on marine life in Britain (Ed. by D.J. Crisp). *J. Anim. Ecol.* **33**, 202–210.

DORRESTEIN, R. (1964) Drijfijswaarnemingen van Nederlandse lichtschepen in Januari–Maart 1963, *De Zee*, **83**, 135–141.

DRINKWAARD, A.C. (1958) The quality of oysters in relation to environmental conditions in the Oosterschelde in 1958. *Ann. biol. Copenh.* **16**, 255–261.

DRINKWAARD, A.C. (1959) The quality of oysters in relation to environmental conditions in the Oosterschelde in 1959. *Ann. biol. Copenh.* **16**, 255–261.

HARMS, J.C. & FAHNESTOCK, R.K. (1965) Stratification, bed forms, and flow phenomena (with an example from the Rio Grande). In: *Primary Sedimentary Structures and their Hydrodynamic Interpretation* (Ed. by G.V. Middleton). *Spec. Publ. Soc. econ. Paleont. Miner., Tulsa*, **12**, 84–115.

KORRINGA, P. (1940) *Experiments and observations on swarming, pelagic life and settling in the European flat oyster Ostrea edulis L.* Doctoral thesis, University of Amsterdam, 249 pp.

KRISTENSEN, I. (1957) Differences in density and growth in a cockle population in the Dutch Wadden Sea. *Archs néerl. Zool.* **12**, 351–453.

LUMBY, J.R. & ATKINSON, G.T. (1929) On the unusual mortality amongst fish during March and April 1929 in the North Sea. *J. Cons. int. Explor. Mer*, **4**, 309–332.

MIDDLETON, G.V. (1978) Mechanics of Sediment Movement. S.E.P.M. Short Course 3, 27 pp.

REINECK, H.-E. & SINGH, I.B. (1973) *Depositional Sedimentary Environments—with Reference to Terrigenous Clastics*, Springer-Verlag, Berlin, 349 pp.

REINECK, H.-E., GUTMAN, W.F. & HERTWECK, G. (1967) Das Schlickgebiet südlich Helgoland als Beispiel recenter Schelfablagerungen. *Senckenberg. Leth.* **48**, 219–275.

REINECK, H.-E. & WUNDERLICH, F. (1968) Classification and origin of flaser and lenticular bedding. *Sedimentology*, **11**, 99–104.

RIJKSWATERSTAAT (1966) *IJsverslag Winter* 1962–1963, Staatsuitgeverij, Den Haag. 85 pp.

SCHÄFER, W. (1962) *Aktuo-Paläontologie nach Studien in der Nordsee.* Kramer, Frankfurt. 666 pp.

TERWINDT, J.H.J. (1967) Mud transport in the Dutch delta area and along the adjacent coastline. *Neth. J. Sea Res.* **3**, 505–531.

TERWINDT, J.H.J. (1977) Mud in the Dutch delta area. *Geologie Mijnb.* **56**, 203–210.

WOODHEAD, P.M.J. (1964) The death of fish and sublittoral fauna in the North Sea and the English Channel during the winter of 1962-1963. In: The effects of the severe winter of 1962–63 on marine life in Britain (Ed. by D.J. Crisp). *J. Anim. Ecol.* **33**, 165–210.

ZIEGELMEIER, E. (1964) Einwirkungen des kalten Winters 1962/63 auf das Makrobenthos im ostteil der Deutschen Bucht. *Helgoländer wiss. Meeresunters.* **10**, 272–282.

Spec. Publs int. Ass. Sediment. (1981) **5**, 161–174

Nearshore processes and shoreline development in St Andrews Bay, Scotland, U.K.

G. FERENTINOS *and* J. McMANUS

Tay Estuary Research Centre, University of Dundee, Dundee, Scotland, U.K.

ABSTRACT

St Andrews Bay is a semi-circular embayment bound by headlands. The present-day sedimentary environments have grown in response to tide- and wave-induced currents and material transfer during and since the Holocene (Flandrian) transgression. The long-shore current pattern is controlled by the shape and orientation of the embayment in relation to wave approach.

Net longshore sand transport in the embayment consists of two 'cell' systems moving in opposite directions and converging towards the head of the embayment. This wave-induced sand transport pattern is implemented by the tidal circulation pattern. The bay-head sand deposits, formed by the convergence of the two cell systems, are further shaped to ebb-delta and back-beach sand environments by estuarine and aeolian processes respectively.

INTRODUCTION

St Andrews Bay is a semi-circular embayment 27 km wide, on the east coast of Scotland (Fig. 1). In the north it is bounded by Deil's Head and in the south by Fife Ness. Two estuaries cut through the bay-head sands; that of the Tay, 50 km long and with high freshwater (198 $m^3 s^{-1}$ mean) and tidal (380·5 $m.m^3$ on ordinary spring tide 5·0 m range) discharge and that of the Eden, 50 km long and of low freshwater (4 $m^3 s^{-1}$ mean) and tidal discharge (8 $m.m^3$ tidal prism).

The purpose of this paper is to describe the present-day sedimentary environments in St Andrews Bay and to discuss their formation in terms of nearshore hydrodynamics.

PALAEOGEOGRAPHICAL SETTING

The entire area was overridden by ice during the Pleistocene glaciations and the effects of subsequent eustatic and isostatic changes of sea-level have left

0141-3600/81/1205-0161 $02.00

their imprint in the morphology of the area. The geographical distribution of the late glacial (Main Perth) and Flandrian (main post-glacial) shorelines, which were formed about 13,500 and 6000 yr BP respectively, suggests that the geometry of the bay was very different from that of today (Fig. 1) (Rice, 1962; Sissons, Smith & Cullingford, 1966; Cullingford, 1972; Chisholm, 1966, 1971).

GEOMORPHOLOGICAL SETTING

Above 30 m the land surface is characterized by mounds of glacial drift, fluvioglacial channels and associated melt-water channels. Below this level are numerous fragments of raised beaches. Near the headlands the raised beaches consist of rounded pebbles and cobbles alternating with layers of comminuted shell fragments and shelly sands forming a terrace at the foot of a prominent cliff (Rice, 1962; Cullingford, 1972). At the head of the bay between St Andrews and Carnoustie, the raised beach deposits vary from silts and clays resting on the sub-Carse peat, and characterizing sheltered depositional environments in the west, to sands resting on bedrock

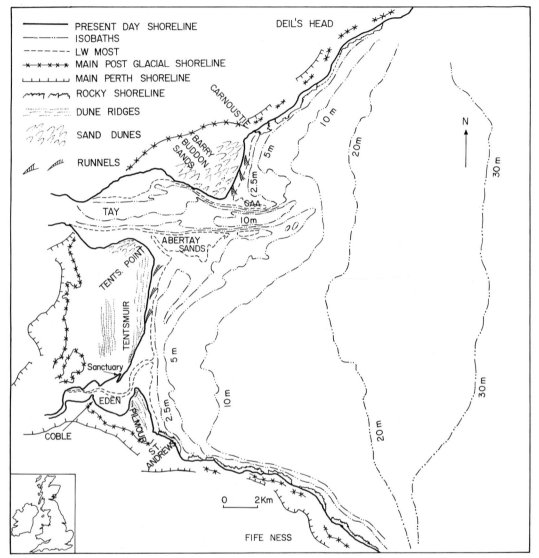

Fig. 1. Location of the study area. Also shown are the late glacial (main Perth) and post-glacial (Flandrian) shorelines (after Rice, 1962; Cullingford, 1972; Chisholm, 1966, 1971), and the post-glacial and present-day features at the head of St Andrews Bay.

or till typifying more exposed environments to the east (Chisholm, 1971; Green, 1974).

PRESENT STUDY

At the head of the bay the post-glacial and present-day sand deposits form three distinct geomorphological units: (1) the sub-littoral zone; (2) the

present and post-glacial beaches and (3) the ebb-delta complexes (Fig. 1).

The sub-littoral zone

The immediate offshore zone (Fig. 1) is characterized by relatively smooth bottom topography. The contours are sub-parallel to the coastline except at the estuary mouths, where ebb-delta deposits push

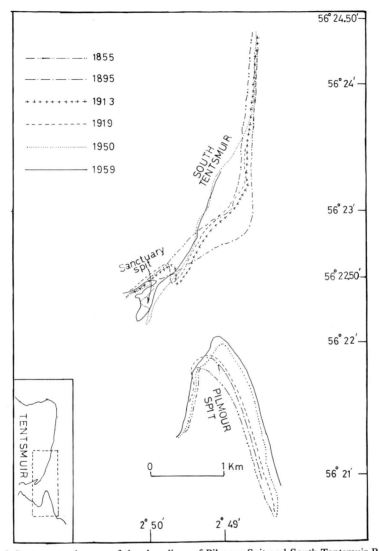

Fig. 2. Long-term changes of the shorelines of Pilmour Spit and South Tentsmuir Beach.

out from the coast. The surface sediments are mainly of very fine sands.

Geophysical surveys in St Andrews Bay reveal the existence of a major reflector beneath the entire sublittoral zone and which is 3–4 m below the sea-bed. A series of boreholes for industrial development exploration have revealed that this reflector is the interface between modern and/or post-glacial deposits and glacial tills. Another reflector identified beneath the southern part of St Andrews Bay at about 20–25 m below sea-bed is probably the interface between bedrock and the overlying glacial deposits (J. Jarvis, personal communication).

Present and post-glacial beaches

The present and post-glacial beach deposits at the head of the bay are divided into three units: the Pilmour-Coble Spits; Tentsmuir Platform and the Barry-Buddon cuspate Foreland (Fig. 1).

(a) *Pilmour and Coble Spits*

These spits are similarly shaped but of differing scale (Fig. 1). The Pilmour sand spit extends 4 km northwards from St Andrews and is formed at an abrupt change in the direction of the old cliff line. The backshore of the spit consists of foredunes

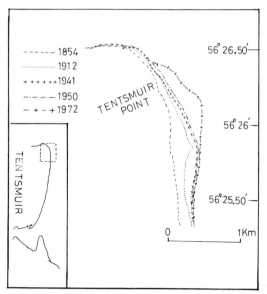

Fig. 3. Long-term changes of the shoreline of Tentsmuir Point.

parallel to the coastline which reveal a northward and eastward pattern of growth of the spit.

The Coble sand spit, 1 km in length, also extends northwards from a change of trend of the former coastline and it seems to be a relic probably formed during higher sea-level.

The study of variations in the position of High Water Mark of Ordinary Spring Tides (H.W.M.O.S.T.) on a sequence of 1-10,560 scale Ordnance Survey maps published between 1855 and 1959 has shown that the seaward margin of the Pilmour sand spit migrated east and north (Fig. 2) as fresh foredunes became attached to the beach line of 1855. From 1885 to 1959 there was a net seaward movement of H.W.M.O.S.T. by 490 m on latitude line 56° 21·45′ N and by 305 m on latitude line 56° 21·30′ N.

(b) *Tentsmuir platform*

This platform is a straight 8 km long beach (Fig. 1) which faces the open sea. The foreshore is dominated by ridge and runnel topography with the runnel openings facing the north-east. The backshore is dominated by foredunes parallel to the present shoreline.

Boreholes and seismic refraction surveys show that the post-glacial/modern deposits are about 18 m thick, with their thickness increasing towards Tents-

muir Point (Green, 1974). These deposits vary from silts and clays resting on a peat layer and characterizing sheltered environments near Morton Lochs, to sands resting on bedrock or till and characterizing more exposed environments near Leuchars (Chisholm, 1971).

The variations in the position of H.W.M.O.S.T. suggest that between 1854 and 1972 continuous net seaward accretion occurred along the central and northern segments of the beach, with the accretion rate increasing northwards (Fig. 3). However, at the southern part of the Tentsmuir (south of latitude 56° 24′ N) shoreline progradation continued until 1895, after which erosion began (Fig. 2).

The spectacular accretion patterns in Tentsmuir beach were first studied by the Royal Commission on Coastal Erosion (1911), which found that between 1855 and 1895, 534 acres (214 hectares) had been gained and only 23 acres (9 hectares) lost. The net seaward accretion along the central and northern sectors of Tentsmuir beach, north of 56° 24·0′ N was apparently continuous from 1855 to 1972, although a current phase of coastal recession became apparent during the mid-1960s along the central segment of Tentsmuir beach. Concrete blocks placed at H.W.M.O.S.T. in 1941 have now been exhumed and rest on the beach face, 60 m seawards of the receding dune ridges. However, concrete blocks along the northern segment (Tentsmuir Point) are over 400 m inland and several low foredunes occur between the blocks and the beach. It is suggested that the materials which were made available by the recession of the shoreline along the southern segment of Tentsmuir beach after 1895, moved southwards into the Eden estuary to provide sediments for the formation of the Sanctuary Split (Fig. 2). The spit appeared between 1855 and 1895 and since then has grown and has migrated south-westwards for about 700 m.

(c) *Barry–Buddon foreland*

The landward margin of the Barry–Buddon foreland is associated with a cliff of the main post-glacial shoreline (Fig. 1). The sandy foreland has developed south of an abrupt change of the old cliffline.

The foreland is covered by parabolic dunes whose flanks are orientated roughly south-west to north-east (Landsberg, 1956). The foreshore is dominated by a ridge and runnel topography with runnels opening towards the south-east.

Boreholes have indicated the presence of a sand layer resting variously on bedrock or late glacial

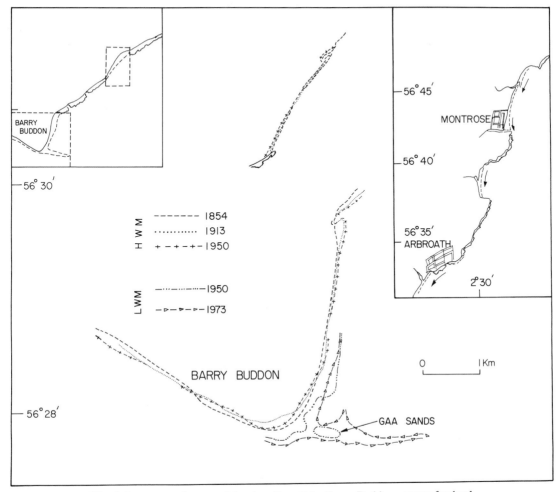

Fig. 4. Long-term changes of the shoreline of the Barry–Buddon cuspate foreland.

deposits. The thickness of the sand layer is about 11 m along the landward margin of the foreland and increases seawards to 17 m.

The variations in the position of H.W.M.O.S.T. show that very little change took place between 1854 and 1850 (Fig. 3). Slight erosion occurred at the head of the foreland and slight deposition at the sides. Similarly, slight erosion of the adjacent beach to the north occurred during the same periods. However, surveys between 1950 and 1973 have shown that considerable accretion has occurred along the low-water line at the head of the foreland and in the Gaa Sands.

Further to the north along the headlands, coastal features suggest that there is a southward longshore drift (Fig. 4).

Ebb-delta complexes

The Tay ebb-delta consists of an ebb-dominated channel, two marginal linear sand-bars, each with a flood-dominated channel, and extensive terminal lobe deposits (Fig. 1). The ebb channel up to 11 m deep is cut into late glacial clays (McManus, Buller & Green, 1980). It shoals gradually seawards and terminates at 6 m on the crest of the arcuate bar.

Geophysical surveys showed that the major sand bodies rest on a flat substrate which is the landward extension of the substrate beneath the sublittoral zone of St Andrews Bay. The thickness of the sand deposits over the bars is about 20 m.

Sediment transport studies using fluorescent tracers and fluxes estimated from current measuring

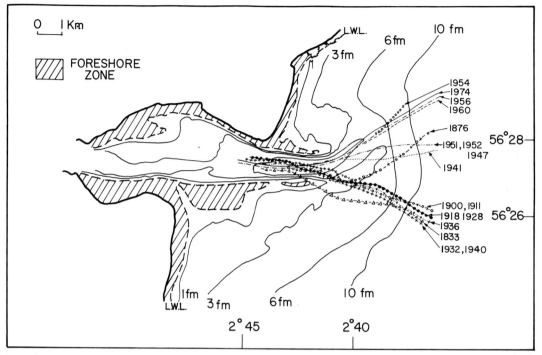

Fig. 5. Long-term variations in the orientation of the axis of the outer channel of the Tay Estuary. L.W.L. is chart datum.

around the Abertay Sands suggest the existence of a clockwise residual circulation (Green, 1974). A similar but anticlockwise sand circulation might be expected to occur around the Gaa Sands linked to the current motions demonstrated by Charlton (1980).

Bathymetric surveys of the mouth of the Tay estuary began in 1833. Most of these were limited to examinations of the channel entrance area, so that changes in that area can be confidently discussed. Comparison of hydrographic surveys carried out between 1833 and 1973 reveals that the axial orientation and the depth of the inner channel have changed little (Fig. 5). However, dramatic changes have taken place in the axial orientation and in the depth of the outer channel (Fig. 5).

In 1833 the axis of the outer channel ran in an east-south-easterly direction, by 1873 it had turned to an east-north-easterly direction, but by 1909 had returned to the east-south-easterly direction. The latter change in channel orientation was followed by an increase in the depth of the outer channel over the arcuate lobe (Fig. 6). Between 1909 and 1940 the orientation of the axis remained unchanged, but in

1941 the axis moved to an easterly orientation and, following dredging in 1944, remained in this direction until 1952. By 1954 the axis had moved to an east-north-easterly direction and it has remained there until today.

As the orientation of the channel axis changed, so too changes took place in the depth of the channel over the area of the arcuate lobe (Fig. 6). In the 1909 and 1911 surveys the depth of the channel was at its greatest (4 fm or 7·32 m) but it had become considerably reduced (3 fm or 5·40 m) by the 1936 and 1940 surveys, as a result of the south-eastward extension of Gaa Sands and an eastward extension of the Abertay Sands. By 1941 the water had formed a new channel by cutting through the south-eastward extension of Gaa Sands. The detached extension of Gaa Sands became linked to the Abertay Sands, as the channel axis swung from east-south-east to east. The new channel had shoaled to less than 3 fm (5·49 m) by 1951 and 1953. By 1954 a new channel had been established again cutting through the Gaa Sands. This ultimately led to the channel axis swinging from the east to east-north-east. It may be relevant that the latest change of the orientation of

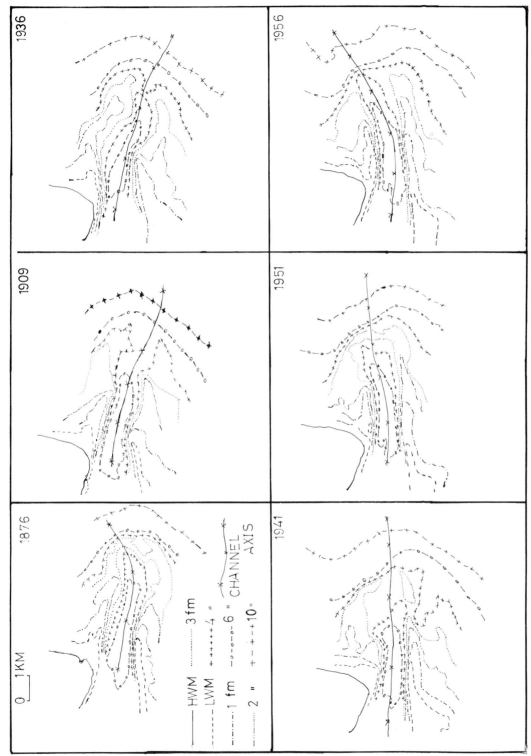

Fig. 6. Sequence of changes of channel arcuation and depth across the terminal lobe of the Tay estuary sand-bar (depth in fathoms).

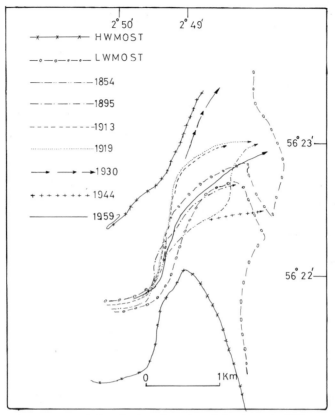

Fig. 7. Long-term variations in the orientation of the channel of Eden estuary.

the channel coincided with the extensive 1953 flooding in the North Sea, caused by extreme surging.

The volume of water entering and leaving the Eden estuary is far less than that of the Tay. This is reflected in the morphology of the Eden mouth. A dominantly northward pattern of longshore drift ensures that the inlet is to the north, where the delta terminates in a convex easterly facing lobe.

The position of the Eden channel has changed frequently since 1855 (Fig. 7). Between 1855 and 1930 it migrated northwards by 880 m but by 1944 it had reoccupied its previous position. The presence of a relict channel in the 1930 position suggests that the southerly movement of the channel was the result of an abrupt avulsion rather than of a gradual migration. Aerial surveys from 1949 to the present indicate that the channel is again moving northwards at a rate of about 10 m yr^{-1}.

OCEANOGRAPHIC SETTING

Wind and waves

Wind records from the Bell Rock lighthouse, 35 km east of Dundee, reveal that during the ten-year period 1960–1970 the prevailing and dominant winds in St Andrews Bay were from the south-west and west respectively. However, not uncommon, and also as strong, were winds from the north-east and south-east.

Analysis of visual observations of wind and wave directions from the North Carr light vessel (2 km off Fife Ness) between 1971 and 1972, confirm the paramount control exerted by wind (Fig. 8). Waves from the south-west were most frequent, followed in turn by those from the south-east and north-east. The former are of little importance to the local

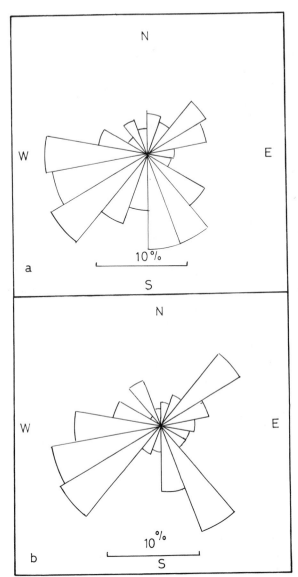

Fig. 8. The relationship between wind direction frequencies: (a) recorded at Bell Rock lighthouse; and (b) the direction of wave crest migration observed from the North Carr Lightship 1972–1973.

commonest period was 6 sec, whereas in spring and summer it was 4 sec.

Nearshore circulation

Wave refraction diagrams constructed for St Andrews Bay relate the offshore wave climate to inshore wave conditions, and indicate the patterns of the longshore currents and the associated sediment transport paths. Waves of 4, 6 and 8 sec periods from the north-east and south-east were refracted to breaking point, using a manual construction technique based on the refraction principles of geometrical optics (Shore Protection Manual, CERC, 1973). The patterns have been partly verified by air photography and also by field observations.

Wave trains approaching from the north-east are initially refracted around Deil's Head and those from the south-east around Fife Ness (Fig. 9). Both are subsequently modified around the Tay ebb-delta deposits.

Waves derived from the north-east induce a strong southward-flowing longshore current along the Angus coastline, and a slight northward-flowing longshore current along the Fife coast (Fig. 9B). Waves from the south-east induce a virtually identical pattern of flow, a strong northward-flowing longshore current along the Fife coastline and a slight southward-flowing current along the Angus coast (Fig. 9A). Refraction around the Tay estuary sand-bars induces longshore currents landwards along the outer flanks of the bars towards the flood channels.

Thus swells from both the south-east and the north-east generate two cells of longshore current circulation flowing in opposite directions. One flows clockwise from Fife Ness to Tentsmuir Point and the other anticlockwise from Deil's Head to Buddon Ness. To this simple pattern are added the local perturbations of westward flows along the seaward margins of the Abertay and Gaa sands.

Tidal currents

Offshore from St Andrews Bay the flood and ebb currents flow in southerly and northerly directions respectively (Fig. 10). Preliminary drogue tracking surveys in St Andrews Bay (Green, 1974) followed by more specific local drogue surveys linked with the use of a 30 × 40 m hydraulic model of the Tay estuary (Charlton, 1980) and detailed current measurements by the authors suggest that the simple

coastal processes because they are locally generated, short-fetched and propagated in an offshore direction. The latter carry most energy shorewards as they are generated over a fetch of up to 750 km. Measurements by a shipboard wave recorder on the North Carr light vessel between June 1969 and August 1970 revealed that heights ranged from 0·3 to 5 m (Draper, 1971). During the autumn and winter the

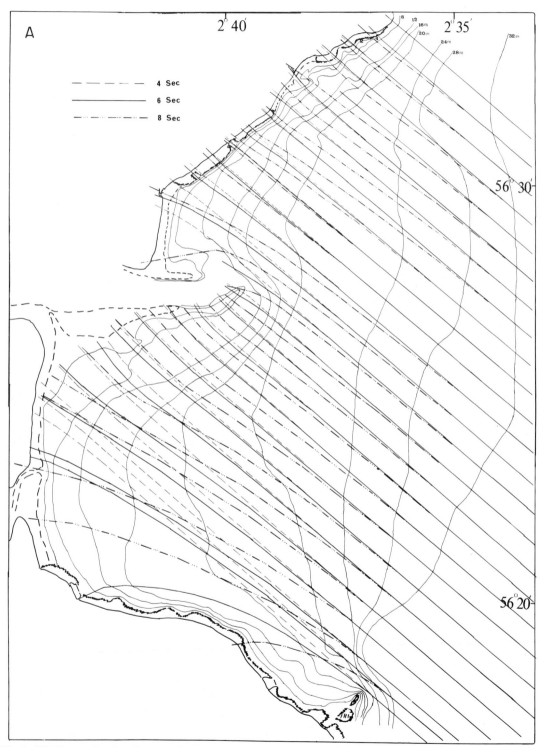

Fig. 9. (A) Wave refraction diagrams for waves of 4, 6 and 8 sec period entering St Andrews Bay from: (a) the south-east, and (b) the north-east at low water.

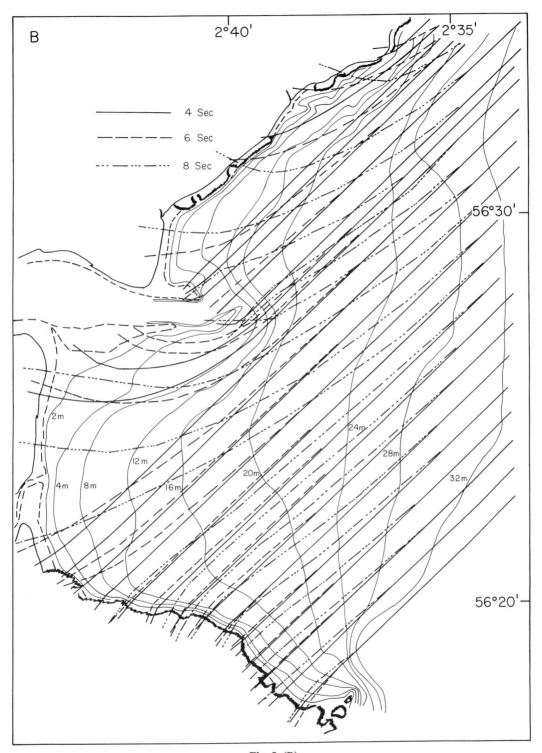

Fig. 9. (B)

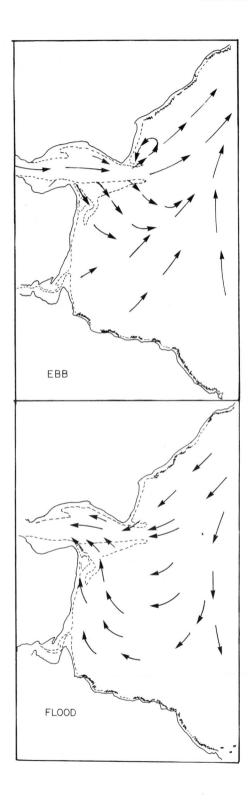

EBB

FLOOD

offshore tidal pattern is modified within the embayments (Fig. 10). The coastal segment of the flood-stream runs south-south-westwards past Arbroath and divides into two streams: one turns west to enter the Tay estuary; the other continues south-south-westwards before forming a clockwise eddy between St Andrews and Abertay Sands. The northward-flowing waters of this eddy sweep the Tentsmuir shoreline and spill over the Abertay sands.

The ebb tide runs eastwards from the Tay estuary entrance. Its momentum leads to the formation of an ebb jet to the north of which is established an anticlockwise eddy. No such eddy is formed to the south of the estuary, where the ebb stream in St Andrews Bay travels north-eastwards.

DISCUSSION

St Andrews Bay is a large semi-circular coastal indentation on a coastline of storm wave and macro-tidal type environment (Davies, 1964). Nearshore hydrodynamics and sedimentary processes in such an area are mainly controlled by wave-induced long-shore currents, and by the coastal tidal water circulation. However, in St Andrews Bay the near-shore dynamics are complicated by the presence of the major Tay estuary and its associated circulation patterns.

Swells from the north-east and south-east are refracted around the headlands and induce net longshore currents, which converge towards the head of the bay. This net longshore current pattern is enhanced by additional refraction of the swells around the Tay ebb-delta. The north-west-moving longshore current extends from Fife Ness to Tents-muir Point. The longshore drift induced by this current provides the sand material for the progradation of Pilmour Spit and Tentsmuir Beach. The net seaward growth of Pilmour Spit and Tentsmuir Beach during the last 120 yr indicates the existence of a continuous sediment supply.

The morphological evolution of the spit suggests that the northward migration is not intermittent, and the eastward-facing coastline has also advanced continuously in that time.

Fig. 10. Mid-flood and mid-ebb tidal current patterns in St Andrews Bay, based on float tracking, current measurement by recording and direct reading meters, and also using a 30×40 m hydraulic model of the Tay estuary and its approaches. Arrows are not vectorial.

The migrating pattern of the Eden channel suggests that the dominant factor in the positioning of the channel is the longshore drift. However, non-periodic catastrophic events, such as extreme fresh-water discharge coupled with extreme spring tides, may lead to rapid relocation of the channel. However, after such a change the longshore drift resumes its role as the main controlling factor.

The morphology of the Eden ebb-delta deposits suggests that the sand carried north by the longshore drift bypasses the Eden estuary and is carried along the Tentsmuir beach towards Tentsmuir Point where motion ceases, giving rise to high local rates of accretion.

The erosion noted since 1895 along a limited stretch of the southern Tentsmuir Beach appears to be associated with a localized refraction pattern in the mouth of the Eden (Eastwood, 1976). At high water, south-westerly longshore currents are induced which transport sediment into the estuary, and are responsible for the formation and migration of Sanctuary Spit along the northern shore of the estuary.

The southward-moving longshore current from Deil's Head to Buddon Ness carried sands towards the Barry–Budden cuspate foreland. No major changes have been observed along the foreland beach but considerable accretion has occurred around the Gaa Sands area suggesting that the southward-transported material is transferred to, and deposited in, the area of the marginal bar.

At the mouth of the Tay estuary the circulation pattern produced by the combined effects of waves and tides implies the existence of two local sediment transport gyres around the Abertay and Gaa Sands. The gyre around Abertay Sands has been verified by fluorescent tracer investigations (Green, 1974). Sediment converging on the mouth of the Tay through longshore current and sediment carried down estuary by ebbing tidal currents may be trapped within these sediment transport gyres to become deposited on the sand-bars. Deposition occurs largely on the gently sloping flanks away from the steep channel sides, but also occurs upon the terminal lobe. Similar sediment transport gyres have been described from Doboy Sound (Oertel & Howard, 1972) Chatham Harbour (Hine, 1975), Merrimack River (Hubbard, 1975), Panamor Island inlet (Byrne, De Alteris & Bulloch, 1974) and Price inlet (Fitzgerald & Fitzgerald, 1977).

The continuous transport of sediment towards the mouth of the Tay, its transfer to the sediment gyres and deposition results in the extension of the marginal bars and shoaling over the terminal lobe. This induces instability to the channel as it restricts the freedom of flow of the tidal waters. In consequence, when the volume of outflowing waters is at its greatest due to the combination of extreme spring tides, perhaps coupled with surging high freshwater discharge, the funnelled water breaks a new channel across the terminal lobe. The shoaling at the entrance of the channel appears to be a continuous process, but the change of direction of the outer channel occurs abruptly in response to the secondary combinations of water circulation.

The present source of sediment supply is not known. Cullingford (1972) has suggested that it may be the post-glacial raised beach deposits which are exposed to wave attack. The volume involved in coastal recession in such areas is totally inadequate for the raised beaches to be the principal source. Some sediment is supplied by the two estuaries and by the many small rivulets cutting into the glacial and late glacial deposits of the coastal plains. Much material may, however, be derived from erosion of unconsolidated sediments on the floor of St Andrews Bay itself.

If we accept that the longshore current pattern described above was in operation over the last 6000 yr, then St Andrews Bay might have been involved during the first stages of its evolution as a barrier beach system. Coarse sediments are deposited along the ocean side of the barrier and silts and clays are deposited in protected inlets, lakes and lagoons. The shoreward migration of the early barrier beaches across the finer materials led to the association observed onshore today, with later barriers migrating on to the face of the earlier barriers. Examination of these relationships is currently in progress.

REFERENCES

BYRNE, R.J., DE ALTERIS, J.T. & BULLOCH, P.A. (1974) Channel stability in tidal inlets. *Proc. 14th Conf. Coastal Engineering, Copenhagen*, Chapter 92. American Society of Civil Engineers.

CHARLTON, J.A. (1980) The tidal circulation and flushing capability of the Outer Tay estuary. *Proc. R. Soc. Edinb.* B **78**, 33–46.

CHISHOLM, J.I. (1966) An association of raised beaches with glacial deposits near Leuchars, Fife. *Bull. geol. Surv. G.B.* **24**, 163–174.

CHISHOLM, J.I. (1971) The stratigraphy of the post-glacial marine transgression in north-east Fife. *Bull. geol. Surv. G.B.* **37**, 91–107.

CULLINGFORD, R.A. (1972) *Late glacial and post-glacial shoreline displacement in the Earn-Tay area and eastern Fife.* Unpublished Ph.D. Thesis. University of Edinburgh.

DAVIES, J.L. (1964) A morphogenic approach to world shorelines. *Z. Geomorph.* **8**, 127–142.

DRAPER, L. (1971) Waves at North Carr Light Vessel, off Fife Ness. *Int. Rep. Instn Ocean. Sci.* **A50**.

EASTWOOD, K.M. (1976) *Some aspects of the sedimentology of the superficial deposits of the Eden estuary.* Unpublished Ph.D. Thesis. University of St Andrews.

FITZGERALD, D.M. & FITZGERALD, S.A. (1977) Factors influencing tidal inlets throat geometry. *Coastal Sediments* 77. American Society of Civil Engineers.

GREEN, C.D. (1974) *Sedimentary and morphological dynamics between St. Andrews Bay and Tayport, Tay Estuary, Scotland.* Unpublished Ph.D. Thesis. University of Dundee.

HINE, A.C. (1975) Bedform distribution and migration patterns on tidal deltas in the Chatham Harbour Estuary, Cape Cod, Massachusetts. In: *Estuarine Research II* (Ed. by E. Cronin), pp. 235–252. Academic Press, London.

HUBBARD, D.K. (1975) Morphology and hydrodynamics of the Merrimack river ebb-tidal delta. In: *Estuarine Research II* (Ed. by E. Cronin), pp. 253–266. Academic Press, London.

LANDSBERG, S.V. (1956) The orientation of dunes in Britain and Denmark in relation to the wind. *Geogr. J.* **122**, 176–189.

McMANUS, J., BULLER, A.T. & GREEN, C.D. (1980) Sediments of the Tay Estuary. VI. Sediments of the lower and outer reaches. *Proc. R. Soc. Edinb.* B **78**, 133–154.

OERTEL, G.F. & HOWARD, J.D. (1972) Water circulation and sedimentation at estuary entrances on the Georgia coast. In: *Shelf Sediment Transport: Process and Pattern* (Ed. by D. J. P. Swift, D. B. Duane and O. H. Pilkey), pp. 461–498. Dowden, Hutchinson & Ross, Stroudsburg, Pennsylvania.

RICE, R.J. (1962) The morphology of the Angus coastal lowlands. *Scott. geogr. Mag.* **78**, 5–14.

SISSONS, J.B., SMITH, D.E. & CULLINGFORD, R.A. (1966) Late-glacial and post-glacial shorelines in the southeast of Scotland. *Trans. Inst. British Geogr.* **39**, 9–18.

Spec. Publs int. Ass. Sediment. (1981) **5**, 175–185

Sand transport in the tidal inlet between Wangerooge and Spiekeroog (W. Germany)

J. HANISCH

Niedersächsisches Landesamt für Bodenforschung/Bundesanstalt für Geowissenschaften und Rohstoffe, Stilleweg 2, 3000 Hannover 51, F.R.G.

ABSTRACT

Based on the evaluation of transport directions and grain-size distributions, a new sand transport model for the tidal inlet between Wangerooge and Spiekeroog (northern Germany) is presented. Only sand with mean diameter > 180 μm forms megaripples and is transported at relatively high rates. The previously assumed shift of sandbars around an arc shaped shoal north of the inlet from west to east is refuted. Instead of this, there are numerous oscillations and internal circulations of the sand in the shoal area before it reaches the beach of Wangerooge. During calm weather periods the arcuate shoal is dominated by ebb-tide currents. Under these conditions it grows higher and widens. During surge conditions the shoal arc is flattened and reduced in radius.

The stability or non-stability of the inlet depends on the ratio of ebb-tide current erosion on the western flank of the inlet channel to the amount of sand accumulating there from the littoral drift. At present, the lack of a sufficient sand supply to the north-western beach of Wangerooge is caused mainly by a decrease of this erosion due to scouring of the Harle bottom at the end of a long groyne.

INTRODUCTION

The area studied is situated on the northern coast of West Germany. An outstanding feature of this coast is a chain of barrier islands, the so-called East Frisian Islands. Spiekeroog and Wangerooge are the two easternmost of these. The large area between the barrier islands and the mainland consists of tidal flats. The tide enters the flats through tidal inlets between the islands. The tidal channel between Wangerooge and Spiekeroog is called 'Harle' (Fig. 1). The tidal range near Wangerooge during neap tide is 2·4 m and 3·3 m during spring tide. The mean tide range is 2·8 m. The highest water level ever recorded was 4·6 m above mean high water level. Winds and storms blow mainly from western directions. Studies on the wave climate in the German Bight have started recently, so detailed wave diagrams are not yet available.

Since Gaye & Walther's study in 1935, it has been

0141–3600/81/1205–0175 $02.00

known that the stability or non-stability of the East Frisian Islands is mainly controlled by the processes in the tidal inlet west of each island. On the basis of an evaluation of annual nautical charts and a few early aerial photographs (Walther, 1969), they established a sand transport model for the East Frisian tidal inlets. They showed that an arc-shaped shoal ('Riffbogen') is formed north of each inlet. It consists of a series of bars with shallow channels between them (Fig. 2). The term 'tidal ebb delta' is not used in this study due to considerable differences between a prograding delta and the tital 'outlet' area treated here, as will be shown.

Gaye & Walther (1935) introduced the following concept: the general eastward littoral drift of sand along the beaches is interrupted at the tidal inlets. The sand from the littoral drift accumulates on bars which lie north of the eastern ends of each island. These bars separate periodically from the beaches and move as a whole around the arcuate shoal from

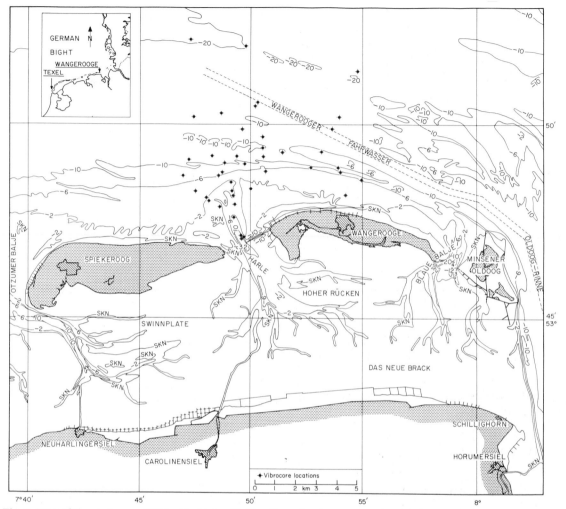

Fig. 1. Map of the study area. SKN = 'Seekarten Null' = Chart datum = mean spring-tide low water level; contours in metres below SKN. Chart from DHI, Hamburg, 1970.

west to east. According to the model they reach the next island to the east after several years and integrate into its beach. In spite of some modifications by later authors (Homeier & Kramer, 1957; Kramer, 1960; Luck, 1975; Nummedal & Penland, 1981), this transport model has been generally accepted. It has formed the sedimentological basis for any coastal protection design.

The beaches and dunes on the north-west part of Wangerooge suffered from considerable erosion during the past 15 years. As no convincing explanation for this was discovered in former expert opinions a new, mainly sedimentological approach was made to find the reasons for these losses.

SEDIMENTOLOGICAL FIELD WORK

At the beginning of our study, several discrepancies in the previous model were detected by the interpretations of new aerial photographs. First we found it impossible to identify individual bars of the seaward part of the inlet system from one year to the next, so we decided to investigate the sedimentary processes and the transport directions there. Our sampling of data covered morphological features, sedimentary structures and grain-size distributions. A large number of undisturbed, oriented samples was taken from the subtidal and intratidal zones. The shipborne devices for the recovery of such

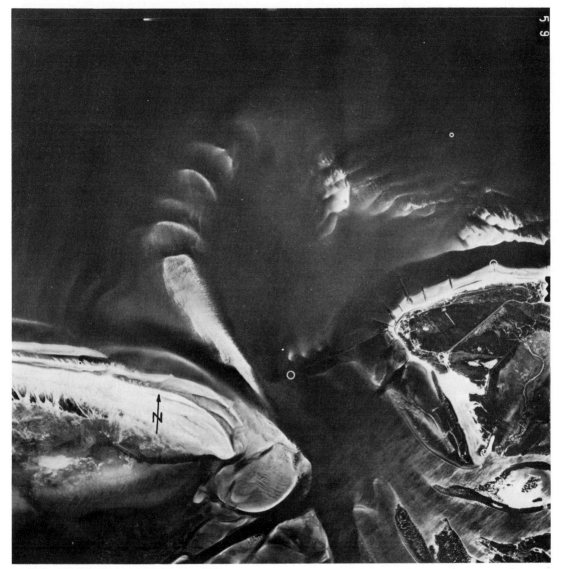

Fig. 2. Aerial photograph of the Wangerooge inlet. Photograph by N. Rüpke, Hamburg, during extreme low water on 23 June 1973 (from Luck & Witte, 1974).

samples were a vibro-corer (Kögler & Veit, 1973) and a submarine box corer (Reineck, 1958). To get unaltered cores from the intratidal area we used Reineck's (1958) box corers and a newly developed manual vibro/flush corer (Hanisch & Husemann, 1979) by which cores $1·5 \times 0·1 \times 0·1$ m can be taken. So-called 'relief peels' were prepared from the cores (Fig. 3) using a new quick-setting polyurethane resin which hardens in wet sand with the moisture as the second component.

TRANSPORT DIRECTIONS

The transport directions of the sand were inferred from the dip-directions of the large-scale cross-bedding and from the morphology of megaripples. Small-ripple cross-bedding has not been considered. Small test pits were dug for examination of the structure of megaripples in the field. This had to be done because sometimes there are features

Fig. 3. Three-dimensional relief peel of a core recovered by the new vibro/flush corer. Scale in centimetres.

morphologically similar to megaripples but which were formed by the erosional action of currents.

The locations of the samples and of the results of direction measurements are presented in Fig. 5. At the eastern end of Spiekeroog ('Robbenplate' = robes' bar) most of the transport directions are toward the south. On the first bar north of it (the so-called 'Tabaksplate') no transport directions could be deduced due to the total absence of megaripples.

Special attention was paid to the sickle-shaped bars in the north-western part of the shoal arc (Fig. 2). Crests of megaripples are visible there in some of the aerial photographs. These are oriented N–S to NE–SW. The examination of them in the field (Fig. 4) showed that the sand transport direction is westward and north-westward. The three-dimensional evaluation of the relief peels from sediment cores down to 1·5 m below the surface confirms the predominance of these transport directions (Fig. 5). These bars are almost completely built up of megaripples which are apparently generated by the ebb-

tide current. Single, sometimes aberrant directions should be explained by the action of wave-induced currents or their interactions with tide currents.

In the eastern part of the arcuate shoal no direct measurements could be made. The indicated transport directions there were deduced from echo-sounder profiles. Several of these were registered along the area of the inlet system (Fig. 5). Figure 6 shows profile no. III which is across the arc. Small megaripples from south to north followed by—with decreasing depth—well-developed larger megaripples can be seen. In spite of the fact that the line was registered during high tide all megaripples are oriented toward the north. So the flood current had not influenced the ebb-current megaripples.

Near the crest of the shoal smaller features are visible. From their symmetrical shape and their dimensions these could represent anti-dunes. If this is true, current velocities of the 'upper flow regime' of at least 3 m sec^{-1} for a water depth of 1 m (assuming a Froude number of $F=1$) would be required according to Kennedy's (1963) formula. On the other

Fig. 4. North–south oriented megaripple crests on a sickle bar at very low water level.

hand, Simons & Richardson (cited in Shen, 1971) pointed out that in natural channels supercritical flow conditions may well develop at $F = 0.5$. So it seems quite possible that during the late phase of the ebb tide there are currents of the 'upper flow regime' in the channels between the sickle bars. Another observation supports this hypothesis: the sickle bars get their shape from erosion along their flanks. As they consist almost entirely of megaripples this erosion must be explained by current velocities corresponding to at least the 'transition zone' between the lower and upper flow regime.

GRAIN-SIZE DISTRIBUTION

The grain sizes in all samples and cores were measured by sieving. The analysis of these data resulted in two groups of sands.

(1) A fraction of very well-sorted sand with a mean diameter of 130–180 μm (Fig. 7). This sand is found in areas (Fig. 8): outside the shoal arc approximately below the 2 m contour line; inside the shoal arc; and on the shoal called 'Tabaksplate'.

(2) A fraction of very well-sorted sand with a mean diameter of 180–240 μm in the areas: at the eastern end of Spiekeroog; on the floor of the main channel (except for the deepest part where clay deposits were found); and on the arc-shaped shoal (except for the 'Tabaksplate'), on the bars as well as in the shallow channels between them (Fig. 8).

In Fig. 7 the mean diameter of the samples is plotted against the sorting coefficient. The two fractions are clearly separated by the 180 μm line. From the evaluation of the bedding types encountered, it can be concluded that this boundary is caused by hydrodynamic differences of the two sand populations. Dillo (1960) and Zanke (1976) found that there is a difference in the behaviour of fine sand and medium-grained sand (i.e. larger or smaller than 200 μm). According to their studies, fine sand is not able to form megaripples at any current velocity (cf. Reineck & Singh, 1973). Simons *et al.* (1965) found that 0.19 mm sand makes 'dunes'. Willis, Coleman &

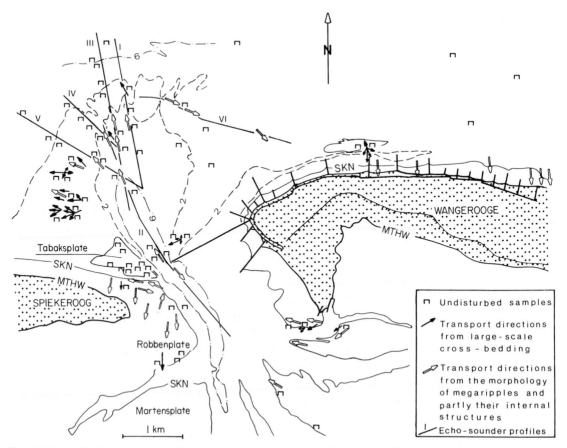

Fig. 5. Map with locations of undisturbed samples and measured transport directions. Contour lines in metres below SKN (see caption to Fig. 1). MTHW=mean high tide water level. After charts from WSA, Wilhelmshaven (1977/78).

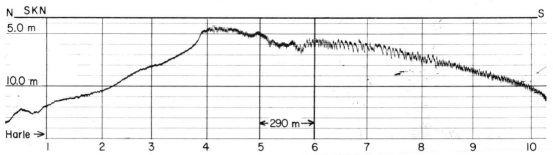

Fig. 6. Echo-sounder profile no. III across the arc-shaped shoal taken during high water level. Very calm sea conditions on 21 August 1978.

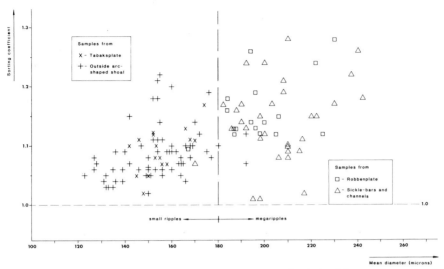

Fig. 7. Mean diameter versus sorting coefficient ($S_0 = \sqrt{Q_3/Q_1}$) of the samples analysed.

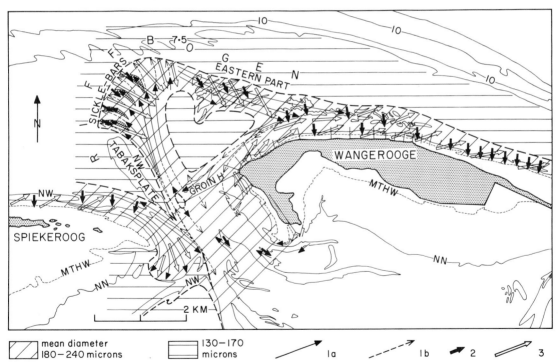

Fig. 8. Grain-size distribution in the Harle inlet and primary transport paths of the sand coarser than 180 μm: (1) transport by tidal currents during calm weather conditions; (1a) by megaripple movement and associated suspension load; (1b) in pure suspension; (2) transport by megaripple movement and associated suspension load created by breakers due to strong south-west to north winds (partly influenced by tidal currents); (3) transport during surge conditions as suspension load (partly influenced by tidal currents).

182   *J. Hanisch*

Ellis (1972) observed that 0·1 mm sand only formed ripples. So the results presented here may well close the gap between these two values at least for relatively shallow water conditions. At greater depths in the Bay of Fundy, Middleton (personal communication) has observed megaripples made by sand of about 150 μm. Therefore, besides current velocity and grain size, the occurrence of megaripples seems to depend on water depth.

CONCLUSIVE SAND TRANSPORT MODEL

From the evaluation of: (1) the measured transport directions, (2) the encountered grain-dize distribution, and (3) a series of current measurements (by WSA, Wilhelmshaven) which have not yet been released for publication, the following sand transport model can be concluded for the Harle inlet (Fig. 8).

There is a general drift of sand from west to east along the beaches of the East Frisian Islands, as has been known for a long time. The driving forces for

this drift are wave-induced currents such as swell currents and oblique rip currents (Reineck, 1963). The mean grain size of the wandering sand is 180–240 μm. The littoral drift of the sand is interrupted at the tidal inlet, where the sand is eroded by the ebb-tide currents and transported northward. This erosion occurs mainly on the western flank of the inlet channel where a cut bank is formed under normal conditions. At present the erosional activity of the ebb-tide current seems to be reduced considerably. The resulting accumulation of sand at the eastern end of Spiekeroog from 1961 to 1973 is documented in Fig. 9. The 'Alte Harle' channel was forced toward the east and a new large bar, 'Robbenplate', formed during this time interval.

The assumed shift of entire sandbars around the arcuate shoal (Gaye & Walther, 1935) is not valid for the western part of the shoal arc. The first large bar north of the eastern end of Spiekeroog consists of sand with a mean diameter of 140–170 μm. The sand of this bar therefore cannot be part of the wandering sands > 180 μm.

At present the (small amount of) sand eroded on

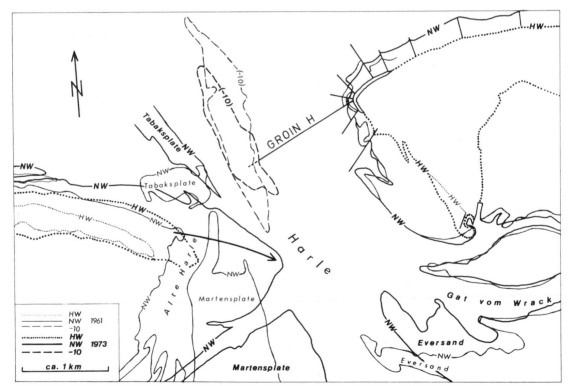

Fig. 9. Dislocations of channels and sandbars in the Harle inlet from 1961 to 1973, after aerial photographs. HW = approximate high water level. NW = approximate low water level.

the western flank of the inlet channel is carried toward the 'Riffbogen' first in suspension (cf. Luck, 1974, fig. 8) and later by moving megaripples together with corresponding suspension load (Fig. 8). When the sand reaches the arc-shaped shoal, differing transport mechanisms take effect. On the bars the megaripple transport continues but in the channels between the sickle bars flow conditions (possibly supercritical) with purely suspension transport are established at least during late ebb-tide phases. As practically no sand coarser than 175 μm was found north of the crest of the shoal (Figs 6 and 7), any transport of the coarser, moving sand fraction must cease directly behind the crest. In this way the arc shoal widens and grows higher during calm weather periods.

During surge conditions the sickle bars are flattened. The sand eroded by wave-induced currents is presumably thrown back to the inner part of the arc. At the same time the arc is reduced in radius. Due to alternating weather conditions the sand is certainly often moved to and fro in this way. By this zig-zag transport it may finally reach the eastern part of the shoal area. According to measurements of currents it is removed there from the influence of the ebb-tide current. It is then driven in groups of megaripples toward the beach of Wangerooge. The driving forces are believed to be the flood-tide currents and wave currents (Fig. 8) or a combination of both. From here some of the sand is transported back to the centre of the Harle mainly along a deep channel which has formed at the ends of the groynes (Fig. 5). It is then eroded at the eastern flank of the Harle channel by the ebb current and thus circulates around the eastern part of the tidal outlet system.

GENERAL CONCLUSIONS

The previous sand transport model is valid only in the eastern part of the arc-shaped shoal. In the western part, the lateral shift of entire bars from Spiekeroog toward the north and north-east is disproved. The shape and lateral displacements of the sickle bars is totally controlled by random erosion. These bars seem to oscillate like bars in a meandering river.

The most important zone for the estimation of stability or non-stability of the Harle channel is the eastern end of Spiekeroog. When, for any reason, the amount of sand eroded there by the ebb-tide current is less than that supplied by the littoral drift the

island starts to spread eastward. The Harle channel is then forced toward the east with all the well-known consequences to the western end of Wangerooge.

At present the erosive force of the ebb-tide current along the eastern end of Spiekeroog is weakened by a deep scour which has formed at the end of a long groyne (Fig. 10). Most of the tidal water flows apparently through this scour. Therefore the current at the western side of the channel is not strong enough to erode the sand accumulating there. The cut bank function is reversed by the groyne to the opposite side of the channel (Fig. 10). The present lack of a sufficient sand supply for the north-western beach of Wangerooge, in the last analysis, is also caused by this lack of erosion. Instead of a satisfactory supply of sand to the 'Riffbogen', which in turn supplies the beach (Krüger, 1937/38), the sand migrates primarily to the tidal flat area (Fig. 8). The formation of the new 'Robbenplate' during the last decades is a striking proof of this view.

REGIONAL ASPECTS

Nummedal & Penland (1981) in a theoretical approach divide the tidal inlets of the German Bight into six morphologic types. According to these authors they depend on the ratio of the tide range to the nearshore wave energy. Within this scheme the Wangerooge inlet would lie in the middle part of the scale (approximately like Nummedal & Penland's fig. 1 : 6). On the other hand, Wangerooge is situated in the area with nearly the highest tidal range of the German Bight. So it seems that the schematic model Nummedal & Penland derived from the eastern United States coast is not fully applicable to the East Frisian inlets.

In a sedimentological study on a Dutch tidal inlet Winkelmolen & Veenstra (1974) came to conclusions similar to those presented in this paper concerning the sorting effects and the transport paths of the sand. The two areas, however, are not completely comparable due to much greater grain-size variation in the Dutch Schiermonnikoog inlet than in the 'Harle'. Moreover, the grain shape and its rollability has not been investigated in our study as was done by Winkelmolen & Venestra.

For the practical aspects of coastal protection measures, much more sedimentological fieldwork should be done in other inlet systems. Otherwise, every protection design remains burdened with unnecessary risks.

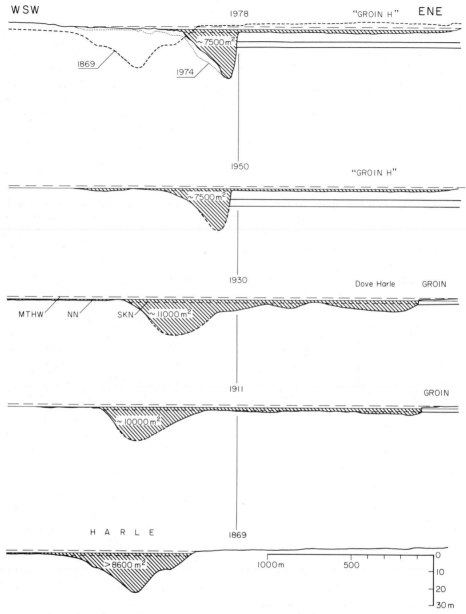

Fig. 10. Development of the Harle channel from 1969 to 1978 in cross-sections. Reversal of the cut bank (between 1911 and 1950) due to the long 'Groin H'.

ACKNOWLEDGMENTS

This study was financed by the Stiftung Volkswagenwerk, Hannover. The author is indebted to Drs Streif, Barckhausen, Kudrass (Hannover) and Ysker (Wilhelmshaven) for their critical help during the study. Without the undaunted engagement of Mr Husemann during the fieldwork (sometimes a little hazardous) this investigation would not have come to a fruitful end. The technical assistance by the WSA, Wilhelmshaven, the Senckenberg Institute, Wilhelmshaven, and the DHI, Hamburg is greatly acknowledged.

REFERENCES

DILLO, H.G. (1960) Sandwanderung in Tideflüssen. *Mitt. Franzius-Inst.* 17, 135–253.

GAYE, J. & WALTHER, F. (1935) Die Wanderung der Sandriffe vor den ostfriesischen Inseln. *Bautechnik,* 41, 1–13.

HANISCH, J. & HUSEMANN, H. (1979) Ein Spül-Stechkasten zur Entnahme ungestörter Sandkerne von 1,5 m Länge. *Senckenberg. Mar.* 11, 47–57.

HOMEIER, H. & KRAMER, J. (1957) Verlagerung der Platen im Riffbogen vor Norderney und ihre Anlandung an den Strand. *Jber. Forschungsstelle Norderney,* 8, 37–60.

KENNEDY, J.F. (1963) The mechanics of dunes and antidunes in erodible-bed channels. *J. Fluid Mech.* 16, 521–544.

KÖGLER, F.-C. & VEIT, K.H. (1973) Entnahme von Kernen aus Lockersedimenten des Schelfgebietes mit dem Vibrohammer-Kerngerät. *Meerestechn.* 4, 91–95.

KRAMER, J. (1960) Zur Frage der Wanderung der ostfriesischen Inseln aufgrund neuerer geologischer Befunde. *Z. geol. Ges.* 112, 529–530.

KRÜGER, W. (1937/38) Riffwanderung vor Wangerooge. *Abh. naturw. Ver. Bremen,* 30, 243–252.

LUCK, G. (1974) Untersuchungen zur Sedimentbewegung mit Hilfe einer Unterwasser-Fernsehanlage. *Jber. Forschungsstelle Norderney,* 25, 55–77.

LUCK, G. (1975) Der Einfluß der Schutzwerke der Ostfriesischen Inseln auf die morphologischen Vorgänge im Bereich der Seegaten und ihrer Einzugsgebiete. *Mitt. Leichtweiss Inst. Braunschweig,* 47, 1–122.

LUCK, G. & WITTE, H.H. (1974) Erfassung morphologischer Vorgänge der ostfriesischen Riffbögen in Luftbildern. *Jber. Forschungsstelle Norderney,* 25, 33–54.

NUMMEDAL, D. & PENLAND, S. (1981). Sediment dispersal in Norderneyer Seegat, West Germany. In: *Holocene Marine Sedimentation in the North Sea Basin* (Ed. by S.-D. Nio et al.). *Spec. Publs int. Ass. Sediment.* 5, 187–211. Blackwell Scientific Publications, Oxford. 524 pp.

REINECK, H.-E. (1958) Kastengreifer und Lotröhre "Schnepfe", Geräte zur Entnahme ungestörter orientierter Meeresgrundproben. *Senckenberg Leth.* 39, 42–48 and 54–56.

REINECK, H.-E. (1963) Sedimentgefüge im Bereich der südlichen Nordsee. *Abh. senckenb. naturforsch. Ges.* 505, 138 pp.

REINECK, H.-E. & SINGH, I.B. (1973) *Depositional Sedimentary Environments.* Springer-Verlag, Berlin. 439 pp.

SHEN, H.W. (1971) *River Mechanics, Vols I and II* (Fort Collins, Colorado).

SIMONS, D.B., RICHARDSON, E.V. & NORDIN, C.F. (1965) Sedimentary structures generated by flow in alluvial channels. *Spec. Publ. Soc. econ. Paleont. Mineral., Tulsa,* 12, 34–52.

WALTHER, F. (1969) Überbilck über die Unterrsuchungen des Wasserbauamtes Norden von 1920 bis 1933 über die Veränderungen der Ostfriesischen Inseln und ihre Ursachen. *Jber. Forschungsstelle Norderney,* 19, 7–30.

WILLIS, J.C., COLEMAN, N.L. & ELLIS, W.M. (1972) Laboratory study of transport of fine sand. *Proc. Am. Soc. civ. Engrs,* 98, 489–501.

WINKELMOLEN, A.M. & VEENSTRA, H.J. (1974) Size and shape sorting in a Dutch tidal inlet. *Sedimentology,* 21, 107–126.

ZANKE, U. (1976) Über den Einfluß von Kornmaterial Strömungen und Wasserständen auf die Kenngrößen von Transportkörpern in offenen Gerinnen. *Mitt. Franzius-Inst.* 44, 1–111.

Spec. Publs int. Ass. Sediment. (1981) **5**, 187–210

Sediment dispersal in Norderneyer Seegat, West Germany

DAG NUMMEDAL *and* SHEA PENLAND

Department of Geology, Louisiana State University, Baton Rouge, Louisiana 70803, U.S.A.

ABSTRACT

The paper presents a sediment dispersal model for a high wave-energy, mesotidal inlet on the southern coast of the North Sea. The model has been derived through the integrated analysis of historical–morphological data, sub- and intertidal bedform distributions, aerial photography, intertidal fluorescent tracer dispersal studies and long-term current velocity time series.

Long-term historical–morphological data permit the reconstruction of Norderney and environs since the year 1350. Over the last 300 years the island has been subject to sustained growth. Annual inlet maps over about three decades demonstrate that this growth is related to the steady migration of bars around the arcuate margin of the ebb-tidal delta from the island of Juist eastward to Norderney.

The pattern of sediment bypassing is not simple. Sand is delivered to the inlet by wave induced longshore transport from the shoreface at Juist and sources further west. As it enters the inlet system, sediment is moved offshore and eastward through a complex set of alternating ebb- and flood-dominated channels. It is delivered to the strongly ebb-dominated main channel through a major marginal flood channel and across large bars (ebb-shields) which form the landward termini of many of the smaller flood ramps. The expanding ebb jet delivers this sediment to two major terminal lobe bar complexes. Once deposited at the terminal lobe the sediment is subject to further eastward movement in the form of large bars. The morphology of the bars demonstrates that they are subject to significant tidal current influence near the updrift (western) margin of the terminal lobe. Tidal influence decreases eastward and the bars become shore-parallel 'swash-bars' before they weld to the island shore about 5 km east of the west end of Norderney.

The overall asymmetry of the Norderneyer Seegat ebb tidal delta, and its pronounced degree of downdrift offset, are deduced to be functions of the strongly eastward-directed momentum of the waves and the residual tidal current. The mechanism of 'wave pumping' is proposed to account for the maintenance of a stable updrift marginal flood channel as contrasted with the highly transient and migratory channels on the downdrift side.

INTRODUCTION

The relative magnitude of wave versus tidal energy has long been recognized as the most crucial factor controlling tidal inlet sand-body geometries. Bruun (1966) considered this when he quantified inlet stability and sediment bypassing mechanisms in terms of the ratio between tidal prism and littoral sediment transport rate. More recently, O'Brien (1976) applied the concept to formulate a 'closure index', defined as the ratio between tidal prism and wave power. Through statistical analysis of inlets

along the coastline of the United States, Jarrett (1976) and Walton & Adams (1976) discovered that both the cross-sectional area of the inlet gorge and the volume of the associated ebb delta increase with the tidal prism and decrease with increasing wave energy. These results are all consistent with the concept that the equilibrium size of an inlet is controlled by a balance between the scouring action of the tidal currents and the infilling by sediment in longshore transport.

These discoveries suggest that there should be a direct link between the ratio of wave energy to

0141-3600/81/1205-0187 $02.00

tide energy and the size, geometry, location, and sedimentary structures of inlet-associated sand bodies. A series of recent papers have investigated this problem. Following the discovery by Price (1955) that coastal embayments generally are wave-

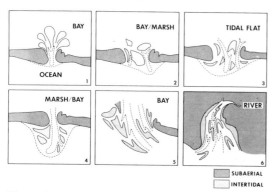

Fig. 1. Schematic display of inlet and sand-body geometries observed in the German and Georgia Bights. Frame 1 identifies the typical flood tidal delta of a wave-dominated inlet into a coastal bay; frames 2 and 3 show the progressive displacement of sand-bodies in a seaward direction as the tidal influence increases; frame 4 identifies the typical ebb-tidal delta of a strongly tide-dominated inlet on the South Carolina or Georgia coast; frame 5 shows the sand-body geometries at the entrance to the Weser estuary where the tide dominance is so strong that barrier islands are about to disappear; and frame 6 schematically displays sand-bodies in Nushagak Bay, Alaska, a large macrotidal coastal embayment. From Nummedal & Fischer (1978).

dominated at the flanks and tide-dominated in the centre, the approach in these studies has been to compare morphology and sedimentation patterns between inlets in different positions within the embayment. The strongly tide-dominated inlets on the southern South Carolina and Georgia coasts (Fitz-Gerald, Nummedal & Kana, 1976; Oertel, 1975) are but one extreme of a spectrum which ranges through transitional inlets in northern South Carolina and northeast Florida (Hubbard, Barwis & Nummedal, 1977; Penland, 1979a) to wave-dominated inlets at North Carolina's Outer Banks (Nummedal *et al.*, 1977). A very similar regional pattern of variability is found in the south-eastern embayment of the North Sea (Nummedal & Fischer, 1978).

Studies of the geometry of inlet sand-bodies within both the German and the Georgia Bights have led to the identification of a continuum of inlet morphologic types, ranging from the completely wave-dominated ones with large flood-tidal deltas and very small ebb-tidal deltas to the completely tide-dominated ones with non-existent flood-tidal deltas and large ebb-tidal deltas (Fig. 1).

The geometries of these sand-bodies, and their internal stratification, reflect the dominant pathways of sand dispersal within the inlet system. Given the fact that a wide range of inlet geometries occurs, there is a corresponding range in inlet sediment dispersal and bypassing mechanisms. Detailed documentation of sand dispersal has been presented

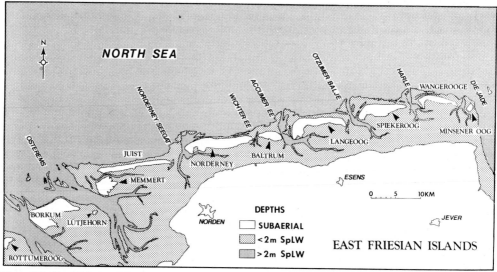

Fig. 2. Map of the East Friesian Islands in the province of Lower Saxony, Republic of Germany. Depth is referred to mean spring low water (SpLW). For location of this area, see Fig. 4.

for only a few inlets. Most of these are located on segments of the American east coast which are subject to rather low wave energy. The primary objective of this paper is to present a sediment dispersal model for an inlet with a high tide range situated in a high-wave-energy environment. The chosen inlet is Norderneyer Seegat on the Friesian coast of West Germany (Fig. 2). The geometry of this inlet would correspond to frame number 3 in Fig. 1. Using the tide range classification proposed by Davies (1964) and inlet models first proposed by Hayes (1975), Norderneyer Seegat falls in the mesotidal inlet category.

PHYSICAL PARAMETERS

The variation in tide range along the open coast of the German Bight is controlled by the North Sea amphidromic system. Classical models of the M_2 tide demonstrate the existence of a counter-clockwise rotation of the tidal wave around an amphidromic point between Jutland and England (Defant, 1958). Iso-range lines are nearly equidistant about this point. Therefore the central German Bight, which is

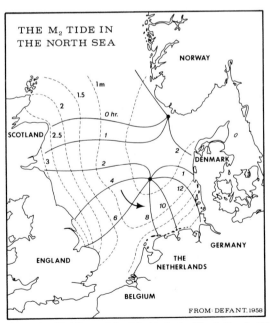

Fig. 3. Co-tidal lines of the M_2 tide in the North Sea (solid lines) and co range lines (dashed). The tide affecting the German coast rotates counter-clockwise about an amphidromic point near the middle of the North Sea. Figure modified after Defant (1958).

further away from the amphidromic point than is Jutland and the north-west corner of the Netherlands, has the greater tide range (Fig. 3). The spring-tide range at Norderneyer Seegat is 2·7 m, the neap range is 2 m, and the mean range is 2·5 m. Tide range increases eastward to 2·9 m at Harle and about 3·8 m at the entrance to the Jade, Weser and Elbe Rivers (Deutsches Hydrographisches Institut, 1979).

Evaluation of the regional variability in nearshore wave climate is difficult because reliable wave records are very scattered and of too short duration to be of much use in studies of the long-term evolution of the coast. Therefore, in order to derive a consistent picture of wave-energy variability, it was decided to utilize the Summary of Synoptic Meteorological Observations (SSMO data) published by the U.S. Naval Weather Service Command (1974). Wave power distributions within pre-established data squares were calculated by a procedure outlined in Nummedal & Stephen (1978). In this same article the authors also discuss in some detail the assumptions and problems associated with the utilization of SSMO data in studies of coastal sedimentation dynamics.

SSMO-derived deep-water wave power values for data squares along the periphery of the North Sea are summarized in Table 1 and Fig. 4. One observes a distinct southward decrease in total wave power. This decrease is clearly fetch-controlled. The dominant prevailing winds at Norderney come from the western quadrant (Luck, 1976a). This wind direction causes the wave power directed along the shore to exceed that directed perpendicularly on to the shore all along the East Friesian coastline. The resultant power for the Bremerhaven data square point to the east-south-east (azimuth = 101°). This resultant vector yields a net eastward longshore power component of $4·44 \times 10^3$ W m^{-1}.

Nearshore wave data, based on *in situ* gauges, are available for Norderney (Niemeyer, 1978) and for Westerland at Sylt (Dette, 1977). Niemeyer's primary objective was to analyse the wave damping across the tidal delta bars. A long-term time series for proper evaluation of the height–frequency distribution of nearshore waves is not yet available. The measurements obtained indicate that the average significant wave height ($H_{1/3}$) exceeds 1·6 m in 10 m of water depth. Dette (1977) presents height-frequency diagrams for a wave-station in 8 m of water off Sylt. These data indicate that significant wave heights exceed 1 m on a daily basis, 4·8 m on

Table 1. Wave power values for SSMO data squares along the margins of the North Sea.
Wave power in units of 10^3 W m^{-1}

Direction	Edinburgh	Grimsby	Rhine Delta	Bremerhaven	Esbjerg	Stavanger
N	2·1	1·5	0·8	0·7	2·1	5·1
NE	0·5	1·5	0·7	0·3	1·0	1·4
E	3·3	1·5	0·5	0·8	1·8	4·5
SE	3·4	0·8	0·3	0·3	1·0	7·8
S	1·8	2·1	0·7	0·5	2·3	3·8
SW	1·7	2·8	2·1	1·5	5·0	5·4
N	3·0	2·3	1·9	2·9	5·5	9·8
NW	3·3	2·7	1·4	2·4	6·3	11·8

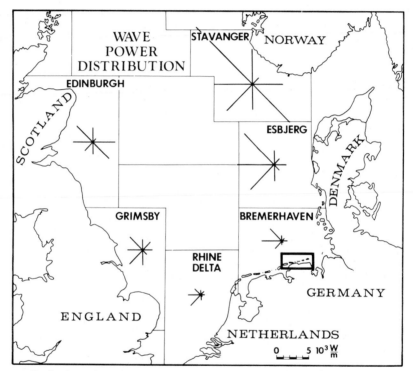

Fig. 4. Wave power distributions along the periphery of the North Sea. Calculations are based on data published by the U.S. Naval Weather Service Command in the Summary of Synoptic Meteorological Observations (SSMO). The length of the bars is proportional to the mean annual wave power from each direction, i.e. the power diagrams are plotted like wind roses (from Nummedal & Fischer, 1978). Dark rectangle identifies area shown in Fig. 2.

an annual basis and (extrapolated) 7·3 m once every 100 years (Fig. 5).

HISTORIC–MORPHOLOGICAL EVOLUTION

The Forschungsstelle Norderney has published a map series at a scale of 1 : 50,000 which reconstructs changes in the morphology of the Friesian coast at intervals of approximately 100 years (Luck, 1975). The map series has been produced under the principal direction of Mr Hans Homeier. According to these historical maps, Norderney's existence can be traced back to the breaching of the barrier island of Buise between AD 1350 and 1400 (Fig. 6). Following the breach, Osterende (later Norderney) began to grow, while Buise decreased in size and ultimately

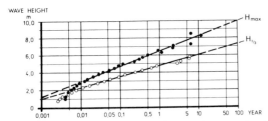

Fig. 5. Wave height–frequency diagram for a station in 8 m of water off Sylt. H_{max} and $H_{1/3}$ refer to maximum and significant wave heights, respectively (from Dette, 1977).

disappeared as a subaerial island by the year 1750. The disappearance of Buise correlated with the eastward growth of Juist.

The size of Norderney (Fig. 7) increased by 117% between AD 1650 and 1960 (Luck, 1975). At the same time Norderneyer Seegat migrated an insignificant distance to the east (Luck, 1975). The inlet appears to be anchored in a major erosional valley which extends northwards from Norddeich (Luck, 1970). This relationship demonstrates conclusively that there exists an efficient pathway for eastward sediment bypassing around Norderneyer Seegat.

It has been pointed out by Oertel (1975, 1977), Nummedal *et al.* (1977), and Hubbard, Oertel & Nummedal (1979) that the pattern of asymmetry in an ebb-tidal delta directly reflects the relative magnitudes and directions of wave-induced and tide-induced sediment transport. Oertel's (1977) schematic representation is reproduced in Fig. 8.

The asymmetric, strongly eastward-directed momentum of the waves combined with an eastward residual tidal current produces a skewed ebb-tidal delta at Norderneyer Seegat. The bathymetry (Fig. 9) closely resembles that of Oertel's (1975) type B inlet.

Through combined use of historical data, bedform orientation maps, tracer dispersal data, and current velocity observations, the remainder of the paper will describe and explain the overall tidal delta asymmetry and the complex interactions of sediment dispersal responsible for the configuration of individual sand-bodies in Norderneyer Seegat.

SEDIMENT TRANSPORT PATTERN

The Nordergründe Bars

The time-sequence maps of tidal delta bars are historical-morphologic data of direct relevance to the understanding of inlet dynamics. These bars are referred to as the 'reef-bow' (in German, *riffbogen*)

by Luck (1975, 1976b). Some of these bars would correspond to delta margin 'swash bars', others would resemble 'spill-over lobes' in the terminology of Hayes *et al.* (1973) and Hayes (1975).

Annual mapping of the bars was carried out between 1926 and 1957 (Homeier & Kramer, 1957). Several annual maps are illustrated in Fig. 10 to document the rates and patterns of morphological change. Fig. 11 presents summary data on the paths of movement of the centres of gravity of selected individual bars between 1926 and 1957. The rate of bar movement is remarkably uniform, averaging 406 m y^{-1}, with a standard deviation of only 85 m y^{-1}. Furthermore, the centre of gravity of individual bars from Nordwestgründe, through Nordergründe and on to the beach at Norderney, all follow essentially the same paths. The inner bars generally move somewhat slower than the more exposed ones and weld to the Norderney shoreline further west. A similar pattern of bar migration has been documented at Wichter Ee and Accumer Ee on the Friesian Coast by Homeier & Luck (1970). This pattern of trajectories is established as a long-term path of dynamic equilibrium between the seaward-directed momentum of the expanding ebb current and the landward- and eastward-directed momentum of the shoaling waves. The residual tidal current on the shelf immediately seaward of Norderneyer Seegat is also oriented to the east; nearshore bedforms reflect this (Ulrich, 1973).

The morphology of the 'reef-bow' bars changes distinctly as one moves eastward from Schluchter. At Nordwestgründe most bars are crescentic in shape; their long axes are oriented NW–SE (Fig. 12A). These bars consist of a high crest on the northeast side, which acts as a flood shield (terminology from Hayes, Anan & Bozeman, 1969; Klein, 1970), and a wide, gently sloping, south-west flank covered with large ebb-oriented sand waves.

Within the Nordergründe complex, the bars are more variable in shape. Near Dovetief, there is evidence of tidal current dominance of bar morphology in the generally cresentic bar outline. Multiple surveys in this region in the summer of 1979, with a precision depth recorder (PDR), however, failed to detect any large-scale bedforms or swash-bar slipfaces. Further east in Nordergründe the bars develop a shore-parallel elongation and a landward-dipping slipface (Fig. 12B). Tidal currents in this area seem to play a subordinate role. These bars correspond to the 'swash bars' in Hayes' (1975) model. Their onshore migration involves an extension of the bar parallel to the beach (Fig. 12C, from Homeier & Kramer, 1957), followed by the classical process of landward migration of a ridge-and-runnel system to form a beach berm (Davis *et al.*, 1972). Similar patterns of swash-bar beach attachment have been documented for North Inlet (Finley, 1976; Nummedal & Humphries, 1978) and Price Inlet (FitzGerald, Hubbard & Nummedal, 1978) in South Carolina and Fort George Inlet in Florida (Penland, 1979b).

The eastward movement of bars from Nordwest-gründe to the beach at Norderney appears to occur

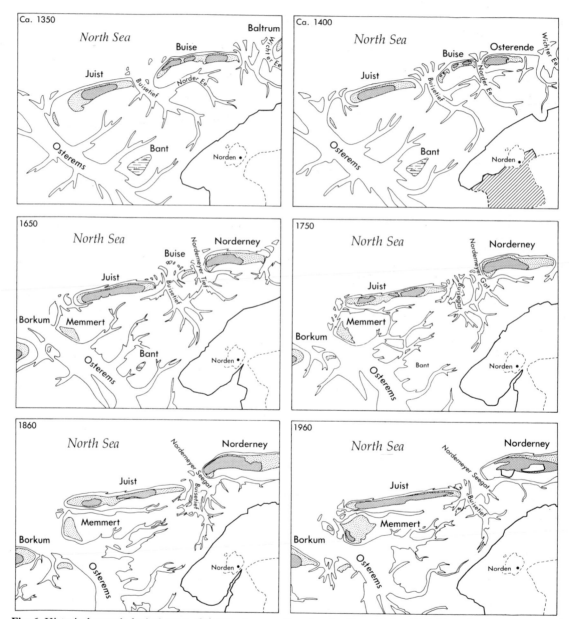

Fig. 6. Historical–morphological maps of the Norderneyer Seegat region from 1350 to 1960. Note the disappearance of Buise between years 1650 and 1750 and the eastward displacement of Busetief in response to the growth of Juist. Norderneyer Seegat has remained stationary (from Luck, 1970).

NORDERNEY

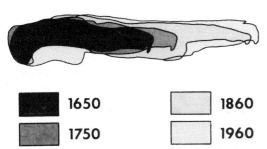

■	**1650**	▨	**1860**
▨	**1750**	▨	**1960**

Fig. 7. Growth of Norderney above the mean high water line since 1650. The figure is based on the 1:50,000 historical map series of the coast of Lower Saxony (see Luck, 1975).

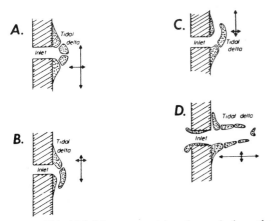

Fig. 8. Ebb-tidal delta asymmetries observed along the Georgia coast, reflecting different relative magnitudes of tidal and longshore currents. Arrow lengths are proportional to relative current magnitudes (from Oertel, 1975).

under the influence of flood- and ebb-segregated tidal current (Nordwestgründe), wave-induced longshore currents (Nordergründe), wave swash (east Nordergründe), and the eastward-directed resultant tidal current on the inner shelf.

The bar complex has steadily grown in size since 1926; however, there is no evidence of bar migration towards the inlet gorge (Homeier & Kramer, 1957). Swash bar growth is generally associated with bar movement towards the gorge in the South Carolina inlets (FitzGerald *et al.*, 1978). The strong longshore component of wave power at Norderney appears to favour bar-bypassing rather than return of sediment to the inlet gorge by wave swash.

Osterriff and Schluchter

Prior to 1650 the inlet east of Juist was Busetief (Fig. 6). With the disappearance of Buise most of Busetief's drainage was diverted into Norderneyer Seegat, greatly diminishing the water exchange through the channels at the east end of Juist, now named Kalfamergat. Recently, however, Kalfamergat has increased in cross-section. Data used for the bathymetric chart of Fig. 9 were compiled prior to 1962. At that time Kalfamergat had a maximum depth of 5 m. The July 1979 survey demonstrated that the channel gorge now is 16 m deep; it has also increased in width. Fig. 13 further demonstrates that the main Kalfameragat channel bifurcates to the north, with the western channel branch feeding the southern lobe of the Osterriff spillover lobe complex, the eastern branch leading into the region of the previously existing Spaniergat. In fact, it is here hypothesized that the infilling of Spaniergat is a direct consequence of the enlargement of Kalfamergat. This, in turn, is related in a complex way to changes in back-barrier drainage.

The seaward depositional lobes related to the Kalfamergat channels constitute Osterriff (Fig. 14). At its western margin near the north shore of Juist, the Osterriff bars assume the characteristic morphology of an ebb-spillover lobe (Fig. 15a). Persistent ebb-oriented sand waves characterize the broad centre of the lobe (Fig. 15b), with the bar crest acting as a flood shield. The seaward margins of the flood shields and the immediately adjacent channels act as conduits for landward sediment transport, due to the momentum flux of obliquely breaking waves (Longuet-Higgins, 1970) and the effect of 'wave-pumping' (Bruun & Kjelstrup, 1977). The tidal component of the current along this outer bar flank is probably also flood-dominant.

Bars at the northern margin of the Osterriff complex are lunate with an ebb-dominated, gently westward-sloping surface and a steep, flood-dominated east flank (Fig. 15c). Large flood-dominated sand waves exist in slightly deeper water behind the bars (Fig. 15d). The spatial arrangement of flood- and ebb-dominated paths in this bar complex is not yet completely determined. However, pathways do exist for alternating landward and seaward tidal sand transport. This tidal flow is superimposed on a net eastward-directed, wave-induced mass movement across the bar complex. The net result is sand movement to the east. The distribution

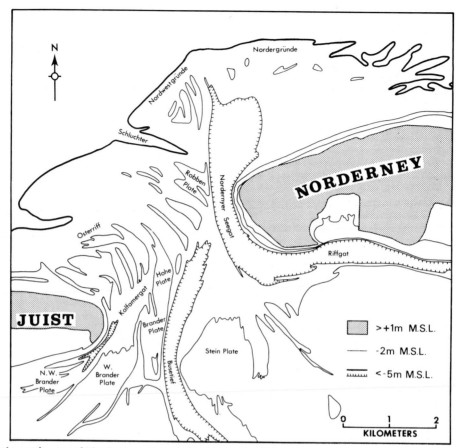

Fig. 9. Bathymetric map of Norderneyer Seegat with labels identifying the major channels and sand shoals (*platen*). This map is based on bathymetric chart no. 5 by the Forschungsstelle Norderney. The original map was published in 1962.

of intertidal bedforms in Osterriff is consistent with this dispersal pattern (Fig. 16).

Sand moving across the Osterriff complex from the shoreface at Juist may reach the Nordergründe swash bars, and thence Norderney, through two separate pathways: (a) across Schluchter to feed the Nordwestgründe bars, and (b) delivery into Norderneyer Seegat and Busetief via Robben Plate, Hohe Plate and Brander Plate; followed by seaward transport through the ebb-dominated main channel of Norderneyer Seegat.

The discussion in this paragraph is limited to path (a). Net sand transport in Schluchter is difficult to ascertain. There are no time-series current data; furthermore, due to rough seas, attempts at profiling failed to reveal any bedforms. There are, however, current data for Spaniergat obtained in 1949, a time when that channel apparently occupied

the same functional role in the Norderneyer dispersal system as Schluchter does today. These data demonstrate flood-dominance under normal conditions, with short periods of ebb-dominance associated with extreme south-westerly winds. The presence of a flood-dominated channel at Schluchter corresponds, in a general sense, to the concept of a marginal flood channel as proposed by Oertel (1972) and Hayes *et al.* (1973) for inlets along the south-east coast of the United States. Because Osterriff should be considered a separate inlet 'nested' within the larger Norderneyer Seegat system, Schluchter occupies the same position relative to Norderneyer Seegat as a typical marginal flood channel in Hayes *et al.*'s (1973) model. Prior to the recent growth of Kalfamergat, Spaniergat probably also played the role of marginal flood channel.

Marginal flood channels develop, in part, in

response to the time-lag between ocean and bay tides. Because of this lag (Nummedal & Humphries, 1978), the ocean tide begins to rise while the main inlet channel is still ebbing. The incoming flood currents are initially forced to avoid the expanding ebb jet and do so by entering the inlet through flood channels at the landward margins of the swash platform. If tides alone were responsible for the ebb-delta morphology, marginal flood channels should form symmetrically around the inlet; they should be about equal in size and conduct the same flood discharge. Nature, however, generally displays great asymmetry in the degree of flood channel development. The bathymetric map of Norderneyer

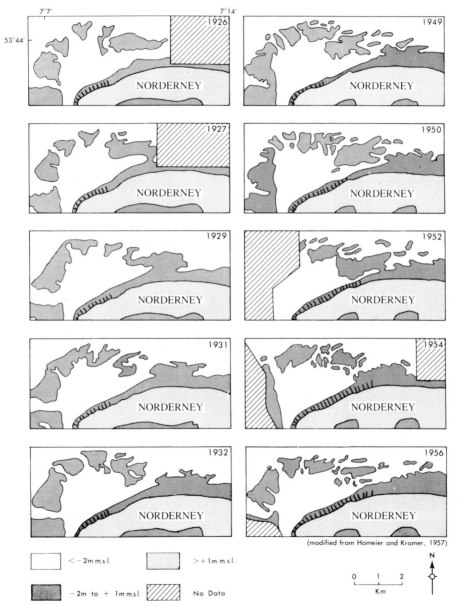

Fig. 10. Morphological changes in the ebb-tidal delta bars at Norderneyer Seegat for two selected time periods illustrating bar welding on to Norderney (from Homeier & Kramer, 1957). For additional information see Luck & Wittle (1973).

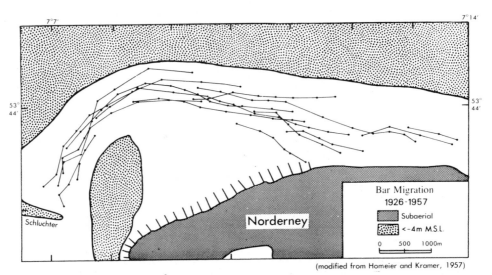

(modified from Homeier and Kramer, 1957)

Fig. 11. Map depicting the annual displacement of the centres of gravity (dot) of individual bars between 1926 and 1957. Bar migration is consistently from left to right (from Homeier & Kramer, 1957).

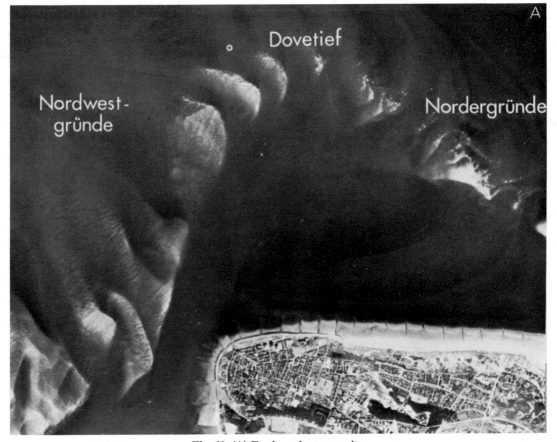

Fig. 12. (A) For legend see opposite.

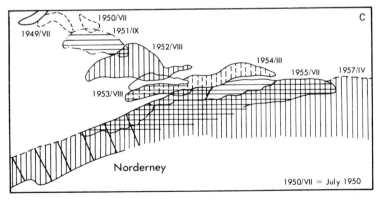

Fig. 12. Illustration of morphological changes associated with the migration of the bars from Nordwestgründe to Norderney. (A) Vertical airphoto (23 June 1973) of Nordwestgründe and central Nordergründe. Note bedforms and the crescentic bar shape demonstrating strong tidal current influence within the western (left) two-thirds of this region. (B) Vertical airphoto (23 June 1973) of the nearshore region of the Nordergründe swash-bar complex. Note landward dipping slipfaces on the bars. These indicate that sediment movement here is controlled by breaking waves. Both airphotos were obtained through the courtesy of the Forschungsstelle Norderney. (C) Map of the beach attachment process of a single large swash bar in the period between 1949 and 1957. Note the gradual bar extension in a shore parallel direction during the attachment process (from Homeier & Kramer, 1957).

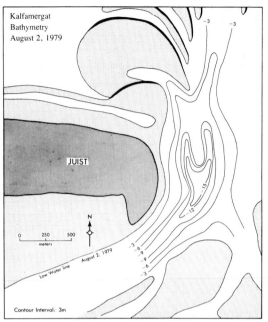

Fig. 13. Bathymetric map of the Kalfamergat region based on multiple PDR surveys in a radial pattern from the beach at Juist.

Seegat (Fig. 9) shows Schluchter to be a deep and large channel. Historical maps demonstrate that Schluchter has always been at that location (Fig. 10) and has been essentially invariant in size. On the east side of the swash platform, on the other hand, marginal flood channels have been rather ephemeral features. Homeier & Kramer (1957) report that a major channel, 3 m deep, once separated the east end of the Nordergründe swash-bar complex from the beach at Norderney. This channel reached its maximum size in 1923, but has subsequently been infilled. Smaller marginal flood channels have periodically opened and filled. Periodic opening and closing of downdrift flood channels is a common process (FitzGerald *et al.*, 1978; Nummedal & Humphries, 1978). Updrift flood channels are always more persistent.

Schluchter, and indeed updrift flood channels on the ebb-tidal deltas investigated on the American east coast, face directly into the dominant waves with a funnel-shaped plan form. This is the most effective orientation for a natural 'wave pump' (Bruun & Kjelstrup, 1977). We propose that a fundamental driving mechanism, hitherto unrecognized, for the landward current through updrift marginal flood channels is the wave pumping.

As a result of this pumping one would expect a persistent mass-flow of the water from Schluchter into Norderneyer Seegat, making the channel flood-dominant during all weather conditions except possibly some storms (Fig. 17).

Nordwestgründe, consisting of complex ebb-dominated spillover lobes, is formed in response to the flood-dominance of Schluchter and the divergent ebb flow out of Norderneyer Seegat. One important source of sediment for this flow is the terminus of Schluchter. Sediment is here effectively fed into the ebb-dominated Norderneyer Seegat system and re-deposited at Nordwestgründe.

Inner part of Seegat

Norderneyer Seegat also contains several shoals (*platen* in German) within, and landward of, the gorge section of the inlet. The main ones are Hohe Plate, Brander Plate and Stein Plate. Two small plates in a general westerly position from Brander Plate are named Brander Plate West and Brander Plate Northwest for identification in the following discussion (Figs 9 and 18a).

A bathymetric profile across the western extension of Brander Plate demonstrates well the transport pattern in this region (Fig. 18a, b). The ridge at P (Fig. 18b) appears to be a zone separating a flood-dominated channel to the west from an ebb-dominated one to the east. The transverse bar crest at Brander Plate reflects even more dramatically the net landward sediment movement across the major part of this feature (Fig. 18c).

The pattern of intertidal bedforms mapped on these plates in general corresponds to the dispersal pattern determined from subtidal bedform evidence. Figure 16 shows intertidal bedform variability and distribution in the inner part of Norderneyer Seegat.

Fluorescent sand tracer was used to provide additional evidence on sediment dispersal directions and rates within the Norderneyer Seegat system. A total of 32 injection stations were used, 23 of which were located on the inlet-associated sand-bodies discussed in this paper. Between 1 and 25 kg of native tagged sand, prepared at Forschungsstelle Norderney by the investigators, was placed in a 1 m diameter circle at each station. The material was allowed to disperse over a 12 h tidal cycle. Surface concentration patterns were determined during night-time low-tide periods by means of hand-held fluorescent lights. All tracer experiments were carried out during 'normal' weather conditions, i.e.

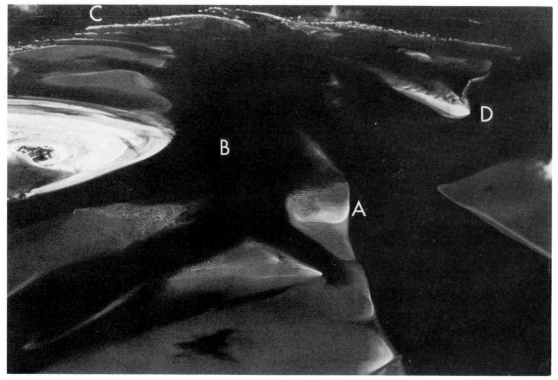

Fig. 14. Oblique airphoto towards the north of Brander Plate (A), Kalfamergat (B), Osterriff (C), and Hohe Plate (D). The Brander Plate region and Osterriff constitute the flood- and ebb-tidal deltas, respectively, with reference to Kalfamergat.

moderate winds from the south-west or north-west quadrant.

Figure 19 presents the results of the tracer experiments. The dispersal patterns at Osterriff are somewhat uncertain due to intense wave-induced diffusion at high water. Further seaward the wave action was too intense for any meaningful sampling. All tracer at Hohe Plate moved to the south and south-east, consistent with the interpretation of this feature as the south-east termination of a flood-transport path. A steep south-east margin, subject to frequent slumping, confirms that Hohe Plate is truncated by the actively eroding ebb-dominated currents of Busetief. Tracer dispersal trends on Brander Plate corroborate a transport path similar to that at Hohe Plate. Stein Plate shows uniform landward transport. The western station shows transport to the south-west, i.e. towards the margin of Busetief. Tracer movement is the only available evidence for sediment dispersal at Stein Plate, because both the intertidal shoal and the shallow subtidal ramps at its northern margin are

completely devoid of any bedforms carrying directional information.

Tracer dispersal at Brander Plate West and North-west presents a picture consistent with their location relative to Kalfamergat. Brander Plate West is, in fact, a flood-tidal delta relative to Kalfamergat. Brander Plate Northwest is part of an ebb-dominated lobe responding to wind-driven flow across the tidal flats behind Juist.

Sediment dispersal in deep channels

The rate of sediment transport (volume per unit time per unit width perpendicular to flow) is proportional to the third (Maddock, 1969) or perhaps a higher power of the current velocity. The strong tidal currents in the deeper parts of the main channels, therefore, are the primary agents of inlet sediment dispersal. The sand-dispersal paths already discussed above have been established in response to these dominant currents. Their effects will here be discussed primarily in the light of bedform evidence.

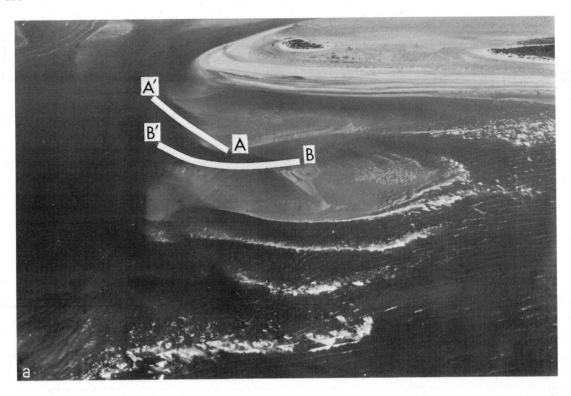

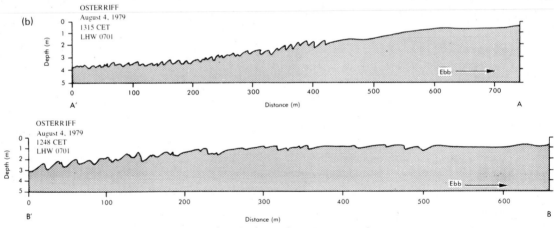

Fig. 15. (a) and (b). For legend see opposite.

Figure 20 presents a summary of selected hydraulics data demonstrating the distribution of ebb- and flood-dominated channels. Riffgat, south of the entrance to Norderney harbour, is strongly ebb-dominated. Kramer (1961) demonstrated this in a current velocity time series obtained in 1960, and it is also evident in the directional asymmetry of large bedforms to the south-east of the harbour entrance (Fig. 21).

The main channel of Norderneyer Seegat also is ebb-dominated. Long-term current velocity time series and large-scale bedform data (Fig. 22) yield a consistent picture. The degree of ebb-dominance, and hence the capacity for net seaward displacement of sediment by tidal currents, rapidly decreases as the ebb jet expands beyond the confined section between Robben Plate and Norderney (Fig. 20). This is accompanied by a rapid decrease in channel depth.

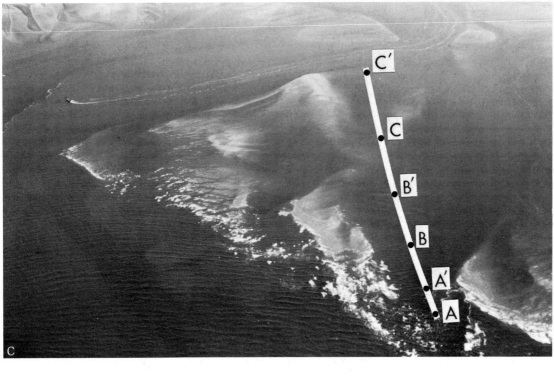

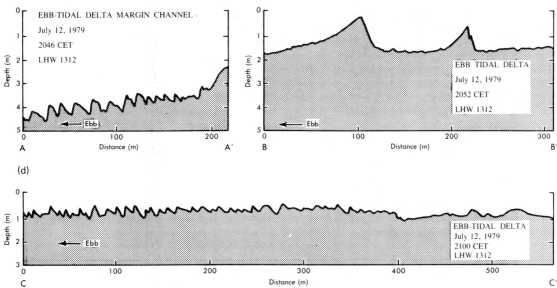

Fig. 15. Morphology and bedforms of the Osterriff bar complex. (a) Ebb spillover lobes near the north shore of Juist. View to the south. White lines indicate locations of PDR profiles shown in (b). (b) Bedform profiles obtained near low water in Osterriff on 4 August 1979. Note ebb-directed asymmetry of all large scale bedforms. (c) Northern margin of the Osterriff complex. View is to the south-east, with Hohe Plate in the background. White line indicates location of PDR profile shown in (d). (d) Selected segments of bedform profile obtained near low water in Osterriff on 13 July 1979.

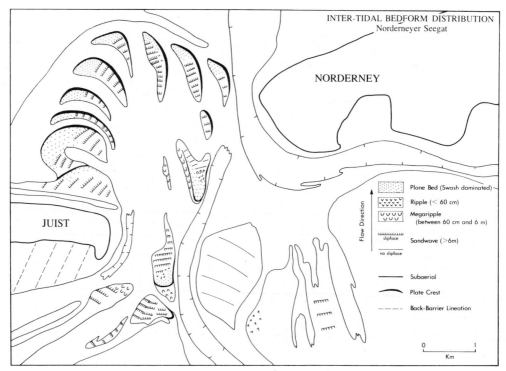

Fig. 16. Map of intertidal bedforms on sand shoals in the central part of Norderneyer Seegat. The map is based on multiple surveys and reconnaissance air flights extending from June to August of 1979. See Fig. 9 for identification of specific shoals. Terminology after Boothroyd & Hubbard (1974).

Busetief is ebb-dominated as well. This is seen most clearly in the asymmetry of the longitudinal profile of Busetief near its confluence with Norderneyer Seegat (Fig. 23). The strong ebb-dominance of Busetief, a long distance landward of the Norderneyer Seegat gorge appears, in part, to reflect the wind-driven mass flux of water across the tidal flats behind Juist (Koch & Luck, 1974). This water is discharged directly into the western tributaries of Busetief and imparts an asymmetry to the tidal prism: the mean ebb prism through Busetief exceeds the flood prism.

SUMMARY INLET DISPERSAL MODEL

All available data have been integrated in the proposed sediment dispersal model for Norderneyer Seegat (Fig. 24). Historical data, demonstrating growth of Norderney without significant migration of Norderney Seegat, clearly prove the existence of efficient sand bypassing (Luck, 1975). Maps of bar locations between 1926 and 1957 (Homeier & Kramer, 1957) further demonstrate that the by-passing is associated with a steady eastward migration of individual bars. This movement is estimated to supply a minimum of 130,000 m³ of sand to the Norderney beach per year. Since there is no reason to assume that the bars trap all, or even a majority, of the total sand in transport along the tidal delta margin, the total annual transport rate past Norderney Seegat is undoubtedly greatly in excess of the above figure.

Nordwestgründe is the updrift end of this bar migration path (Figs 9 and 24). The delivery of sediment from Juist to Nordwestgründe follows complex paths. From the shoreface at Juist, bedload may either move northward and eastward through Osterriff by a back-and-forth motion in alternating flood- and ebb-dominated channels (Fig. 15a, d), or landward through Kalfamergat into its 'flood-tidal deltas' of Brander Plate and Brander Plate West.

Sediment moving through Osterriff has three

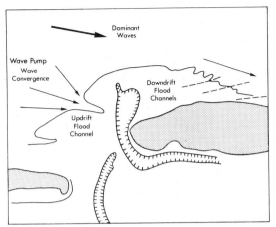

Fig. 17. Schematic illustration of the relative significance of wave pumping on the maintenance of updrift and downdrift marginal flood channels.

ultimate destinations: some is delivered via ebb spill-over lobes into Schluchter (Fig. 15c), some is transported by wave drift via Robben Plate into Norderneyer Seegat, and some is brought into Busetief via Hohe Plate. Because of the wave-pump-induced flood-dominance in Schluchter, sediment delivered there will also be supplied to Norderneyer Seegat. Although the mechanisms are different, the net result of these pathways is that sediment originally derived from the shoreface at Juist will be delivered to the ebb-dominated main channel of Norderneyer Seegat. As the expanding ebb jet decelerates north of the gorge at Norderneyer Seegat, the load is deposited at Nordwestgründe and Nordergründe.

The sediment dispersal pattern here presented correlates quite well with the temporal growth and migration of inlet sand-bodies depicted in Ragutzkie's (1978) map (Fig. 25).

DISCUSSION

Most tidal inlets are found to be 'offset', i.e. the shoreline on the downdrift barrier island is either further seaward (downdrift offset) or further landward (updrift offset) than the updrift shoreline. All the East Friesian inlets are distinctly downdrift offset (Fig. 2). Such an offset is explained to be a consequence of wave refraction around the margin of the ebb-tidal delta platform, causing bar migration and preferential shoreline accretion at the downdrift shore (Hayes, Goldsmith & Hobbs, 1970). It has been demonstrated that some inlets may change

in temporal cycles from downdrift to updrift offset and back (FitzGerald, 1976).

The offset of Norderneyer Seegat appears to conform to the above explanation, because all bar accretion does take place at the downdrift (Norderney) flank of the tidal delta. In addition, there is a historical element to be considered. Luck (1970) traced the origin of Norderneyer Seegat to an initial breach of the earlier island of Buise (Fig. 6). The offset was originally imparted to the system at that time. The sediment dispersal pattern outlined in this paper has contributed to its persistence through time. The large-scale plan-form asymmetry of the Norderneyer Seegat ebb-tidal delta (Fig. 9) clearly distinguishes it from the low-wave energy inlets of the central Georgia Bight (Oertel, 1975; Nummedal *et al.*, 1977). The asymmetry reflects the strong longshore direction of the wave power combined with an eastward residual tidal current on the inner shelf. The western margin (updrift side) of the ebb-tidal delta is oriented nearly perpendicular to the dominant waves; an orientation which both facilitates lateral offshore movement of sediment through the Osterriff spillover lobe complex and landward sediment transport by wave pumping through Schluchter.

Earlier ebb-tidal delta sediment dispersal models assign varying degrees of significance to the wave–current interaction. For example, Dean & Walton (1975) propose transport of water (and implicitly sediment) towards the inlet gorge through the marginal flood channels as a consequence of eddies thought to develop along the sides of the expanding ebb jet.

In contrast to this, Hine (1975), Hubbard (1975) and FitzGerald *et al.* (1976) argue that the landward return of sediment from the terminal lobe of the main ebb channel is a function of shoaling waves and flood-tidal currents. The dispersal of sediment at Norderneyer Seegat clearly supports the latter view. In fact it emphasizes the role of waves even more. The waves account for the overall tidal delta asymmetry, and they explain the difference in stability and degree of development of the updrift and downdrift marginal flood channels.

CONCLUSIONS

Historical data extending about 600 years back in time, extensive annual mapping of bar locations carried out by the staff at Forschungsstelle Norderney, long-term current velocity data, bedform

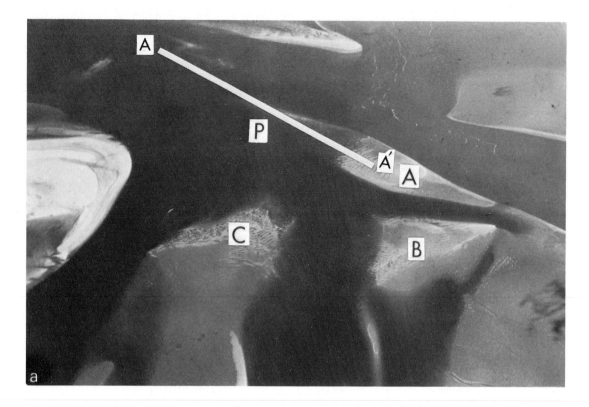

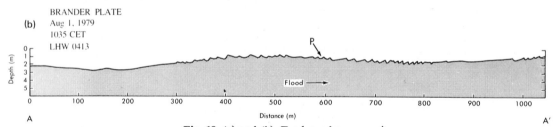

Fig. 18. (a) and (b). For legend see opposite.

profiles and tracer experiments provide a comprehensive data set for the determination of sediment dispersal in Norderneyer Seegat. This data set has revealed the following pathways of sand movement in the inlet.

Sand is supplied from the shoreface at Juist and sources further west. It travels through Osterriff through alternating flood- and ebb-dominated channels and is thence delivered either to the flood-dominated Schluchter or to the margin of bars flanking the west side of the ebb-dominated channels of Norderneyer Seegat or Busetief. The expanding ebb jet delivers the material to Nordwestgründe or

Nordergründe. Transport further to the east, on to the beach at Norderney, takes place through the combined action of wave swash and the eastward-directed residual inner shelf tidal current.

The strong longshore orientation of the momentum flux of the breaking waves imparts an overall asymmetry to this and indeed to all the other inlets in the East Friesian island chain.

Wave pumping is proposed to be an important mechanism in maintaining the effective landward mass flux through updrift marginal flood channels of ebb-tidal deltas situated in a high-wave-energy environment.

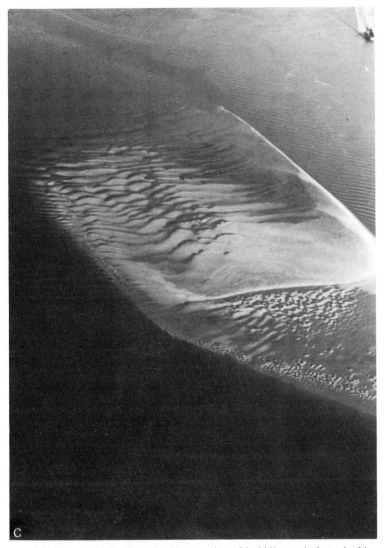

Fig. 18. Morphology and bedforms of the Brander Plate region. (a) Oblique airphoto looking north-east across Brander Plate (a), Brander Plate West (b) and Brander Plate Northwest (c). The east end of the island of Juist is visible on the left side of the photo. Line indicates the location of the PDR profile shown in (b). (b) Bedform profile obtained near low water across the north ramp of Brander Plate. Note the separation between ebb-oriented bedforms left of P and flood-oriented ones to the right. (c) Close-up aerial view of the large transverse bar crest at Brander Plate.

The overall development of an ebb-tidal delta can only be understood by means of a proper integration of the effects of waves and tidal currents.

ACKNOWLEDGMENTS

Funds for this investigation of Norderneyer Seegat were provided by the Office of Naval Research, Coastal Processes Program, through Contract no. N 00014-78-C-0612 to Louisiana State University (Dag Nummedal, Principal Investigator).

The authors would also like to acknowledge the logistics support and ship time provided by the State of Lower Saxony, Germany, through the Forschungsstelle Norderney, Dr Gunter Luck, director. The project would not have been feasible without the assistance, support, and advice received

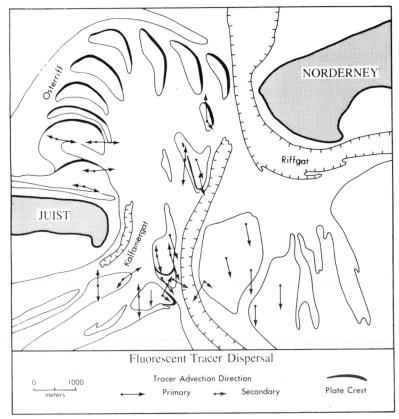

Fig. 19. Directions of dispersal of fluorescent tracer at stations within the central part of the Norderneyer Seegat drainage basin. All tracer experiments were carried out in July 1979, during periods of 'normal' south-west and west winds. Some stations exhibited distinctly bidirectional dispersal, indicated by the presence of a 'secondary' arrow in the diagram.

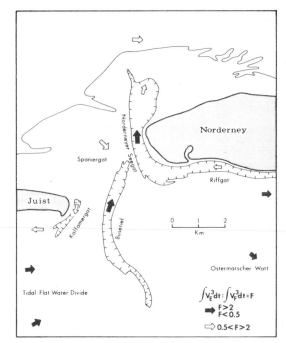

Fig. 20. Map of current dominance in the deep channels of Norderneyer Seegat. Calculations are based on long-term time series current velocity data provided through the courtesy of Forschungsstelle Norderney.

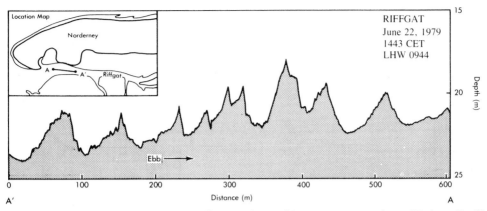

Fig. 21. Profile of large ebb-oriented bedforms in Riffgat, southeast of the entrance to Norderney Harbour. Profiles were obtained with a PDR near low water on 22 June 1979.

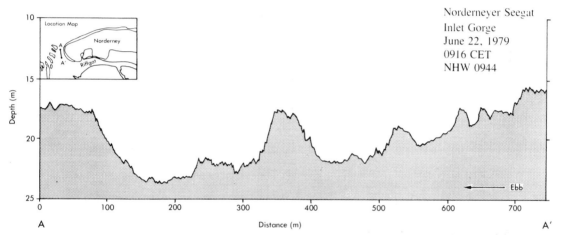

Fig. 22. Profile of bedforms in the Norderneyer Seegat gorge. Small oscillations are due to surface waves.

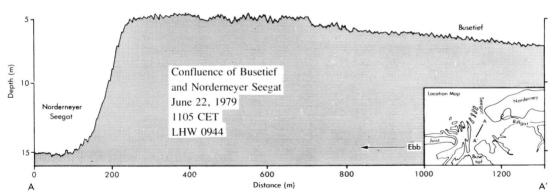

Fig. 23. Profile of Busetief at the confluence with Norderneyer Seegat. The adverse profile to the left followed by a steep bank into Norderney Seegat, suggests a net seaward (left) sediment transport. Small oscillations are due to surface waves.

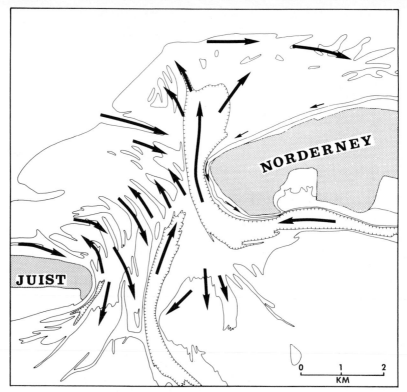

Fig. 24. Integrated sediment dispersal model for Norderneyer Seegat system. This model is based on historic-morphological data, channel and intertidal bedform orientation data, tracer dispersal data and time-series current velocity data. Arrows qualitatively depict net sediment transport directions regardless of which mechanisms operate at any given location.

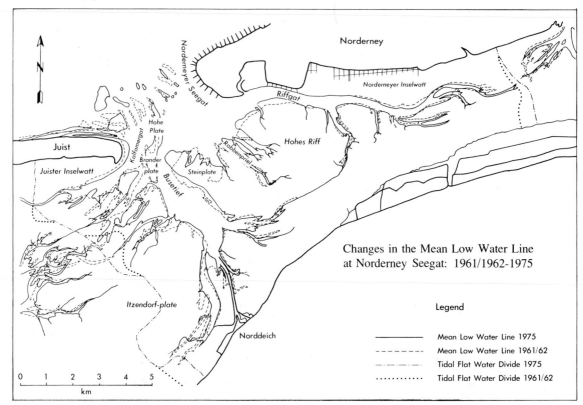

Fig. 25. Historical changes in location of the main channels and shoals within the Norderneyer Seegat system. The map is made from controlled vertical airphoto mosaics obtained in 1961/62 and 1975 (from Ragutzkie, 1978).

from Gunter Luck, Hanz Niemeyer, and the other staff at Forschungsstelle Norderney. Field assistance was provided by Duncan FitzGerald, Amy Maynard, and Mary Penland. Editing was by Mary Penland.

REFERENCES

BOOTHROYD, J.C. & HUBBARD, D.K. (1974) *Bedform development and distribution pattern, Parker and Essex estuaries, Massachusetts.* Misc Paper 1–74. Coastal Engineering Research Center, Ft. Belvoir, Virginia. 39 pp.

BRUUN, P. (1966) *Tidal Inlets and Littoral Drift*, vol. 2. Universitetsforlaget, Oslo. 193 pp.

BRUUN, P. & KJELSTRUP, S. (1977) The wave pump. *Proc. 4th Conf. Port and Ocean Engineering under Arctic Conditions*, pp. 288–308. Memoirs of the University of Newfoundland, St John's.

DAVIES, J.L. (1964) A morphogenetic approach to world shorelines. *Z. Geomorph.* **8**, 27–42.

DAVIS, R.A., FOX, W.T., HAYES, M.O. & BOOTHROYD, J.C. (1972) Comparison of ridge and runnel systems in tidal and non-tidal environments. *J. sedim. Petrol.* **42**, 413–421.

DEAN, R.G. & WALTON, T.L. JR (1975) Sediment transport processes in the vicinity of inlets with special reference to sand trapping. In: *Estuarine Research*, vol. II (Ed. by L. E. Cronin), pp. 129–149. Academic Press, New York.

DEFANT, A. (1958) *Ebb and Flow*. University of Michigan Press, Ann Arbor. 121 pp.

DETTE, H.H. (1977) Ein Vorschlag zur Analyse eines Wellenklimas. *Die Küste*, **31**, 166–180.

Deutsches Hydrographisches Institut (1979) *Hoch- und Niedrigwasserzeiten für die Deutsche Bucht*. Eckardt & Messtorff, Hamburg. 111 pp.

FINLEY, R.J. (1976) Hydraulics and dynamics of North Inlet, South Carolina, 1974–75. *General Investigation of Tidal Inlets*. Rep. no. 10. Coastal Engineering Research Center, Ft. Belvoir, Virginia. 188 pp.

FITZGERALD, D.M. (1976) Ebb-tidal delta of Price Inlet, South Carolina: geomorphology, physical processes, and associated inlet shoreline changes. In: *Terrigenous Clastic Depositional Environments* (Ed. by M. O. Hayes and T. W. Kana), II-143, II-157. Tech. Rep. CRD-11. Department of Geology, University of South Carolina.

FITZGERALD, D.M., HUBBARD, D.K. & NUMMEDAL, D. (1978) Shoreline changes associated with tidal inlets along the South Carolina coast. *Proc. Coastal Zone 1978*, pp. 1973–1994. American Society of Civil Engineers, New York.

FITZGERALD, D.M., NUMMEDAL, D. & KANA, T.W. (1976) Sand circulation pattern of Price Inlet, South Carolina. *Proc. 15th Conf. Coastal Engineering*, II. 1868–1880. American Society of Civil Engineers, New York.

HAYES, M.O. (1975) Morphology of sand accumulation in estuaries: an introduction to the symposium. In: *Estuarine Research*, vol. II (Ed. by L. E. Cronin), pp. 3–22. Academic Press, New York.

HAYES, M.O., ANAN, F.S. & BOZEMAN, R.M. (1969) Sediment dispersal trends in the littoral zone: a problem in paleogeographic reconstruction. In: *Coastal Environments of Northeast Massachusetts and New Hampshire* (Ed. by M. O. Hayes), pp. 290–305. Guidebook, Eastern Section, Society of Economic Paleontologists and Mineralogists, Tulsa.

HAYES, M.O., GOLDSMITH, V. & HOBBS, C.H. III (1970) Offset coastal inlets. *Proc. 12th Conf. Coastal Engineering*, **II**, 1187–1200. American Society of Civil Engineers, New York.

HAYES, M.O., OWENS, E.H., HUBBARD, D.K. & ABELE, R.W. (1973) The investigation of forms and processes in the coastal zone. In: *Coastal Geomorphology*, pp. 11–41. Department of Geology, SUNY Binghamton, New York.

HINE, A.C. (1975) Bedform distribution and migration patterns on tidal deltas in the Chatham Harbor estuary, Cape Cod, Massachusetts. In: *Estuarine Research*, vol. II (Ed. by L. E. Cronin), pp. 235–252. Academic Press, New York.

HOMEIER, H. & KRAMER, J. (1957) Verlagerung der Platen im Riffbogen von Norderney und ihre Anlandung an den Strand. *Jber. Forschungsstelle Norderney*, **8**, 37–60.

HOMEIER, J. & LUCK, G. (1970) Untersuchung morphologischer Gestaltungsvorgänge im Bereich der Accumer Ee als Grundlage für die Beurteilung der Strand und Dünenentwicklung im Westurn und Nordwesten Langeoogs. *Jber. Forschungstelle Norderney*, **22**, 7–42.

HUBBARD, D. K. (1975) Morphology and hydrodynamics of the Merrimack River ebb-tidal delta. In: *Estuarine Research*, vol. II (Ed. by L. E. Cronin), pp. 253–266. Academic Press, New York.

HUBBARD, D.K., BARWIS, J.H. & NUMMEDAL, D. (1977) Sediment transport in four South Carolina inlets. *Proc. Coastal Sediments 1977*, pp. 582–601. American Society of Civil Engineers, New York.

HUBBARD, D.K., OERTEL, G. & NUMMEDAL, D. (1979) The role of waves and tidal currents in the development of tidal-inlet sedimentary structures and sand body geometry: examples from North Carolina, South Carolina, and Georgia. *J. sedim. Petrol.* **49**, 1073–1092.

JARRETT, J.T. (1976) Tidal prism–inlet area relationships. *General Investigations of Tidal Inlets*. Rep. no. 3. Coastal Engineering Research Center, Ft. Belvoir Virginia. 32 pp.

KLEIN, G. DE V. (1970) Depositional and dispersal dynamics of intertidal sandbars. *J. sedim. Petrol.* **40**, 1095–1127.

KOCH, M. & LUCK, G. (1974) Untersuchungen zu den Strömungsverhältnissen auf dem westlichen Juister Watt. *Jber. Forschungsstelle Norderney*, **26**, 29–33.

KRAMER, J. (1961) Strömmessungen im Hafengebiet von Norderney. *Jber. Forschungsstelle Norderney*, **13** 67–94.

LONGUET-HIGGINS, M.S. (1970) Longshore currents generated by obliquely incident sea waves. 1. *J. geophys. Res.* **75**, 6778–6789.

LUCK, G. (1970) *Reisefibel*. Forschungsstelle Norderney. 149 pp.

LUCK, G. (1975) Der Einfluss der Schutzwerke der ostfriesischen Inseln auf die morphologischen Vorgänge im Bereich der Seegaten und ihre Einzugsgebiete. *Mitt. Leichtweiss Inst. Braunschweig*, **47**, 1–122.

LUCK, G. (1976a) Protection of the littoral and seabed against erosion. Fallstudie Norderney. *Jber. Forschungsstelle Norderney*, **27**, 9–78.

LUCK, G. (1976b) Inlet changes of the East Friesian Islands. *Proc. 15th Conf. Coastal Engineering*, II, 1938–1957. American Society of Civil Engineers, New York.

LUCK, G. & WITTLE, H.-H. (1973) Erfassung morphologischer Vorgange der ostfriesischen Riffbogen in Luftbildern. *Jber. Forschungstelle Norderney*, **25**, 33–54.

MADDOCK, J. Jr (1969) The behavior of straight open channels with movable beds. *Prof. Pap. U.S. geol. Surv.* **622**-A. 70 pp.

NIEMEYER, H.D. (1978) Wave climate study in the region of the East Friesian Islands and coast. *Proc. 16th Conf. Coastal Engineering*, I, 134–151. American Society of Civil Engineers, New York.

NUMMEDAL, D. & FISCHER, I.A. (1978) Process-response models for depositional shorelines: the German and the Georgia Bights. *Proc. 16th Conf. Coastal Engineering*, II, 1215–1231. American Society of Civil Engineers, New York.

NUMMEDAL, D. & HUMPHRIES, S.M. (1978) Hydraulics and dynamics of North Inlet, South Carolina, 1975–76. *General Investigations of Tidal Inlets*, Rep. no. 16. Coastal Engineering Research Center, Ft. Belvoir, Virginia. 214 pp.

NUMMEDAL, D., OERTEL, G.F., HUBBARD, D.K. & HINE, A.C. (1977) Tidal inlet variability—Cape Hatteras to Cape Canaveral. *Proc. Coastal Sediments* 1977, pp. 543–562. American Society of Civil Engineers, New York.

NUMMEDAL, D. & STEPHEN, M.F. (1978) Wave climate and littoral sediment transport, northeast Gulf of Alaska. *J. sedim. Petrol.* **48**, 359–371.

O'BRIEN, M.P. (1976) Notes on tidal inlets on sandy shores. *General Investigations of Tidal Inlets*, *Rep. no.* 5, Coastal Engineering Research Center, Ft. Belvoir, Virginia. 26 pp.

OERTEL, G.F. (1972) Sediment transport on estuary entrance shoals and the formation of swash platforms. *J. sedim. Petrol.* **42**, 857–863.

OERTEL, G.F. (1975) Ebb-tidal deltas of Georgia estuaries. In: *Estuarine Research*, vol. II (Ed. by L. E. Cronin), 267–276. Academic Press. New York.

OERTEL, G.F. (1977) Geomorphic cycles in ebb deltas and related pattern of shore erosion and accretion. *J. sedim. Petrol.* **47**, 1121–1131.

PENLAND, S. (1979a) Influence of a jetty system on tidal inlet stability and morphology: Ft. George Inlet, Florida. *Proc. Coastal Structures 1979*, pp. 665–689. American Society of Civil Engineers, New York.

PENLAND, S. (1979b) *The influence of the St John's River jetty system on the morphology, sediment transport pattern and stability of Fort George Inlet, Florida.* Unpublished M.Sc. Thesis, Louisiana State University. 196 pp.

PRICE, W.A. (1955) *Correlation of shoreline type with offshore bottom conditions.* Report for contract N7 no. 48706. Office of Naval Research, Geography Branch. 19 pp +charts.

RAGUTZKIE, G. (1978) Die Wattsedimente im Einzugsbereich des Norderneyer Seegats. *Jber. Forschungsstelle Norderney*, **29**, 175–204.

ULRICH, J. (1973) *Die Verbreitung submariner Riesen- und Grossrippeln in der Deutschen Bucht.* Erganzungsheft zur Deutschen Hydrographischen Zeitschrift Reihe B(4°), Nr 14. Hamburg. 31 pp.

U.S. Naval Weather Service Command (1974) *Summary of Synoptic Meteorological Observations. Western European Coastal Marine Areas.* National Climatic Center, NOAA, Ashville, N.C.

WALTON, T.L. & ADAMS W.D. (1976) Capacity of inlet outer bars to shore sand. *Proc. 15th Conf. Coastal Engineering*, pp. 1919–1937. American Society of Civil Engineers, New York.

Spec. Publs int. Ass. Sediment. (1981) **5**, 211–219

New dates on Holocene sea-level changes in the German Bight

GERHARD LUDWIG*, HELMUT MÜLLER** *and* HANSJÖRG STREIF*

*Niedersächsisches Landesamt für Bodenforschung, Stilleweg 2, D-3000 Hannover 51, F.R.G. and
**Federal Institute for Geosciences and Natural Resources, Stilleweg 2, D-3000 Hannover 51, F.R.G.

ABSTRACT

From 604 shallow boreholes drilled in the German North Sea sector 50 were selected to obtain information about the development of the North Sea basin during the Holocene.

Based on palynological and ^{14}C data, conclusions were drawn as to the development of the facies and the rise in sea-level. Using basic data already published, a graph of the relative sea-level rise was developed. The graph covers the depth zone between 46 and 10 m below German Admiralty Datum and indicates a continuous and steep rise of the sea-level during the period between 8600 and 7100 BP. Additional data from the coastal zone demonstrate a slowly rising sea-level after 6500 BP. Different concepts on dynamics of the sea-level rise and the interpretation of sea-level indicators are compared.

INTRODUCTION

In 1975 and 1976, the Geological State Survey of Lower Saxony, Hannover, cooperated with private companies in prospecting the sand and gravel potential of the German North Sea sector. As a side product, sea-level data were collected which supplement previous data obtained from the offshore zone, tidal flats, and marshlands.

In the course of the field investigations, 604 shallow borings were carried out, of which 291 were done with a vibrocorer. Of these, 50 profiles were evaluated with regard to Holocene sea-level fluctuations. Although the borings reached a maximum depth of only 5 m, they yielded some information about the Tertiary and Pleistocene substratum. The locations of the investigated borings are given in Fig. 1. Detailed lithological descriptions of the profiles and the palynological results are published by Ludwig, Müller & Streif (1979).

THE PRE-HOLOCENE SUBSTRATUM

According to palynological studies, three drill holes reached Tertiary deposits. In borehole 96 (block K1) and 105 (block E18) fine sand and peat were found which represent the seaward continuation of the Pliocene, which is exposed along the 'Rote Kliff' of the island of Sylt (Köster, 1974; Führböter *et al.*, 1976). Borehole 178 (block D 10) is an isolated find of Miocene silty sediments.

Pleistocene deposits were found in 18 boreholes. In most cases it was possible to determine the depositional environment from the lithology and sedimentary structures. However, the stratigraphic classification of the units and their correlation with the stratigraphic sequences in the onshore zone causes some problems.

Boulder clay was found in three boreholes (219, block H6, 302, block L3, A8, block M6). The gravel content of the samples was low and insufficient for an evaluation on the basis of clast counting. As the ice sheets of both the Elsterian and the Saalian glaciations covered this part of the southern North Sea, the stratigraphic classification remains doubtful. A Weichselian age, however, can be excluded because

0141-3600/81/1205–0211 $02.00

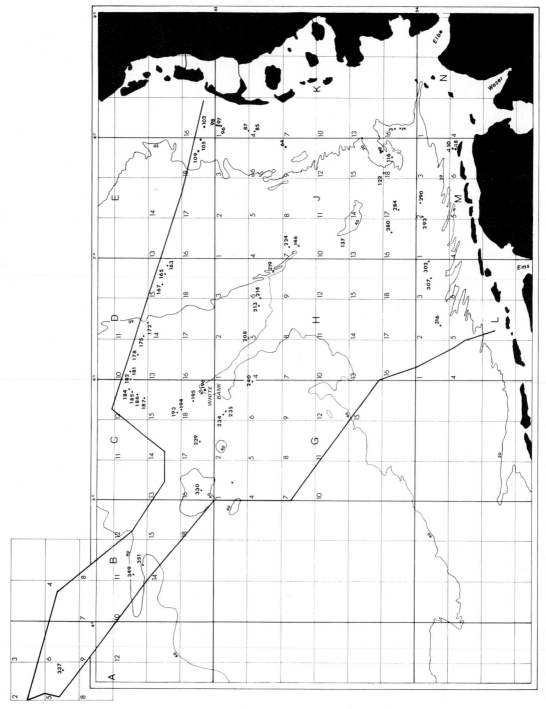

Fig. 1. The German North Sea sector with block numbers and positions and numbers of investigated boreholes. The water depth is given in 20 and 40 m isolines.

the ice margin was in the eastern part of Schleswig-Holstein during the maximum of the most recent glaciation.

Most of the Pleistocene deposits indicate fluvial or glaciofluvial environments and a cold climate. The pollen spectra are rich in *Pinus* or non-arboral pollen. Additionally, they contain numerous Tertiary sporomorphs as well as smaller quantities of Mesozoic and Middle Palaeozoic age. The stratigraphic position of these deposits within the Pleistocene sequences remains doubtful in most cases.

The palynological investigations showed proof of an areal differentiation of these deposits. To the west of the submerged 'Elbe Valley' (Figge, 1977, 1980) the Pleistocene sands contain a relatively high quantity of sporomorphs, late Carboniferous in age. They stem from small coal particles which are incorporated in the Pleistocene sands and which are not present further to the east. The corresponding sediments on the eastern side of the submerged valley contain reworked sporomorphs originating from Neogene, Upper Cretaceous, Devonian and Silurian rocks.

Interglacial sediments of the Eemian were found in two boreholes. Borehole 97 to the west of Sylt (block K1) penetrated freshwater deposits in which a layer of brackish-water sediments is intercalated at a depth of 15·3 m below SKN (SKN is the German Admiralty Datum which corresponds to the mean spring low tide). According to the pollen sequence, the Eemian profile seems to be disturbed. It is not known yet whether this disturbance is due to small-scale solifluction only or to more intensive disturbances and dislocations of the material. Disturbances caused by the drilling method, however, can be excluded. From the floristic development it is obvious that the phase of brackish intercalation did not take place as early as the transition of pollen zone IVa to 'IVb according to Selle (1962); however, it might be considerably younger.

Borehole 316 offshore Borkum (block L2) penetrated clayey and silty deposits at 23·4 m below SKN, which according to diatom analyses can be assigned to a nearshore marine environment.

Late Weichselian deposits were found in boreholes 240 (block G6) and 280 (block J16). They can be palynologically classified in pollen zone IV according to Overbeck (1975), which is the Younger Dryas. A facies analysis for borehole 240 based on pollen and algae demonstrated sediments of a freshwater lake of at least several hectares in size and with a lake level of 38 m below SKN. Previous information about Pleistocene sediments in the German North Sea sector comes from Sindowski (1970).

THE HOLOCENE DEPOSITS

Holocene marine deposits were found in nearly all of the 604 boreholes except for a very few places where erosion is taking place. With regard to the topic of this publication, the interest is mainly focused on non-marine deposits and the non-marine/marine boundary, which can be used for the reconstruction of sea-level changes.

The depth of the samples has been determined in the following way. The water depth, measured with an echo sounder, has been corrected by the Deutsche Hydrographische Institut, Hamburg, and transferred to the German Admiralty Datum SKN, which corresponds to the mean spring low tide. According to the regionally changing tidal amplitude, the SKN is not a horizontal plane, but an undulating surface. At station A10 off Wangerooge SKN is lying 1·65 m below the German Reference Datum NN, which is about mean sea-level. According to the seaward-decreasing tidal amplitude, all other SKN values given in this publication are lying at a smaller vertical

Table 1. Borehole numbers, positions of the borings (shown in Fig. 1), depth interval of the samples below German Admiralty Datum, laboratory numbers, and radiocarbon ages BP

Borehole no.	Position	Depth interval below SKN (m)	Laboratory no. Hv	Radiocarbon age BP
172	55° 19·4′ N 06° 30 0′ E	37·27–37·30	7091	8950± 95
235	54° 56·6′ N 05° 45·0′ E	37·98–38·00 37·09–38·11	7095 7094	8190±140 8485±125
A10	53° 51·4′ N 07° 50·7′ E	21·43–21·59 21·83–22·08 22·42–22·74	8600 8601 8602	7540± 80 7980± 60 7960±205

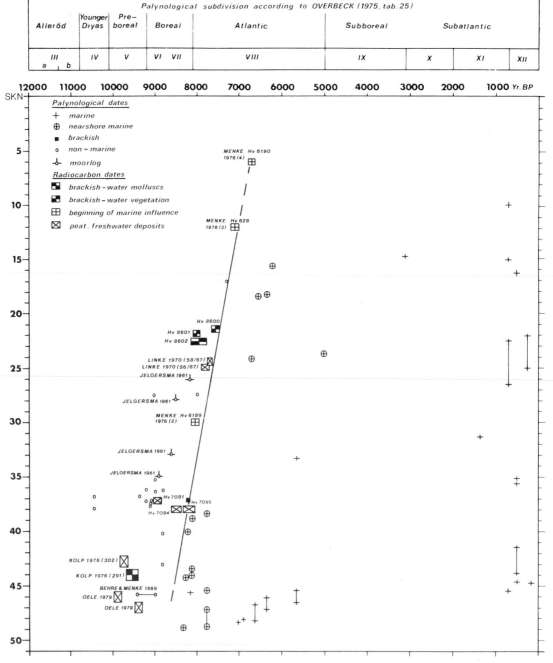

Fig. 2. Time–depth graph of the relative rise in sea-level. The depth is given in metres below German Admiralty Datum (SKN). Palynological subdivision is according to Overbeck (1975) and radiocarbon ages in years BP.

distance below NN. The SKN values plus the vertical distance of the samples from the top of the sedimentary cores give the depths listed in Table 1 and Fig. 2. The depth deviations with respect to SKN can be looked upon as negligible.

Most of the non-marine deposits are described in the Dutch literature as 'lower peat'. Layers of this peat, which are of the same age, were found at different depth below sea-level. In the White Bank region, Preboreal and Boreal peat were found in several boreholes between 38 and 36 m below SKN, whereas it was found 10 m higher in borehole 290 to the north of Langeoog. The same was true for the Atlantic peat which occurs in borehole 290 at 27 m below SKN and at 17 m in borehole 85 west of Sylt. The simultaneous formation of fen peat at different levels indicates that this process was not governed by changes in sea-level. In the area which at present is occupied by the southern North Sea, peat formation was possible under favourable environmental conditions as early as the late glacial. Bog growth started locally and at any elevation above the contemporary sea-level.

The transgressive overlap of brackish or marine deposits on top of fluvial, limnic or semi-terrestrial facies was palynologically dated in 14 boreholes. In nine boreholes the onset of the transgression was connected with erosion and thus there is a time gap between the phases of non-marine deposition and marine sedimentation. An uninterrupted transition from non-marine to brackish or marine environment was observed in only five boreholes (boreholes 102, 172, 196, 235 and 293).

Palynological data

The basic data of the palynological investigations were dealt with by Ludwig *et al.* (1979). They allowed a classification in four facies units such as non-marine, brackish, marine nearshore, and fully marine environments. The term 'marine nearshore environment' was used for marine sediments in which, however, colonies of the freshwater algae *Pediastrum* are represented by as much as 3 % compared with the total pollen. A graphical representation of the palynological results is given in the time–depth graph of Fig. 2. The stratigraphical subdivision of the late glacial and the Holocene on the basis of pollen analyses follows Overbeck (1975, tables 24 and 25).

If we look at the palynological finds in terms of their age, it becomes obvious that the non-marine

samples frequently occur in pollen zones IV to VII according to Overbeck (1975), that is from Younger Dryas to Boreal. A few non-marine samples can also be classified in zone VIII, the Atlantic. The brackish and the nearshore marine samples fall in pollen zones VII and VIII, indicating that brackish and marine influences set in with the Boreal. A few fully marine samples occur in zone VII, but the dominant part of it falls in zone VIII and younger zones.

From the depth distribution of the palynological data in Fig. 2, it becomes obvious that the non-marine finds occur concentrated only in the depth zone between 40 and 25 m below German Admiralty Datum (SKN). In contrast, the marine samples cover the whole depth interval between 50 and 10 m below SKN. According to the palynological studies it is obvious that the marine nearshore sediments form relatively thick layers in the core profiles. This demonstrates a high rate of sedimentation during the early phase of the transgression. It can be seen from the sediment sequences in the boreholes that this rate of sedimentation decreased as the transgression advanced towards the present shoreline.

Radiocarbon data

Radiocarbon datings were carried out in addition to the palynological studies. The radiocarbon data and additional information are gathered in Table 1. One group of dates was collected in the depth zone between 38 and 3 m below SKN. The dated material was Phragmites peat which occurred directly below sediments with marine influence. According to the above-mentioned palynological studies, there is no hiatus between the peat and the sedimentary cover. The samples yielded dates between 8950 ± 95 BP (Hv 7091) and 8190 ± 140 BP (Hv 7095). The latter is the most reliable date with respect to the time of onset of marine influences because it was collected directly below the non-marine/marine boundary.

Another group of dates was obtained between 25 and 23 m below SKN (Hv 8600–8602) from rootlets and stalks of *Phragmites communis* which were separated from a sequence of brackish clayey sediments. According to ecological studies by Scheer (1953), *Phragmites* can exist in the brackwater and tidal zone only between 26 cm below and 72 cm above the mean high tide level. Consequently these dates allow a quite precise determination of the mean high water level between 7980 and 7540 BP.

Previous radiocarbon dates come from different sources. Radiocarbon data obtained from the deepest finds of a so-called 'lower peat' were published by Oele in Jelgersma, Oele & Wiggers (1979). The samples were taken from block F14 and 15 of the Dutch North Sea sector and yielded radiocarbon ages of 9935 ± 55 BP and 9445 ± 80 BP (GrN 5758 and 5759). These results are in a good agreement with the results of Behre & Menke (1969), who made a palynological investigation of Kolp's borehole at station 3 (S-Schill Grund). Kolp (1976, plate 1) published a ^{14}C date of '7783 ± 100 BC' obtained from a 'peat gyttja' taken at $-42 \cdot 8$ m depth from a similar site, station 302 (Oyster Ground). All these dates indicate only that at the time of peat formation the sea-level was below the dated horizon.

On the basis of palynological datings and their comparison with ^{14}C dates from the onshore region, Behre & Menke (1969) assumed that the marine ingression set in at -46 m at about 6500 BC. An age of '7618 ± 160 BC' was published by Kolp (1976, plate 1), derived from a layer of brackish-marine molluscs which he had found at 44 m depth at station 291 (Outer Silver Pit-E). Compared with the above-mentioned peat data, the mollusc shells yielded a higher radiocarbon age for the onset of brackish-marine influences on the sedimentation.

Investigations in the coastal zone yielded additional radiocarbon ages. Menke (1976) dated the initial marine influence in several places in the coastal zone of Schleswig-Holstein. The same dates have been published by Behre, Menke & Streif (1979). Of special interest for this study are the points 2, Marne on the Lower Elbe (Hv 6189: 8075 ± 60 BP); 3, Delve Wallen on the Eider (Hv 628: 7115 ± 90 BP); and 4, Meldorfermoor on the Miele (Hv 6190: 6705 ± 60 BP). Linke (1970) published dates of the marine transgressive overlap in the Scharhörn region, outer Elbe. The most reliable dates come from the boreholes 56/67 and 58/67. Freshwater deposits from 24·90 and 24·35 m below NN yielded radiocarbon ages of 7790 ± 90 BP (Hv 2575) and 7720 ± 65 BP (Hv 2143). All the above-mentioned sea-level data are presented in a time–depth graph, on which is depicted the relative sea-level rise (Fig. 2).

SEA-LEVEL VARIATIONS

Although the above-mentioned data from the southern North Sea offer no new aspects for elucidating the history of the Weichselian late glacial, it seems worthwhile to summarize the present knowledge concerning this subject.

Northern and central North Sea

Jansen (1976) and Jansen *et al.* (1979a, b) published some information on relative sea-level changes in the north-western North Sea. According to these authors, the deepest marks of coastal erosion indicate a maximum fall to about 110 m below present sea-level, which was reached between 20,000 and 15,000 BP. A very important result of this study is the evidence that, at least temporarily, there was a connection between the North Sea and the open sea at the time of the Weichselian maximum and late glacial. This is demonstrated by the glaciomarine sediments of the Upper Hills Deposits, the Fladen Deposits and the early post-glacial Witch Deposits.

This result has to be taken into consideration when making a critical review of the working hypotheses of Valentin (1958), Reinhard (1974) and other authors who have postulated different lake levels and changes in the drainage pattern in the southern North Sea region.

Jansen *et al.* (1979b, p. 182) assume a standstill in the late Weichselian/Holocene sea-level rise at about -45 m, quoting a 'world-wide registered level of stagnation within the period 12,000–9000 BP' (Fairbridge, 1961; Curray, 1965; Mörner, 1971). Field evidence for this assumption comes from the East Bank ridges on the western side of the Dogger Bank (Dingle, 1970; Jansen, 1976; Eisma, Jansen & Van Weering, 1979; Jansen *et al.*, 1979b). These sandy ridges, which reach a crest height of 40–45 m below present sea-level, are interpreted as relict structures of tidal ridges, probably formed in the mouth of an embayment. According to their molluscan and foraminiferal assemblages and a comparison with the Upper Witch Deposits, the East Bank Deposits are Holocene in age, and probably younger than 8700 BP (Jansen *et al.*, 1979a, figs 9–10). Support is also given to this assumption by the results of Kolp (1974, 1976, 1977), who mapped the coastline of a submerged lagoonal (i.e. Haff) system in the Oyster Ground region at about -45 m which dates from 9000 BP Kolp concluded that the sea oscillated for a relatively long period around that -45 m level. However, no radiocarbon data are available which would permit a direct age correlation of the formation of the East Bank ridges with specific events in the dynamics of sea-level fluctuations.

Southern North Sea

The new sea-level data, together with previous data are compiled in Fig. 2 in a time–depth graph of the relative sea-level rise for the southern North Sea region.

The existing working hypotheses (Valentin, 1958; Reinhard, 1974) concerning the Weichselian late glacial melt-water lakes and changes in drainage pattern are mainly based on morphological studies. Jansen *et al.* (1979b, p. 180) stated that 'there are no indications of a large inland lake at the Oyster Ground'. The only proof of a late glacial lake to the south of the White Bank comes from the above-mentioned borehole 240 (block G6). Palynological datings demonstrated a freshwater lake which existed during the Younger Dryas and had a lake level of 38 m below SKN. The configuration of the lake, which had an area of at least several hectares, is not known yet, and nothing can be said about its importance as a local or regional drainage basin. In any case, the late glacial sea-level was below −38 m SKN.

A reliable sea-level curve for the southern North Sea region can be constructed only for the depth zone above the 46 m level (Fig. 2). There we have a population of radiocarbon data points which is sufficiently dense between −46 and −35 m and around −25 m. But also at the remaining depths there are some dates which support the general trend. Since the radiocarbon data of Linke (1970) and Menke (1976) are related to the National Reference Plane and not to the German Admiralty Datum, these points appear approximately 1·5 m too high in the sea-level graph.

Jelgersma (1961) published some palynological results of so-called 'moorlogs' (Nilsson, 1948) for the depth zone between −35 and −27 m which date from 8900 to 8300 BP. These dates have also been used in the sea-level graph of Jelgersma (1979, fig. v-12). Additional information comes from archaeological finds which were dealt with by Louwe Kooijmans (1976, fig. 8).

According to the radiocarbon and palynological dates, there was a steep sea-level rise between 8600 and 7100 BP. The average sea-level rise was about 2·1 m per 100 radiocarbon years. This is in good agreement with the results of Menke (1976) and Jelgersma (1979), who published an average value of 2 m. Menke (1976) assumed a continuation of this steep sea-level rise until 6500 BP to a height of about −5 m. According to a more recent oral communication (Menke, 1979, personal communication) this height has to be revised, and −7 m seems to be more reliable. During this early phase, nearly uninterrupted clastic sedimentary sequences were deposited,

and a undirectional upward and landward shifting of marine and brackish facies zones took place.

For the corresponding time interval, Kolp (1974, 1976, 1977) published a staircase-shaped sea-level curve which is based mostly on field evidence from the Baltic Sea, but which has also been transferred to the North Sea. This is in some contradiction to the line in Fig. 2. Kolp developed a curve with a total of eight steps. In the region south of the Dogger Bank, Kolp (1976, plate 1) demonstrated four terrace levels only, one of which ('−80 m terrace') is below the depth interval which is recorded in his sea-level curve. No counterpart of this −80 m level is described for the Baltic Sea so far. The −60, −45 and −30 m levels of the North Sea region were correlated with the Yoldia, Echeneis and Ancylus terraces respectively. The higher terrace levels (Mastogloia, Cypleus, Litorina I, Litorina II, and Limnaea terraces) do seem not to be distinctly developed or preserved.

The genesis and age are known only for the −45 m terrace (Kolp, 1976, pp. 7, plate 1). It has been demonstrated to be an accumulation level. The borings showed a Holocene sequence which comprises a freshwater peat of Preboreal age, a hiatus, and the onset of marine sedimentation in the Boreal (Behre & Menke, 1969; Kolp, 1976). The onset of the marine sedimentation was observed in several sedimentary cores taken from nearly identical depth. These cores showed repeated facies changes between limnic deposits and brackish-marine sediments. This was interpreted (Kolp, 1976, p. 11) as an indicator for a relatively long stagnation of the sea-level at about −45 m. The other levels are only undated planation surfaces at 80, 60 and 30 m depth below present sea-level, which might have formed during the course of sea-level fluctuations, but which also might be relict structures of pretransgressive morphology. Consequently, it seems difficult to correlate them conclusively with specific events of the late Weichselian/Holocene sea-level history of the North Sea.

The new sea-level data only unsystematically fit the sea-level curve of Kolp (1976, plate 2). Furthermore, it can be assumed that the repeated changes in the vertical trend and the velocity of the sea-level movements, which were described by Kolp, represent the regional sea-level changes of the Baltic Sea. In that case, however, they should no longer be regarded as 'the absolute curve of the eustatic sea-level rise' (Kolp, 1976, plate 2).

The coastal zone

The above-mentioned dates come from non-marine organic deposits which occur at the base of the marine Holocene sequence. They date the onset of the marine ingression. In addition, there are organic layers intercalating the brackish-marine sediments, indicating temporary regressive tendencies. Such intercalated layers occur in the typical cyclic sedimentary sequences of the coastal zone. The formation of intercalated peat layers started at about 6500 BP, when the rate of sea-level rise decreased, and came to an end at about 2600 BP. Thereafter, intercalated peats were formed to a very limited extent only until 1600BP in low-lying moist areas, whereas soils were formed on drier locations.

Although the intercalated peats have been used for the lithostratigraphical subdivision of the Holocene since the very beginning of such studies, a controversial discussion is going on concerning their significance for sea-level fluctuations (Menke, 1976; Behre et al., 1979; Linke, 1979; Streif, 1979).

Without a doubt, the environment of fen peat and lagoonal sediments is most sensitive to water level fluctuations, but vertical changes of sea-level are only one component among a great variety of factors which influence the palaeogeographic tendencies of a retreating or prograding coastline. The whole system of changing water quality, water depth, and bog growth is governed by the ecologically relevant hydrology, which is not necessarily related directly to vertical sea-level fluctuations. It is misleading to interpret each regressive overlap of peat on top of clastic brackish or marine deposits as an indicator of a sinking sea-level. Peat formation in this position is quite possible under the influence of a rising sea-level as long as the bog growth rate is slightly greater than the corresponding rate of sea-level rise. Without a doubt, these conditions existed during the phase of slowly rising sea-level between 6500 and 2600 BP.

Within the general trend of a rising sea-level, a temporary fall in sea-level at about 2000 BP is widely believed to be certain. A regional and synchronous soil formation on top of brackish sediments shows proof of that. At the same time, human settlements were established in the flat marshes up to and even beyond the present coastline (Haarnagel, 1961). There are good grounds for supposing a further phase of falling sea-level at about 2800 BP (Preuss, 1979). In addition to this evidence, Menke (1976) concludes from the vegetational development that there was a large number of minor negative oscillations of the mean high water level between 6500 and 1800 BP (Menke, 1976; Behre et al., 1979). These negative oscillations are, however, controversial.

A new approach to explain the formation of intercalated peat layers and soil horizons within the clastic sequences comes from Linke (1979). He has developed a sea-level curve of the mean high water level without negative oscillations. The curve shows a steep rise in sea-level between 7000 and 4500 BP and a slow rise until 2400 BP. Thereafter, it stagnates until as recently as AD 700 and thereafter again rises continuously to its present height. According to Linke, very stormy phases and storm-free phases were superimposed on these sea-level changes. During the very stormy phases (6000–4500 BP, 3500–3300 BP, and 2400 BP to the present time) marine clastic sedimentation occurred at higher elevations, whereas it took place only at lower levels during storm-free phases. In vertical sedimentary sequences, phases of increased storminess are documented as clastic layers, the storm-free intervals as intercalated peat layers and/or soil horizons. Although the new concept of Linke (1979) offers interesting aspects for the interpretation of a number of sedimentary features they are not conclusive for an explanation of the cyclic Holocene deposits in the coastal zone as a whole.

REFERENCES

BEHRE, K.-E. & MENKE, B. (1969) Pollenanalytische Untersuchungen an einem Bohrkern der südlichen Doggerbank. *Beitr. Meereskde.* **24/25**, 123–129, Dtsch. Akad. Wiss., Berlin.

BEHRE, K.-E., MENKE, B. & STREIF, H. (1979) The Quaternary geological development of the German part of the North Sea. In: *The Quaternary History of the North Sea* (Ed. by E. Oele et al.). *Acta Univ. Ups. Symp. Univ. Ups. Annum Quingentesimum Celebrantis*, **2**, 85–113.

CURRAY, J.R. (1965) Late Quaternary history, continental shelves of the United States. In: *The Quaternary of the United States* (Ed. by H.E. Wright and D.G. Frey), pp. 723–735. Princeton University Press.

DINGLE, R.V. (1970) Quaternary sediments and erosional features off the north Yorkshire coast, western North Sea. *Mar. Geol.* **9**, M17–M22.

EISMA, D., JANSEN, J.H.F. & VAN WEERING, T.C.E. (1979) Sea-floor morphology and recent sediment movement in the North Sea. In: *The Quaternary History of the North Sea* (Ed. by E. Oele et al.). *Acta Univ. Ups. Symp. Univ. Ups. Annum Quingentesimum Celebrantis*, **2**, 217–231.

FAIRBRIDGE, R.W. (1961) Eustatic changes in sea level. In: *Physics and Chemistry of the Earth* (Ed. by L.H. Ahrens et al.), **4**, 99–185. Pergamon Press, London.

FIGGE, K. (1974) Sediment distribution mapping in the German Bight (North Sea). *Mém. Inst. Géol. Bassin Aquitaine*, **7**, 253–257.

FIGGE, K. (1980) Das Elbe–Urstromtal im Bereich der Deutschen Bucht (Nordsee). *Eiszeitalter Gegenw.* **30**, 203–211.

FÜHRBÖTER, A., KÖSTER, R., KRAMER, J., SCHWITTERS, J. & SINDERN, J. (1976) Beurteilung der Sandvorspülung für die künftige Stranderhaltung am Westrand der Insel Sylt. *Die Küste*, **29**, 23–95.

HAARNAGEL, W. (1961) Die Marschen im deutschen Küstengebiet der Nordsee und ihre Besiedlung. *Ber Landeskde.* **27**, 204–219.

JANSEN, J.H.F. (1976) Late Pleistocene and Holocene history of the northern North Sea, based on acoustic reflection records. *Neth. J. Sea Res.* **10**, 1–43.

JANSEN, J.H.F., DOPPERT, J.W.C., HOOGENDOORN-TOERING, K., DE JONG, J. & SPAINK, G. (1979a) Late Pleistocene and Holocene deposits in the Witch and Fladen Ground area, northern North Sea. *Neth. J. Sea Res.* **13**, 1–39.

JANSEN, J.H.F., VAN WEERING, T.C.E. & EISMA, D. (1979b) Late Quaternary sedimentation in the North Sea. In: *The Quaternary History of the North Sea* (Ed. by E. Oele *et al.*). *Acta Univ. Ups. Symp. Univ. Ups. Annum Quingentesimum Celebrantis*, **2**, 175–187.

JELGERSMA, S. (1961) Holocene sea level changes in the Netherlands. *Meded. geol. Sticht. C*, **VI**, No. 7, 100 pp.

JELGERSMA, S. (1979) Sea-level changes in the North Sea basin. In: *The Quaternary History of the North Sea* (Ed. by E. Oele *et al.*). *Acta Univ. Ups. Symp. Univ. Ups. Annum Quingentesimum Celebrantis*, **2**, 233–248.

JELGERSMA, S., OELE, E. & WIGGERS, A.J. (1979) Depositional history and coastal development in the Netherlands and the adjacent North Sea since the Eemian. In: *The Quaternary History of the North Sea* (Ed. by E. Oele *et al.*). *Acta Univ. Ups. Symp. Univ. Ups. Annum Quingentesimum Celebrantis*, **2**, 115–142.

KÖSTER, R. (1974) Geologie des Seegrundes vor den Nordfriesischen Inseln Sylt und Amrum. *Meyniana*, **24**, 27–41.

KOLP, O. (1974) Submarine Uferterrassen in der südlichen Ost- und Nordsee als Marken eines stufenweise erfolgten holozänen Meeresspiegelanstiegs. *Balthica*, **5**, 11–40.

KOLP, O. (1976) Submarine Uferterrassen der südlichen Ost- und Nordsee als Marken des holozänen Meeresanstieges und der Überflutungsphasen der Ostsee. *Petermanns Geogr. Mitt.* **1**, 1–23.

KOLP, O. (1977) Die Beziehungen zwischen dem eustatischen Meeresanstieg, submarinen Terrassen und den Entwicklungsphasen im Holozän der Ostsee (Diskussion einer eustatischen Kurve). *Z. geol. Wiss.* **7**, 853–870.

LINKE, G. (1970) Über die geologischen Verhätnisse im Gebiet Neuwerk/Scharhörn. *Hamburger Küstenforsch.* **17**, 17–58.

LINKE, G. (1979) Ergebnisse geologischer Untersuchungen im Küstenbereich südlich Cuxhaven—ein Beitrag zur Diskussion holozäner Fragen. *Probl. Küstenforsch.* **13**, 39–83.

LUDWIG, G., MÜLLER, H. & STREIF, H. (1979) Neuere Daten zum holozänen Meeresspiegelanstieg im Bereich der Deutschen Bucht. *Geol. Jb.* **D32**, 3–22.

MENKE, B. (1976) Befunde und Überlegungen zum nacheiszeitlichen Meeresspiegelanstieg (Dithmarschen und Eiderstedt, Schleswig-Holstein). *Probl. Küstenforsch.* **11**, 145–161.

MÖRNER, N.A. (1971) The Holocene Eustatic sea level problem. *Geol. Mijnb.* **50**, 699–702.

NILSSON, T. (1948) Versuch einer Anknüpfung der postglazialen Entwicklung des norddeutschen und niederländischen Flachlandes an die pollenfloristische Zonengliederung Südskandinaviens. *Meddn. Lunds geol.-miner. Instn.* **112**, 79.

OVERBECK, F. (1975) Botanisch geologische Moorkunde unter besonderer Berücksichtigung der Moore Nordwestdeutschlands als Quellen zur Vegetations, Klima- und Siedlungsgeschichte. Neumünster, Wachholtz. 719 pp.

PREUSS, H. (1979) Die holozäne Entwicklung der Nordseeküste im Gebiet der östlichen Wesermarsch. *Geol. Jb.* **A53**, 3–85.

REINHARD, H. (1974) Genese des Nordseeraumes im Quartär. *Fennia*, **129**, 95 pp.

SCHEER, K. (1953) Die Bedeutung von Phragmites communis Trin. für die Fragen der Küstenbildung. *Probl. Küstenforsch.* **5**, 15–25.

SELLE, W. (1962) Geologische und vegetationskundliche Untersuchungen an einigen wichtigen Vorkommen des letzten Interglazials in Nordwestdeutschland. *Geol. Jb.* **79**, 295–352.

SINDOWSKI, K.-H. (1970) Das Quartär im Untergrund der Deutschen Bucht (Nordsee). *Eiszeitalter Gegenw.* **21**, 33–46.

STREIF, H. (1979) Cyclic formation of coastal deposits and their indications of vertical sea-level changes. *Oceanis*, **5**, 303–306.

VALENTIN, H. (1958) Die Grenze der letzten Vereisung im Nordseeraum. *Verh. 30. Geographentag Hamburg 1955*, 359–366.

Spec. Publs int. Ass. Sediment. (1981) **5**, 221–227

Sedimentary events during Flandrian sea-level rise in the south-west corner of the North Sea

B. D'OLIER

Department of Geology, City of London Polytechnic,
Walburgh House, Bigland Street, London E1 2NG, U.K.

ABSTRACT

By the extensive use of reflection seismic profiler, side-scan sonar, echo-sounder, corer and grab sampler a detailed picture of the initial surface topography over which the Flandrian transgressive sea migrated has been drawn up. The positions of several of the tidal sand ridges are shown to be directly related to this model of a partly modified drowned landscape. Their present-day locations can be expressed in terms of transitory island sites, coalesced beaches, confluent channel bars, and changing coastline positions.

The presence of a tectonically stable zone from the Thames Estuary to the Belgian coast is suggested as a factor to be considered when comparing apparent Flandrian sea-level rise data from different parts of the southern North Sea.

INTRODUCTION

That the present southern North Sea floor represents an old land surface only partially altered by recent marine activity has been evident to many workers in the area. Edelman, 1933; Jelgersma, 1961, etc. This postulation is certainly correct for that part of it which includes the approaches to the Thames Estuary and adjacent parts of the North Sea between 51° 10′ N and 52° N and 1° E to 2 10′ E. D'Olier (1972, 1975) has presented evidence showing that the Cretaceous Chalk and Eocene sands and clays of that area still have clearly incised within them the channels, terraces and related features of a pre-Holocene drainage system. Additionally glacially incised features such as tunnel valleys can also be recognized. The Flandrian transgressive sea therefore flooded across a surface with a considerable relief, invading valleys and converting them temporally to tidal channels. As the transgression continued, the valley sides and eventually in some places the watershed areas were covered, while a tidal

system similar to that operating today became the dominant agent affecting sedimentary processes. Today the primary sedimentary depositional forms are elongate tidal sand ridges, and whilst these vary in size and in position relative to the present coastline, in some cases recently investigated they can be shown to relate to transient coastlines and to features of the pre-Flandrian morphology. In addition it is noted that caution should be exercised when considering data relating to Flandrian sea-level rise from separate areas in the southern North Sea. It is suggested that a zone of relative tectonic stability might well persist from the Thames Estuary to Belgium (D'Olier, 1975). If further work confirms this stable zone then differences of apparent sea-level rise in various parts of the southern North Sea may be explained.

METHODS

Several hundred kilometres of seismic profiler track using a variety of sound sources have been run. Sound sources include Huntec and EG and G Boomer, EG and G 'low power' sparker, Kelvin

Hughes and EG and G side-scan sonar and various precision echo-sounder equipment. In addition some short gravity and vibro-cores, a shell and auger core, and many sea-bed surface sediment samples have been taken.

PRESENTATION AND DISCUSSION OF DATA

It is evident from a study of the bathymetric charts of the area supplemented, where there is a moderate to thick sediment cover, by data obtained by seismic profiler, that the Flandrian transgressive sea first entered the southern North Sea through the Dover Straits. Leading from the Dover Straits northwards between the Kent coast and Goodwin Sands to the west and the South Falls Sandbank to the east is a valley deeply incised into the upper chalk. The valley bottom lies at approximately 54 m below chart datum and on its sides are a series of terrace levels. These terraces can be clearly traced on the western side where they are cut into the chalk, but are less clear on the eastern side where the softer Eocene sands and clays have been partially modified by more recent marine action. Initially this proto-Rhine/Thames valley became a tidal inlet. Fluvial material was sorted and reworked, some of it to form beaches, while later, as the sea invaded the flat terraces on the edge of the valley, these beach ridges formed the cores for offshore bars. Zenkovitch (1957) suggests just such a mechanism for bar origin. Both the Goodwin Sands and the South Falls sandbar owe their origin to this early stage of the transgression. Seismic profiling reveals some evidence in places for this though the acoustic contrast between the early 'beach' material and the bulk of the material added at a later date is only slight and only locally apparent.

As the transgression continued the valley of the proto-Thames which ran west to east as a tributary of the proto-Rhine became in its turn a tidal inlet (Fig. 4). This valley is now largely infilled with sediment. Sufficient seismic profile evidence has accrued to indicate that this main channel at least in part became an ebb-dominated tidal inlet with sand-bars forming and growing over fluvial coarse sand and flint lag gravels (Fig. 2) while other sections indicate a flood-dominant component (Fig. 3). It would seem that the ebb and flood currents took different pathways within this large river channel/tidal inlet much as they do at the present day in confined tidal inlets (Robinson, 1960).

These features have subsequently been covered by sheet sands brought in and deposited over them by the tidal current system that developed when the northern and southern parts of the North Sea were joined and the standing wave oscillation of the northern part became dominant.

At later stages of the transgression various elements of the pre-transgressive morphology have exerted an influence upon sedimentation. This is apparent when considering some of the larger 'inner' estuarine sandbanks such as Long Sand, Sunk Sand and the Barrows (Fig. 1). However, consideration here will be given to three sand banks in the Thames Estuary approaches, the Kentish Knock, Galloper and the Unnamed Bank (Fig. 1).

The Kentish Knock

This sandbank is constructed largely of fine- to medium-grained, well-sorted sands. It lies at the present day on a 'high' of Eocene London clay, the latter being immediately overlain by a coarse sand and flint gravel layer which fills in irregularities within the surface and locally adds an additional feature to the relief. To the south and south-east a series of terrace steps descend into the position of the buried valley of the Thames whilst immediately to the east, to the west in the Knock Deep, and to the north are the infilled sites of tributary stream channels that once flowed into the proto-Thames (Fig. 4).

The site of the bank was therefore almost completely surrounded by the low ground of the previous valley systems and became the location of a series of coalescing beaches of coarse sand and gravel as transgression took place. These deposits acted as the core for the later growth of the sandbar. Growth was complex, but seismic profiles indicate that both the flood- and the ebb-tidal currents were instrumental in this later stage of formation.

The Galloper

This bank lies on the south-eastern corner of a wide, partially peneplaned surface of London clay (Fig. 1). This surface, which lies at approximately 25 m below Chart Datum (C.D.), is bounded to the east at approximately 2° 5′ E by a deep, where present-day water depths of between 35 and 60 m are to be found. To the north similar depths are encountered, while to the west around 1° 45′ E are two deeply incised north–south-trending valleys where depths up to 60 m below C.D. are found. These two

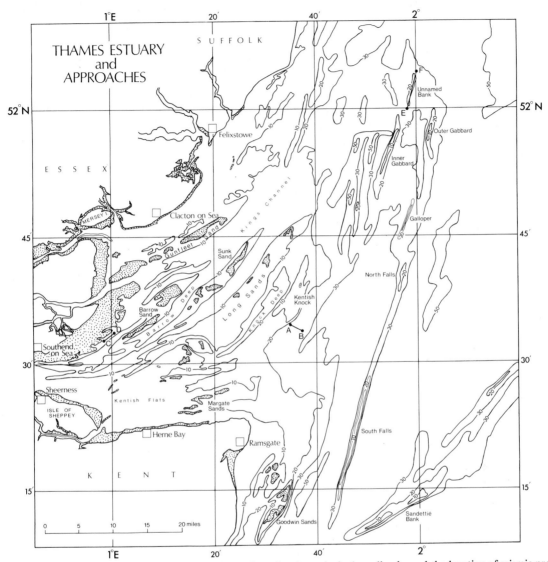

Fig. 1. Map of the Thames Estuary and its approaches, showing principal sandbanks and the location of seismic profiles referred to in the text.

valves are believed to have been formed subglacially and are similar in form to the tunnel valleys of East Anglia and many parts of the North Sea (Woodland, 1970). Also situated on this shallower surface are three other sandbars, the Inner and Outer Gabbards and the Unnamed Bank.

Seismic profiling has revealed the shape of the London clay surface morphology underlying the Galloper sandbar. This shows an elongate core of London clay some 6 m higher than the surrounding surface and around which the sandbar is located

(Fig. 5B). The present-day sandbar surface morphology (Fig. 5A) reflects the shape of this underlying core, the two shallowest sections of approximately 5 m or less in depth being situated over the highest parts of the core, which are around 24 m below Chart Datum (Fig. 5). A small swatchway which crosses the central area of the sandbank at 1° 57′ 40″ E, 51° 46′ 30″ N reflects a depression in the London clay core at that same position. This relationship between present sandbank surface morphology and underlying morphology has been noted

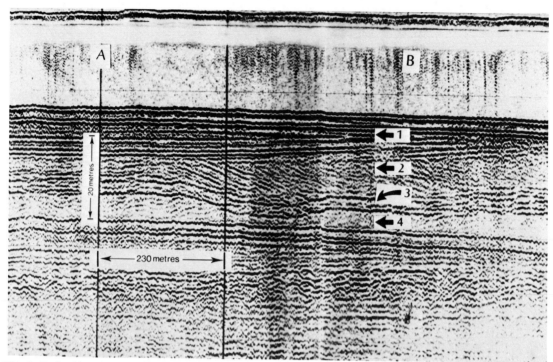

Fig. 2. Reflection seismic profile A–B of an ebb-dominated tidal channel sand within the proto-Thames buried channel. Layer 1, sea sand derived from the north-east; layer 2, ebb-dominated channel bar sand; layer 3, coarse sand and gravel lag; layer 4, London clay.

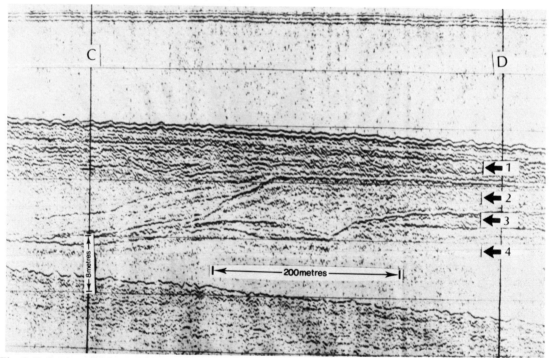

Fig. 3. Reflection seismic profile C–D of a flood-dominated channel bar situated in the proto-Thames buried channel. Layers 1–4 as for Fig. 2.

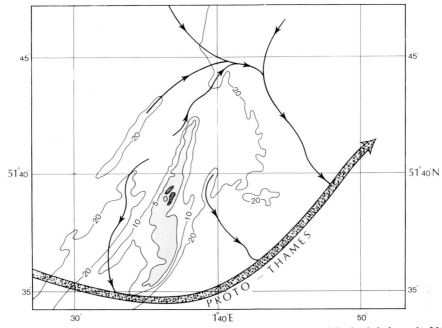

Fig. 4. Kentish Knock Sandbank showing positions of pre-Holocene buried or partially buried channels. N.B. Modern submarine contours in metres below chart datum.

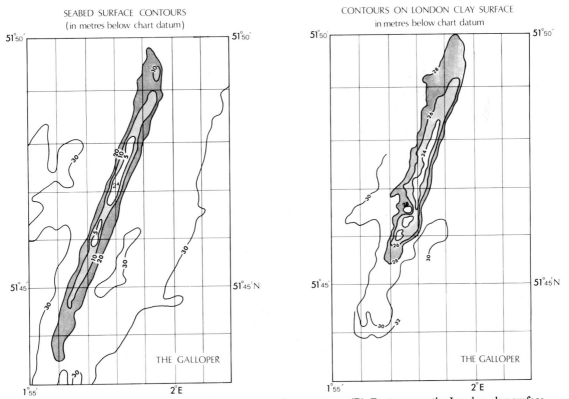

Fig. 5. Galloper Sandbank. (A) Present-day sea-bed surface contours. (B) Contours on the London clay surface.

226   *B. D'Olier*

by the author in many places within the sandbank complex of the Thames Estuary approaches. Where the ebb and flood tides are of closely equal strength and duration, as in the area of the Galloper sandbar, then the relationship is undistorted; where the ebb or flood tide dominates then there is often an offset reflecting that dominance.

The Unnamed Bank

This bank is the most northerly of the four sandbars situated on the relatively shallow partially peneplaned surface of London clay (Fig. 1). Its position relative to the deeper water to the east seems to epitomize a relationship shown by all four of these banks. They all lie in shallow water of approximately 25 m depth, close to deeper water either to the east or to the west. They all lie parallel or closely parallel to the edge of the shallow platform, and all tend to lie in echelon and only slightly overlapping. The Unnamed Bank lies close to and sometimes overlapping a high knoll of London clay (Fig. 6). It would seem that the initial stages in the formation of this bank might have involved a small beach bar

which advanced across the relatively flat surface from the east until it met with the small knoll around which it became stabilized. In addition another beach bar advancing from the deep water to the west might have compounded the beach bar core. Where each of the four banks are situated would seem to depend on where this initial core material came to rest which in turn depended on the position of London clay knolls, the amount of beach bar material being moved and the relative competency of wave action and the early tidal action to move it.

The view that some of the sandbanks of the southern North Sea might have a core is contrary to the view of Houbolt (1968). This is perhaps explained when one considers that he did not run closely spaced transverse and longitudinal seismic profiles and that he did not run profiles over or close to the shallowest sections of the sandbanks where the core, if present, is most likely to be located.

Tectonic stable zone

Boswell (1915) noted how deposits in Essex and Suffolk, ranging in age from Cretaceous to Pleisto-

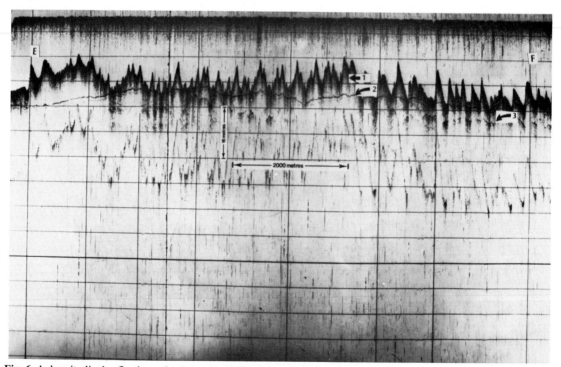

Fig. 6. A longitudinal reflection seismic profile E–F of the Unnamed Bank showing a London clay knoll underlying recent marine sands. Layer 1, Recent marine fine sand; layer 2, surface of London clay; layer 3, reflector in the London clay.

cene, thinned over an 'axis of instability', the axial trace of which trended north-west to south-east. The effects of this axis he noted ranged from the region of Mersea Island in the south-west to beyond Felix- stowe in the north-east. Seismic profiling has sub- stantiated Boswell's view, showing the chalk and the Eocene sands and clays folded into a monoclinal structure with its steep limb facing to the south-west and extending out from the Essex coast at least as far as the Kentish Knock Sandbank. This feature is sometimes faulted on the steep limb. Recent work indicates that this axis of folding and faulting might well continue intermittently across the southern North Sea to Belgium. Rossiter (1972) showed that the sea-level at present was rising at Southend, which is situated to the south-west of the axis of instability and at the entrance to the River Thames (Fig. 1), at the rate of 3·4 mm yr^{-1}, whilst at Felixstowe situated on the axis and some 65 km to the north-east, the sea level appeared to be rising at only 1·6 mm yr^{-1}, indicating a differential subsidence at the rate of 1·8 mm in the Southend area. Devoy (1979) also showed differential subsidence to be operating in the lower reaches of the River Thames. Bridgland (thesis work in progress) also shows evidence as to how the River Thames during the Pleistocene has pro- gressively moved its channel southwards from the area of Clacton-on-Sea to latitude 51° 30′ N before the recent submergence, possibly under the same influence of differential subsidence. It might appear that Boswell's 'axis of instability' is still operating and, if this is the case, features of this kind should be located and their influence taken into account when attempts are being made to evaluate rates of Flandrian sea-level rise in the various countries surrounding the southern North Sea.

CONCLUSIONS

Evidence is presented to show that several tidal sand ridges in the British sector of the southern North Sea owe their position and origin to small elevations of the pre-Flandrian morphology and to beach bar cores. In addition the influence of deep- seated tectonic activity is suggested to account for variable rates of apparent sea-level rise in different parts of the southern North Sea.

REFERENCES

BOSWELL, P.G.H. (1915) Stratigraphy and petrology of the Lower Eocene deposits of the N.E. part of the London Basin. *Q. Jl geol. Soc. Lond.* **71**, 536–591.

DEVOY, R.J.N. (1979) The Thames Estuary: coastal development and the problems of Flandrian land/sea level changes in southern Britain. Abstr. 2. Theme 1 (Holocene sea-level fluctuation). *International Meeting on Holocene Marine Sedimentation in the North Sea Basin.* Texel, the Netherlands.

D'OLIER, B. (1972) Subsidence and sea-level rise in the Thames Estuary. *Phil. Trans. R. Soc.* A **272**, 21–130.

D'OLIER, B. (1975) Some aspects of Late Pleistocene– Holocene drainage of the River Thames in the eastern part of the London Basin. *Phil. Trans. R. Soc.* A **279**, 269–277.

EDELMAN, C.H. (1933) *Petrologische Provinces in het Nederlandsche.* Kwartair, Amsterdam.

HOUBOLT, J.J.H.C. (1968) Recent sediments in the Southern Bight of the North Sea. *Geol. Mijnb.* **47**, 245–273.

JELGERSMA, S. (1961) *Holocene sea-level changes in the Netherlands.* Ph.D. Thesis, Leiden, Mededel. *Geol. Stricht.* C, 6–7.

ROBINSON, A.H.W. (1960) Ebb-flood channel systems in sandy bays and estuaries. *J. Geogr.* **45**, 183–199.

ROSSITER, J.R. (1972) Sea-level observations and their secular variation. *Phil. Trans. R. Soc.* A **272**, 131–139.

WOODLAND, A.W. (1970) The buried tunnel-valleys of East Anglia. *Proc. Yorks. geol. Soc.* **37**, 521–578.

ZENKOVITCH, V.P. (1957) Coastlines. *Tr. Inst. Okeanol. Akad. Nauk. U.S.S.R.* **21**, 240 pp.

Spec. Publs int. Ass. Sediment. (1981) **5**, 229-237

An early Holocene tidal flat in the Southern Bight

D. EISMA,* W. G. MOOK† *and* C. LABAN‡

* *Netherlands Institute for Sea Research, Texel, the Netherlands,*
† *Physical Laboratory, State University, Groningen, the Netherlands and*
‡ *Geological Survey of the Netherlands, Haarlem, the Netherlands*

ABSTRACT

Salinity, temperature and age have been determined for *Cardium edule* shells, collected in the Southern Bight of the North Sea, from their isotopic composition. The results indicate that before *c.* 8100 BP (probably between *c.* 10,000 and 8000 BP) a brackish-water tidal-flat area existed in the present Deep Water Channel (chlorinity *c.* 13‰). Between *c.* 8000 and 7000 BP most of the Southern Bight was a tidal flat, and after *c.* 7000 BP most became fully marine, except for the Thames estuary (which remained an estuary) and an area off the southern Dutch coast that remained brackish until after 6000 BP (probably up to *c.* 5500 BP). The data for the *Cardium edule* shells are in agreement with the curve of sea-level rise in the Southern North Sea, except for two samples. The chlorinity distribution in the Southern Bight between *c* 9300 and 5900 BP clearly shows the influence of the freshwater outflow of the Rhine and Meuse. The growth-temperatures indicate that *Cardium edule* populations have been able to adapt themselves to the prevailing temperature conditions. The sharp increase in yearly average air temperatures in Central England between *c.* 10,000 and 7000 BP coincides with the period during which the Southern Bight was flooded and warm water from the south could reach the Southern North Sea. The *Cardeum edule* shells were transported to some extent but transport occurred only over relatively short distances: the present distribution of the early Holocene *Cardium edule* shells largely reflects the conditions of 9300-6000 years ago.

INTRODUCTION

Brown- or bluish-coloured shells of *Cardium edule* and *Macoma balthica* have been found in the North Sea on the Dogger Bank, in the Oyster Grounds off the Frisian Islands and along the Dutch North Sea coast (Davis, 1925; Pratje, 1929; Stride, 1959; Veenstra, 1965; Eisma, 1965). They were also found in large numbers during a grab-sampling programme, carried out in the Southern Bight in 1964-7. At present both species live abundantly on the tidal flats and in brackish water areas around the Southern North Sea, but offshore are confined to a narrow belt along the coast (Eisma, 1966; Tebble, 1966). It was therefore assumed that the shells found further offshore date from an earlier period when the

Southern North Sea was more brackish. A quick inspection of some *Cardium edule* samples collected in the Southern Bight, using the average number of ribs on the shell (Eisma, 1965; Eisma, Mook & Das, 1976), indicated that the shells had indeed been formed at a lower salinity than occurs at present. In order to gain some insight into the conditions in the Southern Bight during that period, the (average) salinity and temperature during the period of shell growth, as well as their (geological) age, were determined for the *Cardium edule* samples. This species was chosen instead of *Macoma balthica* because the average rib-number would give an independent check on the isotopic determination of salinity (Mook, 1971; Eisma *et al.*, 1976). Chlorinity is taken as a measure of salinity.

0141–3600/81/1205–0229 $02.00

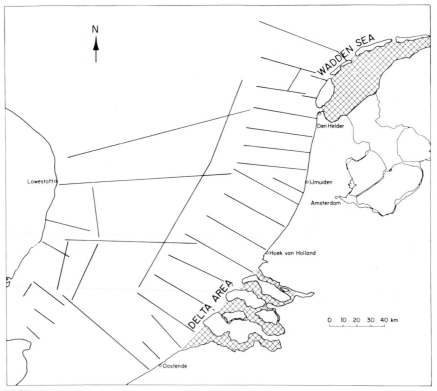

Fig. 1. Traverses where grab samples were collected in the Southern North Sea by Dutch research vessels in 1964–7. The coastal areas with very extensive sampling are cross-hatched.

METHODS

The samples were collected in the Southern Bight in 1964–7 (Fig. 1) with a Van Veen grab that samples the top 15 cm of the bottom sediment. After collecting a subsample from the upper 5 cm for grain-size analysis, the sample was washed through a 2 mm sieve. The material > 2 mm (shells and sometimes gravel) was collected, washed with tap water and dried at room temperature. The *Cardium edule* and *Macoma balthica* shells were removed and counted. For each sample that contained more than 20 *Cardium edule* shells the average rib-number of these shells was determined as a measure of salinity (Eisma, 1965). Then the oxygen and carbon isotope composition of these samples was determined by drilling small portions of a large number of shells in each sample. Several samples were divided into two equal portions of which the average isotopic composition was determined. These duplicates differed by less than a few tenths of a permil. In a few cases single specimens from one sample showed

differences of a few permil. The conventional C-14 age ($t_{\frac{1}{2}}$ = 5568 yr fractionation correction to $\delta^{13}C$ = -25% versus **PDB**) was determined for a number of randomly selected shells in a selected number of samples covering the Southern Bight. From the oxygen and carbon isotope ratios the temperature and salinity during the growth of the shell were determined using the method given in Mook (1971) and Eisma *et al.* (1976). The conventional C-14 ages were corrected for an apparent age of surface ocean water of 400 yr and a lower salinity, taking the C-14 content in river water as 85%, similar to average groundwater. The corrected ages were obtained by comparing the measured C-14 content of the sample, a^{14}, with a calculated C-14 content of the original brackish water, a_0^{14}, in which the shell was formed. The latter follows from a mixing equation using marine values of $a_m^{14} = 100\%$, $\delta^{13} = +1.5\%$ and $Cl_m = 19.3\%$ and freshwater values of $a_f^{14} = 85\%$, $\delta^{13} = -12.5\%$ and $Cl_f = 0\%$:

$$a_0^{14} = \frac{[\Sigma C_f a_f^{14} - \Sigma C_m a_m^{14}]\, Cl + \Sigma C_m a_m^{14} Cl_f - \Sigma C_f a_f^{14} Cl_m}{[\Sigma C_f - \Sigma C_m]\, Cl + \Sigma C_m Cl_f - \Sigma C_f Cl_m} \quad (1)$$

Table 1. Isotopic data of those samples which were dated by C-14: Cl is deduced from δ^{13} and δ^{18}; a_0^{14} is calculated according to eq. 1; a_{0c}^{14} implies the standard correction for isotopic fractionation; $T = -8033 \ln a^{14}/a_{0c}^{14}$, where a^{14} is the measured ^{14}C content; T_c is the conventional C^{14} age

Sample no.	δ^{13} ‰	δ^{18} ‰	Cl ‰	a_0^{14} %	a_{0c}^{14} %	a^{14} %	T y	T_c y	GrN-
Z 250	−2·10	−1·60	13·2	95·3	90·4	38·8	6790	7610±80	8023
283	−4·45	−0·94	11·2	93·7					
354	−2·28	−1·48	13·1	95·2	90·9	28·3	9370	10,130±60	7875
531	−2·00	−1·65	13·3	95·3	90·9	32·5	8260	9030±60	7876
552	−0·83	−0·13	15·7	97·2	92·5	34·3	7960	8585±80	7548
722	−4·16	−1·57	11·1	93·6	89·7	35·7	7400	8265±80	7549
839	−0·27	+0·36	16·5	97·8	93·0	34·9	7870	8450±75	7550
938	−2·13	−0·91	13·7	95·6	91·2	36·0	7460	8200±50	8025
1135	−1·02	−0·02	15·4	97·0	92·3	33·9	8040	8680±50	8026
1302	−3·27	−2·74	11·2	93·7					
1304	−4·52	−4·37	8·6	91·7					
1306	−3·56	−3·69	10·2	92·9	88·8	42·5	5910	6870±90	8024
1309	−3·53	−2·94	10·7	93·3					
1310	−4·86	−3·78	8·8	91·8					
3016	−1·29	−0·66	14·7	96·4	91·8	35·1	7720	8420±50	7879
Ye 1434	−3·50	−2·10	11·3	93·8	89·8	34·4	7700	8570±80	7877
1461	−0·10	+0·71	16·9	98·1	93·2	40·7	6650	7215±45	7878
1463	−0·20	+0·57	16·7	98·0	93·1	28·3	9560	10,150±90	7551
1464	−0·16	+0·70	16·8	98·1	93·2	30·4	8990	9570±55	9040
1465	−0·11	+0·65	16·9	98·1					

(cf. Mook, 1970, p. 185). ΣC denotes the total carbon content of the water, while the subscripts m and f refer to the marine and river-water components. We have arbitrarily chosen $\Sigma C_f = EC_m$. After correcting a_0^{14} for isotope fractionation ($\delta^{13} = -25$ ‰) to a_{0c}^{14} the corrected age is calculated from $T = T = -8033 \ln a^{14}/a_{0c}^{14}$. The results are compiled in Table 1.

For the evaluation of mollusc shell transport during and after the early Holocene transgression a comparison was made with the distribution of Eocene and Plio-Pleistocene fossils present in the grab samples. Determination of the fossils was carried out by G. Spaink (Netherlands Geological Survey).

SALINITY, TEMPERATURE AND AGE

The relation between δ^{13} and δ^{18} is shown in Fig. 2, which gives the salinity of the water and its temperature during shell growth. The empirical curve on the left agrees with a $\Sigma C_f/\Sigma C_m$ ratio = 1 and a δ^{18} value of about -10 ‰ similar to the present Rhine value. The salinity thus found (expressed as chlorinity) agrees well with the salinity determined from the average number of ribs (Fig. 3), although the isotopic method has a greater precision.

The distribution of the isotope salinity is given in Fig. 4(A). It clearly shows the influence of the outflow of the Rhine and Meuse as well as a relatively low salinity for most of the Southern Bight as compared with the present salinity. The C-14 ages (corrected for salinity), however, show that the shells collected in the different parts of the Southern Bight are not of the same age (Fig. 4B).

The oldest samples occur in the Deep Water Channel in the centre of the Southern Bight (9370 and 8260 BP) with chlorinities of 13·1–13·3‰, and off Walcheren (9650 and 8990 BP) with chlorinities of 16·7–16·9‰. The samples collected in the centre of the Southern Bight lie within the area that, according to Netherlands Geological Survey data on peat, was already marine at the beginning of the Holocene. Those collected off Walcheren lie outside this area and, as will be seen below, are also anomalous in other respects.

The younger samples, collected all over the Southern Bight, date from between 8040 and 5919 y BP, indicating that almost the entire Southern Bight became brackish during that period. This agrees very well with the data obtained previously for this area. As can be seen in Fig. 4(B), the ages of the *Cardium edule* shells are younger than those of early Holocene peat layers collected nearby (peat data from Jelgersma, 1961, Kirby & Oele, 1975, and

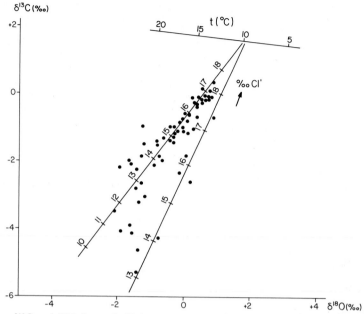

Fig. 2. Relation between δ¹³C and δ¹⁸O for early Holocene *Cardium edule* from the Southern Bight of the North Sea.

Netherlands Geological Survey). In the Thames estuary estuarine conditions were first established around 8900 BP and the area has remained estuarine up to the present time (Greensmith & Tucker, 1973). Along the present Belgian–Dutch coast a marine influence was first felt between 6700 and 6200 BP, when a coastal lake was gradually transformed into a brackish lagoon and tidal flats (Van Straaten,

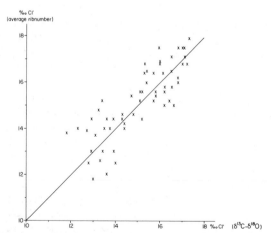

Fig. 3. Relation between chlorinity determined for *Cardium edule* shells from the Southern Bight of the North Sea from the average number of ribs on the shells and from their stable carbon and oxygen isotopes.

1965). Offshore early Holocene tidal flat deposits are still present along the Dutch coast below a surface layer of younger sands of 3–4 m thickness (Oele, 1971).

For most samples the relation between the age of the shells and the water depth at which they were found forms a curve that on average follows the curve of Holocene sea-level rise given by Louwe Kooijmans (1976) and Jelgersma (1979) for the Souther Bight, but lies about 8 m lower (Fig. 5). At present *Cardium edule* lives in the Waddensea on the tidal flats, mainly in the intertidal zone, at chlorinities of 10–16‰, whereas in the coastal North Sea it lives below the zone of strong wave action at 5–15 m waterdepth at chlorinities of 16–18‰. Assuming that the early Holocene *Cardium edule* had a similar distribution, the chlorinity of most samples would indicate an intertidal environment. Dead shells of molluscs that live on tidal flats are easily washed from the flats into gullies and channels, where they are deposited (and often accumulated) at least several metres below sea-level. In the present Waddensea, *Cardium edule* shells can be found in quite large numbers down to c. 50 m water depth in the deepest parts of the tidal inlets. Horizontally, however, transport is confined to one tidal area, involving transport along relatively short distances. The present depth at which the shells were found,

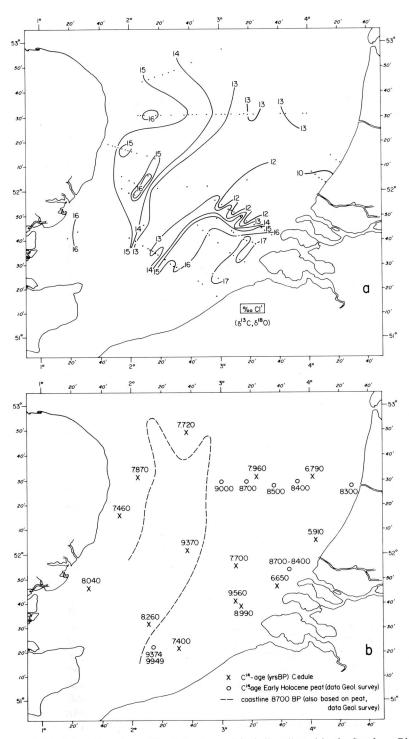

Fig. 4. Distribution of chlorinity (A) and age (B) of *Cardium edule* shells collected in the Southern Bight of the North Sea.

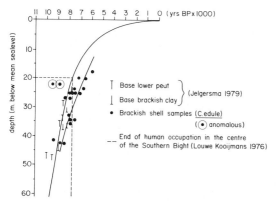

Fig. 5. Relation between the age of *Cardium edule* shells and the water depth at which they were found in the Southern Bight of the North Sea, compared with the curve of sea-level rise during the Holocene and with the period of human occupation in the central part of the Southern Bight.

therefore, is in agreement with an intertidal life-position of the *Cardium edule* shells so that most of the Southern Bight between *c.* 8000 and 7000 BP can be regarded as a tidal flat area, gradually changing into a shallow sea. Off the southern Dutch coast tidal flat conditions remained up to *c.* 5900 BP and probably up to *c.* 5500 BP, when the area along the present Dutch coast gradually became fully marine (Van Straaten, 1965).

Two of the oldest *Cardium edule* samples (nos Ye 1463 and Ye 1464+1465, both collected off Walcheren) are anomalous: they are found at too shallow a depth or their age is too high. This can be due to an anomaly in the carbon isotopic composition, to reworking of older material and (or) to transport of shells. Reworking of older shell material cannot easily explain the anomaly: this would imply either the presence of old early Holocene moraine deposits dating from *c.* 9000 BP at *c.* 60 m depth in that area, whereas these are known not to be present from shallow borings and seismic profiles (unpublished data, Netherlands Geological Survey), or horizontal transport over a considerable distance. Transport of shells was not important during the early Holocene, as will be seen below. A post-depositional change in the carbon isotopic composition, reducing the C-14 content and thus increasing the apparent age of the samples, is unlikely. The salinity, as determined for the stable oxygen and carbon isotopes (16·7–16·9 ‰ Cl′) agrees well with the salinity determined from the average rib-number (17·1–17·5 ‰ Cl′). This is in contradiction

with a post-depositional isotopic change. Furthermore we have no indications for a higher apparent age of the ocean water during the time of deposition. Therefore these two dates are not understood.

The growth temperature of the early Holocene *Cardium edule* shells was *c.* 10 °C (Fig. 2). For recent shells it is 13·5–14·0 °C and for Eemian *Cardium edule* shells more than 15 °C, possibly up to 17 °C (Eisma *et al.*, 1976). This indicates that *Cardium edule* populations are able to adapt shell growth to the prevailing temperature conditions and are not dependent on a fixed temperature to start shell formation. This is also indicated by the higher growth temperature during the Eemian, which is in accordance with the somewhat higher temperatures inferred for that period generally (Van Straaten, 1956). The data of Lamb, Lewis & Woodroffe (1966) show a sharp increase in air temperature in Central England between 10,000 BP and 7000 BP (yearly average increasing from −1 to +12 °C). This is the period when the Southern North Sea was flooded and warmer water could reach the Southern North Sea from the South through the Strait of Dover.

TRANSPORT OF SHELLS

The *Cardium edule* shells were not found in life position but mixed with sand, recent and fossil shells and sometimes gravel. The shells have evidently been reworked and transported to some extent. Fine grained material is known to be transported in suspension through the entire North Sea and material from various sources may be thoroughly mixed (Eisma, 1981), but there are several indications that during the early Holocene transport of mollusc shells has only taken place over relatively short distances:

(*a*) With a few exceptions the data found for the *Cardium edule* shells agree well with the other data for the Southern Bight such as the curve of sea-level rise and the extension of the early Holocene transgression. Such an agreement would not have been found if the shells had been transported over large distances and shell material of different origin had become extensively mixed.

(*b*) The variability in the number of ribs on the *Cardium edule* shells from the Southern Bight is of the same order as for the populations at present living in the Waddensea and the coastal North Sea. In recent populations the number of ribs varies ±2 around the average rib number, with an occasional

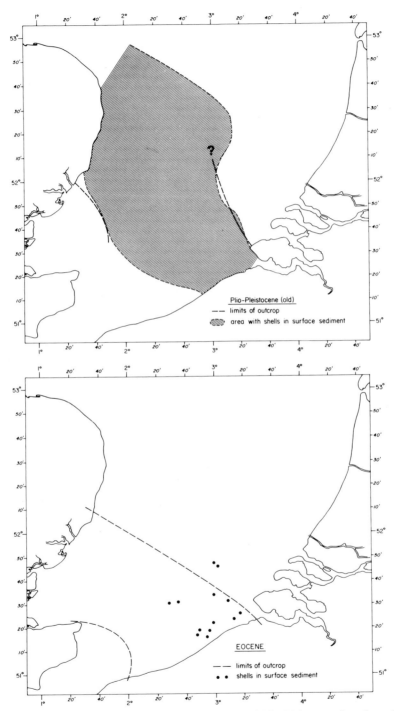

Fig. 6. Relation between the distribution of the outcrops of Eocene and Plio-Pleistocene deposits and the distribution of fossil mollusc shells from these periods in the surface sediment of the Southern Bight.

shell having a rib number that differs more than 2 from the average. The same variability was present in the early Holocene samples. This indicates that mixing of shell populations that grew up at different salinities has been small but does not exclude the possibility that the shell populations are transported over large distances as a whole, without large-scale mixing or sorting.

(*c*) Fossil Eocene and Plio-Pleistocene mollusc shells occur in the same grab samples as the *Cardium edule* shells. Their distribution in the upper 15 cm of the bottom sands in the Southern Bight is given in Fig. 6 and compared with the distribution of Eocene and Plio-Pleistocene deposits on the mainland and below the Recent sea sands that form the present seafloor (Netherlands Geological Survey data). The areas in the surface sediment where the fossils dating from these periods are found almost coincide with the outcrop areas, indicating that shell transport has not been very large during and after the early Holocene transgression.

On the basis of these arguments it can be concluded that the present distribution of the early Holocene *Cardium edule* shells largely reflects the situation of *c.* 9500–6000 y ago. There is, however, sedimentological evidence for large-scale sand transport that cannot easily be reconciled with this conclusion. In a large part of the Southern Bight older Holocene sands with a cold water mollusc fauna are separated by a strong acoustical reflector from overlying younger Holocene sands containing a warm water (Lusitanian) mollusc fauna (Laban & Schüttenhelm, 1981). The reflector is very clear in acoustical records but cannot be correlated with a lithological horizon in sediment cores. The older and younger sands can only be divided on the basis of their mollusc fauna, indicating that a considerable amount of sand with warm water molluscs has been deposited on top of the older sands during the Holocene. Transport was presumably from south to north as is indicated by the gradual decrease in grain size of the top sands in that direction. The warm water fauna consist of very small mollusc shells (*Angulus pygmaeus*) or fragments and were moved with the sand, but the larger and heavier shells, like those of *Cardium edule* and the Eocene and Plio-Pleistocene fossils, may have remained behind when the finer material was gradually moved to the north. The larger molluscs would thus behave like gravel and coarse sands, which show a patchy distribution in the Southern North Sea and have not been transported very much. To explain how

the early Holocene *Cardium edule* shells can still be found in the surface sands above the acoustical reflector, mixed with the warm water fauna, large-scale reworking of the older bottom sands during the period of northward transport has to be assumed. This, however, is likely to have happened, the acoustic reflector indicating the depth to which reworking has taken place.

CONCLUSIONS

Analysis of the salinity and temperature conditions as determined from early Holocene *Cardium edule* shells collected in the Southern Bight indicate that:

(*a*) before *c.* 8100 BP (probably between *c.* 10,000 and 8000 BP) a brackish water area, presumably a tidal flat, existed in the present Deep Water Channel (salinity *c.* 13 ‰);

(*b*) between *c.* 8000 and 7000 BP most of the Southern Bight was a brackish-water, tidal-flat area;

(*c*) after *c.* 7000 BP most of the Southern Bight became fully marine, except for the Thames Estuary (which remained an estuary) and an area off the southern part of the Dutch coast, that remained brackish until after 6000 BP (probably up to *c.* 5500 BP);

(*d*) the influence of freshwater outflow from the Rhine and Meuse rivers is clearly visible in the salinity distribution;

(*e*) the growth-temperature of the early Holocene *Cardium edule* shells was *c.* 10 °C, while for Recent *Cardium edule* it is 13·5–14·0 °C and for Eemian specimens 15 °C up to possibly 17 °C. These temperatures reflect the prevailing temperature conditions in the water, to which the *Cardium edule* populations were adapted. The sharp increase in yearly average air temperatures in Central England between *c.* 10,000 and *c.* 7000 BP coincides with the period when the Southern Bight was flooded and warmer water from the south could have reached the Southern North Sea;

(*f*) the data from the *Cardium edule* shells show good agreement with the curve of sea-level rise in this area, except for two samples;

(*g*) the *Cardium edule* shells were not found in life positions and had been transported to some extent, but transport of mollusc shells during or after the early Holocene transgression occurred only over relatively short distances. The present distribution of early Holocene *Cardium edule* shells therefore largely reflects the conditions of 9500–6000 y ago.

ACKNOWLEDGMENTS

We are indebted to the Royal Netherlands Navy for enabling us to carry out part of the sampling programme with H.M. *Fret*. We also like to thank skipper and crew of the HD 211 and the R.V. *Ephyra* for their help on board. To Mr G. Spaink of the Geological Survey of the Netherlands we owe the determination of the Eocene and Plio-Pleistocene fossils.

REFERENCES

DAVIS, F.M. (1925) Quantitative studies on the fauna of the sea bottom. 2. Results of the investigations in the southern North Sea, 1921–1924. *Fisheries Invest. Ser. II*, **8**, 1–50. M.A.F.F., London.

EISMA, D. (1965) Shell characteristics of *Cardium edule* L. as indicators of salinity. *Neth. J. Sea Res.* **2**, 493–540.

EISMA, D. (1966) The distribution of benthic marine molluscs off the main dutch coast. *Neth. J. Sea Res.* **3**, 107–163.

EISMA, D. (1981) Supply and deposition of suspended matter in the North Sea. In: *Holocene and Marine Sedimentation in the North Sea* (Ed. by S.-D. Nio et al.). *Spec. Publs int. Ass. Sediment.* **5**, 415–428. Blackwell Scientific Publications, Oxford. 524 pp.

EISMA, D., MOOK, W.G. & DAS, H.A. (1976) Shell characteristics, isotopic composition and trace element contents of some euryhaline molluscs as indicators of salinity. *Palaeogr. Palaeoclim. Palaeoecol.* **19**, 39–62.

GREENSMITH, J.T. & TUCKER, E.V. (1973) Holocene transgressions and regressions on the Essex coast, Outer Thames estuary. *Geol. Mijnb.* **52**, 193–202.

JELGERSMA, S. (1961) Holocene sealevel changes in the Netherlands. *Meded. geol. Sticht.* C–VI, 1–101.

JELGERSMA, S. (1979). Sea level changes in the North Sea basin. In: *The Quaternary History of the North Sea* (Ed. by E. Oele, R. T. E. Schüttenhelm and A. J.

Wiggers). *Acta Univ. Upsaliensis, Symp. Univ. Ups. Annum Quinqentesimum Celebr.* **2**, 233–248.

KIRBY, R. & OELE, E. (1975) The geological history of the Sandettie-Fairy Bank area, southern North Sea. *Phil. Trans. R. Soc.* A **279**, 257–267.

LABAN, C. & SCHÜTTENHELM, R.T.E. (1981) Some new evidence on the origin of the Zeeland ridges. In: *Holocene Marine Sedimentation in the North Sea Basin* (Ed. by S.-D. Nio et al.). *Spec. Publs int. Ass. Sediment.* **5**, 239–245. Blackwell Scientific Publications, Oxford. 524 pp.

LAMB, H.H., LEWIS, R.P.W. & WOODROFFE, A. (1966) Atmospheric circulation and the main climatic variables between 8000 and 0 B.C.: meteorological evidence. *Proc. Int. Symp. World Climate from 8000 to 0 B.C.*, pp. 174–217.

LOUWE KOOIJMANS, L.P. (1976) Prähistorische Besiedlung im Rhein-Maas-Deltagebiet und die Bestimmung ehemaliger Wasserhöhen. *Probl. Küstenforsch. südlichen Nordseegebeieit*, **11**, 119–143.

MOOK, W.G. (1970) Stable carbon and oxygen isotopes of natural waters in the Netherlands. *Proc. I.A.E.A. Conf. Use of Isotopes in Hydrology, Vienna*, pp. 1963–190.

MOOK, W.G. (1971) Paleotemperatures and chlorinities from stable carbon and oxygen isotopes in shell carbonate. *Palaeogeogr. Palaeoclim. Palaeoecol.* **9**, 245–263.

OELE, E. (1971) The Quaternary geology of the southern area of the Dutch part of the North Sea. *Geol. Mijnb.* **50**, 461–474.

PRATJE, O. (1929) Subfossile Seichtwassermuscheln auf der Doggerbank und in der südlichen Nordsee. *Zentbl. Miner. Geol. Paläont.* B 56–61.

VAN STRAATEN, L.M.J.U. (1956) Composition of shell beds formed in tidal flat environment in the Netherlands and in the Bay of Arachon (France). *Geol. Mijnb.* **18**, 209–226.

VAN STRAATEN, L.M.J.U. (1965) Coastal barrier deposits in South- and North-Holland. *Meded. geol. Sticht.* **17**, 41–75.

STRIDE, A.H. (1959) On the origin of the Dogger Bank in the North Sea. *Geol. Mag.* **96**, 33–44.

TEBBLE, N. (1966) British bivalve seashells. *Brit. Mus. nat. hist. London*, 212 pp.

VEENSTRA, H.J. (1965) Geology of the Dogger Bank area. *Mar. Geol.* **3**, 245–262.

Spec. Publs int. Ass. Sediment. (1981) **5**, 239–245

Some new evidence on the origin of the Zeeland ridges

CEES LABAN *and* RUUD T. E. SCHÜTTENHELM

Department of Marine Geology, Geological Survey of the Netherlands,
Spaarne 17, Haarlem, the Netherlands

ABSTRACT

Survey campaigns in an area off the Zeeland coast in the south-eastern part of the Southern Bight with sub-bottom profilers and coring and drilling equipment resulted in relatively detailed information on the sub-seabed conditions.

Our data shed some new light on the origin of at least some of the ridges, investigated before by Baak (1936) and Houbolt (1968). Several of the modern Zeeland ridges seem to have been formed essentially by sand accumulation around small pre-existing sediment bodies. One of these so-called 'initial ridges' has been dated as early Atlantic. The sand which makes up the bulk of the modern ridges contains several mollusc species that are no longer living—or are very rare now— in the Southern Bight. This suggests that the ridge formation was essentially completed some time ago, before the deposition of up to 2000 years old, slightly clayey, lee-side deposits.

INTRODUCTION

In 1975 a detailed survey programme started off the Zeeland coast, investigating the area largely with the SONIA sub-bottom profiler (a 3·5 kHz pinger) and various kinds of sampling equipment. The survey forms part of a geological mapping programme for the North Sea area off the south-western Netherlands, carried out by the Geological Survey of the Netherlands and the North Sea Directorate of the Ministry of Transport and Public Works. The area of the present study is situated in the south-easternmost part of the Southern Bight.

Water depths are generally between 5 and 28 m. In the eastern part of the area SW–NE-striking, elongated ridges called the Zeeland ridges occur more or less parallel to the present coast. These ridges, situated 3–7 km apart, show elevations of up to 15 m above the surrounding seabed (Fig. 1). The whole area, except for a nearshore zone of 5–14 km, is covered with sandwaves. The height of these sandwaves varies from 2 to 8 m (Fig. 2). The surface

sediment consists of medium sand with fine sand in the nearshore area (Schüttenhelm, 1980).

PREVIOUS WORK

The origin of the ridges in the Southern Bight of the North Sea has been very much in debate. Baak (1936) compared the position of the so-called Old Dunes in western Holland—ill-defined barriers with dunes on top—with that of the Zeeland ridges. He considered the Zeeland ridges as drowned extensions of the barrier chain. At present, we know that the onshore barriers were formed in two different phases (Jelgersma *et al.*, 1970). The first phase is dated as late Atlantic and early Sub-boreal (5300–4700 y BP); the second phase as early Sub-boreal (4700–3800 y BP). Dune formation started before 4100 y BP and continued up to Roman times.

Houbolt (1968) concluded from sparker records and shallow cores that the Zeeland ridges were in part erosional forms and were not formed by submarine or subaerial sand accumulation only. One of his arguments was the absence of flat-based ridges, indicative for lateral movement of these

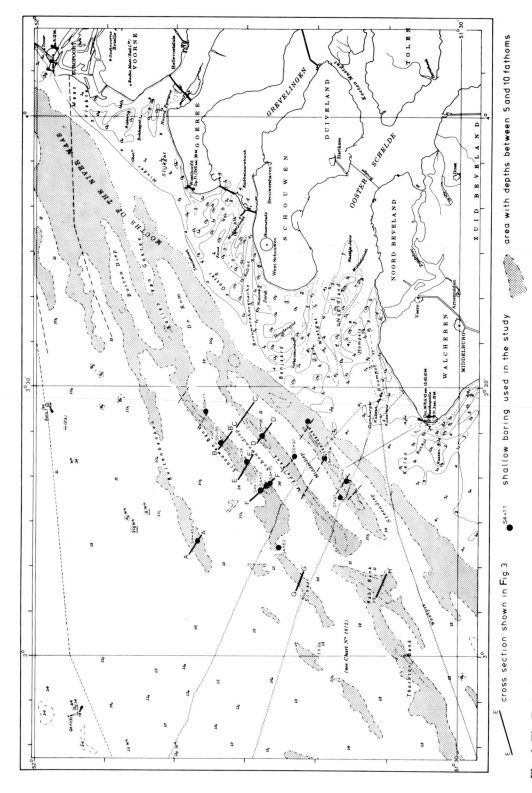

Fig. 1. The Zeeland ridges off the south-western Dutch coast. Depths are in fathoms (and feet) below LLWS. Areas with water depths between 5 and 10 fathoms are shaded. Positions of the cross-sections over the Zeeland ridges are indicated by thick black lines.

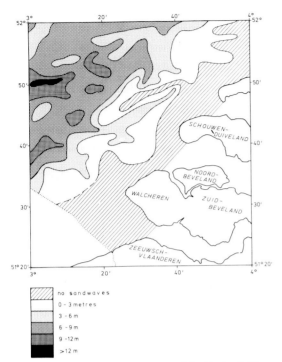

Fig. 2. Sandwave height in the area of the Zeeland ridges.

THE ZEELAND RIDGES:
A SUMMARY OF OUR RESULTS

A study of high-resolution seismic records, cores and other sub-bottom samples revealed the presence of several more or less continuous reflectors between which sediments with characteristic macrofossil associations occur. A strong reflector—the 'horizontal' reflector of Fig. 3—has been recorded during our surveys over almost the entire area. The position of this reflector ranges from just below to some metres below the base of the sandwaves. This reflector also continues under the Zeeland ridges (see Fig. 3).

Within some of the ridges reflectors were observed indicating the presence of small sedimentary bodies on top of the 'horizontal' reflector, provisionally named 'initial ridges'. In one of these 'initial ridges' in the Schouwenbank a core (S5-34) was taken consisting of bluish-grey, clayey, very fine sand covered by 0·4 m of yellowish brown medium sand. Pollen analysis (de Jong, 1979) done on samples from the top of the clayey sand and at 0·8 m revealed respectively an Atlantic age (Dutch pollen zone III, high content of *Ulmus*) and an early Atlantic age (Dutch pollen zone III, high content of *Corylus* and relatively low content of *Quercus* and *Alnus*).

From the 'initial ridge' reflectors run towards the top of the modern ridge. At the north-western flank these reflectors rest directly on the 'horizontal' reflector. Such internal structures dipping in the same direction as the present slopes have been observed in most of the records perpendicular to the ridges.

Besides the core mentioned above, 16 contra-flush/airlift borings were carried out down to 10 m below seabed with the Geodoff MKII, a drilling device developed at the Geological Survey. The borings are located near the top and on the flanks of the ridges, and samples consist mainly of medium sand with a few per cent of mollusc shells. The sand does not contain appreciable amounts of pollen, diatoms or ostracods. Information on the stratigraphic position of these deposits was obtained in recent years by studies by Spaink (1972, 1973) and Spaink & Sliggers (1978a, b, c, 1979a, b, c) of the mollusc content (see below).

The underlying Pleistocene sediments generally consist of medium to coarse sand with some marine molluscs. The sands form a part of the Eem Formation (Zagwijn & Van Staalduinen, 1975) and they are of Eemian age. These sands can be distinguished

ridges. Instead, on a schematized cross-section he showed markers rising from below within the ridges. Another argument was that he found bluish-grey sand and clayey sand covered with yellowish brown sand in some cores from both sides of the Middel-bank, one of the Zeeland ridges. He concluded that, given the absence of a transition zone, the nature of the contact was erosional. Houbolt supposed that the bluish-grey deposits in the lower part of the Zeeland ridges were fossil and the upper parts probably recent.

Moreover, in some places, amongst others on the Thorntonbank, the truncation of reflectors rising from below was observed at about the same depth as a break and a change in the mollusc content. Finally, Houbolt compared the orientation of the Zeeland ridges with those of the ridges of the Hinder Group situated to the south-west. He suggested that the Zeeland ridges are the older ones because they seem to be truncated by those of the Hinder Group.

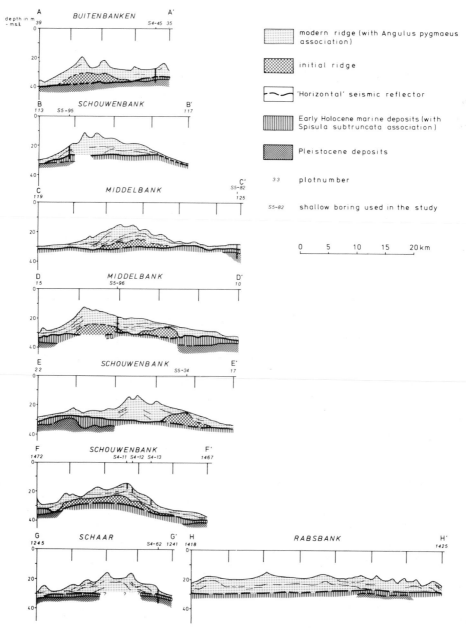

Fig. 3. Cross-sections over some of the Zeeland ridges. Locations are indicated on Fig. 1.

Table 1. The main constituents of the *Angulus pygmaeus* mollusc association (after Spaink, 1973). Some 200 species occur in this association. A = abundant, B = common, C = rather rare

Emarginula conica Lamarck	B	*Odostomia conoidea* (Brocchi)	B
Calliostoma zyziphinum (Lamarck)	B	*Odostomia eulimoides* (Hanley)	A
Gibbula cineraria (Linné)	B	*Odostomia insculpta* (Montagu)	C
Gibbula tumida (Montagu)	A	*Odostomia plicata* (Montagu)	C
Cantharidus montagui (W. Wood)	B	*Odostomia turrita* (Hanley)	A
Skenea cuttleriana Clarck	C	*Odostomia unidentata* (Montagu)	B
Skenea divisa (Fleming)	B	*Nucula nucleus* (Linné)	B
Skenea pusilla Jeffreys	A	*Nucula sulcata* Brown	B
Phaseanella pullus (Linné)	A	*Nucula turgida* Leckenby & Marshall	C
Lacuna crassior (Montagu)	B	*Arca lactea* Linné	B
Cingula islandica (Friele)	B	*Glycymeris* spec.	C
Cingula semicostata	A	*Limopsis aurita* (Brocchi)	B
Cingula semistriata (Montagu)	B	*Modiolus phaseolinus* (Philippi)	B
Alvania cancellata (Da Costa)	C	*Crenella decussata* Montagu	C
Rissoa alderi Jeffreys	A	*Musculus discors* (Linné)	B
Rissoa inconspicua Alder	A	*Chlamys opercularis* (Linné)	B
Rissoa parva (Da Costa)	B	*Chlamys varius* (Linné)	B
Rissoa parva interrupta (Da Costa)	C	*Lima sulcata* Brown	B
Tornus subcarinatus (Montagu)	B	*Astarte montagui* Dilwyn	B
Skeneopsis planorbis Fabricius	B	*Astarte triangularis* (Montagu)	A
Caecum glabrum (Montagu)	B	*Altenaeum nortoni* Spaink	B
Cerithiopsis tubercularis (Montagu)	C	*Diplodonta rotundata* (Montagu)	B
Triphora perversa (Linné)	B	*Phacoides borealis* (Linné)	C
Epitonium clathratulus (Kanmacher)	B	*Kellia suborbicularis* (Montagu)	B
Aclis ascaris (Turton)	C	*Lepton nitidum* Turton	B
Aclis supranitida (Wood)	B	*Lepton squamosum* (Montagu)	B
Eulima bilineata (Alder)	C	*Montacuta substriata* (Montagu)	C
Chrysallida indistincta (Montagu)	C	*Cardium ovale* (Sowerby)	B
Chrysallida obtusa (Brown)	B	*Cardium papillosum* Poli	B
Chrysallida spiralis (Montagu)	A	*Cardium scabrum* Philippi	B
Trivia sp.	B	*Venus ovata* Pennant	A
Trophon truncatus (Ström)	B	*Angulus donacinus* (Linné)	C
Ocenebra erinacea (Linné)	B	*Angulus pygmaeus* (Lovén)	A
Nassarius incrassatus (Ström)	B	*Sphenia binghami* Turton	A
Philbertia linearis (Montagu)	C	*Thracia phaseolina* (Lamarck)	A

from the overlying marine Holocene sands with the *Spisula subtruncata* and *Angulus pygmaeus* associations by the occurrence of characteristic mollusc species like *Divaricella divaricata* (Linné) and *Corbula gibba* (Olivi). Medium to coarse sand of the Kreftenheye Formation occurs locally in the westernmost part, in the area of the Buitenbanken and Schouwenbank. These fluvial sands were deposited by the rivers Rhine and Meuse during the Eemian and Weichselian stages. The top of the Pleistocene deposits is situated at about 25 m below mean sealevel (MSL) near the coast. It dips gently to about 40 m below MSL in the Buitenbanken area.

A great number of shallow cores (2–5 m) were taken between the Zeeland ridges and the coast in a surface layer of fine silty sand with intercalated clay laminae. The surface layer ranges in thickness from 0·5 to 4·5 m. Pollen analysis (de Jong, 1972) resulted in the recognition of pollen assemblages character-

istic for the period from 2000 to 400 years ago (part of Dutch pollen zone Vb). High percentages in some cores of cereals like rye (*Secale*; indicative for the period after *c.* AD 700) and the presence of cornflower (*Centaurea cyanus*: after *c.* AD 1300) permitted an even more precise idea on the age of these deposits. They reflect relatively quiet conditions in a sheltered area, which means that the ridges pre-date the deposits mentioned above.

THE FAUNAL EVIDENCE

Apart from a recent, also living, mollusc association, Spaink (1973) distinguished two mollusc associations in the marine Holocene sediments in the Southern Bight, i.e. the association of *Angulus pygmaeus* and that of *Spisula subtruncata*. Species lists are represented in Tables 1 and 2 respectively.

Table 2. The *Spisula subtruncata* mollusc association (after Spaink, 1973). A—abundant, B—common, C—rare

Littorina littorea (Linné)	B
Littorina saxatilis (Olivi)	B
Hydrobia ulvae (Pennant)	B
Polinices polianus (Della Chiaje)	A
Buccinum undatum Linné	C
Retusa alba (Kanmacher)	B
Mytilus edulis Linné	A
Anomia squamula Linné	A
Ostrea edulis Linné	B
Mysella bidentata (Montagu)	A
Montacuta ferruginosa (Montagu)	A
Cardium edule Linné	A
Cardium lamarcki Reeve	B
Venus striatula (Da Costa)	B
Spisula elliptica (Brown)	A
Spisula subtruncata (Da Costa)	A
Mactra corallina (Linné)	B
Donax vittatus (Da Costa)	A
Scrobicularia plana (Da Costa)	B
Abra alba (Wood)	A
Abra prismatica (Montagu)	C
Macoma balthica (Linné)	A
Angulus fabulus (Gmelin)	A
Tellina tenuis Da Costa	B
Ensis arcuatus (Jeffreys)	B
Ensis ensis (Linné)	B
Barnea candida (Linné)	A

The latter association, which strongly resembles the living mollusc fauna of nearshore areas in the southern North Sea, occurs in sediments below the 'horizontal' reflector. Lithological evidence suggests that these deposits are to a large extent reworked fluvial sands and coastal sands deposited by the rivers Rhine and Meuse during and before pre-Boreal times. Hereafter reworking in a shallow marine environment was caused by the rapidly rising sea-level. The top of the deposits is situated at a depth of 45 m in the western part of the area and at about 30 m in the easternmost part. Sub-bottom profiler records show that the *Spisula subtruncata* sands are overlain by sands with the *Angulus pygmaeus* association or by 'initial-ridge'-like sediment bodies (see above). In places, a transitional interval separates both mollusc associations.

Angulus pygmaeus is one of the most characteristic though not the most abundant species of the *Angulus pygmaeus* mollusc association. The species is not known from pre- and early Holocene deposits in the southern North Sea (Spaink, 1973). At present only a few living specimens are found very locally in seabed areas with medium sand or coarser deposits in the Southern Bight. This means that the widespread

occurrence of *Angulus pygmaeus* in this area has to be considered as a thing of the past. In the eastern part of the Southern Bight the *Angulus pygmaeus* association is found north to about 52° N. *Angulus pygmaeus* shells are only locally found in surface samples in the area between 52° N and about 53° 20′ N, but generally without most of the associated species. *Angulus pygmaeus* specimens have not been found in borings north of 53° 20′ N in the Dutch sector. Today many of the associated species are living in the English Channel and other shelf areas farther to the south (Spaink, 1973). Sand from the uppermost metres in borings on the Zeeland ridges and the sandwave area to the west generally contain this association (Fig. 3). The association is absent, however, in some shallow cores, e.g. in S5-97 and S5-38 on the Steenbanken. The recent mollusc association observed in the upper part of these cores includes specimens of *Petricola pholadiformis*, which was brought into the North Sea around AD 1900. The *Angulus pygmaeus* association is also absent in marine nearshore sands deposited in historical times and to a large extent derived from eroded coastal deposits.

FINAL REMARKS AND CONCLUSIONS

The evidence mentioned above suggests that the *Angulus pygmaeus* association was brought together with surface sediment into the Southern Bight from the south some time after the submergence of the Straits of Dover. The presence of numerous chert and chalk particles derived from Cretaceous rocks in that area support this view as well as the overall decrease in grain size in the surface sand layer in the eastern part of the Southern Bight from the south-west to the north-east. The Zeeland ridges and the nearby sandwave field essentially consist of *Angulus pygmaeus* sands. It may be concluded that several of the Zeeland ridges are built on top of or are leaning against a core consisting of older deposits at least in part of (early) Atlantic age. These older deposits, provisionally named 'initial ridges', appear to be elongated in shape. Some of the Zeeland ridges seem to lack an older core, however, The Zeeland ridges pre-date leeside deposits that are up to 2000 years old.

As regards the processes that caused the inferred transport of large amounts of sand into the southeastern Southern Bight, one can merely speculate.

The present authors favour transport mechanisms induced by asymmetrical tidal flow. Perhaps part of the water brought from the south into the Southern Bight by tidal currents spilled at lower sea-level stands over the drowned former land bridge between Norfolk and the Dutch island of Texel and did not flow back to the south. This land bridge submerged when the sea-level rose to about -30 m below the present level, i.e. around 8300 years ago (Jansen, Van Weering & Eisma, 1979).

ACKNOWLEDGMENTS

The discussions with, and good advice of, Messrs G. Spaink, B. C. Sliggers and J. de Jong are gratefully acknowledged. Mr A. Walkeuter carefully composed the figures, and Miss E. Y. Lamboo typed the manuscript.

REFERENCES

BAAK, J.A. (1936) *Regional Petrology of the Southern North Sea.* Veenman & Zonen, Wageningen. 127 pp.

HOUBOLT, J.J.H.C. (1968) Recent sediments in the Southern Bight of the North Sea. *Geologie Mijnb.* **47**, 245–273.

JANSEN, J.H.F., VAN WEERING, T.C.E. & EISMA, D. (1979) Late Quaternary sedimentation in the North Sea. In: *The Quaternary History of the North Sea* (Ed. by E. Oele, R.T.E. Schüttenhelm and A.J. Wiggers), pp. 175–187. *Acta Univ. Ups. Symp. Univ. Ups. Annum Quingentesimum Celebrantis,* **2**.

JELGERSMA, S., DE JONG, J., ZAGWIJN, W.H. & VAN REGTEREN ALTENA, J.F. (1970) The coastal dunes of the western Netherlands; geology, vegetational history and archaeology. *Meded. Rijks geol. Dienst. Nieuwe Ser.* **21**, 93–167.

DE JONG, J. (1972) Pollenanalytisch onderzoek van een aantal kernen, afkomstig van het gebied voor de kust van Goeree. *Rijks Geol. Dienst, Afd. Palaeobot., Int. Rep.* 660, 1 p.

DE JONG, J. (1979) Pollenanlytisch onderzoek van Noordzeeboring 74H32. *Rijks Geol. Dienst, Afd. Palaeobot., Int. Rep.* 853, 2 pp.

SCHÜTTENHELM, R.T.E. (1980) The superficial geology of the Dutch sector of the North Sea. *Mar. Geol.* **34**, M27–37.

SPAINK, G. (1972) Noordzee, boring 72H25. *Rijks Geol. Dienst, Afd. Macropalaeontol., Int. Rep.* 674, 3 pp.

SPAINK, G. (1973) De 'Fauna van Angulus pygmaeus' en de 'Fauna van Spisula subtruncata' in de Zuidelijke Noordzeekom. *Rijks Geol. Dienst, Afd. Macropalaeontol., Int. Rep.* 578, 9 pp.

SPAINK, G. & SLIGGERS, B.C. (1978a) Molluskenonderzoek van de Noordzeeboring 74GS38. *Rijks Geol. Dienst, Afd. Macropalaeontol., Int. Rep.* 1087, 8 pp.

SPAINK, G. & SLIGGERS, B.C. (1978b) Molluskenonderzoek van de Noordzeeboring 77MK63. *Rijks Geol. Dienst, Afd. Macropalaeontol., Int. Rep.* 1142, 7 pp.

SPAINK, G. & SLIGGERS, B.C. (1978c) Molluskenonderzoek van de Noordzeeboring 75H59. *Rijks Geol. Dienst, Afd. Macropalaeontol., Int. Rep.* 1283, 4 pp.

SPAINK, G. & SLIGGERS, B.C. (1979a) Molluskenonderzoek van de Noordzeeboring 72GD23, vak S4-11. *Rijks Geol. Dienst, Afd. Macropalaeontol., Int. Rep.* 1316, 5 pp.

SPAINK, G. & SLIGGERS, B.C. (1979b) Molluskenonderzoek van de Noordzeeboring 76MK5, vak S5. *Rijks Geol. Dienst, Afd. Macropalaeontol., Int. Rep.* 1276, 5 pp.

SPAINK, G. & SLIGGERS, B.C. (1979c) Molluskenonderzoek van Noordzeeboring 76MK08. *Rijks Geol. Dienst, Afd. Macropalaeontol., Int. Rep.* 1335, 5 pp.

ZAGWIJN, W.H. & VAN STAALDUINEN, C.J. (eds) (1975) Toelichting bij geologische overzichtskaarten van Nederland. *Rijks Geologische Dienst, Haarlem,* 134 pp.

Spec. Publs int. Ass. Sediment. (1981) **5**, 247–256

Predicted sand-wave formation and decay on a large offshore tidal-current sand-sheet

MICHAEL A. JOHNSON, ARTHUR H. STRIDE, ROBERT H. BELDERSON
and NEIL H. KENYON

Institute of Oceanographic Sciences, Wormley, Godalming, Surrey, U.K.

ABSTRACT

The conditions required for generation of small sand-waves (rather than sand-ripples) in flumes have been applied to the extensive field of sand-waves, occurring on a depositional sand-sheet in the Southern Bight of the North Sea, for mean spring tides and extreme spring tides at both summer and winter water temperatures and viscosities. The variation in the position of the predicted boundaries of sand-wave formation (from an initially rippled bed) is striking. The tidal currents at near-minimum winter water temperatures would be unable (on their own) to generate sand-waves on almost any part of the sand-sheet. At near-maximum summer water temperatures the tidal currents of mean springs would (on their own) generate sand-waves up to 40 km further across the sand-sheet, and very extreme spring tides to a still further distance of about 40 km. Smaller boundary migrations would be predicted for the combined action of the tidal and non-tidal currents. Across a large part of the sand-sheet the small sand-waves are active over a long enough period to have grown into large sand-waves (with small ones on them). However, in an outer zone the small sand-waves will be largely flattened by bioturbation and storm waves during periods of weak tidal currents, and so will not be able to develop into large ones.

The overall form, grain size and internal structure of the sand-sheet will reflect both the periodic and episodic building and destruction of the small sand-waves in the outer zone, the presence and variability of smaller on larger sand-waves in the inner zone, and the geographical differences in the shape of the large sand-waves.

INTRODUCTION

Sand-sheets are one of the two main depositional facies of continental-shelf areas with strong offshore tidal currents (the other being the tidal sand-banks). They have been recognized in several predominantly tidal seas. One of the best known is located to the west of Holland, north of a bed-load parting in the Southern Bight of the North Sea (Fig. 1). This sheet has an area of several thousand km² and a thickness of up to 10 m in the south, decreasing to 4–6 m further north over much of the area (Laban, 1977). The median grain size of the sands decreases progressively northwards from about 0·45 mm at the bed-load parting in the south, with the sands ultimately giving

way to muddy sands and muds (Jarke, 1956). The sorting of the sands is very good, with less than 3% clay and silt and little very fine sand. It is best for the fine sand range (0·125–0·25 mm) (Volpel, 1959). Almost all the material is quartz.

The surface of much of the sand-sheet is shown by extensive side-scan sonar coverage to have both large and small sand-waves, with presumed super-imposed ripples. The sand-ripples have a height and wavelength of up to 5 and 60 cm respectively. The small sand-waves are up to 2 m high. In the south, at the bed-load parting, the sand-waves are symmetrical in cross-section. Northwards they evolve progressively into a 'catback' profile (van Veen, 1936) and then further north again into more clearly

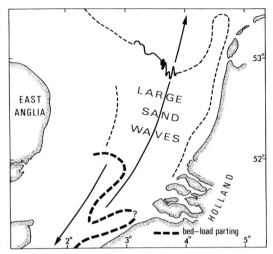

Fig. 1. The sand sheet of the eastern half of the Southern Bight of the North Sea, seaward of Holland, extends from a bed-load parting in the south to the northern side of the figure. Much of the sheet has a surface of large sand waves within the limits shown. Arrows indicate directions of net sand transport. The western half of the seabight is largely occupied by the sand-bank facies.

asymmetrical forms, with the northerly facing slope being the steeper one. Thus, the obvious asymmetrical profile of most of the large sand waves indicates a net northerly sand transport. The more southerly of the asymmetric sand-waves have northerly facing lee slopes with a tangential contact with the bottom of the adjacent trough, while those further north have a more abrupt change in angle at the base of their steeper slopes. The height of the large sand waves generally decreases northwards to a limiting value of about 2–3 m (Stride, 1970; McCave, 1971). Beyond this line sonographs show that there is a zone of small sand waves about 1 m high (but not generally recognizable on echo-sounder records); beyond this again a rippled sand-floor is expected (Johnson & Stride, 1969).

The sand-sheet is associated with tidal currents whose peak strength gradually decreases northwards except for a local 'high' west of the island of Texel. Laterally there is a steeper increase of current speed towards the East Anglian coast. The associated northwards net transport of sand is in keeping with the peak northwards flowing (ebb) tidal current being slightly stronger (by up to 10 cm sec^{-1} near surface) than the peak southwards flowing flood current (Stride, 1963). Computer simulations have recently confirmed that net peak bed shear stress due to the principal tidal constituent M_2 plus its first harmonic M_4 is directed northwards (Pingree & Griffiths, 1979). Another simulation integrated sand-transport rate assuming a cubic dependence on current speed, over the computed twice daily and spring/neap current speed cycles. It reproduced the expected northwards sand transport over the sand sheet, with net erosion in the south and net deposition in the north (Sündermann & Krohn, 1977).

The present paper examines the conditions under which small sand-waves (rather than ripples alone) are the stable bedform in controlled experiments in flumes, in the absence of analytical theory or adequate data sets from rivers or the subtidal seafloor. Intertidal sand-waves will not be considered. It then draws conclusions about the generation of small sand waves on this sand sheet and discusses the resulting internal structure of the sheet.

MINIMUM BED SHEAR STRESS FOR GENERATION OF SMALL SAND-WAVES IN FLUMES

There is as yet no detailed theory of either ripples or sand-waves. Therefore recourse has to be had to observations and dimensional analysis. The latter indicates (Yalin, 1972) that water viscosity is of basic importance for ripples. They are therefore features of the flow near the bed, and independent of flow depth provided this exceeds 12 cm or so. For sand-waves fluid viscosity is of negligible influence but the turbulence in the flow is essential.

The available flume data are almost all for uni-directional steady flows, with flow and bedforms allowed to reach near-equilibrium. Bed shear stress has been calculated as the product of density, gravitational acceleration, mean depth and mean water-surface slope, less the correction for sidewall friction, estimated by the usual method (Vanoni & Brooks, 1957) where the correction has not already been applied in the data publication. The commonly used 'shear velocity' is the square root of bed shear stress divided by water density.

The height of sand waves that can form in a flume is severely limited by the water depth, to about $\frac{1}{3}$–$\frac{1}{2}$ of mean water depth and is therefore often not much greater than heights of ripples (which can reach only 5 cm). However, in all data sets used it is fairly certain that sand waves were validly distinguished from ripples, both by their much longer wavelength (greater than 60 cm) and by profile and plan differences. Nomenclature is confusing, since the hydraulics literature almost always uses 'dune' rather than sand-wave, and sometimes uses it to include sand-ripple (e.g. Vanoni & Brooks, 1957; Vanoni & Hwang, 1967) and also uses 'sand-wave' in the sense of a very long, flat-topped accumulation of sand moving relatively rapidly down a flume. Data sets involving such 'sand-waves' were excluded and this meaning of 'sand-wave' will not be used again in this paper. Only data sets with mean flow depth above 12 cm have been used here, since for these the Froude numbers were less than 0·5 and it is fairly certain that formation and building of sand-waves was not inhibited by surface-wave interference.

The data were initially plotted in terms of Shields variables (Fig. 2: defined in its caption). Dimensional analysis (Yalin, 1972) shows that for steady flow the ripple/sand-wave boundary should be a relation between these two variables only, on the assumption of 'geometrically similar' sediment, discussed later. The Reynolds number takes account of differing water viscosity (due to differing water temperature which ranged from 8 to 63°C) in different flume runs. fig. 2 shows that the grain shear Reynolds number is the main influence on the boundary between the ripple/sand-wave existence regions. Yalin (1972, fig. 7.23) shows a transition zone between Reynolds number values 8 and 24, but including bed states of ripples on sand waves. The line drawn on Fig. 2 passes through all but one of the weakest flow occurrences of sand-waves, and almost through Hill, Robinson & Srinivasan's (1971) two points for stated ripple/sand-wave boundary, so that some reported ripple occurrences lie on the sand-wave (right) side of the line. Where ripples were present on sand-waves the present paper takes the bed state as sand-waves, since the ripples must have formed subsequently. Such ripples on sand-waves can form for higher values of total bed shear stress and grain-shear Reynolds number than on a flat bed since the ripples are in equilibrium with the smaller local bed shear stress.

The disregarded sand-wave point corresponds to a run (no. 47 with '0·47 mm' sand of Guy, Simons & Richardson, 1966) for which there seems to be some error in the published data. Its tabulated depth and slope give lower values of bed shear stress than for ripple runs with the same material, and the tabulated sand-wave dimensions and movement speed are close to those for the following run, which had twice the bed shear stress. Fig. 2 is drawn on a linear scale of grain shear Reynolds number to show the lower

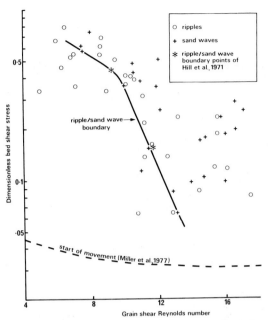

Fig. 2. Plot of dimensionless bed shear stress (bed shear stress divided by the product of gravitational acceleration, grain diameter and sediment density minus water density) and grain-shear Reynolds number (grain diameter multiplied by the square root of bed shear stress divided by water density, all divided by kinematic water viscosity), of flume data sets (Annambhotla, 1969; Barton & Lin, 1955; Chabert & Chauvin, 1963; Crickmore, 1967; Franco, 1968; Guy *et al.*, 1966; Hill *et al.*, 1971; Pratt, 1970; Stein, 1965; Taylor, 1971; Vanoni & Brooks, 1957; Vanoni & Hwang, 1967) showing sand-ripples, or sand-waves with or without superimposed sand-ripples, together with two approximate data sets for observed ripple/sand-wave instability (Hill *et al.*, 1971). The linear grain shear Reynolds number scale emphasizes the boundary area. Many points well removed from it have been omitted. The boundary shown joins the sand-wave points with lowest flows, with one exception discussed in the text.

sand-wave boundary region in detail and omits many ripple or sand-wave points on the expected sides of the boundary. These would have appeared on a conventional Shields diagram (with both scales logarithmic and 1000:1 or so range) which would also have made the overlap of ripple and sand-wave areas less conspicuous. The overlap is fortunately less in the upper part (smaller grain sizes) of Fig. 2, which is the part required for the North Sea application.

The overlap seems too large to be accounted for by measurement inaccuracies and lack of exact equilibrium in the flume runs, and a major cause

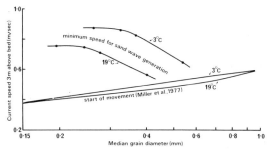

Fig. 3. The minimum current speed at 3 m above the bed required for generation of small sand-waves from an initially rippled bed for two water temperatures (and for the start of sand movement on a flat bed). Salinity is 35 parts per thousand, immersed sand density is 1·62. The curves are derived from Fig. 2.

must be the lack of similarity between the sediments used. For exact 'geometrical similarity' (Yalin, 1972) between different sediments, their grain-shape characteristics are required to be the same, the grain-size distribution curves to have the same shape, and the values of gradation (geometric standard deviation: or some other non-dimensional measure of dispersion coefficient could be used) to be the same. Yalin (1972) assumes the grains all to have the same density, but if grain densities vary, presumably similarity requires their probability distribution curves to have the same shape and standard deviation. Sediment transport rate studies do not seem to have shown any important effects of lack of similarity, provided that the grains were subangular or subrounded, their size distribution was approximately lognormal, their gradation had a value in the usual range for river sands of that median diameter, and that a sediment contained no more than a few per cent of grains of widely differing density. However, these sediment characteristics, especially the last two, may be more critical for an instability phenomenon like initial sand-wave formation, which depends strongly on the effective roughness presented to the flow by the sediment grains and ripples.

In the flume data sets used, sediment geometric standard deviation varies between 1·26 and 1·67, except for the value 1·11 for Pratt's (1970) sand. For the other sands gradation variations had no clear effect on the positions of the plotted points on Fig. 2. Pratt's relevant flume bed-state phases are 2B 'three-dimensional irregular ripples' and 3 'development of larger, longer bedforms by superimposition of one on another'. Pratt (1970, fig. 9) found a decrease of bed shear stress as flow speed increased through

phase 2B. This would cause decreases of the Shields variables so that some points from the start of phase 3 (sand waves) plot in Fig. 2 to the left of points from phases 2A and 2B (ripples).

Chabert & Chauvin's (1963) data points plot well to the right of those of other investigators. Thus, use of their data led to Bonnefille's (1964) conclusion that the ripple/sand-wave boundary could be taken as grain shear Reynolds number equal to 15. One reason may be their use of Einstein's (1942) technique for sidewall friction correction which involves estimation of a Strickler coefficient for the sidewalls rather than taking the walls as hydraulically smooth as in Vanoni & Brooks' method. The effect of this could not be checked as the depth, slope, velocity and temperature values were not given. Also no materials smaller in median diameter than 0·45 mm sand and 0·397 mm crushed gravel were used.

For the equivalent expression of the lower sand-wave boundary in terms of velocity and median grain diameter the density and sea temperatures chosen refer to the southern North Sea sand-sheet, where sea temperatures of 3 and 19°C correspond respectively to near-minimum winter and near-maximum summer temperatures. Then for temperatures 3 and 19°C, and grain density 2·65 (immersed density 1·62 since salt water density is about 1·03), the corresponding lower sand-wave boundaries are plotted on Fig. 3. Water salinity, taken as 35 parts per thousand, increases water viscosity for a given temperature by about 5% at 19°C and 3% at 3°C, but for given salinity the viscosity at 3°C is about 53% more than at 19°C. Thus, the seasonal changes of temperature have much more effect than would a change from salt water to fresh water.

If instead of the lowest-flow sand-wave points the highest-flow ripple points had been taken as defining the ripple/sand-wave boundary, the curves of minimum speed for sand-wave generation would have had speeds about 10% higher than in Fig. 3 for median grain sizes up to 0·3 mm, and about 15% (for temperature 3°) and 20% (for temperature 19°) higher for median size 0·4 mm. Speeds using Vanoni's (1975) ripple/sand-wave boundary converted into the present variables would have been approximately mid-way.

The assumptions to note in applying such flume data to sand-waves in tidal seas are as follows: (a) unidirectional steady-current data can be used for the tidal case using the maximum tidal current speed during the period considered; (b) near-bottom tidal current speeds can be predicted from known near-surface tidal current speeds; (c) the shelf and flume sediments are 'geometrically similar'; (d) an equivalent bed roughness height of 6 cm can be used for relating bed shear stress and mean flow velocity; (e) the data for flume depths (0·12–0·5 m) are relevant to deeper water of rivers and the sea.

Regarding assumption (a), the tidal periods constitute additional variables and an accurate solution would involve them. However, it can be argued that the rapid increase (approximately a cube-law) of sand transport rate with the excess of current speed above a threshold value means that sand-wave building and movement by tidal currents occur primarily at and near the times of maximum current. Also, the tidal periodicity will often cause near-crest reversals of sand-wave asymmetry between ebb and flood currents to be superimposed, but will not affect the question of sand-wave building. Observed wavelength/depth and height/depth ratios of developed subtidal sand waves in the Southern Bight of the North Sea and near Japan are stated (Yalin & Price, 1976) to be consistent with those (Yalin, 1972) of sand waves under a steady current with the tidal maximum speed, although they take longer to build.

For assumption (b), velocity profiles over the whole water column are found not to be continuations of the near-bed logarithmic profile, as they are in flumes and shallow rivers. The ratio of current speed 3 m above the bed to current speed given by tidal current charts and tables (which can be regarded as an average speed over the uppermost 10 m) is taken as 0·75 for the water depth of about 30 m. The quantity 0·75, used in the absence of adequate theory or observations, is an approximate average of values for existing current meter recordings of that ratio for the sand sheet. It corresponds to a van Veen type power law with index 1/8.

For assumption (c), the lack of 'geometrical similarity' between the sand-sheet and flume sediments probably introduces little more uncertainty than that already discussed for the different flume sediments, since both flume and sheet sediments are largely river-derived and neither contains appreciable silt or clay.

For assumption (d) an equivalent bed roughness height of 6 cm ($z_0 = 0·2$ cm in the usual logarithmic velocity profile formula) over a bed of fine or medium sand with well-developed ripples is an approximate average of equivalent roughness heights derived from flume data and from near-bed velocity profiles over sandy rippled seafloors.

For assumption (e) it seems fairly certain, from Bagnold's (1956) and Yalin's (1972) discussions, that sand-ripples depend on viscous forces in the steep velocity gradients adjacent to the bed and the sediment grains, and require a viscous sublayer over the initially flat bed. Sediment movement is likely to be initiated mainly by 'streaks' of high-speed flow within the sublayer, associated with turbulent 'bursts' in the flow outside the sublayer (Jackson, 1976). The small irregularities in the sand bed then grow into sand ripples, particularly when their down-current slopes become slip faces with flow separation behind them (Bagnold, 1956). If bed shear stress increases sufficiently, the turbulence presumably breaks up any remaining viscous sublayer over parts of the ripples, destroys the ripples and then sand waves begin to build. With the use of bed shear stress, all the statements so far in this paragraph, and the ripple/sand-wave change, are independent of flow depth, provided this exceeds 12 cm or so. Use of flow velocity instead of bed shear stress would involve use of a third dimensionless quantity, e.g. bed friction factor, or equivalently dimensionless velocity, grain size and depth as by Southard (1971). There does not yet seem to be sufficient data over the wide range of depths to carry out such a three-dimensional analysis for the ripple/sand wave-boundary.

For a given median grain size a given bed shear stress corresponds to a higher current speed the lower the temperature, as the viscosity and sublayer thickness are higher. For a given temperature sand-wave formation requires a higher current speed the smaller the median grain size (also for very coarse sizes). An illustration is Reineck & Singh's (1967) finding of sand-waves in medium sand and ripples in fine sand for approximately the same current speeds in a tidal channel. The temperature and grain size effects on the threshold of sediment movement and the formation of ripples are much smaller, so that the lower the temperature or median grain size the wider the current speed-range giving ripples (Fig. 3). The 'minimum speed for sand wave generation' curves on Fig. 3 must curve upwards to the right of the portions shown, to stay above the movement threshold curves.

The depth of flow does, of course, largely determine the maximum size to which sand-waves can build. However, whether they attain maximum (equilibrium) size depends not on whether the current exceeds the minimum for sand-wave formation, but by how much and for how long it exceeds this minimum, assuming sufficient sand is present.

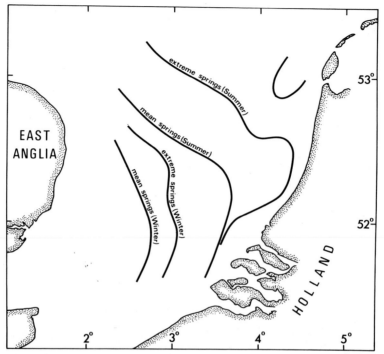

Fig. 4. The predicted outer limit of generation of small sand-waves about 1 m high in the Southern Bight of the North Sea for extreme and mean spring tidal currents (equivalent to tidal range coefficients at Brest of 120 and 95, respectively) and two temperature conditions using the upper two curves of Fig. 3 (but without non-tidal water movements). The detached portion of the northernmost line is due to a local current speed 'high'.

GENERATION OF SAND-WAVES IN THE SOUTHERN BIGHT OF THE NORTH SEA

The above discussion shows that sand-waves, initially small, of course, will be formed from an initially rippled bed by weaker currents at summer water temperatures than at winter water temperatures. This is expressed in Fig. 4 as four predicted outer boundaries of (initially) small sand-waves for the sand-sheet west of Holland. These boundaries are for mean spring tidal currents and for extreme spring tidal currents for sea temperatures of 3 or 19°C. In the absence of harmonic analyses of near-bed current meter records for positions on the sand sheet, from which tidal currents for other times could be computed, the extreme tidal current speeds (reached perhaps once a century on average) were taken as 120/95 of mean springs current speeds. This is the ratio of coefficients from the French Tide Tables, which strictly are proportional to tidal range at Brest, but will apply to within a few per cent to tidal currents in the study area. The median grain sizes

that were used (McCave, 1971) are based mainly on Netherlands Institute of Sea Research data and agree quite well with Jarke's (1956) map of sand sizes.

The seasonal variation in position of the predicted boundaries is striking, and its magnitude does not depend strongly on the assumptions or the data analysis. Mean spring tidal currents at winter water temperatures would be unable, on their own, to generate even small sand-waves on almost any part of the sand sheet. Even extreme tidal currents at winter temperatures would be effective in building sand-waves on only a slightly larger area than the mean spring tides.

In contrast, the mean spring tides at summer water temperatures are predicted to generate small sand-waves up to about 40 km north of the innermost boundary shown above. Furthermore, extreme tides associated with summer water temperatures would (even when unaided by non-tidal currents) generate small sand-waves within the whole area where large sand-waves are found in the Southern Bight of the North Sea, a distance of about 40 km beyond the mean spring-tide boundary for summer sea temperatures. Displacements in the east–west direction are smaller because of the steeper tidal current speed gradient.

Intermediate sea temperatures and tidal current strengths will give intermediate positions of the outer boundary for generation of sand-waves less than 1 m high. The frequency with which the outer boundary will be located at any particular position will be a maximum at its mean position and will decrease northwards and southwards from there, reflecting the full range of tidal variation in combination with the seasonal variation of water temperatures.

In practice, of course, superimposed on these tidal (and therefore regular) effects will be the effect of occasions of storm-wave incidence and unidirectional current contributions. When aiding the tidal currents these will cause increased sand transport, particularly during winter, so helping to offset the effect of low water temperatures. (The statistical distribution of tidal current speeds is almost the same during summer and winter half-years.) The magnitude of the non-tidal currents during storms can exceed 0·6 m sec^{-1}, even at 2 m above the bed, off the North Sea coast of Germany (Gienapp, 1973). However, well away from coasts some values of non-tidal currents (obtained by subtracting predicted tidal currents from observed currents) in storms are only 36 cm sec^{-1} in about 38 m of water (Caston, 1976) and 23 cm sec^{-1} in about 90 m of water (Howarth, 1975) at 4·6 and 15 m above the bed with maximum hourly mean wind speeds of about 25 and 18 m sec^{-1} respectively. Caston suggests that at his measuring site, between two sand-banks, funnelling and shift of current axes between the banks increased the value considerably. This can also happen near coasts for some wind directions. On the open shelf at about 3 m above the bed in a water depth of 30 m the typical maximum every few years may be 25 cm sec^{-1}, decreasing slightly with increase in water depth.

The more numerous occasions when sand-waves will be built and moved on the southern part of the sand-sheet (due to the stronger currents there) than on the northern part, is probably responsible for the greater size of the sand-waves in the south and for their progressive decrease in height to the north. However, the height of the sand-waves must also be dependent on the availability of suitable grain sizes. A contributory factor is probably a progressive northwards increase in the ratio of suspension transport rate to bed-load transport rate (McCave, 1971, who applied a theory of Kennedy, 1969). Further north, in the outermost zone where small sand-waves

occur on their own, the growth of sand-waves is likely to require tidal and non-tidal currents combined. However, it is possible that once small sand-waves have been formed by the combined water movements, they can be maintained (as can sand-ripples, e.g. Sutherland, 1967) by a lower tidal current speed than is required to form them in the first place. This is because the increased bed roughness they cause gives sufficient bed shear stress for sediment movement near sand-wave crests. Such sediment movement causes the formation of small subsidiary crests on ebb and flood tides.

The relatively abrupt change from the zone of large sand-waves, with small ones on them, to the outlying zone of small ones on their own is not thought to indicate a fundamental difference in origin for large and small sand-waves. No such difference is indicated by dimensional analysis (Yalin, 1972) nor can it be explained by the ratio of suspension to bed-load transport rates, as this can only change gradually northwards, as current speed and median grain size only change gradually and water depth changes little. On the contrary it is suggested to be a growth problem. In the outer zone the small sand-waves are probably made and then at least partially destroyed again before they can grow into large enough ones to withstand future storms. They lie in a region where tidal currents acting on their own will only construct them at about ten-year intervals, say, and where storm waves and bioturbation will tend to destroy or largely flatten them more frequently. Even the additional unidirectional currents will still be too limited in time to enable them to grow much larger.

REDUCTION IN HEIGHT OF SMALL AND LARGE SAND-WAVES DURING STORMS

Repeated echo-sounder surveys, with good navigational control, at the eastern edge of the sand sheet near to Holland show that the height of large sand waves decreased in winter and increased in summer, and that there were corresponding changes in height of the associated small sand-waves. Lowering of crests between two successive surveys was found to occur when there had been a period of waves with calculated near-bed oscillatory water speeds (in water depth of 18–23 m) exceeding about 0·5 m sec^{-1} (Terwindt, 1971). Similar changes were reported from shallower water off the U.S.A. (Ludwick, 1972). Near-bed oscillatory water movements are likely to assist in sediment transport and sand-wave building when the current at the time is stronger. However, when the current becomes weak, e.g. every six hours or so in twice-daily tidal currents, the wave movements may be stronger and then the water movement directions approximately reverses every few seconds. If this happens on either slope of a sand-wave, the water velocity will have downslope components as often as upslope, but sediment grains can move downslope more readily than upslope, so lowering the sand-wave crest in time.

A more widespread echo-sounder survey with less repetition and poorer navigational control led to the belief that there was a band of floor along the whole of the eastern and northern edges of the field of large sand waves where there were substantial height changes (McCave, 1971). However, some at least of those supposed changes may be due to the admitted navigational errors of up to 2 miles, as the northernmost sand-waves are known from accurate surveys to be patchily developed (Caston & Stride, 1973). The survey dates given by McCave (1971) were 4–22 November, during which the only wind speed of force 8 or above recorded at Dutch lightvessels (K.N.M.I., 1972) was force 8 at Terschellinger Bank at midnight on the 6th/7th. There were several observations of force 7 winds during the 6th, 7th and between the 13th and 17th but these winds were off the land. Accordingly the lightvessel wave observations lead to no calculated near-bed oscillatory speeds exceeding 0·13 m sec^{-1}. (The linear wave theory used for these calculations is known to be accurate to within 20%.) The second most marked non-tidal south-going current of the year, which occurred on 14–16 November (from the tabulations in K.N.M.I., 1972) at Texel Light Vessel at 18 m depth (mean water depth away from the immediate surroundings was about 30 m), does not seem to be relevant either. This current reversal might have been expected to lead to some reversal of the asymmetry of the small sand-waves rather than to smooth them out. Clearly, new surveys with good navigational control are required to substantiate the proposed changes in sand-wave height along their northern boundary.

Preservation of much of the sand sheet is inevitable, as the currents lack the ability to move all the sand away from the sand-wave zone. However, the southern boundary of the sand sheet may gradually move somewhat northwards. In the long term, though, it is anticipated that the preservation even of this portion may be partly achieved because of the continued sinking of the North Sea floor and the general increase in basin dimensions due to coast erosion. Together these should tend to lessen the peak tidal current strength progressively and thus restrict the tidal currents' ability to erode and transport sand. However, tidal current speed reduction would need to be verified by computations of tidal currents for estimated future dimensions, since (Johnson & Belderson, 1969) an increase of depth can even increase tidal current speed if it causes near-resonance, and also it expands the spatial pattern of current velocity variation along the basin. Also increase of depth and width could increase the effect of Atlantic tides. The large sand-waves themselves are seen as being both an indicator of net sand transport, as well as approaching ever closer to being a deposit. In the majority of cases they will probably not be preserved intact, but the internal structure of the lower part of them is likely to be preserved in the deposit.

INTERNAL STRUCTURE OF THE SAND SHEET

Short cores taken within the zone of large sand-waves (with small sand-waves on them) show the presence of northerly dipping cross-bedding (Houbolt, 1968), as might be expected. The measured lee slopes of the large sand-waves imply a northwards decrease in angle of dip from about 11° to 5° (with angles of up to 20° expected in the symmetrical sand waves at the bed-load parting further south). As these values are much less than those observed in the cores, the core data would appear to confirm that the almost ubiquitous small sand-waves (which commonly have avalanche lee slopes) found on top of the large ones are responsible for most of the cross-bedding.

Within the deposits in the zone of small sand waves on their own both a dominant northward dip and a possible subsidiary southward dip direction might be expected. This is because the occasional near-bed, south-flowing, unidirectional currents that are known from current meter records, and have been computed for particular storms (Davies, 1976), may temporarily reverse the normal ebb-current dominance. These small sand-waves could acquire reversed asymmetry or be completely rebuilt or partially smoothed by waves, much less slowly than the large ones further south. Cross-bedding dips should be up to about 30°, but there are likely to be numerous erosional episodes, and also periods when ripple-bedding is formed by weaker currents. Beyond this zone the sands are expected to be flat or ripple-bedded. In an overall longitudinal section there should be some alternations of the structures found in adjacent zones. However, this is not expected to apply to the large sand-waves to any extent. This is because their bulk is so great that occasional episodes of abnormal current flow will be unlikely to modify them appreciably, except for their crests, which in any case are unlikely to be preserved.

REFERENCES

ANNAMBHOTLA, V.S.S. (1969) *Statistical properties of bed forms in alluvial channels in relation to flow resistance.* Ph.D. Thesis, University of Iowa.

BAGNOLD, R.A. (1956) The flow of cohesionless grains in fluids. *Phil. Trans. R. Soc.* A **249**, 235–297.

BARTON, J.R. & LIN, P.N. (1955) *A Study of the Sediment Transport in Alluvial Channels.* Rep. 55 JRB 2. Civil Engineering Department, Colorado A & M College, Fort Collins.

BONNEFILLE, R. (1964) Étude d'un critère de debut d'apparition des rides et des dunes fluviales. *Bull. Cent. Rech. Essais Chatou,* **11**, 18–22.

CASTON, V.N.D. (1976) A wind-driven near-bottom current in the Southern North Sea. *Est. Coastal Mar. Sci.* **4**, 23–32.

CASTON, V.N.D. & STRIDE, A.H. (1973) Influence of older relief on the location of sand waves in a part of the Southern North Sea. *Est. Coastal Mar. Sci.* **1**, 379–386.

CHABERT, J. & CHAUVIN, J.-L. (1963) Formation des dunes et des rides dans les modèles fluviaux. *Bull. Cent. Rech. Essais Chatou,* **4**, 31–52.

CRICKMORE, J.J. (1967) Measurement of sand transport in rivers with special reference to tracer methods. *Sedimentology,* **8**, 175–228.

DAVIES, A.M. (1976) *Application of a Fine Mesh Numerical Model of the North Sea to the Calculation of Storm Surge Elevations and Currents.* Rep. 28. Institute of Oceanographic Sciences.

EINSTEIN, H.A. (1942) Formulas for the transport of bed-load. *Trans. Am. Soc. civ. Engrs* **107**, 561–577.

FRANCO, J.J. (1968) Effects of water temperature on bed-load movement. *J. WatWays Harb. Div. Am. Soc. civ. Engrs* **94** (WW3), 343–352.

GIENAPP, H. (1973) Strömungen während der Stormflut vom 2. November 1965 in der Deutschen Bucht und ihre Bedeutung für den Sediment transport. *Senckenberg. Mar.* **5**, 135–151.

GUY, H.P., SIMONS, D.B. & RICHARDSON, E.V. (1966) Summary of alluvial channel data from flume experiments, 1956–61. *Prof. Pap. U.S. geol. Surv.* **462-I**.

HILL, H.M., ROBINSON, A.J. & SRINIVASAN, V.S. (1971) On the occurrence of bed forms in alluvial channels. *Proc. 14th Congr int. Ass. hydr. Res.,* Paris, **3**, 91–100.

HOUBOLT, J.J.H.C. (1968) Recent sediments in the Southern Bight of the North Sea. *Geologie Mijnb.* **47**, 245–273.

HOWARTH, M.J. (1975) Current surges in the St George's Channel. *Est. Coastal Mar. Sci.* **3**, 57–70.

JACKSON, R.G. (1976) Large-scale ripples of the Lower Wabash River. *Sedimentology,* **23**, 593–623.

JARKE, J. (1956) Eine neue Bodenkarte der südlichen Nordsee. *Dt. hydrogr. Z.* **9**, 1–9.

JOHNSON, M.A. & BELDERSON, R.H. (1969) The tidal origin of some vertical sedimentary changes in epicontinental seas. *J. Geol.* **77**, 353–357.

JOHNSON, M.A. & STRIDE, A.H. (1969) Geological significance of North Sea sand transport rates. *Nature,* **224**, 1016–1017.

KENNEDY, J.F. (1969) The formation of ripples, dunes and antidunes. *A. Rev. Fluid Mech.* pp. 147–168.

K.N.M.l. (Kon. Ned. Met. Inst.) (1972) *Meteorologische en oceanografische waarnemingen verricht aan boord van Nederlandse lichtschepen in de Noordzee,* **20**, 1968.

LABAN, C. (1977) Sand occurrences in the North Sea, part 1 (P and Q blocks). *Geol. Surv. Netherlands, Int. Rep.* OP 7514.

LUDWICK, J.C. (1972) Migration of tidal sand waves in Chesapeake Bay entrance. In: *Shelf Sediment Transport* (Ed. by D. J. P. Swift, D. B. Duane and O. H. Pilkey), pp. 377–410. Dowden, Hutchinson & Ross, Stroudsburg, Pennsylvania.

McCAVE, I.N. (1971) Sand waves in the North Sea off the coast of Holland. *Mar. Geol.* **10**, 199–225.

MILLER, M.C., McCAVE, I.N. & KOMAR, P.D. (1977) Threshold of sediment motion under unidirectional currents. *Sedimentology,* **24**, 507–527.

PINGREE, R.D. & GRIFFITHS, D.K. (1979) Sand transport paths around the British Isles resulting from M_2 and M_4 tidal interactions. *J. mar. biol. Ass. U.K.* **59**, 497–513.

PRATT, C.J. (1970) *Summary of Experimental Data for Flume Tests over 0·49 mm Sand.* Rep. CE/9/70. Department of Civil Engineering, University of Southampton, England.

REINECK, H.E. & SINGH, I.B. (1967) Primary sedimentary structures in the recent sediments of the Jade, North Sea. *Mar. Geol.* **5**, 227–235.

SOUTHARD, J.B. (1971) Representation of bed configurations in depth-velocity-size diagrams. *J. sedim. Petrol.* **41**, 903–915.

STEIN, R.A. (1965) Laboratory studies of total land and apparent bed load. *J. geophys. Res.* **70**, 1831-1842.

STRIDE, A.H. (1963) Current-swept sea floors near the southern half of Great Britain. *Q. Jl geol. Soc. Lond.* **119**, 175–199.

STRIDE, A.H. (1970) Shape and size trends for sand waves in a depositional zone of the North Sea. *Geol. Mag.* **107**, 469–477.

SUNDERMANN, J. & KROHN, J. (1977) Numerical simulation of tidal caused sand transport in coastal waters. *Proc. 17th Congr. int. Ass. hydrogr. Res.* **1**, 173–181.

SUTHERLAND, A.J. (1967) Proposed mechanism for sediment entrainment by turbulent flow. *J. geophys. Res.* **82**, 6183–6194.

TAYLOR, B.D. (1971) *Temperature Effects in Alluvial Streams.* Rep. KH-R-27, W. M. Keck Laboratory, California Institute of Technology.

TERWINDT, J.H.J. (1971) Sand waves in the Southern Bight of the North Sea. *Mar. Geol.* **10**, 51–67.

VANONI, V.A. (1975) Closure to discussion of: factors determining bed forms of alluvial streams. *J. Hydr. Div. Am. Soc. civ. Engrs* **101**, 1435–1440.

VANONI, V.A. & BROOKS, N.H. (1957) *Laboratory Studies of the Roughness and Suspended Load of Alluvial Streams.* Rep. E68. Sedimentation Laboratory, California Institute of Technology.

VANONI, V.A. & HWANG, L.S. (1967) Relation between bed forms and friction in streams. *J. Hydr. Div. Am. Soc. civ. Engrs* **93**, 121–144.

VEEN, J VAN (1936) *Onderzoekingen in der Hoofden in verband met de gestaldenheidin der Nederlandsche Kust.* Algemeene Landsdrukkerij, S-Gravenhage.

VOLPEL, VON F. (1959) Studie über der Verhalten weitwandernder Flachseesande in der südlichen Nordsee. *Dt. hydrogr. Z.* **12**, 64–76.

YALIN, M.S. (1972) *Mechanics of Sediment Transport.* Pergamon Press, Oxford. 290 pp.

YALIN, M.S. & PRICE, W.A. (1976) Time growth of tidal dunes in a physical model. *Proc. Symp. Modelling Techniques*, pp. 936–944. Engineering Division, American Society of Civil Engineers, San Francisco.

Spec. Publs int. Ass. Sediment. (1981) **5**, 257–268

Offshore tidal sand-banks as indicators of net sand transport and as potential deposits

NEIL H. KENYON, ROBERT H. BELDERSON, ARTHUR H. STRIDE
and MICHAEL A. JOHNSON

Institute of Oceanographic Sciences, Wormley, Godalming, Surrey GU8 5UB, U.K.

ABSTRACT

Modern tidal sand-banks are generally not quite longitudinal bedforms, but are aligned obliquely to the regional direction of peak tidal flow and to the resulting net sand transport direction by as much as 20°. The crest lines of most sand-banks are rotated in an anticlockwise sense with respect to the regional peak tidal flow direction. In such cases the local sand transport direction on the banks, as indicated by sand-waves, veers to the right on approaching the crest of the bank from either side. Where the regional net sand transport direction is known, the steeper side of the sand-bank is found to face obliquely down the transport path. Thus, as with sand-waves, the asymmetry of the bank may be used as an indicator of net sand transport direction, providing the bank's sense of offset with respect to the peak tidal flow is known. Sand-banks in areas where the peak speeds of ebb and flood tidal currents are equal tend to have symmetrical profiles. New predictions of net sand transport directions for the Southern Bight of the North Sea are in agreement with earlier predictions. The predicted direction in the entrance to the White Sea (U.S.S.R.) is northwards, in the Yellow Sea it is seaward out of Korea Bay but it is landward into the Approaches to Seoul.

Actively maintained sand-banks are distinguished from moribund ones formed at lower sea-levels. The moribund sand-banks have gentler slopes and sand-waves are less frequently found upon them. With continuing transgression such moribund sand-banks could be wholly or partly preserved beneath a mud blanket. Any preserved internal structure will probably take the form of small-scale high-angle cross-stratification, whose dip is almost opposite to that of the regional net sand transport direction, as well as to the large-scale low-angle interfaces parallel to the steep face of the bank.

INTRODUCTION

Tidal sand-banks (the 'tidal current ridges' but not the 'shoreface connected sand ridges' of some authors) are one of the two main depositional facies of the offshore tidal environment, the other being the offshore sand-sheet facies. They are the largest bedforms of shallow, strongly tidal seas, and can, in extreme cases, reach up to 120 km long and 30 km wide. They tend to occur in groups, either well away from the coast or in estuaries, or else they are found as solitary near-coastal and banner banks in the lee of headlands, islands or submerged rock shoals. The tidal currents passing over modern sand-banks are usually stronger than those associated with the modern offshore sand-sheet facies, where the grain size of the sand is in equilibrium with peak current strength. The sand within the banks must therefore either be trapped by transport from seawards into an estuary, lie in the lee of an obstacle, be supplied to a region faster than it can be removed, or have been originally so abundant that there has not yet been time for all of it to be transported away.

Some tidal sand-banks, particularly near the coast and in estuaries, are strongly V- or S-shaped in plan view. The shapes of these have frequently been related in the literature to ebb- or flood-dominant

0141-3600/81/1205-0257 $02.00

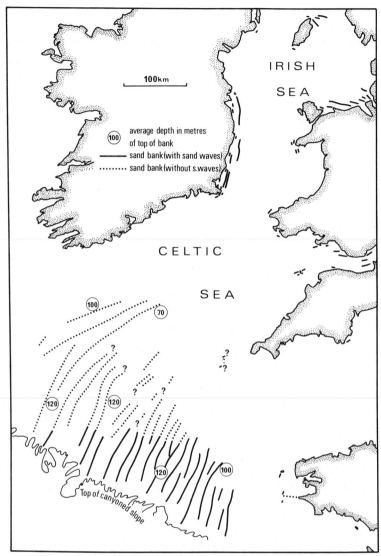

Fig. 1. Offshore tidal sand-banks of the Celtic and Irish Seas. The sand-banks in deep water of the Celtic Sea have relatively gentle slopes and are considered to be moribund (largely predating modern tidal conditions). However, the sand-waves on some of them suggest that they are still subject to some constructional activity by modern currents.

channel systems. This paper will exclude such complex bedforms and confine itself to the more or less straight-crested parts of offshore tidal sand-banks.

The straight or slightly sinuous-crested sand-banks have generally been considered to extend parallel to the peak tidal currents (Off, 1963). However, current observations made by Smith (1969) showed a near-coastal sand-bank to be oblique to peak tidal flow. The sand-bank could thus be thought of as being somewhat analogous to a sand-wave. There were

doubts about the relevance of this example because of uncertainties about the relationship of the bank to other relief. Subsequently, an analytical model (Huthnance, 1973) showed that an oblique orientation could account for the apparent net sand circulation around a group of sand-banks east of Norfolk, U.K., as implied by asymmetries of sand-waves found on the banks. More recent current measurements (McCave, 1979) showed peak flow to be oblique to the North Hinder sand-bank, so leading

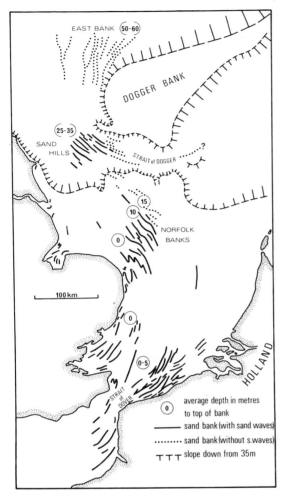

Fig. 2. Offshore tidal sand-banks of the south-western part of the North Sea. The East Bank group, the Sand Hills group and the outermost of the Norfolk Banks are considered to be moribund. However, the sand-waves on much of the Sand Hills group suggest that they are still subject to some constructional activity by modern currents.

him to question whether this relationship might not be more common.

All readily available data have now been assembled for a comparison of regional peak tidal flow directions and net sand transport directions near and on the active Holocene tidal sand-banks of northern Europe. Apart from published information, use has been made for seas around the British Isles of unpublished side-scan sonar records of the Institute of Oceanographic Sciences and *Seasat* side-scan radar images (by courtesy of the European Space Agency).

The regional net sand-transport direction is determined by the stronger of the peak ebb or flood currents and by sand-wave asymmetries away from the immediate vicinity of the sand-bank.

A compilation of all major tidal sand-banks for the seas around the southern half of the British Isles (Figs 1 and 2) includes not only the numerous actively maintained modern (late Holocene) sand-banks but also large numbers of sand-banks first formed at lower sea-levels and now considered to be moribund. Such moribund sand-banks offer favourable sites for preservation of this important facies.

OBLIQUITY OF SAND-BANK AXES TO REGIONAL PEAK TIDAL FLOW

The long axes of many tidal sand-banks are shown on close inspection to be oblique to the regional peak tidal flow. Around the British Isles the orientation of the sand-bank axis is rotated relative to the peak tidal flow direction in an anticlockwise sense in 49 known cases, and in a clockwise sense in only three known cases. (Sand-banks related to headlands are excluded since those tied to anticlockwise eddies are offset in a clockwise sense and those tied to clockwise eddies are offset in an anticlockwise sense.) The pattern is generally consistent along most of the length of a sand-bank except near major bends in the bank axis, and is also consistent within a group of neighbouring banks.

The usual angle between the axis of the sand-bank and the regional peak tidal-flow direction (and the resulting regional net sand transport direction) is between about 7° and 15°, but may range from 0° to 20° or more for extreme values. The directions of regional peak tidal flow and sand transport well away from the sand-banks were based on long-term near-bottom and the many near-surface current measurements, and widespread data from sand-wave, sand-ribbon and wreck-mark orientations. (Near-bottom and near-surface tidal current directions of strong tidal currents are predicted by theory and shown by most observations to be aligned within 5° or less of each other, i.e. much less than the offset of many of the sand-banks.)

Most tidal sand-banks have asymmetric cross-sectional profiles. Where such is the case the steeper slope is observed to be on the side of the bank that faces obliquely in the general direction of regional net sand-transport. Thus, for the purpose of predicting regional net sand-transport paths the

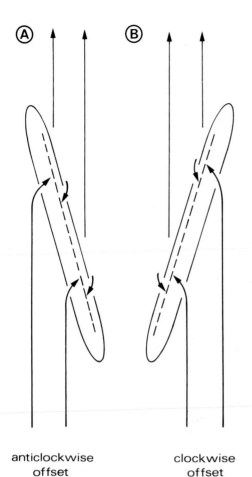

anticlockwise   clockwise
offset   offset

Fig. 3. Two classes of active asymmetrical tidal sand-banks found on the open shelf. The axial offset is either (A) anticlockwise, or (B) clockwise with respect to the regional direction of peak tidal flow. The arrows indicate the direction of net sand transport, which veers towards the right on approaching the sand-bank crest in case (A) and towards the left in case (B). The sand-bank crest is shown by a dashed line. The steeper face of the bank is relatively protected from the regionally stronger (up-page) flow, but more exposed to the regionally weaker reverse (down-page) flow, which thus locally becomes the stronger (as indicated by the pairs of short arrows). These models have been used in later figures both to improve on existing sand-transport maps and to predict net sand-transport direction in new regions.

asymmetric cross-section of a sand-bank may in general be used as an indicator in much the same way as a highly oblique sand-wave, except that one must also know the orientation of the bank relative to the regional peak tidal flow. This is consistent with the observed internal structure of at least some sand-banks, where large-scale growth or migration

surfaces lie parallel to the steeper face of the bank (e.g. Houbolt, 1968). The long-term migration of some of the Norfolk sand-banks in the direction faced by their steeper side was demonstrated by Caston (1972).

LOCAL SAND-TRANSPORT DIRECTIONS ON BANKS

The modern sand-banks under discussion all have large sand-waves upon them. The sand-waves on the upper slopes of the sand-bank are usually orientated obliquely to the bank crest and have their steeper slopes facing towards the crest from either side of the bank (e.g. Houbolt, 1968; Caston & Stride, 1970; Caston, 1972). The sharp crestlines of such sand-banks are thus maintained by sand-wave movement. However, so long as the sand-banks were thought to be parallel to the regional peak tidal flow, it was necessary, in order to explain this sand-wave 'circulation' system on the banks, to have recourse to mutually evasive ebb–flood channel systems (as occur in some sand-choked estuaries) or to separate a channel into ebb-dominance on one side and flood-dominance along the other side of the channel. But, as described above, the sand-banks lie slightly oblique to the peak tidal flow, so that the explanation for the opposed sense of sand-wave migration on either side of the crest is found in the relatively greater exposure of one side and protection of the other side of the bank (together with a narrow band of adjacent floor) during peak flood flow, and vice versa during peak ebb flow. Thus, the degree of obstruction of the sand-bank to the tidal flow is sufficient to allow the local net sand transport direction on the lee (steeper) face of the bank to be opposite to the net regional direction.

In rather more detail the observations show that the sand-waves, as they approach the crest of the sand-bank, do not maintain their crest-lines at approximately normal to the regional peak tidal flow, but bend around to become more parallel to the crest of the bank. In the vicinity of the British Isles the sense of this veering over those banks that are offset in an anticlockwise sense from the peak tidal flow is towards the right as the sand-wave approaches the bank crest (Fig. 3, model A). On the few offshore sand-banks with axes offset in a clockwise sense from the peak tidal flow, the sand-wave crests are found to veer towards the left on approaching the bank crest (Fig. 3, model B).

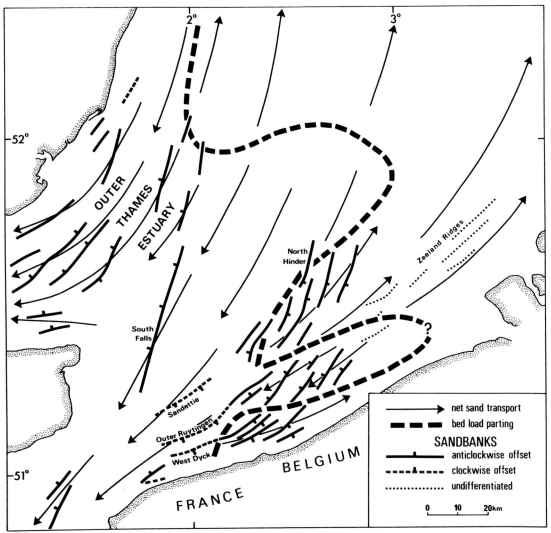

Fig. 4. Offshore tidal sand-banks and net sand-transport directions in the Southern Bight of the North Sea. The steeper sides of the sand-banks are indicated with a tick. The sinuous bed-load parting has been more closely defined than previously by application of the hypothesis set out in Fig. 3. This figure is based on Institute of Oceanographic Sciences data, alignment of flow-parallel streaks relative to many bank axes on *Seasat* side-scan radar imagery, and data from British Admiralty charts, Talbot & Harvey (1981), Houbolt (1968), McCave (1971), D'Olier (1974), Terwindt (1971), Burton (1977), and G. F. Caston (1979).

At least part of this veering of the sand-wave crests towards the sand-bank crest is attributable to variations in the cross-bank component of the current during the tidal cycle. During much of the tidal cycle the cross-bank component will represent a greater proportion of the total flow than it does at peak flow, while the bank itself will also present a

greater obstruction to the flow the greater the angle of approach of the flow towards the crest. The wider the tidal envelope, the greater will be this effect. This may account, for instance, for the observed parallelism of the sand-wave crests to the bank crest on Cultivator Shoal, George's Bank, U.S.A. (Uchupi, 1968), where the tidal current envelope is wide.

SAND-BANKS AS INDICATORS OF REGIONAL NET SAND TRANSPORT DIRECTION

The 'banner' banks that occur in the lee of headlands, islands or submerged rock shoals indicate, by their location, the net sand transport direction in the vicinity of the associated relief feature. They might perhaps best be considered (in the case of the headland-associated banks) as detached submarine spits, or (in the case of the island and rock shoal-associated banks) as giant sand shadows. As this paper is mainly concerned with the more open-shelf tidal current sand-banks, these banner banks will not be discussed further here.

Provided the clockwise or anticlockwise sense of axial offset of the offshore sand-bank crest with respect to the regional peak tidal flow and the asymmetry of the bank (in profile) is known, then the regional sand transport direction can be predicted. If the alignment of peak tidal flow is not known accurately enough, then either the direction faced by the steeper side of sand-wave profiles on the sides of the bank, or the sense of veering of sand-wave crests in plan view, may be used to determine whether the sense of axial offset of the bank is clockwise or anticlockwise relative to the peak tidal flow. This hypothesis is summarized diagrammatically in models A and B of Fig. 3. Predictions of regional net sand transport direction using tidal sand-banks in this way are made below for sand-bank areas in the Southern Bight of the North Sea, the entrance to the White Sea (U.S.S.R.) and the eastern Yellow Sea (west of Korea).

Southern Bight of the North Sea

Sand-wave profiles and the observed sense of veering of their crests on the Norfolk sand-banks indicate that these banks are offset in an anticlockwise sense with respect to the regional peak tidal flow (Fig. 3, model A). Since their steeper slopes face to the north-east then a northwards regional sand transport path is predicted, consistent with other regional evidence (Stride, 1973).

In the southernmost part of the North Sea the already known sand transport directions and those now predicted from sand-bank asymmetry are in agreement. These are shown in a new map of net sand-transport directions (Fig. 4). The sand-banks of the outer Thames Estuary are dominantly steeper on their north-western flanks and have an anticlockwise offset to the regional peak tidal flow, as determined from peak tidal current directions (Talbot & Harvey, 1981) and sand-wave data (D'Olier, 1974). By application of model A of Fig. 3 a net westerly sand-transport direction is predicted. This is consistent with the earlier described net sand-transport direction (Stride, 1973) and with bed shear stress computations (Pingree & Griffiths, 1979), and also with the long-term transport of sand implied by the abundance of sand, as banks, choking the mouth of the Thames Estuary.

The South Falls and the Sandettie sand-banks, which lie obliquely across the sand-transport paths heading in towards the Strait of Dover from the east, are offset in opposite senses with respect to the regional peak tidal flow. A detailed survey shows that the South Falls is offset anticlockwise (G. F. Caston, 1979) and is steeper on its west side, whereas the Sandettie is offset clockwise with respect to regional peak tidal flow (Burton, 1977; and unpublished sector scanning sonar records showing sand-wave asymmetry) and is steeper on its south side. Thus, they lie at an angle of 40° to each other instead of parallel, as might otherwise be expected. The Sandettie, the Outer Ruytingen and the West Dyck are a parallel group of banks that, on present evidence, are the only ones (except for sand-banks tied to anticlockwise eddies in the lee of headlands) known to have clockwise offsets of their crests relative to the peak tidal flow (Fig. 3, model B). The relationship of their crest lines to the crest orientations of the large sand-waves upon them and to streaks on the water surface over them (presumed to be parallel to the peak tidal flow) is clearly seen from a *Seasat* side-scan radar image of patterns of water disturbance at the sea surface (Fig. 5).

The other sand-banks near the Belgian coast, for which there are data, are offset in the more usual anticlockwise sense relative to the peak tidal flow (Houbolt, 1968; *Seasat* side-scan radar images; and tidal current data on British Admiralty charts). There are three groups of these sand-banks. It is predicted from model A of Fig. 3 that each adjacent group has an oppositely directed net sand-transport direction. The groups are thus separated from one another by a sinuous bed-load parting. This bed-load parting is in the same general position as that already proposed (Stride, 1963a; McCave, 1971; Pingree & Griffiths, 1979), but is now seen as being much more complex than previously realized. The proposal of an easterly net sand-transport path near the Belgian coast is supported by an easterly shift in the position

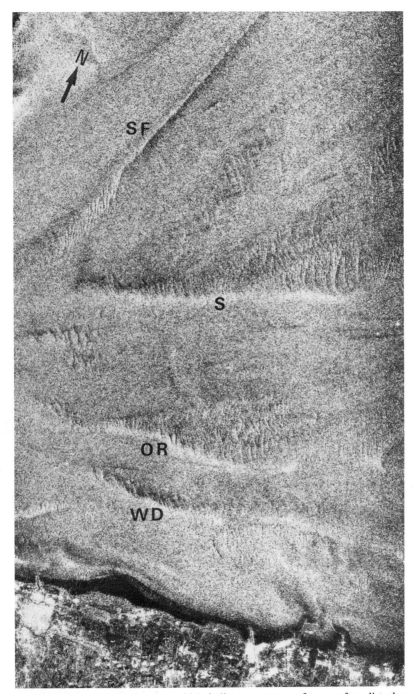

Fig. 5. A portion of *Seasat* side-scan radar imagery which indicates patterns of sea surface disturbance in the south-westernmost corner of the North Sea. The disturbances define the crests of the South Falls (SF), Sandettie (S), Outer Ruytingen (OR) and West Dyck (WD) sand-banks, as well as the crests of numerous large sand-waves. On some of the banks sand-wave crests are seen to make an angle with the sand-bank crest. In particular, note the veering of sand-wave crests towards the crest of Sandettie. A train of sand-waves is seen extending from the southern end of the South Falls. Longitudinal streaks on the water surface, presumed to be flow-parallel, are at an angle of about 20° to the sand-bank axes.

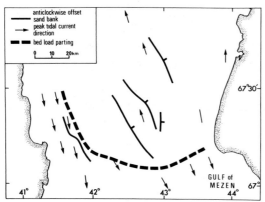

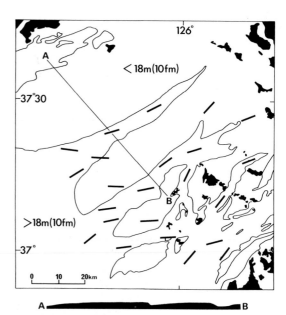

Fig. 6. Offshore tidal sand-banks, the directions of strongest tidal flow, and a predicted bed-load parting in the 'White Sea Funnel' at the entrance to the White Sea, northern U.S.S.R. The steeper sides of the sand-banks are indicated by a tick.

Fig. 7. Exceptionally wide tidal sand-banks (as indicated by the 18 m contour and by profile A–B) in the Approaches to Seoul, eastern Yellow Sea. The alignment of peak tidal flow (from Kang, 1969) is indicated by the black bars. Profile A–B shows the sand-banks to be steeper towards the south-east. The predicted direction of net sand-transport is to the north-east.

of some sand-banks, noted from a comparison of navigational charts of various ages (Van Cauwenberghe, 1971). As with sand-wave profiles (Stride, 1963a), there is evidence that sand-bank profiles at or near the bed-load parting are symmetrical. The Zeeland Ridges off the Dutch coast have not been considered as net sand-transport path indicators because, although they are covered by active sand waves, these banks have older relief within them (Laban & Schüttenhelm, 1981).

White Sea entrance

Sufficient data are available from a group of four parallel sand-banks in the 'White Sea Funnel' to help clarify net sand-transport paths in this area (Fig. 6). Two current measurements (British Admiralty chart 2270) at stations about 5 km from the nearest sand-banks indicate that they are offset by about 10° in an anticlockwise sense from the peak tidal current direction. The three easternmost sand-banks have steeper slopes facing to the east. Thus, by applying model A of Fig. 3 it is possible to predict a sand transport path to the north. This agrees with the direction proposed by Chakhotin, Medvedev & Longinov (1972) on the basis of a shortage of sand in the Gulf of Mezen. If a southerly sand transport into the Gulf of Mezen predicted by Belderson, Johnson & Stride (1978) is correct, then there should be a bed-load parting just to the south of the sand-banks. No obvious asymmetry for the southwesternmost sand-bank can be ascertained from the chart,

which may therefore be close to a bed-load parting between the northerly path proposed here and the southerly near-coastal path already proposed (Belderson *et al.* 1978). Within the White Sea itself, the Western Solovetskaya Salma sand-bank appears (from Chakhotin, 1977, fig. 4b) to be offset by about 9° in an anticlockwise sense from peak tidal flow as indicated by sand-wave orientations.

Yellow Sea

Extensive areas of sand-banks in the eastern Yellow Sea are shown on available navigational charts (based mainly on nineteenth-century surveys). In the southern part of the Approaches to Seoul there are numerous relatively small sand-banks tied to rocky outcrops and islands. In the northern part of this area (Fig. 7) there are very wide (up to 30 km), flat-topped banks separated by narrow channels. The steeper slopes of the banks face south-eastwards. Measurements of peak tidal currents (of between $2\frac{3}{4}$ and $5\frac{3}{4}$ knots) at about twenty stations amongst these banks (Kang, 1969) show that the sand-bank axes are

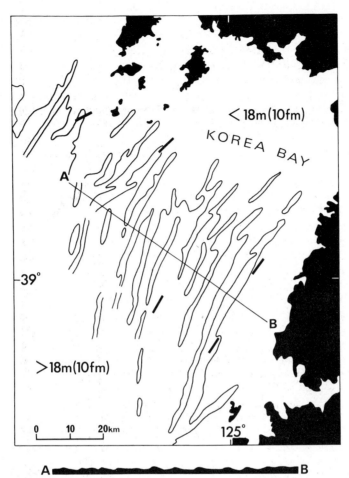

Fig. 8. Narrow tidal sand-banks (as defined by the 18 m contour and profile A–B) in Korea Bay, eastern Yellow Sea. The alignment of peak tidal flow is indicated by the black bars. Profile A–B shows the sand-banks to be steeper towards the north-west. The predicted net sand-transport direction is to the south-south-west.

consistently offset in an anticlockwise sense with respect to direction of peak tidal flow. Thus, one can predict from model A of Fig. 3 that the overall sand transport direction is north-easterly, in towards the coast.

In Korea Bay, off the western coast of North Korea, the sand-banks (Fig. 8) are much narrower than those in the Approaches to Seoul. Off (1963) noted that they are up to 90 km long and between 0·5 and 3 km wide. From the British Admiralty chart 1257 the profiles of the outer banks appear to be either symmetrical, or steeper on their western sides. The few peak current directions on the chart indicate that the bank axes are offset in an anticlockwise sense

with respect to the direction of peak tidal flow. Thus there is just enough information to predict that there should be a net transport of sand seaward to the south-west.

The contrasting widths of the sand-banks (Figs 7 and 8) in these two areas of the Yellow Sea offer some support for the sand transport directions proposed here. The broader banks are found where the predicted net sand transport direction is landward (thus trapping the sand in shallow water so that they must grow laterally) while the narrower banks occur where the predicted direction is seaward (allowing the sand-banks to develop in deeper and less laterally confined waters).

MORIBUND TIDAL SAND-BANKS
AS POTENTIAL DEPOSITS

Various criteria have been used to distinguish between sand-banks maintained by the modern (late Holocene) tidal current regime and those formed at times of lower sea-level (Figs 1 and 2). The actively maintained tidal sand-banks are found where the peak near-surface mean springs tidal currents generally attain well over 1 knot (50 cm sec^{-1}); they have large sand-waves present upon them; the bank crests (when not shallow enough to be flattened by wave action) tend to be sharp, with the steeper of their two slopes approaching a maximum angle of 6°; they are usually separated by gravel floors; and their crests are relatively shallow, often approaching sea-level, at which stage the effects of wave action may plane them off. When confined in estuaries, with a plentiful supply of sand, they have broad, flat tops, frequently exposed at low tide. In contrast, the moribund sand-banks are found where the tidal currents now reach a peak near-surface mean springs speed of less than about 1 knot (50 cm sec^{-1}); they do not have large sand-waves upon them; their crests have rounded profiles and their slopes are only 1° or so; they are separated by sandy or muddy floors; and their crests are in relatively deep water.

No absolute demarcation can be made between the actively maintained and the moribund tidal sand-banks. The term 'moribund' itself implies a variable (though low) level of activity. For example, in the outer Celtic Sea there is an extensive group of moribund sand-banks (Fig. 1) presumably first formed at lower sea-levels (Stride, 1963b; Bouysse *et al.*, 1976) that now have gently dipping slopes. Those in the region of weakest currents (the northernmost ones) have rounded crests and an absence of sand-waves, but when followed towards the shelf edge and south-eastwards their crests become sharper and sand-waves occur upon them as they come increasingly under the influence of stronger currents. Thus, there is a transition from the moribund state to a more active state, both longitudinally along the length of individual sand-banks, and laterally from bank to bank within the group. The level of activity will change in time, depending on variations in water movements.

In the North Sea there are at least two groups of moribund tidal sand-banks (Fig. 2). The East Bank group, located north west of the Dogger Bank, are thought to be related to a sea-level 40 m below that at present (Jansen, 1976) and have no sand-waves on their flanks. The 'Sand Hills' group were previously thought to be a kind of transverse bedform (Dingle, 1965). However, these are now known from new Institute of Oceanographic Sciences, Hydrographic Department and *Seasat* data to be nearly longitudinal to the present-day tidal currents. They are very similar to the transitional, southeasternmost, banks of the Celtic Sea in that they have low angles of slope and yet are partially covered by sand-waves. They were probably initiated at lower sea-levels, when stronger currents flowed in the relatively deep passage south of the Dogger Bank ('Strait of Dogger').

The existence of moribund sand-banks indicates that preservation of some offshore tidal sand-banks is inevitable, particularly during a major and rapid marine transgression. Thus they could be buried beneath a blanket of mud, and are likely to occur in such a stratigraphic setting in ancient deposits. In contrast, the live, flat surfaced tidal sand-banks found in estuaries should, with sufficient sand supply, grow laterally until they tend to merge into a single sand-sheet overlain by tidal flats with migrating tidal channels.

Some tidal sand-banks have been shown to grow by deposition of low-angle layers parallel to their steeper side, with relatively small-scale high-angle cross-stratification associated with sand-waves occurring on that slope (Houbolt, 1968). Since the sand-waves on the steep 'lee' slope of the sand-bank themselves have steep lee slopes facing in a direction opposed to that of the regional net sand transport direction, the cross-stratification resulting from them will give a wrong impression of the regional net sand transport direction if observed in an ancient offshore tidal sand-bank deposit. In the case of the estuarine sand-sheet resulting from the lateral amalgamation of flat-topped sand-banks, the internal structure should differ substantially from that of the offshore sand-sheet facies.

DISCUSSION

The predictions, using the hypothesis summarized in Fig. 3, of sand transport paths made above for the Southern Bight of the North Sea are confirmed by evidence from other methods, while those made for the White and Yellow Seas remain to be tested. This approach, if further verified, substantially increases the total area of a strongly tidal sea for which net sand-transport directions can be described.

In this paper the sand-banks are used as indicators of regional net sand transport only. How much they move, and in precisely which direction, whether directly down the sand-transport path, or with some component transverse to the regional peak tidal flow, needs to be examined. Many other aspects concerning these bedforms also remain to be clarified, for example, net gain and loss of sand to the banks and how this is achieved, and variations in bank shape, particularly at either end.

Discussion on the mechanisms controlling the variable amount and sense of offset of tidal sand-banks with respect to peak tidal flow will not be attempted in any detail here. However, account needs to be taken of the facts: (*a*) that all the active tidal sand-banks so far studied exhibit upon them the pattern of sand-wave asymmetry first described for the Norfolk Banks (Houbolt, 1968; Caston & Stride, 1970; Caston, 1972); (*b*) that most tidal sand-banks so far observed are offset in an anticlockwise sense from the regional peak tidal flow, and consequently from the regional net sand-transport path; and (*c*) that laboratory rotating flows, modelling rotating-earth shelf currents, are found (Greenspan, 1968) to develop spiral-flow vortices, aligned either a few degrees to the right (larger vortices) or 14°–15° to the left (2–3 times smaller vortices) of the current direction: this suggests a possible mechanism for building sand-banks parallel to the vortex axes.

The sense of axial offset of a sand-bank with respect to the regional peak tidal flow may be related in some way to the sense of rotation of the tidal current vector and the associated lag in the rate of picking up, transport (especially in suspension) and setting down of sand after a change in local current speed (Stride, 1974). Many of the sand-banks which are offset in an anticlockwise sense with peak tidal flow are in fact situated in a region where the tidal current vectors rotate clockwise, while the three sand-banks which are offset in a clockwise sense are situated in a region where the tidal current vectors rotate anticlockwise (Sager & Sammler, 1975). However, the group of neighbouring banks to the east of the latter appears to be an exception to such a proposed relationship. On the other hand, should the sand-bank orientation be more directly influenced by Coriolis force (which in any case has an influence on the sense of rotation of tidal current vectors), then in the southern hemisphere the preferred sense of axial offset to the peak tidal flow could be in a clockwise sense. This also remains to be tested. Evidence, largely from ancient deposits, of a differing sense of deflexion of various bedforms in the northern and southern hemispheres has been put forward by Smirnov & Khramov (1975).

It has been assumed throughout this paper that net sand-transport paths are orientated parallel to the flow of the strongest tidal current flow direction and at right angles to the mean crest orientation of large sand-waves. The longitudinal tidal sand-banks are here shown generally to be somewhat obliquely orientated bedforms. Further evidence of sand transport at oblique angles to the strongest tidal current direction should be sought. Indeed there is some evidence, from *Seasat* side-scan radar images of the outer Bristol Channel, that the overall crest orientations within some fields of large sand-waves are not always exactly normal to the peak tidal flow. Coriolis force may also offer some explanation for this.

ACKNOWLEDGMENTS

The authors wish to thank the Fisheries Laboratory, Lowestoft (and in particular Dr F. R. Harden-Jones), for the use of the sector scanning sonar on board R.V. *Clione*, and the European Space Agency (and in particular Dr T. D. Allan) for making available side-scan radar imagery from *Seasat*.

REFERENCES

Belderson, R.H., Johnson, M.A. & Stride, A.H. (1978) Bed-load partings and convergences at the entrance to the White Sea, U.S.S.R., and between Cape Cod and Georges Bank, U.S.A. *Mar. Geol.* **28**, 65–75.

Bouysse, P., Horn, R., Lapierre, F. & Le Lann, F. (1976) Etude des grands bancs de sable du Sud-est de la mer Celtique. *Mar. Geol.* **20**, 251–275.

Burton, B.W. (1977) An investigation of a sandwave field at the south-western end of Sandettie Bank, Dover Strait. *Int. hydrogr. Rev.* **54**, 45–59.

Caston, G.F. (1979) Wreck marks: indicators of net sand transport. *Mar. Geol.* **33**, 193–204.

Caston, V.N.D. (1972) Linear sand banks in the southern North Sea. *Sedimentology*, **18**, 63–78.

Caston, V.N.D. & Stride, A.H. (1970) Tidal sand movement between some linear sand banks in the North Sea off northeast Norfolk. *Mar. Geol.* **9**, M38–M42.

Chakhotin, P.S. (1977) Some results of a study of the tidal sand waves in the White Sea. *Oceanology*, **17**, 182–188.

Chakhotin, P.S., Medvedev, V.S. & Longinov, V.V. (1972) Sand ridges and waves on the shelf of tidal seas. *Oceanology*, **12**, 386–394.

Dingle, R.V. (1965) Sand waves in the North Sea mapped by continuous reflection profiling. *Mar. Geol.* **3**, 391–400.

D'OLIER, B. (1974) Past and present sedimentation in the Thames Estuary, England. *Mem. Inst. Geol. Bassin d'Aquitaine,* **7,** 287–290.

GREENSPAN, H.P. (1968) *The Theory of Rotating Fluids.* Cambridge University Press. 327 pp.

HOUBOLT, J.J.H.C. (1968) Recent sediments in the Southern Bight of the North Sea. *Geologie Mijnb.* **47,** 245–273.

HUTHNANCE, J.M. (1973) Tidal current asymmetries over the Norfolk Sandbanks. *Est. Coastal Mar. Sci.* **1,** 89–99.

JANSEN, J.H.F. (1976) Late Pleistocene and Holocene history of the northern North Sea, based on acoustic reflection records. *Neth. J. Sea Res.* **10,** 1–43.

KANG, Y.C. (1969) Tidal current in vicinity of Yeon Pyeong-Do (West coast of Korea). *Technical Reports for the year 1969,* pp. 98–122. Hydrographic Office, Republic of Korea.

LABAN, C. & SCHÜTTENHELM, R.T. (1981) Some new evidence on the origin of the Zeeland Ridges. In: *Holocene Marine Sedimentation in the North Sea Basin* (Ed. by S.-D. Nio *et al.*). *Spec. Publs int. Ass. Sediment.* **5,** 239–245. Blackwell Scientific Publications, Oxford. 524 pp.

McCAVE, I.N. (1971) Sand waves in the North Sea off the coast of Holland. *Mar. Geol.* **10,** 199–225.

McCAVE, I.N. (1979) Tidal currents at the North Hinder Lightship, southern North Sea: flow directions and turbulence in relation to maintenance of sand banks. *Mar. Geol.* **31,** 101–114.

OFF, T. (1963) Rhythmic linear sand bodies caused by tidal currents. *Bull. Am. Ass. Petrol. Geol.* **47,** 324–341.

PINGREE, R.D. & GRIFFITHS, D.K. (1979) Sand transport paths around the British Isles resulting from M_2 and M_4 tidal interactions. *J. mar. biol. Ass. U.K.* **59,** 497–513.

SAGER, G. & SAMMLER, R. (1975) *Atlas der Gezeitenströme für die Nordsee, den Kanal und die Irische See.* See-hydrographischer Dienst der Deutschen Demokratischen Republik, Rostock. 58 pp.

SMIRNOV, L.S. & KHRAMOV, A.N. (1975) Coriolis force and the texture of the sandstone-siltstone rocks *vis-à-vis* the paleomagnetic latitudes. *Izv. Earth Phys.* **3,** 66–79 (translation).

SMITH, J.D. (1969) Geomorphology of a sand ridge. *J. Geol.* **77,** 39–55.

STRIDE, A.H. (1963a) Current-swept sea floors near the southern half of Great Britain. *Q. Jl geol. Soc. Lond.* **119,** 175–199.

STRIDE, A.H. (1963b) North-east trending ridges of the Celtic Sea. *Proc. Ussher Soc.* **1,** 62–63.

STRIDE, A.H. (1973) Sediment transport by the North Sea. In: *North Sea Science* (Ed. by E.D. Goldberg), pp. 101–130. MIT Press, Cambridge.

STRIDE, A.H. (1974) Indications of long term, tidal control of net sand loss or gain by European coasts. *Est. Coastal Mar. Sci.* **2,** 27–36.

TALBOT, J.W. & HARVEY, B.R. (1981) Investigation of the dispersal of sewage sludge in the Thames Estuary. *Fish. Res. Tech. Rep.* MAFF Direct. Fish. Res., Lowestoft.

TERWINDT, J.H.J. (1971) Sand waves in the southern bight of the North Sea. *Mar. Geol.* **10,** 51–67.

UCHUPI, E. (1968) Atlantic continental shelf and slope of the United States—physiography. *Prof. Pap. U.S. geol. Surv.* **529C,** 30 pp.

VAN CAUWENBERGHE, C. (1971) Hydrografische analyse van de Vlaamse banken langs de Belgisch-Franse kust. *Ingenieurstijdingen,* **20,** 141–149.

Spec. Publs int. Ass. Sediment . (1981) **5**, 269–281

Sediment transport measurements in the Sizewell–Dunwich Banks area, East Anglia, U.K.

BARBARA J. LEES

*Institute of Oceanographic Sciences, Taunton TA*1 *2DW, U.K.*

ABSTRACT

Sediment transport measurements have been made of the sand fraction of both bedload and suspended load in the Sizewell–Dunwich area, East Anglia, U.K., as part of a programme studying the processes of sediment movement near linear sandbanks.

The bedload has been measured using a fluorescent coated sand. Grain-size analyses of pumped samples, taken at times of maximum flow, have shown distributions overlapping those of the top few centimetres of box core samples of the sea-bed. Therefore a coated sand, slightly coarser than the mean of that found on the sea-bed, was used. Three-quarters of a tonne of wetted fluorescent sand was pumped as a slurry down a pipe directly to the sea-bed. Surveys were carried out at 1, 3, 5, 6, 51, 165 and 230 days after injection, with a mean of 150 grab samples being obtained each time. The tracer cloud ranged from 450 to 1050 m long and 50 to 300 m wide. Depth of burial measurements were made using box and vibrocores. By D 231 the mean measured depth was $16\cdot8$ cm. The centroid movement shows that in the short term the tracer distribution was dependent on the tidal flow pattern, but in the longer term owed more to wave conditions. Calculated bedload sediment transport rates are relatively low. The maximum rate was $0\cdot040$ g cm^{-1} sec^{-1} before the tracer had achieved equilibrium with the background sand, slowing to $0\cdot012$ g cm^{-1} sec^{-1} after equilibrium appeared to have been reached. During a period when there were high waves, caused by severe predominantly north-east gales, the rate rose to $0\cdot015$ g cm^{-1} sec^{-1} with the centroid having moved towards the crest of the banks.

The suspended load was measured for one tidal cycle during spring tides at each of five stations. Calculations show that in marked contrast to bedload rates, the suspended sand transport rate was high, with sediment concentrations of 423 mg l^{-1} at $1\cdot75$ m above the sea-bed and 1892 mg l^{-1} at 15 cm above the sea-bed at Station PS 4. An instantaneous rate of $30\cdot819$ g cm^{-1} sec^{-1} for the total water column has been measured at maximum flow during a spring ebb tide, and of $12\cdot630$ g cm^{-1} sec^{-1} during the flood tide. The calculated net suspended sediment transport rate is $5\cdot660$ g cm^{-1} sec^{-1} but the high rate at this site in the ebb (NNE) direction at this station is probably not typical.

INTRODUCTION

This paper describes the early results of experiments performed during 1978 and 1979 to measure sediment transport rates of the sand fraction, both bedload and suspended load, in the Sizewell–Dunwich Banks area of East Anglia (Fig. 1).

There are many ways of calculating sediment transport rates and directions from a knowledge of current velocities and grain sizes. For example, Heathershaw & Hammond (1979) have used formulae

0141–3600/81/1205–0269 $02.00
© 1981 International Association of Sedimentologists

developed by Einstein (1950), Yalin (1963) and Bagnold (1963), in the modified form due to Gadd, Lavelle & Swift (1978), to predict net bedload transport rates, and those of Engelund & Hansen (1967) and Ackers & White (1973) for total load rates. They have compared these with observed transport rates measured in Swansea Bay, South Wales, using radioactive tracers and have found that in this particular case the modified form of Bagnold's (1963) formula gave the best estimate of the bedload rates.

To test the generality of this conclusion further

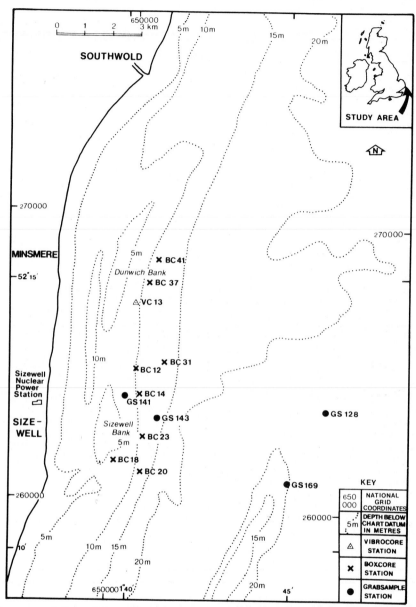

Fig. 1. Location map of Sizewell–Dunwich Banks showing grab sample, box core and vibrocore stations mentioned in text.

work of this type was carried out in an area differing in sedimentological characteristics, tidal flow, wave regime and shoreline geometry from Swansea Bay, and the area selected was that of the Sizewell–Dunwich Banks.

The sedimentary characteristics of the area are well suited to sand transport rate measurements as

the sand lies on a platform of alluvial clay in the north, and coarser sediments of the Norwich Crag Series in the south (Lees, 1980). This platform slopes gently from the shore to a depth of 15 m over a distance of 5 km. For a depth of at least 1 m the bank comprises unconsolidated material, mainly fine to very fine sand. Geophysical work shows a maxi-

Table 1. Grain-size analysis of selected grab and box core samples

Type of sample and number	Percentage of mud	Percentage of sand	Percentage of gravel	Mean of sand fraction (ϕ)	Standard deviation of sand fraction (ϕ)
Grab sample					
128	Nil	94·2	5·8	1·06	0·71
141	8·7	91·3	Nil	3·09	0·46
143	Nil	100·0	Nil	2·55	0·26
169	Nil	100·0	Nil	2·12	0·58
Box core					
12	5·8	94·2	Nil	3·38	0·64
14	1·0	99·0	Nil	2·89	0·32
18	0·3	99·7	Nil	2·52	0·47
20	0·3	99·7	Nil	2·19	0·48
23	1·0	99·0	Nil	2·74	0·33
31	2·0	98·0	Nil	3·02	0·43
37	1·0	99·0	Nil	2·92	0·43
41	1·6	98·4	Nil	3·01	0·38

mum sand thickness of 9·2 m in the north with the bank rising to within 3·5 m of the sea-surface, the base of the sand being clearly identifiable. It becomes less distinct towards the south where the bank overlies sandy and shelly sediments. Grab samples, box and vibrocores have enabled textural measurements to be made. Mean grain size over the whole surface area, and in the 20 vibrocores, has been assessed according to the Wentworth Scale, using a Grain Size Comparator Disc (Kirby, 1973) in the field. Visual field estimates were confirmed in the laboratory when the top 3 cm of eight box cores and four grab samples were sieved at $\frac{1}{4}\phi$ intervals (Table 1). Fig. 2 shows the sediment distribution mapped in 1975.

Tidal flow is parallel to the bank and coastline, which are aligned in a NNE to SSW direction. Maximum velocities reach 150 cm sec^{-1} with only a short slack water period. The tidal range is 1·1 m during neap tides and 1·9 m during maximum springs.

BEDLOAD TRANSPORT MEASUREMENTS

Experimental procedure

Very few workers have attempted to use fluorescent tracer for quantitative work in the offshore zone. The earliest fluorescent tracer studies were by Medvedev & Aibulatov (1956) and Boldyrev (1956) in the USSR, cited in Nelson & Coakley (1974). Jolliffe (1963) carried out work on the Lowestoft Sandbank, an area to the north of the Sizewell–Dunwich Banks, and Vernon (1966) used fluorescent tracer off southern California. Brattelund & Bruun (1974) employed this technique when examining scour round the bases of the legs of an oil rig platform in the North Sea. Quantitative sediment transport measurements, using fluorescent dyed sand, have been made on beaches and in surf zones by many workers (e.g. Bruun, 1968; Komar, 1969; Ingle & Gorsline, 1973). Radioactive tracer techniques have been employed frequently offshore (e.g. Courtois & Monaco, 1969; Heathershaw & Carr, 1977). Radioactive labelling has the advantage of ease in monitoring, as no direct samples are necessary, although there is a potential health risk. There is also the difficulty of measuring the depth to which the tracer becomes mixed in the sediment. However, it was not possible to use radioactive tracer techniques in the Sizewell–Dunwich area because of the health physics monitoring requirements at the nearby Sizewell Nuclear Power Station (Fig. 1). Therefore fluorescent tracer was used, and several new techniques were developed to cope with the associated problems (Lees, 1979).

The injection site was chosen at the sandy southeast corner of the Dunwich Bank where the structure is reasonably well understood (Fig. 2).

The first requirement was that of matching the tracer to the sand on the sea-bed. This is not as critical as it would be for radioactive labelling because, provided that recovery rates are high enough, subsequent sea-bed samples can be sieved and the sediment transport rates calculated for each fraction (de Vries, 1973). Nevertheless, the tracer must be reasonably similar in grain-size distribution to the background sand. Grain-size analyses for samples from the sea-bed and of sediment in suspension at

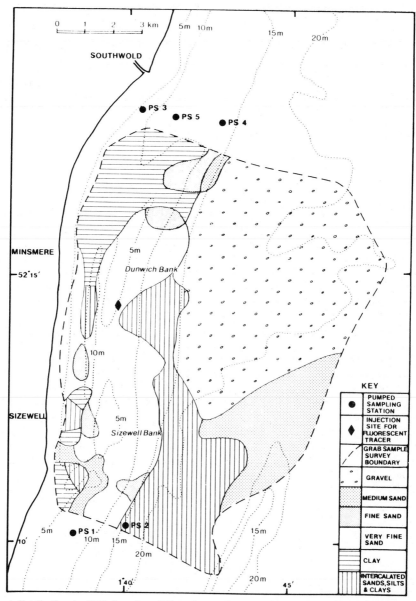

Fig. 2. Sediment distribution in 1975, based on 251 grab samples spaced at approximately 500 m intervals, plus nine box cores. Stations for sediment transport measurements also shown. For details of sampling see Lees (1980).

times of maximum flow show that the material in suspension is very close in grain-size distribution to that on the sea-bed (Fig. 3). In other words, at times of maximum flow all size fractions to be found on the sea-bed can also become suspended in the water column. Therefore a readily available slightly coarser sand, whose grain-size distribution is also shown in Fig. 3, was chosen for dyeing, with the aim

of keeping as much of it as possible on the sea-bed and so restricting transport to bedload processes. The subsequent recovery rates would suggest that we have largely achieved this. The sand was commercially produced by Feslente (BIS) Ltd, using a red fluorescent dye which was fixed from aqueous solution.

Prior to injection a background survey of all the

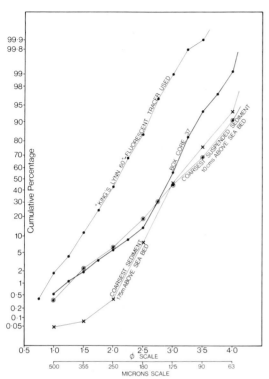

Fig. 3. Grain-size analyses of sea-bed, suspended sediment and fluorescent tracer samples.

area likely to be traversed by the tracer was carried out and no material which could be confused with the coated sand was found.

The second requirement was one of getting a large quantity of the tracer sand quickly and directly on to the sea-bed, with none escaping into the water column, in an area where diving is impracticable because of the bad visibility. This was achieved by pumping a slurry, comprising the tracer plus sea-water, down a pipe directly to the sea-bed at slack water (Lees, 1979).

Previous work and reference to the appropriate literature would suggest that the mobile layer can be at least 10 cm thick (Courtois & Monaco, 1969; Heathershaw & Carr, 1977). In general, calculation of sediment transport rates requires the total amount of tracer injected to be accounted for (de Vries, 1973). Therefore a grab is preferable to sticky cards (Nelson & Coakley, 1974) as the latter samples only the surface layer of grains and recovery is minimal on a rippled surface. The Shipek grab used cuts a semicylindrical sample from the sea-bed of maximum depth of 15 cm. Each sample from the post-injection surveys was split into two sub-samples. One part was immediately spread out thinly and examined in the dark under ultraviolet light so that the tracer cloud could be delineated on site. The other sub-sample was returned to the laboratory for washing,

Table 2. Depth of mixing measurements

Date	Box core number	Maximum depth (m)	Mean tracer concentration (grains kg^{-1})
23 August 1978. Injection day + 2	60	0·003	412
	61	0·003	1414
	63	0·010	177
	64	0·008	1966
	65	0·004	1856
	66	0·019	1172
	67	0·018	1060
	68	0·010	2050
	69	0·016	12,977
	71	0·005	1979
	72	0·010	2085
	75	0·010	424
5 April 1979. Injection day + 226	Vibrocore VC 13	0·22	1861
10 April 1979. Injection day + 231 (Day 231)	S 1	0·22	1844
	S 2	0·18	882
	S 3	0·20	566
	S 4	0·12	409
	S 5	0·12	492
	S 7	0·16	931
	S 8	0·18	2984

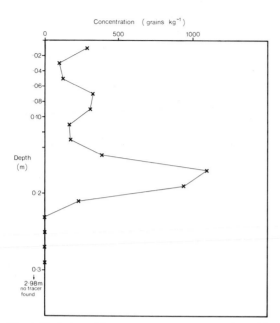

Fig. 4. Variation of fluorescent tracer concentration with depth in vibrocore VC 13.

drying, subdividing further if necessary, and accurate counting of the tracer grains.

Seven surveys were undertaken, each comprising on average 150 samples. It quickly became apparent that the bulk of the tracer was not moving over large distances, and therefore the original grid spacing of 500 m was reduced to approximately 10 m. The necessary accurate position fixing was obtained with a Del Norte Trisponder. This equipment has a resolution of 1 m, with a range repeat accuracy of ±3 m. The readings for each sample were corrected for the offset of the grab from the Trisponder master on board ship.

Of the seven surveys, the first four were in August 1978, soon after the injection, the fifth in October 1978, the sixth in February 1979, and the final one in April 1979 when the tracer was recovered with concentrations of no more than 10^2 grains kg^{-1}. The four-month gap between Surveys 5 and 6 was due to adverse weather conditions. Originally a survey had been planned for December.

The last requirement was to measure the depth to which the tracer had become mixed with the sea-bed sediment. As both box and vibrocoring techniques provide relatively undisturbed samples of the sea-bed, cores for depth analysis were taken in the area of the tracer cloud in August 1978 and April 1979. Subsamples of the box cores were made using 6 cm

diameter plastic tubing. All the samples were halved lengthways and the sediment removed in 0·1 cm thick horizontal sections for the first box cores, and in 2 cm sections for the later box cores and the vibro-core, until no more tracer was found. This enabled the depths to which it had penetrated to be measured. The sections containing tracer were washed, dried, the number of fluorescent grains counted and the mean concentration in grains kg^{-1} calculated for each core (Table 2). Additionally the typical variation of concentration with depth is shown for the vibro-core in Fig. 4.

RESULTS

The stations for the field readings form an irregular grid which was regularized by interpolation, using a computer program, in order to facilitate calculation of the centroid. Graphic printouts in the form of concentration contour maps were produced from the concentrations measured in the laboratory. Manually smoothed versions of these are shown as Fig. 5(A, B). The data from the first two surveys have been combined, and also those from the second two.

The first three contour maps, covering tracer dispersion up to 51 days after the injection, show elongation of the tracer cloud on the sea-bed parallel to the bank, reflecting the almost rectilinear tidal flow pattern. In spite of strong currents of up to 150 cm sec^{-1} measured near the injection site, the recorded tracer sand did not move far. After Day (D) 51, lateral spreading became apparent, particularly towards the bank in a westerly (shorewards) direction. Table 3 shows the maximum excursions for the different surveys, using a concentration of 10 grains kg^{-1} as the criterion for the boundary.

These concentration results were converted to sediment transport rates using the method described by Heathershaw & Carr (1977), based on similar methods used for radioactive tracer by Courtois & Monaco (1969) and other workers. The method involves calculating the centroid, or centre of gravity of each tracer cloud, knowing the depth to which the tracer has become mixed and combining the two. The x and y coordinates of the centroid of each cloud are given by

$$\bar{x} = \frac{\sum\limits_{i=1}^{M} C_i x_i}{\sum\limits_{i=1}^{M} C_i} \quad \text{and} \quad \bar{y} = \frac{\sum\limits_{i=1}^{M} C_i y_i}{\sum\limits_{i=1}^{M} C_i} \qquad (1)$$

where C_i is the concentration of the tracer at a

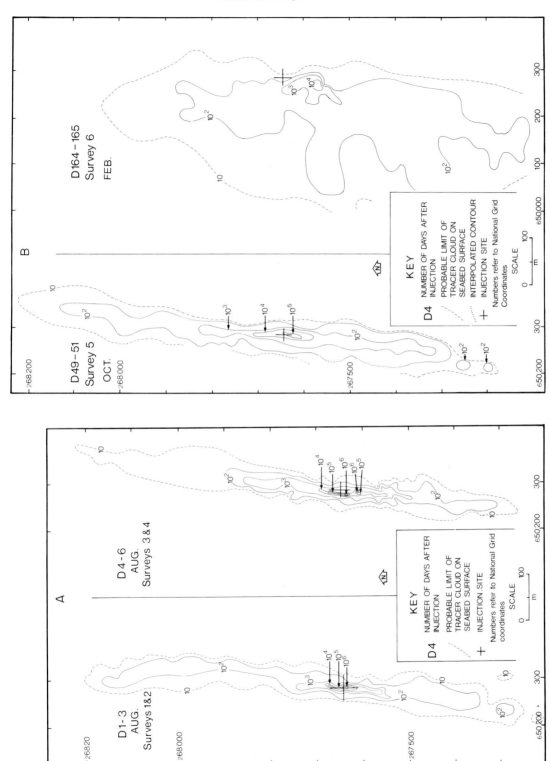

Fig. 5. (A) Contours of fluorescent tracer concentrations in grains kg⁻¹ for Surveys 1 + 2 and Surveys 3 + 4. (B) Contours of fluorescent tracer concentrations in grains kg⁻¹ for Surveys 5 and 6.

measured sampling point, x_i and y_i are its co-ordinates, and M is the number of sample points.

Table 2 shows that the measured depth of mixing and the concentration of tracer in the cores varies in both sets of observations, with the greatest depths and highest concentrations near the injection site, as would be expected. It has been assumed that the depth of burial increases exponentially with time, and that by D 231 the system is close to equilibrium. The mean depth from the seven box cores taken at the end of the experiment is 16·86 cm and therefore the equilibrium depth has been taken as 17·0 cm. The computed depths of mixing for the various surveys are then as shown in Table 4.

Sediment transport rates have been calculated using the relationship

$$q_{sb} = \rho VE, \qquad (2)$$

where q_{sb} is the sediment transport rate per unit width, E is the depth of mixing, and V is the velocity of the centroid. The bulk density of the sediment ρ, was taken as 2·03 g cm^{-3} (Terzaghi & Peck, 1967).

The method depends on the recovery rate of the tracer being high, and calculations indicate that at least 60% of the tracer can be accounted for in each survey. Table 4 shows the results. For each survey the distance and bearing of the centroid from the injection site is given, the centroid velocity, or drift rate, and also the mean rate of daily transport since the injection. Table 5 gives the same information in a slightly different form, where the sediment transport rates are calculated from survey to survey rather than from the original injection time (see also Fig. 6).

SUSPENDED LOAD TRANSPORT MEASUREMENTS

Experimental procedure

The apparatus used for measuring the suspended sediment load was similar to that described by Crickmore & Aked (1975). Four Braystoke rotors at heights of 25, 45, 85 and 175 cm above the sea-bed measured the current speeds, whilst the six sediment sampling nozzles were at the same heights, with two additional ones at 10 cm and 15 cm above the sea-bed. The pump measured the volume of water passing through the filtering system, whilst nylon filters with a 40 μm mesh trapped the sand-sized

Table 3. Maximum observed excursion of fluorescent tracer cloud during each survey. Concentration 10 grains kg^{-1} used as boundary

Survey number	Maximum elongation (m)	Maximum width (m)
1+2	968	92
3+4	984	92
5	1072	112
6	c. 1070	c. 320
7	c. 360	c. 180

Table 4. Drift rates and net sediment transport rates for each survey, referred to injection site

Survey number	Days after injection (mean)	Distance from injection site (m)	Bearing from true north (°)	Centroid velocity or drift rate (m day^{-1})	Estimated depth of mixing (m)	q_{sb} (g cm^{-1} sec^{-1})
1+2	2	8·06	10	4·03	0·013	0·012
3+4	5	14·32	171	2·86	0·023	0·016
5	50	33·24	190	0·66	0·112	0·017
6	164·5	55·07	244	0·33	0·165	0·013

Table 5. Drift rates and sediment transport rates between surveys

Survey numbers	Number of days between	Distance travelled (m)	Drift rate (m day^{-1})	Estimated depth of mixing (m)	q_{sb} (g cm^{-1} sec^{-1})
1+2 from injection site	2	8·06	4·03	0·013	0·012
3+4 from 1+2	3	22·09	7·36	0·023	0·040
5 from 3+4	45	20·25	0·45	0·112	0·012
6 from 5	114·5	44·41	0·39	0·165	0·015

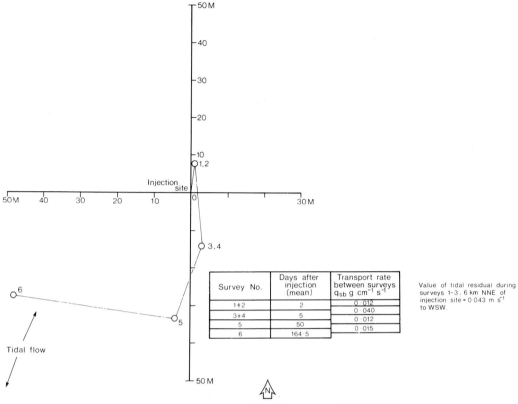

Fig. 6. Movement of tracer centroid observed over a period of 165 days.

Survey No.	Days after injection (mean)	Transport rate between surveys q_{sb} g cm^{-1} s^{-1}
1+2	2	0·012
3+4	5	0·040
5	50	0·012
6	164·5	0·015

Value of tidal residual during surveys 1-3, 6 km NNE of injection site = 0·043 m s^{-1} to WSW.

material. The samples were returned to the laboratory, then dried and sieved, primarily to separate the sand fraction from the silt and clay. This was necessary because, as sand accumulated on the filter, it began to act as a filter itself, trapping increasingly smaller particles.

Five stations were occupied, each for the duration of a tidal cycle, during spring tides. Their locations are shown in Fig. 2. Because there is always a large amount of sand in suspension in the Sizewell–Dunwich area, the filters trapped an adequate amount of sediment for analysis quickly, often becoming completely blocked in under a minute. Time is needed between sampling levels to flush the system to remove the previous level's sediment. Profiles were therefore usually completed in between 10 and 20 min. Complete profiles were taken at half-hourly intervals. The longest uninterrupted data series was obtained at PS 4, located approximately 7 km NNE of the injection site for the bedload experiment (Fig. 2) and therefore the results from that station will be presented in full. On the basis of earlier experimental

data it is assumed that the sea-bed at that site is rippled sand with a wavelength of about 6 cm.

RESULTS

Velocity profiles were averaged over 10 min intervals and then fitted to the Karman–Prandtl logarithmic profiles using the equation

$$u = \frac{U_*}{k} \cdot \ln\frac{z}{z_0}, \qquad (3)$$

where u is the velocity at a height above the sea-bed z, U_* is the friction velocity, z_0 is the roughness length (the height at which the velocity equals zero) and k is von Karman's constant, generally accepted as equal to 0·4.

Of 53 readings made at PS 4, 50 fit Karman–Prandtl logarithmic profiles with a 95% confidence level, and 32 at a 98% confidence level. Such high correlation coefficients do not necessarily mean that the profiles are strictly logarithmic, but Fig. 7 shows that only those measured near slack water depart much from this ideal. A possible reason for this departure is discussed below.

From the relationships shown by equation (3), the friction velocities and roughness lengths were also calculated for each 10 min group of profiles. Fig. 8 is a plot of friction velocity through the tidal cycle, with a value at maximum tidal flow of just over 5 cm sec^{-1} during both ebb and flood. Concentrations of sediment in mg l^{-1} can be calculated from the known volume of water pumped and known weights of sand trapped on the filters.

The resulting sand concentrations were fitted to Rouse concentration profiles, using the relationship

$$\frac{C_z}{C_a} = \left(\frac{h-z}{z} \cdot \frac{a}{h-a}\right)^{w_s/kU_*}, \qquad (4)$$

where C_z is the concentration at height z, C_a is the reference concentration, h is the total water depth (1300 cm), and a is the height of the reference concentration (100 cm). The Stokes' settling velocity W_s is 0·35 cm sec^{-1} for a mean grain size of approximately 72 μm/3·8ϕ and the density of quartz sand taken as 2·65 g cm^{-3}.

Of the 21 profiles sampled, two-thirds fit Rouse profiles at the 95 % confidence level, and half of these at the 99 % level. The ones which fit least well tend to be those measured as the tide was slackening during the ebb flow, or when accelerating after turning, during both flood and ebb.

Figure 9 shows the variation of the reference concentration with time, calculated at a height of 100 cm above the sea-bed, with a maximum of 320 mg l^{-1} during the flood and 680 mg l^{-1} during the ebb.

Having fitted the velocity profiles to the Karman–Prandtl logarithmic form, and the concentrations to Rouse profiles, the suspended sediment transport rate, q_{ss}, may then be calculated by numerically integrating the product of equations (3) and (4) over the entire flow depth, i.e.

$$q_{ss} = \frac{C_a U_*}{k}\left(\frac{a}{h-a}\right)^{w_s/kU_*} \int_{z_0}^{h}\left(\frac{h-z}{z}\right)^{w_s/kU_*} . \ln\frac{z}{z_0} . dz$$

$$(5)$$

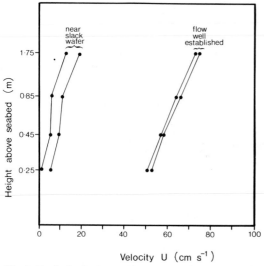

Fig. 7. Typical velocity profiles for Station PS 4.

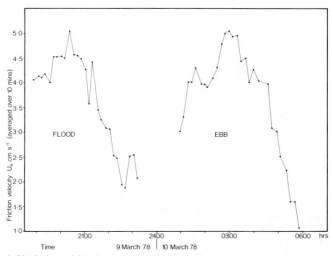

Fig. 8. Variation of friction velocity with time during one tidal cycle at Station PS 4.

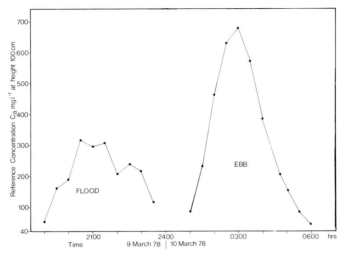

Fig. 9. Variation of reference concentration (at height 100 cm) with time during one tidal cycle at Station PS 4.

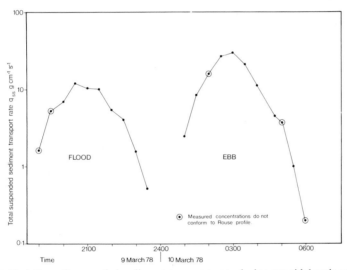

Fig. 10. Variation of suspended sediment transport rate during one tidal cycle at Station PS 4.

where z_0 is the roughness length and the integration increment dz is equal to 0.01 cm.

The variation of the total suspended sediment transport rate during one spring tidal cycle is shown in Fig. 10, with a maximum value of 12.630 g cm^{-1} sec^{-1} during the flood, and 30.819 g cm^{-1} sec^{-1} during the ebb. The times when the concentrations do not fit Rouse profiles are also indicated. Integrating flood and ebb values gives an overall ebb residual of 5.660 g cm^{-1} sec^{-1} for one tidal cycle.

DISCUSSION

Sediment may be moved on or above the sea-bed in two modes, firstly as bedload in which grains roll or saltate along the sea-bed, perhaps at heights of up to a few grain diameters, and secondly as suspended load in which grains are transported within the body of the fluid at some distance (many grain diameters) from the boundary.

For both modes of transport it has been assumed

for the purposes of the transport rate calculations that all particles have the same density, that of quartz sand (2·65 g cm⁻³). Direct comparisons can be made between suspended and bedload measurements for particles exceeding 40 μm mean diameter as long as it is borne in mind that the experimental time-scales differ.

The results of the bedload transport experiment are considered first.

The initial tracer dispersion shown in Fig. 5(A) reflects the tidal flow pattern, with elongation in the directions of the tidal currents, with very little lateral dispersion. The tracer injection was made at high water slack and therefore by D2 the centroid is slightly east of north of the injection site, having moved first with the ebb tide (Fig. 6). The transport rate of 0·012 g cm⁻¹ sec⁻¹ was not particularly high at this stage, being a resultant of movement first with the ebb, and then with an equal number of flood and ebb tides. As the tracer gradually attained equilibrium with the background sediment, the rate changed from 0·040 g cm⁻¹ sec⁻¹ between D 3 and D 5, to 0·012 g cm⁻¹ sec⁻¹ between D 5 and D 50 and the centroid had moved to 33 m south of the injection site. This time taken to reach equilibrium is consistent with the value of at least 7 days suggested by Heathershaw & Carr (1977).

Wave records from a frequency-modulated pressure recorder located at Dunwich (Fig. 1) show that between D 50 and D 164·5 gales gave rise to a maximum significant wave height of 2·5 m, with a mean zero crossing period of 9·9 sec. The wind direction at the time measured at Gorleston Meteorological Station was NE/ENE. The effect of wave action on the sea-bed was shown by the rate of transport increasing to 0·015 g cm⁻¹ sec⁻¹, integrated over the period 10–12 October 1978 to 2–3 February 1979, and the change in direction of centroid movement to a point with a bearing of 241° and 55 m distance from the injection site (Fig. 6).

A seventh survey took place in April 1979 (D 229), but only isolated patches of tracer in low concentrations, less than 10^2 grains kg⁻¹, were found.

When the suspended load transport rates are considered it is seen that, in contrast to the bedload rates, they were high, and at PS 4 the maximum ebb rate exceeded that on the flood by as much as 18·19 g cm⁻¹ sec⁻¹. The dominance of ebb transport rates over flood rates, also shown at PS 1 and PS 3, is the result of the ebb velocities being greater than those on the flood by about 10 cm sec⁻¹, although this produces only a small increase in the friction

velocity. The non-linear relationship between velocity and concentration accounts for the concentrations during the ebb exceeding those on the flood by a factor of more than two. Also the ebb tide exceeded the flood by approximately 30 min in length, which would again mean that more suspended sediment would be transported during the ebb. At PS 2 and PS 5 the transport rates were approximately equal on both ebb and flood.

When the tidal flow was well established, both velocity and concentration profiles could be fitted to the accepted theoretical patterns with a high degree of confidence. The fit was less good as slack water times were approached. The velocity profiles departed from the logarithmic form by becoming concave downwards, mainly due to the values at the height of 85 cm being less than might be expected (Fig. 7). This may be the result of sediment settling into the sampling area from above, with the density of the suspension affecting the velocity field. Near slack water it is noticeable that the fit of the concentration profiles to the Rouse equation is not as good, the concentrations near the top of the rig being too high. The phenomenon is shown particularly as low water slack is approached, when the last set of concentration measurements made before slack water gives as negative correlation.

In conclusion, the experiments have shown that in the Sizewell–Dunwich area the dominant mode of transport is likely to be by suspension. The figures cannot, of course, be compared directly, partly because of the different time-scales, but also because the suspended rates given are instantaneous whereas the bedload rates are net values, although the order of magnitude difference may be estimated. However, the tracer experiment indicated that there was relatively little movement along the sea-bed, with a maximum calculated excursion of the centroid from the injection site of only 55 m after 165 days. The maximum tracer cloud length was just over 1 km. The maximum net transport rate, after equilibrium had been reached, was 0·015 g cm⁻¹ sec⁻¹. In contrast a maximum instantaneous suspended sediment transport rate of 30·819 g cm⁻¹ sec⁻¹ for the total water column has been measured on a spring ebb tide, with sediment concentrations as high as 423 mg l⁻¹ of sand at 1·75 m above the sea-bed and 1892 mg l⁻¹ at 15 cm above. An order of magnitude for the net suspended sediment transport rate is indicated by the figure of 5·660 g cm⁻¹ sec⁻¹ in the ebb direction for this one spring tide. Such a high value may well be balanced by strong flood residuals in other parts of the area. The rate is also likely to be considerably less during neap tides.

Future work includes the analysis of tidal current data obtained in the area over 4·5 yr. The various sediment transport formulae mentioned above will be assessed for their ability to predict sediment transport rates comparable to those already measured.

ACKNOWLEDGMENTS

This work was supported financially by the Department of the Environment. The author would like to thank colleagues at the Institute of Oceanographic Sciences, Taunton, for their help and cooperation at all stages of the work. I am indebted to the Ministry of Agriculture, Fisheries and Food, Lowestoft, for the loan of their box corer, designed by Tennant, and for the use of their vessel R.V. *Tellina*; also to the Marine Biological Association, Plymouth, for the use of R.V. *Sarsia*; and to the Natural Environment Research Council's Research Vessel Services for the use of R.V. *Edward Forbes*. May I also thank Decca Survey Ltd for the charter of their vessel M.V. *Decca Mariner*.

REFERENCES

ACKERS, P. & WHITE, W.R. (1973) Sediment transport: new approach and analysis. *Proc. Am. Soc. civ. Engrs, J. Hydr. Div.* HY11, 2041–2060.

BAGNOLD, R.A. (1963) Mechanics of marine sedimentation. In: *The Sea* (Ed. by M.N. Hill), pp. 507–582. Interscience Publications, New York.

BOLDYREV, V.L. (1956) Tests with tracer sand to investigate silting. *Proc. Ass. Mar. Proj., Min. Merchant Marine (USSR)*, 3, 63–71. Cited in Nelson & Coakley (1974).

BRATTELUND, E. & BRUUN, P. (1974) Tracer tests in the middle North Sea. *Proc. 14th Conf. Coastal Engng* Copenhagen, 2, 978–989.

BRUUN, P. (1968) Quantitative tracing of littoral drift. *Proc. 11th Conf. Coastal Engng* London, 1, 322–328.

COURTOIS, G. & MONACO, A. (1969) Radioactive methods for the quantitative determination of coastal drift rate. *Mar. Geol.* 7, 183–206.

CRICKMORE, M.J. & AKED, R.F. (1975) Pump samplers for measuring sand transport in tidal waters. *Proc. Conf. Inst. Electron. Radio Eng.* 32, 311–326.

EINSTEIN, H. A. (1950) The bed load function for sediment transportation in open channel flows. *Soil Conserv. Serv. Tech. Bull., US Dep. Agric. No.* 1026.

ENGELUND, F. & HANSEN, E. (1967) *A Monograph on Sediment Transport in Alluvial Streams*. Teknisk Vorlag, Copenhagen.

GADD, P.E., LAVELLE, J.W. & SWIFT, D.J.P. (1978) Estimates of sand transport on the New York Shelf using near-bottom current meter observations. *J. sedim. Petrol.* 48, 239–252.

HEATHERSHAW, A.D. & CARR, A.P. (1977) Measurements of sediment transport rates using radioactive tracers. *Coastal Sediments '77, Am. Soc. civ. Engrs, Charleston, S.C., U.S.A.* pp. 399–416.

HEATHERSHAW, A.D. & HAMMOND, F.D.C. (1979) Offshore sediment movement and its relation to observed tidal current and wave data. In: *Swansea Bay (Sker) Project. Topic Report: 6. Inst. Oceanogr. Sci. Rep. No.* 92.

INGLE, J.C.J. & GORSLINE, D.S. (1973) Use of fluorescent tracers in the nearshore environment. In: *Tracer Techniques in Sediment Transport. Tech. Rep. Ser. No.* 145, pp. 125–150. International Atomic Agency, Vienna.

JOLLIFFE, I.P. (1963) A study of sand movements on the Lowestoft Sandbank using fluorescent tracers. *Geogr. J.* 129, 480–493.

KIRBY, R. (1973) UCS Grain size comparator disc. *Mar. Geol.* 14, M 11pM 14.

KOMAR, P.D. (1969) *The longshore transport of sand on beaches*. Ph D. Thesis. University of California, San Diego.

LEES, B. J. (1979) A new technique for injecting fluorescent sand tracer in sediment transport experiments in a shallow water marine environment. *Mar. Geol.* 33, M 95–M 98.

LEES, B.J. (1980) Introduction and geological background to the Sizewell–Dunwich Banks field study. *Topic Report* 1. *Inst. Oceanogr. Sci. Rep. No.* 88.

MEDVEDEV, V.C. & AIBULATOV, N.A. (1956) The use of marked sand for the study of the transport of marine detritus. *Izv. Akad. Nauk SSSR. Geogr. Ser.* 4, 99–102 (in Russian). Cited in Nelson & Coakley (1974).

NELSON, D.E. & COAKLEY, J.P. (1974) Techniques for tracing sediment movement. *Sci. Ser. No.* 32. Inland Waters Directorate, Canada, Center for Inland Waters, Burlington, Ontario.

TERZAGHI, K. & PECK, R.B. (1967) *Soil Mechanics in Engineering Practice*. Wiley, New York. 729 pp.

VERNON, J.W. (1966) *Shelf sediment transport system*. Ph.D. dissertation, University of Southern California.

VRIES, M. DE (1973) Applicability of fluorescent tracers. In: *Tracer Techniques in Sediment Transport. Tech. Rep. Ser. No.* 145, pp. 105–124. International Atomic Agency, Vienna.

YALIN, M.S. (1963) An expression for bed load transportation. *Proc. Am. Soc. civ. Engrs, J. Hydr. Div.* pp. 221–250.

Spec. Publs int. Ass. Sediment. (1981) **5**, 283–301

Sediment response to waves and currents, North Yorkshire Shelf, North Sea

C. F. JAGO*

Wellcome Marine Laboratory, Robin Hood's Bay, Yorkshire, U.K.

ABSTRACT

The response of shelf sediments off North Yorkshire to the contemporary hydraulic regime has been examined by: (1) measuring the concentrations and *in situ* grain size distributions of sediments put into suspension by tidal currents, (2) following the dispersal of sea-bed and sea-surface drifters, and (3) examining the surficial sediment distribution from the shoreline out to 50 m depth.

Both tidal currents and waves can entrain sand-size grains, and the sediment distribution reflects the present-day hydraulic regime. Sediments, mostly silt-size, are suspended near the bottom by tidal currents, both concentration and mean grain size fluctuate during the tidal cycle, reaching maxima at peak currents on spring tides. Grains suspended by tidal currents reach neither the surface nor mid-depths except in winter, when waves reinforce tidal currents, stir the bottom and attack an eroding shoreline.

Sea-bed drifters identify a residual bottom drift that is shoreward. Sea-surface drifters indicate that surface water moves in response to winds, with a seawards component of drift when winds are offshore. Since surface concentrations of suspended sediment are high during the winter months, seawards movement of suspended sediment will occur if winds are offshore. But during the summer months suspended sediment, confined near the bottom, will not move seawards against the residual bottom drift. Drifters can move northwards along the coast under certain conditions, but their long-term movement is to the south. Sediments suspended by tidal currents, although redeposited at slack water, will have a net southerly drift.

The surficial bottom sediments are aligned in well-defined zones.

(1) A nearshore sediment-free wave cut platform that extends from the foreshore to 20–40 m.

(2) Well sorted sands characterized by a seaward-fining textural gradient that is established by shoaling waves; these extend to 30·m but occur only in certain areas.

(3) Well sorted sands characterized by a seaward-coarsening textural gradient that is established by tidal currents (which increase in deeper water) and/or residual currents (shorewards on the bottom); these extend to 44 m.

(4) Sandy gravels and sandy muds at depths greater than 44 m. Muds encroach on the shoreface during the summer months. Although deepwater patches of mud may locally interrupt it, the textural trend is one of increasing mean grain size seawards. This trend may be a valid analogue for shelf sediments deposited in ancient tidal seas.

INTRODUCTION

This study forms part of a programme designed to examine the feasibility of disposing large quantities (350,000 metric tonnes per annum) of fine sand, silt

*Present address: Marine Science Laboratories, Menai Bridge, Gwynedd, Wales, U.K.

0141–3600/81/1205–0283 $02.00

and clay-size material into the sea off North Yorkshire. This material would derive from the extraction of potash deposits in the Whitby area. Since the coastal area of North Yorkshire is of importance as a shellfish and demersal fishery, there is concern as to the eventual fate of the discharged sediment.

The results described here stem from three aspects

of this programme: (1) measurements of the concentrations and grain-size distributions of sediments put into suspension by tidal currents; (2) the dispersal of sea-bed and sea-surface drifters released near the proposed discharge point; (3) an examination of the present distribution of bottom sediments from the shoreline out to 50 m depth.

The North Yorkshire coast

The coastline of North Yorkshire (Fig. 1) consists of mostly high cliffs, exceeding 190 m at Ravenscar, but decreasing to the north. Drift reaches sea-level in the bays, notably in parts of Robin Hood's Bay and around Whitby. The cliffs are fronted by a wave-cut platform—up to 300 m or more wide at mean low tide—which consists of two parts: an upper part, cut during the Flandrian transgression and sloping at 6°; a lower part, cut during the Eemian interglacial and sloping at 1° (Agar, 1960). With sea-level static during the past millennium, the wave-cut platform has continued to erode downwards (0·1–2·9 cm yr^{-1}) and the cliff landwards (Robinson, 1977). Surveys have shown that the average rate of cliff recession is 9·2 m per century where the cliff and foreshore are of bare shale, 28·0 m per century for the glacial drift (Agar, 1960).

Tidal currents are essentially rectilinear, with the main tidal axis parallel to the coast. The tidal range is large: 4·60 m for mean springs, 2·23 m for mean neaps. Wave data for this part of the North Sea are not available, but data from the North Carr Light Vessel (just off the Forth Estuary) are probably applicable to the North Yorkshire coast. During 1969/70, the most common waves were those with a significant height (H_s) of 0·5–0·6 m and a zero-crossing period (T_z) of 3·5–4·5 sec. The maximum height was 7·7 m, the maximum period 12 sec (Draper, 1971).

Methods

Water samples (2 litre) and current measurements (using Plessey direct-reading current meters) were taken at 30 min intervals over tidal cycles at stations off Maw Wyke (the proposed discharge point) during spring and neap tides (during calm weather in summertime). These stations were 1·6 and 3·2 km offshore in water depths of 27 and 45 m (Fig. 1). Samples and data were collected from the surface, mid-depth and about 1 m from the bottom.

Water samples were analysed for suspended sedi-

ment in the laboratory as soon as possible after collection using a Model A Coulter Counter. No attempt was made to deflocculate aggregates and, in fact, the inorganic particle size distributions were found to be quite stable. Counts were made of particles between 1 μm (10 phi units) and 125 μm (3 phi units) at approximately 1 phi unit intervals. Smaller particles were estimated by extrapolation of cumulative probability curves. Details of the counting procedure are given in the Appendix.

The drifter studies employed three types of plastic drifter: the standard Woodhead sea-bed drifter (released close to the bottom from a weighted cage); the same type modified to float at the surface; the same suspended 3 m below the surface from a cork float. Some 400 drifters were released in 1970/71, these supplementing earlier releases of 600 drifters from the same points by Newton (1973).

Surficial sediments were sampled by Dietz-Lafond snapper grab from 1000 locations at 0·4 km intervals along transects 0·8 km apart, the grid being fixed according to Decca Main Chain coordinates (Fig. 1). The northern sector was sampled twice—in October 1970 and May 1971—the southern sector once in October 1971. Samples were analysed for grain size using standard techniques—sieving (at 0·25 phi unit intervals) and pipette analysis (after deflocculation with sodium hexametaphosphate).

SEDIMENT SUSPENSION BY TIDAL CURRENTS

The tidal variations of current velocity and total flocculated suspended sediment for a spring tide at the inshore station are shown in Fig. 2. Data from this and other stations are summarized in Table 1. Mid-depth data are not included since in every case mid-depth characteristics were virtually identical to surface characteristics.

Tidal currents were quite rapid, reaching 135 cm sec^{-1} at the surface and 100 cm sec^{-1} near the bottom on a spring tide; 90 cm sec^{-1} at the surface, 60 cm sec^{-1} near the bottom on a neap. Suspended sediment concentrations were essentially the same at the inshore and offshore stations. Surface (and mid-depth) concentrations varied significantly neither between tides nor during the tidal cycle. Near-bottom concentrations, by contrast, were up to 500% greater than surface concentrations and almost 200% greater on springs than on neaps. While the small fluctuations of surface concentration appeared to be

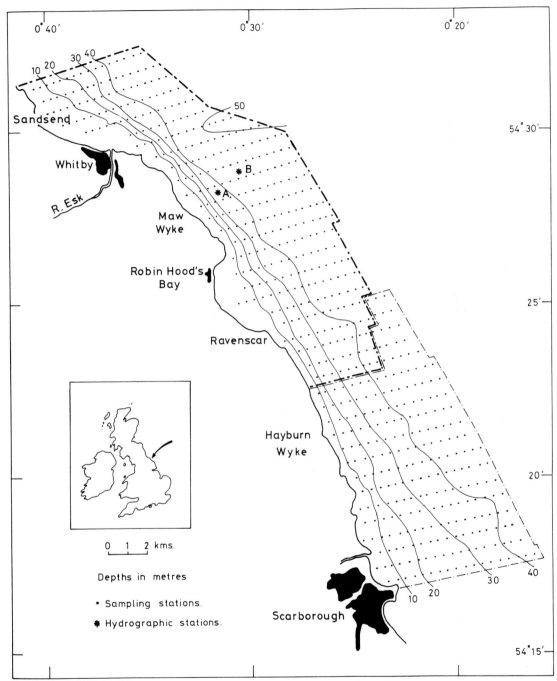

Fig. 1. The North Yorkshire shelf. Bathymetry and sampling stations.

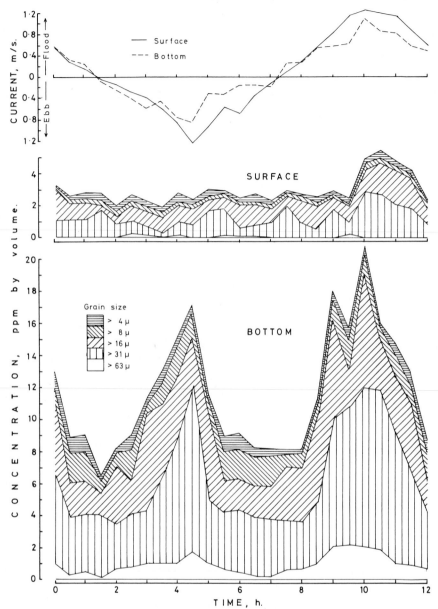

Fig. 2. Current velocity and suspended sediment variations over a spring tide, station A.

random, near-bottom concentrations were obviously dependent on current velocities, the greatest concentrations coinciding with peak currents (though with a 30 min time lag on neaps).

The surface grain-size distributions were relatively uniform for all tides, their modes ranging from 45 to 20 μm. Most of the grains were coarse and medium silt with some finer sizes, but clay-size grains were almost totally absent. The small fluctuations that did occur resulted from adjustments to the coarse and medium silt grades (63–16 μm).

Near-bottom grain-size distributions were more variable, though their modes again ranged from 45 down to 20 μm. Clay-size grains were as uncommon near the bottom as at the surface. But grains coarser than 8 μm increased considerably near the bottom.

Table 1. Summary of maximum current speeds and suspended sediment concentrations at stations **A** and **B**. Range of sediment concentrations in parentheses

	Tide		Maximum concentration ppm	
Range m	Maximum current cm sec⁻¹			
	Surface	Bottom	Surface	Bottom
A Inshore station				
5·5	135	100	5·5 (3·5)	21·5 (15·0)
3·4	90	60	3·5 (1·5)	13·5 (8·5)
B Offshore station				
5·1	120	92	5·0 (2·5)	20·5 (15·5)
2·8	84	56	3·5 (1·0)	13·5 (9·5)

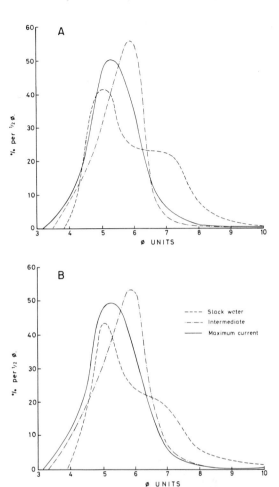

Fig. 4. Adjustments to the particle spectra of near-bottom sediment suspensions during: (A) accelerating, (B) decelerating currents. Data at 1 phi unit intervals.

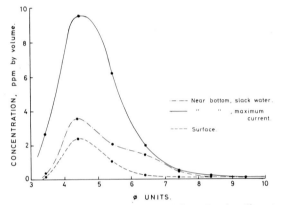

Fig. 3. Typical particle spectra of flocculated sediment suspensions. Data at 1 phi unit intervals.

Concentrations of silt were 3–5 times and very fine sand 10 times greater near the bottom than at the surface, the greatest increase coinciding with peak currents on spring tides. Concentrations of grains coarser than 8 μm varied with current velocity, whereas concentrations of grains finer than 8 μm varied as randomly as at the surface.

Typical particle-size counts are shown as particle spectra in Fig. 3, where the equivalent diameter of particle volume is plotted against the concentration in ppm (Sheldon & Parsons, 1967a, b). These spectra are very similar to the distributions of unsorted mixtures of flocs and single grains described by Kranck (1973, 1975). She argues that in a stable distribution particles finer than the mode settle as flocs while particles coarser than the mode settle as single grains. If this is so, we can assume that particles coarser than about 44 μm settle as single grains. However, the situation will be complicated if the flocs are unstable. Such could be the case here,

especially during spring tides. Some aggregates coarser than 8 μm would be suspended during springs but not during neaps. These aggregates, since they would remain on the bottom during neaps, might undergo post-depositional modifications of floc size, so that when resuspended during springs they would be unstable.

The grain-size distributions were not usually lognormal since they invariably exhibited a fine tail and were truncated at their coarse ends. Sengupta (1975) has shown that lognormality in suspended sediment grain-size distributions is an ephemeral response to certain critical controls—current velocity and height above the bottom. His conclusions are borne out here. Lognormal distributions were

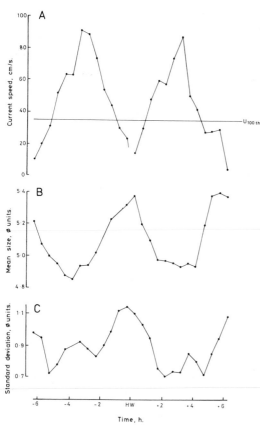

Fig. 5. Variations of: (A) current speed, (B) mean grain size, (C) standard deviation of suspended grain-size distributions on a spring tide.

generated near the bottom only by intermediate current velocities of about 40 cm sec^{-1} an hour after slack water (Fig. 4). Near the bottom, both mean grain size and standard deviation described broadly cyclic trends, more so on springs than on neaps. The mean became coarser as the current increased, finer as it decreased (Fig. 5). The variation of standard deviation was more complex. Sorting improved with increasing frequencies of silt but deteriorated with increasing frequencies of fine sand. As the current velocity increased after slack tide, more and more of the modal grain sizes were picked up, so that the suspension became better sorted. But as the velocity increased further, and grains coarser than the mode were suspended, sorting deteriorated. This progression recurred in reverse as the current slackened. Hence the best sorting was produced by currents of intermediate speed (Fig. 4).

DRIFTER MOVEMENTS

Recovery rates were good, particularly of sea-bed drifters: 50–70% were returned and, of these, 60–90% within a month of release (Figs 6 and 7). The arrangement of releases allows for several comparisons of drifter movement: (1) between sea-bed, sea-surface and 3 m drifters; (2) between drifters released from the same point but separated by half a tidal cycle (drops 1A and 1B); (3) between drifters released from three points at the same time (drops 2A, 2B and 2C); (4) between drifters released on similar tides but under different meteorological conditions (drops 1B and 2A).

Onshore/offshore movement

After all releases, considerable numbers of sea-bed drifters came ashore very quickly. Even those released 4·8 km offshore began to beach within ten days; this represents a net shoreward motion of at least 0·6 cm sec^{-1}. The greatest frequency of recovery followed onshore winds and heavy swell, but large numbers came ashore even after strong offshore winds and little swell. A net shoreward movement of sea-bed drifters from these points was also found by Newton (1973).

The sea-surface drifters showed a contrary movement. Far fewer of them reached the shore and, after some drops, none at all. Clearly, there can be no persistent shoreward drift of surface water. A closer look at the weekly frequencies of recoveries suggests that the sea-surface drifters moved in response to wind stresses at the surface. No sea-surface drifter was recovered when the prevailing winds were offshore during the month following release, but some were recovered if the prevailing winds were onshore following release. None of the 3 m drifters was recovered, so these presumably moved seawards.

Longshore movement

The short-term longshore movement of sea-bed drifters was more variable than the onshore/offshore movement. After some drops, recoveries within one month were confined to the shore north of the release points, after other drops to the shore south of the release points (Figs 6 and 7). This short-term movement might be related to local wind direction, but not at all obviously. However, the long-term longshore movement of sea-bed drifters was to the south. The longer a sea-bed drifter remained in the water,

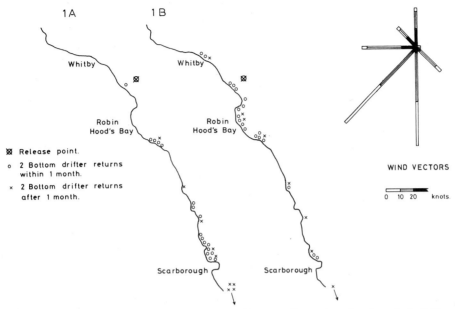

Fig. 6. Sea-bed drifter returns from release points 1·6 km offshore. Drifters released at low slack (1A) and high slack waters (1B).

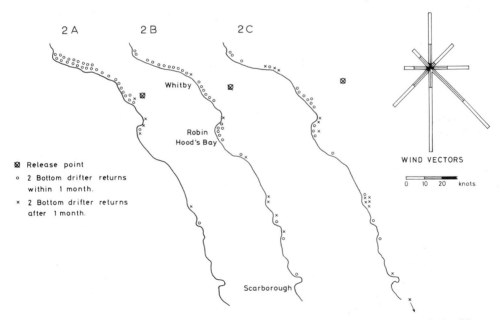

Fig. 7. Sea-bed drifter returns from release points 1·6 km (2A), 3·2 km (2B) and 4·8 km (2C) offshore.

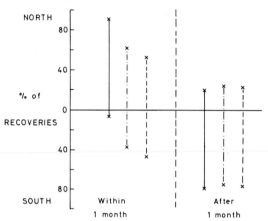

Fig. 8. Long-term southerly movement of drifters.

and the further offshore it was released, the more likely it was ultimately to be beached to the south (Fig. 8). This applied also to sea-surface drifters. A pronounced south-going movement of sea-bed drifters off this coast is reported by Ramster & Jones (1978), although contrary results were obtained by Newton (1973), whose sea-bed drifters were beached to the north of the release points (but owing to their rapid recovery few of Newton's drifters reflect a long-term drift).

Good evidence that the sea-bed drifters do meaningfully trace water movements was provided by drops 1A and 1B. These drifters were released from the inshore station, 1A at low water slack, 1B at the following high water slack. This should have resulted in the drifters from 1A being several kilometres south of those from 1B at any given time. That this was indeed so is shown by the distribution of the returns (Fig. 6). Most of the drifters were stranded during the third week from release, and at that time those from 1B had reached Robin Hood's Bay while those from 1A were 10 km further south towards Scarborough.

Comparison with current meter results

From data kindly supplied by the Fisheries Laboratory, M.A.F.F., Lowestoft, residual drift

values have been calculated for six moored current meter stations deployed for five days in January/ February 1970 (Fig. 9). All meters recorded a residual drift whose longshore component was to the SSE (1·8–7·8 cm sec⁻¹). Four of the five meters near the surface or mid-depth recorded a seaward component of drift (2·2–6·7 cm sec⁻¹), but the one near-bottom meter recorded a shoreward component (2·8 cm sec⁻¹).

These results must be viewed with caution. Current meters have large speed errors when used in shallow water stirred by surface waves (Hammond & Collins, 1979). And even a small directional error can indicate a spurious residual at an angle to the main tidal axis.

SURFICIAL SEDIMENT DISTRIBUTION

Sands are present on the foreshore in the sheltered bays but elsewhere the cliffs are fronted by a wave-cut platform which extends seawards to depths of 20–40 m. Well-sorted sands, with a minor admixture of fines, occur beyond the platform to depths of 40–44 m. In deeper water, gravels and—locally—muds become increasingly abundant. Most of the sediments are therefore sandy, with admixtures of less than 30% gravel and less than 50% mud (Fig. 10).

The gravel distribution is a patchy one. Sediments containing more than 30% gravel-size grains are not found shoreward of 35 m, and are mostly restricted to depths greater than 45 m (Fig. 11).

Muddy sediments occur mostly in deep water, but their distribution with depth varied markedly from one survey to the next. The October 1970 survey (carried out after the first autumn gales) showed muddy sediments at depths mostly beyond 40–45 m. But both the May 1971 survey and the October 1971 survey (completed before the first autumn gales) proved muddy sediments closer to the coast at 20–30 m (Fig. 12). Indeed, in summertime, muds are locally deposited on the wave-cut platform and foreshore. This temporal variation clearly reflects seasonal changes in wave climate.

Mean grain-size contours trend approximately parallel to the bathymetric contours, so the sediment distribution reflects, at least partially, the present-day hydraulic regime. Broadly, the trend is one of increasing mean grain size seawards, although patches of mud in deep water may locally interrupt it. The seaward-coarsening gradient is highlighted in Fig. 13

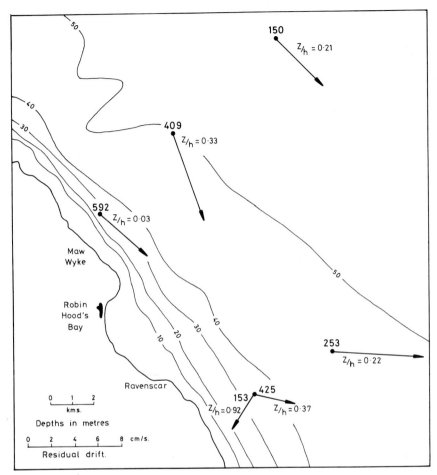

Fig. 9. Residual drift at moored current meter stations over 5 days (from M.A.F.F. data). z=depth of meter, h=total water depth.

where grains finer than 63 μm have been omitted and the truncated grain-size distributions normalized.

There are important variations within this trend. In some places, in particular off Scarborough and, to a limited degree, in Robin Hood's Bay and near Sandsend, the textural gradient within the shoreface sands is a seaward-fining one, extending to a maximum depth of 32 m off Scarborough and much less elsewhere (Fig. 13). The trend reverses in deeper water. But generally, a progressive seaward coarsening characterizes the whole sand prism from its landward margin out to about 44 m (Fig. 14). At depths greater than 43–45 m, the seaward coarsening continues but with significant wanderings (Fig. 13). This is a consequence not only of an increasing but variable frequency of the gravel fraction but also a continued seaward coarsening of the sand fraction.

Allowing for local variations, the surficial sediments are therefore arranged in zones more or less parallel to the coastline:

(1) A nearshore sediment-free wave-cut platform extending from the foreshore to 20–40 m.

(2) Well-sorted sands displaying a progressive seaward fining of mean grain size, but occurring only in some areas.

(3) Well-sorted sands displaying a progressive seaward coarsening of mean grain size and extending to no more than 44 m.

(4) A heterogeneous mixture of gravels, sands and muds in deeper water. Except where muds are present, these sediments are coarser than those in shallower water.

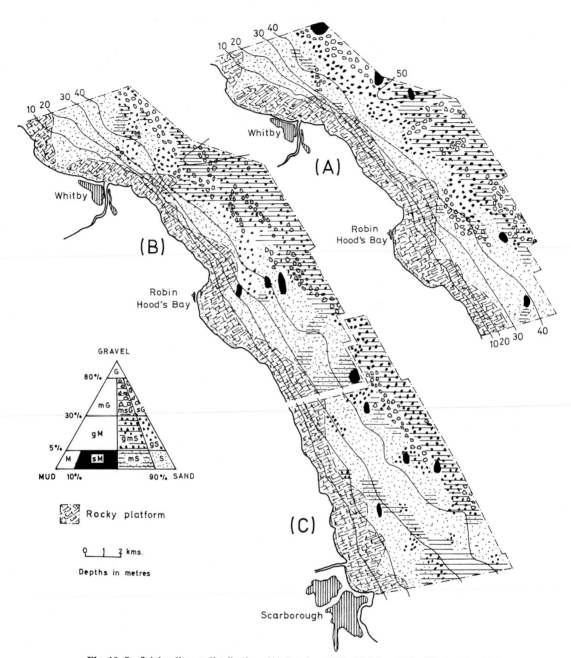

Fig. 10. Surficial sediment distribution. (A) October 1970. (B) May 1971. (C) October 1971.

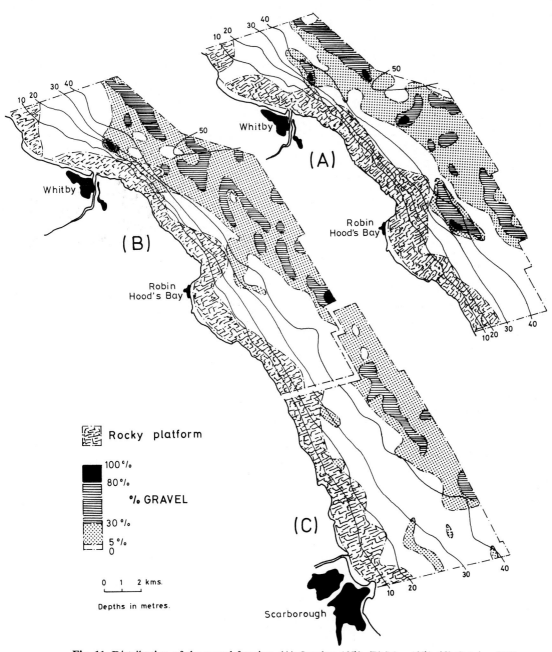

Fig. 11. Distribution of the gravel fraction. (A) October 1970. (B) May 1971. (C) October 1971.

C. F. Jago

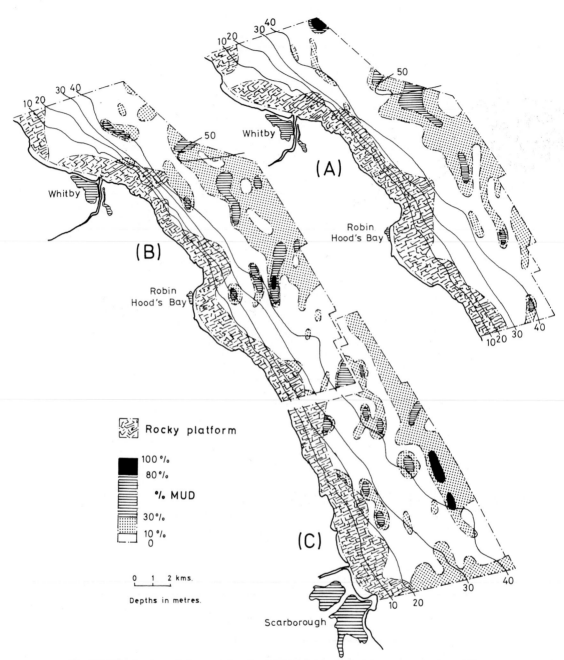

Fig. 12. Distribution of the mud fraction. (A) October 1970. (B) May 1971. (C) October 1971.

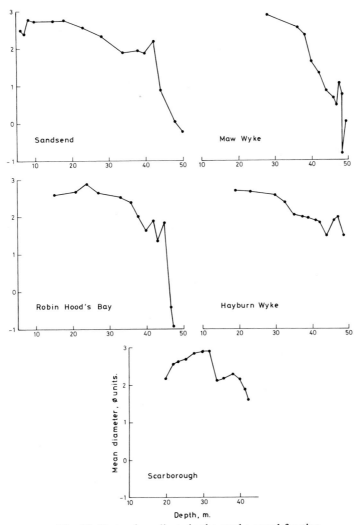

Fig. 13. Textural gradients in the sand+gravel fraction.

DISCUSSION

Tidal currents do not raise suspended grains far from the bottom and affect neither mid-depth nor surface concentrations. Nevertheless, Newton & Gray (1972) report mid-depth concentrations an order of magnitude greater in winter than in summer; and the mid-depth concentrations correlate significantly with sea state (i.e. wave height), Esk river flow and tidal range. Waves can raise suspended sediment concentrations by reinforcing the tidal currents and by direct stirring of the bottom deposits —e.g. a wave of height 1·5 m and period 7 sec can entrain sand-size grains down to 30 m depth (Fig. 15

is based on the threshold criteria of Komar & Miller, 1975). Furthermore, waves hasten shoreline erosion, particularly during spring tides, by attacking the vulnerable drift cliffs of Whitby and Robin Hood's Bay, so much so that the suspended sediment concentration can be 1000% greater at the water's edge in Robin Hood's Bay than at a point 0·8 km offshore (Moore, 1972).

The drifter observations and current-meter data suggest that the residual drift off the North Yorkshire coast has components that are alongshore (to the SSE), shorewards near the bottom, seawards at mid-depth and surface. These results are supported by: Newton (1973), whose drogue studies identified

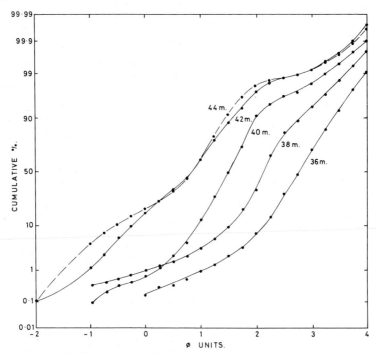

Fig. 14. Grain-size distributions of seaward-coarsening sands.

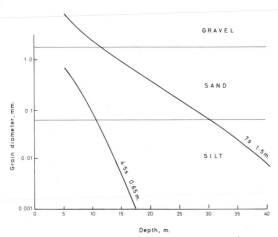

Fig. 15. Grain diameters at threshold for the most common waves ($H_s=0.65$ m, $T_z=4.5$ sec) and wintertime waves ($H_s=1.5$ m, $T_z=7$ sec).

a residual drift of 3.2–23.6 cm sec^{-1} predominantly southwards along the coast, although occasionally reversed: Ramster & Jones (1978), whose sea-bed drifters give comparable results: Ramster (1977), working from extensive moored current data off Flamborough Head (just to the south of Scarbo-

rough), who estimates a shoreward bottom residual drift of 0.3–0.8 cm sec^{-1} for the summer months—cf. 0.6 cm sec^{-1} (drifters) and 2.8 cm sec^{-1} (current meters) estimated above.

A number of hydrographic conditions might generate a shoreward bottom drift. Wave drift cannot be solely responsible since the shoreward residual is most strongly developed in summertime (Ramster, 1977; Ramster & Jones, 1978). Offshore winds can cause upwelling and a net shoreward bottom return flow (Fonds & Eisma, 1967); off this coast the prevailing winds—westerlies—are offshore. Less dense, low-salinity surface outflow may instigate a net shoreward bottom drift (Bumpus, 1965; Halliwell, 1973). Preferential heating of surface coastal waters might result in a seaward surface flow and compensatory shoreward bottom flow.

The residual drift patterns will determine the dispersal of sediment in suspension. Suspended sediment must move alongshore to the south. In summertime, when surface concentrations are low, bottom drift will tend to confine suspended sediment to the shoreline. In wintertime, when surface concentrations are boosted by the effects of waves, the seaward surface drift can carry sediment away from the coast.

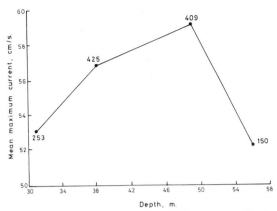

Fig. 16. Variation of maximum tidal currents with distance from the shore. Hourly means from M.A.F.F. data (station positions are shown in Fig. 9).

Does the distribution of bottom sediments reflect the prevailing hydraulic conditions? The nearshore rocky platform and the seaward-fining sands are clearly wave-controlled. While smaller waves, if more common, do not affect sands beyond about 11 m, the larger, albeit less frequent, waves of winter can entrain sand-size grains at 30 m (Fig. 15). It is noteworthy that McCave (1971) estimates that, at least for the southern North Sea, the onset of low wave effectiveness occurs at 30 m.

Beyond the zone of wave sorting, the textural gradient is seaward coarsening. Aloisi *et al.* (1977) have marked a comparable trend close to the French coast of the English Channel, and they relate it to a seaward-increasing tidal energy. Current data for the Yorkshire shelf are very sparse and we are limited to the M.A.F.F. data referred to above. But these data do show that between 30 and 50 m the maximum tidal currents increase in deeper water (Fig. 16). These data relate to neap tides, so we would expect the effect to be stepped up during springs. The texture of the sand then reflects the maximum tidal energy available on the bottom. That the tidal currents are competent to entrain the bottom sediments for much of the spring tidal cycle is clear from Fig. 5(A). U_{100th}, the threshold velocity 1 m above the bed, has been estimated for the sand at station B, by assuming a drag coefficient of $3 \cdot 0 \times 10^{-3}$ (Sternberg, 1972), and superimposed on the spring tide near-bed current velocity profile. Clearly, sand must be moving on the bottom for a good part of the spring cycle (and for part of the neap cycle too). Such limited data as we have suggest that the seaward coarsening of the sands is controlled by tidal currents (cf. Swift, 1970).

Thus the drag exerted on the bed contains two components. The one (generated by waves) diminishing, the other (generated by currents) increasing, seawards (to about 50 m), so that a point occurs at which wave dominance gives way to current dominance. Hence we can recognize two units—wave-sorted and current-sorted—within the sands (Fig. 17). Of course, currents influence the former and waves the latter, but the textural gradient identifies the dominant control in each case. That the transition from wave to current dominance does not appear at a uniform depth illustrates the complexity of wave/current interaction. The transition depth, about 30 m, seems to be dependent on offshore bottom slope and the width of the zone of shoaling waves.

Sandy and muddy gravels generally occur in deeper water, where transport may be limited to storms when wave effects boost tidal currents sufficiently to lift the bottom drag above threshold.

Another factor could be important: the shoreward bottom drift. Though this may be a small component (< 1 cm sec^{-1}) of the total tidal flow (> 100 cm sec^{-1}), it must deflect the transport paths of grains entrained by tidal currents. Given that, by selective sorting, finer grains will outrun coarser ones down the energy gradient from source (the sandy gravels at 45 m+) towards shore, we might anticipate a progressive fining of bottom sediments from shelf to shoreface up to the point where sorting by waves becomes paramount.

A shoreward residual drift of water does not necessarily imply a commensurate drift of bottom sediment since only the faster currents entrain sediment. An estimate of residual sediment movement has been attempted using the data at hydrographic station B (Fig. 1). Transport occurs only when the threshold velocity is exceeded; the transport volume is then

$$Q_i = \alpha \int_i (u - u_{th})^3 \, dt$$

where u is the measured current speed, u_{th} is the threshold speed, α is a constant of proportionality, and t_i is the duration of the transport event (Swift *et al.*, 1976). Without knowing α, we can calculate the rate of transport at any time relative to the total transport per tide or per day. This has been done for successive time intervals during a spring tide and a neap tide at station B. The normalized resultant, or

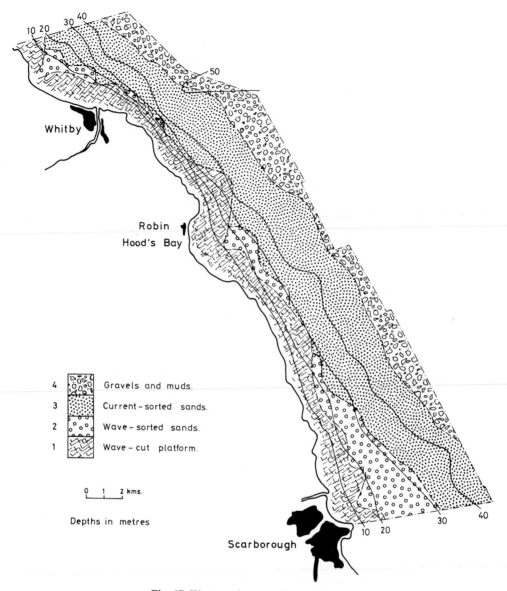

Fig. 17. Wave- and current-dominated areas.

vector sum, of all values has then been calculated. This gives a neap tide residual sediment transport (0.12×10^4 day^{-1}) which is less than 5% of the spring tide transport (2.57×10^4 day^{-1}). On both sides, the residual transport is to the south with a shoreward component. If we take $\alpha = 4.48 \times 10^{-5}$ g cm^{-4} sec^2 (Gadd, Lavelle & Swift, 1978), then the residual transport rate is 5.7×10^{-1} g cm^{-1} sec^{-1} on the spring tide, 2.6×10^{-2} g cm^{-1} sec^{-1} on the neap, with shore-

ward components of 2.3×10^{-1} and 1.1×10^{-2} g cm^{-1} sec^{-1}, respectively (Fig. 18). It would be rash to make too much of these figures, but they do suggest that a net shoreward drift of sand-size grains is feasible. Suspended sediment near the bottom also has a residual drift with a shoreward component (Fig. 18, from the suspended sediment/current velocity data).

Whatever the cause of the shoreward bottom drift, it must be intermittent and partly compensated by a

CONCLUSIONS

The distribution of shelf sediments off North Yorkshire results from transport by both waves and tidal currents. Sediments, mainly silt-size, suspended near the bottom by tidal currents, fluctuate in concentration and grain-size characteristics during the tidal cycle, reaching maxima at peak current speeds on spring tides. Suspended grains do not reach either the surface or mid-depths except in winter when waves reinforce tidal currents, stir the bottom and attack an eroding shoreline. Sea-bed drifters identify a shoreward residual bottom drift (which is not simply wave-controlled), so that near-bed seawards escape of suspended sediment cannot occur. Sea-surface drifters indicate that the movement of surface water is wind-controlled, so surface seawards escape of suspended sediments can occur in response to offshore winds. Since surface concentrations are high during the winter months, significant seawards movement of suspended sediment is then possible. Drifters identify a long-term residual drift southwards along the coast although short-term reversals seem to be not uncommon. Sediments suspended by tidal currents, even though redeposited at slack water, will have a net southerly drift.

Surficial bottom sediments are arranged in well-defined zones as we move seawards.

(1) A nearshore sediment-free wave-cut platform which extends from the foreshore to 20–40 m.

(2) Well-sorted sands characterized by a seaward-fining textural gradient that is established by shoaling waves; these are developed only where the bottom slope is small and the zone of shoaling waves wide, out to a depth of 30 m. While the most frequently occurring waves can entrain sands at 11 m depths and less, wintertime waves can do so at 30 m or more.

(3) Well-sorted sands characterized by a seaward-coarsening textural gradient that is established by tidal currents and/or residual currents; these sands extend to 44 m. Maximum tidal currents increase seawards, and the sand texture reflects the maximum tidal energy available near the bottom. A residual shoreward bottom drift of sediment might also, by selective sorting, produce the observed textural gradient.

(4) Sandy gravels, muddy sandy gravels, and sandy muds in deeper water (the source area for the nearshore sands). Muds encroach on the shoreface during the summer months when wave action is reduced and the shoreward bottom drift most developed.

The surficial sediment distribution therefore largely

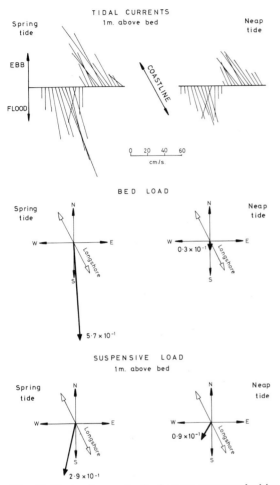

Fig. 18. Vector time series for bottom current velocities and estimated residual bed and suspension transport in g cm⁻¹ sec⁻¹ at station B.

seaward movement of material during certain periods—otherwise the offshore zone would be depleted of sediment. Such a seaward drift could arise during storms, as on the Atlantic shelf of North America where onshore winds generate strong surface currents and a downwelling regime (Swift, Duane & McKinney, 1973). Although our limited data do not support his hypothesis, Stride (1974) postulates a seaward bottom drift for the North Yorkshire shelf, due to the combined effects of a time lag in sediment suspension by, and an anti-clockwise rotation of, the tidal currents so that the maximum transport occurs just after peak flood tide.

reflects the prevailing hydraulic conditions, as well as the effects of the post-glacial marine transgression. European tidal seas may well be particularly suitable for comparison with ancient epeiric seas (which were surely tidal), so we might postulate that the predominant seaward-coarsening textural gradient of the Yorkshire shelf would be a valid analogue for some shelf deposits of the past. True, the North Sea is atypical in that it is floored with glacial sediments that must be the primary source from which the present-day sediment distribution has developed. But any epeiric sea, transgressing over coarse basal sediments, should generate the seawards-coarsening textural gradient, given an adequate sand supply. Such a gradient would be diametrically opposite to that developed on wave-dominated shelves.

ACKNOWLEDGMENTS

The study was financed by Messrs Cremer and Warner (consultant chemical engineers). Mr S. R. Jones, Fisheries Laboratory, M.A.F.F., Lowestoft, kindly supplied the moored current meter data, and Dr L. Draper, I.O.S., Wormley, the North Carr Light Vessel wave data. I am grateful to Dr I. N. McCave for constructive criticism of an early draft of the paper, to Mr J. E. Whale for his expert seamanship, and to Mr F. Dewes who drafted the figures.

REFERENCES

AGAR, R. (1960) Post-glacial erosion of the North Yorkshire coast from the Tees estuary to Ravenscar. *Proc. Yorks. geol. Soc.* **32**, 408–425.

ALOISI, J.C. et al. (1977) Essai de modélisation de la sédimentation actuelle sur les plateaux continentaux français. *Bull. Soc. géol. Fr.* **XIX**, 183–195.

BRUN-COTTAN, J.C. (1971) Etude de la granulométrie des particules marines mesurés effectuées avec un compteur Coulter. *Cah. océanogr.* **XXIII**, 193–205.

BUMPUS, D.F. (1965) Residual drift along the bottom on the continental shelf in the Middle Atlantic shelf area. *Limnol. Oceanog.*, Suppl. to **3**, 48–53.

DRAPER, L. (1971) Waves at North Carr Light Vessel, off Fife Ness. *N.I.O. Internal Rep.* No. A50, 5 pp.

FONDS, M. & EISMA, D. (1967). Upwelling water as a possible cause of red plankton bloom along the Dutch coast. *Neth. J. Sea Res.* **3**, 458–463.

GADD, P.E., LAVELLE, J.W. & SWIFT, D.J.P. (1978) Calculations of sand transport on the New York Shelf using near-bottom current meter observations. *J. sedim. Petrol.* **48**, 239–252.

HALLIWELL, A.R. (1973) Residual drift near the sea bed in Liverpool Bay: an observational study. *Geophys. J. R. astr. Soc.* **32**, 439–458.

HAMMOND, T.M. & COLLINS, M.B. (1979) Flume studies of the response of various current meter rotor/propellers to combinations of unidirectional and oscillary flow. *Dt. hydrogr. Z.* **32**, 39–58.

KOMAR, P.D. & MILLER, M.C. (1975) On the comparison between the threshold of sediment motion under waves and unidirectional currents with a discussion of the practical evaluation of the threshold. *J. sedim. Petrol.* **45**, 362–367.

KRANCK, K. (1973) Flocculation of suspended sediment in the sea. *Nature*, **246**, 348–350.

KRANCK, K. (1975) Sediment deposition from flocculated suspensions. *Sedimentology*, **22**, 111–123.

McCAVE, I.N. (1971) Wave effectiveness at the sea bed and its relationship to bed-forms and deposition of mud. *J. sedim. Petrol.* **41**, 89–96.

McCAVE, I.N. (1975) Vertical flux of particles in the ocean. *Deep Sea Res.* **22**, 491–502.

MOORE, P.G. (1972) Particulate matter in the sublittoral zone of an exposed coast and its ecological significance with special reference to the fauna inhabiting Kelp holdfasts. *J. exp. mar. biol. Ecol.* **10**, 59–80.

NEWTON, A.J. (1973) *Marine biological investigation concerning a proposed industrial effluent.* Unpublished Thesis, University of Leeds. 370 pp.

NEWTON, A.J. & GRAY, J.S. (1972) Seasonal variation of the suspended solid matter off the coast of North Yorkshire. *J. mar. biol. Ass. U.K.* **52**, 33–47.

RAMSTER, J.W. (1977) Residual drift regimes off the north-east coast of England during 1976. *I.C.E.S.* CM 1977/C, **8**, 6 pp. MAFF, Fisheries Laboratory, Lowestoft.

RAMSTER, J.W. & JONES, S.R. (1978) Residual trends from Woodhead sea-bed drifters released off the northeast coast of England in 1976. *I.C.E.S.* CM 1978/C, **29**, 6 pp. MAFF, Fisheries Laboratory, Lowestoft.

ROBINSON, L.A. (1977) Erosive processes on the shore platform of northeast Yorkshire, England. *Mar. Geol.* **23**, 339–361.

SENGUPTA, SUPRIYA (1975) Size-sorting during suspension transportation—lognormality and other characteristics. *Sedimentology*, **22**, 257–273.

SHELDON, R.W. (1968) Sedimentation in the estuary of the River Crouch, Essex, England. *Limnol Oceanog.* **13**, 72–83.

SHELDON, R.W. & PARSONS, T.R. (1967a) *A Practical Manual on the Use of the Coulter Counter in Marine Science.* Coulter Electronics, Toronto. 66 pp.

SHELDON, R.W. & PARSONS, T.R. (1967b) A continuous size spectrum for particulate matter in the sea. *J. Fish. Res. Bd, Can.* **24**, 909–915.

STERNBERG, R.W. (1972) Predicting initial motion and bedload transport of sediment in the marine environment. In: *Shelf Sediment Transport: Process and Pattern* (Ed. by D.J. Swift et al.), pp. 61–82. Dowden, Hutchinson & Ross, Stroudsburg, Pennsylvania.

STRIDE, A.H. (1974) Indications of long term, tidal control of net sand loss or gain by European coasts. *Est. Coastal Mar. Sci.* **2**, 27–36.

SWIFT, D.J.P. (1970) Quaternary shelves and the return to grade. *Mar. Geol.* **8**, 5–30.

Swift, D.J.P., Duane, D.B. & McKinney, T.F. (1973) Ridge and swale topography of the Middle Atlantic Bight, North America: secular response to the Holocene hydraulic regime. *Mar. Geol.* **15**, 227–247.

Swift, D.J.P. *et al.* (1976) Morphologic evolution and coastal sand transport, New York–New Jersey shelf. In: *Middle Atlantic Continental Shelf and the New York Bight. Spec. Symp. Am. Soc. Limnol. Oceanogr.* **2**, 69–89.

APPENDIX

Particle size analysis using the Coulter Counter

The principal difficulty encountered was in the separation of inorganic particles from plankton and bacteria, since chemical removal of the organic material was likely to modify the degree of flocculation of the inorganic particles. Therefore the inorganic particles of a particular diameter were counted indirectly by: measuring the total natural particulate population; allowing the inorganic particles coarser than the required diameter to settle out; measuring the total particulate population left in suspension; computing the total inorganic particulate population finer than the required diameter by difference; repeating the procedure for successive diameters.

Such a procedure assumes that the organic particles have specific gravities close to unity so that they remain suspended for some time, even after death. Settling times for the inorganic particles were estimated from Stokes' Law, which of course assumes that particles are spherical. This is probably a reasonable assumption (McCave, 1975). Estimating the particle density posed a problem. An estimate of the sediment density was made by determining the total inorganic volume concentration from the Coulter Counter and the total weight concentration from filtering through a 0·45 μm Millipore filter (assuming negligible weight for the organic material). This gave a mean density of 1·6 g ml^{-1}. The effective mean density of the flocs *in situ* would tend to be: greater than 1·6 g ml^{-1}, as filter weight would be less than true weight owing to loss of water from hydrated minerals during drying; less than 1·6 g ml^{-1} since particle aggregates enclose significant amounts of water.

Thoroughly flocculated suspensions will consist of particles of different densities (Kranck, 1975). Aggregates of different sizes are likely to contain variable amounts of organic material and water, larger flocs tending to have lower wet bulk densities than smaller flocs (McCave, 1975). As discussed above, particles larger than 44 μm probably occur as single grains rather than as aggregates, so that the large flocs of high organic content and low density described by McCave probably do not occur. Brun-Cottan (1971) estimates a maximum mean particle density of 1·44 g ml^{-1} for flocculated particles from 1 to 64 μm, assuming a primary particle density of 2·5 g ml^{-1}. Sheldon (1968) estimates particle densities of 1·5–1·8 g ml^{-1} for flocculated sediment of 8–15 μm diameter suspended in waters of the Crouch estuary. A density of 1·6 g ml^{-1} was used in the present calculations of settling velocities for all diameters.

Spec. Publs int. Ass. Sediment. (1981) **5**, 303–322

Holocene sedimentation in the North-western North Sea

RODERICK OWENS

Institute of Geological Sciences, Murchison House, West Mains Road,
Edinburgh EH9 3LA, Scotland, U.K.

ABSTRACT

Maps of superficial sediments have been prepared from textural analyses of samples taken at 495 sample stations between latitudes 56° and 58° N; longitudes 2° W and 0°. The areal and vertical relationships of the textural variables provide evidence for a degree of bathymetric control of their distribution. While sand is ubiquitous, gravelly sediments occur mainly on topographic highs in shallow waters, and muddy sediments generally in water depths greater than 120 m.

Cluster and trend-surface analyses of the data indicate a dynamic sedimentary environment in which the coarser terrigenous and biogenic carbonate components of the sediments are derived by erosion from topographic highs formed of glacigenic sediments. The dispersion directions of the sediments are controlled by the tidal stream. Tidal currents in the study area are weak, except in shallow waters near coasts, and there is evidence that their dispersive role is aided by storm-wave-induced oscillatory currents. Relative topography is as important a factor as absolute water depth in controlling the distribution of facies. Shallow geophysical data confirm these observations. They also reveal anomalously located bedforms, such as large sand waves, which may be the result of exceptionally severe storms. The abundance of biogenic carbonate in the sediments and its generally relict appearance may result from continuous slow accumulation following the Holocene transgression.

Use of the terms relict and palimpsest to describe the sediments mapped may be inappropriate until more information on temporal variations in the sedimentary environment is available.

INTRODUCTION

The study area is a portion of the United Kingdom continental shelf between latitudes 56° and 58° N; longitudes 2° W and 0° (Fig. 1). In common with other high-latitude shelves, this area was glaciated during the Pleistocene with the consequence that the uppermost sedimentary deposits are relict glacigenic sediments. These sediments were reworked by transgressing seas and the subsequent marine environment. The post-glacial rise in sea-level starved the area of further significant input of terrigenous sediment. Thus, the superficial sediments should exhibit attributes of glacial and post-glacial processes and sedimentation.

The Institute of Geological Sciences commenced geophysical and sampling reconnaissance surveys of the area in 1969. A detailed study of the complex

0141/3600/81/1205-0303 $02.00
© 1981 International Association of Sedimentologists

distribution of the superficial sediments was begun in 1973 and is reported in Owens (1977a, 1980). Independently, these sediments were also mapped by Jansen (1976) and Jansen, van Weering & Eisma (1979).

FIELD AND LABORATORY TECHNIQUES

Sample stations were selected following an analysis of shallow seismic data and a total of 495 sites were occupied (Fig. 2). Shipek grab and gravity or vibro-corer samples were taken at each station. Grab samples were analysed by sieving and pipette techniques (Carver, 1971). Carbonate content was determined gravimetrically after hand picking the gravel fraction and leaching the finer fractions with dilute hydrochloric acid.

Colour was determined by reference to Munsell colour charts and the earlier (pre-1973) use of

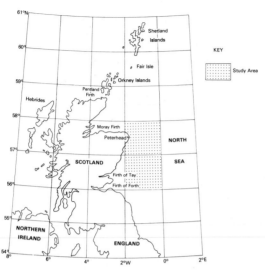

Fig. 1. Location map. The study area is between latitudes 56° and 58° N; longitudes 2° W and 0°.

subjective colour classifications corrected by hindcasting. Summary statistics of grain-size distributions were calculated by the method of moments using the computer program of Schlee & Webster (1967). Contour maps and trend surface analyses were made using the SACM (Batcha & Reese, 1964) and SURFACE II (Sampson, 1978) computer programs.

GEOLOGICAL SETTING

The study area lies to the east of the Scottish basement block. An offshore sedimentary succession from Palaeozoic to Tertiary subcrops beneath Quaternary sediments (Eden, Holmes & Fannin, 1978).

Total inundation of the area by ice at some time during the Pleistocene glaciations is inferred from the presence of erratics of Scandinavian origin in glacial deposits on the east coast of Scotland (Flint, 1957; Johnstone, 1966). The average thickness of Quaternary sediments over most of the study area is 30 m but in the north·east, on the western flank of the Central Graben, it increases to approximately 300 m (Holmes, 1977).

The uppermost glacial sediments were deposited between 22,000 and 18,000 y BP when an ice front formed a terminal moraine in the south-west of the study area. Ground moraine and glacio-

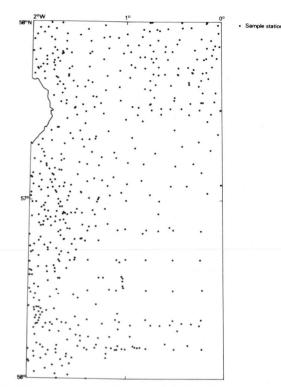

Fig. 2. Sample stations. All were sampled by Shipek grab between 1969 and 1977.

marine sediments were laid down as the Wee Bankie Beds in a zone between latitudes 56° 10′ N and 56° 48′ N in the region of longitude 2° W (Thomson, 1977). East of this zone and north to 57° 20′ N, the shallow arctic water marine sands of the Marr Bank Beds were deposited in open water. Penecontemporaneously, shallow water marine clays with dropstones were deposited in the north as the Upper Swatchway Beds (Holmes, 1977; Thomson, 1977; Eden *et al.*, 1978).

Deglaciation and the accompanying eustatic rise in sea-level resulted in a transgressive phase of littoral and shallow-water marine erosion in areas of largely unconsolidated deposits previously sub-aerially exposed or exhumed from beneath glaciers. Reworked sediments were transported and deposited elsewhere leaving a lag armour. 'Modern' oceanographic conditions probably became established in the area at 5000–6000 BP when sea-level stood at 5 m below the present level (Owens, 1977a), although actual water depth was probably greater than at present because of isostatic depression.

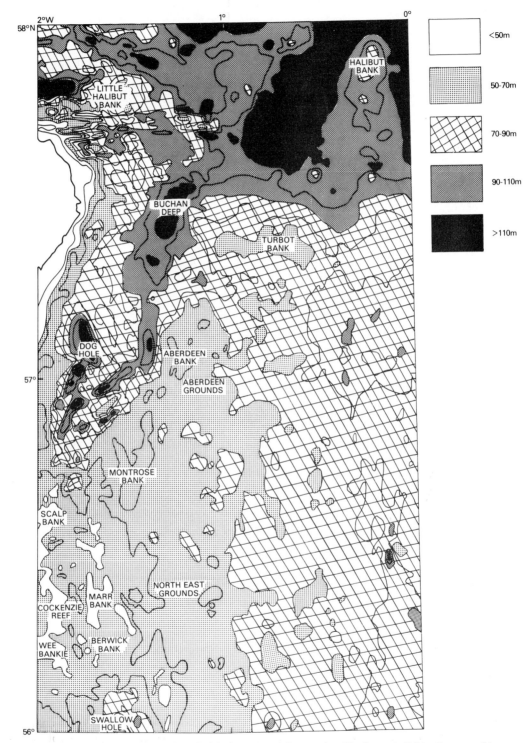

Fig. 3. Bathymetry. The high relief in the west is underlain by a morainic complex. To the east of this, the area of low relief is underlain by a marine erosion surface. Locations after Holmes (1977).

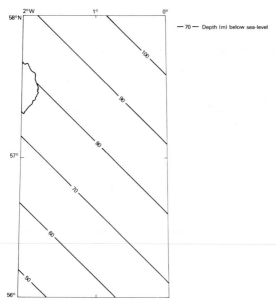

Fig. 4. Bathymetry—linear trend surface. The regional trend is north-east and slopes at 1 m in 3·9 km. The linear surface accounts for 50 % of sums of squares.

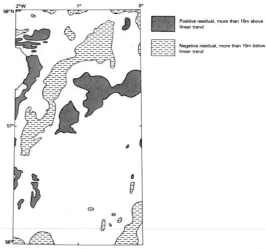

Fig. 5. Bathymetry—residuals from linear trend surface. The residuals are obtained by subtracting the linear trend from the bathymetry.

PHYSIOGRAPHY

Morphology

The bathymetric map (Fig. 3) was prepared from IGS echo sounder, shallow seismic and sample station records, with the addition of Hydrographic Office soundings data.

Two topographic provinces are indicated, one of high and one of low relief. High relief occurs mainly in the west and corresponds with the moraine and channel system of Thomson & Eden (1977). The banks shoal to 40 m in a mean water depth of 60 m and, north of 57° 40′, the north-east–south-west aligned channels deepen to 150 m. In the north a series of deep east–west lineated channels extend eastwards from the Moray Firth. There is an extensive area of low topographic relief in the east and south-east, with a mean water depth of 70 m.

A first-degree trend surface, calculated from the bathymetric data, slopes north-east at 1 m in 3·9 km and has a fit of 50% (Fig. 4). The direction and degree of slope correspond well with that of the sub-Marr Bank Beds erosion surface (Holmes, 1977, fig. 6), suggesting that this feature has controlled the pattern of sedimentation during the late Pleistocene and the Holocene. Subtraction of the planar trend from the original bathymetric surface eliminates this underlying control and reveals residuals about the planar surface (Fig. 5). Positive residuals (areas 10 m above the planar surface) are largely confined to the northern half of the area with little expression of the morainic banks in the south-west. Negative residuals (areas 10 m below the planar surface) are similarly confined, although a long sinusoidal one extends from the north-east to the south-west. Elsewhere, there are small scattered negative residuals at the north and south margins of the area.

Current systems

Tides in the North Sea are predominantly semi-diurnal. The principal tidal wave enters from the north, moves south and reflects off the southern coast of the North Sea basin. Tidal currents (mean spring tides) range from less than 0·38 m sec^{-1} in the south-east of the study area to more than 1·00 m sec^{-1} at the coastal zone in the north-west (Fig. 6). Current directions over most of the area are NNE–SSW and are generally parallel to the isotachs. Near-bottom current values were computed, using the velocity profile of van Veen (1938) with an exponent of 1/5·3, to produce the zone of minimum critical erosion velocities on Fig. 6 from a Hjulström diagram, as modified by Sundborg (1956). This predicts mean spring tidal near-bottom current velocities sufficient for erosion and transport of unconsolidated non-cohesive sediments up to

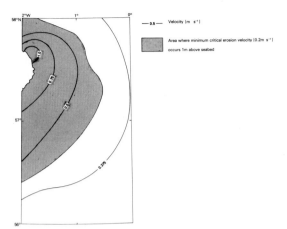

Fig. 6. Maximum surface tidal currents, mean spring tides. Current flow is generally parallel to the isotachs. The area of minimum critical velocity was obtained by extrapolating surface currents to 1 m above sea-bed by means of the parabolic decay curve of van Veen (1938). (Adapted from 'Tidal Currents for the North Sea, The English Channel and Irish Sea', Marine Hydrographic Service, German Democratic Republic, Rostock.)

medium sand grade over 40% of the area. Current velocities sufficient for transport only extend to 70% of the area.

The wave climate in the study area may be compared to that of the Sevenstones Light (open to oceanic waves) for which Draper (1967) indicates storm-wave-induced peak oscillatory current speeds in excess of 0.26 m sec^{-1} at water depths down to 100 m on approximately three days of the year. Ewing (1973) calculated that oscillatory currents with speeds of 0.62 m sec^{-1} may be experienced in water depths down to 100 m on the continental shelf during Beaufort force 10 storm conditions, with current speeds in excess of 0.48 m sec^{-1} down to depths of 120 m. He also observes that these speeds may be exceeded by a factor of 3.2 approximately every 30 min.

If these data for oscillatory currents are applied to the study area, then since accelerating currents produce more bottom stress than unidirectional steady flows (Komar, Neudeck & Kulm, 1972), it appears likely that unconsolidated, non-cohesive sediment at least up to medium sand grade can be eroded from the seabed during storms of force 10 and above. The occasional peak current values may also result in erosion of gravel-grade particles. Storms are common in the North Sea during winter months and it is likely that a combination of storm wave-induced oscillatory currents and tidal currents will be able to produce significant net sediment transport. Low-velocity tidal currents thus aided may be able to control the direction of transport simply by their constancy (Kenyon & Stride, 1970).

SEDIMENT TEXTURAL ANALYSES

Thickness of superficial sediments

The term 'superficial sediments' refers to sea-bed deposits which have been sampled by means of a Shipek grab. Shallow cores show that sediment thickness varies from less than 0.05 m to greater than 1.00 m. At over 80% of sample stations it is less than 0.50 m thick. This variability of distribution makes realistic areal correlation difficult but, generally, the thinnest sediment covers occurs on topographic highs and the thickest in lows. A regional trend of thickening towards the north-east is observed (Owens, 1980).

Sediment colour

The mixture of qualitative and semi-quantitative colour assessment required the grouping of sediment colours into three reasonably consistent classes ('yellow', 'grey' and 'green') prior to mapping.

The origins of colour in marine sediments are reviewed by Pantin (1969), Stanley (1969) and Swift & Boehmer (1972). Their findings indicate that 'yellow' sediment is stained by ferric compounds. Adsorption of ferrous and chlorophyll-related compounds could cause 'green' coloration. 'Grey' sediment may reflect a colour state transitional to those of 'yellow' and 'green'. An oxidizing environment is indicated by the presence of 'yellow' sediment whilst 'green' and 'grey' imply a reducing environment within the sediments. The latter case is supported by shipboard observations of colour changes following the recovery of samples.

The mapped distribution of sediment colour infers a relationship with bathymetry or topography (Fig. 7). 'Yellow' sediments are confined to shallow near-shore waters and to the submarine banks. In deep waters 'green' is the dominant sediment colour whilst 'grey' sediment is mainly confined to sediments at intermediate water depths (70–90 m). Isolated small patches of mixed colour groups suggest transition zones or areas where bottom conditions vary considerably over small distances.

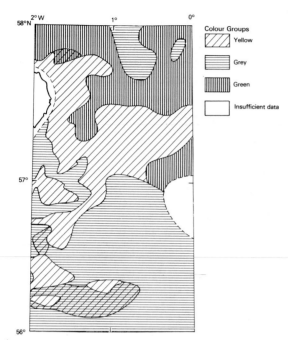

Fig. 7. Sediment colour. The colour classes are an integration of qualitative and semi-quantitative assessments of sediment colours.

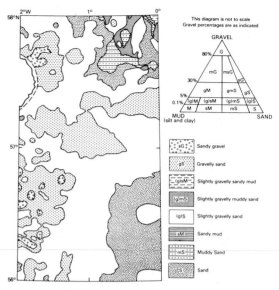

Fig. 8. Sediment distribution—Folk classification. A modified Folk classification is used with the trace gravel level at 0·1 %.

Folk classification analysis

Gravel, sand and mud (silt plus clay) sample percentage data are more readily comprehensive as 'facies' in terms of the Folk (1954) classification. Considerations of sample size and of the hydraulic significance of some of the gravel-grade biogenic carbonate makes the use of a 0·01 % gravel 'trace' threshold unrealistic. Instead, a value of 0·1 % was (arbitrarily) adopted, although even this may also be unrealistic. A practice of overlaying contoured maps of gravel percentages and sand/mud ratios to produce the Folk classes was employed, since experience showed that gradients thus preserved in the data produce the most realistic portrayal of the sediment distribution.

The mapped distribution of Folk sediment classes (Fig. 8) shows slightly gravelly sand to be the most common. This is because in many areas the dominantly sandy superficial sediments contain an admixture of biogenic carbonate gravel. Remapping on a carbonate free' basis virtually eliminates this class and demonstrates the dominance of sand. Little change is seen in the areas of gravelly sand (gS) and sandy gravel (sG), suggesting that their gravel content is dominantly lithic.

Correlation between areas of gravelly sediment and the glacigenic submarine banks (Fig. 3) is evident. Gradients fining downslope off the banks are also indicated. A significant area of muddy sediments in the north-east corresponds with an extensive topographic low in which water depths range from 100 to 130 m.

Summary statistics

Sample mean grain size

The mapped distribution of this statistic reinforces the observations made on correlation between sample grade and water depth (Fig. 9). Banks in waters shallower than 50 m are approximately delineated by the 0·0 phi contour. Sediments finer than 3·0 phi are mapped in waters deeper than 110 m. The 2·0 phi contour in the south separates the complex coarse sediments of the west from the area of uniformly sandy sediment in the east which corresponds with a region of low, featureless topography.

The distribution of sorting values follows closely that of mean grain size. Poorest sorting is confined to the submarine banks in shallow waters. Areas of well-sorted sediments are restricted to grain sizes around 2·0 phi and occur adjacent to the coast south of Peterhead and south of the area of muddy sediments.

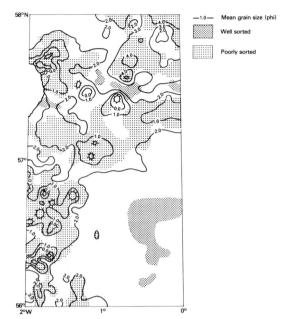

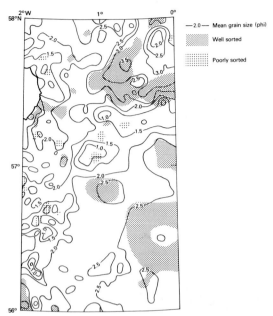

Fig. 9. Sample mean grain size. Areas of well-sorted sediment have phi sorting values less than 0·5; poorly sorted sediments have phi sorting values greater than 1·0.

Fig. 11. Sand mean grain size. Areas of well-sorted sediment have phi sorting values less than 0·5; poorly sorted sediments have phi sorting values greater than 1·0.

The correlation between mean grain size and water depth is further evaluated in Fig. 10. This reveals a non-linear relationship, although sample mean grain size does decrease with increasing water depth, particularly below 90 m in the case of sediment finer than 2·0 phi. One possible reason for the absence of a clear trend may be the assumption of normally distributed data. Such an assumption is clearly invalid in the case of coarse polymodal gravelly sediments. Palimpsest sediments, which are mixtures of relict and modern materials (Swift, Stanley & Curry, 1971) can be polymodal, particularly with modern inputs of biogenic carbonate.

Sand mean grain size

The distribution of this parameter is similar to that of the total sample, reflecting the dominance of sand-grade sediments (Fig. 11). Sand coarser than 2·0 phi is confined mainly to the west, with the exception of a restricted zone extending across the centre of the area. Submarine banks are clearly delineated by sand coarser than 1·5 phi. The south-east and north-east of the study area is dominated by sand finer than 2·0 phi, fining to less than 3·0 phi in the area of muddy sediment.

Sand-sorting values generally lie between 1·0 and

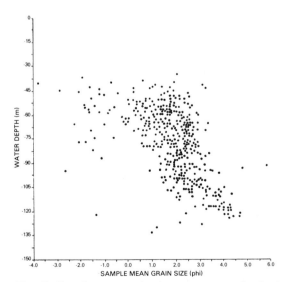

Fig. 10. Sample mean grain size versus water depth. A general trend to fining with increasing water depth is evident, although only distinct with grain sizes finer than 2·0 phi.

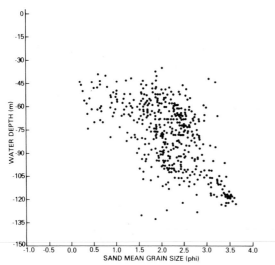

Fig. 12. Sand mean grain size versus water depth. The fining trend with increasing water depth is better defined than with the total sample but is obviously not simply linear. (Scale of x-axis × 2 compared to Fig. 10.)

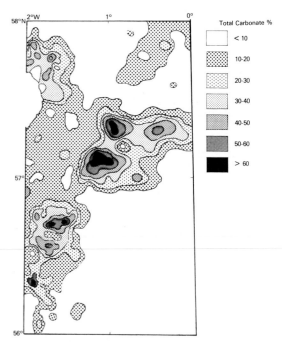

Fig. 13. Carbonate distribution (total sample) showing that high values generally correspond with the morainic banks.

0·5 (moderately well sorted). A restricted area of well-sorted sand south of the Peterhead coast corresponds with a similarly located area of well-sorted total sediment. A plot of sand mean grain size against water depth shows a near linear decrease in size with increasing water depth (Fig. 12). This corresponds with the 'classical' view of a pattern of seaward fining of sediments and supports a conclusion that sand is the equilibrium sediment in the area.

SEDIMENT COMPOSITION

Lithic fraction

Gravel-grade lithic clasts range up to 76% of the sample with a mean value of 4%. Metamorphic, igneous and well-indurated fine-grained sandstones are the most common lithologies encountered. Limestone, chalk and flints are rare. The concentrated occurrence of lithic gravels, the identification of glacial striae on occasional clasts and their poorly sorted grain-size distribution indicate that they are derived by erosion of the morainic banks (Owens, 1977a, 1980).

A limited petrographic study of the sand fraction of 32 randomly selected samples shows that it consists of mono- and polycrystalline quartz, feldspars, micas, heavy minerals, glauconite and igneous, metamorphic and sedimentary rock fragments. The .

sand composition, particularly its high content of rock fragments, polycrystalline grains and heavy minerals, and the angularity of many of the grains (Owens, 1977a, 1980) suggests direct derivation from glacial sediments (Trumbull, 1972).

Carbonate fraction

Petrographic analysis has not disclosed any significant presence of detrital carbonate; carbonate content here refers to biogenic carbonate content.

Gravel and sand-grade carbonate is composed dominantly of molluscan remains with subordinate barnacles, echinoids and foraminifera. This is the foramol assemblage typical of high-latitude carbonates (Lees & Buller, 1972). In samples where biogenic carbonate is a significant constituent (greater than 20%) it is commonly iron-stained, rounded and bored, suggesting it is relict (Swift *et al.*, 1971).

Sample carbonate content may exceed 70% on the shallower banks but decreases with increasing water depth (Fig. 13), declining rapidly east of the banks where the 20% contour follows the 50–60 m zone. Low values correspond with two distinct areas. The

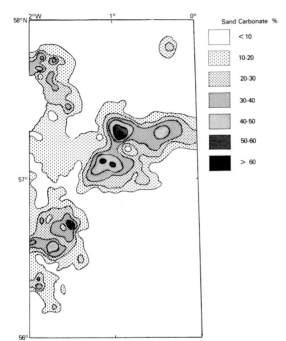

Fig. 14. Carbonate distribution (sand fraction). Comparison with Fig. 13 shows that most of the carbonate is found in the sand fraction.

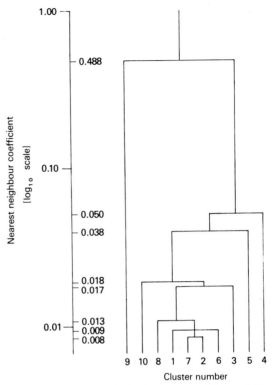

Fig. 15. Cluster analysis dendrogram. The variables analysed for each sample were the half-phi variables for the total sample, carbonate and lithic gravel, sand carbonate, mud carbonate and mud content.

zone south of latitude 57° N and east of longitude 1° 30′ W, mapped as greater than 90% sand, corresponds with a carbonate content of less than 10%. Similarly, in the north the area of low carbonate corresponds with a zone of muddy sands in the depth range 90–110 m.

Comparison of the distribution of total carbonate (Fig. 13) with that of sediment colour (Fig. 7) shows a correlation between 'yellow' sediment and areas of high carbonate content. The degree of correlation is considerably improved on comparison of the distribution of sand-grade carbonate (Fig. 14) with that of the 'yellow' sediments. Here, the fit is almost exact, indicating that the 'yellow' sediment colour is due largely to the presence of iron-stained carbonate sand. Petrographic examination reveals that the subordinate lithic component of these sediments is similarly stained.

The areal distribution of sand-grade carbonate overlaps that of gravel-grade carbonate by a considerable extent. This, and a consideration of the size distribution curves, indicates production and dispersion of sand-grade carbonate on the banks by *in situ* degradation of gravel-grade carbonate

through a combination of physical and biological processes (Owens, 1980).

FACIES ANALYSIS

Allocation of samples to facies involves assessment of up to 50 variables from 495 samples and necessitated the use of statistical techniques. Cluster analysis was selected because the technique is independent of sample frequency distribution and does not require the experimental sample to be representative of the general population or carefully randomized (Parks, 1966). Bias due to redundant variables is eliminated by principal component analysis prior to clustering. Ten basic cluster types were identified by cluster analysis of half-phi variables (total sample and gravel carbonate) and carbonate percentages. These were then grouped into six facies following a shape analysis of grain-

Table 1. Cluster summary data

Cluster no.	Cluster component percentages							Cluster statistics (mean phi sizes)				
	G	S	M	G carb.	S carb.	M carb.	Total carb.	Total sample	Gravel total	Gravel carb.	Gravel lithic	Sand total
1	0·6	93·6	5·8	0·4	5·4	1·5	7·3	2·6	−3·0	−2·3	−4·0	2·6
2	0·7	98·2	1·1	0·3	4·0	0·3	4·6	2·4	−2·3	−1·5	−2·9	2·5
3	7·4	91·9	0·7	4·1	20·9	0·3	25·3	1·3	−2·4	−1·9	−3·1	1·6
4	0·2	94·3	5·5	0·2	4·9	1·3	6·4	3·2	−2·6	−2·6	—	3·3
5	0·1	73·1	26·8	0·1	6·7	6·7	13·5	3·3	−2·1	−2·1	—	3·3
6	0·5	98·2	1·3	0·5	4·0	0·3	4·8	2·6	−2·2	−2·2	−1·3	2·6
7	11·0	87·8	1·3	3·5	10·9	0·4	14·8	1·5	−2·9	−2·3	−3·2	2·1
8	0·1	98·6	1·3	0·1	2·3	0·3	2·7	2·8	−1·8	−1·9	−1·7	2·8
9	3·6	93·9	2·5	1·6	3·9	0·2	5·7	2·3	−3·8	−4·0	−3·6	2·5
10	0·0	97·2	2·8	0·0	4·3	0·7	5·0	3·0	—	—	—	3·0

Table 2. Facies summary data

Facies	Component clusters and number in each cluster	Characteristics (percentages given are approximate)
Gravelly–coarse	3 (120)	High carbonate (25 %); sand- and gravel-grade carbonate in equilibrium: lithic gravel has three modes at -5ϕ, -3ϕ and -1ϕ; sand fraction is medium grade and moderately sorted
Gravelly–sand transitional	7 (168)	Medium carbonate (15 %); sand- and fine fraction of gravel-grade carbonate in equilibrium; lithic gravel has three modes at -5ϕ, -3ϕ and -1ϕ; sand fraction is fine grade and moderately sorted
Sandy	2 (9) 6 (18) 8 (13) 9 (3)	Low carbonate (6 %); sand forms 98 % of samples, is fine, and moderately to well sorted
Sandy–sensitive	1 (36)	Low carbonate (8 %); sand forms 94 %, mud 6 %; sand is fine and well sorted, with a frequency distribution identical to the main sandy facies
Sandy–muddy transitional	4 (5) 10 (15)	Low carbonate (6 %); sand forms 94–97 %, mud 3–6 %; sand is fine–very fine and well sorted
Muddy	5 (39)	Medium carbonate (14 %); sand forms 73 %, mud 27 %; sand is very fine and well–very well sorted

size distribution curves and the dendrogram (Fig. 15, Tables 1 and 2). If the particle size distributions of the facies are used as a measure of the extent of influence of currents in the depositional environment (Swinchatt, 1965), they appear to form a dynamic sedimentary system covering the spectrum of activity from erosion to deposition.

Comparison of the facies distribution (Fig. 16) with the bathymetry (Fig. 3) and the distribution of topographic residuals (Fig. 5) reveals a set of correlations indicative of complex relationships. There is evidence, for example, that the facies mapped relate not only to absolute water depth but also to relative topography. This is best illustrated by the large

spread of the gravelly–coarse facies in the centre of the map, in the region of the Turbot and Aberdeen Banks. These banks have low relief, except on their northern and north-western flanks where they slope into deeper waters. Northerly currents encountering these topographic highs will accelerate across their profile, while there may also be an increase in wave activity in their vicinity. Thus, areas of greatest positive topographic anomaly should correspond with facies indicative of an environment with a high degree of hydraulic activity. This may account for the relative absence of the gravelly–coarse facies from the morainic banks of the south-west and for its presence on other isolated banks in the north. It

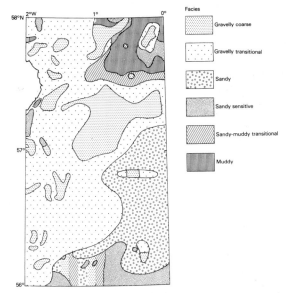

Fig. 16. Facies distribution. The environments represented are: (1) erosional (gravelly–coarse facies); (2) transportational (gravelly–sand transitional and sandy facies); (3) depositional (sandy–sensitive, sandy–muddy transitional and muddy facies).

may also explain the distribution of depositional facies that appear to have an anomalous disposition. For example, the occurrence of the sandy–sensitive facies in the south-west and south-east is clearly related to depressions and negative topographic residual features. However, in the north-east it covers an elevated area south and east of the Halibut Bank, where northerly currents will decelerate following an acceleration across the Halibut Bank.

The distribution of the gravelly–sand transitional and sandy facies is also explicable in terms of a combination of the effects of bathymetric and relative topographic controls. Both occur on relatively flat featureless areas. Differentiation between these facies appears related to absolute water depth; the gravelly–sand transitional facies occurs in water shallower than 70 m, whereas the sandy facies is encountered in depths of 70–90 m. The model of absolute and relative bathymetric control on facies fails to account for the almost complete absence of depositional facies in the Buchan Deep–Dog Hole linear complex of bathymetric lows. This may reflect the coarseness of the sampling grid in this area.

The facies distribution reveals a complex situation, where facies are correlated not only with

absolute water depth but also with topography and degree of exposure. Summarizing, the environments that the facies appear to represent are erosional (gravelly–coarse), transportational (gravelly–sand transitional and sandy) and depositional (muddy, sandy muddy transitional and sandy–sensitive). Sorting trends reinforce the impression of a dynamic system.

DISPERSION PATTERNS

Sedimentary evidence

The earlier interpretation of facies indicating areas dominated by erosional and depositional conditions implies that dispersion paths should exist between these areas. Contoured maps of textural parameters are extremely 'noisy', with considerable apparently random variation in sample characteristics over very short distances. Accordingly, trend surface analysis was used to detect any major trends (Gorsline & Grant, 1972) underlying the areal distribution of the mean grain size of the total sample and the main sample components (lithic and biogenic carbonate gravel and total sand). Surfaces of degrees 1–6 were calculated for the whole area and, to improve their fit, for restricted areas. Fit varied from almost zero, in the case of total gravel for the entire area, to 50% in the case of one of the restricted sand areas.

The identification of possible dispersion paths from the trend surfaces depends on the assumption of a process–response model. Fining of sediment in the direction of declining current strength and improved sorting with duration of transport are expected. Dispersion paths should be indicated by the presence of linear maxima of sediment grain size and the direction(s) of dispersion by that of minimum gradient(s). Differences in particle characteristics may cause varying responses to similar levels of hydraulic activity, resulting in dispersion paths controlled not only by the processes, but also by the response of the particles to the processes.

The presence of relict glacial sediments in an area where erosional zones are identified implies that residual lag sediments may also occur. Poor sorting is recognized in association with the morainic banks, considered to be a major source of sediment in the area. Similarly, a depositional zone at the distal end of a dispersion path may be expected to contain well-sorted sediments.

Dispersion paths identified from the trend surface analyses are summarized in Fig. 17. All fractions

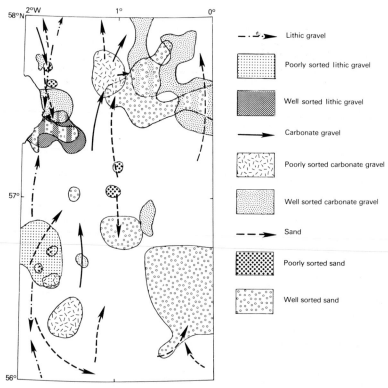

Fig. 17. Sediment dispersion paths. The paths are identified from trend surface analysis of particle size data.

analysed exhibit a similarly aligned north–south pattern with dispersion to either direction, generally parallel to the tidal currents (Fig. 6). The proximal ends of dispersion paths correlate with areas of poorly sorted sediment on the morainic banks, identifying them as sources of sand and lithic gravel. Near the coast, a zone of well-sorted lithic gravel- and sand-grade sediment corresponds with the distal ends of the lithic gravel and sand dispersion paths. This area also coincides with a sand-transport bed-load convergence zone mapped by Kenyon & Stride (1970).

In the eastern half of the study area dispersion trends are less clear. The zone of fine muddy sediment in the north-east corresponds with areas of well-sorted sand and carbonate gravel, and with the distal ends of well-defined sand dispersion paths. Similarly, in the south-east, a zone of well-sorted sand coincides with the ends of other, less well defined dispersion paths.

In the south-west a deviation from the general north–south alignment is seen in the anticlockwise curvature of a sand dispersion path. This deviation

occurs in an area of low-velocity tidal currents (Fig. 6), possibly subject to the influence of randomly oriented wind-drift residual currents (Dooley, 1974).

An input of carbonate gravel from the Moray Firth corresponds with one of the sand transport paths mapped in Kenyon & Stride (1970). It is supported by evidence of a 'tongue' of biogenic carbonate extending across the floor of the Moray Firth in a north-west–south-east direction (Owens, 1977b). This is the only evidence for an input of allogenic sediment to the area.

Geophysical evidence

The data consist mostly of Kelvin-Hughes MS47 transit sonar records, supplemented by echo-sounder and Huntec deep-tow boomer records. Bedforms were identified following Belderson, Kenyon & Stubbs (1972) and were placed in the following categories.

(1) Lineations—sand ribbons, elongated sand and/ or gravel patches and lineated mounds of coarse or tough sediment emergent through a thin sand

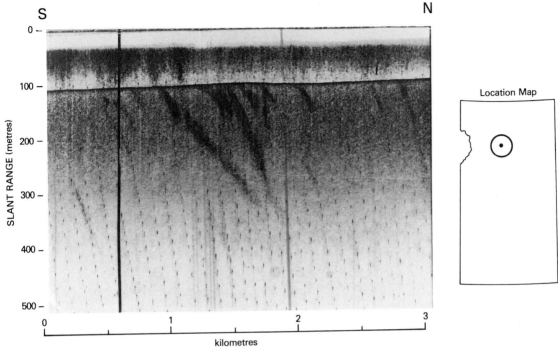

Fig. 18. Lineations. A side-scan sonar record of coarse or hard substrate emerging as a low ridge through a thin sand cover.

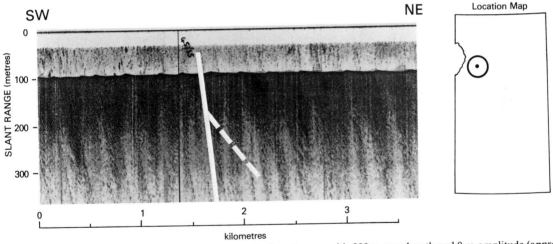

Fig. 19. Large sandwaves. A side-scan sonar record of sandwaves with 200 m wavelength and 8 m amplitude (approximately). The trend of the superimposed megaripples (broken line) is oblique to that of the sandwaves (continuous line).

cover (e.g. Fig. 18). They are aligned parallel to the sediment transport direction.

(2) Waveforms—classified, following McCave (1971) as megaripples (wavelength less than 30 m) and sandwaves (wavelength greater than 30 m). Transport direction can only be resolved in the case

of the larger features (Fig. 19) and is at right angles to the strike of the bedforms.

The sediment transport directions inferred from these features (Fig. 20) show considerable agreement with the results of Kenyon & Stride (1970), Stride (1973) and the sedimentary evidence of this

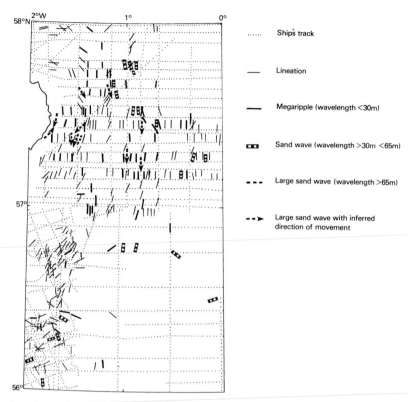

Fig. 20. Sediment transport directions inferred from the alignments of lineations and perpendiculars to the directions of strike of waveforms.

study. They have a dominant north–south orientation, with deviations generally conforming with coastal alignment, but with a less organized appearance in the south-west. Lineations are most common in the north and west of the area and correspond with zones of positive topographic features. Waveforms are sparse and the occurrence of megaripples and small sandwaves appears unrelated to particular features or zones. Large sandwaves occur only in the north, with average wavelengths of 200 m and heights up to 17 m. The bedload convergence zone of Kenyon & Stride (1970) is clearly indicated by convergent asymmetry of the sandwaves (Fig. 21).

Comparison of the mapped distribution of these features with the tidal currents (Figs 20 and 6) supports the sedimentary evidence that directions of transport are controlled by the tidal streams. The locations of features also reflect control by the strength of the tidal currents. Lineations, generally located on topographic highs identified as positive residuals, occur mainly within the area of minimum critical erosion velocity at the sea-bed (Fig. 6). Since

these features indicate erosional conditions, or areas 'starved' of sediment supply, their presence on positive topographic residuals (Fig. 5), and absence elsewhere in areas of similarly weak tidal currents implies that the direction and intensity of sedimentation are here controlled by an interaction of relative topography, tidal currents and (possibly) storm-wave-induced oscillatory currents.

Large-scale sandwaves are generally found in the shallow water near coastal zones, where tidal currents range from 0·6 to 1·3 m sec^{-1}. They also occur 50 km east of the coast in waters approximately 80 m deep in an area of low-speed tidal currents (0·3– 0·5 m sec^{-1}). Sandwaves of this magnitude (amplitude approximately 8 m; wavelength 160–270 m) are normally associated with high current speeds, generally in shallow waters, where sand supply is abundant (Kenyon & Stride, 1970; Belderson *et al.*, 1971; Caston, 1974). Their occurrence here is anomalous and may indicate that they are relict. Evidence of possibly relict tidal sand ridges in the southern North Sea is given in Jansen (1976), who

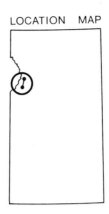

LOCATION MAP

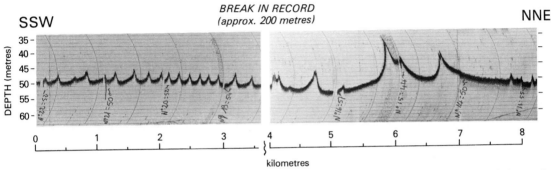

SSW   BREAK IN RECORD   NNE
                       (approx. 200 metres)

Fig. 21. Bedload convergence. This echo-sounder record of sandwaves shows convergent asymmetry. (N.B. vertical exaggeration × 50. The lee side of the largest sandwave has slopes of 4·2° on the upper section, 0·3° on the lower section and 1·5° overall.)

suggests that they formed between 12,000 and 9000 y BP when the sea-level was 35–45 m below the present level. The absence of megaripples from the flanks of the ridges is used as part of the evidence. The presence of megaripples on these sandwaves, as seen in Fig. 19, may indicate that they are active features (McCave, 1971; Langhorne, 1973). However, the absence of an obvious present-day mechanism to account for the presence of the sandwaves may still mean that they are relict, but not necessarily many thousands of years old. They could be caused by strong currents generated during the occasional severe storms experienced in the area (see below).

The geophysical evidence also suggests that there are zones of deposition within the study area. The bedload convergence of Kenyon & Stride (1970) has already been mentioned. Less obvious examples of depositional zones can be found; in the north-east an apparently featureless area corresponds with the zone of deposition of fine sandy and muddy sediment. This environment is likely to contain only

small-scale bedform features not detectable by the seismic techniques employed.

A discontinuous extension of this featureless area can be followed to the south-west, through Buchan Deep and Dog Hole, to the complex of deeps north of the Scalp and Montrose Banks. This possibly provides the evidence of depositional conditions which could not be obtained from the textural data.

Another featureless zone in the south-east has also been interpreted, on textural evidence, as a zone of deposition. However, the facies here (dominantly sandy with some sandy–sensitive) are representative of a transportational regime with some restricted deposition. It is possible that this area is one of metastable conditions where weak tidal currents allow deposition in a depth zone subject to fairly regular disturbance by wave-induced oscillatory motion.

East of Turbot Bank (Fig. 3) a zone of strongly lineated sea-bed occurs in water depths similar to the area of possible metastable deposition. The difference between these areas may be due to a broad

Table 3. Radiocarbon dates of biogenic carbonate

Sample no.	Position Lat. (N)	Long. (W)	Water depth (m)	Radiocarbon age (y BP)	Nature of dated material	Comment
DF 7	56° 26′	01° 53′	60	10,865 ± 160	Balanid fragments	Worn, abraded and strongly iron-stained
DF 13	56° 47′	01° 17′	68	6370 ± 160	*Acanthiocardia echinata*, single shell	Worn, abraded and strongly iron-stained
SF 43	57° 05′	01° 12′	62	3070 ± 80	Bulk sample, dominantly *Mollusca*	Worn, abraded and mostly iron-stained
SF 44	57° 02′	01° 10′	61	4270 ± 90	Bulk sample, dominantly *Balanus*	Fragmentary, worn and commonly iron-stained
SF 46	57° 08′	01° 08′	63	4220 ± 100	Bulk sand-grade sample	Badly worn and commonly iron-stained
SF 215	57° 21′	01° 39′	65	6170 ± 100	Bulk sand-grade sample	Badly worn and commonly iron-stained

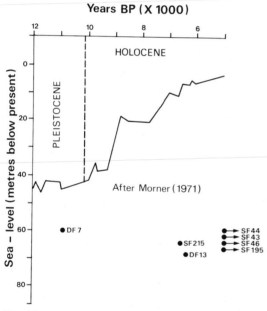

Fig. 22. Sea-level curve. Dated samples are positioned according to the age and the water depth in which they were found.

positive topographic residual feature. If this interpretation is correct, then marginal conditions of dynamic stability are indicated in this and other similar locations within the study area, emphasizing the importance of relative topography in facies control.

AGE OF DEPOSITS

The biogenic carbonate component is a significant part of the apparently palimpsest sediments mapped in the area. It is probably the only clearly identifiable post-glacial input and is of importance in the differentiation of relict from modern sediment.

Much of the biogenic carbonate sampled has a classically 'relict' texture, a combination of iron oxide stain, roundness and biological degradation which leads to an expectation of considerable age partially substantiated by the radiocarbon ages of some samples (Table 3). A feature of these dates is their lack of consistency, even between results from single species. When plotted against a sea-level curve for the area, the samples do not correlate with any low stand of sea-level which could have caused the sub-aerial exposure commonly invoked to account for iron-stained sediments (Fig. 22). This lack of fit could be explained by post-depositional transport of the shells. The observed high degree of rounding, together with evidence of mixing with 'fresh' material (Owens, 1980), supports this view. However, the distribution of bathymetric highs or positive topographic residuals (Figs 3 and 5), total carbonate, sand carbonate (Figs 13 and 14) and sediment colour (Fig. 7) are so intimately associated that a causal relationship is evident. The topographic highs and associated gravelly sediments are considered zones of high biological productivity (Belderson *et al.*, 1971). Since the amount of biogenic carbonate present at any given location is a function of productivity, dilution by terrigenous sediment and transport, conditions on the highs may be favourable for a slow, long-term accumulation of carbonate sediments.

The production of sand-grade carbonate by *in situ* attrition of gravel-grade carbonate implies considerable agitation at the sea-bed. The shallowness of the submarine banks renders the sediments on them vulnerable to the action of wave-induced oscillatory currents and, since the highest degree of shell abrasion occurs when they are mixed with gravelly sands (Driscoll, 1967), conditions well suited to rounding of shells and shell fragments are indicated (Owens, 1980). Exposure of the carbonate particles at the sea-bed renders them liable to biological attack, such as boring by sponges, annelids and bryozoa. Thus slow rates of accumulation or long residence times should favour biological degradation of the shell material.

The iron-stained appearance, commonly considered diagnostic of relict sediment (Swift *et al.*, 1971), can also be explained by halmyrolysis. Sediments exposed to frequent agitation and aeration at the sea-bed in an area of little or no deposition are in an oxidizing environment where iron present may be deposited as ferric compounds, particularly iron oxides (Swift & Boehmer, 1972). The current-agitated coarse carbonate-rich sediments of the banks appear to form such an environment. Evidence to support this view is found in the distribution of the mineral glauconite. This occurs in two mutually exclusive forms in the study area; a fresh green form and an altered brown form. In thin-section the altered form is seen to range from a brown rim surrounding a green core to totally brown (Owens, 1980). The brown compound is identified as limonite, formed by the oxidation of glauconite (McRae, 1972). The altered form of glauconite correlates with topographic highs, 'yellow' sediments, high carbonate values and current sculpted lineations on the sea-bed. These observations support a hypothesis that iron staining could be the product of modern processes and not necessarily the result of sub-aerial weathering in a former environment (Swift & Boehmer, 1972).

The areas of carbonate-rich sediment are interpreted as representing a long-term dynamic balance between carbonate productivity in the favourable environments of the current-swept banks, and erosional losses following physical and biological degradation of the shelly debris. Thus, although these sediments are old, they need not be relict *sensu stricto*. In the absence of data on process rates, for both accumulation and degradation of carbonate (Farrow, Cucci & Scoffin, 1978), differentiation between truly relict deposits of carbonate and long-term accumulations is impossible without a considerable amount of radiometric or other dating information. Even this approach may be inconclu-

sive. Shallow burial could result in shells from the same sample giving identical radiometric ages, but having contrasting 'fresh' and 'relict' appearances. Dates compatible with this hypothesis have been obtained from samples of carbonate sediment from the west coast of Scotland (H. Allen, 1979, personal communication).

DISCUSSION

Sediment distribution in the area is subject to several levels of control. Topographic highs, the source of the lithic gravel fraction, are most exposed to the effects of tidal and wave-induced oscillatory currents. Hence the lithic gravels remain unburied as part of the zonal equilibrium sediment, the gravelly-coarse facies. The remaining coarse components, carbonate gravel and sand, may be interpreted as modern input to a palimpsest sediment mixture.

Areas of low topographic relief shallower than 70 m are generally covered by sediments of the gravelly-sand transitional facies. Its composition (Table 2) suggests it represents the gravelly-coarse facies diluted by an admixture of fine, dominantly quartzose sand eroded from the glacigenic sediments of the banks. In deeper waters, between 70 and 90 m, a decrease in hydraulic activity leads to a dominance of fine sands in the sandy facies. In both cases these facies represent metastable deposits, possibly subject to erosion and redistribution by the action of storm-wave-induced oscillatory currents. At depths below 90 m the depositional facies (sandy-sensitive, sandy-muddy transitional and muddy) represent a further decline in levels of hydraulic activity at the sea-bed. Carbonate content in the latter two facies reverses the trend of decline with increasing water depth because of an input of foraminiferal tests.

The apparently anomalous location of sea-bed lineations in areas of weak tidal currents has been explained in terms of localized increases in levels of hydraulic activity on topographic highs. Alternatively, this may be a particular expression of the more general case that relatively weak tidal currents are capable of causing net sediment transport when aided by the erosive effects of storm-wave-induced oscillatory currents.

The possibility that the anomalous sandwaves, mapped in the centre of the study area, may be the 'relict' product of an exceptional storm casts doubt on the synoptic significance of the sediments mapped. Storms appear to form a regular part of the shelf processes on a geological time-scale (Lewis, 1979). They can raise transport rates by a factor of 10 (Johnson & Stride, 1969) and, with extreme conditions, possibly by several orders of magnitude (Owens, 1980). Apparently gradual geological processes may have a 'catastrophic fine structure' (Swift *et al.*, 1971), with the result that the effects of storms may be the chief, if not the only, events recorded in the sediments (Ager, 1974). The likely effects of exceptional storms (singular events) further complicate matters since their long return periods and probably exceptionally effective processes have to be envisaged against a background of knowledge of the 'normal' processes operating within a given area (Hayes, 1967). Thus, at any point within the area, the sediments as mapped can represent three conditions.

(1) Equilibrium—modern sediment in equilibrium with its hydraulic environment. This includes palimpsest sediments.

(2) Disequilibrium—sediments, exhumed or transported by the effects of a singular event, that are out of equilibrium with the normal hydraulic regime of their present environment. These sediments may appear relict on the short time-scale of recent observation, but could be palimpsest on the longer (geological) time-scale (Owens, 1977a).

(3) Transient—sediment in transit as part, for example, of a sandwave. These sediments are likely to have a short, but finite, residence time as part of the sediment body.

Accordingly, the sediments mapped within the area may be considered both palimpsest and relict over several time-scales. The lack of detailed long-term data on meteorological and oceanographic conditions in the area makes meaningless the conventional definition of modern conditions: 'having no significant difference from present configurations' (McManus, 1975). It also calls into question the simple assumption that any lag conglomerate underlying the superficial sediments is the Holocene transgressive lag. It could equally well represent localized lags developed by the last singular storm event, the last winter's most severe storm or, simply, the last storm prior to sampling. Combinations of these circumstances are also possible.

CONCLUSIONS

The areal and vertical relationships of textural variables has revealed a degree of bathymetric con-

trol on their distribution. While sand is ubiquitous, gravelly sediments occur mainly on topographic highs in shallow waters, and muddy sediments generally in depths below 120 m. A dynamic sedimentary environment is indicated in which the coarser terrigenous and carbonate components of the sediments are derived from topographic highs formed of glacigenic sediments.

The directions of sediment dispersion are controlled by the tidal stream. Since tidal currents within the area are generally of low velocity, it is inferred that their dispersive action is initiated and aided by storm-wave-induced oscillatory currents. Relative topography is as important as absolute water depth in controlling the consequent facies distribution. Supporting evidence from geophysical data also reveals anomalously located large sandwaves, possibly caused by the effects of exceptionally severe storms.

The abundance of apparently relict carbonate on topographic highs may be the result of slow accumulation and long residence times in current-swept environments of little or no terrigenous sedimentation. Consequent long-term exposure on the sea-bed could account for physical and biological degradation of the shelly sediment. Current action may cause sufficient aeration to give an iron-stained appearance from deposits of ferric compounds. Support for this view is gained from the mapped distribution of an altered form of glauconite.

The use of the terms 'relict' and 'palimpsest' to describe any of these sediments may be misleading until more information on spatial and temporal variations in the sedimentary environment is available. Similarly, the definition of continental shelves as either 'tide-dominated' or 'storm-dominated' (e.g. Johnson, 1978) may be inappropriate when the relative significance of tides and storms and their influence in combination on shelf sediments is not fully understood.

ACKNOWLEDGMENTS

I thank Dr Bryan Lovell of Edinburgh University and Drs Nigel Fannin, Daniel Evans and Mr Alexander Skinner of the Institute of Geological Sciences for their careful criticism of the manuscript. The study formed a part of the author's Ph.D. thesis which was jointly supervised by Dr Bryan Lovell and Messrs Robert Eden and Dennis Ardus, Institute of Geological Sciences. The paper is published with the permission of the Director, Institute of Geological Sciences.

REFERENCES

AGER, D.V. (1974) Storm deposits in the Jurassic of the Moroccan High Atlas. *Palaeogeogr. Palaeoclim. Palaeoecol.* **15**, 83–93.

BATCHA, J.P. & REESE, J.R. (1964) Surface determination and automatic contouring for mineral exploration extraction and processing. *Colo. Sch. Mines Q.* **59**, 1–4.

BELDERSON, R.H., KENYON, N.H. & STRIDE, A.H. (1971) Holocene sediments on the continental shelf west of the British Isles. In: *ICSU/SCOR Working Party 31 Symposium, Cambridge,* 1970, *The Geology of the East Atlantic Continental Margin.* 2. *Europe* (Ed. by F.M. Delaney). *Rep. Inst. Geol. Sci. No.* 70/14, pp. 157–170.

BELDERSON, R.H., KENYON, N.H. & STUBBS, A.P. (1972) *Sonographs of the Sea Floor: a Picture Atlas.* Elsevier, Amsterdam.

CARVER, R.E. (1971) *Procedures in Sedimentary Petrology.* Wiley, New York.

CASTON, V.N.D. (1974) Bathymetry of the northern North Sea. Knowledge is vital for offshore oil. *Offshore,* February, 76–84.

DOOLEY, H.D. (1974) Hypotheses concerning the circulation of the northern North Sea. *J. Cons. int. Explor. Mer.* **36**, 54–61.

DRAPER, L. (1967) Wave activity at the seabed around northwestern Europe. *Mar. Geol.* **5**, 133–140.

DRISCOLL, E.G. (1967) Experimental field study of shell abrasion. *J. sedim. Petrol.* **37**, 1117–1123.

EDEN, R.A., HOLMES, R. & FANNIN, N.G.T. (1978) Depositional environment of offshore Quaternary deposits of the Continental Shelf around Scotland. *Rep. Inst. Geol. Sci. No.* 77/15.

EWING, J.A. (1973) Wave-induced bottom currents on the outer shelf. *Mar. Geol.* **15**, M31–M35.

FARROW, G.E., CUCCI, M. & SCOFFIN, T.P. (1978) Calcareous sediments on the nearshore continental shelf of western Scotland. *Proc. R. Soc. Edin.* **76 B**, 55–76.

FLINT, R.F. (1957) *Glacial and Pleistocene Geology.* Chapman & Hall, London.

FOLK, R.L. (1954) Sedimentary rock nomenclature. *J. Geol.* **62**, 345–351.

GORSLINE, D.S. & GRANT, D.J. (1972) Sediment textural patterns. In: *Shelf Sediment Transport: Process and Pattern* (Ed. by D.J.P. Swift, D.B. Duane and O.H. Pilkey), pp. 575–600. Dowden, Hutchinson & Ross, Stroudsburg, Pennsylvania.

HAYES, M.O. (1967) Hurricane as geologic agents: case studies of hurricane Carla, 1961 and Cindy, 1963. *Rep. Inv.* **61**, 1–56. Texas University Bureau of Economic Geology.

HOLMES, R.H. (1977) Quaternary deposits of the central North Sea. 5. The Quaternary geology of the UK sector of the North Sea between 56° and 58° N. *Rep. Inst. geol. Sci. No.* 77/14.

JANSEN, J.H.F. (1976) Late Pleistocene and Holocene history of the northern North Sea, based on acoustic reflection records. *Neth. J. Sea Res.* **10**, 1–43.

JANSEN, J.H.F., VAN WEERING, TJ. C.E. & EISMA, D. (1979) Late Quaternary Sedimentation in the North Sea. In: *The Quaternary History of the North Sea* (Ed. by E. Oele, R.T.E. Schüttenhelm and A.J. Wiggers). *Acta. Univ. Ups. Symp. Univ. Ups. Annum Quingentesimum Celebrantis,* **2**, 175–187.

JOHNSON, H.D. (1978) Shallow siliclastic seas. In: *Sedimentary Environments and Facies* (Ed. by H.G. Reading), pp. 207–258. Blackwell Scientific Publications, Oxford.

JOHNSON, M.A. & STRIDE, A.H. (1969) Geological significance of North Sea sand transport rates. *Nature,* **224**, 1016–1017.

JOHNSTONE, G.S. (1966) *The Grampian Highlands.* HMSO, Edinburgh.

KENYON, N.H. & STRIDE, A.H. (1970) The tide-swept continental shelf sediments between the Shetland Isles and France. *Sedimentology,* **14**, 159–173.

KOMAR, P.D., NEUDECK, R.H. & KULM, L.D. (1972) Observations and significance of deep-water oscillatory ripple marks on the Oregon continental shelf. In: *Shelf Sediment Transport: Process and Pattern* (Ed. by D.J.P. Swift, D.B. Duane and O.H. Pilkey), pp. 601–619. Dowden, Hutchinson & Ross, Stroudsburg, Pennsylvania.

LANGHORNE, D.N. (1973) A sandwave field in the Outer Thames Estuary, Great Britain. *Mar. Geol.* **14**, 129–143.

LEES, A. & BULLER, A.T. (1972) Modern temperate water and warm water shelf carbonates contrasted. *Mar. Geol.* **13**, M67–M73.

LEWIS, K.B. (1979) A storm dominated inner shelf, Western Cook Strait, New Zealand. *Mar. Geol.* **31**, 31–43.

McCAVE, I.N. (1971) Wave effectiveness at the seabed and its relationship to bed-forms and deposition of mud. *J. Sedim. Petrol.* **41**, 89–96.

McMANUS, D.A. (1975) Modern versus relict sediment on the continental shelf. *Bull. geol. Soc. Am.* **86**, 1154–1160.

McRAE, S.G. (1972) Glauconite. *Earth Sci. Rev.* **8**, 397–440.

OWENS, R. (1977a) Quaternary deposits of the central North Sea. 4. Preliminary report on the superficial sediments of the central North Sea. *Rep. Inst. geol. Sci. No.* 77/13.

OWENS, R. (1977b) Cruise Report—John Murray 77JM02, Leg 2. *Internal Rep.* 77/15. IGS Continental Shelf Northern Unit.

OWENS, R. (1980) *Holocene sedimentation in the central North Sea.* Unpublished Ph.D. Thesis, University of Edinburgh.

PANTIN, H.M. (1969) The appearance and origin of colours in muddy marine sediments around New Zealand. *N.Z. J. Geol. Geophys.* **12**, 51–66.

PARKS, J.M. (1966) Cluster analysis applied to multivariate geologic problems. *J. Geol.* **74**, 703–714.

SAMPSON, R.J. (1978) *Surface II Graphics System.* Kansas Geological Survey Series on Spatial Analysis. Kansas Geological Survey, Kansas.

SCHLEE, J. & WEBSTER, J. (1967) A computer program for grain size data. *Sedimentology,* **8**, 45–53.

STANLEY, D.J. (1969) Atlantic continental shelf and slope of the United States—colour of marine sediments. *Prof. Pap. U.S. geol. Surv.* 529-*D*.

STRIDE, A.H. (1973) Sediment transport by the North Sea. In: *North Sea Science* (Ed. by E.D. Goldberg), pp. 101–130. MIT Press, London.

SUNDBORG, A. (1956) The River Klaralven, a study in fluvial processes. *Geogr. Annlr* **38**, 125–316.

SWIFT, D.J.P. & BOEHMER, W.R. (1972) Brown and gray sands on the Virginia shelf: color as a function of grain size. *Bull. geol. Soc. Am.* **83**, 877–884.

SWIFT, D.J.P., STANLEY, D.J. & CURRY, J.R. (1971) Relict sediments on continental shelves: a reconsideration. *J. Geol.* **69**, 322–346.

SWINCHATT, J.P. (1965) Significance of constituent composition, texture and skeletal breakdown in some Recent carbonate sediments. *J. sedim. Petrol.* **35**, 71–90.

THOMSON, M.E. (1977) IGS studies of the geology of the Firth of Forth and its approaches. *Rep. Inst. Geol. Sci. No.* 77/17.

THOMSON, M.E. & EDEN, R.A. (1977) The Quaternary sequence in the west-central North Sea. *Rep. Inst. Geol. Sci. No.* 77/12.

TRUMBULL, J.V.A. (1972) Atlantic continental shelf and slope of the United States—sand size fraction of bottom sediments, New Jersey to Nova Scotia. *Prof. Pap. U.S. geol. Surv.* 529-*K*.

VAN VEEN, J. (1938) Water movement in the Straits of Dover. *J. Cons. int. Explor. Mer*, **13**, 7–38.

Spec. Publs int. Ass. Sediment. (1981) **5**, 323–334

Interglacial and Holocene sedimentation in the northern North Sea: an example of Eemian deposits in the Tartan Field

J. H. FRED JANSEN *and* ANTOINETTE M. HENSEY*

Netherlands Institute for Sea Research, Texel, the Netherlands

ABSTRACT

Since the temperate marine Holocene deposits in the northern North Sea are not thicker than 1 m, it may be expected that temperate interglacial deposits will also be thin. Consequently interglacial episodes will be hard to recognize. This is illustrated by an Eemian deposit in the Tartan Field, in which Foraminifera and pollen analyses show only limited climatic amelioration. The Eemian deposit is part of the Swatchway Beds. This implies that the Weichselian deposits in the northern North Sea, instead of providing the greater part of the Quaternary succession, cover a more limited interval.

INTRODUCTION

A major problem in understanding Quaternary sedimentation processes in the central and northern North Sea is the lack of a generally accepted stratigraphy which is founded on reliable absolute or relative dates. Extensive seismic and shallow-seismic reflection surveys in the last decade resulted in a well-established framework of seismic units in the northern North Sea basin (Van Weering, Jansen & Eisma, 1973; Van Weering, 1975; Jansen, 1976; Holmes, 1977; Thomson & Eden, 1977) but there is no consensus about the ages of these units.

Hughes *et al.* (1977) found only shallow Arctic fossil assemblages in samples from the central and north-western North Sea representing the greater part of the Quaternary deposits, and six radiocarbon datings, covering this part, revealed Middle and late Weichselian ages (Holmes, 1977). Consequently Holmes attributed a disproportionate part of the Quaternary sediments to the Weichselian glacial stage, and Eden, Holmes & Fannin (1977) proposed a model to explain this by repeated deep scouring

during the glacial stages. Jansen, who did more detailed work on the upper layers, concluded older ages of these layers by comparison with Holmes' interpretation. A few of our cores demonstrated temperate, fully marine conditions, pointing to an interstadial or interglacial origin, and cast doubt on the reliability of the radiocarbon dates mentioned above (Jansen, 1976; Jansen *et al.*, 1979). Gregory & Harland (1978) also reported the presence of interstadial or interglacial sediments. Samples from a location where the Quaternary is *c.* 350 m thick show two strong climatic ameliorations at *c.* 110 and 170 m depth, of which the latter was considered to be of Eemian age. Recent reviews of the Quaternary history of the North Sea are given by Jansen, Van Weering & Eisma (1979), Oele & Schüttenhelm (1979), Eisma, Jansen & Van Weering (1979) and Caston (1979).

Inasmuch as the known glacial deposits in the northern North Sea basin are indistinguishable from one another by their fossil contents, the chronological attribution will rest on the identification of interglacial deposits. The only uncontested interglacial deposits are of Holocene age: the Upper Witch Deposits in the north-western North Sea, the East Bank Deposit in the central North Sea north-west of

*Present address: Creagboy Cottage, Clare Galway, County Galway, Ireland.

0141–3600/81/1205–0323 $02.00
© 1981 International Association of Sedimentologists

Dogger Bank, the Forth Beds in the Firth of Forth approaches, and the superficial deposits in the Norwegian Channel and the Skagerrak–Kattegat area (Van Weering *et al.*, 1973; Jansen, 1976; Thomson & Eden, 1977; Jansen *et al.*, 1979; Van Weering, 1981). It can be expected that there will be many resemblances between Holocene and older interglacial sedimentary processes, and so a comparison of inferred climatic ameliorations with the Holocene sedimentary record will be useful for the recognition and understanding of the interglacials of the North Sea.

In the present paper data from four borings from the Tartan Field in the northern North Sea are reported, and the evidence of warmer climatic episodes in the northern North Sea is discussed in the light of the Holocene sedimentary history.

HOLOCENE SEDIMENTARY HISTORY

In the central and northern North Sea the last glaciation was followed by the development of several deposits (Jansen, 1976; Jansen *et al.*, 1979). The uppermost deposits, named Witch Deposits, are at maximum 15 m thick, and their formation started probably around 15,000 BP, when the North Sea reached its lowest relative sea-level of approximately −110 m. During the subsequent transgression sandy clays were deposited: the Lower Witch Deposits, which form the major part of the Witch Deposits. Their fossil content (Mollusca, Foraminifera, pollen) mirrors a transition form Arctic marine to temperate marine conditions of Holocene age. These Lower Witch Deposits pass gradually into the Upper Witch Deposits, which consist of a veneer of fine sands not more than 0·5 m thick, and were formed in a temperate marine environment comparable to the present one. Pollen analyses showed that the shift from clay to sand deposition took place between 8700 and 8400 BP, and that the sedimentation of the sands of the Upper Witch Deposits ended before 8400 BP. At that time the relative sea-level was about −45 to −40 m. Large quantities of sand were mobile, as is demonstrated north-west of Dogger Bank by the presence of a system of relict tidal sand ridges and an associated sand deposit up to 10 m thick—the East Bank Deposit. After 8400 BP no sedimentation of importance took place in the central and northern North Sea. A number of radiocarbon dates of carbonate and organic carbon from two cores of unmentioned locations in the Fladen Ground area

(Erlenkeuser, 1978, fig. 3) showed that the upper 15 cm of the sediment are subject to reworking, which causes admixture with recent organic material. This is also in agreement with dates by Johnson & Elkins (1979), although their geological inferences are thought to be incorrect (Jansen, 1980). Younger Holocene deposits are only present in the Norwegian Channel and in the Skagerrak–Kattegat area, where sedimentation still occurs (Van Weering, 1981).

After the last glaciation planktonic Foraminifera, mainly *Globigerina* sp., were first recorded in the northern North Sea during the formation of the Lower Witch Deposits, and reached considerable numbers after 9000 BP, in company with benthonic Foraminifera of southern origin (Jansen *et al.*, 1979). During this period the water depth in the Witch and Fladen Ground area increased from c. 30 to 90 m. In the present North Sea tests of *Globigerina* sp. are only found at water depths exceeding 60–70 m (Jarke, 1961). Apparently these depths are necessary to allow the entrance of planktonic Foraminifera into the North Sea, and possibly these planktonic forms need thermally stratified waters. Besides sufficient water depths, necessary for exchange with Atlantic water, the entrance of benthonic species of southern origin is also dependent on the northward migration of the North Atlantic polar front, which reached the Norwegian coast at 60° N after about 13,500 BP and prior to 12,600 BP (Ruddiman, Sancetta & McIntyre, 1977; Mangerud, 1977). The same is true for *Orbulina universa*, a preferentially subtropical planktonic species which at present is absent in Arctic and Subarctic waters (Bé, 1977). Although *O. universa* occurs in the actual northern North Sea (Jarke, 1961), it was not recorded in the samples from the Upper Witch Deposits, which were formed after 8700 BP but before 8400 BP (Jansen *et al.*, 1979). So *O. universa* found its way into the North Sea only after 8700 BP.

MATERIAL AND ANALYTICAL METHODS

Four foundation borings from the Tartan Field in block 15/16 of the U.K. sector, drilled at distances of about 50–100 m, provided an opportunity to examine their Foraminifera and pollen content (Fig. 1, Table 1). They were carried out in 1976 for Texaco North Sea U.K. Company of London by McClelland Engineers on board the M/V *Ferder*, using rotary drilling methods and a sand-gel slurry as a stabilizing

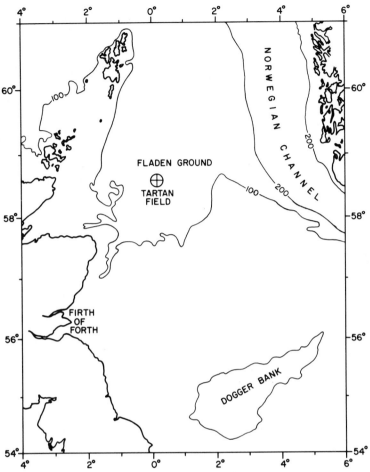

Fig. 1. Map showing the position of the Tartan Field.

Table 1. Locations, water depths, and lengths of the borings

	Latitude	Longitude	Water depth m	Length m
TP1	58° 22′ 12·3″ N	0° 04′ 22·2″ E	140·2	96·5
TP2	58° 22′ 11·1″ N	0° 04′ 24·4″ E	140·2	31·7
TP3	58° 22′ 10·7″ N	0° 04′ 28·1″ E	139·6	122·4
TP4	58° 22′ 13·3″ N	0° 04′ 28·7″ E	140·5	46·0

drilling fluid. The samples were recovered with a wire-liner sampler operated through the bore of the drill string after the drill bit had been advanced to the desired sampling depth. Samples were obtained either by latching the wire-line sampler into the bit and pushing a thin-wall tube into the sediment with the drill string, or by driving a thin-wall tube into the sediment using a sliding weight. A number of these samples were put at our disposal for examination. It appeared from the soil tests that the mechanical properties of the four borings were very similar at equivalent depths. Therefore we consider them together in this publication as one single boring. However, it must be realized that the stratigraphic relation between samples recovered from nearly the same depths in different borings is ultimately uncertain.

Fortunately one of the shallow seismic profiles of Jansen (1976) was recorded almost exactly across the location where the Tartan platform was to be built. This allowed us to correlate the reflectors separating the seismic units, with breaks in mechanical soil properties in the borings. The positions of the seismic units are indicated in Fig. 3. For the geological implications of these units see the discussion below.

The description of silty clay at the top of the borings fits the known character of the Lower Witch Deposits (Jansen et al., 1979). The water contents were determined as 40–50%, the cohesive shear strength is smaller than 1·0 N cm^{-2}. The sample at 0·5 m belongs to this deposit, and the 4·0 m sample of TP3 is probably situated at the boundary with the underlying Fladen Deposits. The Fladen Deposits contain more water (50–60%) and the cohesive shear strength is somewhat larger, up to 1·5 kg cm^{-2}. The reflector between the Fladen Deposits and the proglacial Hills Deposit lies at 8·5 m, and at this depth a break is present in the curve of the water content, which decreases downwards to c. 40%. The shear strength increases to 2·5 N cm^{-2}. The base of the proglacial Hills Deposits is marked by breaks in the profiles of the shear strength as well as the water content. The shear strength gradient becomes larger, and the water content is smaller than 40%, decreasing to values of approximately 20% at 24 m and deeper.

Below the proglacial Hills Deposits the sediments most probably consist of Swatchway Beds and Aberdeen Ground Beds (see Holmes, 1977), as was concluded from a deep tow boomer record by the Institute of Geological Sciences at Edinburgh, passing about 4 km east of the Tartan Field (N. G. T. Fannin, personal communication). Since at 46 m depth the consistency in the borings changes from normally consolidated to heavily over-consolidated (cohesive shear strengths of 10 and >30 N cm^{-2} respectively), this level is interpreted as representing the Prominent Seismic Reflector separating the Swatchway Beds and the over-consolidated Aberdeen Ground Beds (Holmes, 1977; Harland et al., 1978). Below 60 m depth no stratigraphical information is available because this is the maximum penetration of the boomer record.

For the determinations of the content of silt and clay subsamples were treated with hydrogen peroxide and hydrochloric acid to remove organic matter and carbonate. Then the material was dry-sieved with a 63 μm sieve.

The Foraminifera samples were treated following the method described by Knudsen (1978), using a sieve with a mesh diameter of 100 μm. Usually 100 specimens of benthonic forms were counted from each sample, and the rest of the slides were examined for rare species.

In this publication the results of the pollen investigations by J. de Jong are summarized and the geological interpretation is discussed. The laboratory methods and the palynological data and their implications will be reported in a separate paper by De Jong & Jansen (1981).

FORAMINIFERA

The examinations of the Foraminifera have concentrated on the benthonic forms, because of the usually low percentages of planktonic species. The total contribution of plankton is also recorded since it appeared to be an important indicator of the influence of Atlantic Ocean water in the North Sea (Jansen et al., 1979). Special attention is paid to the proportion of benthonic species of southern origin, like *Bulimina marginata*, *Cassidulina laevigata* var. *carinata*, *Fissurina lucida* and *Protelphidium anglicum* (Fig. 2). In the present North Sea these species are close to their northern limit of distribution (Murray, 1971), and it is thought that sediments in which they are found, even in low percentages, indicate the proximity of North Atlantic central water. (For a more extensive discussion see Murray, 1971, and Harland et al., 1978.) On the other hand species, which in this paper are called of northern origin, are in the North Sea close to their southern limit of distribution. Furthermore the term Arctic is applied to Foraminifera that have a distinct maximum abundance in Arctic waters, or are even absent in the present North Sea. Following Harland et al. (1978) we distinguish two major assemblages, Arctic and Mixed, indicating the absence or presence of species of southern origin respectively. For the environmental character of the Foraminifera and the interpretations of the assemblages we consulted the following review publications: Jarke (1961), Feyling-Hanssen et al. (1971), Murray (1971), Boltovskoy & Wright (1976), Bé (1977), and also the papers of Kihle (1971), Bartlett & Molinsky (1972), Løfaldli (1973), Aarseth et al. (1975), Knudsen (1976), Vilks

& Rashid (1976), Hughes *et al.* (1977), Knudsen (1977, 1978), Harland *et al.* (1978), and Gregory & Harland (1978). The faunal diversity in Fig. 2 is calculated as defined by Walton, indicating the number of ranked species in a sample that account for 95% of the total assemblage (Walton, 1964, in Feyling-Hanssen *et al.*, 1971).

Zone X

This zone is represented by a single sample at 121·9 m depth, and is characterized by dominant *Elphidium albiumbilicatum* (43%) and *E. excavatum* f. *clavata* (36%), while *E. ustulatum, Cassidulina crassa* and *Fissurina lucida* are common accessory species. *E. albiumbilicatum* is a normal North Sea form preferring shallow waters of lowered salinity. The association is interpreted as a Mixed shallow-water association with a dominance of species of northern origin.

Zone IX and zone VII

In these zones *E. excavatum* f. *clavata* is highly dominant (>85%), accessory species are *E. albiumbilicatum, C. crassa* and *Protelphidium anglicum*. The faunal diversity is very low (1–5) and the samples are very poor in specimens (<2 g^{-1} dry sediment). This indicates high Arctic circumstances in shallow hyposaline waters of about 10–20 m deep (Knudsen, 1976).

Zone VIII

This interval shows a relatively high contribution of southern species (*c.* 5 and 20%) and a higher faunal diversity (⩾8). Also remarkable is the *c.* 15% of plankton, mainly *Globigerina* sp., in the upper sample, which indicates a rather good connection to the Atlantic Ocean, and related to this a fairly high relative sea-level. The latter conclusion also agrees with the large amounts of *C. laevigata* var. *carinata* and *Islandiella teretis*, both outer shelf species. After the Weichselian glaciation comparable constituents of planktonic and southern forms did not occur until about 9000 BP in the northern North Sea (Jansen *et al.*, 1979). Consequently the Mixed assemblages of zone VIII are considered as fully marine assemblages from an interstadial or part of an interglacial period.

Zone VI

The foraminiferal assemblages of the complex zone VI are dominated by *Elphidium excavatum* f. *clavata*; the northern species *E. asklundi*, the Arctic species *E. subarcticum* and *E. bartletti*, the shallow-water species *E. albiumbilicatum*, and the shallow arctic *Protelphidium orbiculare* are all frequent. The percentages of southern species are <2%. This faunal composition indicates a marine Arctic shallow-water environment. Since, except at 48·8 m, *E. excavatum* f. *clavata* is less abundant and the samples are more rich in specimens compared to the zones IX and VII, the environment was probably less extreme.

At 47·4 m and at *c.* 42–38·3 m the zone is interrupted twice by associations with a higher contribution of *E. excavatum* f. *clavata* in relation to Arctic species, and with up to 8% *Globigerina* sp. This indicates amelioration of the palaeoceanographical conditions and higher relative sea-levels, but not as warm and as high as in zone VIII. The sample from boring TP4 at 42·8 m is possibly not in the right stratigraphic position between the others from TP1, and may belong somewhat higher in the column.

Zone V

In zone V *Elphidium excavatum* f. *clavata* is dominant (26–63%) and *C. crassa* enters for the first time in large amounts into the borings (17–30%). A number of Arctic species and species of northern origin like *P. orbiculare, Bucella frigida, E. subarcticum*, and *E. ustulatum* are common, but so also are some southern species, especially in the upper part, which are present in amounts of up to 18%. The presence of these southern Foraminifera is accompanied with the occurrence of up to 17% planktonic forms and faunal diversities of 6–15. The benthonic association is of the Mixed type and indicates nearly normal marine circumstances, comparable to the sedimentary circumstances below the top of the Lower Witch Deposits which was formed at the same locality during early Holocene times after 9000 BP (Jansen *et al.*, 1979). In the 27·4 m sample 12% of the 17% plankton consists of *Orbulina universa*, which suggests an oceanic circulation with the position of the North Atlantic polar front very similar to the present situation. This places the nearly normal marine conditions of zone V in an interstadial and probably partly in an interglacial period. The presence of *E. ustulatum* in all samples from this zone is worth noting. This species, of

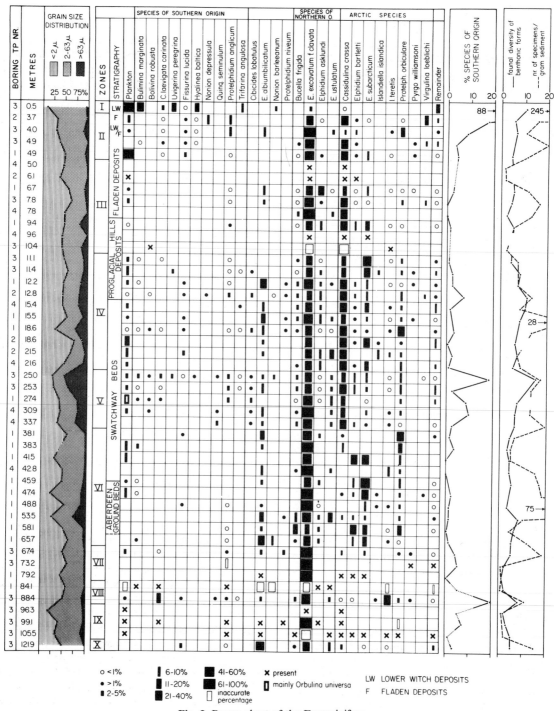

Fig. 2. Range chart of the Foraminifera.

northern origin, has never been recorded from Holocene faunas (Gregory & Bridge, 1979), and this suggests that the zone is at least early Weichselian, probably Eemian or older.

Zone IV

E. excavatum f. *clavata* and *C. crassa* dominate in this section; together they cover 45–64% of the benthonic Foraminifera. *E. subarcticum* is also abundant in varying percentages up to 48%, and other northern and Arctic forms are common, as well as the shallow-water species *E. albiumbilicatum* and *P. orbiculare*. The Arctic species exceed the species of northern origin in frequency, and the assemblage is considered as a high Arctic shallow-water assemblage. However, this picture seems to contradict the considerable amount of *Globigerina* sp., the common presence of southern species, and the comparatively high faunal diversity (5–12). These conflicting components are attributed to admixture by glacial processes of material from older deposits in the North Sea or from the surrounding mainland. A specimen of the Tertiary form *Pseudohastigerina micra* was found at 11·1 m.

Zone III

Zone III shows the same dominance of *E. excavatum* f. *clavata* and *C. crassa* (together 46–96%); of the accessory species *E. bartletti*, *E. subarcticum* and *B. frigida* are common. In the lower part the Arctic forms dominate the Subarctic, but in the upper part the Subarctic Foraminifera become the major element. The low faunal diversity (2–5) and the small numbers of southern species (not more than 2%) indicate high Arctic conditions, which were most severe during the deposition of the lower part. The sea was probably deeper than at the time of zone IV.

Zone II

Besides the abundant *E. excavatum* f. *clavata* and *C. crassa* (together 56–87%) this zone is characterized by the increasing influence of southern species (4–17%) and the smaller contribution of Arctic forms. In this interval the chronological succession is obscured, because the TP2 sample at 3·7 m probably belongs somewhat lower in the stratigraphy and the TP1 sample at 4·9 m somewhat higher. The assemblage is of the Mixed type, with a dominance of species of northern origin and to a less

extent of Arctic forms. The sample from boring TP3 at 4 m is situated at the boundary between the Fladen Deposits and the Lower Witch Deposits, the level which is assumed from seismic evidence to represent the lowest relative sea level of the last Weichselian substage (Jansen *et al.*, 1979). In the sample *E. excavatum* f. *clavata* is very high compared to *C. crassa* (63 and 2% respectively), indicating shallow circumstances (Knudsen, 1976). The next most important species, accounting for 28% of the specimens, are *P. orbiculare*, *P. anglicum*, and *E. albiumbilicatum*. Both are abundant in shallow waters, the latter two especially in areas of lowered salinity. The agreement between this faunal evidence and the seismic interpretation is conspicuous, although it may be a coincidence.

Zone I

In the foraminiferal fauna of zone I the southern species make up 86% of the association. The 63% planktonic species, mainly *Globigerina* sp., points to a good connection with the Atlantic Ocean waters at the time of sedimentation. The benthonic assemblage, which is characteristic of the present northern North Sea and of the top of the Lower Witch Deposits, is regarded as Holocene.

POLLEN

In broad outline the palynological analysis of the borings leads to the differentiation of two principal groups, based on the contributions of herbaceous and *Ericaceae* pollen and of the thermophilic tree pollen. Group I, with a low percentage of herbs and *Ericaceae* and a high percentage of thermophilic trees, is generally regarded as belonging to an interglacial period. Group II, containing a relatively high amount of herbs and *Ericaceae* and few thermophilic trees, is attributed to a glacial period, although parts of interstadials cannot be excluded. Furthermore group III is distinguished as a subgroup of group II with smaller contributions of herbs and *Ericaceae*. This group is interpreted as representing part of an interstadial, or possibly the beginning, or less probably the end, of an interglacial. Environmental conditions are also indicated by the contribution of cysts of *Hystrichosphaeridae* (dinoflagellates), reflecting the degree of marine influence, of colonies of *Pediastrum* sp. (freshwater green algae), which point to continental influence, and reworked

Tertiary and older pollen, which indicate the action of glacial processes. The samples are divided into nine zones (A–K) (Fig. 3).

Zone K and zone G

The pollen of these zones is assigned to group III, although the content of thermophilic tree pollen is relatively small. The assemblages resemble the known assemblages of early Weichselian times, but an older interstadial or interglacial age is also within the range of possibility. The spectrum of the thermophilic trees in the lowest sample at 121·9 m reveals somewhat higher temperatures than the rest of the samples.

Zone H

This sample, in which a high contribution of reworked Tertiary palynomorphs is observed, belongs to the glacial group II.

Zone F

Zone F is attributed to group III, the thermophilic tree pollen contribution is fairly large and the influence of reworked forms is very low. Therefore, compared to zones K and G, the interval is thought to represent a warmer climatic period.

Zone E

Zone E contains pollen spectra of glacial origin as defined by group II. They show a large contribution of reworked forms and indicate a glacial climate, which was more severe than the cold climatic intervals of the early Weichselian around 60,000 BP (see Zagwijn, 1975, also Jelgersma, Oele & Wiggers, 1979).

Zone D

The assemblages of this zone, ascribed to group III, do not really have an interglacial character, but are also different from the known early Weichselian spectra. Although they show low percentages of herbs and Ericaceae and little influence of reworked pollen, the contribution of thermophilic tree pollen is rather small. Nevertheless the associations record a warmer climate than prevailed during the Brørup and Amersfoort interstadials (about 65,000 and 67,000 BP respectively, Zagwijn, 1975). Consequently

it is concluded that zone D is, at least partly, of Eemian age or older.

Zone C and zone B

Both zone C and B have pollen assemblages belonging to group II, but they are distinguished by the amounts of Tertiary and other reworked pollen. Zone B, with many reworked specimens, represents a glacial marine environment, especially in the lower part, 12·2–18·6 m. On the other hand zone C, containing fewer reworked forms, reflects less extreme glacial circumstances.

Zone A

Above the sample at 4·0 m with a transitional pollen assemblage, the top sample belongs to group I. The spectrum is clearly recognizable as of early Holocene origin.

DISCUSSION

Inasmuch as the examined material is not from continuously cored sections, and consequently there is no sedimentological information available of the complete succession, the data should be checked for downhole contamination by drilling. However, several climatic ameliorations in the foraminiferal as well as in the pollen record were observed in different borings, where they occur at the same depths (see the zones V, VIII, D and F). For that reason contamination cannot be invoked to explain the occurrence of these intervals, and the quality of the samples can be relied on.

The combined Foraminifera and pollen results demonstrate an early Holocene age for the top sample (Fig. 3).

The foraminiferal zones II to IV and the pollen zones B, and C partly, are interpreted as belonging to the Middle Weichselian glacial period, lying roughly between 55,000 and 15,000 BP (Zagwijn, 1975). The two records are in good agreement though in detail there are differences. This is not surprising because the responses of even closely situated marine and continental environments to climatic changes did not necessarily occur simultaneously. In the interval under discussion the pollen assemblages appear to have changed prior to the foraminiferal faunas. The maximum of glacial influence is observed between about 20 and 12 m, in zone IV and in the lower part of zone B, and mirrors the maximum of the Weichselian glaciation (*c.* 18,000 BP).

Below, pollen zone D corresponds to the foraminiferal zone V and the upper amelioration of the top of zone VI. The Foraminifera data reveal lower sea-water temperatures than existed during early Holocene times around 8500 BP, and a position for the North Atlantic polar front very similar to the present. A minimum early Weichselian age was suggested. However, the continental climate, as shown by the pollen, was warmer than at that time. This argument, together with the very glacial zone E (see below), places a large part of the zones D, V, and of the top of zone VI in the Eemian interglacial. The top parts of zone V and zone C possibly represent the early Weichselian period.

The 'Prominent Seismic Reflector' marks an unconformity with evidence of channelling (Holmes, 1977; Harland *et al.*, 1978), and it is to be expected that a hiatus occurs at this level. It is not known whether the samples at 45·9 m are situated above or below the reflector, so no attempt is made to explain their contents.

Pollen zone E and the foraminiferal sample at 48·8 m are of a glacial nature. Since, especially on palynological grounds, the temperatures were too low for the early Weichselian cold intervals, and combined with the warm zone D above, there is no choice but a Saalian or older age for zone E. The amelioration at 47·4 m in the foraminiferal record was not recognized in the pollen because it was not examined.

Below zone E the interpretation becomes problematic owing to the poor correspondence between the results of the foraminiferal and the pollen analyses. This discrepancy could be caused by several possible effects. For instance, as the transitions in the different records are not usually simultaneous, the wider spacing between the samples in the lower part of the borings will more easily obscure the interrelations. Another possible cause is the lack of a seismic stratigraphy below 60 m depth. Because of this it is impossible to detect possible gaps in the sedimentary column, and if they occur at crucial places they may produce unwelcome misinterpretations. Therefore we must, for the zones below 55 m, confine ourself to the separate conclusions of the foraminiferal and pollen records.

The results from the upper 55 m of the borings corroborate the chronological interpretation of the seismic data by Jansen (1976) and Jansen *et al.* (1979). The Swatchway Beds appear to be of Eemian to

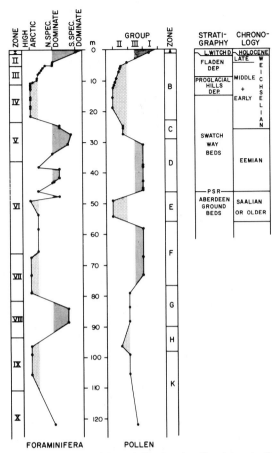

Fig. 3. Summary of foraminiferal and pollen data, stratigraphy and chronological interpretation. PSR—Prominent Seismic Reflector.

Middle Weichselian age. The glaciomarine Proglacial Hills Deposits, assigned to the Weichselian maximum around 18,000 BP, include the upper part of the interval of maximum glacial influence between *c.* 12 and 20 m. The amelioration of the marine environment during the development of the Fladen Deposits is also expressed in the corresponding assemblages of the Tartan Field borings. Finally the succession merges into the uppermost sample of early Holocene age at the top of the Lower Witch Deposits.

The Tartan Field borings are correlated with a 182 m long boring, SLN 75/33, taken 65 km further south-eastwards, using the seismic interpretation by Holmes (1977, fig. 7). From Foraminifera and dinoflagellate cysts in this boring four episodes of climatic ameliorations were defined (Gregory & Harland, 1978; Harland *et al.*, 1978). Two are situated above and two below the Prominent Seismic Reflector (Fig. 3). The relatively warm periods of foraminiferal zones I plus II, and V plus the top of VI are connected with the upper (above *c.* 10 m) and the greater part of the lower episode (*c.* 25–55 m) respectively, and the arctic zones III and IV with the interval between *c.* 10 and 25 m. The latter interval also comprises part of the second episode (*c.* 15–19 m), which still belongs to the combined Fladen and Hills Deposits (or to the Witch Ground Beds in Holmes, 1977), and is accordingly not equal to zone V. It is suggested here that the occurrence of southern species in this interval is due to glacial reworking as in zone IV, and that no associated climatic amelioration has taken place. The two episodes below the reflector are not recognized in the Tartan Field borings. High numbers of *Elphidium* cf. *williamsoni*, as reported by Gregory & Harland (1978), are not found, nor other abundant southern species. Possibly the sample of zone X, with 43% *E. albiumbilicatum*, denotes a correlation with one of the episodes, since both species indicate southern influences and prefer shallow hyposaline environments. Based upon the above age interpretations the supposed Eemian episode at *c.* 170 m (Harland *et al.*, 1978) is of Holsteinian or older age.

Because of the evidence of Eemian deposits in the Swatchway Beds, which were thought to be younger than 25,000 years, the chronology of the seismic units in the central and north-western North Sea by Holmes (1977) needs to be revised. The question arises why no interglacial faunas with high numbers of planktonic and southern Foraminifera, like those known from Holocene times, have been found.

If the older interglacials are comparable to the Holocene, records of full interglacial environments will be very scarce in the central and northern North Sea. This is illustrated by the Eemian interval of the Tartan Field borings, in which the assemblages of both pollen and Foraminifera indicate an improvement in conditions, but still different from the fully Holocene amelioration. As is demonstrated by the post-glacial sequence, the contribution of Foraminifera of southern origin is highly indicative of the eustatic and climatic events in the area. However, if their presence is not in agreement with the image of the entire assemblage, including planktonic forms, they are more likely to have been produced under the influence of glacial rather than interglacial processes.

ACKNOWLEDGMENTS

We wish to thank Texaco North Sea U.K. Company, London, for access to the material from the Tartan Field borings, and Dr N. G. T. Fannin (Institute of Geological Sciences, Edinburgh) for providing us with the samples. We would also like to acknowledge Dr J. W. Chr. Doppert (Geological Survey of the Netherlands, Haarlem) for his helpful discussions. Our gratitude is extended to Mrs R. Gieles for the grain-size analyses, Mrs J. Hart and Mrs L. Everhardus for typing the manuscript, and Messrs R. P. D. Aggenbach, H. Hobbelink and B. Verschuur for drawing the figures.

REFERENCES

AARSETH, I., BJERKLI, K., BJÖRKLUND, K.R., BÖE, D., HOLM, J.P., LORENZEN-STYR, T.J., MYHRE, L.A., UGLAND, E.S. & THIEDE, J. (1975) Late Quaternary sediments from Korsfjorden, western Norway. *Sarsia*, **58**, 43–66.

BARTLETT, G.A. & MOLINSKI, L. (1972) Foraminifera and the Holocene history of the Gulf of St. Lawrence. *Can. J. Earth Sci.* **9**, 1204–1215.

BÉ, A.W.H. (1977) An ecological, zoogeographic and taxonomic review of recent planktonic foraminifera. In: *Oceanic Micropaleontology*, 1 (Ed. by A.T.S. Ramsay), pp. 1–100. Academic Press, London.

BOLTOVSKOY, E. & WRIGHT, R. (1976). *Recent Foraminifera*, pp. 1–515. Junk, The Hague.

CASTON, V.N.D. (1979) The Quaternary sediments of the North Sea. In: *The North-west European Shelf Seas: the sea bed and the sea in motion, I. Geology and sedimentology* (Ed. by F.T. Banner, M.B. Collins and K.S. Massie), pp. 195–270. Elsevier, Amsterdam.

DE JONG, J. & JANSEN, J.H.F. (1981) A palynological study of Late Quaternary deposits in the northern North Sea. In preparation.

EDEN, R.A., HOLMES, R. & FANNIN, N.G.T. (1977) Quaternary deposits of the central North Sea, 6. Depositional environment of offshore Quaternary deposits of the continental shelf around Scotland. *Rep. Inst. geol. Sci.* 77/15, pp. 1–18.

EISMA, D., JANSEN, J.H.F. & VAN WEERING, TJ.C.E. (1979) Sea-floor morphology and recent sediment movement in the North Sea. In: *The Quaternary History of the North Sea* (Ed. by E. Oele, R.T.E. Schüttenhelm and A.J. Wiggers). *Acta Univ. Ups. Symp. Univ. Ups. Annum Quingentesimum Celebrantis*, **2**, 217–231.

ERLENKEUSER, H. (1978) The use of radiocarbon in estuarine research. In: *Biogeochemistry of Estuarine Sediments*, pp. 140–153. Proc. UNESCO/SCOR workshop, Melreux, Belgium, 29 November–3 December 1976. UNESCO, Paris.

FEYLING-HANSSEN, R.W., JØRGENSEN, J.A., KNUDSEN, K.L. & ANDERSON, A.-L.L. (1971) Late Quaternary foraminifera from Vendsyssel, Denmark and Sandness, Norway. *Bull. geol. Soc. Denmark*, **21**, 67–317.

GREGORY, D. & BRIDGE, V.A. (1979) On the Quaternary foraminiferal species *Elphidium*? *ustulatum* Todd 1957: its stratigraphic and paleoecological implications. *J. Foraminiferal Res.* **9**, 70–75.

GREGORY, D. & HARLAND, R. (1978) The late Quaternary climatostratigraphy of IGS borehole SLN 75/33 and its application to the palaeoceanography of the north-central North Sea. *Scott. J. Geol.* **14**, 147–155.

HARLAND, R., GREGORY, D.M., HUGHES, M.J. & WILKINSON, I.P. (1978) A late Quaternary bio- and climatostratigraphy for marine sediments in the north-central part of the North Sea. *Boreas*, **7**, 91–96.

HOLMES, R. (1977) Quaternary deposits of the central North Sea, 5. The Quaternary geology of the UK sector of the North Sea between 56° and 58° N. *Rep. Inst. geol. Sci.* 77/14, pp. 1–50.

HUGHES, M.J., GREGORY, D.M., HARLAND, R. & WILKINSON, I.P. (1977) Late Quaternary foraminifera and dinoflagellate cysts from boreholes in the UK sector of the North Sea between 56° and 58° N. *Rep. Inst. geol. Sci.* 77/14, pp. 36–46.

JANSEN, J.H.F. (1976) Late Pleistocene and Holocene history of the northern North Sea, based on acoustic reflection records. *Neth. J. Sea Res.* **10**, 1–43.

JANSEN, J.H.F. (1980) Holocene deposits in the northern North Sea: evidence for dynamic control of their mineral and chemical composition?—a comment. *Geol. Mijnb.* **59**, 179–180.

JANSEN, J.H.F., DOPPERT, J.W.C., HOOGENDOORN-TOERING, K., DE JONG, J. & SPAINK, G. (1979) Late Pleistocene and Holocene deposits in the Witch and Fladen Ground area, northern North Sea. *Neth. J. Sea Res.* **13**, 1–39.

JANSEN, J.H.F., VAN WEERING, Tj.C.E. & EISMA, D. (1979) Late Quaternary sedimentation in the North Sea. In: *The Quaternary History of the North Sea* (Ed. by E. Oele, R.T.E. Schüttenhelm and A.J. Wiggers). *Acta Univ. Ups. Symp. Univ. Ups. Annum Quingentesimum Celebrantis*, **2**, 175–187.

JARKE, J. (1961) Die Beziehungen zwischen hydrographischen Verhältnissen, Faziesentwicklung und Foraminiferenverbreitung in der heutigen Nordsee als Vorbild für die Verhältnisse während der Miocän Zeit. *Meyniana*, **10**, 21–36.

JELGERSMA, S., OELE, E. & WIGGERS, A.J. (1979) Depositional history and coastal development in the Netherlands and the adjacent North Sea since the Eemian. In: *The Quaternary History of the North Sea* (Ed. by E. Oele, R.T.E. Schüttenhelm and A.J. Wiggers). *Acta Univ. Ups. Symp. Univ. Ups. Annum Quingentesimum Celebrantis*, **2**, 115–142.

JOHNSON, T.C. & ELKINS, S.R. (1979) Holocene deposits of the northern North Sea: evidence for dynamic control of their mineral and chemical composition. *Geol. Mijnb.* **58**, 353–366.

KIHLE, R. (1971) Foraminifera in five sediment cores in a profile across the Norwegian Channel south of Mandal. *Norsk geol. Tidsskr.* **51**, 261–286.

KNUDSEN, K.L. (1976) Foraminifer faunas in Weichselian stadial and interstadial deposits of the Skearumhede boring, Jutland, Denmark. *Spec. Publ. Mar. Sedim.* **1**, 431–449.

KNUDSEN, K.L. (1977) Foraminiferal faunas of the Quaternary Hostrup Clay from northern Jutland, Denmark. *Boreas*, **6**, 229–245.

KNUDSEN, K.L. (1978) Middle and Late Weichselian marine deposits at Nørre Lyngby, northern Jutland, Denmark, and their foraminiferal faunas. *Danm. geol. Unders.* **112**, 1–44.

LØFALDLI, M. (1973) Foraminiferal biostratigraphy of Late Quaternary deposits from the Frigg Field and Booster Station. *NTNF's Kontinentalsokkelkontor Publ.* **18**, 1–82.

MANGERUD, J. (1977) Late Weichselian marine sediments containing shells, foraminifera, and pollen, at Ågotnes, western Norway. *Norsk geol. Tidsskr.* **57**, 23–54.

MURRAY, J.W. (1971) *An Atlas of British Recent Foraminiferids*, pp. 1–244. Heinemann, London.

OELE, E. & SCHÜTTENHELM, R.T.E. (1979) Development of the North Sea after the Saalian glaciation. In: *The Quaternary History of the North Sea* (Ed. by E. Oele, R.T.E. Schüttenhelm and A.J. Wiggers). *Acta Univ. Ups. Symp. Univ. Ups. Annum Quingentesimum Celebrantis*, **2**, 191–216.

RUDDIMAN, W.F., SANCETTA, D.C. & MCINTYRE, A. (1977) Glacial/interglacial response rate of subpolar North Atlantic waters to climatic change: the record in oceanic sediments. *Phil. Trans. R. Soc. B* **280**, 119–142.

THOMSON, M.E. & EDEN, R.A. (1977) Quaternary deposits of the central North Sea, 3. The Quaternary sequence in the west-central North Sea. *Rep. Inst. geol. Sci.* 77/12, pp. 1–18.

VAN WEERING, Tj.C.E. (1975) Late Quaternary history of the Skagerrak; an interpretation of acoustical profiles. *Geol. Mijnb.* **54**, 130–145.

VAN WEERING, Tj.C.E. (1981) Recent sediments and sediment transport in the northern North Sea: surface sediments of the Skagerrak. In: *Holocene Marine Sedimentation in the North sea Basin* (Ed. by S.-D. Nio et al.). *Spec. Publs int. Ass. Sediment.* **5**, 335–359. Blackwell Scientific Publications, Oxford. 524 pp.

VAN WEERING, T., JANSEN, J.H.F. & EISMA, D. (1973) Acoustic reflection profiles of the Norwegian Channel between Oslo and Bergen. *Neth. J. Sea Res.* **6**, 241–263.

VILKS, G. & RASHID, M.A. (1976) Post-glacial paleo-oceanography of Emerald Basin, Scotian shelf. *Can. J. Earth Sci.* **13**, 1256–1267.

WALTON, W.R. (1964) Recent foraminiferal ecology and paleoecology. In: *Approaches to Paleoecology* (Ed. by J. Imbrie and N.D. Newell), pp. 151–237. Wiley, New York.

ZAGWIJN, W.H. (1975) Indeling van het Kwartair op grond van veranderingen in vegetatie en klimaat. In: *Toelichting bij Geologische Overzichtskaarten van Nederland* (Ed. by W.H. Zagwijn and C.J. van Staalduinen), pp. 109–114. Rijks Geologische Dienst, Haarlem.

APPENDIX

Bolivina robusta Brady, 1884
Bucella frigida (Cushman, 1922)
Bulimina marginata d'Orbigny, 1826
Cassidulina crassa d'Orbigny, 1839
Cassidulina laevigata d'Orbigny var. *carinata* Cushman, 1922
Cibicides lobatulus (Walker & Jacob, 1798)
Elphidium albiumbilicatum (Weiss, 1954)
Elphidium asklundi Brotzen, 1943
Elphidium bartletti Cushman, 1933
Elphidium excavatum (Terquem) f. *clavata* Cushman, 1930
Elphidium subarcticum Cushman, 1944
Elphidium ustulatum Todd, 1957
Fissurina lucida (Williamson, 1848)
Hyalinea baltica (Schroeter, 1783)

Islandiella islandica (Nørvang, 1945)
Islandiella teretis (Tappan, 1951)
Nonion barleeanum (Williamson, 1858)
Nonion depressula (Walker & Jacob, 1798)
Orbulina universa d'Orbigny, 1839
Protelphidium anglicum Murray, 1965
Protelphidium niveum (Lafrenz, 1963)
Protelphidium orbiculare (Brady, 1881)
Pseudohastigerina micra (Cole, 1927)
Pyrgo williamsoni (Silvestri, 1923)
Quinqueloculina seminulum (Linné, 1758)
Trifarina angulosa (Williamson, 1858)
Uvigerina peregrina Cushman, 1923
Virgulina loeblichi Feyling-Hanssen, 1954

Spec. Publs int. Ass. Sediment. (1981) **5**, 335–359

Recent sediments and sediment transport in the northern North Sea: surface sediments of the Skagerrak

TJEERD C. E. van WEERING

Netherlands Institute for Sea Research, P.O. Box 59, Texel, the Netherlands

ABSTRACT

Bottom sediments from the Skagerrak were collected on localities chosen on the basis of their acoustical characteristics. The grain-size distribution, lithology and clay mineralogy have been studied, as well as carbonate content and content of organic carbon. These data have been related to the present-day current pattern. The Skagerrak obviously acts as a sink for recent sediments that are transported by traction currents along the bottom and in suspension, or by a combination of these. The main depot centres for very fine sand are situated north-west of Skagen and in the entrance of the Kattegat, whereas silt and clay-size materials are found in the deeper parts and off the Norwegian and Swedish coasts. Locally till-like sediments occur.

INTRODUCTION

The Skagerrak forms the deepest part (about 700 m) of the North Sea, and is the deepest of a series of basins which together form the Norwegian Channel, a 900 km long, rather narrow (80–90 km), elongated basin along the southern and south-western coast of Norway (Fig. 1). The bathymetry of the Skagerrak and Norwegian Channel was first studied by O. Holtedahl (1940) as part of a study of Norwegian coastal waters. He also studied details of the seafloor morphology with an echo sounder (O. Holtedahl, 1964), indicating that in the Skagerrak the floor of the central deep basin is flat with a slope towards Norway that is relatively irregular and steep, while the southern slope of the Skagerrak has a convex form with a very gradual inclination and some marked incisions.

Only recently continuous seismic profiling in the Skagerrak indicated the existence of presumably Quaternary sediments of strongly varying thickness, that wedge out towards the Norwegian subcontinent and which overlie uncomformably the eroded Mesozoic strata that form the bottom of the basin

(Sellevol & Aalstad, 1971). There is also a strong difference in thickness of the Quaternary sediments in the western part of the Skagerrak as compared with the central and eastern parts. This is probably the result of the occurrence of hiatuses in the Quaternary record, which is a rather general phenomenon in areas which have been influenced by glaciers during the Pleistocene. During the last glaciation the Scandinavian glaciers that reached as far south as Denmark most probably covered the Skagerrak completely during the maximum extension of the ice sheet.

Post-glacial sedimentation in the Skagerrak is, at least along the southern flank of the Skagerrak, strongly influenced by synsedimentary tectonic movements in the underground, which most likely occur along previously existing fault lines in the pre-Quaternary underground (van Weering, 1981).

During an acoustical reflection profiling survey carried out during 1973–1976 it was found (van Weering, Eisma & Jansen, 1973; van Weering, 1975, 1981) that a number of different sedimentary units of presumably late-Pleistocene and Holocene age can be distinguished over large areas in the Skagerrak

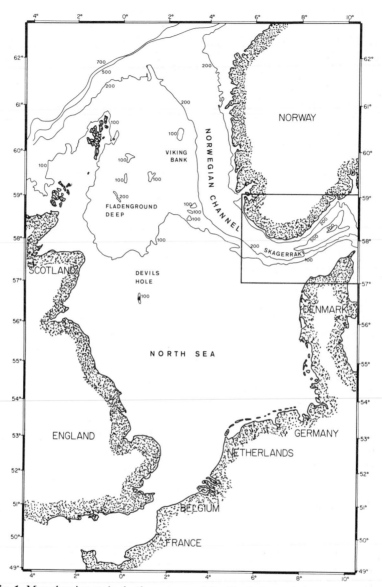

Fig. 1. Map showing main depth contours of the North Sea and the area of survey.

and Norwegian Channel on the basis of their acoustic characteristics. From the distribution and character of these units a hypothesis of the sedimentary history was tentatively made. In this hypothesis a rapid deposition of sediments of variable grain size (due to the rapid retreat of the Scandinavian glaciers followed by partial reworking of previously deposited glacial sediments under the influence of rising sea-level) was postulated. The rise of sea-level would have resulted in a decrease of

wave disturbance and turbulence and probably also in a diminishing of current velocities close to the bottom. When the present-day circulation pattern was established about 8000 BP (Eisma, Jansen & van Weering, 1979), fine material was transported from the central and southern North Sea and subsequently deposited in quiet water areas such as the Skagerrak and the Norwegian Channel. This resulted in a gradual decrease in grain size of the bottom sediments since that time. At present, transport of fine sediments from southern sources such as the Channel, the rivers Rhine, Meuse, Thames, Weser and Elbe, coastal erosion and sea-floor erosion still goes on under the influence of the northward directed resultant current (forming part of the general anticlockwise circulation in the North Sea) along the Danish coast. However, apart from the data published by Pratje (1951) explaining the coarse, poorly sorted sediments along the southern flank of the Skagerrak and the western flank of the Norwegian Channel as terminal moraines, the work of Lange (1956) and the more recent work of Holtedahl & Bjerkli (1975), little is known from the sediment distribution and character of bottom sediments from the Skagerrak. An indication of expected bottom sediments in an area where sand-waves occur along the southern slope of the Skager-rak was given by Stride & Chesterman (1973), though no samples were taken.

In order to test the postulated sedimentary history, sampling of the bottom sediments of the Skagerrak was carried out on a regional scale. The sampling localities were selected on the basis of the acoustical profiles obtained previously. A number of grab samples and shallow sediment cores were taken in 1975 in the eastern part of the Skagerrak and in 1976 and 1977 in the central and western parts of the Skagerrak. Additional sampling was carrried out in 1978 (Table 1).

Apart from the cores collected in 1975, which were taken with a 2 m long gravity corer, all cores were obtained with a 6 m long piston corer, while in 1978 an 8 m long core was obtained in the entrance to the Kattegat.

This study deals with the interpretation of grain-size data, content of organic carbon, and carbonate content of about 115 grab samples as well as with the results of textural and mineralogical determinations of a selected number of grab samples. The results of the analyses of the cores will be published separately.

ACOUSTICAL INVESTIGATIONS

The investigations in the Norwegian Channel and the Skagerrak as mentioned above indicates the presence of thick accumulations of presumably late and post-glacial sediments which can be divided into an upper, rather transparent deposit almost without internal reflections (unit 1, presumably of Holocene age) and a lower deposit (unit 2, presumably late glacial to early Holocene in age) which is characterized by the presence of several strong re-flectors of varying intensity and which is generally well stratified. In the lower part of this unit acousti-cally transparent layers occur separated by strong reflectors, while scattered hyperbolic reflectors are also found. Unit 2 sediments were interpreted as being of mixed glacial-marine and proglacial origin.

The thickness of these deposits is rather variable; a maximum of 127 m for units 1 + 2 is found along the southern slope of the Skagerrak, where a thick sediment wedge is present, while in the deepest part of the Skagerrak the thickness is about 20 m. Along the Norwegian coast and locally along the Swedish west coast there are strong differences in thickness (Holtedahl & Bjerkli, 1975); in the entrance to the Kattegat the thickness is of the order of 160 m (Flodén, 1973; van Weering, 1981).

These deposits rest partly (in the western Skagerrak and locally along the Norwegian and Swedish coast) on top of a deposit which contains internal hyperbolic reflectors (probably representing stones and boulders) and no internal stratification. This deposit has a clearly recognizable, irregular, un-dulating top reflector and was interpreted as a glacial drift deposit (unit 3). Where this unit is absent, the underground consists of hard rock of presumably Cretaceous and Lower Tertiary age along the southern slope and of even older sediments along the Norwegian side of the Skagerrak. Towards the east the sediments rest directly on granitic rocks of Pre-cambrian age. Only locally, in the deepest part of the Skagerrak, another deposit can be recognized below unit 1 and unit 2 sediments. This deposit is most probably of pre-Weichselian age and fills a depression in the Mesozoic hardrock that forms the pre-Quaternary underground here.

The surface sediments of the Skagerrak generally belong to unit 1, while scattered outcrops of unit 2 sediments occur as well, mainly in front of the morainic ridges that form the sea-bottom close to the Norwegian coast and the Swedish west coast. In the south-western part of the Skagerrak unit 2 sediments are also present at the sea-bottom.

In general, unit 1 sediments wedge out towards the north and decrease gradually in grain size, while they are very thick in the (north-)eastern part of the Skagerrak. Shadow zones, i.e. zones where the penetration of sound in sediment is prevented and an acoustic mask is formed, are found close ($\sim$ 1.5–5 m) below the surface in the south-eastern part in

Table 1. Locations and water depths of the sample stations in the Skagerrak (V = grab sample, G = gravity core, P = piston core)

Station	Latitude (°N)	Longitude (°E)	Water depth (m)	Station	Latitude (°N)	Longitude (°E)	Water depth (m)
V75-14	57° 46'	10° 48'	35	69	57° 23' 30"	8° 39'	30
15	57° 48'	11° 04'	43	70	57° 22'	8° 53'	25
16	57° 49'	11° 14'	64	71	57° 11'	8° 0'	45
17	57° 50'	11° 22' 30"	61	72	57° 18'	8° 01'	42
18	58° 03'	11° 13' 30"	66	73	57° 29'	7° 51'	134
19	58° 01' 30"	11° 02'	123	G75-23	58° 11' 30"	9° 08'	540
20	58° 03'	10° 50'	202	24	58° 10'	9° 08'	620
21	58° 01'	10° 34'	160	30	57° 50' 30"	8° 45'	425
22	58° 02' 30"	10° 16'	93	31	57° 43' 30"	8° 29'	295
23	57° 54' 40"	10° 15'	82	32	57° 43'	8° 32'	272
24	57° 52' 30"	9° 58'	58	34	57° 35' 40"	7° 49'	255
25	57° 59'	9° 56'	88	35	57° 37' 20"	7° 47'	340
26	58° 02' 30"	9° 55'	149	36	57° 53' 40"	7° 32' 30"	400
27	58° 07'	9° 54' 30"	285	V76-28	57° 40'	6° 56'	308
28	58° 09'	9° 53' 30"	382	29	57° 36' 40"	6° 48'	270
29	58° 13'	9° 52'	500	38	57° 48' 40"	6° 09' 20"	273
30	58° 03'	9° 22'	530	39	57° 45'	6° 04' 20"	240
31	57° 59' 30"	9° 28'	390	40	57° 41' 20"	6° 21' 50"	154
32	57° 56'	9° 27'	155	41	57° 37' 20"	5° 58' 0"	144
33	57° 51'	9° 33'	62	42	57° 35' 40"	6° 45' 30"	213
34	58° 19'	9° 52'	570	43	57° 32' 40"	6° 46'	176
35	58° 27'	9° 52'	450	44	57° 27' 10"	6° 46' 10"	121
36	58° 29'	10° 15'	420	45	57° 18' 30"	6° 39'	77
37	58° 30' 30"	10° 36'	164	46	57° 19' 30"	6° 37'	75
38	58° 32' 30"	10° 55'	70	47	57° 36'	6° 32' 30"	155
39	58° 38' 30"	10° 36'	129	48	57° 32'	8° 46'	51
40	58° 44'	10° 44'	99	49	57° 32'	8° 16'	107
41	58° 45' 30"	10° 24'	148	V77- 7	58° 0' 40"	6° 34'	380
42	58° 47' 30"	10° 05'	233	8	57° 56' 20"	6° 49' 40"	402
43	58° 49'	9° 41' 30"	185	9	57° 43' 0"	5° 45' 10"	136
44	58° 48 30"	9° 42' 30"	165	10	57° 38' 20"	5° 41' 40"	77
45	58° 47' 30"	9° 45'	350	11	57° 31' 50"	5° 59'	87
46	58° 37'	9° 36'	470	12	57° 30' 50"	6° 21'	100
47	58° 21' 30"	9° 30'	615	13	57° 39'	6° 28'	177
48	58° 06'	8° 52'	515	14	57° 51'	7° 07'	427
49	58° 01'	8° 59'	560	15	57° 52' 30"	7° 14' 30"	423
50	57° 53'	9° 06'	280	16	57° 41' 30"	7° 21'	333
51	57° 52' 30"	9° 08'	240	17	57° 37'	7° 31' 30"	324
52	57° 51'	9° 10'	152	18	57° 42' 30"	8° 15' 40"	300
53	57° 45'	9° 13'	66	19	58° 18'	10° 16'	433
54	57° 39'	9° 20'	32	20	58° 12'	10° 13'	278
55	57° 28'	8° 56'	26	22	58° 12' 30"	10° 40' 30"	242
56	57° 34'	8° 53	47	23	58° 12' 30"	11° 03'	122
57	57° 28	8° 49'	90	24	57° 54'	11° 19'	63
58	57° 44' 30"	8° 48'	172	25	57° 55'	11° 02'	97
59	57° 48'	8° 45'	350	26	57° 46'	8° 31'	305
60	57° 58'	8° 37'	515	27	57° 20'	7° 21'	60
61	58° 0'	8° 37'	550	28	57° 11' 20"	7° 23' 30"	55
62	57° 56'	8° 26'	490	P78- 1	57° 52' 20"	11° 06' 30"	60
63	57° 47' 30"	8° 28'	440	4	58° 29'	10° 05'	537
65	57° 40' 30"	8° 29'	202	P77- 6	= V77-7		380
66	57° 36'	8° 32'	113	9	= V77-13		177
67	57° 28 30"	8° 37'	55	18	= V77-26		305
68	57° 30' 30"	8° 35'	62				

unit 1 sediments. This indicates the presence of biogenic gas in the underground, which is probably related to the rapid sedimentation of rather fine sediments containing relatively high amounts of organic matter (Vilks *et al.*, 1974). In fact during coring operations in 1977 a part of a core taken in this area was extruded and gas escaped from the core (core P77-17).

Glacial morphology is still present on part of the seafloor in the western part of the Skagerrak where local sedimentation has not been sufficient for a complete burial of pre-Holocene sediments and relief. This is illustrated by the presence of iceberg grooves on side-scan sonar profiles (Belderson & Wilson, 1973). Locally the effect of partial reworking of morainic sediments can be noted in the distribution of the sediments.

METHODS

Grab samples were obtained with a van Veen grab sampler. On board a subsample of about 1 l was taken from each sample and the remainder was wet sieved over sieves of 3 and 2 mm. The coarse material was collected for further study.

Grain-size analysis was carried out using the standard sieve and pipette method (Folk, 1968), employing whole phi intervals. For the dispersion of the sediments sodium diphosphate was used, in combination with an ultrasonic treatment.

The carbonate content of the surface samples was measured by the Scheibler method, a gasometric technique which is comparable to the method described by Hülsemann (1966).

For the determination of organic carbon phosphoric acid was used to remove the inorganic carbon, and after adding $K_2S_2O_8$ to the remainder the amount of carbon dioxide developed was measured.

The method of the analysis of the clay minerals is dealt with later. The textural and mineralogical character and composition of the sand fraction of a number of selected samples were determined employing the method of Shepard & Moore (1954), as adapted by Trumbull (1972) for the eastern continental shelf of the U.S.A. The amount of the major components of the sand fraction was estimated on the basis of Terry & Chilingar's (1955) technique. Initially these results have been compared with the results of counting 300–500 grains. When it was found that only minor differences were encountered,

the remaining samples were studied by means of the method of Terry & Chilingar only.

RESULTS

Grain-size determinations

After determination of the grain-size distribution, cumulative curves were drawn and a number of parameters studied comprising the distribution of ϕ_{50}, the sand content, the distribution of the various sand fractions, the clay content and the distribution of gravel (material > 2000 μm). The total number of samples used for the data presented in the maps (Figs 2–8) included 115 grab samples, with in addition the top 5 cm of several piston cores.

The cumulative curves of these samples indicate the presence of several mechanisms involved in the present-day distribution of the surface sediments in the Skagerrak. The bottom sediments can be divided into three main populations, while another sediment population is found more scattered (Fig. 2).

A rather well-sorted sand population with a silt fraction restricted mainly to the size adjacent to the sand is found in the shallower southern part of the Skagerrak. In this area sand waves of composite character occur locally (Stride & Chesterman, 1973; van Weering, 1975). According to Klovan (1966) curves of a type similar to those found here reflect sediment transport by currents. In fact current measurements carried out in this area during the Joint Skagerrak Expedition in 1966 reveal maximum current velocities of up to 40 cm sec^{-1} at 5 m above the bottom (see section on hydrography). These currents prevent settling of fine particles and can probably also remove part of the silt and fine sand fractions.

The second population consists of a mixture of a coarse, rather well-sorted sandy part and a moderately sorted silt and clay part. Similar curves were described for shallow-water deposits by Klovan (1966) who attributed them to transport by bottom-currents. Schlee (1973) also found this type of curve in finer grained deposits around several basins in the Gulf of Maine, the Scotian Shelf and adjacent coastal areas. Visher (1965, 1969) attributed this type of grain-size curve to a combination of (1) settling out of suspension and (2) a contribution by traction currents. In the Skagerrak these sediments are found in a zone adjacent to the north-western Danish coast, in the entrance to the Kattegat and in

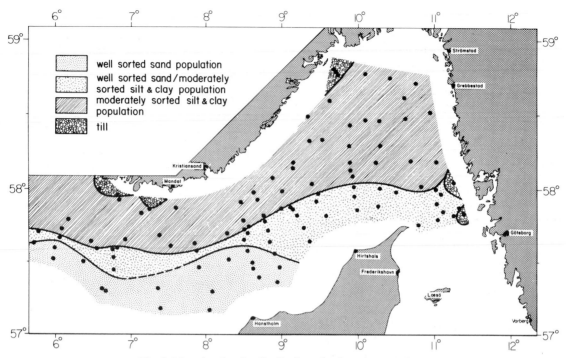

Fig. 2. Map showing the distribution of sediment populations.

a narrow zone along the southern slope of the basin. The results of the sorting action of the bottom-currents are well illustrated by the gradual decrease of the median diameter and of the total sand content in a north-easterly direction as well as by the distribution of the separate size fractions of the sand fraction (Figs 5–8).

The third main type of grain-size curve and sediment population is found in the northern and north-eastern part of the Skagerrak in the deeper areas, where moderately sorted silts and clay with a considerable range in grain size are found. Though a very small sand component may be present, most of the sediment consists of clay and silt. This type of sediment is most likely formed by settling out of suspension and has, after deposition, undergone no reworking by bottom-currents. Curves of almost identical shape were described by Schlee (1973) from deposits in several small basins in the Gulf of Maine where at present pelagic sedimentation occurs, and from a few sheltered bays and lagoons.

Another type of grain-size curve probably reflects till-like sediments, as the sediments contain a considerable amount of coarse particles in the sand range and are poorly sorted. These sediments are found in areas where glacial relics and morainic deposits occur: locally off the Swedish west coast, along the Norwegian coast (for example the station west of Mandal), and in the western part of the Skagerrak. These size distributions probably reflect the outwash character of the sediments caused by bottom-currents in the late Pleistocene or early Holocene.

Curves of similar shape were also ascribed to the occurrence of morainic deposits in the Gulf of Maine and in isolated areas south of Long Island (Pratt & Schlee, 1969; Schlee, 1973), and in the northern North Sea (Jansen, van Weering & Eisma, 1979).

Median grain size (ϕ_{50}) (Fig. 3)

The distribution of ϕ_{50} values indicates that the sediments with the coarsest median grain sizes occur in a zone along the southern slope of the Skagerrak and in the adjacent parts of the North Sea, merging gradually into finer grades towards the north and north-east. Locally, where partly winnowed glacial deposits form the seafloor this picture is interrupted, as off the Swedish and Norwegian coasts. It is clear that the finest sediments in the Skagerrak are found

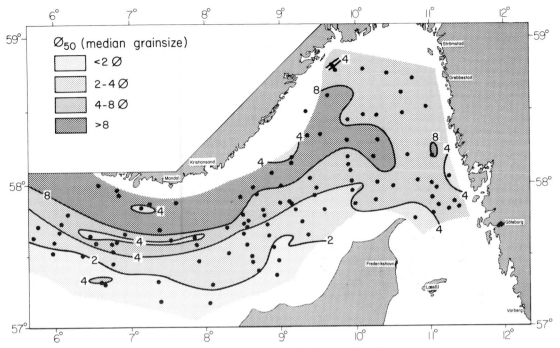

Fig. 3. Map showing the distribution of Q_{50} values.

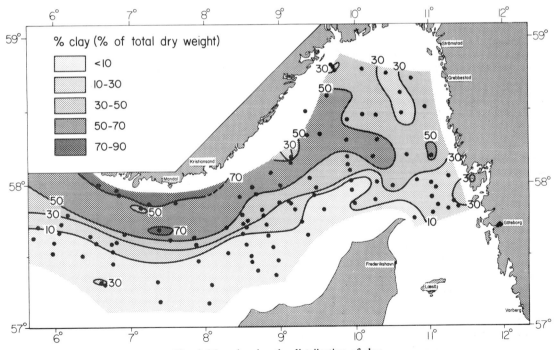

Fig. 4. Map showing the distribution of clay.

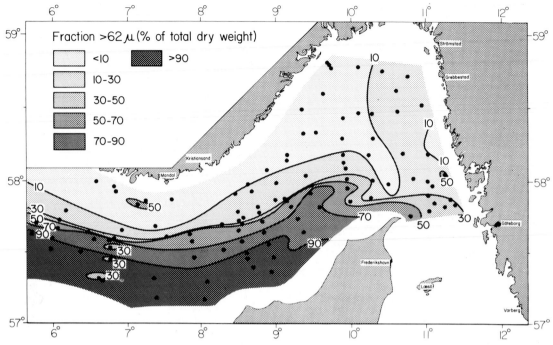

Fig. 5. Map showing the distribution of sand (63–2000 μm).

in a wide zone along the Norwegian mainland and in the deepest part of the basin.

As sampling close to the Norwegian coast was not possible, the extension of the belt of morainic sediments which is found off Mandal and also further north could not be established. It is probable, however, that near the Norwegian coast coarser sediments form a (semi-) continuous deposit which has partly been winnowed out, leaving a coarse lag deposit on top as demonstrated by Holtedahl & Bjerkli (1975) for an area off Jomfruland. This effect is also found near Mandal, where local areas with a value of ϕ_{50} of less than 4 are found.

Clay fraction (< 2 μm) (Fig. 4)

The percentages of clay in the Skagerrak increase from the Kattegat to the northern Skagerrak, where it forms about 30–50% of the surface sediments. Along the southern coast of Norway the bottom sediments contain locally 70% of clay-sized material.

The clay mineral composition of a number of samples (tops of cores P78-1, P78-4, P77-18, P77-9, P77-6) has been analysed by X-ray diffraction, using a pre-treatment which consisted of a separation into

two fractions (2–$\frac{1}{2}$ μm, and < $\frac{1}{2}$ μm respectively) by a centrifugation method in which an anti-whirl apparatus as described by de Jonge (1979) was used. Carbonate was removed in a sodium acetate–acetic acid buffer (pH = 5) and organic carbon by oxidation with bromine, as this method has shown to be the most effective without affecting the clay minerals (van Langeveld, van der Gaast & Eisma, 1978). The samples were then washed with a 1 N CaCl$_2$ solution and mounted on ceramic slabs.

Relative abundances were determined by measuring and comparing peak heights. Comparison of the components in the two fractions shows that in the fraction < $\frac{1}{2}$ μm montmorillonite occurs most frequently, followed in order of decreasing abundance by illite, kaolinite and chlorite. In one sample (P77-18), the montmorillonite and illite occur in almost equal quantities.

In the fraction $\frac{1}{2}$–2 μm vermiculite, illite and kaolinite are the most abundant minerals, while montmorillonite and chlorite are only present in small amounts. The vermiculite content is highest in P77-6, followed by P77-9 and P77-18, but is almost completely lacking in samples P78-1 and P78-4.

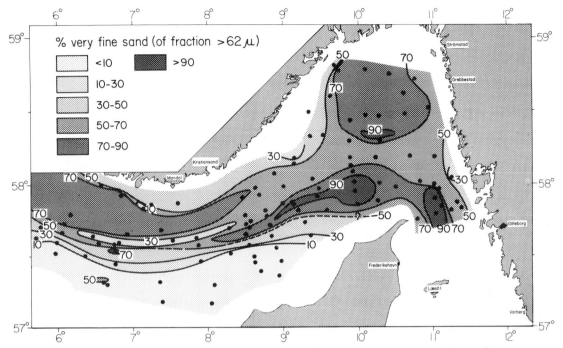

Fig. 6. Map showing the distribution of the very fine sand fraction (as part of the fraction > 62 μm).

Sand fraction (63–2000 μm) (Fig. 5)

There is a gradual decrease in sand content from the southern side of the Skagerrak, via a broad zone with values of 70–90 % sand, towards the entrance of the Kattegat where the seafloor consists of 30–50 % sand. From the northern tip of Denmark towards the Swedish coast there is also a marked decrease in grain size, which points to a sediment transport along the Danish coast and a gradual settling of the components. This trend is interrupted along the Swedish coast, where morainic deposits occasionally crop out on the seafloor and more coarse components are present, resulting in a higher sand content.

The decrease in sand content along the southern and south-eastern flank in a direction perpendicular to the depth contours of the basin is rather rapid. There is, most probably, a relation with the Jutland current which has a direction more or less parallel to the depth contours and reaches maximum current velocities which (at least) prevents settling in the shallower parts of the Skagerrak.

The local presence of glacial deposits is indicated by scattered deviations of the previously described pattern. This applies mainly to an area west of 7° E.L. where irregular glacial morphology is still present

with a hummocky seafloor and a number of small, low-lying areas where locally finer sediments (and thus smaller amounts of sand) are found. The reverse can be noted in some elevated areas and locally along the Swedish west coast.

The distribution of the very fine (63–125 μm) sand fraction (as part of the fraction > 62 μm, Fig. 6) shows that, apart from the shallower area along the southern flank of the Skagerrak where considerable amounts of fine and medium sand also occur, the greater part of the sand fraction consists of very fine sand. There are two areas where the bulk of the very fine sand fraction settles; notably north-west of Skagen and in the central part of the area between Denmark and Sweden. In both areas the amount of very fine sand is more than 90% of the total fraction > 62 μm. In the northern part of the Skagerrak the total sand fraction (which is very low) consists almost exclusively of very fine sand (Fig. 6).

In a narrow zone along the southern flank of the Skagerrak the very fine sand fraction forms only 10–30% of the sand fraction, in contrast with the areas north (50–70%) and south (30–50%) of it. This is caused by the presence in the sub-bottom of late glacial silty clays containing gravel and coarse

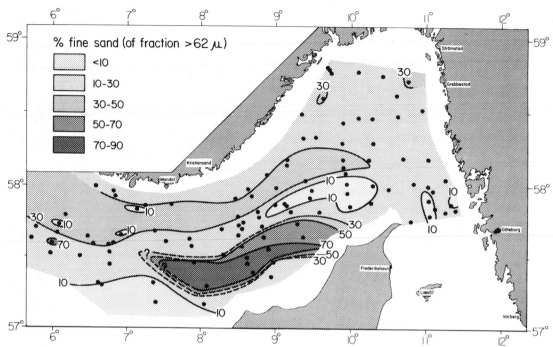

Fig. 7. Map showing the distribution of the fine sand fraction (as part of the fraction > 62 μm).

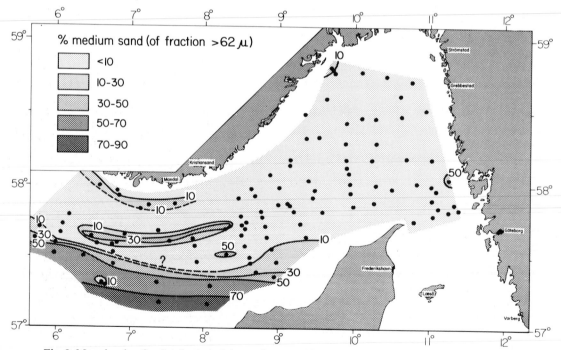

Fig. 8. Map showing the distribution of the medium sand fraction (as part of the fraction > 62 μm).

sand, which are covered only by a thin Holocene sediment blanket. The effects of the glacial relief can be noticed as well in considerable grain-size differences over short distances, mainly indicating the occurrence of isolated patches of morainic deposits.

Fine (Fig. 7) and medium (Fig. 8) sand-size particles are present usually in minor amounts, but it is obvious that the fraction 63–250 μm accounts for the bulk of the sand. Only in the shallower areas along the Danish side of the Skagerrak considerable amounts of medium sand are present, while, as was mentioned before, this is also the case in the narrow zone parallel to the depth contours along the southern flank.

Gravel fraction (> 2000 μm)

Gravel is present on the seafloor in rather considerable amounts but its distribution is patchy. The distribution depends mainly on: (a) the presence of glacial deposits on the seafloor, (b) the amount of reworking of these deposits, (c) the formation of lag deposits on elevated areas versus rapid burial in low lying areas, and (d) the total rate of sedimentation since the post-glacial. Well-polished, rounded to well rounded very fine, fine and medium gravel with an admixture of some subrounded to subangular pebbles are found off the north-western coast of Denmark in depths of 25–110 m. Though the greater part of the gravel consists of granitic, metamorphic or to a small extent porphyric rocks, sedimentary rocks are also found as pebbles of chert, chalk and limestone (locally up to 30% of the total gravel fraction), and sandstone (about 10%). A characteristic brownish coating is present on the greater part of the gravel, with some overgrowth of serpula and bryozoa.

Biogenic components are found at most stations and form the greatest part of the fractions > 8 and > 16 mm, especially at the deeper stations. The biogenic components mainly consist of various kinds of mollusc shells, locally with encrustations of serpulas, bryozoa and the like, as well as remains of echinoids. A number of the mollusc shells have been intensely burrowed and give a worn, retransported impression. Other biogenic components include polychaete tubes, wood, rounded peat fragments and leaf fragments (though the latter are very rare). Close to the Swedish coast subangular to subrounded gravel in a clayey matrix is found, the gravel being composed mainly of granitic rocks, with only a small admixture ($\sim$ 10%) of partly rounded limestones

and about 10% quartzitic sandstones. A small number of the pebbles here show striations on the surface. In the 2–4 mm fraction limonitic clay pebbles occur abundantly (mainly fragmented). At this station (V75-18) almost no encrustations were found.

Sampling off Jomfruland in water depths of 165–185 m revealed the presence of angular to subangular, medium to fine gravel in a clay matrix. The gravel contains no chert; limestone is only present in small amounts (less than 10%). Glacial striae are present on a number of pebbles. In the fractions 2–4 and 4–8 mm mollusc remains form about 20–25% of the total.

In the western part of the Skagerrak a belt of morainic sediments is present rather close to the Norwegian coast from Egersund to Mandal, with a possible connection towards Kristiansand and further north. The gravel along the Norwegian coast is mainly angular to subangular and contains granitic pebbles, greenstones, slates, no chert and up to 7% of chalk and limestone. Occasionally porphyries are found, most of them showing signs of glacial abrasion and glacial striae. In the finer gravel fractions some sedimentary rocks occur as well. Biogenic components are almost completely absent and no encrustations are found.

On the opposite side, along the south-western flank of the Skagerrak the gravel is composed of angular to subangular rock fragments of which many show the presence of glacial striae. In the finer gravel fraction an increasing amount (up to 30%) of (slightly) better rounded glauconitic chalk and chert is found. There is a good correlation between the degree of roundness and the grain size, which is specially true for the (softer) sedimentary rocks. The amount of biogenic components, although generally rather low, increases with decreasing grain size. In some samples, rich in chalk fragments, the fossil microfauna was determined (samples V75-16, V77-7, V76-29 fraction > 16 mm, V76-44, V76-42 both fraction > 8 mm). This yielded (J. Smit, personal communication) a lower Danian age (pseudobulloides zone) for most of the chalk, indicating that the Lower Tertiary sediments and probably also the Upper Maastrichtian sediments which form the pre-Quaternary underground have been actively eroded.

Lithology of the sand fraction

The textural and mineralogical characteristics of the sand fraction (63–2000 μm) of the bottom sediments of the Skagerrak have been studied using an adapted version (Trumbull, 1972) of the coarse fraction analysis developed originally by Shepard & Moore (1954). The components of the sand fraction are of detrital, authigenic or biogenic origin. The detrital components include quartz (and felspar), rock fragments, mica and glauconite, and other dark minerals. The authigenic components are pyrite and probably some glauconite, whereas the biogenic components include mainly remains of benthonic and planktonic foraminifera, molluscs and to a lesser extent echinoids, ostracods, sponges (spicules), fish, wood, and leaves.

Detrital components

(1) *Quartz and feldspar*

In the sediments belonging to the well-sorted sand population, quartz and feldspar are the dominating minerals, forming 95–100% of the very fine and fine sand fraction with a decrease in the coarser fractions. The quartz group minerals in this population are well-polished clear grains with an admixture of frosted grains. The coarse and medium sand fractions contain well-rounded grains, while in the very fine and fine sand fractions most grains are rounded and (for a small part) subrounded. The samples from the western part of the Skagerrak on the whole are somewhat less rounded, the coarser fractions being rounded and the very fine and fine sand fraction showing subrounded to subangular forms. In a number of samples a reddish-brown iron coating is found in depressions on the grains. This is found more frequently in the shallower samples.

The sediments of the well-sorted sand–moderately sorted silt and clay population contain a comparable though somewhat lower amount of quartz in the very fine and fine sand fractions in the zone adjacent to the previous, with a clear, sharp decrease towards the north-west, where in the fine sand fraction quartz accounts for only up to 20% of the detrital components. In the medium and coarse sand fractions there is also a notable decrease in quartz content, varying from around 50% for samples in a zone close to the well-sorted sand population to 30% or less in the north-eastern part. This suggests a gradual transition into the moderately sorted silt and clay populations. Towards deeper water there is also a rapid decrease in quartz content in both the fine and medium sand fractions. The grains are generally well polished, while the amount of frosted grains is variable (up to 30%).

The grains in the coarse and medium sand fraction are rounded and for a small part well rounded, whereas in the very fine and fine sand fraction the roundness is slightly less. Most grains in these fractions are subrounded to rounded. Only a very small part of the grains have an iron oxide coating.

Where the moderately sorted silt and clay population is found, the sediments contain only small amounts of quartz, mainly in the very fine sand fraction (locally up to 40%, but at most stations less than 20%) as subrounded to subangular grains. The fine and medium sand fractions contain up to 10% of quartz, apart from the stations where morainic material has been (or is) winnowed. Here enrichment of the detrital fraction causes locally a quartz and feldspar content of up to 40%.

(2) *Rock fragments*

Rock fragments are found only in low percentages (up to 5%) in the very fine and fine sand fraction of the well-sorted sand population. In the medium and coarse fraction values seldom exceed 10%. The rock fragments occur as well-rounded to rounded grains with small admixtures of subangular or subrounded grains. Only in the western part of the Skagerrak the roundness is slightly less, and here locally subangular to subrounded grains form the main part of this component, with a small percentage of angular rock fragments occurring as well. In a number of samples well-rounded chalk fragments are notable, while also granitic, metamorphic and sedimentary rock including chert is found. Part of the granitic rocks show the presence of abraded biotite within the grain.

In the well-sorted sand to moderately sorted silt and clay sediments rock fragments form a minor constituent. They are mostly present in the samples adjacent to the zone of the well-sorted sand population. Here values of 1–5% in the medium and coarse fraction are found, whereas in the very fine and fine sand fractions rock fragments are rare and form only up to 1% of the sediments. The moderately sorted silt and clay population contains rock fragments only exceptionally in all fractions. Where morainic sediments in the neighbourhood have been (or are) winnowed, rock fragments are found in very high percentages (20–90%) of the medium and coarser sand fractions and then have a subangular to angular form.

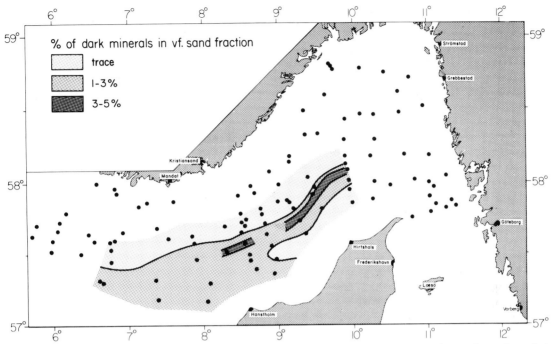

Fig. 9. Map showing the occurrence of dark minerals in the very fine sand fraction along the southern slope of the Skagerrak.

(3) *Dark minerals* (*Fig.* 9)

The distribution of the dark minerals (by X-ray diffraction shown to be mainly hornblende), which appear as elongate, well-polished grains with rounded and abraded edges, is rather limited. Dark minerals are mainly present in the very fine and fine sand fraction along the southern and south-eastern slope of the Skagerrak. In the well-sorted sand population dark minerals generally form 1–3% of the components, though maximum values of 5% are found in the very fine sand fraction. These values are found in a narrow zone close to the southern slope of the Skagerrak and extend well into the area where well-sorted sands and moderately sorted silt and clay sediments are found. This zone is situated along the north-western margin of the very fine sand accumulation north-west of Hirtshals (Fig. 6). In the fine sand fraction of the well-sorted sand to moderately sorted silt, dark minerals reach maximum values of 3%, with a rather rapid decrease towards the north-east. The roundness of the dark minerals is slightly less towards the north-east. The amount of dark minerals is rather negligible in the moderately sorted silt and clay sediments (the dark minerals occurring mainly in the very fine sand fraction). In the neighbourhood of morainic sediments percentages of up to 5% are found.

(4) *Mica*

Micas are completely absent in the well-sorted sand population along the southern side of the Skagerrak or occur only in trace amounts (< 1%) (Fig. 10). They occur as transparent, green or brown flakes (in order of decreasing abundance) with well-rounded edges. Where the well-sorted sands pass into the well-sorted sand to moderately sorted silt and clay population the values of mica are lowest in the very fine sand fraction (< 1%) and show a gradual increase towards the north-east where values of 1–3% occur. In the fine sand fraction the amount of mica is highest (up to 5%) at the deeper stations, whereas towards the north-east values of up to 3% are found, indicating a slight increase in this direction. In the medium sand fraction mica occurs locally as well, though in small amounts. In the sediments of the moderately sorted silt and clay population micas are found almost exclusively in the very fine sand fraction. In the deepest parts of the

23-2

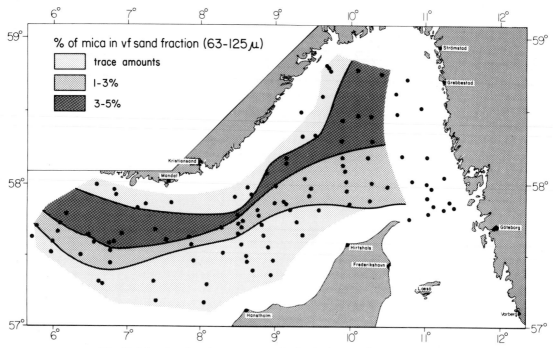

Fig. 10. Map showing the occurrence of mica in the very fine sand fraction.

basin the micas form 3–5 %. whereas in a zone along the Norwegian coast only traces (< 1 %) occur. An exceptional value of about 7 % was found west of Grebbestad (station V 75-40). In the fine sand and coarser fractions mica is mostly absent.

(5) Glauconite

Glauconite is not found in abundance. It occurs as either clear green, slightly angular grains or as dull, dark brownish-green well-rounded grains (probably foram moulds). In the well-sorted sand population it forms generally less than 1 % of the sand fraction. In some samples slightly higher values (∼ 2 %) are found in the fine sand fraction. In the well-sorted sand, moderately sorted silt and clay population the glauconite locally forms up to 5 % of the sand, while in the very fine sand fraction it may form up to 3 %, indicating a slight increase in occurrence. In addition to the above-mentioned types part of the glauconite occurs as well-rounded–rounded dull dark greenish grains. In the remainder of the area glauconite is almost absent apart from those stations which were taken in the vicinity of a morainic deposit. Here glauconite occurs as clear, rounded to subrounded grains, forming up to 1 % of the sand fraction.

(6) Pyrite

Detrital pyrite in the form of partly rounded yellow grains was found in only a small number of the sediments belonging to the well-sorted sand population and only occurs as a trace in the surface sediments.

Authigenic components

(1) Glauconite

Authigenic glauconite may be present in the form of foram moulds in only small amounts. The clear, bright green grains which are found may be derived partly from eroded Upper Cretaceous to Lower Tertiary chalks, as rock fragments eroded from these rocks contain glauconite. Therefore no conclusion concerning their origin can be drawn based on these occurrences, since the foram moulds may also be (partly) of reworked origin.

(2) Pyrite

Pyrite is found as pyritized faecal pellets, foram infill or burrow, or as (dark) reddish-brown aggregates. It forms only a minor constituent, occurs only sporadically ($< 1\%$) in the well-sorted sand population, but is more widespread in the adjoining sediment population and reaches, in the form of pyritized pellets, maximum values of up to 5% in the sediments belonging to the moderately sorted silt and clay population. Aggregates also occur in these sediments, and as no preference for any size class can be observed, this is probably an *in situ* product of early diagenesis.

Faecal pellets and biogenic components

In the well-sorted sand population faecal pellets and biogenic components occur mainly in the medium sand and coarser fractions (occasionally forming the bulk ($\sim 90\%$) of these fractions). However, they occur mostly in small amounts ($< 3\%$ or even $< 1\%$). Biogenic components are (in order of decreasing abundance): echinoid fragments, mollusc remains, tests of benthonic foraminifera, ostracod valves and fish remains. Wood fragments or plant remains are very rare. In one sample a rounded peat fragment was found. Planktonic foraminifera and sponge spicules were also found rarely. In the well-sorted sand to moderately sorted silt and clay population faecal pellets and biogenic components are found widespread and in considerable amounts, increasing in abundance towards the medium sand and coaser grain sizes which consist locally for 95% of faecal pellets, indicating extensive biological reworking. Benthonic foraminifera (which locally comprise up to 1000 per g dry sediment), echinoid fragments and ostracods form the greater part of the biogenic fraction, while mollusc shells, tests of planktonic foraminifera, fish remains and sponge spicules occur in minor quantities.

In the moderately sorted silt and clay population faecal pellets and biogenic components generally form the bulk of the sand fraction. As a general rule the amount of biogenic components increases towards the medium sand range and coarser size grades. The main biogenic components are benthonic foraminifera and mollusc fragments followed by ostracod valves and echinoid fragments. Tests of planktonic foraminifera form only a very small part of the biogenic material ($< 1\%$ of the total number of foraminiferal tests (Qvale, 1980), while fish remains, wood and sponge spicules are found in varying orders of abundance as trace constituents.

Carbonate content

The amount of carbonate (Fig. 11) is rather variable though there is a clear relation with the presence of the fine-grained deposits (Fig. 4). In the western part of the Skagerrak and locally off the Swedish west coast the sediments contain more than 15% of carbonate. Sediments containing more than 15% of carbonate are present as far north as Egersund, the areal distribution being closely related to both the depth and the presence of fine sediments. This is probably partly influenced by the inflowing Atlantic water, as in the more northerly part of the Norwegian Channel, where the Atlantic water is transported southwards along the western side of the Norwegian Channel (Helland Hansen, 1907) the amount of carbonate is much higher and locally exceeds 30% (Van Weering, unpublished data).

The lowest carbonate values occur along the southern flank of the Skagerrak where late glacial sediments crop out. Intermediate values (below 10%) are found at the entrance of the Kattegat and along the Swedish and Norwegian coasts. The greatest part of the basin, however, contains between 10 and 15% of carbonate, which is considerably less than in other parts of the Norwegian Channel and may be the result of a combination of greater sediment influx to the Skagerrak and/or slight dissolution of the carbonate (Alexanderson, 1978, 1979). In fact a number of tests of foraminifera from this area showed extensive dissolution effects. Whether this is only a local phenomenon or occurs on a regional scale throughout the Norwegian Channel is not yet clear, though on account of the otoliths from surface samples collected during 1976, Gaemers (1978) concluded that dissolution occurs in the entire Norwegian Channel area. However, part of his otoliths were of (sub) fossil origin and further research is necessary.

Content of organic carbon

The differences in organic carbon content of the surface samples are only small. Values of 3% or more are found in the central part of the basin where the sediments belong to the moderately sorted silt and clay population, while in contrast the shallow southern part of the Skagerrak sediments only contain less than 1% of organic carbon (Fig. 12). When the relation between carbon and clay content and the carbon:silt ratio are considered, it is seen that there is a direct connection between the two in population

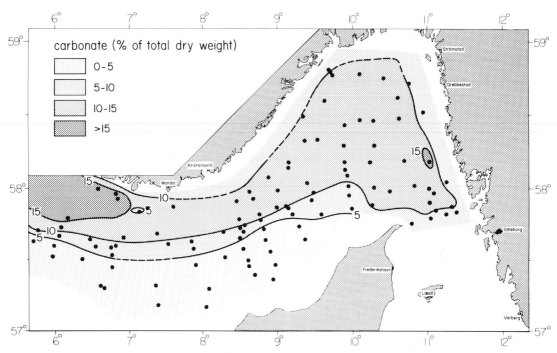

Fig. 11. Map showing the amount of carbonate in the surface sediments.

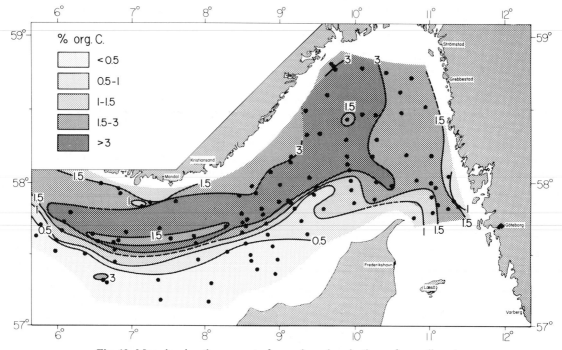

Fig. 12. Map showing the amount of organic carbon in the surface sediments.

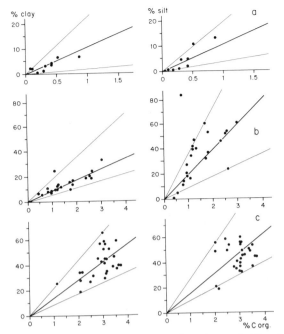

Fig. 13. Relation between % organic carbon and silt and clay content for bottom deposits from the Skagerrak; from top to bottom the well sorted sand population, the well-sorted sand to moderately sorted silt and the moderately sorted silt and clay population.

1, while in populations 2 and 3 the spread is greater (Fig. 13). However, the regression lines show a decrease in slope in populations 2 and 3 which may indicate that the organic carbon is occurring adsorbed to the finest fractions without any preference in population 3, and with a slight preference for clay in population 2. Few data on the occurrence of organic carbon in North Sea sediments have been published. The values found in the Skagerrak are consistent with those of Olausson for the eastern part of the Skagerrak (Olausson, 1975), but are considerably higher than those encountered in the northern part of the Norwegian Channel (where in the deepest part values only slightly higher than 1 % were found).

This may be attributed to the fact that the Skagerrak acts as a sink for fine material which has been transported from the southern North Sea and gradually settles out of suspension, as is also indicated by the distrubution of the sand, silt and clay fractions. This also fits well with observations by Eisma (1981) concerning the amount and distribution of suspended material in the water column and the contents of organic matter.

The somewhat lower values in a narrow zone along the southern flank of the Skagerrak (values 1½–3 %) may be caused by the presence of late glacial sediments in this area. It probably reflects a somewhat lower productivity in the post-glacial period, or a decomposition of organic matter.

HYDROGRAPHY

The general circulation pattern in the Skagerrak is well known, though strong differences may occur over short periods due to the combined effects of winds and changes in the atmospheric pressure over the North Sea (Dietrich, 1951). The older oceanographic data have been summarized by Svansson (1975); recently the work of Larsson & Rohde (1979) has contributed to a more detailed knowledge of the hydrography.

The surface current system in the Skagerrak forms part of the anticlockwise rotating system in the North Sea which, along the north-western Danish coast, gives rise to a current of north-easterly direction, the Jutland current. This current mixes in the Kattegat entrance with the outflowing Baltic current, and continues along the Swedish and Norwegian coasts as the Norwegian Coastal Current. It follows the south-western coast of Norway and then flows northwards until it joins the Atlantic current in the northern Atlantic Ocean.

Within the Skagerrak there are indications of a closed horizontal circulation (Engström, 1967; Lindquist, 1970); recently the existence of a deep countercurrent below the coastal current has been postulated (Dahl, 1978).

The North Sea water consists partly of water that has entered between the Orkney and Shetland Islands, whereas another part enters via the Straits of Dover and the Channel (Lee, 1970; Dooley, 1974; Dooley, Martin & Payne, 1976). Recent research by Kautsky (1977) has shown that a limited amount flows in north of Shetland as well. Helland Hansen noted already as early as 1907 that another inflow exists along the western side of the Norwegian Channel, where Atlantic water follows the 200 m depth contour in a southerly direction.

As the Atlantic water clearly has different charactersistics and the Norwegian Coastal Current is diluted with fresh water from rivers and from the Baltic, both water masses are conventionally separated by the 35°/oo isohaline (Helland Hansen

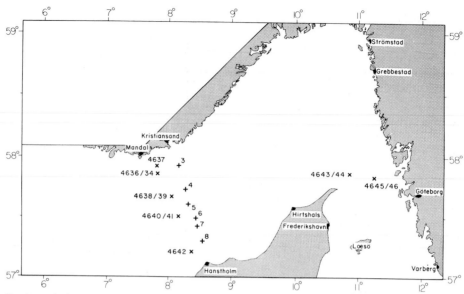

Fig. 14. Stations during joint oceanographic Skagerrak expedition (mentioned in the text) and stations of Larsson & Rohde (1979).

& Nansen, 1909; Saëlen, 1959; Furnes & Saëlen, 1977) and can be recognized easily.

Surface currents

There is only limited information available concerning the surface current pattern. The data lists of the Joint Skagerrak Expedition (Anonymous, 1966) and observations by Rohde (1973) and Larsson & Rohde (1979) form the best source. The current velocities close to the Norwegian coast (south of Mandal, station 4635, Fig. 14) are between 14 and 57·5 cm sec^{-1}, with current directions towards the south-west (mostly between 212° and 250°). On the opposite side of the Skagerrak (station 4641) strong variations in current velocity (between below 10 and 68 cm sec^{-1}) were encountered, the current direction here being mainly towards the north-east.

Close to the Danish coast (off Hanstholm, station 4642) the currents are rather strong (> 80 cm sec^{-1}) and show a main flow towards the north-east, while occasionally a flow towards the north-west is present.

Just north of Skagen strong differences in velocity occur, the currents ranging between 120·5 and 25·7 cm sec^{-1} with a direction between north-east and east (station 4643).

Off the Swedish coast west of Göteborg the currents show directions between north and north-west and between east and south. The velocities here are around 70 cm sec^{-1}. The measurements of Rohde (1973) in this area give more details, while the recent work of Larsson & Rohde (1979) in a section perpendicular to the Swedish coast at 58° 18′ N shows that in the centre of the Skagerrak (station 7 at 475 m water depth) the surface currents are at a maximum of 38 cm sec^{-1} but are mostly around 20 cm sec^{-1}.

Bottom currents

During the Joint Skagerrak Expedition current measurements were carried out at 5 m above the bottom along the Mandal–Hanstholm section. More recently Larsson & Rohde (1979, see also Figs 14 and 15) measured current velocities at 15–20 m above the bottom. These current velocity values therefore cannot be related directly to sediment transport studies. Yet, for lack of better data part of their results will be used as well. In the present study the values of the bottom current velocities exceeding 10 cm sec^{-1} were taken. From these values the mean velocity and direction were calculated as giving the best relation to sediment transport and sediment distribution at the sea-bottom. Half-hourly means were given in the tables of the expedition (Anonymous, 1969). At station 4634 (water depth 516 m) 22 measurements (> 10 cm sec^{-1}) give a mean

Table 2. Mean direction and velocity calculations for station 4640 (only values > 10 cm sec^{-1} considered), also showing the maximum and minimum current velocities per day.

Day	Mean direction (°)	Mean velocity cm sec^{-1}	Max. velocity cm sec^{-1}	Min. velocity cm sec^{-1}
June				
24	42	19·4	26	10·1
25	52	17·8	25·9	10·4
26	52	17·4	24·6	10
27	54	18·3	30·1	10·5
28	46	18·4	23·9	11·5
29	49	19·9	25·9	12·9
30	50	18·4	31·5	10
July				
01	46	20·1	29·5	14·6
02	49	23·5	28·8	16
03	53	21·1	30·1	11
04	52	20·7	24·4	11·1
05	49	21·7	28·8	17·7
06	52	25·5	30·8	21·1
07	53	22·3	27·8	16·4
08	52	27·4	36·6	20
09	49	36·4	40·8	30
10	49	31·6	35·3	25·3

velocity of 11·4 cm sec^{-1}, and a maximum current velocity of 13·8 cm sec^{-1}; the mean direction was towards the south-west (214°). At station 3 of Larsson & Rohde currents up to 20·8 cm sec^{-1}, with variable directions towards the WSW, ESE, and another to the NNW were encountered. However, a great number of observations shows currents well below 10 cm sec^{-1}. At station 4639 measurements over a period of 2 days at 285 m water depth clearly showed two opposite directions. A maximum current velocity of 21·3 cm sec^{-1} was measured, with a mean of 13 cm sec^{-1}, in a direction towards 72°. The other direction was towards 180° with a mean velocity of 14·4 cm sec^{-1}. These results are confirmed by the observations of Larsson & Rohde (station 4, in the direct neighbourhood) who measured a maximum value of 32·3 cm sec^{-1} towards 75°. The main part of their observations show current velocities between 10 and 20 cm sec^{-1} in directions between 20° and 115°.

Results of the measurements carried out continuously from 23 June to 10 July 1966 at station 4640 (water depth 125 m) are given in Table 2, which shows calculated daily values for mean velocity and direction. It is evident that the current velocities are

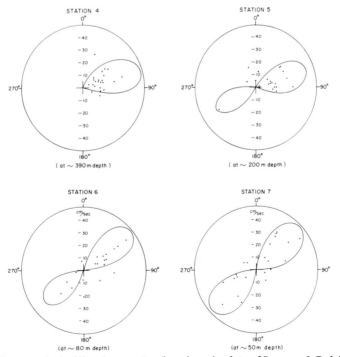

Fig. 15. Current velocity/direction rosettes (based on the data of Larsson & Rohde, 1979).

sufficient for the erosion of sandy deposits and only occasionally will be so low as to allow settling of fine particles. The direction of the currents is uniform and indicates that the Jutland current here flows towards 51°.

Off Hanstholm (station 7 of Larsson & Rohde) the current velocities are at a maximum of 45 cm sec⁻¹, with a direction towards 55°. Here two directions are generally dominant, one towards the north-east, the other towards the south-west (see Fig. 15).

In the south-eastern and eastern parts of the Skagerrak the currents are rather variable, depending partly on the effect of winds, the Baltic outflow and the Atlantic inflow (Svansson, 1975). Current velocities at station 4644 (40 m below the surface) showed variations between 0 and 63·6 cm sec⁻¹, indicating transport to the east, the south and the north-west. Close to the Swedish coast the currents show strong variations in both direction and velocity, the general direction being towards the NNW. Larsson & Rohde show that off the Swedish coast current velocities (5–20 m above the bottom) are strongest close to the coast; at their station 2 (water depth 80 m) a maximum current velocity of 36·3 cm sec⁻¹ was measured while farther seaward (at 290 m water depth), most velocities are below 10 cm sec⁻¹.

DISCUSSION

During the last glaciation the Weichselian ice sheet in Denmark reached a boundary known as the Main stationary line or C-line (Madsen, 1928). The maximum extension of the ice in the North Sea is not well known, although Pratje (1951) concluded that remnants of terminal moraines occur along the western side of the Norwegian Channel and south of the Skagerrak as a series of stony grounds and banks. Although different authors have varying opinions as to the southern limit of the ice sheet (Pratje, 1951, Valentin, 1957, Veenstra, 1970 and Jansen, 1979), general agreement exists that the Skagerrak was completely covered with ice during the period of maximum extension of the ice sheet.

In that period the ice divide in Norway was situated a considerable distance east and south of the present watershed, and ice flow directions suggest a confluence of Norwegian and Swedish ice in the eastern part of the Skagerrak and a diffluence in the area south of Lista and Lindesnes (Vorren, 1977). This author has also calculated the thickness of the

ice cap in the period of maximum glaciation, indicating that the Skagerrak was completely ice filled in the deepest part; this may have caused considerable erosion of the bottom. This fits well with my own observations that in the deepest part of the Skagerrak the sedimentary cover on top of hard rock is only very thin (20–30 m) and consists presumably of late and post-glacial sediments (van Weering et al., 1973; van Weering, 1975). On the contrary, the diffluent pattern of the ice flow lines south of Lista and Lindesnes in combination with the limited catchment area suggests that the ice sheet in this region was less thick. Although the ice sheet here was probably grounded during its maximum extension (van Weering, 1975), it has eroded the underlying deposits only partly. In fact airgun profiles across this part of the Skagerrak show a thickness of the sediments on top of Mesozoic and younger hard rock of 100 m close to the Norwegian coast and of 225 m in the centre of the Skagerrak. Only locally hard rock of Cretaceous and Tertiary age crops out on the sea-floor along the southern side of the Skagerrak.

This accounts well for the occurrence of chalk fragments of Upper Maastrichtian and Lower Danian age, together with chert and angular to sub-angular rock fragments showing glacial striae along the south-western flank of the Skagerrak. The angularity of the gravel illustrates that reworking here has taken place only on a limited scale. Moreover, the presence of these deposits at or near the sea-bottom indicates that sedimentation since then (if any) here has been very low, in contrast to other parts of the Skagerrak (which is corroborated by the acoustic data) or that even erosion by bottom-currents may have occurred. When the ice had retreated to the Norwegian and Swedish coasts at about 13,000 BP, the Lista moraine was deposited along a part of the Norwegian coast, while in the meantime the Halland coastal moraine (shortly followed by the Gothenburg moraine) was formed off the Swedish west coast (Berglund, 1976). These moraines are probably reflected in the patchy distribution of subrounded gravel off the Swedish west coast, and in the occurrence of angular to subangular gravel in the stations south of Mandal and off Jomfruland.

This also indicates that the morainic deposits of the Swedish west coast which are found at 65 m water depth have been abraded and have undergone reworking during the post-glacial rise in sea-level. Off Mandal and Jomfruland, where the water depth is much greater, the gravel has been unchanged as it

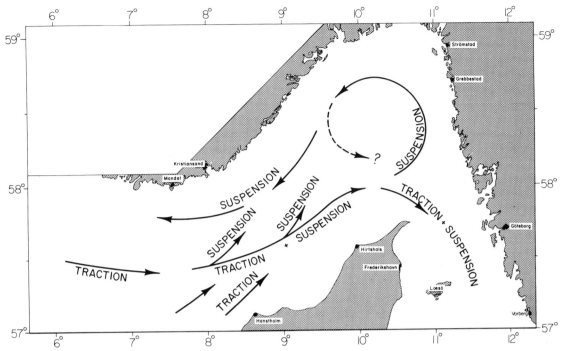

Fig. 16. Map showing the bottom transport and sediment distribution mechanisms.

consists primarily of angular and subangular fragments in a clay matrix.

The gravel off the north-west coast of Denmark has apparently been reworked considerably, as it consists of well-polished, rounded to well-rounded rock fragments. The high content of chert, chalk limestone and sandstone in these samples points to an additional source rock area which has probably been the bottom of the Skagerrak. Although theoretically this gravel may partly originate from an older glaciation (and in that case has undergone repeated reworking) it has most probably been influenced and reworked during an early phase of the post-glacial rise in sea-level. Then the bottom-currents were much stronger and wave turbulence might have influenced the bottom as well.

The distribution of surface sediments and sediment populations, as depicted in Figs 2–8, can be explained very well by assuming a sediment transport and deposition under influence of traction currents for the well-sorted sand population (Fig. 16). The fairly high maximum current velocities measured by Larsson & Rohde (1979) and during the Joint Skagerrak Expedition show that the current velocities which are required for the erosion of very fine to medium sand deposits by Banner (1979), based on

a recent modification of the graph (by Postma, 1967) which relates median diameter to current velocity, are exceeded. These velocities are also high enough to keep silt and clay-sized particles in suspension and prevent settling. Therefore it is suggested that sediment transport along the bottom as bedload in combination with the bypassing of fines has resulted in the deposits of the well-sorted sand population. This interpretation is strengthened by the almost complete absence of mica along the southern flank, and the presence of the rounded and subrounded form of the quartz grains, as well as by the occurrence of granitic rock fragments containing abraded biotite flakes. The presence of rounded chalk particles and rounded and well-polished dark minerals also suggests a transport as bedload, requiring bottom-current velocities that prevent fine particles from settling.

Micas, because of their platy nature, are especially sensitive to only slight changes in the energy conditions, the absence of mica indicating that winnowing or bypassing of fines are important processes (Doyle, Cleary & Pilkey, 1967), or that no mica is being contributed. Off Nigeria (Adegoke & Stanley, 1972) and along the coast of Israel (Pomeranblum, 1966) it has been shown that mica behaves as a hydraulic equivalent of silts and that winnowing processes are effectively transporting micas to a lower energy environment. The distribution of mica in the Skagerrak strongly indicates a sedimentation mechanism due to a gradual decrease in energy towards the north-east and north. This is reflected in the distribution patterns of the sand fractions, as well as in the distribution of the dark minerals. The zone where the dark minerals are enriched in the very fine sand fraction probably indicates a rapid change in energy conditions.

Due to a gradual decrease in energy towards the north-east and east the very fine sand fraction is deposited together with the silt fraction and some clay as the well-sorted sand to moderately sorted silt population. A further decrease in bottom-current velocity towards the north, in combination with a gradual settling of the suspension fraction, has resulted in the deposition of the moderately sorted silt and clay population.

Few data exist about the carbonate content of bottom sediments from the North Sea, apart from those published by Jansen (1979) for the north-western North Sea and some measurements on widely spaced cores from the Norwegian Channel by Moyes *et al.* (1974). My own (unpublished) data for the Norwegian Channel indicate that the Skagerrak bottom sediments contain far less carbonate than those in the northern part of the Norwegian Channel (at maximum 15% compared to values up to 30–50%).

The carbonate dissolution in the eastern, coastal part of the Skagerrak as discussed by Alexandersson (1978, 1979) may have led to the relatively low values of carbonate found there. The process of maceration which he described may occur on a more regional scale, as in a number of samples (from the deeper stations) the mollusc shells were found to be very weak and easily ground to powder. However, this is not the case for all stations. For example, along the southern part of the Skagerrak mechanical abrasion and boring activity play an important role in destruction. The importance and regional occurrence of the maceration process is not yet clear and so far I have not detected the remains of the destructive diagenesis in the sediments. In particular no lime mud has been noted, though Alexandersson stated that this lime mud, a result of the decomposition process, may become easily suspended (Alexandersson, 1979). Therefore it should be found in the northern part of the Skagerrak as well.

Whether the carbonate dissolution as noted by

Gaemers (1978) is of recent or older age is not yet clear, and no conclusion for the occurrence of the process of destructive diagenesis in the northern part of the Norwegian Channel can be drawn at this stage.

The relatively high content of organic carbon in the moderately sorted silt and clay population cannot be explained by high primary productivity in the Skagerrak. This is not significantly different from the North Sea basin as a whole. As only a minor fraction of the organic matter produced in the photic zone reaches the sediment surface of shelves or oceans (cf. de Vooys, 1979), a possible explanation is that the material is swept into the deeper Skagerrak from the area south of it. This area is much shallower (< 100 m) so that a large percentage of the organic matter produced in the photic zone (about 30 m) will reach the bottom before being decomposed. In the eastern part of the Skagerrak the combination of rapid sedimentation and a high content of organic matter in the sediments may have led to the formation of bacterial gas in the bottom sediments by sulphate reduction. This results in somewhat lower percentages of organic carbon in the bottom sediments in this area, as compared to the northern and central parts of the Skagerrak, and in the presence of shadow zones in the sub-bottom (van Weering, 1975). This can also explain the extrusion of cores taken in this area as well as the escape of gas from cores and the presence of gas cracks in sediments obtained from this area. Moreover, in a paper by Jörgensen *et al.* (1981) it was pointed out that in sediments along the southern slope of the Skagerrak pyrite is found. This is in good agreement with my own observations that in the moderately sorted silt and clay population are considerable amounts of pyritized pellets and pyritized burrows. A high amount of organic matter for sediments in suspension was found in the Skagerrak by Eisma (personal communication). His observations also point to a gradual settling of suspended particles in the Skagerrak in accordance with the surface and bottom currents. The previously discussed sediment composition and distribution, as well as the acoustic profiles obtained in this area (van Weering *et al.*, 1973; van Weering, 1975) can be best explained by a sediment dispersal mechanism as illustrated in Fig. 16. Nevertheless more data on near-bottom currents and sediment transport are needed. For example the use of sediment traps and continuous recording current meters may provide a more detailed picture of the area.

CONCLUSIONS

The transport of sediments in the Skagerrak mainly takes place in suspension, while along the southern side of the Skagerrak traction currents are also of importance. There are two areas where the bulk of the sand fraction settles, notably north-west of Skagen and in the entrance to the Kattegat, which shows that there is a clear relation between sediment distribution and the present-day hydrographic data. The bottom sediments are a combination of relict (Emery, 1968), reworked and modern sediments, the distribution of which coincides well with the acoustical characteristics and sediment thickness observed previously. The amount of carbonate in the bottom sediment is low compared to other parts of the Norwegian Channel, which may be caused by carbonate dissolution. In contrast, the amount of organic carbon is considerable, which points to an additional supply apart from the organic matter escaped from the photic zone. This source area is probably the shallower southern North Sea, which is also the source area of the inorganic particles of the recent sediments found in the Skagerrak.

The Skagerrak obviously acts as a sink for fine-grained materials derived from the southern North Sea.

ACKNOWLEDGMENTS

This study was supported by the Netherlands Organization for the Advancement of Science, ZWO. Technical support and ship's time was provided by the Netherlands Institute for Sea Research. Permission to carry out the research was obtained from the Norwegian Petroleum Directorate and through the Danish and Swedish Ministries of Foreign Affairs. Mr Hamburger of the Dutch Ministry of Foreign Affairs is thanked for his help in obtaining the above-mentioned permissions.

I thank all my colleagues from the department of Marine Geology at the NIOZ for their encouragement, help and criticism, and above all Jack Schilling for his continuous efforts to obtain the best sampling results. Thanks are also due to captain and crew of the RV *Aurelia*.

H. Hobbelink and B. Aggenbach made the drawings, typing was done by Mrs L. Everhardus, I thank them as well.

Critical reading of the manuscript by Professor Postma and L. M. J. U. van Straaten and by Dr D. Eisma is also appreciated.

REFERENCES

ADEGOKE, O.S. & STANLEY, D.J. (1972) Mica and shell as indicators of energy level and depositional regime on the Nigerian Shelf. *Mar. Geol.* **13**, M61–M66.

ALEXANDERSSON, E.T. (1978) Destructive diagenesis of carbonate sediments in the eastern Skagerrak, North Sea. *Geology*, **6**, 324–327.

ALEXANDERSSON, E.T. (1979) Marine maceration of skeletal carbonates in the Skagerrak, North Sea. *Sedimentology*, **26**, 845–852.

ANONYMOUS (1969) *ICES Oceanographic Data Lists, Joint Skagerrak Expedition 1966. Vol. 4: Current Measurements.* Turbidity Records, 151 pp.

BANNER, F.T. (1979) Sediments of the North-Western European Shelf. In: *The Northwest European Shelf Seas: the sea bed and the sea in motion. I. Geology and Sedimentology* (Ed. by F. T. Banner, M. B. Collins and K. S. Massie), **24A**, 271–300. Elsevier, New York.

BELDERSON, R.H. & WILSON, J.B. (1973) Iceberg plough marks in the vicinity of the Norwegian trough. *Norsk geol. Tidsskr.* **53**, 323–328.

BERGLUND, B.E. (1976) *The deglaciation of southern Sweden. Presentation of a research project and a tentative radio carbon chronology.* Rep. 10, Department of Quaternary Geology, University of Lund, 67 pp.

DAHL, F.E. (1978) On the existence of a deep countercurrent to the Norwegian coastal current in Skagerrak. *Tellus*, **30**, 552–556.

DIETRICH, G. (1951) Oberflächenströmungen im Kattegat, im Sund, und in der Beltsee. *Dt. hydrogr. Z.* **4**, 129–150.

DOOLEY, H. D. (1974) Hypotheses concerning the circulation of the Northern North Sea. *J. Cons. int. Explor. Mer.* **36**, 54–61.

DOOLEY, H.D., MARTIN, J.H.A. & PAYNE, R. (1976) Flow across the continental slope off Northern Scotland. *Deep Sea Res.* **23**, 875–880.

DOYLE, L.J., CLEARY, W.J. & PILKEY, O.H. (1968) Mica, its use in determining shelf depositional regimes. *Mar. Geol.* **6**, 381–389.

EISMA, D. (1981) Supply and deposition of suspended matter in the North Sea. In: *Holocene Marine Sedimentation in the North Sea Basin* (Ed. by S.-D. Nio et al.). *Spec. Publs int. Ass. Sediment.* **5**, 415–428. Blackwell Scientific Publications, Oxford. 524 pp.

EISMA, D., JANSEN, J.H.F. & VAN WEERING, TJ.C.E. (1979) Sea floor morphology and recent sediment movement in the North Sea. In: *The Quaternary of the North Sea* (Ed. by E. Oele, R. T. E. Schüttenhelm and A. J. Wiggers), pp. 217–231. Acta Univ. Ups. Symp. Univ. Ups. Annum Quingentesimum Celebrantis, Uppsala.

EMERY, K.O. (1968) Relict sediments on continental shelves of the world. *Bull. Am. Ass. Petrol. Geol.* **52**, 445–464.

ENGSTRÖM, S.G. (1967) Laying out surface drifters in the eastern North Sea and Skagerrak in the summer of 1966. *Medd, Havsfiskelab.* **33**, 8 pp.

FLODÉN, T. (1973) Notes on the bedrock of the eastern Skagerrak with remarks on the Pleistocene deposits. *Stockholm Contr. Geol.* **XXIV**, 79–102.

FOLK, R.L. (1968) *Petrology of Sedimentary Rocks.* University of Texas, 170 pp.

FURNES, G.K. & SAËLEN, O.H. (1977) *Currents and hydrography in the Norwegian coastal current off Utsira during Jonsdap-76.* Rep. 2/77, The Norwegian Coastal Current Project.

GAEMERS, P.A.M. (1978) Late Quaternary and recent otoliths from the seas around southern Norway. *Meded. Werkgr. Tert. Kwart. Geol.* **15**, 101–117.

HELLAND HANSEN, H. (1907) Current measurements in Norwegian fjords, the Norwegian Sea and the North Sea in 1906. *Bergens Mus. Årb.* **15**, 1–61.

HELLAND HANSEN, H. & NANSEN, F. (1909) The Norwegian Sea. *Rep. Norw. Fishery mar. Invest.* **II-2**, 360 pp.

HOLTEDAHL, H. & BJERKLI, K. (1975) Pleistocene and recent sediments of the Norwegian Continental Shelf (62°–71 °N) and the Norwegian Channel area. *Norg. geol. Unders.* **316**, 241–253.

HOLTEDAHL, O. (1940) The submarine relief off the Norwegian coast. *Det Norske Videnskaps Academi i Oslo*, 1–41.

HOLTEDAHL, O. (1964) Echosoundings in the Skagerrak. *Norg. geol. Unders.* **223**, 139–160.

HÜLSEMANN, J. (1966) On the routine analysis of carbonates in unconsolidated sediments. *J. sedim. Petrol.* **36**, 622–625.

JANSEN, J.H.F. (1979) *Late Quaternary history of the northern North Sea.* Thesis, University of Groningen.

JANSEN, J.H.F. VAN WEERING, TJ.C.E. & EISMA, D. (1979) Late Quaternary sedimentation in the North Sea. In: *The Quaternary History of the North Sea* (Ed. by E. Oele, R. T. E. Schüttenhelm and A. J. Wiggers), pp. 175–187. Acta Univ. Ups. Symp. Univ. Ups. Annum Quingentesimum Celebrantis, 2, Uppsala.

DE JONGE, V.N. (1979) Quantitative separation of benthic diatoms from sediments using density gradient centrifugation in the colloidal Silica Ludox-TM. *Mar. Biol.* **51**, 267–278.

JØRGENSEN, P., ERLENKEUSER, H., LANGE, H., NAGY, J., RUMOHR, J. & WERNER, F. (1981) Sedimentological and stratigraphical studies of two cores from the Skagerrak. In: *Holocene Marine Sedimentation in the North Sea Basin* (Ed. by S.-D. Nio *et al.*). Spec. Publs int. Ass. Sediment. 5, 397–414. Blackwell Scientific Publications, Oxford. 524 pp.

KAUTSKY, H. (1977) Strömungen in der Nordsee. *Umschau*, **77**, 672–673.

KLOVAN, J.E. (1966) The use of factor analysis in determining depositional environments from grainsize distributions. *J. sedim. Petrol.* **36**, 115–125.

LANGE, W. (1956) Grundproben aus Skagerrak und Kattegat, microfaunistisch und sedimenpetrografisch untersucht. *Meijniana*, **5**, 51–86.

VAN LANGEVELD, A.P., VAN DER GAAST, S.J. & EISMA, D. (1978) A comparison of the effectiveness of eight methods for the removal or organic matter from clay. *Clay Clay Miner.* **26**, 361–364.

LARSSON, A.M. & ROHDE, J. (1979) *Hydrographical and chemical observations in the Skagerrak 1975–1977.* Report 29. Oceanografiska Institutionen, Götreborg University.

LEE, A. (1970) The current and watermasses of the North Sea. *Oceanogr. Mar. Biol. Ann. Rev.* 1970, no. 8, pp. 33–71.

LINDQUIST, A. (1970) Zur Verbreitung der Fisheier und Fishlarven im Skagerrak in den Monaten Mai und Juni. *Inst. Mar. Res., Lysekil. Ser. Biol.* **19**, 92 pp.

MADSEN, V. (1928) Summary of the geology of Denmark. *Danm. geol. Unders.* **V-4**, 14–47.

MOYES, J., GAYET, J., PUJOL, C. & PUJOS-LAMY, A. (1974) *Etude stratigraphique et sedimentologique.* Orgon 1 (Mer de Norvège).

OLAUSSON, E. (1975) Man made effects in sediments from Kattegat and Skagerrak. *Geol. För. Stockh. Förh*, **97**, S60, 3–13.

POMERANCBLUM, M. (1966) The distribution of heavy minerals and their hydraulic equivalents in sediments of the Mediterranean continental shelf of Israel. *J. sedim. Petrol.* **36**, 162–174.

POSTMA, H. (1967) Sediment transport and sedimentation in the estuarine environment. In: *Estuaries* (Ed. by Lauff), pp. 158–179. Am. Ass. Adv. Sci., Washington.

PRATJE, O. (1951) Die Deutung der Steingründe in der Nordsee als Endmoränen. *Dt. hydrogr. Z.* **4**, 106–114.

PRATT, R.M. & SCHLEE, J. (1969) Glaciation on the continental margin off New England. *Bull. geol. Soc. Am.* **80**, 2335–2341.

QVALE, G. (1980) Foraminifer fordelingen I overflate sedimenter fra Skagerrak og nordlige del av Nordsjoen. In: *Abstracts 14th Nordiske Geologiske Vintermøte*, p. 58. Bergen.

ROHDE, J. (1973) *Sediment transport and accumulation at the Skagerrak-Kattegat border.* Rep. 8. Oceanografiska Institutionen, University of Göteborg.

SAËLEN, O.H. (1959) Studies in the Norwegian Atlantic current. Part I: the Sogne fjord section. *Geofys. Publ.* **XX**, 1–28.

SCHLEE, J.S. (1973) Atlantic continental shelf and slope of the United States—sediment texture of the northeastern part. *Prof. Pap. U.S. geol. Surv.* **529-L**, 64 pp.

SELLEVOL, M.A. & AALSTAD, I. (1971) Magnetic measurements and seismic profiling in the Skagerrak. *Mar. geophys. Res.* **I**, 284–302.

SHEPARD, F.P. & MOORE, D.G. (1954) Sedimentary environments differentiated by coarse-fraction studies. *Bull. Am. Ass. Petrol. Geol.* **38**, 1792–1802.

STRIDE, A.H. & CHESTERMAN, W.D. (1973) Sedimentation by nontidal currents around Northern Denmark. *Mar. Geol.* **15**, M53-M58.

SVANSSON, A. (1975) *Physical and chemical oceanography of the Skagerrak and the Kattegat. I. Open sea conditions.* Rep. 1, Fishery Board of Sweden, Institute of Marine Research.

TERRY, R.D. & CHILINGAR, G.V. (1955) Summary of 'Concerning some additional aids in studying sedimentary formations by Shvetslov, M.S.' *J. sedim. Petrol.* **25**, 229–234.

TRUMBULL, J.V.A. (1972) Atlantic continental shelf and slope of the U.S.A. Sand size fraction of bottom sediments, New Jersey to Nova Scotia. *Prof. Pap. U.S. geol. Surv.*, **529-K**.

VALENTIN, H. (1957) Die Grenze der letzten Vereisung im Nordseeraum. *Abh. d. Geographentg* (Hamburg 1955) **30**, 359–372.

VEENSTRA, H.J. (1970) Quaternary North Sea coasts. *Quaternaria*, **XXII**, 169–184.

VILKS, G. et al. (1974) Methane in recent sediments of the Labrador Shelf. *Can. J. Earth Sci.* **11**, 1427–1434.

VISHER, G.S. (1965) Fluvial processes as interpreted from

ancient and recent fluvial deposits. In: *Primary Sedimentary Structures and their Hydrodynamic Interpretation* (Ed. by E. V. Middleton). *Spec. Publ. Soc. econ. Paleont. Miner.*, *Tulsa*, **12**, 265 pp.

VISHER, G.S. (1969) Grainsize distribution and depositional processes. *J. sedim. Petrol.* **39**, 1074–1106.

DE VOOYS, C.G. (1979) Primary production in aquatic environments. In: *The Global Carbon Cycle* SCOPE 13 (Ed. by B. Bolin, E. T. Degens, S. Kempe and P. Ketner). Wiley, London.

VORREN, T. O. (1971) Weicheslian ice movement in South Norway and adjacent areas. *Boreas*, **6**, 247–257.

VAN WEERING, TJ.C.E. (1975) Late Quaternary history of the Skagerrak; an interpretation of acoustical profiles. *Geol. Mijnb.* **54**, 130–145.

VAN WEERING, TJ.C.E. (1981) Shallow seismic and acoustic reflection profiles from the Skagerrak; implications for recent sedimentation. (in preparation).

VAN WEERING, TJ.C.E., EISMA, D. & JANSEN, J.H.F. (1973) Acoustic reflection profiling of the Norwegian Channel between Oslo and Bergen. *Neth. J. Sea Res.* **6**, 241–263.

Spec. Publs int. Ass. Sediment. (1981) **5**, 361–383

Sediment transport in the Middle Atlantic Bight of North America: synopsis of recent observations

DONALD J. P. SWIFT, ROBERT A. YOUNG, THOMAS L. CLARKE,
CHRISTOPHER E. VINCENT*, ALAN NIEDORODA† *and* BARRY LESHT‡

*Atlantic Oceanographic and Meteorological Laboratories, National Oceanic and Atmospheric Administration, 15 Rickenbacker Causeway, Virginia Key, Miami, Florida 33149, U.S.A., *School of Environmental Sciences, University of East Anglia, Norwich NR4 7TC, U.K., †Woodward-Clyde Consultants, 7330 Westview Drive, Houston, Texas 77055, U.S.A. and ‡RER/203, Argonne National Laboratory, Argonne, Illinois 60439, U.S.A.*

ABSTRACT

The Middle Atlantic Bight of the North American continental shelf is a north-east-trending arcuate water body measuring approximately 600 by 180 km. The most energetic water motions are storm-driven flows, except near estuary mouths and in the Georges Bank–Nantucket Shoals sector, where tidal flows are dominant. The flow climate is complex, and a three-year time series is not sufficient to develop stable statistics. The shelf surface is a gently sloping plain, traversed by shelf valleys and associated landforms. Intervalley surfaces bear a sand-ridge topography formed during erosional retreat of the shoreface in the course of the Holocene transgression. Erosional shoreface retreat has veneered the shelf surface with a sand sheet up to 10 m thick. The ridges are moulded into the sand sheet, whose basal gravel is exposed in swales between the ridges. To the north, fine sediment is accumulating down-current from the eroding high of Nantucket Shoals–Georges Bank.

Sediment transport by storm flows is highly episodic and is seasonal in nature. In a 135-day observation of near-bottom flow, over 95 % of the computed transport occurred during a single winter storm. Wave energy is best predictor for concentration in the boundary layer. Concentration–wave energy curves at different sites have similar patterns: after a threshold value has been exceeded, the concentration increases linearly with increasing wave energy.

Studies of sediment transport on the Long Island shoreface reveal the presence of two contrasting shoreface provinces. An upper shoreface zone is subject to nearly continuous alongshore transport by the wave-driven littoral current, and to onshore transport in response to landward-oriented wave orbital currents. Below approximately 10 m, the lower shoreface is subjected to alongshore tide- and wind-driven currents. These are not normally capable of transporting sediment. However, during a north-easter storm the wind-driven component intensified and was associated with downwelling. If wave orbital velocity is intense enough during storms, these flows will entrain sediment all the way down the shoreface. Transport by downwelling storm currents is alongshore and offshore, since on the lower shoreface wave orbital currents lack landward asymmetry.

Studies of sand ridges on the Maryland coast indicate that there is an onshore–offshore continuum of morphologic parameters. Along a shore-normal transect, grain size and topography are systematically related: grain size becomes finer across the ridge and is finest on the seaward flank. The relationship is believed to be indicative of a phase lag between bottom shear stress during storm flows and bottom morphology. Shear stress is greatest on the upcurrent flank rather than at the crest, inducing ridge growth. Tide-built sand ridges on Nantucket Shoals are morphologically similar to the ridges built by storm currents on the Maryland coast, and like the latter they are oriented obliquely to flow.

INTRODUCTION

Environmental pressures on the Middle Atlantic Bight of the North American continental shelf have led to several large-scale studies designed to assess the impact of proposed petroleum development and of waste disposal practices on the environment. These studies, which have devoted as much attention to sediment transport dynamics as to sedimentary deposits, have resulted in some basic insights into continental shelf processes. In this paper, we will first summarize information concerning shelf circulation and sediment distribution in order to portray the setting. We will then report on continuing studies of sediment transport and summarize studies recently completed. We believe that many of these observations are pertinent to other shelves, including the North Sea and, as we report them, we will point out similarities and differences between the two shelf areas.

HYDRAULIC REGIME

Storm- and tide-dominated regimes

The Middle Atlantic Bight of the North American continental shelf is a north-east-trending shelf sector extending from Cape Hatteras, North Carolina, to Cape Cod, Massachusetts, a distance of approximately 600 km (Fig. 1). At its maximum extent off New York, the shelf is about 180 km wide. The shelf break climbs from about 150 m at Georges Bank to 50 m at Cape Hatteras.

To the north, on Nantucket Shoals, tidal currents dominate the hydraulic regime, and tidal currents also dominate the succession of large estuaries that divide the coast of the Middle Atlantic Bight into compartments: Long Island Sound, Hudson Estuary, and Delaware and Chesapeake Bays (Fig. 2). Tidal influence extends on the shelf 5–10 km seaward of the mouths of these estuaries, and several kilometres distance seaward of the many tidal inlets of the barrier island chains between estuaries.

Elsewhere in the Middle Atlantic Bight, storm-driven currents dominate the hydraulic regime. Figure 3 presents spectral analyses from current meter stations at four depths in the Middle Atlantic Bight prepared by Beardsley, Boicourt & Hansen (1976). Diurnal, inertial, and semi-diurnal peaks are visible. The diurnal and semi-diurnal peaks are most intense at site A, a shallow-water station near a tidal inlet. The inertial peak is most intense at sites C and D, near and over the shelf edge. Such inertial currents are responses to meteorological transients: fast-moving fronts or strong veering winds which rotate clockwise with near-inertial frequency and excite an inertial response. However, these peaks occur as spikes on the side of a 'red' spectrum: a spectrum whose kinetic energy density increases with decreasing frequency. Most of the energy in the curves is contained to the left, or low-frequency side of a shoulder in the curve at a period of 4 days. Inertial shelf currents occur primarily in the summertime when stratification insulates the flow from the bottom (Mayer, Hansen & Ortman, 1979). The wind stress spectrum is similarly a 'red' curve with a shoulder at 4 days.

In Fig. 3, the current records obtained at shelf sites B and C are too short for the computed spectra to indicate locations and magnitudes of lower-frequency peaks, hence the question mark. Some of the energy may be due to a leakage of energy from the deeper ocean on to the continental shelf. At site D over the slope, much of the energy is apparently associated with a 30-day peak that may be related to propagation of Rossby waves up the continental rise, Gulf Stream meanders, or formation of warm-core eddies.

More recent studies of the Middle Atlantic shelf flow field (Mayer *et al.*, 1979; Butman, Noble & Folger, 1979) shed more light on the nature of wind-induced flows. Mayer *et al.* (1979) analysed a three-year time series from mid-shelf stations. They concluded that the event-dominated nature of the regime tended to obscure the flow climate even in a time series of this duration. Energetic wind-driven transient current events rendered it impossible to determine a clear seasonal pattern in the sequence of monthly mean flows. During 1973–1976 such a transient reversed the normally south-westerly near-bottom flow for three months. However, a clear seasonal pattern occurs in the distribution of meso-frequency fluctuations. Storm-induced transient currents of 3–10 days duration appear prominently in winter records.

Beardsley *et al.* (1976) have presented a simple conceptual model of wind-forced events which will be summarized here together with additional information from Beardsley & Butman (1974), Boicourt & Hacker (1976), and Butman *et al.* (1979). During the winter, intense low-pressure disturbances form over the south-eastern United States and propagate up the eastern seaboard with characteristic periods of 2–4 days. The response of the Bight

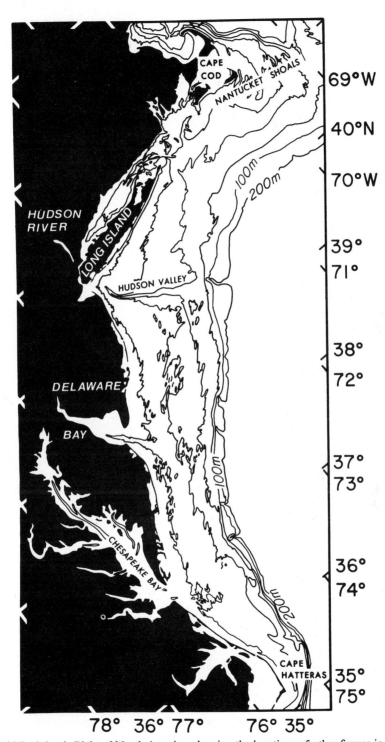

Fig. 1. Map of the Middle Atlantic Bight of North America, showing the locations of other figures in this paper.

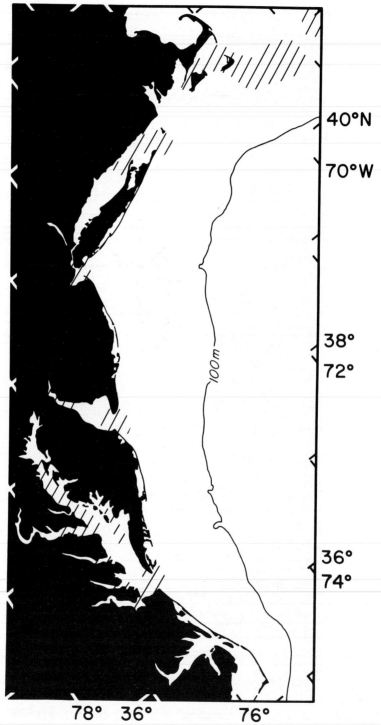

40°N
70°W
38°
72°
36°
74°
78° 36° 76°

Fig. 2. Tidal currents in excess of 25 cm sec⁻¹ in the Middle Atlantic Bight. Data from Redfield (1958).

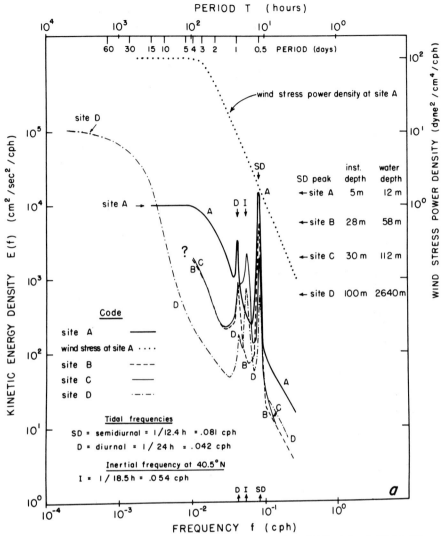

Fig. 3. Spectral analyses of current meter stations at four depths in the Middle Atlantic Bight. Measurements were made by Savonius rotor current meters moored at mid-depth. From Beardsley *et al.* (1976).

depends on storm trajectories. Storms passing landward of the Bight may create brief intense currents but there is little net displacement of water. However, storms which pass directly over the Bight may generate strong north-east winds paralleling the coast over the entire Bight. During such periods, Ekman transport to the right of the wind results in coastal set-up of up to 1 m, causing an onshore pressure gradient that is in rough geostrophic balance with a strong alongshore flow. Such a 'scale matching' storm will result in a coherent, in-phase, 'slab-like' flow over the entire Bight. Typical maximum daily

mean speeds of 40 cm sec^{-1} can produce alongshore fluid particle excursions of 40–80 km during the several days of the storm. Such events are imposed on a background of a low-frequency alongshelf flow of 5–20 cm sec^{-1} and cross-shelf tidal flow of 5–10 cm sec^{-1}.

The wave climate

Wind data from the Chief of Naval Operations (summarized in Emery & Uchupi, 1972, fig. 211) suggest that in the northern half of the Middle

Atlantic Bight during the winter, the largest storm waves will come from the north-east. In the southern portion, the largest waves will be associated with south-westerly winds more frequently than with north-easterly winds. In the summer, the predominant wave direction for both areas will be from the south-west, but storm waves will be from the north.

The frequency of observation of large waves increases northward along the Middle Atlantic Bight. During the winter, waves larger than 3·5 m are observed 7% of the time in deep water along the southern end of the Middle Atlantic Bight, while they are observed 12% of the time off Long Island and New England (Office of Climatology and Division of Oceanography, U.S. Navy, chart 137).

The lowest wave month in the Middle Atlantic Bight is typically August (mean significant wave height of 0·6 m at Atlantic City), with June and July close behind. The highest wave month is September (mean significant wave height of 1·35 m at Atlantic City), closely followed by the winter months of January, February and March. During the winter, however, the waves are locally generated by north-eastern storms and are markedly shorter than the long-period waves generated by offshore tropical storms during the fall (9·3 sec for September at Atlantic City).

In summary, an examination of wave observations along the Atlantic coast of the United States indicates that the influence of gravity waves will be felt over most of the shelf during the highest wave month (September). During the other high wave months (winter), wave orbital currents frequently exceed threshold for fine sand shoreward of 50 m (inner two-thirds of shelf) and occur in combination with strong wind-driven currents. The summer months are the lowest wave months on the average, and average wave conditions are only effective out to 30 or 40 m water depths.

MORPHOLOGY AND BOTTOM SEDIMENT DISTRIBUTION

Seaward of the shoreface, the Atlantic shelf surface is a gently sloping plane, with gradients of a degree or less. The major morphologic divisions of the shelf are the shelf valley complexes and the intervalley plains. These surfaces have been shaped primarily by nearshore marine rather than by subaerial processes during the Holocene transgression

(Swift et al., 1972a). Shelf valley complexes extend seaward from the major estuaries that separate the coastal compartments. These features are moulded into the upper surface of estuarine deposits which partially or completely fill the subaerial river valleys. Shelf valleys per se occur because the original river valley has not been completely buried (Hudson Valley, Fig. 1), or because the retreating estuary mouth has cut a scour channel into the upper surface of the estuarine deposits (Delaware Valley, Fig. 1). They may be accompanied by shoal retreat massifs; levee-like highs marking the retreat paths of littoral drift depositional centres at estuary margins (Swift et al., 1972b; Swift, 1973). The Delaware Shelf Valley and its associated massif can be traced back into, respectively, the scour trench at the mouth of Delaware Bay and the bay mouth sand shoals fed by littoral drift from the New Jersey coast (Swift, 1974). Shelf valleys may terminate in mid-shelf or shelf edge deltas, and some may be traced to submarine canyons.

The intervalley plains are smooth, or are moulded into sand ridges and swales. The ridges trend obliquely to the shoreline. They are characterized by heights of 10 m and spacings of 2–5 km, and may be traced into the shoreface where they currently appear to be forming in response to storm flows (Duane et al., 1972; Swift & Field, 1976). A similar array of sand ridges on Nantucket Shoals and Georges Banks are apparently responses to tidal flows.

The distribution of bottom sediments in the Middle Atlantic Bight has been summarized by Schlee (1973), Milliman, Pilkey & Ross (1972), Trumbull (1972), and Hollister (1973). The conclusions of these authors are presented in graphic form in Fig. 4. Over 85% of the floor of the Middle Atlantic Bight is covered by sand. The major exception to this generalization is the 'mud patch' of the southern New England shelf, south of Nantucket Shoals, an elliptical zone of silt and clay 85 km wide (Twichell, McClennen & Butman, 1981). Fine-grained deposits also occur in the Hudson Shelf Valley and in the lee of Assateague Shoals on the Maryland shelf. Thin mud lenses a few tens or hundred of metres in extent are common within 20 km of the beach. They occur in the swales between sand ridges, or in the troughs between sand waves and megaripples. They tend to grow in extent during the summer months and shrink or disappear during the winter (Swift et al., 1972a; Freeland, Swift & Young, 1978; Olsen et al., 1981).

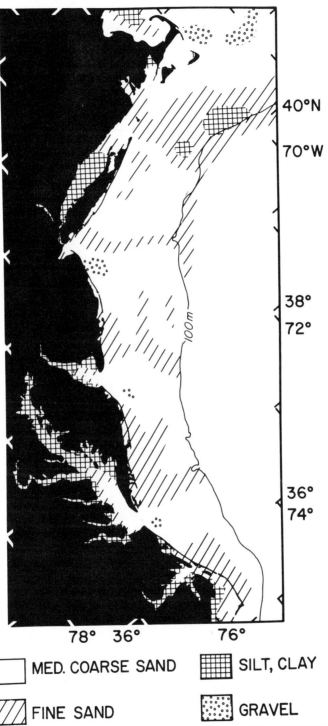

MED. COARSE SAND   SILT, CLAY

FINE SAND   GRAVEL

Fig. 4. Distribution of bottom sediments in the Middle Atlantic Bight. From Schlee (1973), Milliman *et al.* (1972), Trumbull (1972), and Hollister (1973).

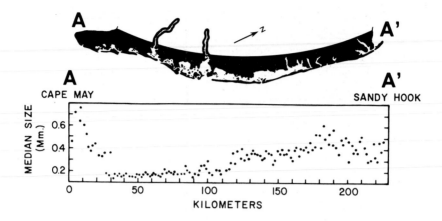

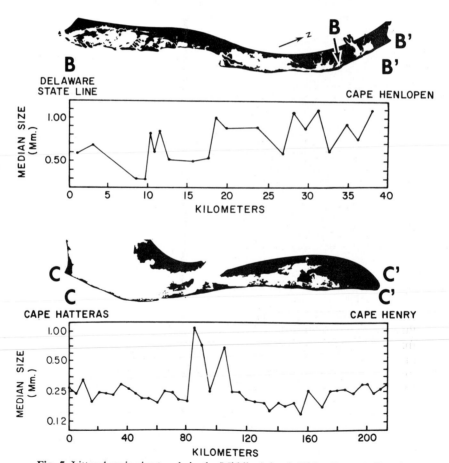

Fig. 5. Littoral grain-size trends in the Middle Atlantic Bight. From Swift (1976).

Gravel deposits are most extensive on the tide-swept upper surfaces of Georges Bank and Nantucket Shoals. They also occur in the troughs between sand ridges on the Long Island shelf (McKinney & Friedman, 1970), New Jersey shelf (Stubblefield *et al.*, 1975), Delmarva shelf (Swift & Field, 1981), and on the Virginia–North Carolina shelf (Swift *et al.*, 1977).

These deposits occur in a characteristic vertical sequence as well as in a characteristic horizontal pattern (Swift *et al.*, 1972a; Sheridan, Dill & Kraft, 1974). Gravel everywhere rests on a surface of Pleistocene or early Holocene age; it is a basal gravel or lag deposit generated as the transgressing Holocene shoreface cut through its own barrier island and lagoon deposits. In this respect it is comparable to the basal gravel on the shelf around the British Isles, described by Belderson & Stride. (1966). The sands occur as a discontinuous sheet ranging from less than 1 m to over 10 m in thickness. In the ridge topography of the intervalley plains, the swales between ridges form windows in which the basal gravel is exposed. Within the shelf valley complexes, the sand sheet thickens to 15 or 20 m to fill the shelf river valleys cut during the Pleistocene low stand of the sea.

The creation of this sand sheet is clearly associated with the erosional retreat of the shoreface, the sloping surface between 0 and 15 m water depth. Erosional shoreface retreat is seen as a response to rising sea-level; the agents of erosion are wave-driven currents in the surf zone and wind-driven currents immediately seaward of the surf (Swift *et al.*, 1970, 1978b).

The characteristic grain-size pattern of the successive shelf sectors has been imprinted by nearshore processes during the transgression. As a consequence of the prevailing north-easterly storm wave approach, littoral drift trends south-westerly between estuaries (except very near the mouths, where the drift moves into the estuary from both sides). Grain size decreases down-drift along each coastal compartment between estuary mouths, as far to the south-west as the Virginia–North Carolina Boundary, where the trend reverses (Fig. 5). The shelf grain-size distribution seen in Fig. 4 may be at least in part a consequence of the landward transgression of a littoral drift system acting as a grain-size fractionating machine. The sand sheet deposited during shoreface retreat at the eroding north end of each littoral drift system is coarse, thin, and locally lacking, exposing gravel. At the southern aggrading end, much fine sand has leaked out of the retreating surf zone on to the shelf, resulting in a relatively thick, fine sand sheet. Sand wave patterns and local grain-size gradients suggest that south-westerly, coast-parallel transport seen along each coastal compartment also occurs on the inner shelf seaward of the littoral drift, in response to wind-riven storm flows (Swift *et al.*, 1970, 1978a, fig. 5, 1979, fig. 8).

CALCULATIONS AND OBSERVATIONS OF SEDIMENT TRANSPORT

Observations on the Inner Long Island Shelf

The response of the shelf floor in the Middle Atlantic Bight to the flow regime described on previous pages has been studied by Lavelle *et al.* (1976; 1978a, b, 1981), Gadd, Lavelle & Swift (1978), and Butman *et al.* (1979). These studies have revealed the intermittent nature of sediment transport in the Middle Atlantic Bight: flows capable of transporting fine sand occur once every 10–12 days, with durations of several hours to several days (e.g. Fig. 6).

Transport on the shelf floor is markedly seasonal in nature. Storms are only slightly more frequent in the period November–March than in the period April–October, but the associated winds and the geostrophic currents induced by them are markedly more intense.

In Fig. 7, curves are plotted for the monthly percentage surplus over the 16 cm sec^{-1} threshold, using the data of Mayer *et al.* (1979), Beardsley *et al.* (1976), Lavelle *et al.* (1978a), as well as unpublished data. Both the inner and outer shelf curves reveal a November–March maximum. As a consequence of the dependence of sediment transport on the cube of fluid velocity, extreme winter storms dominate transport on the middle Atlantic shelf. During the 135-day observation period of Lavelle *et al.* (1976, 1978b), over 95% of the total transport occurred during a single major two-day storm. The importance of major storms in shelf sediment transport has been previously noted by Smith & Hopkins (1972).

During summertime observations by Butman *et al.* (1979) variations in suspended matter occurred, but there was no obvious bottom scour and ripples did not form. Increases in suspended matter concentration could be correlated only in part with observed currents, and may have been due in part to biological activity or advection.

Calculations of inner shelf sediment transport rates by Gadd *et al.* (1978), using the formulae of

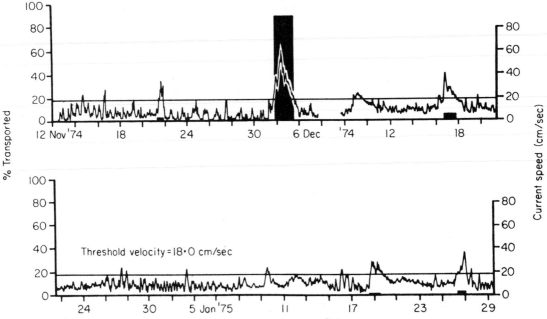

Fig. 6. Near-bottom current velocities on the Long Island inner shelf during the fall and winter of 1974. Measurement by Anderaa current meter moored 2 m off the seafloor. From Lavelle *et al.* (1978).

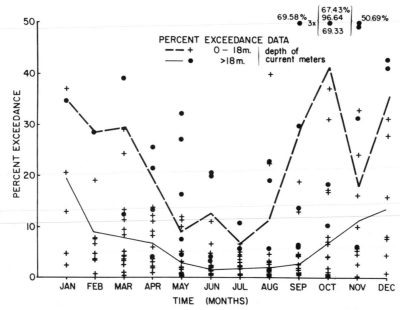

Fig. 7. Percentage surplus by month of 16 cm sec^{-1} threshold in response to mean flow at 16 stations on the New York–New Jersey continental shelf. Dashed lines are mean values of plotted samples for each of two sample sets. From Vincent *et al.* (1981).

Bagnold (1963), Einstein (1950), and Yalin (1963) and a drag coefficient of 3×10^{-3}, yield values ranging from $2 \cdot 8 \times 10^{-3}$ to $2 \cdot 2 \times 10^{-1} \, \mathrm{g \, cm^{-1} \, sec^{-1}}$. These calculations do not take into account the role of high-frequency, wave-induced bottom motions in entraining sediment into the mean flow, yet the rates are comparable to those computed by Lavelle *et al.* (1978) on the basis of radioisotope tracer studies. The analysis of Lavelle *et al.* (1978) yielded lower bounds for the average sediment mass flux of $3 \cdot 2 \times 10^{-3} \, \mathrm{g \, cm^{-1} \, sec^{-1}}$ and $1 \cdot 7 \times 10^{-1} \, \mathrm{g \, cm^{-1} \, sec^{-1}}$ for typical and extreme winter storm activity.

Sediment resuspension by storm waves

Simultaneous observations of fluid motion and sediment concentration by Lavelle *et al.* (1978), (see also Fig. 8), and Lesht *et al.* (1981) suggest that wave power is a good predictor of sediment concentration in the bottom boundary layer. Figure 9 presents the relationship between these two variables during the storm of 20 October 1976 (see also the effect of the storm in Fig. 8). As the storm intensifies, the concentration–wave energy values rise along the lower half of a hysteresis loop; during the decline of the storm, the values descend along the upper half. Higher values of wave power are needed to resuspend sediment from the bottom than to keep it in motion once it has been entrained in the flow. Concentration–wave power curves at different sites have similar patterns: after a threshold value has been exceeded the concentration increases linearly with increasing wave power (straight lines, Fig. 9). However, the height of the curve above the origin varies from station to station. The controlling variable appears to be the mud content of the muddy sand bottom at the site examined. Lesht *et al.* conclude that bottom wave motions stir the sand bed to a depth proportional to the root mean square velocity. After the threshold has been exceeded, the suspended sediment concentration is equal to a constant plus an increment proportional to the admixture of fine sediment in the bottom, times the rms wave velocity:

$$C = C_0, \quad \sigma_\omega < \sigma_{\mathrm{Th}}$$
$$C = KF(\sigma_\omega - \sigma_{\mathrm{Th}}) + C_0, \quad \sigma_\omega \geqslant \sigma_{\mathrm{Th}}$$

where F is the percentage of fine sediment in the bottom, C_0 is the ambient concentration, and σ_ω is the standard deviation of wave orbital velocity (cm sec^{-1}).

Analysis of the near-bottom concentration and velocity data for coherence indicates that the response of the bottom to flow is relatively slow (of the order of a few seconds) and near-bottom concentration and fluid velocity values are not coherent at surface wave frequencies (Lavelle *et al.*, 1978; Clarke *et al.*, 1981). As a consequence, somewhat more esoteric relationships emerge. During storms concentration and velocity are coherent in the 100–150 sec period band due to the beating of waves of similar frequency. During the waning periods of storms, coherence at surface wave periods occurs as a graded suspension develops in the bottom nepheloid layer and oscillates up and down at surface wave frequencies past the nephelometer.

Sediment transport on the shoreface

Recent observations indicate something of the complex nature of the fluid velocity and sediment transport fields over the shoreface of the Middle Atlantic Bight. The sector in question is the more steeply dipping, innermost sector of the shelf floor, from the beach to a water depth of 15 m some 3–5 km from the beach. Calculations by Gadd *et al.* (1978) indicate that across this nearshore zone, sediment transport rates tend to decrease as depth and distance from shore increase. The calculations of Gadd *et al.* (1978) are based on Savonius rotor meters, which do not clearly distinguish between the slowly varying components of flow and high frequency motions due to surface wind waves; the nearshore transport gradient that they observed is probably caused by the increasing frequency of occurrence of wave orbital currents as the bottom shoals towards the beach.

A comparable decrease in transport with increasing depth does not occur between the outermost nearshore stations of Gadd *et al.* (1978) and Lavelle *et al.* (1976, 1978), and the outer shelf stations of Butman *et al.* (1978). The intensity of the geostrophic mean flow component induced by storms is independent of depth, and during many storms waves are able to stir the seafloor at all shelf depths. The main difference between the nearshore stations and the outer shelf stations sppears to be that 'fair-weather' wave conditions sometimes induce transport at the former, while only storms affect the rest of the shelf surface. A comparison of monthly percentage surpluses (Fig. 7) indicates that the 16 cm sec^{-1} threshold is exceeded more frequently in shallow water (less than 18 m). Anomalies on the shallow water surplus curve are due to data collected

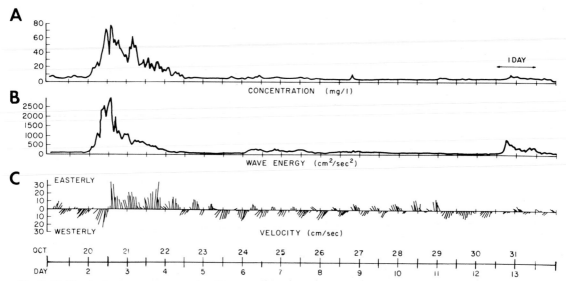

Fig. 8. Fluid velocity, wave energy, and sediment concentration, measured 100 cm off the bottom at the 10 m isobath on the Long Island inner shelf. Measurements were made by means of a Marsh-McBirney electromagnetic current meter and a Monitek nephelometer-transmissometer, mounted 100 cm off the seafloor. Modified from Lavelle *et al.* (1978).

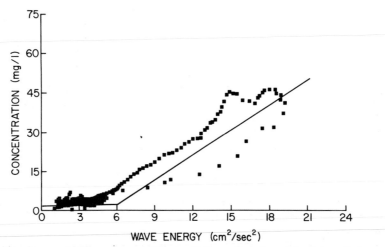

Fig. 9. Relationship of concentration measured 100 cm off the seafloor by a Monitek nephelometer-transmissometer to rms wave energy measured by an electromagnetic current meter. Data collected 100 cm off the seafloor. From Clarke *et al.* (1981).

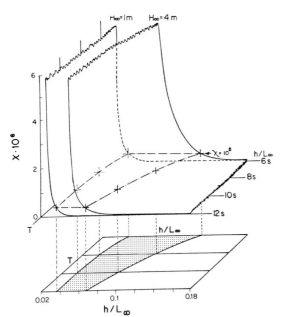

Fig. 10. Three dimensional plot of sediment transport parameter (χ) against depth to wavelength ratios (h/L_∞) and wave period (T) from 1 and 4 m deep-water wave heights (H). The horizontal projection of the zone of maximum inflection is plotted in h/L_∞ versus T space. From Niedoroda (1980).

in the proximity of tidal inlets (September–November), and data collected from very nearshore (10 m water depth; April–May). Radioisotope studies (Lavelle *et al.*, 1976, 1978) reveal a diffusive pattern of fair-weather transport in shallow water, in which labelled sediments are smeared out over areas of 100 m or less in radius by peak tidal flows during periods of wave activity. These patterns contrast with the highly elongate, shore-parallel patterns that develop during major storms. Such patterns may extend up to 1500 m.

A more detailed analysis by several of us (Niedoroda & Swift, 1981) of the Long Island shoreface shows that shoreface sediment transport occurs in two contrasting zones. Calculations of sediment transport by asymmetric wave orbital currents for field conditions were undertaken, using a formulation based on Bagnold (1963). If the threshold grain entrainment velocity is small compared to the maximum near-bottom orbital velocity the average mass flux of bed load per unit width of wave crest χ_s can be approximated by:

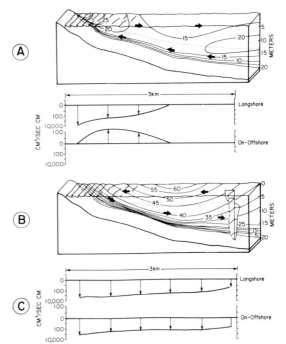

Fig. 11. (A) Coastal flow field at Tiana Beach, Long Island, 26 August 1976. Block diagram at top illustrates eastward longshore current component and shore-normal component (arrows are approximately 10 cm sec^{-1}). Figure near right side of block diagram (b) shows profiles of longshore (dashed) and shore-normal (solid) current components measured at Shelton Spar. The two diagrams below show computed bed-load flux. (B) Coastal flow field on 24 August. Transport calculations are for composite of mean flow fields of 24 August and wave regime of 26 August. Modified from Niedoroda (1980).

$$\chi_s = \int_t^{t+T} Ku_s^3\, dt = \frac{9}{8} K \frac{(\pi H)^4}{LT^3} \frac{1}{\sinh^6 kh} + \frac{189}{512}$$
$$K \frac{(\pi H)^6}{L^3 T^3} \frac{1}{\sinh^{12} kh},$$

where t is the time, T is the wave period, K is a coefficient of proportionality, H is the wave height, L is the wavelength, h is the local depth, k is the wavenumber, and u_s is the second-order Stokes horizontal wave orbital component (Niedoroda, 1980). The calculations indicate that the relative mass flux parameter (χ_s) for a given wavelength has small values at large depths, but undergoes a sharp increase as the depth becomes smaller (Fig. 10). For the wave climate of the Long Island coast, the transition occurs over the depth range of 5–10 m. The

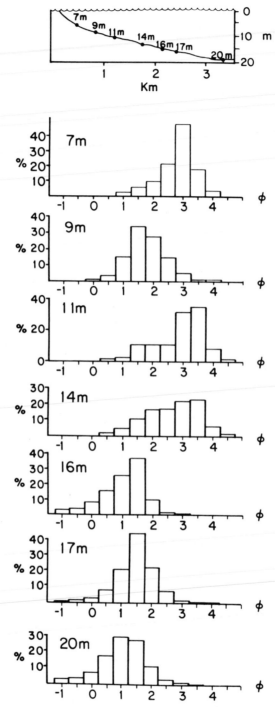

Fig. 12. Grain-size frequency distributions from successive levels on the Long Island shoreface at Tiana Beach, showing differing responses of bottom to differing hydraulic regimes.

diagram indicates that at depths less than approximately 10 m, wave orbital asymmetry is strong and therefore wave-driven sediment transport is intense. At depths greater than 10 m, wave orbital velocity may still be high, but orbital asymmetry is weak, and so therefore is wave-driven net sediment transport. Waves on the shallow inner shelf below this depth play an important role in entraining the sediment to be transported by the mean flow, but they are not in themselves transporting agents.

Observations of transport on the upper and lower shoreface of the Long Island shoreface have been recently described by Niedoroda & Swift (1981). The data were collected along a 3 km, shore-normal transect at Tiana Beach, Long Island south shore (Fig. 1). Current velocities were measured at the inshore end of the transect by means of electro-magnetic current meters mounted 1 m off the bottom in approximately 3 m of water. Data at the offshore end were collected by means of a 'Shelton Spar', a vertical array of electromagnetic current meters (see Scott & Csanady, 1976). Data in between were collected by means of boat stations using a Bendix Q-15 current meter. See Niedoroda (1980), for details of the experimental design.

Observations were made during part of August 1976. During this period, two different regimes were observed, which are believed to be the two contrasting regimes which characterize the shoreface. Figure 11 (A) presents the situation on 26 August. There was no wind on that day. Swells with heights of 1·5 m and a period of 13 sec made an 8° breaker angle opening to the west. The wave regime created a 40 cm sec^{-1} easterly coast-parallel current that extended as far seaward as the 10 m isobath. Beyond 10 m a weak, upwelling, westerly tidal flow attained a velocity of 20 cm sec^{-1}.

Sediment transport induced by the combined regime of wave orbital currents and slowly varying currents has been calculated using the formulation of Madsen & Grant (1976). On the 26th, the regime of shoaling and breaking waves on the upper shoreface resulted in onshore and alongshore sediment transport, but the weak flow over the lower shoreface was incapable of inducing significant sediment movement. This pattern of fluid motion and sediment transport appears to be typical of the summer period of fair weather, as well as those intervals during the winter between storm periods.

During an earlier period (24 August), mild storm conditions were observed (Fig. 11B). The wind on 24 August had changed during the night from the nearly calm conditions of the previous day to a 5·5 m sec^{-1} wind from the east-north-east. The resulting landward Ekman drift of the surface layer resulted in a well-defined coastal jet (Csanady, 1977), with a maximum surface velocity of 60 cm sec^{-1}. The regime was downwelling, with offshore bottom flow. Waves, however, remained small: significant wave height was 65 cm and significant wave period was 9·2 sec. Calculated sediment transport during this single summer storm observation was limited to the upper shoreface, as in the case of the fair weather observations, despite the 60 cm sec^{-1} velocity. It is apparent that the strong mean flows by themselves are not capable of significant sediment transport.

Since the summer wave regime was too weak to cause significant sediment transport on the shoreface at the time when the most interesting flow event was measured, the wave conditions of 25–26 August have combined with the current conditions of 26 August. These long swells originated at a great distance from the study area and could have arrived a day earlier without significantly changing the wind-driven flow over the shoreface. The resulting calculations, undertaken in 4 h steps, show that sediment transport spread from the seaward edge of the surf across the entire shoreface as the coast jet strengthened and developed. The final stage is shown in Fig. 11 (B). Interestingly, the magnitude of the offshore component of bed-load transport is the same as the alongshore component.

The response of the seafloor to the flow in the spatially contrasting regions of wave-dominated upper shoreface and mean-flow-dominated lower shoreface is revealed by grain-size frequency distributions from the study area (Fig. 12). Size distributions out to 12 m primarily are fine and well sorted and become increasingly fine grained with depth. These materials are best understood as surf fallout, reworked and graded by the landward-directed forces of asymmetrical wave orbital currents (Swift, 1976, pp. 264–267). A coarse admixture is present in most of these upper shoreface samples and dominates the 7 m sample. It re-emerges as the dominant mode on the lower shoreface seaward of 12 m water depth. It is interpreted here as a lag, developed in response to the alongshore and offshore mean flows associated with north-easter storms. Similar studies of the North Carolina coast (Swift *et al.*, 1971) and of the Netherlands coast (Van Straaten, 1965, in Swift, 1976) indicate that the two facies bear a systematic stratigraphic relationship: a seaward-thinning

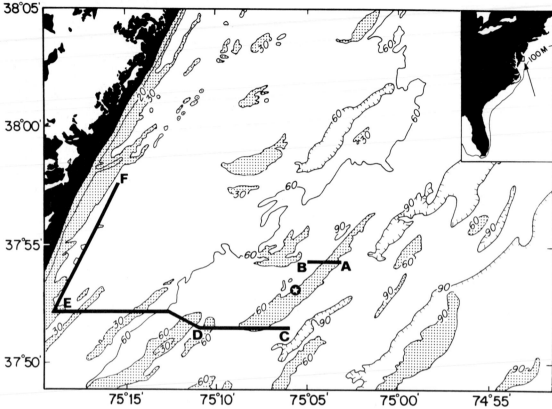

Fig. 13. Portion of the Maryland inner shelf from which the statistics of Table 1 were drawn. Heavy lines are lines of section in Fig. 14; star indicates location of current meter. From Swift & Field (1981).

Table 1. Characteristics of shoreface, nearshore and offshore ridges

Parameter	Shoreface ridge (12 ridges)	Nearshore ridge (eight ridges)	Offshore ridge (15 ridges)
Mean slope (°)	1·5	1·0	0·5
Steepest slope (°)	2·5	2·0	7·0
Mean asymmetry (landward slope: seaward slope)	1:1	1:2	1:5
Mean aspect ratio (landward slope: seaward slope)	9:1	6:1	3:1
Maximum cross-sectional area ($\times 10^3$ m^2)	87	187	481

blanket of surf zone fallout pinches out at about 10 m, exposing a lag developed on the eroding surface of older strata.

An analysis of Moody (1964, in Swift, 1976) of a 33-year time series of bathymetric transects of the Delaware coast has revealed how variation in the importance of storm and fair weather regimes through time has driven the erosional retreat of the Delaware shoreface. Moody reports that the shoreface steepens over a period of years towards the ideal wave-graded profile. The steepening process is terminated by a major storm. The blanket of wave-graded surf fallout is stripped off, the shoreface gradient is reduced, and a significant landward translation of the profile occurs. Shoreface erosion during the retreat process is largely balanced by aggradation of the adjacent inner shelf.

STUDIES OF LARGE-SCALE BEDFORMS

Storm-built sand ridges, Maryland coast

Recent observations of storm flows and seafloor response on the Maryland coast (Fig. 13) have led to a reappraisal of theories for the maintenance of sand ridges (Swift & Field, 1981). Important clues to ridge genesis are afforded by regional patterns of ridge morphology, by the relationship between grain size and bottom morphology, by analysis of topographic time series, and by limited flow observations.

Morphometric analysis of sand ridges on the Maryland inner shelf indicates that ridge characteristics change systematically as ridges are examined at successively greater distances from shore (Table 1). It is possible to erect three ridge classes. *Shoreface ridges* are defined by contours that are finger-like extensions of the coast-parallel contours of the shoreface. *Nearshore* and *offshore ridges* are defined by closed contours. Nearshore ridges are shallower than 15 m; offshore ridges are deeper than this value.

Measurements of ridge parameters indicate that the class mean for maximum ridge slope diminishes from 1·5 for shoreface ridges to 1·0 for nearshore ridges to 0·5 for offshore ridges. Shoreface ridges have steeper landward slopes on their landward ends, and are symmetrical or have steeper seaward

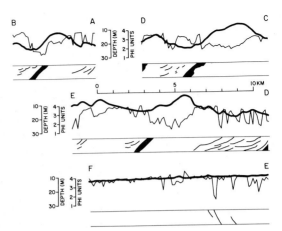

Fig. 14. Bathymetric profiles (heavy lines), grain-size profiles (lighter lines). See Fig. 13 for location. Schematic representation of side-scan sonar record is presented beneath curves. Black zones are coarse sediment and shell hash at base of ridge's landward flank. Thin lines are sand wave crestlines. From Swift & Field (1981).

slopes at their seaward ends. Seaward asymmetry increases for nearshore and offshore ridges with distance from shore. Cross-sectional area increases. The ridges are moulded into the lag sand deposit that formed as the shoreface underwent erosional retreat in response to Holocene sea-level rise, and the onshore–offshore morphologic sequence of

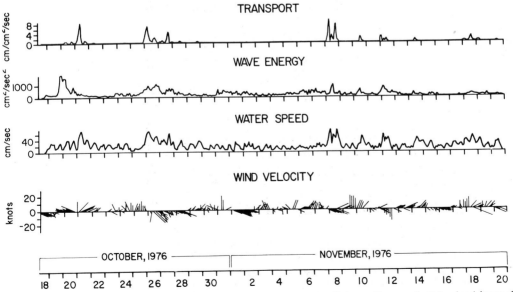

Fig. 15. Data from electromagnetic current meter positioned 100 cm off the seafloor on the Maryland inner shelf. See star in Fig. 13 for location. From Swift & Field (1981).

ridges must therefore constitute an evolutionary sequence.

Grain-size gradients associated with the Maryland and Delaware ridge topographies have been noted by Moody, 1964 (in Swift, 1976) and by Palmer, Wilson & Greala (1975) (in Swift *et al.*, 1978b). On shoreface and nearshore ridges, grain size and topographic variation normal to shore are both sinusoidal but are 90° out of phase. Grain size is coarsest not on the ridge crest where the bottom is highest, but on the landward flank. A transect of grab samples and side-scan sonar records show that this relationship extends for 20 km seaward through the zones of nearshore and offshore ridges (Fig. 14). A 43-year time series of morphologic change on the outer-

most ridge of this transect shows that the landward-facing flank has undergone erosion, while the seaward-facing flank has undergone aggradation. Near-bottom current measurements indicate that during a 1 month period in the fall of 1976, sustained storm flows in excess of 40 cm sec^{-1} occurred seven times (Fig. 15). Sediment transport, calculated using a Bagnold-type equation described by Gadd *et al.* (1978) and the threshold criteria described by Madsen & Grant (1976), reveal a mean transport direction southward and slightly offshore relative to the trend of the ridge crest.

The observed relationship between grain size and bathymetry constitutes strong evidence that the ridges are initiated as responses to the strong mean

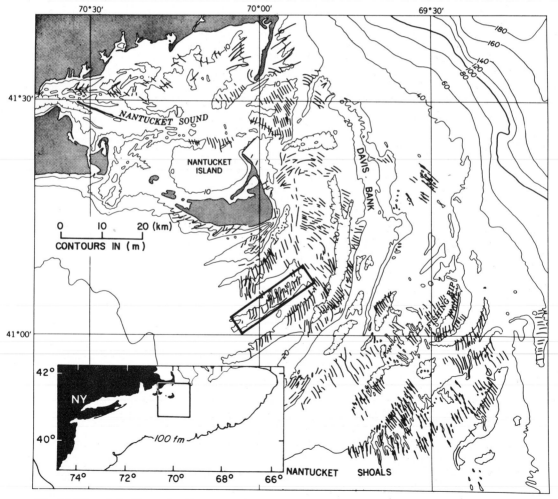

Fig. 16. Map of the Nantucket Shoals sector of the Atlantic continental shelf. Modified from Uchupi (1968).

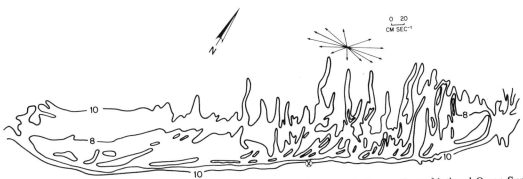

Fig. 17. Map of the Nantucket study area; showing portion 10 fathoms or shallower. From National Ocean Survey bathymetric map. Tidal ellipse is based on a 12·5 h portion of electromagnetic current meter deployed 100 cm from the seafloor. From Mann & Swift (1981).

flows of the sort observed on the lower shoreface (Fig. 11B). In the stability analysis of Smith (1970), bedforms growing in response to such sustained flows would experience maximum shear stress on the up-current flanks. There would be a positive shear stress gradient up this flank to the point of maximum stress, and therefore erosion. Fines would be winnowed out, leaving a coarse residue. The fine sediment would be transported to the crest and downslope side. In this sector, downcurrent from the point of maximum shear stress, the shear stress gradient would be negative and deposition would occur. Thus any initial bottom high would tend to grow.

Smith's model does not directly specify the scale and spacing of the sand ridges or their characteristic oblique orientation with respect to the shoreline. Conceivably, the spacing of initial bottom irregularities that trigger ridge growth may be a response to upper shoreface processes such as those responsible for 'migrating sand humps' (Bruun, 1954; Dolan, 1971). The oblique angle of the ridge can be tentatively attributed to a modification of flow-substrate interactions by the shoreface slope. In the shallower nearshore water bottom-wave orbital-velocity currents are more intense and more efficient in entraining sediment into the mean flow. The inshore end of the ridge will tend to migrate faster for this reason. As the ridge grows, constriction of the overlying water column and consequent flow acceleration will be more intense at the inshore end, further increasing the migration rate of this end. Furthermore, ridge height is less at this end than offshore. There is less volume of sand in the ridge per unit length of crestline, and the migration rate of the inshore end will be less for this third reason.

A high migration rate at the inshore end of the ridge would skew obliquely with respect to the orientation of the coast. However, the extreme regularity of ridge orientations over large areas of the Middle Atlantic shelf (Duane *et al.*, 1972) suggests that a further factor is responsible: the relationship between the rate of ridge migration and the rate of shoreface retreat. If these are subequal (probably of the order of 1 km per 1000 years) then the shoreface will have retreated out from under any given ridge segment before it has migrated very far alongshore. Once on the inner shelf floor, where wave-assisted sediment transport is greatly diminished, the ridge segment would be 'frozen' into place. According to this model, long inner shelf ridges such as those of Fig. 13 would be trails left by the movement of a ridge-generating perturbation of the upper shoreface.

Tide-built sand ridges, Nantucket shoals

A second major class of large-scale bedforms occurs on Nantucket Shoals (Fig. 16). It has been suggested that the array of shoal-transverse ridges is analogous to a ridge sequence that extends seaward from the East Anglian coast in the southern bight of the North Sea (Swift, 1975a). The distribution of ridges with respect to Nantucket Island in Fig. 17 indicates that, as in the case of the Maryland Coast ridges, the Nantucket ridges must be an evolutionary sequence, formed as the south-west coast of ancestral Nantucket Island retreated along the crest of the shoals to its present position (Swift, 1976).

A current meter deployed for 1 month near a

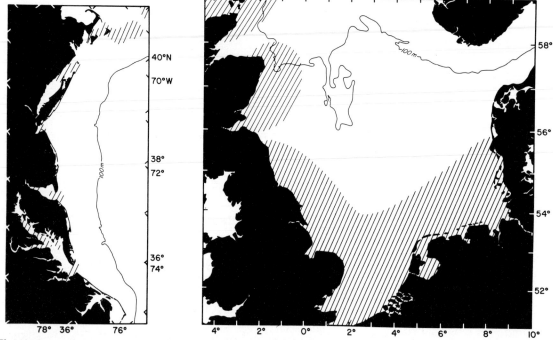

Fig. 18. Comparison of Middle Atlantic Bight and North Sea. Ruled areas experience tidal velocities in excess of 25 cm sec⁻¹. Data from Redfield (1958) and Houbolt (1968).

ridge crest reveals that the tide-built Nantucket ridges, like the storm-built Maryland ridges, are oriented obliquely to flow (Fig. 17). A side-scan sonar survey indicates that sand wave asymmetry reverses across the ridge, so that the ridge constitutes a sand circulation cell, or closed loop in the sand transport path. An array of large (5 m high) sand waves lies on the gentle landward flank of this ridge. Sand wave crests tend to lie nearly normal to the ridge crest at their deep ends but curve into near parallelism with the ridge crest as the latter is approached. These relationships suggest that mean flow at the ridge crest is nearly crest-normal, perhaps as a response of the flow to the rule of conservation of vorticity. Grain-size profiles indicate that Smith's (1970) model for bedform genesis may be applicable here also.

Huthnance (1972) has presented a model for the maintenance of tidal sand ridges oriented obliquely with respect to the tidal ellipse. The reason for the orientation is not explained by the model. Perhaps, as suggested for the Maryland coast ridges, orientation is fixed during the nearshore formative period.

MIDDLE ATLANTIC BIGHT VERSUS THE NORTH SEA

The North Sea and the Middle Atlantic Bight of North America exhibit obvious differences of shape and hydraulic regime; the Middle Atlantic Bight is a crescentic water body while the North Sea is a box-like re-entrant; strong tidal flows are more widespread in the latter than the former (Fig. 18). However, the sediment transport observations summarized in this paper and the observations presented in companion papers in this volume show that there are many similarities in the respective sedimentary regimes. In both cases, the sedimentary regime is autochthonous, in that sediments have been derived primarily from the process of erosional shoreface retreat rather than from river discharge (see Belderson & Stride (1966) for a discussion of the generation of this facies during a transgression of the British Isles; Swift *et al.* (1972b) for an analysis of the role of erosional shoreface retreat in the Middle Atlantic Bight). On both shelves, a shoreface zone of wave-graded sand can be identified. Nearshore flow patterns are seen as critical to the release of the debris of nearshore erosion to offshore transport pathways (this paper; Jago, 1981). On both shelves, grain-size gradients indicative of transport paths for sands released by the shoreface retreat process may be defined (Kenyon & Stride, 1970; Swift *et al.*, 1977, fig. 6). On both shelves, some of the sand released by erosional shoreface retreat is trapped in sand ridges that may be maintained by either storm currents (Swift *et al.*, 1978b; examples for both shelves) or tidal currents (this paper; Kenyon *et al.*, 1981). On both shelves, deposits of fine sediments are locally spreading over older sands (Twichell *et al.*, 1981; Creutzberg & Postma, 1979).

ACKNOWLEDGMENTS

The paper was prepared in the course of a study of sediment transport in the New York Bight funded by the Office of Marine Pollution Assessment, National Oceanic and Atmospheric Administration. Some of the data were kindly provided by Drs J. Scott and G. Csanady from the COBOLT Project of Brookhaven National Laboratory, funded by the Department of Energy. Portions of the research were funded by ONR contract N00014-76-C-0145.

REFERENCES

BAGNOLD, R.A. (1963) Mechanics of marine sedimentation. In: *The Sea* (Ed. by M.N. Hill), pp. 507–582. Interscience Publications, New York.

BEARDSLEY, R.C., BOICOURT, W.C. & HANSEN, D.V. (1976) Physical oceanography of the Middle Atlantic Bight. *Spec. Symp. Am. Soc. Limnol. Oceanogr.* **2**, 20–33.

BEARDSLEY, R.C. & BUTMAN, B. (1974) Circulation on the New England continental shelf: response to strong winter storms. *Geophys. Res. Lett.* **1**, 181–184.

BELDERSON, R.H. & STRIDE, A.H. (1966) Tidal current fashioning of a basal bed. *Mar. Geol.* **4**, 237–257.

BOICOURT, W.C.P. & HACKER, P.W. (1976) Circulation of the Atlantic continental shelf of the United States, Cape May to Cape Hatteras. *Mem. Soc. R. Sci. Liège*, **10**, 187–200.

BRUUN, P. (1954) Migrating sand waves or sand humps, with special reference to investigations carried out on the Danish North Sea coast. In: *Fifth Conf. Coastal Engineering* (Ed. by J.W. Johnson), pp. 269–295. Grenoble, France.

BUTMAN, B., NOBLE, M. & FOLGER, D.W. (1979) Long-term observations of bottom current and bottom sediment movement on the Mid-Atlantic continental shelf. *J. geophys. Res.* **84**, 1187–1205.

CLARKE, T.L., FREELAND, G.L., LESHT, B., SWIFT, D.J.P. & YOUNG, R.L. (1981) Sediment resuspension by wave orbital motion on the Long Island inner shelf. *Geophys. Res. Lett.* (in preparation).

CREUTZBERG, E. & POSTMA, H. (1979) An experimental approach to the distribution of mud in the Southern North Sea. *Neth. J. Sea Res.* **13**, 99–116.

CSANADY, G.T. (1977) The coastal jet conceptual model in the dynamics of shallow seas. In: *The Sea* (Ed. by E.D. Goldberg), pp. 117–144. Wiley, New York.

DOLAN, R. (1971) Coastal landforms: crescentic and rhythmic. *Bull. geol. Soc. Am.* **82**, 177–180.

DUANE, D.B., FIELD, M.E., MIESBERGER, E.P., SWIFT, D.J.P. & WILLIAMS, S.J. (1972) Linear shoals on the Atlantic Continental Shelf, Florida to Long Island. In: *Shelf Sediment Transport, Process and Pattern* (Ed. by D.J.P. Swift, D.B. Duane and O.H. Pilkey), pp. 477–498. Dowden, Hutchinson & Ross, Stroudsburg, Pennsylvania. 656 pp.

EINSTEIN, H.A. (1950) The bedload function for sediment transportation in open channels. *Tech. Bull. U.S. Dep. Agric.* **1026**, 71 pp.

EMERY, K.O. & UCHUPI, E. (1972) Western North Atlantic Ocean: Topography, rocks, structure, water, life and sediments. *Mem. Am. Ass. Petrol. Geol.* **17**, 253 pp.

FREELAND, G.L., SWIFT, D.J.P. & YOUNG, R.A. (1979) Mud deposits near the New York Bight dumpsites: origin and behavior. In: *Proc. SEPM Symp. Geological Aspects of Ocean Waste Disposal* (Ed. by G. Gross, and H. Palmer), June 12–66. Society of Economic Mineralogists and Paleontologists, Tulsa. 668 pp.

GADD, P.E., LAVELLE, J.W. & SWIFT, D.J.P. (1978) Calculations of sand transport on the New York shelf using near-bottom current meter observations. *J. sedim. Petrol.* **88**, 239–252.

HOLLISTER, C.D. (1973) Atlantic continental shelf and slope of the United States: texture of the sediments from northern New Jersey to southern Florida. *Prof. Pap. U.S. geol. Surv.* **529**, 23 pp.

HOUBOLT, J.H.C. (1968) Recent sediment in the southern Bight of the North Sea. *Geol. Mijnb.* **47**, 245–273.

HUTHNANCE, J.M. (1972) Tidal current asymmetries over the Norfolk sandbanks. *Est. Coastal Mar. Sci.* **1**, 89–99.

JAGO, C.F. (1981) Sediment responses to waves and currents, North Yorkshire Shelf, North Sea. In: *Holocene Marine Sedimentation in the North Sea Basin* (Ed. by S.-D. Nio *et al.*). *Spec. Publs int. Ass. Sediment.* **5**, 283–301. Blackwell Scientific Publications, Oxford. 524 pp.

KENYON, N.H. *et al.* (1981) Offshore tidal sand-banks as indicators of net sand transport and as potential deposits. In: *Holocene Marine Sedimentation in the North Sea Basin* (Ed. by S.-D. Nio *et al.*). *Spec. Publs int. Ass. Sediment.* **5**, 257–268. Blackwell Scientific Publications, Oxford. 524 pp.

KENYON, N.H. & STRIDE, A.H. (1970) The tide-swept continental shelf sediments between the Shetland Isles and France. *Sedimentology*, **14**, 159–173.

LAVELLE, J.W., BRASHEAR, H.R., CASE, F.N., CHARNELL, R.L., GADD, P.E., HAFF, K.W., HAN, G.A., KUNSELMAN, C.A., MAYER, D.A., STUBBLEFIELD, W.L. & SWIFT, D.J.P. (1976) Preliminary results of coincident current meter and sediment transport observations for wintertime conditions, Long Island inner shelf. *Geophys. Res. Lett.* **3**, 97–100.

LAVELLE, J.W., SWIFT, D.J.P., GADD, P.E., STUBBLE-FIELD, W.L., CASE, F.N., BRASHEAR, H.R. & HAFF, K.W. (1978b) Fair weather and sand storm transport on the Long Island, New York, inner shelf. *Sedimentology*, **25**, 823–842.

LAVELLE, J.W., YOUNG, R.A., SWIFT, D.J.P. & CLARKE, T.L. (1978a) Near-bottom sediment concentration and fluid velocity measurements on the inner continental shelf, New York. *J. geophys. Res.* **83**, 6052–6061.

LESHT, B., CLARKE, T.L., YOUNG, R.A. & SWIFT,, D.J.P. (1981) Sediment resuspension by wave orbital pattern: data from the Long Island inner shelf. *Geophys. Res. Lett.* (in press).

McKINNEY, T.F. & FRIEDMAN, G.M. (1970) Continental shelf sediments of Long Island, New York. *J. sedim. Petrol.* **40**, 213–248.

MADSEN, O.S. & GRANT, W.D. (1976) Quantitative description of sediment transport by waves. *Proc. 15th Coastal Eng. Conf.* pp. 1093–1112. American Society of Civil Engineers.

MANN, R.G. & SWIFT, D.J.P. (1981) Size classes of flow-transverse bedforms in a subtidal environment, Nantucket Shoals, North Atlantic Shelf. *Geomar. Letts* (in press).

MAYER, D.E., HANSEN, D.V. & ORTMAN, D.A. (1979) Long-term current and temperature observations on the middle Atlantic shelf. *J. geophys. Res.* **84**, 1776–1792.

MILLIMAN, J.D., PILKEY, O.H. & ROSS, D.A. (1972) Sediments of the continental margin off the eastern United States. *Bull. geol. Soc. Am.* **83**, 1315–1334.

MOODY, D.W. (1964) *Coastal morphology and processes in relation to the development of submarine sand ridges off Bethany Beach, Delaware.* Ph.D. thesis, Johns Hopkins University. 167 pp.

NIEDORODA, A.W. (1980) Shoreface-surf zone sediment exchange processes and shoreface dynamics. *NOAA Tech. Memo. OMPA–1*, 89 pp.

NIEDORODA, A. & SWIFT, D.J.P. (1981) The role of wave orbital currents versus wind-driven currents in maintaining the shoreface: observations from the Long Island inner shelf. *Geophys. Res. Lett.* (in press).

OLSEN, C.R., BISCAYE, P.E., SIMPSON, H.J., TRIER, R.M., KOSTYK, N., BOPP, R.F., LI, Y.H. & FEELY, H.W. (1981) Reactor-released radionuclides and fine-grained sediment transport and accumulation patterns in Barnegat Bay, New Jersey and Adjacent Shelf Waters. *J. sedim. Petrol.* (in press).

PALMER, H.D., WILSON, D.G. & GREALA, J.R. (1975) *Sediment response to inner shelf hydraulic regimes: progress report February 1975–November 1975.* Unpublished report, Westinghouse Ocean Research Laboratory, Annapolis.

REDFIELD, A.C. (1958) The influence of the continental shelf on tides of the Atlantic Coast. *J. Mar. Res.* **17**, 432–448.

SCHLEE, J.S. (1973) Atlantic continental shelf and slope of the United States: sediment texture of the northeastern part. *Prof. Pap. U.S. geol. Surv.* **529L**, 64 pp.

SCOTT, J.T. & CSANADY, G.T. (1976) Nearshore currents off Long Island. *J. geophys. Res.* **81**, 5403–5409.

SHERIDAN, R.E., DILL, C.E. Jr & KRAFT, J.C. (1974) Holocene sedimentary environment of the Atlantic Inner Shelf off Delaware. *Bull. geol. Soc. Am.* **85**, 1319–1328.

SMITH, J.D. (1970) Stability of a sand bed subjected to a shear flow of low Froude Number. *J. geophys. Res.* **75**, 5928–5940.

SMITH, J.D. & HOPKINS, T.S. (1972) Sediment transport on the continental shelf off Washington and Oregon in the light of recent current meter measurements. In: *Shelf Sediment Transport, Process and Pattern* (Ed. by D.J.P. Swift, D.B. Duane and O.H. Pilkey). Dowden, Hutchinson & Ross, Stroudsburg, Pennsylvania. 656 pp.

STUBBLEFIELD, W.L., LAVELLE, J.W., SWIFT, D.J.P. & McKINNEY, T.F. (1975) Sediment response to the present hydraulic regime on the central New Jersey shelf. *J. sedim. Petrol.* **45**, 337–358.

SWIFT, D.J.P. (1973) Delaware Shelf Valley: Estuary retreat path, not drowned river valley. *Bull. geol. Soc. Am.* **84**, 2743–2748.

SWIFT, D.J.P. (1975a) Barrier Island genesis: evidence from the central Atlantic shelf, eastern U.S.A. *Sedim. Geol.* **14**, 1–43.

SWIFT, D.J.P. (1976) Tidal sand ridges and shoal retreat massifs. *Mar. Geol.* **18**, 105–113.

SWIFT, D.J.P. (1976) Coastal sedimentation. In: *Marine Sediment Transport and Environmental Management* (Ed. by D.J. Stanley and D.J.P. Swift), pp. 255–310. Wiley, New York. 602 pp.

SWIFT, D.J.P. & FIELD, M. (1981) Evolution of a classic sand wave field, Maryland sector, North American inner shelf. *Sedimentology*, **28**, 461–482.

SWIFT, D.J.P., FREELAND, G.L. & YOUNG, R.A. (1979) Time and space distributions of megaripples and associated bedforms, Middle Atlantic Bight, North American Atlantic shelf. *Sedimentology*, **26**, 289–406.

SWIFT, D.J.P., HOLLIDAY, B., AVIGNONE, N. & SHIDELER, G. (1972a) Anatomy of a shoreface ridge system, False Cape, Virginia. *Mar. Geol.* **12**, 59–84.

SWIFT, D.J.P., KOFOED, J.W., SAULSBURY, F.P. & SEARS, P. (1972b) Holocene evolution of the shelf surface, south and central Atlantic shelf of North America. In: *Shelf Sediment Transport: process and pattern* (Ed. by D.J.P. Swift, D.B. Duane and O.H. Pilkey), pp. 499–574. Dowden, Hutchinson & Ross, Stroudsburg, Pennsylvania. 656 pp.

SWIFT, D.J.P., SANFORD, R., DILL, C.E. JR & AVIGNONE, N. (1971) Textural differentiation on the shoreface during erosional retreat of an unconsolidated coast, Cape Henry to Cape Hatteras western North Atlantic shelf. *Sedimentology*, **16**, 221–250.

SWIFT, D.J.P., NELSON, T., McHONE, J., HOLLIDAY, B., PALMER, H. & SHIDLER, G. (1977) Holocene evolution of the inner shelf of southern Virgina. *J. sedim. Petrol.* **47**, 1454–1474.

SWIFT, D.J.P., PARKER, G., LANFREDI, N.W., PERILLO, G. & FIGGE, K. (1978b) Shoreface connected sand ridges on American and European shelves: a comparison. *Est. Coastal Mar. Res.* **7**, 257–273.

SWIFT, D.J.P., SEARS, P.C., BOHLKE, B. & HUNT, R. (1978a) Evolution of a shoal retreat massif, North Carolina shelf: inferences from areal geology. *Mar. Geol.* **27**, 19–72.

SWIFT, D.J.P., SHIDLER, G.L., AVIGNONE, N.F. & HOLLIDAY, B.W. (1970) Holocene transgressive sand sheet of the Middle Atlantic Bight: a model for generation by shoreface erosion. *Prog. geol. Soc. Am.* **2**, 757–759.

TRUMBULL, V.A.J. (1972) Atlantic continental shelf and slope of the United States—sand sized fraction of bottom sediments, New Jersey to Nova Scotia. *Prof. Pap. U.S. geol. Surv.* **529K**, 45 pp.

TWICHELL, D.C., MCCLEMMEN, C.E. & BUTMAN, B. (1981) Fine grained sediment accumulation, southern New England shelf. *J. sedim. Petrol.* (in press).

UCHUPI, E. (1968) Atlantic continental shelf and slope of the United States—physiography. *Prof. Pap. U.S. geol. Surv.* **529C**, 30 pp.

VAN STRAATEN, L.M.J.U. (1965) Coastal barrier reports in south and north Holland—in particular the area around Schevenengen and Ijmiden. *Meded. geol. Sticht*, **71**, 41–75.

VINCENT, C., SWIFT, D.J.P., YOUNG, R.A. & HILLARD, B. (1981) Sediment transport climatology of the New York–New Jersey continental shelf. In: *Shelf Sediment Transport* (Ed. by C. Nittrover and G. Allen) (in press).

YALIN, M.S. (1963) An expression for bedload transportation. *J. Hydraul. Div., Proc. Am. Soc. civ. Engrs* **88**, 221–250.

Spec. Publs int. Ass. Sediment. (1981) **5**, 385–396

Formation of storm deposits by wind-forced currents in the Gulf of Mexico and the North Sea*

ROBERT A. MORTON

Bureau of Economic Geology, The University of Texas at Austin, Austin, Texas 78712, U.S.A.

ABSTRACT

Textural variations, sedimentary structures, and vertical sequences preserved in shelf storm beds from the Gulf of Mexico are remarkably similar to published examples from the North Sea. In both areas, interlaminated and graded sands and muds characterize proximal and distal deposits. Proximal deposits, which are dominantly composed of sand, are first influenced by strong unidirectional currents, which may later give way to storm waves. Advective processes involving lateral sediment transport are also responsible for finer grained distal beds deposited in deeper water below wave base.

The formation of graded shelf deposits by storm runoff from coastal lagoons and adjacent barriers was first proposed by Hayes (1967). Since then, many sedimentologists studying ancient shelf deposits have appealed to storm-surge ebb as a mechanism for explaining size sorting and vertical succession of sedimentary structures found in shallow marine sand and mud sequences. However, re evaluation of topographic and geomorphic data for the type area (Padre Island and Laguna Madre) in conjunction with flood data for Hurricane Carla indicates that storm-surge ebb, as defined by Hayes, was probably not responsible for the graded bed deposited during Carla.

Judging from nearshore current data for recent storms, large-scale bottom flows produced or augmented by wind forcing are the most likely explanation for graded shelf deposits. Highest current velocities on the shelf, of the order of 2 m sec⁻¹, are recorded shortly after the greatest wind stress is applied. These wind-driven currents occur near or within the bottom boundary layer and are directed alongshore or offshore. In storm-dominated basins such as the Gulf of Mexico, breaking waves and orbital velocities are important for assisting in sediment entrainment, but unidirectional bottom flows produced by wind forcing are probably the most important mechanism for shelf sediment transport. In other areas, such as the North Sea, which is characterized by substantially higher wave activity and stronger tidal currents than is the Gulf of Mexico, wind forcing can temporarily alter the tidal component and significantly increase the bottom current velocities.

INTRODUCTION

Storm-related processes are known to be responsible for graded sequences and well-sorted sand beds that are deposited in shallow marine environments at depths generally less than 40 m and within approximately 45 km of the shoreline (Reineck & Singh, 1972). These episodic high-energy events of brief duration account for much of the sedimentological record preserved in some modern and ancient shelf sediments, and yet the exact nature of the physical processes involved in their formation is still poorly understood. As a result, nearshore graded beds have been attributed to various origins, including density currents generated by storm-surge ebb (Hayes, 1967), clouds of suspended sediment formed by waves (Reineck & Singh, 1972, 1973), and *in situ* sorting of sand and shell by wave-induced pressure fluctuations (Powers & Kinsman, 1953).

In recent years, the development and deployment of instrumented arrays in the shelf environment have permitted the acquisition of near-bottom current

* Publication authorized by the Director, Bureau of Economic Geology, The University of Texas at Austin.

0141-3600/81/1205-0385 $02.00

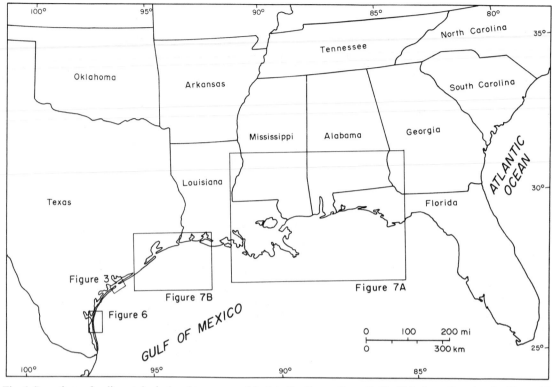

Fig. 1. Locations of sedimentological and oceanographic data for the northern Gulf of Mexico referred to in this study.

data during storms (Murray, 1970; Gienapp, 1973; Caston, 1976; Forristall, Hamilton & Cardone, 1977; Smith, 1978; Swift, Freeland & Young, 1979). These data indicate that nearshore shelf sedimentation is largely related to strong unidirectional bottom currents produced or augmented by wind forcing. Pronounced riverine discharge and astronomical tides influence shelf currents and sediment transport in many areas, but wind-induced bottom currents can form on most shelves regardless of pre-existing conditions.

Despite evidence favouring wind-forced bottom-return flows developed at the height of the storm (Murray, 1970; Gienapp, 1973; Forristall *et al.*, 1977), some workers continue to explain the observed textural gradations and sedimentary structures by wave activity or waning currents after the storm diminishes (Reineck & Singh, 1972; Walker, 1979). Several workers studying ancient clastic and carbonate shelf sediments (Brenner & Davies, 1973; Goldring & Bridges, 1973; Kelling & Mullin, 1975; Scott *et al.*, 1978) have also appealed to these post-

storm processes collectively referred to as storm-surge ebb.

Hayes (1967) introduced the concept of storm-surge ebb as an explanation for graded and well-sorted sandy sediments that were cored on the Texas inner shelf following Hurricane Carla. One purpose of the present study was to test Hayes' storm-surge ebb hypothesis by using meteorological and hydrographic data from Hurricane Carla, as well as morphological and topographic data from the type area. The study also examined near-bottom hydrographic measurements published for the Gulf of Mexico (Murray, 1970; Forristall *et al.*, 1977) and the North Sea (Gienapp, 1973; Caston, 1976). These data confirm the importance of wind-induced currents in shelf sediment transport during storms. The data also establish more precisely the timing and direction of the strongest currents. Reconstructions of modern storm deposits from the Gulf of Mexico (Fig. 1) and the North Sea (Fig. 2) are also presented to establish the thickness, textural variability, lateral continuity, and spatial distribution of storm beds.

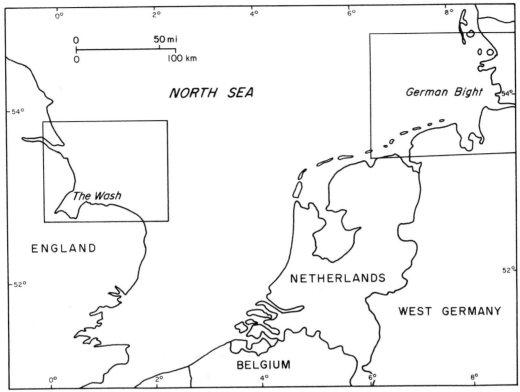

Fig. 2. Field locations for the North Sea data. The German Bight area was studied by Gadow & Reineck (1969) and Gienapp (1973); the south-western North Sea area was studied by Caston (1976).

MODERN STORM DEPOSITS

Storm beds commonly comprise two lithologic units, sand and mud, that are sometimes but not always graded. The base of each graded or interlaminated sequence is dominated by physical structures and usually marked by an erosional contact (Hayes, 1967; Reineck & Singh, 1972; Howard & Reineck, 1972). Overlying sediments exhibit varying degrees of bioturbation; when present, burrowing is usually concentrated in the uppermost fine-grained units (Reineck & Singh, 1973). Rapid deposition of storm beds is commonly indicated by homogeneity of mud and lack of burrowing.

Modern storm beds are typically a few centimetres to a few tens of centimetres thick (Hayes, 1967; Reineck & Singh, 1972). This is considerably less than some ancient examples that are up to 1 m thick. The apparent discrepancy in thickness is probably due to coring locations of nearshore sediments (Figs 3 and 4; Gadow & Reineck, 1969) that

rarely include sites from the shoreface or proximal depositional setting. Alternatively, the greater thickness of some ancient storm beds may reflect selective preservation of storm events with recurrence intervals that exceed time spans represented in cores of modern sediments.

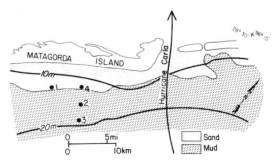

Fig. 3. Inner shelf bathymetry and position of vibracores in relation to the track of Hurricane Carla. Surface sediment distribution generalized from McGowen & Morton (1979). Location shown in Fig. 1.

Gulf of Mexico

Graded or texturally sorted deposits from the Gulf of Mexico (Fig. 1) are easily identified because shelf sediments are predominantly muds seaward of the lower shoreface (Figs 3 and 4). The lateral extent and gross textural changes within individual storm deposits can be observed by correlating distinct bedding sequences in vibracores from the inner shelf (Fig. 3) where Hurricane Carla crossed the Texas coast in September 1961. Although the cores were taken more than seventeen years after Carla, the persistent graded bed that forms the uppermost sedimentation unit (Fig. 4) is interpreted as the Carla deposit because the shelf off Matagorda Island has not been greatly influenced by storms since 1961.

The uppermost bed, which varies in thickness from 20 to 25 cm, grades laterally from nearly featureless sand to interlaminated mud and sandy mud and finally to homogeneous mud in an offshore direction (Fig. 4). This particular unit can be traced along depositional strike over a distance of 6 km.

A coarse-graded bed and sand layer, which was also deposited during Hurricane Carla off central Padre Island in up to 36 m of water (Figs 1 and 6), was mapped and described by Hayes (1967) in what is now a classical study of the role of hurricanes in coastal sedimentation. This Carla deposit, which contrasted with underlying homogeneous mud sampled before the storm, covered nearly 800 km² and extended offshore more than 24 km; it attained a maximum thickness of about 9 cm (Hayes, 1967).

By any standard, Carla was an extreme storm, which partly explains the lack of data on sea conditions. Near the storm centre (Fig. 3), open coast surge was about 4 m above sea-level and maximum

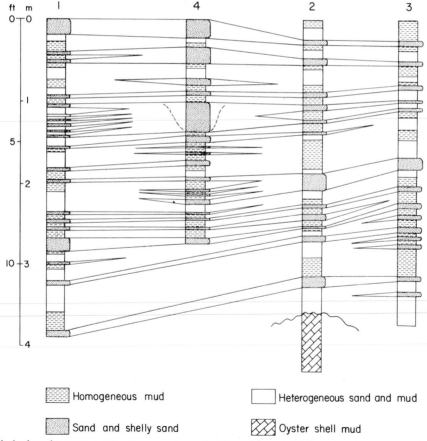

Fig. 4. Lithologic units preserved in vibracores from the Texas inner shelf. Core locations shown in Fig. 3. Proximal units are best illustrated in core 4. Except for the uppermost sand in core 1, the deposits in cores 1, 2, and 3 represent distal facies.

winds of 65 m sec^{-1} were directed onshore. Along central Padre Island (Fig. 6), peak winds of 30 m sec^{-1} were directed offshore, and surge heights were only about 2·5 m above mean sea-level (United States Army Corps of Engineers, 1962).

North Sea

The widespread occurrence of sand waves in the North Sea results in a depositional style that limits the formation and recognition of storm deposits to certain areas. Gadow & Reineck (1969) and Reineck & Singh (1972) described graded and laminated storm deposits from the German Bight (Fig. 2) where strong offshore bottom currents accompanying winter storms had previously been recorded (Gienapp, 1973). The short cores containing laminated sand and mud couplets were taken up to 50 km from the shoreline in water depths ranging from 14 to 24 m. At these depths, surface sediments were fine-grained sandy silt and clayey silt (Gadow & Reineck, 1969). Individual distal storm beds were slightly less than 10 cm thick and thinned progressively offshore.

Gienapp (1973) provided detailed descriptions of storm conditions that preceded sampling by Gadow & Reineck (1969). The storm peaked shortly before 2 November 1965, when estimated wave heights were 6 m, peak wind gusts exceeded 30 m sec^{-1}, and storm surge was 1·6 m above normal high tide (Gienapp, 1973).

Proximal and distal units

Core data from the Texas inner shelf (Fig. 4; Hayes, 1967) and the North Sea (Gadow & Reineck, 1969) are admittedly sparse, but they are adequate for showing that proximal storm deposits are composed mostly of sand; in contrast, overlying muds are subordinate in comparison to total storm bed thickness.

Vertical grain size changes within proximal beds (Fig. 5) are related primarily to sediment composition. The basal shell and rock fragment gravel grades upward into very fine sand, which is predominantly quartz. The sand, in turn, grades into the overlying clay-rich mud. Textural changes within the sand itself are usually minor, because this size fraction was derived from well-sorted beach and shoreface sediments before final deposition. These thick beds

Fig. 5. Graded bed preserved in core 4 (Fig. 4) at a depth of 90 cm. Core width equals 7·6 cm.

(Fig. 4) also exhibit the greatest lateral variability and discontinuity. The lack of precise correlation among most of the storm beds suggests that subtle changes in seafloor morphology control the spatial distribution of the coarsest sediments, which may be concentrated in shallow scour channels (Fig. 4).

The number and thickness of discrete sand layers and the total thickness of individual storm beds decreases between proximal and distal deposits (Fig. 4; Gadow & Reineck, 1969). Because of coring locations and poor recovery in clean sands, studies of modern storm beds (Hayes, 1967; Gadow & Reineck, 1969; Figs 3 and 4) have mainly described physical characteristics of distal deposits. In the coarsest distal units, which usually consist of fine sand, basal sequences are delineated by erosional

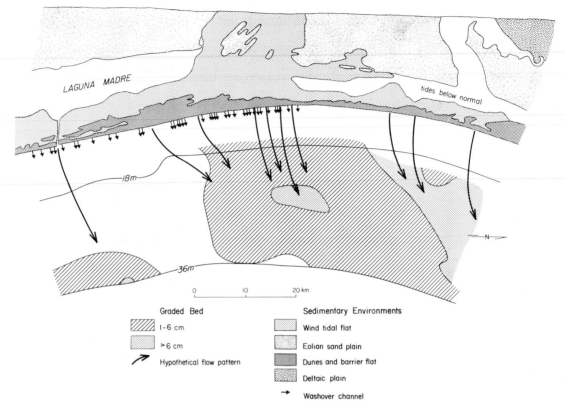

Fig. 6. Geomorphic features and physical characteristics of a graded bed, central Padre Island. Bed thickness and hypothetical flow patterns from Hayes (1967). Sedimentary environments modified from Brown *et al.* (1977). Location shown in Fig. 1.

contacts occasionally overlain by thin shell lags (Reineck & Singh, 1973). Parallel laminations and small-scale ripple cross-laminations are the most common stratification types found in the fine sand layers. Overlying mud layers can be extensively bioturbated or homogeneous (Fig. 4), depending on rates of sedimentation and recolonization by benthic fauna (Reineck & Singh, 1973).

Grain size of storm deposits diminishes in an offshore direction. These lateral variations are expressed as textural gradations of sand into mud, as decreases in the size and amount of comminuted shells, and as pinch-outs of individual sand layers (Fig. 4). The correlations also suggest that thin alternating layers of sand and mud can comprise a single storm deposit. Such contemporaneous cyclic deposition may be caused by pulsations in current velocity, variations in sediment concentration, or selective sorting in the bottom boundary layer as described by Stow & Bowen (1980).

ORIGINS OF BOTTOM CURRENTS

Storm-surge ebb

Since it was first proposed, storm-surge ebb (Hayes, 1967) has been cited as an explanation for ancient storm deposits (Brenner & Davies, 1973; Goldring & Bridges, 1973; Kelling & Mullin, 1975; Scott *et al.*, 1978; Walker, 1979). These interpretations assume that storm surges, in bays and lagoons, are capable of generating strong gravity-induced currents that flow seaward through washover channels (Fig. 6) as the strength of the storm diminishes. According to this hypothesis, the surface currents become sediment-laden density currents as they cross the shoreface and inner shelf.

A more recent compilation of data presented herein indicates that storm runoff from Laguna Madre (Fig. 6) was probably negligible and, therefore, cannot be considered a reasonable explanation for the Carla deposit. The lines of evidence against storm-surge ebb are as follows. (1) The thickest and most extensive area of the graded bed was found offshore from a barrier segment where well-developed and stabilized fore-island dunes prevented overwash and where washover channels were absent. Furthermore, closely spaced washover channels are found south of the main storm deposit (Fig. 6). (2) Water levels in Laguna Madre were actually below normal (U.S. Army Corps of Engineers, 1962) as a result of wind circulation. (3) Composite wind directions (U.S. Army Corps of Engineers, 1962) and drift bottle data (Chew, Drennan & Demoran, 1962) clearly show that nearshore current directions were southwestward, not northward as envisaged by Hayes (1967). (4) The broad sub-aerial plain that forms the back-barrier environment of Padre Island (Fig. 6) limited the hydraulic head and offered considerable frictional resistance to flow, thus greatly retarding water returning to the Gulf. This is substantiated by post-Carla photographs (Hayes, 1967, p. 22) that show no evidence of strong currents in washover channels along central Padre Island.

Open coast surge during Carla was greatest (3·7 m) where the storm centre crossed the coast (Fig. 3). In the same area, flood depths in adjacent bays and lagoons were nearly 6·5 m above sea-level (U.S. Army Corps of Engineers, 1962). Numerous deep washover channels and extreme surge heights in this area would have provided the best opportunity to develop storm-surge ebb as originally proposed (Hayes, 1967). However, field studies and examination of post-Carla photographs (Morton, 1979) indicate that storm-surge ebb was subordinate to wind-driven currents and washovers that were flood oriented and directed onshore. As floodwaters receded, currents flowing through washover channels may have contributed to the offshore transportation of fine-grained sediment in suspension (Shideler, 1978), but their overall influence on the inner shelf was negligible. These currents were probably diffused in, and became indistinguishable from, the longshore currents beyond the surf zone.

Wind-forced currents

Strong bottom currents that form in response to wind forcing (Csanady, 1976) are a plausible explanation for the laminated and graded shelf deposits described by Hayes (1967) and Reineck & Singh (1972). The conservation of mass provided by these bottom flows was predicted from theoretical considerations (Reid, 1957) and documented by field data from the Western Atlantic (see Swift *et al.*, 1979, for references), the North Sea (Gienapp, 1973; Caston, 1976), the northern Gulf of Mexico (Murray, 1970, 1975; Smith, 1977; Forristall *et al.*, 1977), and elsewhere. Swift (Swift, Heron & Dill, 1969; Swift *et al.*, 1979) was one of the first advocates of wind-forced currents and their ability to transport sediment on continental shelves. The influence of modern shelf studies is seen in the excellent discussion of storm bed origins presented by Brenchley, Newall & Stanistreet (1979).

Depending on wind velocity and duration, water depth, and pre-existing conditions, a variety of relationships can exist between wind and water. The magnitude, direction, and duration of bottom currents depend on such things as shelf slope, shoreline orientation, geographic position relative to the storm centre, and storm characteristics. The latter group of conditions includes surge height, wind velocities and direction, forward speed of the storm, and its angle of approach to the shoreline. The last two parameters concern only those storms with organized wind circulation such as tropical cyclones.

Although field data collected during storms are sparse, they generally show a single-layer or a two-layer system. In the single-layer system, the entire water column generally moves alongshore in the same direction as the wind. In the two-layer system, bottom water masses generally move offshore in a direction opposite to that of the wind. When wind blows onshore at a relatively high angle or nearly perpendicular to the coast, large volumes of water are transported onshore in the surface layers, while bottom currents flow seaward. These bottom-return flows, as well as the longshore currents, provide effective mechanisms for transporting sand across the shoreface and inner shelf. Examples of these two systems are recorded in field data from the Gulf of Mexico (Murray, 1970; Forristall *et al.*, 1977; Smith, 1977, 1978) and the North Sea (Gienapp, 1973; Caston, 1976).

Gulf of Mexico

Wind-driven currents are especially important along storm-dominated coasts such as the Gulf of Mexico, where currents caused by astronomical tides and density contrasts are minor. Murray (1970) presented field measurements for Hurricane Camille (Fig. 7A) that crossed the Mississippi Gulf coast in August 1969. Murray's data, taken off the Florida

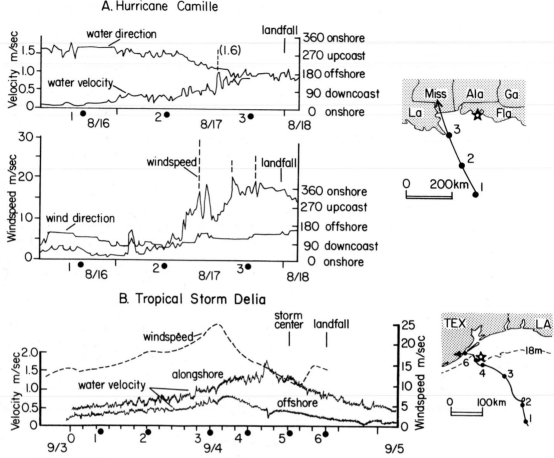

Fig. 7. (A) Wind and current velocity profiles during Hurricane Camille (16–18 August 1969). Current velocities recorded 1·5 m off the bottom in 6·3 m of water and 360 m from the shoreline. Data from Murray (1970). Location shown in Fig. 1. (B) Current velocity profiles during Tropical Storm Delia (3–5 September 1973). Current velocities recorded 3 m off the bottom in 20 m of water and 40 km from the shoreline. Data from Forristall *et al.* (1977). Location shown in Fig. 1.

coast, showed that winds gradually shifted from east to south-east as the storm moved onshore. Corresponding data for the water column showed that currents 1·5 m above the seafloor flowed westward and then turned southward in response to changes in wind direction and intensity. The shift from a single-layer to a two-layer system in this example is attributed to a change in wind direction with respect to shoreline orientation. Maximum velocities of 1·6 m sec^{-1} (Fig. 7A) were recorded before landfall when the wind was blowing north-westerly (on-shore) and bottom currents were flowing in a southerly direction (offshore). Velocities for these bottom-return flows are higher than expected con-

sidering that measurements were made more than 160 km from the storm centre.

Other records of near-bottom conditions in the Gulf of Mexico were reported for tropical storm Delia (Forristall *et al.*, 1977) that made landfall during September 1973 near Galveston Island on the upper Texas coast (Fig. 7B). As the storm passed the recording platform located 40 km offshore, currents directed seaward were between 0·50 and 0·75 m sec^{-1}, while currents flowing alongshore at depths of about 18 m were nearly 2 m sec^{-1} (Fig. 7B). For a brief period, velocities of the organized unidirectional currents that flowed alongshore were greater than orbital velocities of even the highest waves (Forristall *et al.*, 1977). These strong currents occurred about 4 h after highest wind speeds (Fig. 7B) accompanied storm intensification and a change in wind direction (Forristall *et al.*, 1977) from north-easterly (offshore) to southerly (onshore). Simultaneous tide measurements at Galveston (U.S. Army Corps of Engineers, unpublished data) show that water levels rose rapidly as wind direction and intensity changed and the storm approached the coast. This storm surge (1·3 m) preceded maximum current velocities; however, it did not peak (2 m) until 30 h after storm landfall (Fig. 7B). Together, these data prove that bottom current velocities during Delia depended primarily on wind stresses, not on storm-surge ebb.

Neither wind velocities nor storm surge associated with Delia were particularly impressive, and yet unusually high current velocities were generated in the bottom-boundary layer under those conditions. This suggests that bottom current velocities near storm centres could exceed 2 m sec^{-1} for extreme storms such as Carla and Camille.

These and other examples from the Gulf of Mexico (Smith, 1978) demonstrate that maximum bottom velocities occur shortly after maximum wind stress is applied but before passage of the storm's centre and before landfall. The timing of these events also suggests that storm-surge ebb has little if anything to do with strong bottom currents and deposition of coarse graded beds.

The fact that the highest current velocities occur after the maximum wind velocities is not necessarily evidence that return flows are generated by release of water impounded along the shoreline, as implied by Walker (1979). In fact, the time difference is probably a hysteretical effect compounded, in some cases, by the offshore locations of current measurements.

North Sea

Even though tidal currents are considerably greater in the North Sea than in the Gulf of Mexico, the importance of wind forcing cannot be completely ignored, as shown by simultaneous recordings of wind and bottom current velocities and directions at several locations in the North Sea (Fig. 2). Perhaps the best collection of regional synoptic data was published by Gienapp (1973) for a winter storm in 1965 that affected the German Bight. Although the data include tidal effects as well as those of wind forcing, they represent conditions in the German Bight where the physical setting and meteorological forces combined to produce bottom currents that temporarily overwhelmed the tidal currents. Offshore current velocities measured 2 m above the seafloor were up to 1·5 m sec^{-1}. Highest bottom velocities were recorded shortly after the strongest winds (27 m sec^{-1}) and occurred either in shallow water or where shelf slopes are steep. At two stations where simultaneous measurements were made, maximum bottom currents were slightly greater than near-surface currents. The opposite is normally true in purely wind-driven systems because of internal and bottom friction. This suggests that the tidal component probably was responsible for the anomalous increase in velocity with depth.

Because of the westerly wind direction and the concave shape of the shoreline, an alongshore single-layer system was developed in the southern part of Gienapp's study area (Fig. 2), while a two-layer offshore bottom flow developed in the zone of convergence to the north. Unidirectional currents in excess of 1 m sec^{-1} lasted for more than 6 h, an indication that attendant water masses were capable of transporting water particles and suspended sediment more than 20 km during that period.

Caston (1976) also demonstrated the importance of wind forcing by reviewing published data and presenting wind and bottom current measurements from the south-western North Sea (Fig. 2). These data corroborated the presence of wind-driven currents superimposed on the tidal regime. The conditions reported by Caston represent an intermediate case in the spectrum of wind- and tide-dominated currents. In his example, wind forces augmented tidal currents moving in the same direction as the wind and reduced tidal currents moving in a direction opposite to that of the wind. An important aspect of this example is that wind-driven currents did not prevail over tidal currents; they simply interfered constructively and destructively, depending on the phase of the tides. Despite the relatively high wind velocities (26 m sec^{-1}) and shallow depths (35 m), maximum currents recorded 4·6 m above the seafloor ranged only from 0·6 to 1·0 m sec^{-1} (Caston, 1976). These moderate velocities were products of wind blowing alongshore; therefore, higher bottom velocities might have been encountered if the wind had blown directly onshore. By calculating expected current velocities, Caston (1976) was able to estimate the wind-driven component, which represented about 2% of the surface wind velocity. The data published by Caston (1976)

SED 5

indicate that near-bottom currents flowing alongshore in a southerly direction were part of a single-layer system, whereas the northerly flowing bottom currents were most likely part of a two-layer system characterized by southerly flowing surface currents. However, data supporting or refuting this interpretation are not available.

DISCUSSION

Several authors have suggested that strong unidirectional bottom currents are analogous to large-scale rip currents or coastal jets (Swift et al., 1979). The formation of these jet-like flow patterns may be augmented by shelf bathymetry. For example, in the German Bight, currents are locally concentrated and increased by tidal channels emerging from nearshore sand flats. In other areas, such as the Gulf of Mexico, nearshore bathymetry is fairly smooth, but cross-shelf flow is locally channelized, as indicated by the areal distribution of individual shell-bearing beds (Fig. 4).

The flow patterns may be manifested by the shape of the storm deposits. Indeed, maps of storm beds (Hayes, 1967; Gadow & Reineck, 1969), cross-sections from cores (Fig. 4), and outcrops (Hobday & Morton, 1980) suggest that the thickest storm deposits have lobate geometries that thin locally in alongshore and offshore directions.

On a broader scale, comparison of the Carla deposits off Padre Island (9 cm, Fig. 6) and Matagorda Island (25 cm, Fig. 4) indicates that storm beds are thickest in the vicinity of maximum storm influence, but textural fining patterns are principally controlled by water depth and distance from the coast.

Nearshore bottom gradients may also control bottom velocities and sediment dispersal. For example, the steeper shelves off Matagorda and Padre Islands ($1 \cdot 3$–$1 \cdot 5$ m km^{-1}, Figs 3 and 6) and within the German Bight ($0 \cdot 75$ m km^{-1}; Reineck & Singh, 1972) are more conducive to development of stronger currents than is the Mississippi Sound shelf, where Hurricanes Camille (1969) and Frederick (1979) encountered extremely low nearshore slopes ($0 \cdot 24$ m km^{-1}).

The anticipated bottom velocities and the recurrence of extreme storms of historical record, as well as the number of graded beds in shallow core of Holocene marine sediments (Fig. 4), indicate that the frequency of storm bed deposition could easily

exceed the rate of once every few thousand years estimated by Brenchley et al. (1979). In fact, the frequency of storm deposition preserved in the regressive shelf sequence (Fig. 3) is on the order of once every few hundred years based on the similarity of oyster shells recorded in core 2 (Fig. 4) with those from the same area dated at 8000 y BP by Curray (1960). Furthermore, meteorological data for the Gulf of Mexico suggest that storms capable of generating strong bottom currents occur on the order of once every few years (Hayes, 1967).

Hydrodynamic interpretation

Direct correlation between storm deposits and specific processes has not been achieved. However in recent years, hummocky cross-stratification has been suggested as evidence for storm waves (Harms et al., 1975; Walker, 1979). Although widely recognized in association with ancient shelf sediments (Walker, 1979), this stratification type has not been reported from modern sediments because of the narrowness of cores relative to the large low-relief bedforms that produce the hummocky structures. Thus, details concerning the sedimentological record from modern storm beds are presently incomplete. To overcome this limitation, an idealized sequence and hydrodynamic interpretation for storm deposits was developed from sedimentary structures preserved in sandstones of the Cretaceous Washita Group (Hobday & Morton, 1980). The erosional base of the sandstone bed is directly overlain by a thin layer of high-spired gastropods with preferred offshore orientation. The overlying basal sands, which are massive to parallel laminated, grade upward into hummocky cross-stratification, which in turn grades into ripple cross-lamination. Each successive sequence is usually thinner than the preceding one except for the overlying mudstones, which vary in thickness.

This sequence of structures, which is similar to those reported by Brenchley et al. (1979) and Walker (1979), first records the influence of strong, unidirectional bottom currents that were gradually modified by orbital velocities of storm waves. Both the large, three-dimensional bedforms that produce hummocky cross-stratification, and its stratigraphic position in the vertical sequence, suggest that this unique type of bedding is the joint product of storm waves and bottom currents (Walker, 1979). As the organized currents weakened, waves became even more influential until, finally, they dominated the mode of deposition, as evidenced by oscillation ripples at the top of the sand unit. Elsewhere, unidirectional currents persisted, as indicated by current ripples. The overlying laminated mudstones were deposited rapidly from suspension, as evidenced by parallel laminations and the lack of bioturbation.

Interlaminated and graded sands and muds should form the bulk of nearshore sediments deposited on storm-dominated shelves. The coarsest detritus is derived from adjacent barriers and strandplains (Brenner & Davies, 1973; Brenchley *et al.*, 1979) or delta-front environments (Johnson, 1975; Walker, 1979; Hobday & Morton, 1980). Along these progradational (regressive) shorelines where storm deposists have the best chance of being preserved, elevations of aeolian dunes and other surficial features commonly exceed storm-surge heights, thus preventing or retarding washover and subsequent runoff. These boundary conditions do not prevent the development of strong bottom currents; in fact, their formation is enhanced because wind-driven water impounded against the shoreline is forced to flow alongshore or offshore. In this manner, unidirectional bottom currents are developed which are similar to those interpreted from modern and ancient storm beds.

ACKNOWLEDGMENTS

Appreciation is expressed to personnel with the United States Geological Survey Marine Geology Branch in Corpus Christi, Texas, and the crew of the University of Texas RV *Longhorn*, who were responsible for obtaining the vibracores. The coring cruise was conducted while the author was on teaching leave at the University of Texas Marine Science Institute. D. K. Hobday and L. F. Brown, Jr read the manuscript and made helpful comments.

REFERENCES

BRENCHLEY, P.J., NEWALL, G. & STANISTREET, I.G. (1979) A storm surge origin for sandstone beds in an epicontinental platform sequence, Ordovician Norway. *Sedim. Geol.* **22**, 185–217.

BRENNER, R.L. & DAVIES, D.K. (1973) Storm-generated coquinoid sandstone: genesis of high energy marine sediments from the Upper Jurassic of Wyoming and Montana. *Bull. geol. Soc. Am.* **84**, 1685–1698.

BROWN, L.F. Jr, McGOWEN, J.H., EVANS, T.J., GROAT, C.G. & FISHER, W.L. (1977) *Environmental Geologic Atlas of the Texas Coastal Zone—Kingsville Area.* The University of Texas at Austin, Bureau of Economic Geology. 131 pp.

CASTON, V.N.D. (1976) A wind-driven near bottom current in the southern North Sea. *Est. Coastal Mar. Sci.* **4**, 23–32.

CHEW, F., DRENNAN, K.L. & DEMORAN, W.J. (1962) Drift-bottle return in the wake of hurricane Carla, 1961. *J. geophys. Res.* **67**, 2773–2776.

CSANADY, G.T. (1976) Wind-driven and thermohaline circulation over continental shelves. In: *Effects of Energy-Related Activities on the Atlantic Continental Shelf* (Ed. by B. Manowitz). *BNL Rep.* 50484, pp. 31–47.

CURRAY, J.R. (1960) Sediments and history of Holocene transgression, continental shelf, northwest Gulf of Mexico. In: *Recent Sediments, Northwestern Gulf of Mexico*, pp. 221–266. American Association of Petroleum Geologists, Tulsa.

FORRISTALL, G.Z., HAMILTON, R.C. & CARDONE, V.J. (1977) Continental shelf currents in Tropical Storm Delia: observations and theory. *J. Phys. Oceanogr.* **7**, 532–546.

GADOW, S. & REINECK, H.E. (1969) Ablandiger sandtransport bei sturmfluten. *Senckenberg. Mar.* **1**, 63–78.

GIENAPP, H. (1973) Strömungen während der Sturmflut vom 2, November 1965 in der Deutschen Bucht und ihre Bedeutung für den Sedimenttransport. *Senckenberg. Mar.* **5**, 135–151.

GOLDRING, R. & BRIDGES, P. (1973) Sublittoral sheet sandstones. *J. sedim. Petrol.* **43**, 736–747.

HARMS, J.C., SOUTHARD, J.B., SPEARING, D.R. & WALKER, R.G. (1975) Depositional environments as interpreted from primary sedimentary structures and stratification sequences. *Short Course Soc. Econ. Paleont. Miner. Tulsa*, **2**, 161 pp.

HAYES, M.O. (1967) Hurricanes as geological agents: case studies of hurricane Carla, 1961, and Cindy, 1963. *Rep. Inv.* **61**, 54 pp. The University of Texas at Austin, Bureau of Economic Geology.

HOBDAY, D.K. & MORTON, R.A. (1980) Lower Cretaceous shelf storm deposits, North Texas. *Abstr. Bull. Am. Ass. Petrol. Geol.* **64**, 723.

HOWARD, J.D. & REINECK, H.E. (1972) Georgia coastal region, Sapelo Island, U.S.A., physical and biogenic sedimentary structures of the nearshore shelf. *Senckenberg. Mar.* **4**, 81–123.

JOHNSON, H.D. (1975) Shallow marine sand bar sequences: an example from the late Precambrian of North Norway. *Sedimentology*, **24**, 245–270.

KELLING, G. & MULLIN, P. (1975) Graded limestone and limestone-quartzite couplets: possible storm deposits from the Moroccan carboniferous. *Sedim. Geol.* **13**, 161–190.

McGOWEN, J.H. & MORTON, R.A. (1979) *Sediment Distribution, Bathymetry, Faults, and Salt Diapirs, Submerged Lands of Texas.* The University of Texas at Austin, Bureau of Economic Geology. 31 pp.

MORTON, R.A. (1979) Subaerial storm deposits formed on barrier flats by wind-driven currents. *Sedim. Geol.* **24**, 105–122.

MURRAY, S.P. (1970) Bottom currents near the coast during hurricane Camille. *J. geophys. Res.* **75**, 4579–4582.

MURRAY, S.P. (1975) Trajectories and speeds of wind-driven currents near the coast. *J. Phys. Oceanogr.* **5**, 347–360.

POWERS, M.C. & KINSMAN, B. (1953) Shell accumulations in underwater sediments and their relation to the thickness of the traction zone. *J. sedim. Petrol.* **23**, 229–234.

REID, R.O. (1957) Modification of the quadratic bottom-stress law for turbulent channel flow in the presence of surface wind stress. *Tech. Memo. Beach Eros. B. U.S.* **93**, 33 pp.

REINECK, H.E. & SINGH, I.B. (1972) Genesis of laminated sand and graded rhythmites in storm-sand layers of shelf mud. *Sedimentology*, **18**, 123–128.

REINECK, H.E. & SINGH, I.B. (1973) *Depositional Sedimentary Environments*. Springer-Verlag, New York. 439 pp.

SCOTT, R.W., FEE, D., MAGEE, R. & LAALI, H. (1978) Epeiric depositional models for the Lower Cretaceous Washita Group, North-Central Texas. *Rep. Inv.* **94**, 23 pp. The University of Texas at Austin, Bureau of Economic Geology.

SHIDELER, G.L. (1978) A sediment-dispersal model for the South Texas continental shelf, northwest Gulf of Mexico. *Mar. Geol.* **26**, 289–313.

SMITH, N.P. (1977) Near-bottom cross-shelf currents in the northwestern Gulf of Mexico: a response to wind forcing. *J. Phys. Oceanogr.* **7**, 615–620.

SMITH, N.P. (1978) Longshore currents on the fringe of Hurricane Anita. *J. geophys. Res.* **83**, 6047–6051.

STOW, D.A.V. & BOWEN, A.J. (1980) A physical model for the transport and sorting of fine-grained sediment by turbidity currents. *Sedimentology*, **27**, 31–46.

SWIFT, D.J.P., HERON, S.D. Jr & DILL, C.E. Jr (1969) The Carolina Cretaceous: petrographic reconnaissance of a graded shelf. *J. sedim. Petrol.* **39**, 18–33.

SWIFT, D.J.P., FREELAND, G.L. & YOUNG, R.A. (1979) Time and space distribution of megaripples and associated bedforms, Middle Atlantic Bight, North American Atlantic shelf. *Sedimentology*, **26**, 389–406.

UNITED STATES ARMY CORPS OF ENGINEERS (1962) *Report on Hurricane Carla*, 9–12 *Sept*. 1961. Galveston District Corps of Engineers, 29 pp.

WALKER, R.G. (1979) Shallow marine sands. In: *Facies Models* (Ed. by R.G. Walker). *Geosci. Canada Repr. Ser.* **1**, 75–89.

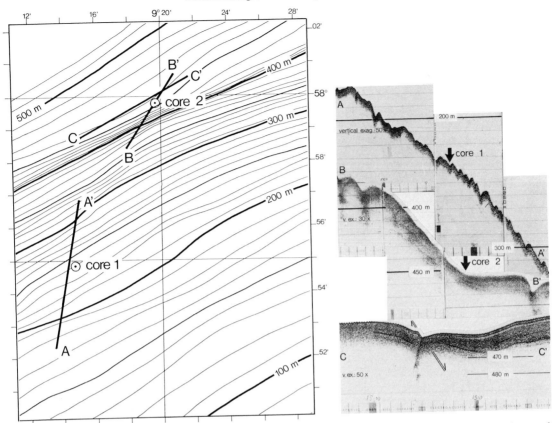

Fig. 2. Location of piston cores and echosounder profiles in the central Skagerrak (see Fig. 1). The echosounder profiles (18 kHz) are shown to the right.

dicates a relatively low supply of sediment from Norway and Western Sweden whereas a large amount of sediment from Denmark and the North Sea covers the southern flank of the basin.

The well-stratified sediments of the lower part of the flank show a general downslope decrease of thickness. Continuing across the flat-bottomed basin the layering becomes less distinct and vanishes towards the Norwegian mainland. Most of the sediments on this side are trapped in small basins near the coast (van Weering & Paauwe, 1979).

In contrast to this general trend there is an area of thick fine-grained Holocene sediments south of Larvik. This area may have a sedimentological influence on the central basin by means of submarine canyon sediment transport.

The occasionally high near-bottom current velocities lead to the suggestion of a temporary downslope bed load transport which is superimposed on the general sediment discharge from suspension. Thus the question arises whether the interference of two

types of sedimentation process may be detected from sediment analysis and, if so, how its variation in Holocene history may influence the sedimentation.

In this respect two piston cores in 239 and 454 m depth from the southern slope of the Skagerrak were taken. The location of the cores (Figs 1 and 2) is based on a survey with the 18 kHz sediment echosounder of RV *Poseidon* (Kiel). The accoustical penetration was up to 40 m, depending on sediment composition. Positioning was made by Decca Navigation System Mark 21 at 3–5 min interval.

The slope shows a general convex outline with inclinations of 0·7° in 150 m depth, 1·3° in 250 m (core 1), 2·2° in 350 m and a maximum of 5·2° in 450 m depth (core 2). A downslope cascade morphology is interpreted as an effect of sediment creeping (Fig. 2, A–A′) that has built up, parallel to the slope, a series of elongated sediment traps for near-bottom transported material.

These traps are buried under varying degrees of sediment cover, suggesting that the process of disrupting the sediment cover was active from early Holocene until recent times (Fig. 2, C–C′).

In an area 10–20 nm SSW of core 1 repeated side-scan sonar surveys revealed sand wave fields in 40–70 m depth changing in space and time and thus indicating the occurrence of recent strong unidirectional bottom currents to NNE (30° (Fig. 1, arrow) (Kujpers & Rumohr, in preparation).

Possibly core 1 was more affected by the coring process than core 2. The uppermost centimetres have been pushed off during penetration. Moreover, the sediment is partially affected by small vertical cracks (Fig. 11), probably due to degassing effects. On the other hand, severe disturbance of the core can be ruled out, because of the preservation of subtile organic structures continuing across the cracks and of high Pb 210 content, indicating the surface layers.

The cores were studied with respect to water content, shear strength (cone penetration), grain-size distribution, mineral composition, sedimentary structures using X-ray radiography, organic and carbonate content, foraminiferal stratigraphy and radio carbon age determination.

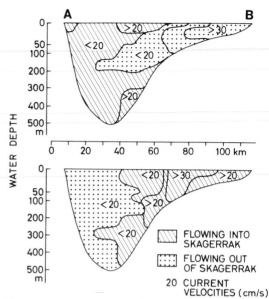

Fig. 3. Water movement in two observed situations in a vertical section along line A–B in Fig. 1 (data from Larsson & Rodhe, 1979).

GRAIN SIZE AND GEOTECHNICAL DATA

In core 1 there is a variation in sand content between 7 and 65 % (Fig. 4), and the amount of sand increases from 4·1 m upwards. A close examination revealed that the grain-size distribution for material finer than 63 μm is almost independent of the sand content. This appears from a presentation of the grain-size data in a triangular diagram (Fig. 5). Our interpretation is that these sediments are composed of two modes deposited from suspension load and bed load respectively.

The grain-size distributions for these two modes are shown in Fig. 6, and the observed coarsening upward is explained as a result of increasing bed load transport to this area during the latter part of the Holocene. Material coarser than 63 μm (sand) was separated from the finer material in 32 samples from this core. Ninety-eight per cent of the sand was in the fraction 63–125 μm and the remaining 2 % consisted mainly of mica. Consequently, there are no particles with an equivalent spherical diameter above 125 μm. Transport of this material as bed load requires current bottom velocities of about 20 cm sec⁻¹. As demonstrated by Larsson & Rodhe (1979), velocities like this are frequently observed 5–20 m above the

bottom in the Skagerrak down to 300 m depth or more. Consequently, there is an agreement between observed current velocities and the proposed bottom transport of sand material. Moreover, observation of sand-silt material in different bedforms as continuous wavy, lenticular and flaser (Fig. 12) indicate clearly that bed load transport occurred periodically.

Recent studies of particle size spectra of non-living suspended matter in the North Sea and Baltic Sea (Eisma & Gieskes, 1977; Eisma & Kalf, 1978; Eisma, 1976; Eisma, 1981; Dera et al., 1974) showed that this material consists of aggregates with diameters rarely exceeding 60 μm. Therefore it is unlikely that the sand fraction was transported in suspension, and we feel that the proposed model (Fig. 6) is reasonable.

MINERALOGY

The variation in mineralogy with grain size is illustrated in Fig. 7. The content of phyllosilicates rapidly decreases with increasing particle size. The fractions 20–63 and 63–125 μm mainly contain quartz and small amounts of feldspar and calcite. The fraction larger than 125 μm contains large amounts of dioctahedral mica and some chlorite,

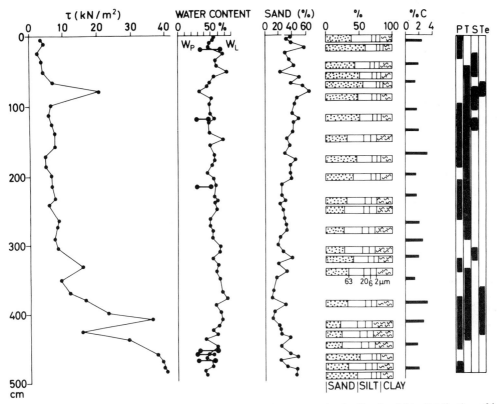

Fig. 4. Geotechnical and textural data from core 1 taken at 239 m water depth. On the right: distribution of biogenic structures (for explanation see Fig. 12).

quartz and calcite. As pointed out in connection with the grain-size distribution, the phyllosilicates in this fraction belong hydrodynamically to the fractions 20–63 and 63–125 μm, but are separated from these fractions during sieving. Calcite has its maximum concentration in the fractions 2–6 and 6–20 μm. By determining the amount of CO_2 released during treatment with 0·1 N-HCl the amount of calcite in each fraction was calculated (Table 1). This table together with the diffractograms illustrates the fact that calcite has its highest concentration in the fraction 2–20 μm. Most of this calcite is probably produced by Quaternary erosion of fine-grained Mesozoic and Tertiary calcareous sediments in the North Sea area.

The clay fraction has a large influence upon the properties of these sediments. It was separated from each 25 cm interval in both cores. In fact all the diffractograms from one core, within experimental error, were identical. The main conclusion from this study is that no change in clay mineralogy as a func-

tion of depth in the core could be observed. Consequently the material from different sources must be well mixed and the contribution from the various sources must have been more or less constant throughout the late Holocene.

The diffractograms (Fig. 8) show that the main constituents are illite, smectite, kaolinite, chlorite and quartz. In addition, small amounts of feldspar were observed. Treatment with NaTPB revealed that part of the illite is of a trioctahedral nature.

When comparing the diffractograms from the two localities it is observed (Fig. 8) that the clay fraction from core 1 has a weaker 14 Å reflection and a stronger 7 Å reflection than clays from core 2. We conclude that the clay fraction deposited at shallow depth (239) contains more kaolinite and less smectite than the material deposited at greater depth (454 m). Diffractograms of material finer than 0·2 μm showed that this fraction contains mainly smectite. Hence it may be concluded that kaolinite and illite are coarser than smectite, and small aggregates with high

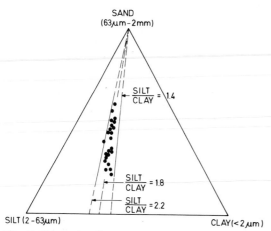

Fig. 5. Triangular diagram illustrating the variations in total mechanical composition for samples from core 1.

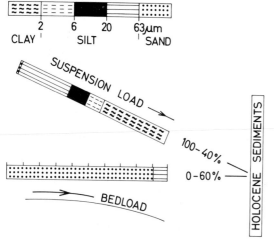

Fig. 6. Mechanical composition of suspension load and bed load at 239 m water depth based on core 1.

smectite/kaolinite ratios will have a greater chance of reaching deep water than larger aggregates containing more kaolinite.

Geotechnical data and texture

Water contents, Atterberg limits and shear strength values are shown in Figs 4 and 9.

The difference between liquid limit (W_L) and plastic limit (W_P) is called the plasticity index:

$$I_P = W_L - W_P.$$

Skempton (1953) has defined activity as the ratio of the plasticity index and the clay content:

$$\text{activity} = I_P/\% \text{ clay.}$$

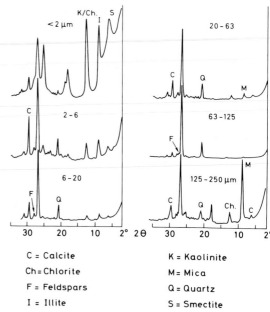

Fig. 7. Mineralogical composition for different size fractions

C = Calcite   K = Kaolinite
Ch = Chlorite   M = Mica
F = Feldspars   Q = Quartz
I = Illite   S = Smectite

Table 1. Amount of calcite in different fractions

Core	Depth (cm)	percentage calcite (μm)					
		< 2	2–6	6–20	20–63	63–125	> 125
1	116–120	9·9	20·2	18·7	10·2	6·8	10·1
2	148–151	6·1	18·1	17·6	11·7	8·1	6·4
2	398–400	8·4	17·5	18·8	11·8	8·2	8·0

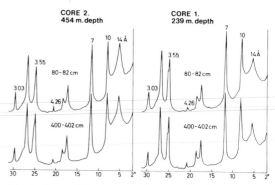

Fig. 8. X-ray diffractograms for clay fractions separated from the two cores studied.

When plotting the plasticity index versus the clay content (Fig. 10) the slope of a line from the origin to the sample point gives the activity. Salt content has a strong influence upon the activity (Bjerrum,

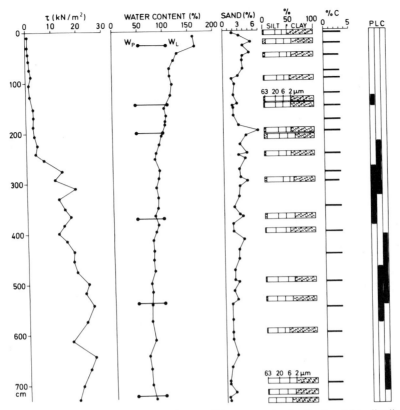

Fig. 9. Geotechnical and textural data from core 2 taken at 454 m water depth. On the right: distribution of biogenic structures (for explanation see Fig. 12).

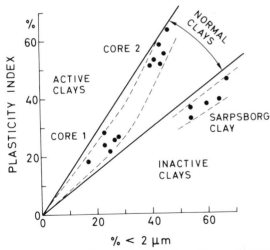

Fig. 10. Activity diagram for clays from Skagerrak compared to clays from Sarpsborg in the Oslofjord area (Bjerrum, 1954).

1954). All the samples from the Skagerrak fall within the normal clays group, while the most active clays from the Oslofjord area fall within the inactive clays group.

All the samples plotted in Fig. 10 have a similar salt content, thus differences in activity cannot be related to this parameter. Kaolinite is a very inactive mineral, and consequently, the high activity observed for the Skagerrak clays must be attributed to the fine-grained smectite. The material from core 2 has a slightly higher activity than the material from core 1. This is probably due to a small difference in the smectite/kaolinite ratio as illustrated by the X-ray data. The activity data strongly support the previous conclusion that there is no change in composition as a function of depth in these cores. This is important for interpreting the data on shear strength.

Shear strength

There is a drastic change in shear strength around 2·5 m depth in core 2 (Fig. 9). Above this depth the shear strength is slowly increasing with increasing depth and this is followed by a gradual decrease in water content. Below this depth the shear strength is much higher than above, and it is gradually increasing with increasing depth. This increase in shear strength is followed by a very small change in water content. Since the mineralogical and mechanical composition and Atterberg limits are not significantly different above and below 2·5 m, the observed change in strength cannot be due to a change in composition. This could be explained as a result of a period with erosion, and as a consequence a large hiatus would be expected around 2·5 m depth. There are, however, no indications of such an erosional period, and in particular the foraminiferal assemblages are identical above and below this level. Since only a small change in water content accompanies this large change in shear strength, a factor in addition to reduction in water content is required to explain the observed change in shear strength. We propose the following models.

(1) Above 2·5 m there is continuous consolidation resulting in frequent distortion of particle–particle contacts. Due to this the clay remains soft. As the water content is approaching 90% the movement slows down and firm particle–particle bonds are established. Since the shear strength due to this process is increasing more rapidly than the shear stress due to increasing sediment thickness, the consolidation will stop, and consequently, the water content remains constant for a long time.

(2) Overconsolidation in the two cores can also be caused by horizontal compaction (Plessmann, 1966) in the course of sediment creep and distortion on a slope 1·3° at core 1 and 5·2° at core 2. Even without much change in water content this process will result in a better grain to grain contact. Sedimentation continued after these movements with higher rates, as indicated by ^{14}C dates and change in bioturbation structures in core 2.

Looking to the sedimentary structures in Figs 11 and 12, one could argue that below 2·5 m depth different types of bedding (flaser-lenticular) and biogenic structures occur, which may affect the shear strength properties. However, an increase in shear strength of more than 100% is certainly not possible. The first model has in addition the advantage of continuity. It may be easily checked by examination of further cores in that region.

Sedimentary structures

General remarks

The sedimentary structures of the two cores have been evaluated from X-ray radiographs. The results are summarized in graphs at the same scale as the other parameters in Figs 4 and 9. As the diameter of the cores is relatively small (5 cm) in relation to the scale of many structures, the generalization and classification necessary for obtaining a suitable log graph may therefore imply some errors.

In both cores, bioturbation dominates the sedimentary structures. However, bedding features are preserved in varying amounts. Pyritic structures are important only in core 2.

Bedding structures

The more important primary structures preserved in the cores can be classified as follows:

(a) Horizontal rhythmic parallel lamination. Continuous, undisturbed bedding of very small alternating clay/silt layers is present in a few small sections (Fig. 11c), but may be important just for this reason, indicating stratigraphic events which are equivalent to periods of reduced (or absent) bottom fauna.

(b) Wavy-lenticular flaser bedding. In spite of the partial disturbance, this bedding type was recognized in several sections of both cores (Fig. 11d). The light lenses of coarser material (probably coarse silt/very fine sand) are interpreted as relics of a very small current ripple bedding, analogous to the much larger lenticular bedding described by Reineck, Singh & Wunderlich (1968). The sections with this bedding type are therefore interpreted as periods of higher activity of bottom currents.

(c) Flaser-lenticular bedding. At the top of core 1, a sequence 25 cm thick of alternating sand/mud layers occurs, displaying a typical flaser-lenticular structure in the sense of Reineck et al. (1968). They indicate strong bottom currents, able to carry bed load. Being the only occurrence of this bedding type in the core, a significant change in environmental conditions taking place in the geological present may be assumed. As the thickness of this sequence far exceeds the depth of todays burrowing activity the sequence will not be destroyed in the future by bioturbation. In contrast, it seems possible that similar (but thinner) layers in deeper parts of the profile were eliminated by bioturbation. The change to conditions with stronger bottom currents could therefore have occurred either gradually or abruptly.

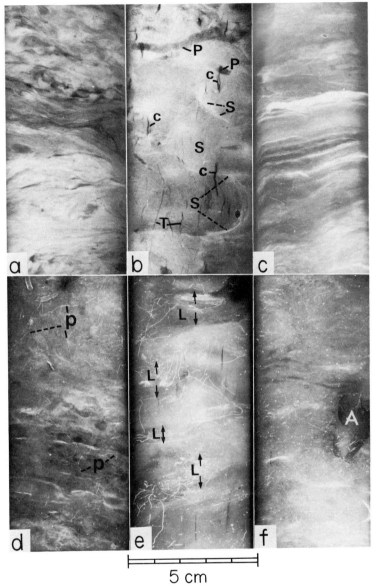

5 cm

Fig. 11. X-ray radiographs from sediment cores. (a) Core 1, 1–12 cm, flaser-lenticular bedding, in the upper part strongly bioturbated (deformational structure). (b) Core 1, 43–54 cm, bioturbation dominated by Scolicia burrows (S). P = Planolites tubes, T = hollow Trichichnus tubes, C = degassing (?) cracks. (c) Core 2, 32–43 cm, horizontal rhythmic parallel lamination (light sand/silt layers alternating with dark mud layers). (d) Core 2, 411–421 cm, upper part intensively bioturbated (deformational structures, lower part with non-rhythmic horizontal lamination, partly destroyed by bioturbation. Small light lines (p) are pyritic structures. (e) Core 2, 487–498 cm, pseudo-stratification due to Lophoctenium (?)-spreiten (L). Arrows indicate vertical extension of spreiten burrows. Small light dots and white lines are pyrite structures, partially concentrated in spreiten. (f) Core 2, 570–581 cm, relicts of horizontal lamination and/or biogenic structures. White dots are pyrite spheres.

The steady increase of sand content in the profile (Fig. 4) favours the first possibility.

Bioturbation

In general, only a part of the bioturbation structures present in the radiographs can be defined as 'Lebensspuren', a larger number of the structures being biogenic deformational structures in the sense of Schäfer (1962).

The biogenic structures were analysed in terms of the palaeoichnological nomenclature. The genus categories according to Häntzschel (1975) were used, in spite of possible errors of determination due to the 'two-dimensional' and fragmentary way of observation in the radiograph. Nevertheless, we believe that the application of this nomenclature is the adequate method to describe biogenic structures in marine sediments as far as the generating animals are largely unknown. In the following, the more important of the observed structures are described, leading to characteristic associations of lebensspuren as denoted in the graphs of Figs 4 and 9.

(a) *Planolites*. Here a variety of different sizes and shapes of cylindrical tubes are comprised, although it is felt that this category may be too large for being of great value for environmental analysis. The specimens are, however, too rarely distributed to give satisfying results from the statistical point of view when split further. Most of the tubes are filled with finer material than the surroundings, yielding darker structures in the X-ray negatives (Fig. 11d). In core 1 the specimens are widely distributed, while core 2 contains a considerable amount of individuals only in the section between 3 and 4 m. This occurrence is significantly related to the lenticular bedding type.

(b) *Trichichnus*. Numerous small, hollow tubes in various marine sediments (Seibold *et al.*, 1973; Wetzel, 1979) are in agreement with the morphologic characteristic of *Trichichnus* defined by Frey (1970). In core 1, abundant hollow tubes of winded shape and a few tenths of a millimetre in diameter belong to this type. In core 2 it is less frequent, but a part of the pyritized structures belongs to this group.

(c) *Chondrites*. The specimens occurring in core 2 show a rather uniform morphology. Their tubes have a relatively small diameter (0·5–0·8 mm) and are often hollow without too much branching. They may best be related to the *Chondrites* type D defined by Wetzel (1979). According to the general bathymetric distribution of Chondrites, the location of core 1 is probably too shallow for the ecological needs of the generating organism.

(d) *Lophoctenium*. Following Wetzel (1979), we tentatively attribute these feeding burrows with irregular character and little internal structures (Fig. 11e), but relatively large extension and thicknesses of more than 1 cm, to the ichnogenus *Lophoctenium*, described from ancient sediments.

(e) *Scolicia*. In the upper part of core 1 *Scolicia* burrows with typical concentric structures occur, which are known to be generated by irregular echinoids (Fig. 11). The occurrence of this together with an increasing sand content towards the top of this core seems to be in agreement with the ecological relationships of these type of burrows (Wetzel, 1979).

(f) *Vertical spreiten burrows*. In this category, different types of vertical spreiten burrows, including *Teichichnus* (Seilacher, 1955) are comprised. To describe them as a *Teichichnus* association seems to be justified because of their occurrence within one definite core section, in the lower part of core 1. Generally, vertical spreiten burrows are considered as indicating phases of higher accumulation rates (Goldring, 1964), and therefore the occurrence of these burrows is in agreement with the location of core 1 rather than core 2.

Pyritic structures

Thin linear, often string-like pyritic forms are abundant in core 2 (Fig. 11d–f), but nearly absent in core 1. They are recognized frequently as pseudomorphic structures of *Chondrites* and *Trichichnus*. In addition to these elongated structures, isolated pyrite spherulites (Love, 1963) are frequent, in some sections even abundant (Fig. 11f). Occasionally, pyrite formation appears to be favoured by the backfill material of the *Lophoctenium* burrows (Fig. 11e), string-like forms being strongly concentrated within them. The upper limit of pyrite formation is sharply defined at 90 cm sediment depth in core 2. A second limit at 4 m separates very thin and small pyrite structures from longer and thicker forms (up to 10 cm vertical extension and 0·5 mm in diameter) in the lower section.

Foraminiferal stratigraphy

The foraminiferal assemblages occurring in the two cores (Figs 13 and 14) are generally composed

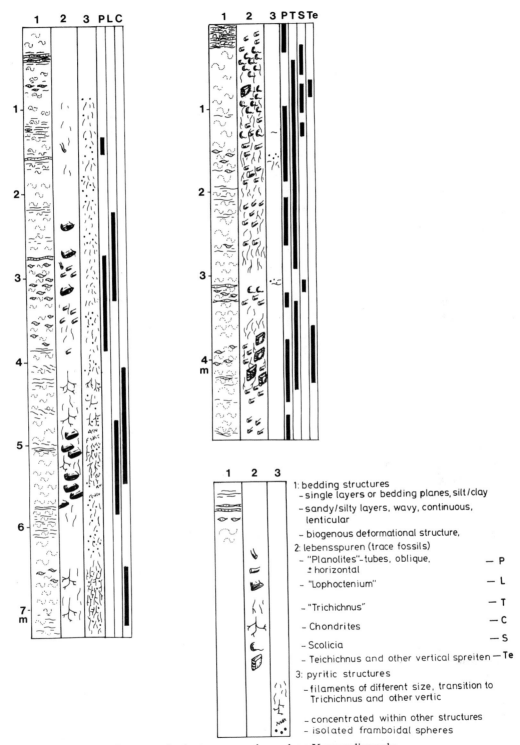

Fig. 12. Structures in the two cores observed on X-ray radiographs.

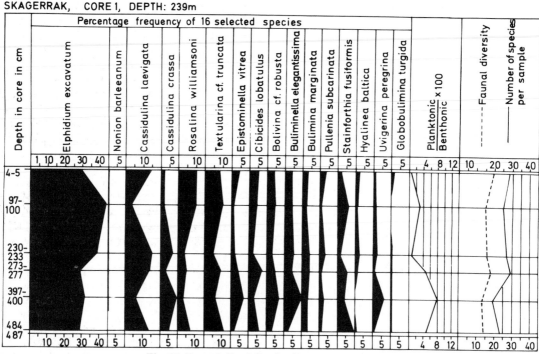

Fig. 13. Foraminiferal distribution chart for core 1.

of species with boreo-arctic, boreal, and boreo-lusitanian distribution. Although the relative frequency of the individual species shows greater or smaller changes, the general aspect of the assemblages remains the same throughout the cores, which are referred to the Flandiran. Quantitatively important species with a boreal or boreo-lusitanian main distribution are: *Nonion barleeanum*, *Cassidulina laevigata*, *Bolivina* cf. *robusta*, *Buliminella elegantissima*, *Bulimina marginata*, *Stainforthia fusiformis*, *Hyalinea baltica*, *Uvigerina peregrina*, and *Globobulimina turgida*.

The faunal diversity is high in both cores and shows an increase slightly from the bottom to the top; it is 14–19 in core 1, and 11–23 in core 2. The ratio between planktonic and benthonic specimens varies from 0·003 to 0·08 in core 1, and from 0·022 to 0·18 in core 2. These very low values indicate the reduced influence of Atlantic waters in this area.

The laboratory treatment and analysis of the material is in accordance with the standard procedure for quaternary samples, described by Feyling-Hanssen (1964).

Faunal comparisons

The two cores show marked differences with regard to the quantitative distribution of several species which must be explained by environmental differences at the two stations.

In core 1 the dominant species *Elphidium excavatum* accounts for 29–44 % of the fauna, while in core 2 it forms only 0·5–9 %. This arctic–boreal species is quite common in some Flandrian and recent assemblages described from the shallower parts of the Skagerrak, e.g. by Lange (1956) and Fält (1977).

Nonion barleeanum occurs at <1 % in core 1, while in core 2 it is represented at 6–31 %. The considerably higher frequency in the deeper core is in accordance with the general tendency of *N. barleeanum* to increase with increasing depth. This tendency appears from papers concerning the North Sea (Lange, 1956) and the Barents Sea (Jarke, 1960; Lorange, 1977).

Bolivina cf. *robusta* has immigrated into the Skagerrak during the Flandrian as indicated by Lange (1956). In this region the species is frequent at depths from 300 to 500 m. In the Oslofjord it is dominant at depths greater than 100 m (Risdal,

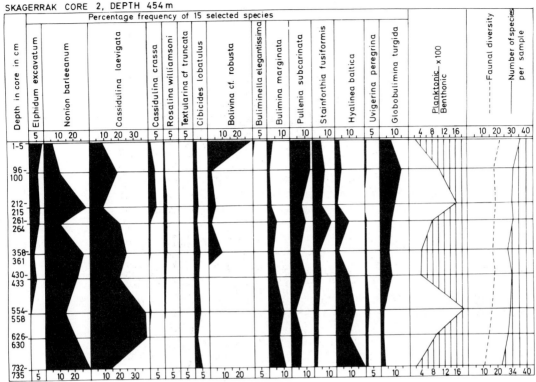

SKAGERRAK CORE 2, DEPTH 454 m

Fig. 14. Foraminiferal distribution chart for core 2.

1964). In the present material *B*, cf. *robusta* is present throughout core 1, where it accounts for 3–7% of the assemblages. In the lower part of core 2, from 735 to 430 cm, the species is missing (with the exception of a single specimen). In the upper part of the core it is represented by 2–28%.

Other important differences between the two cores are: the higher frequency of *Cassidulina laevigata*, *Pullenia subcarinata*, *Hyalinea baltica*, and *Globobulimina turgida* in core 2; and the higher frequency of *Cassidulina crassa*, *Rosalina williamsoni*, *Textularina* cf. *truncata*, and *Bulliminella elegantissima*, in core 1.

Age and correlation

A foraminiferal sequence similar to that of core 2 was described from a nearby station at a water depth of 453 m by Lange (1956). The length of his core was 925 cm, which he subdivided in to a lower Cassidulina–Nonion zone (925–330 cm) characterized by the dominance of *Cassidulina laevigata* and *Nonion barleeanum*, and an upper Bolivina zone (330–0 cm)

characterized by *Bolivina* cf. *robusta* showing strong dominance at three horizons.

The similarity of the foraminiferal sequences found in the two cores warrant correlation of the lower 735–400 cm of our core 2 with the Cassidulina–Nonion zone and the upper 400–0 cm with the Bolivina zone. Core 1 contains *Bolivina* cf. *robusta* throughout its whole length and, therefore, it must be correlated with the Bolivina zone. From these correlations it follows that the Bolivina zone increases in thickness from deeper to shallower waters on the southern slope of the Norwegian Channel. The most probable explanation of this increase is the higher sedimentation rate in the shallower southern areas which are more influenced by the north-east currents that follow the coast of northern Jutland and are responsible for considerable mass transport.

The Cassidulina–Nonion zone of Lange (1956) and the same unit in our core 2 seems to be younger than zone C (*B. margimata, U. peregrina, T. bocki*) described by Fält (1977) from the Skagerrak and the Uvigerina–Cassidulina zone described by Nagy & Ofstad (1980) from the northern part of the Norwe-

27

gian Channel. The foraminiferal assemblages occurring in the latter two zones are regarded as transitional between the cold late Weichselian, and the temperate Middle Flandrian faunas, and are referred mainly to Preboreal and Boreal time. Consequently, the Cassidulina–Nonion zone and the overlying Bolivina zone must be referred to the Atlantic, Subboreal and Sub-atlantic periods.

Dating and sedimentation rates

A preliminary chronostratigraphy of the present cores was established by radiocarbon dating. Analysis for ^{14}C and stable carbon-isotope composition was performed on CO_2 prepared from the organic fraction of the sediment samples by wet oxidation (Suess & Erlenkeuser, 1975; Erlenkeuser, 1979). The radiocarbon dates reported here are based on the 'conventional' ^{14}C half life (5568 yr) and 95% of the NBS oxalic acid ^{14}C standard activity (Stuiver & Polach, 1977). The ^{14}C dates were not normalized with respect to δ ^{13}C variations. Normalization (Stuiver & Polach, 1977) would increase the ^{14}C age by about 65 yr.

δ ^{13}C was measured against a laboratory reference, which was calibrated against the NBS 20 ^{13}C/^{12}C standard. The laboratory reference has a δ ^{13}C = $+1.8‰$ on the PDB scale. This figure replaces a previously used value, taken from the literature, of $-1.06‰$ (Erlenkeuser, Suess & Willkomm, 1974; Suess & Erlenkeuser, 1975; Erlenkeuser, Metzner & Willkomm, 1975). The present δ ^{13}C results are presented as per mille on the PDB scale.

Results of dating

The results of the carbon isotope measurements are presented in Table 2 and Fig. 15. Core 1 shows a nearly uniform ^{14}C age of about 2600 yr BP throughout the core. The radiocarbon dates suggest very rapid accretion of the sediment at a rate greater than 12.7 mm yr^{-1} (at the 90% confidence level). A similarly rapid rate of sedimentation (greater than 5.2 mm yr^{-1} at the 90% confidence level) is found for the upper section of core 2 while the part below accumulated at a rate of roughly 1.4 mm yr^{-1}.

The stable carbon isotope composition is about δ ^{13}C = $-21‰$. This value is typical for many marine sediments (Erlenkeuser, 1978b) and is the same within a range of about $\pm 1‰$ as found for the marine organic matter of other cores from the North

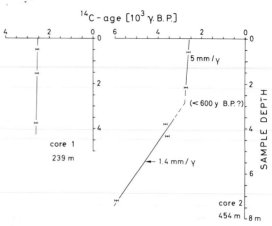

Fig. 15. Radiocarbon dates of the organic fraction versus sample depth in the cores.

Table 2. Conventional radiocarbon dates of the organic fraction and stable carbon isotope composition

Lab. no.	Depth in the core (cm)	^{14}C-age (yr BP)$\pm 1\sigma$	δ ^{13}C (0/00 PDB)
\multicolumn{4}{c}{Core 1 (239 m water depth)}			
KI-1606 0·04	44–54	2550±80	−20·8
0·01	150–160	2560±95	−21·2
0·02	370–380	2620±80	−20·4
0·03	485–495	nd	−20·8
\multicolumn{4}{c}{Core 2 (454 m water depth)}			
KI-1607 0·03	53–63	2610±85	−21·4
0·01	210–220	2710±85	−20·5
0·04	374–382	3660±90	−20·2
0·05	427–435	3570±90	−21·9
0·02	711–720	5910±90	−21·4

Sea and the western Baltic (Erlenkeuser et al., 1974, 1975; Suess & Erlenkeuser, 1975).

Discussion of age

The radiocarbon ages based on the bulk organic fraction of shallow-water marine sediments are generally found to be affected by older sedimentary particles that have been resuspended and redeposited by the action of waves and bottom currents. The ^{14}C dates are frequently much older than the depositional age of the sediment (Erlenkeuser et al., 1974; Suess & Erlenkeuser, 1975; Erlenkeuser, 1978a, 1979). It appeared from these studies that the large number of small-scale processes involved in the interaction of the sea and the sea bottom results in a fairly constant and reproducible radiocarbon age of the freshly arriving sedimentary particles at a given location even through longer periods ($\sim 10^3$ yr) of the depositional history (Erlenkeuser, 1978a, 1979). This initial ^{14}C age of the sediment hence appears suitable to define the starting point of a radiocarbon-based depositional chronostratigraphy.

(1) The initial age of the sediment is represented by the ^{14}C age of the sediment surface and may be estimated by extrapolating the sedimentation line upward to the top of the sediment pile (Suess & Erlenkeuser, 1975). For the present cores, the surface ^{14}C age is about 2600 yr BP. Similar values have been observed for other cores from the North Sea, e.g. about 3000 yr for sediments from the tidal flats of Nordstrand Island, Schleswig-Holstein, Germany; about 2000 yr for two cores from the Fladenground area, northern North Sea; and 2500 yr for a core from the muddy grounds of the eastern middle North Sea. For the latter core, the young depositional age of the sediment surface was proved by the ^{210}Pb ^{137}Cs distribution. A somewhat lower ^{14}C age of the top sediment, about 1000 yr, has been reported for the Baltic (Erlenkeuser, 1978a, 1979).

(2) The stable carbon isotope ratio of the organic fraction of the recent North Sea deposits does not provide evidence for a significant contribution of terrigenous plant litter. The high surface ^{14}C age is therefore unlikely to result from erosion of Holocene-aged peat deposits in the North Sea area.

(3) Nearly identical ^{14}C age values are found for the upper 2 or 3 m section of core 2 and the upper 4 m section of core 1. Both these intervals of rapid sedimentation coincide with a reduced shear strength (see Fig. 9). Assuming identical age of these core sections—as is suggested by the radiocarbon dates and the foraminiferal stratigraphy—the shallower lying core 1 appears to have accumulated more rapidly than the deeper lying core 2. This conclusion is also supported by the different thickness of the *Bolivina* cf. *robusta* zone in the two cores (Figs 13 and 14). The higher rate of sedimentation in core 1 is apparently accounted for by a rapid supply of a sand-sized particle fraction as revealed by the grain-size distribution of core 1. If the recent rapid sediment growth observed at both locations is not the result of a local and/or episodic

event, then it points to a possibly high mass transport of particulate matter from the more energetic environment of the shallow water zone of the Skagerrak towards the abyssal depths of the Norwegian Channel.

(4) Subtracting the ^{14}C age of the sediment surface, the radiocarbon dates of the deeper samples of core 2 provide an estimate of the depositional age of 1000 yr BP at a sample depth of 4 m and of 3200 yr BP at 7·9 m. The 4 m depth corresponds roughly to the base of the *Bolivina* cf. *robusta* zone.

The sedimentation rates suggested by the ^{14}C dates of the upper sections of both cores appear strikingly high. In contrast to these results, the ^{210}Pb distribution yielded much lower rates of sediment accumulation. Excess ^{210}Pb was clearly identified only above the 14 cm sample depth in core 1 and above 20 cm in core 2. However, a first weakly pronounced appearance of excess radiolead was possibly found at 28 cm in core 1 and at 43 cm in core 2. The ^{210}Pb analyses suggest accumulation rates of roughly 1 to 2 mm yr^{-1} and 2–4 mm yr^{-1} for cores 1 and 2, respectively. These rates represent the average over the last 100 or 150 years of the depositional history. They suggest that in recent times core 1 accumulated at a lower rate than core 2, while the ^{14}C results indicate the opposite relation of sedimentation rates for the deeper core sections.

The differences between the sedimentation rates suggested by the two dating methods appear rather large, and the processes producing them are not well understood yet. The apparent changes of sedimentation rates may or may not be real. In the course of the climatic history of the Holocene, the sedimentary source material and the depositional environment, particularly in marginal seas, are likely to have undergone long-term changes which have possibly affected the initial ^{14}C age of the sediment as well. The North Sea sediments should have been largely influenced by older materials remobilized by the great number of heavy storm floods, which occurred between AD 1200 and AD 1750 (Lamb, 1977, table 13.3). Not only could this climatic deterioration have caused an increased supply of sedimentary matter to the Skagerrak area, thus giving rise to an increased rate of sedimentation, but could also have led to a gradually increasing initial ^{14}C age of the freshly arriving particulate load, thus leaving an even steeper ^{14}C age distribution in the sediment than the

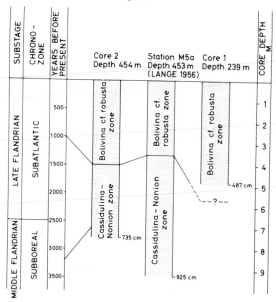

Fig. 16. Tentative correlation of cores from the Skagerrak.

faster rate of sediment growth alone would account for.

As shown by the foraminiferal stratigraphy (Fig. 16), core 2 belongs completely to the Flandrian (735 cm), and is probably younger than the Boreal chronozone. It indicates that the average sedimentation rate in the core is higher than 0·92 mm yr^{-1}. In this connection it must be noted that the nearby core of Lange (1956) with a length of 925 cm shows an average Flandrian sedimentation rate higher than 1·15 mm yr^{-1}.

If we assume that the base of the *Bolivina* cf. *robusta* zone is contemporaneous in the area, then we can conclude that the sedimentation rate in the upper part of the cores increases from the core of Lange (3·4 mm yr^{-1}) through our core No. 2 (4 mm yr^{-1}) to core No. 1 (>4·8 mm yr^{-1}). It is, however, clear that exact sedimentation rates cannot be calculated without more reliable age determinations.

FINAL DISCUSSION AND CONCLUSION

The investigation on the two cores revealed the following major trends, changes and events in the profiles:

(1) an upward increase of sand content at constant silt/clay ratio;

(2) a strong decrease in shear strength, starting at 400 cm below the surface in core 1 and at 250 cm in core 2;

(3) an exceptional layer of flaser-lenticular bedding at the top of core 1;

(4) a general upward decrease of definite biogenous structures and, simultaneously with an increase of deformational structures in core 2, above 300 cm depth;

(5) a stratigraphic division of core 2 according to *Bolivina* cf. *robusta* while *B.* cf. *robusta* is found throughout core 1;

(6) a sudden increase of sedimentation rate in core 2 at 250 cm depth and a generally higher accumulation rate in core 1.

The sand content is gradually increasing in both cores and indicates an increasing influence of bedload carrying currents. This is associated with an increased rate of sedimentation in the upper part of core 2 and may also account for the even higher supply of sandy material observed in core 1.

The different thickness of the *B. robusta* zone in the cores reveals a higher sedimentation rate in the area of core 1. This is in accordance with the ^{14}C results.

The change in lebensspuren characteristics in core 2 may be explained as a consequence of an increased sedimentation rate which disfavoured the activity of spreiten borrowing sediment feeders (Seilacher, 1967). A major problem is the explanation of the break in shear strength observed in the two cores.

This observation does not parallel with similar changes of the water content, clay mineralogy and grain-size distribution. As many shear strength profiles of long cores have shown (Inderbitzen, 1974; Silva *et al.*, 1976), no inherent mechanism in the three-variable system, sedimentation rate/overburden/consolidation state, has been found to produce a consolidation front in the profile.

However, the change in shear strength might be related to a sudden change of the sedimentation rate in core 2; but this cannot explain overconsolidation in the deeper parts of the cores. Therefore other causes have to be considered.

As mentioned above, the formation of furrows parallel to the slope under varying thicknesses of sediment cover are interpreted as the effect of slow downward sediment creeping rather than of a single mass slide. Further down the slope the sediment layer with increased shear strength in core 2 could then be interpreted as a result of horizontal compaction by an earlier phase of active mass movements as Plessmann (1966) has shown in Mesozoic and Palaeozoic sediments. The increased sedimentation rate above this layer could be attributed to an increased availability of fine-grained material due to an increased roughness and enlarged surface exposed to the bottom currents.

An increased current activity as deduced from the sand content may have contributed as well to the high sedimentation rate. The increased sedimentation rate may also explain the reduction of shear strength above the sub-bottom reflector at the site of core 2 and in core 1. Analogous phenomena have been reported by Silva *et al.* (1976).

An increase in frequency of bed-load carrying currents in the Skagerrak may be caused by climatic changes in the younger Holocene. There are many indications by recent investigations that a general increase of storm frequency took place in the past thousand or two thousand years (Lamb, 1977).

As the Jutland current is controlled also by meteorological conditions (Larsson & Rodhe, 1979), these conditions could probably have induced a higher activity of the Jutland current and downslope transport of sand. The occurrence of flaser–lenticular bedding at the top of core 1 is considered as a clear illustration of these conditions. However, further investigations are necessary to test the validity of this model and its regional extension or corresponding systems respectively.

ACKNOWLEDGMENTS

P. Jørgensen would like to express thanks to Deutscher Akademischer Austauschdienst (DAAD) for stipend during his stay in Kiel. We all want to thank the captain and crew on R.V. *Poseidon* for their help during cruises.

We appreciate helpful discussions with Professors E. Seibold and J. Thiede.

REFERENCES

BJERRUM, L. (1954) Geotechnical properties of Norwegian marine clays. *Norwegian Geotechn. Inst. Publ.* **4**, 1–21.

DERA, J., HAPTER, R., PRANDKE, H., WOZNIAK, B., GOHS, L., KAISER, W., RÜTUNG, W. & ZALEWSKI, S. (1974) Untersuchungen zur Wechselwirkung zwischen den optischen, physikalischen, biologischen und chemischen Umweltfaktoren in der Ostsee. *Geod. Geoph. Veröff.* **13**, 1–100.

EISMA, D. (1976) Deeltjesgrootte van dood gesuspendeerd materiaal in de Zuidelijke Bocht van de Nordzee. *Rep. Neth. Inst. Sea Res.* p. 19.

EISMA, D. (1981) Sedimentatie van gesuspendeerd materiaal in de Noordzee. (in preparation).

EISMA, D. & GIESKES, W.W.L. (1977) Particle size spectra of non-living suspended matter in the Southern North Sea. *Rep. Neth. Inst. Sea Res. 1977*, p. 7.

EISMA, D. & KALF, J. (1978) Suspended matter between Norway and Shetland and in the Sognefjord. *Rep. Neth. Inst. Sea Res. 1978*, p. 13.

ERLENKEUSER, H. (1978a) The use of radiocarbon in estuarine research. In: *Biogeochemistry of Estuarine Sediments. Proc. UNESCO/SCOR workshop, Melreux, Belgium, December 1976*, pp. 140–153. UNESCO, Paris, 293 pp.

ERLENKEUSER, H. (1978b) Stable carbon isotope characteristics of organic sedimentary source materials entering the estuarine zone. In: *Biogeochemistry of Estuarine Sediments. Proc. UNESCO/SCOR workshop, Melreux, Belgium, December 1976*, pp. 199-206. UNESCO, Paris, 293 pp.

ERLENKEUSER, H. (1979) Environmental effects on radiocarbon in coastal marine sediments. In: *Radiocarbon Dating, Proc. Ninth int. Radiocarbon Conference, Los Angeles and La Jolla, 1976* (Ed. by R. Berger and H. E. Suess), pp. 453–469. University of California Press, Berkeley. 787 pp.

ERLENKEUSER, H., METZNER, H. & WILLKOMM, H. (1975) University of Kiel radiocarbon measurements: VIII. *Radiocarbon* **17**, 276–300.

ERLENKEUSER, H., SUESS, E. & WILLKOMM, H. (1974) Industrialization affects heavy metal and carbon isotope concentrations in recent Baltic Sea sediments. *Geochim. Cosmochim. Acta*, **38**, 823–842.

FÄLT, L.M. (1977) Gränsen Pleistocen/Holocen i marina sediment utanför svenska västkusten. Chalmers Tekn. Högsk. *Göteb. Univ. Geol. Inst., Publ.* A **14**, 39 pp.

FEYLING-HANSSEN, R.W. (1964) Foraminifera in Late Quarternary deposits from the Oslofjord area. *Norges geol. Unders.* **225**, 383 pp.

FREY, R.W. (1970) Trace fossils of Fort Hays Limestone Member of Niobrara Chalk (Upper Cretaceous), west-central Kansas. *Paleontol. Contrib. Art. Univ. Kansas*, **53**, 1–41.

GOLDRING, R. (1964) Trace-fossils and the sedimentary surface in shallow-water marine sediments. *Develop. Sediment.* **1**, 136–143.

HÄNTZSCHEL, W. (1975) Trace fossils and problematica. In: *Treatise on Invertebrate Paleontology*, Part W, *Miscellanea*, suppl. 1, 2nd edn (Ed. by C. Teichert). Boulder, Colorado, and Lawrence, Kansas. 269 pp.

INDERBITZEN, A.L. (1974) *Deep-sea Sediments.* Plenum Press, New York.

JARKE, J. (1960) Beitrag zur Kenntnis der Foraminiferenfauna der mittleren und westlichen Barents-See. *Int. Revue ges. Hydrobiol. Hydrogr.* **45**, 581–654.

KAUTSKY, H. (1977) Strömungen in der Nordsee. Radioisotope aus Wiederaufbevertungsanlagen signalisieren den Weg. *Umschau*, **77**, 672–673.

LAMB, H.H. (1977) *Climate Present, Past and Future.* Vol. 2. *Climatic History and the Future.* Methuen, London. 835 pp.

LANGE, W. (1956) Grundproben aus Skagerrak und Kattegat, mikrofaunistisch und stratigraphisch untersucht. *Meyniana*, **5**, 51–86.

LARSSON, A.-M. & RODHE, J. (1979) Hydrographical and chemical observations in the Skagerrak 1975–1977. *Rep. 29.* Oceanografiska Institutionen, Gøteborgs Universitetet.

LORANGE, K. (1977) *En mikropaleontologisk undersøkelse av kvartære sedimenter i nordvestre del av Barentshavet.* Degree thesis, University of Oslo.

LOVE, L.G. (1963) Pyrite spheres in sediments. In: *Biochemistry of Sulphur Isotopes* (Ed. by M. L. Jensen). *Proc. Nat. Sci. Found. Symp.* pp. 121–143. Yale, 1962.

NAGY, J. & OFSTAD, K. (1980) Quaternary foraminifera and sediments in the Norwegian Channel. *Boreas*, **9**, 39–52.

PLESSMANN, W. (1966) Diagenetische und kompressive Verformung in der Oberkreide des Harz-Nordrandes sowie im Flysch von San Remo. *Neues Jb. Geol. Paläont.* **8**, 480–493.

REINECK, H.-E., SINGH, I.B. & WUNDERLICH, F. (1968) Classification and origin of flaser and lenticular bedding. *Sedimentology*, **11**, 99–104.

RISDAL, D. (1964) Foraminiferfaunaens relasjon til dybdeforholdene i Oslofjorden, med an diskusjon av de senkvartære foraminifersoner. *Norg. geol. Unders.* **226**, 142 pp.

SCHÄFER, W. (1962) *Aktuo-Paläontologie.* Kramer, Frankfurt. 666 pp.

SEIBOLD, E., DIESTER-HAASS, L. FÜTTERER, D., LANGE, H., MÜLLER, P. & WERNER, F. (1973) Holocene sediments and sedimentary processes in the Iranian part of the Persian Gulf. In: *The Persian Gulf* (Ed. by B. H. Purser), pp. 57–80. Springer Verlag, Berlin.

SEILACHER, A. (1955) Spuren und Lebensweise der Trilobiten; Spuren und Fazies im Unterkambrium. In: *Beitr. zur Kenntnis des Kambriums in der Salt Range (Pakistan)* (Ed. by O. H. Schindewolf and A. Seilacher). *Akad. Wiss. Lit. Mainz, Math.-naturwiss. Kl., Abh., Jg.* **10**, 342–399.

SEILACHER, A. (1967) Bathymetry of trace fossils. *Mar. Geol.* **5**, 413–428.

SILVA, A.J., HOLLISTER, C., LAINE, E.P. & BEVERLY, B.E. (1976) Geotechnical properties of deep sea sediments: Bermuda Rise. *Mar. Geotech.* **3**, 195–232.

SKEMPTON, A.W. (1953) The colloidal 'activity' of clays. *Proc. 3rd Int. Conf. Soil Mech. Found. Engin.*, Zürich, **1**, 57.

STUIVER, M. & POLACH, H.A. (1977) Discussion: reporting of ^{14}C data. *Radiocarbon*, **19**, 355–363.

SUESS, E. & ERLENKEUSER, H. (1975) History of metal pollution and carbon input in Baltic Sea sediments. *Meyniana*, **27**, 63–75.

VAN WEERING, T.C.E. (1975) Late Quaternary history of the Skagerrak; an interpretation of acoustic profiles. *Geologie Mijnb.* **54**, 130–145.

VAN WEERING, T.C.E., JANSEN, J.H.F. & EISMA, D. (1973) Acoustic reflection profiles of the Norwegian Channel between Oslo and Bergen. *Neth. J. Sea Res.* **6**, 241–263.

VAN WEERING, T.C.E. & PAAUWE, E.F. (1979) A coarse fraction analysis and scanning electron microscope observations on four sediment cores from the Norwegian Channel. *Int. Meeting Holocene Marine Sedimentation in the North Sea Basin* (Abstr.), p. 56.

WETZEL, A. (1979) *Bioturbation in spätquartären Sedimenten von NW-Africa.* Dissertation, University of Kiel. 111 pp.

Spec. Publs int. Ass. Sediment. (1981) **5**, 415–428

Supply and deposition of suspended matter in the North Sea

D. EISMA

Netherlands Institute for Sea Research, Texel, the Netherlands

ABSTRACT

Suspended matter in the North Sea is supplied from the northern Atlantic Ocean, the Channel, the Baltic, rivers, coastal erosion, seafloor erosion, primary production and the atmosphere. It is moved through the North Sea with a general anti clockwise circulation, is concentrated near the coasts of the southern North Sea and is deposited in a number of small areas: river estuaries, tidal flats, small areas south-east of Helgoland and south-west of Dogger Bank and in the Skagerrak–Kattegat and possibly also in the Elbe Rinne. The total amount of suspended matter yearly supplied to the North Sea is estimated to be at least approximately 34×10^6 ton yr^{-1}, to which an unknown amount should be added supplied by seafloor erosion and from the Channel through the coastal waters of the Dover–Calais Straits. Approximately $11 \cdot 4 \times 10^6$ ton yr^{-1} is supplied from the Atlantic Ocean and more than $22 \cdot 5 \times 10^6$ ton yr^{-1} from sources in the Channel and the North Sea. Outflow into the Northern Atlantic is less than $14 \cdot 4 \times 10^6$ ton yr^{-1}. From the deposition rate in the areas of sedimentation it is estimated that $21–31 \cdot 5 \times 10^6$ ton yr^{-1} is being deposited, whereas $c.\ 2 \times 10^6$ ton is yearly dumped on land. The total of estimated outflow, deposition and dumping is $37 \cdot 5–48 \times 10^6$ ton yr^{-1}. Although the total estimated supply comes within the range of estimated outflow + deposition, the figures that have been used can only be regarded as broad estimates. Uncertainties chiefly concern the amount of seafloor erosion, the sedimentation rates in the areas of deposition and the transport of fine-grained material in suspension near the bottom.

INTRODUCTION

The general distribution of suspended matter in the North Sea is characterized by high concentrations in the coastal waters of the Southern North Sea and low values offshore in the Skagerrak and north of Dogger Bank (Fig. 1). Actual concentrations range from more than 100 mg l^{-1} in nearshore waters off Belgium–Holland and East Anglia to less than $0 \cdot 2 \text{ mg l}^{-1}$ in the northern North Sea and the Norwegian Channel (Eisma, 1976; Eisma & Kalf, 1978, 1979). This would suggest that the main sources of suspended matter are the rivers that enter the southern North Sea and the coastal erosion in that area. However, as was pointed out by Postma (1978), the circulation pattern in the southern North Sea tends to concentrate suspended matter in the near-shore areas because residual currents carry bottom water to the shore (Dietrich, 1955; Ramster, 1965).

Further concentration takes place in the estuaries and in the Wadden Sea where suspended matter is accumulated (Postma & Kalle, 1955; Van Straaten and Van Kuenen, 1957; Postma, 1961), but also some suspended material will follow the general anti-clockwise circulation in the North Sea and will reach the Skagerrak or the Norwegian Trough, eventually flowing out into the Atlantic Ocean. Due to these complications and to a general lack of sufficient data, the origin and dispersion of suspended matter in the North Sea is not fully understood, but some estimate of the quantities involved can be obtained by considering the budget of supply, deposition and outflow into the Atlantic Ocean. This was tried by McCave (1973), but on the basis of the data available at that time the estimated supply could not be brought into agreement with the estimated deposition. Since then new data have become available that give a better balance. A

0141-3600/81/1205-0415 $02.00

416　　　　　　　　　　　　　　　　　　*D. Eisma*

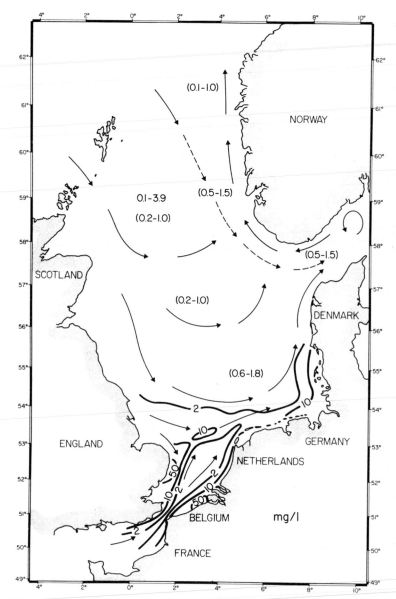

Fig. 1. General distribution of suspended matter and water circulation in the North Sea. Heavy lines and numbers: suspended matter concentrations (mg, 1⁻¹). The numbers between brackets were obtained during only a few cruises made in 1977–1979. The others are based on data from the literature as well. Arrows, general direction of flow: broken arrows, subsurface flow.

balanced budget, besides being of interest for the understanding of suspended-matter dispersal in the North Sea, is of practical value for estimating the origin of material deposited in estuaries and other shoaling areas, as well as for estimating the fate of pollutants associated with the suspended particles.

SUPPLY

Suspended matter in the North Sea, where measured, was found to consist largely of particles smaller than 60–70 μm (Lee & Folkard, 1969; McCave, 1979; Eisma & Kalf, 1978, 1979 and unpublished data NIOZ). During storms larger particles can be temporarily in suspension in large quantities in the shallow parts of the southern North Sea where the bottom is sandy: Coulter Counter measurements indicate a coarsening of the suspended matter in these areas when the weather becomes rough, especially near the bottom, but this coarser material drops out soon after the weather calms down. This temporarily and locally stirred up material is not included in the budget calculations carried out in this paper. Also not included is the material (either sand or finer particles) moving in suspension or saltation within 0·5 m above the bottom.

Suspended matter is supplied to the North Sea from the Atlantic Ocean, the Channel and the Baltic, from rivers and the atmosphere, from coastal erosion and sea-floor erosion, and from primary production. Atlantic water flows in around Scotland and around Shetland (Fig. 2). The inflow around Scotland goes southward through the northern and central North Sea, partly following the Scottish–English coast. The inflow around Shetland dips below the northward-flowing surface water and continues southward, following the western side of the Norwegian Channel into the Skagerrak. Here it mixes with southern North Sea water flowing in along the north coast of Denmark and with inflow from the Baltic. Atlantic water enters also through the Channel. It decreases in salinity from 35·3‰ at the western entrance to just above 35·0‰ near the Dover–Calais Straits due to admixture of fresh water from the Channel coasts (Dietrich, 1950). The Baltic inflow has a salinity below 35‰ and in the Skagerrak mixes with North Sea water and Atlantic water. River water enters the North Sea chiefly in the south where the large rivers (Rhine, Thames, Humber and Elbe) are located. Rainfall in the North Sea is almost entirely balanced by evaporation (Otto, 1976).

The suspended matter concentrations in the inflowing Atlantic water are of the order of 0·1–0·2 mg l^{-1} with an average near 0·2 mg l^{-1} (Eisma & Kalf, 1978 and unpublished data NIOZ). This is slightly higher than the amounts found by Jacobs & Ewing (1961) in the North Atlantic Ocean (<0·1–0·25 mg l^{-1}) so that probably some material is already picked up from the continental shelf before the oceanic water enters the North Sea. At a yearly inflow of 51,000 km^3 from the Atlantic (Otto, 1976), the amount of suspended matter supplied is of the order of 10×10^6 ton (dry weight). Suspended-matter concentrations in the Channel water entering the Dover–Calais Straits are 1·2–2·7 mg l^{-1} (average 1·9 mg l^{-1}) at salinities above 35‰ (Eisma & Kalf, 1979) but higher in water of lower salinity. Lower-salinity water is present in the Dover–Calais Straits during periods of northerly winds when North Sea water is blown into the Channel. Also in the coastal waters of the Dover–Calais Straits suspended-matter concentrations may be higher (up to 11 mg l^{-1}) usually with somewhat higher concentrations on the English side. This material may be supplied by the rivers discharging into the Channel and/or by erosion of the Channel coasts and the seafloor, but can also be supplied from the North Sea during northerly winds and by tidal dispersion. At a water inflow into the North Sea through the Dover–Calais Straits of 4900±725 km^3 yr^{-1} (Prandle, 1978) and an average suspended-matter concentration of 1·9 mg l^{-1} in the water of >35‰ salinity, the yearly supply through the Dover–Calais Straits is at least 8–11·5×10^6 tons (average 10×10^6 tons) with an additional unknown quantity being supplied through the coastal waters. Taking the coastal transport as estimated by Terwindt (1967) and McCave (1973), the supply becomes *c.* 11·5–15×10^6 ton yr^{-1}. This is much higher than previous estimates (McCave, 1973) due to the more reliable data that have become available on water inflow and suspended-matter content.

In the eastern Channel the suspended-matter concentrations are much higher than in the adjacent Atlantic Ocean: the material is supplied by sources within the Channel (coastal runoff and rivers, coastal erosion, seafloor erosion). That the large sandbanks in the eastern Channel may be an important source is demonstrated by suspended-matter measurements made at the north-eastern end of Varne Bank. Comparing size distributions of suspensions from this area with those of suspensions from the eastern Channel shows a large addition of particles of

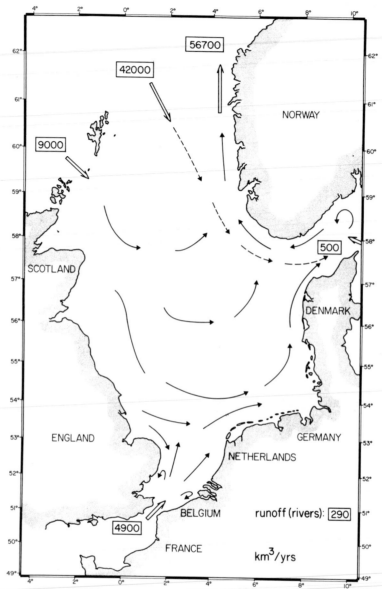

Fig. 2. Inflow of water into the North Sea (numbers indicate km³ yr⁻¹) and the general water circulation, indicated by arrows (general direction of flow) and broken arrows (subsurface flow).

9–35 μm, presumably by uptake from the bottom (Eisma & Kalf, 1979, fig. 3). Further north, in the central part of the Southern Bight, the particle concentrations in these sizes are much lower again, but there is still an important admixture of particles of 15–30 μm. If this uptake is a continuous process several million tons of material may be supplied annually to the North Sea from the Varne Bank

area. From these data it is understandable how in the Channel the suspended-matter concentrations can increase about one order of magnitude going from the Atlantic Ocean to the Dover–Calais Straits.

The inflow from the Baltic has a suspended-matter concentration of c. 1 mg l⁻¹, as was estimated by McCave (1973) on the basis of data from Brogmus

(1952) and was measured by Eisma & Kalf (in preparation) during a cruise in July 1978. At a yearly inflow of 500 km³ (Otto, 1976) this gives a supply of 0.5×10^6 ton yr⁻¹. Rivers supply c. 4.5×10^6 ton yr⁻¹: c. 1.7×10^6 ton yr⁻¹ comes at present from the Rhine and Meuse (Terwindt, personal communication), which has become much lower during the past years because the main outlets have been closed; from the Ems comes 0.07×10^6 ton yr⁻¹ (Hinrich, 1974) and from Weser, Elbe and Thames + Humber 0.35, 0.86 and 1.47×10^6 ton yr⁻¹ respectively (McCave, 1973; Hinrich, 1975). Supply from the atmosphere was estimated to be c. 1.6×10^6 ton yr⁻¹ by a group of meteorologists (McCave, 1973).

Coastal erosion occurs chiefly at the Holderness cliffs and the East Anglian cliffs and supplies a total of 0.7×10^6 ton yr⁻¹ (McCave, 1973). Erosion of the seafloor is much less known. By comparing sea charts from 1894/95 and 1970/71 Gossé (1977) estimated that up to 2.4×10^6 ton yr⁻¹ may be supplied from the Flemish Banks by erosion. This confirmed earlier estimates by Morra *et al.* (1961) which were based on a comparison of sounding charts dating from 1823 to 1958. Old clay layers and silty deposits are being eroded: the eroded material is lost from the area and is assumed to have been removed in suspension. Also off East Anglia there are indications of bottom erosion: outcrops of old clay deposits (Sheldon, 1968), lag deposits containing Plio-Pleistocene shells (Eisma & Spaink, unpublished data) and a high turbidity (Lee & Folkard, 1969). Another area where seafloor erosion may be important lies north and east of Cape Skagen (Rohde, 1973): data on suspended sediment, current velocities and bottom deposits indicate that the intermittent effect of deposition and erosion has a sorting effect, with only coarser material being finally deposited in the shallow areas and the finer material being deposited in the Kattegat. Erosion of the seafloor may therefore supply large quantities of suspended matter. The yearly supply is at least several million tons and probably is of the order of 5×10^6 ton yr⁻¹.

Primary production, although large (c. 38.5×10^6 ton dry weight per year in the southern North Sea alone; Postma, 1973), gives only a very temporary supply since almost all organic material is mineralized. In suspended matter collected in the Southern Bight during the winter in the absence of living plankton, the organic content (loss on ignition at 520 °C) is 10–20%, about the same content as in the suspended material supplied by the river Rhine. In bottom sediments the organic content drops to less than 5% even in fine-grained deposits: 50% to >75% of the organic material in suspension is removed during or after deposition by decomposition or utilization by bottom fauna. The organic matter, found in suspension during the winter, probably has a mixed origin, being derived from bottom deposits, coastal sediments, river supply, primary production and bottom-dwelling organisms. The organic matter is tied up in aggregates where it glues inorganic particles together: entirely organic particles are rare when there is no plankton growth. The origin of the organic matter in suspension is therefore primarily related to the formation of aggregates and not to primary production. Bottom fauna probably plays a large role in aggregate formation, so that a dominant portion of the organic matter in suspension may be supplied by bottom fauna through the formation of faeces and pseudofaeces that subsequently fall apart into aggregates. In this way up to several million tons

Table 1. Supply

	ton yr⁻¹ (dry weight) ($\times 10^6$)
Atlantic Ocean	10
Channel	10 (+ ?)
Baltic	0.5
Rivers	4.5
Rhine + Meuse 1.7×10^6 (Terwindt, personal communication)	
Ems 0.07×10^6 (Hinrich, 1974)	
Weser 0.95×10^6 } (Hinrich, 1975)	
Elbe 0.86×10^6	
Thames + Humber 1.47×10^6 (McCave, 1973)	
$\overline{4.5 \times 10^6}$	
Atmosphere (meteorological group, McCave, 1973)	1.6
Coastal erosion	0.7
East Anglia + Holderness (McCave, 1973)	
Seafloor erosion	$\simeq 5 (+ ?)$
Flemish Banks: up to 2.4×10^6 (Gossé, 1977)	
Varne Bank- $\simeq 1–2 \times 10^6$	
Off East Anglia?	
Skagerrak off Denmark?	
Primary production	$\simeq 1$
Total	$\simeq 34 (+ ?)$

of organic matter may be supplied annually, which is mostly mineralized again when the suspended matter is deposited, the net contribution being small. The hard parts of plankton (e.g. diatom frustules) are rare in recent bottom sediments in the North Sea (Outer Silver Pit, German Bight, Elbe Rinne, Skagerrak and Norwegian Channel, Eisma & Van Weering, unpublished data) and are hardly ever seen in suspension during the winter, even in the rather productive southern North Sea (Eisma & Kalf, 1979). The total net contribution from primary production is therefore small, of the order of, at most, a few per cent, i.e. in the order of 1×10^6 ton yr^{-1}.

The supply from all sources together is in the order of 34×10^6 ton yr^{-1} (Table 1) plus an unknown amount supplied from the Channel through the coastal waters of the Dover–Calais Straits and supplied from seafloor erosion.

OUTFLOW INTO THE NORTH ATLANTIC OCEAN (NORWEGIAN SEA)

Inflow into the North Sea from the Atlantic as well as outflow from the North Sea into the Norwegian Sea are concentrated in the area between Norway and Shetland (Fig. 2). Suspended-matter concentrations measured in the inflow and outflow in May 1977 in this area (Eisma & Kalf, 1978) and in the Norwegian Channel south of Bergen in July 1979 (Eisma & Kalf, in preparation) indicated that there is no significant difference between the inflow

and outflow concentrations, although in both cases the outflow concentrations were slightly higher. On the basis of these data the excess leaving the North Sea over what comes in from the Atlantic is at most 3×10^6 ton yr^{-1}: only c. 10 % or less of the amount of material supplied from other sources than the Atlantic flows out of the North Sea. Actually more material from other sources may leave the North Sea, but this is compensated for by material coming in from the Atlantic that remains behind.

DEPOSITION

A systematic overall survey of the bottom deposits in the North Sea has not been made. The principal sediment distribution maps that have been published are based on the Fisheries Charts and make a rough distinction between gravel, coarse sand, sand, fine sand, sandy mud and mud. They show a patchwork of sediment types (Lüders, 1939; Veenstra, 1971; Jansen et al. 1979). Pratje (1949), who carried out the most extensive bottom sampling programme in the North Sea, published a map which is rather similar to the Fisheries Charts, on which a distinction is made between gravel, sand and mud. Bruns (1958) gives a bottom sediment map based on unknown data. For the southern North Sea a sediment distribution map has been made by Jarke (1956) based on the samples collected by Pratje (1949, 1952). Fine-grained deposits are divided into 'Mehlsand' (15·6–125 μm) and 'Schluff und Ton' (<15·6 μm). On the map a distinction is also made between 'Mehlsand' with >20 % 'Schluff und Ton' and 'Schluff und Ton' with >20 % 'Mehlsand'.

Earlier work in this area was carried out by Borley (1923). The German Hydrographic Institute at Hamburg made a more detailed survey of bottom deposits in the German sector of the North Sea (Figge, 1974) and the results have been partly published (Ludwig & Figge, 1979). The State Geological Survey of the Netherlands prepared a sediment distribution map of the Dutch sector (Schüttenhelm, 1980). Smaller studies have been carried out in many areas of the North Sea and by now give a comprehensive picture of the types and distribution of the bottom deposits of the North Sea as a whole.

Mud deposits or mud-containing deposits are known from the following areas (the numbers between brackets refer to the numbers in Fig. 4): (a) northern (1) and central North Sea (2): the Fladen Grounds, the Devils Hole area and smaller

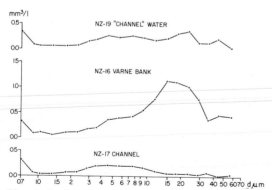

Fig. 3. Particle-size distributions (determined with a Coulter Counter) of suspended matter from the eastern Channel, of the northern part of Varne Bank in the Dover–Calais Straits, and in the Channel water in the Southern Bight (after Eisma & Kalf, 1979).

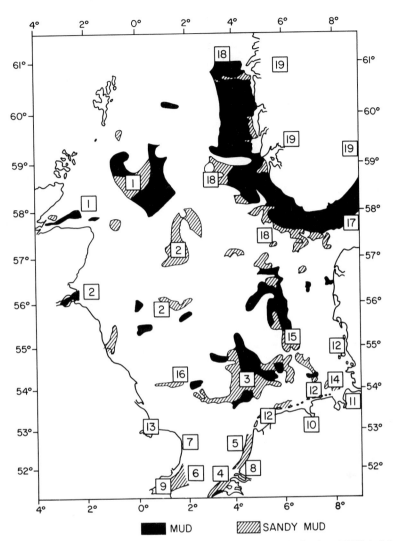

Fig. 4. Distribution of mud and sandy mud in the North Sea, largely as given by Lüders (1939) in McCave (1973) and with additional data for the Southern Bight from Eisma (1966), Gullentops *et al.* (1977), Joyce (1979) and McCave (1979). Explanation of the numbers is given in the text on pp. 422–3.

patches along the Scottish and English coasts; (b) the Oyster Grounds (3); (c) the Flemish Banks off the Belgian–Dutch coast (4); (d) off the Dutch coast up to Den Helder and off the eastern Wadden Islands (5); (e) an area off the Thames estuary and the East Anglian coast (6) and small areas off Yarmouth and Lowestoft (7); (f) river estuaries (and harbours): Rhine–Meuse–Scheldt (8), Thames (9), Ems (10), Elbe–Weser (11); (g) the Wadden Sea (12) and the Wash (13); (h) the German Bight (14) and the 'Elbe Rinne' (15); (i) the Outer Silver Pit area (16); (j) the

Skagerrak (17) and Norwegian Channel (18); (k) the Norwegian (and some Swedish) fjords (19). The definition of 'mud' differs with different authors. For all practical purposes mud can be regarded as synonymous with suspended matter, and mud deposits or mud admixtures can be considered to consist of material $< 60–70\ \mu m$ that has settled out of suspension.

Fladen Grounds, Devils Hole area, Oyster Grounds

North of Dogger Bank (in the Devils Hole area and the Fladen Grounds) the muddy deposits that form the sea-floor are old Holocene dating from before 8400 BP (Jansen, 1976; Jansen et al., 1979) with only a younger influx of carbonate material and organic carbon (Erlenkeuser, 1978). Johnson & Elkins (1979) give for the Fladen Ground a sedimentation rate of 0·3–0·5 cm/100 yr based on radiocarbon datings, which is equivalent to a deposition of c. $0·1 \times 10^6$ ton yr^{-1}. South of Dogger Bank in the Oyster Grounds there is a large area with sandy mud containing up to 24% of material of $< 60 \mu m$ (Creutzberg & Postma, 1979). This deposit has a thickness of 0·5–1·5 m (locally up to 2·0 m) and lies on top of old Holocene tidal flat deposits and older Pleistocene sediments (Oele, 1969 and Netherlands Geological Survey, unpublished data). The deposit is almost entirely reworked and contains a reworked tidal flat (mollusc) fauna. Data on current velocities and wave action (Postma, 1978; Creutzberg & Postma, 1979) indicate that fine-grained material will not be removed from this area, in contrast to the more sandy deposits in the shallower parts of the southern North Sea, where fine-grained material will be winnowed out. The distribution of suspended matter and the residual current pattern in the southern North Sea indicate that only small amounts of suspended matter will reach this area. If it is assumed that the mud admixture has been deposited out of suspension and not supplied by local reworking of the older deposits, the average deposition rate since c. 8000 BP, when the area became fully marine, is in the order of $0·2–0·3 \times 10^6$ ton yr^{-1}.

On the smaller patches of muddy sediment off the Scottish–English coast no data are available to the author: if recent deposition occurs in these areas, it will be related to local erosion of the coast and the seafloor and to local circulation, because suspended-matter concentrations in the rivers and in the northern and central North Sea are very low.

Southern Bight

Smaller areas with muddy admixtures are located in the Southern Bight. On the Flemish Banks areas of erosion and deposition intermingle but there is a net erosion in this area (Gossé, 1977). A small area of mud deposits along the Belgian coast off Ostend–Kadzand coincides with the area where suspended matter is trapped by local flow conditions and the maximum current velocity at 1 m above the bottom does not exceed 50 cm sec^{-1} (Gullentops et al., 1977). The area is shallow (5–20 m water depth) and there is much reworking by waves, tidal currents and bottom fauna. A ^{210}Pb dating of Petit (1977) gives a sedimentation rate of c.6 cm/100 yr, but the ^{210}Pb distribution in the bottom profile can better be explained by reworking of the upper 6 cm of the deposit than by deposition of this amount during the past 100 yr.

Along the Belgian–Dutch coast up to Den Helder and further to the east off the islands of Ameland and Schiermonnikoog there is an admixture of mud (1–5%, locally up to >20%) in nearshore sands at predominantly 4–14 m water depth (Van Straaten, 1961, 1965; Eisma, 1966). This mud is present in thin laminae and is deposited during slack tide (Van Straaten, 1961) or during prolonged periods of calm weather: temporary mud deposits have been observed offshore near Camperduin and Petten during calm periods where normally the bottom is sandy. These temporary mud deposits are eroded again when wave action increases but some mud is buried under encroaching sand. There will also be deposition through organisms that consume the organic matter in the mud. The deposition of mud in a narrow belt along the coast is probably due to the high concentration of suspended matter near shore and to the high concentrations of bottom-dwelling organisms in the zone with mud laminae. Exposure to wave action is important: a muddy admixture is absent in the surf zone and small or absent off the islands of Texel, Vlieland and Terschelling although suspended matter concentrations in the water are not lower. It is not likely that appreciable quantities of mud are deposited in the nearshore area on a yearly basis. There is considerable sediment movement along the Dutch coast, resulting in alternating erosion or deposition on the beach with a net overall retreat of the beach up to 2 m yr^{-1} along the southern and northern parts, and a net progression of the beach of up to 1 m yr^{-1}, or zero displacement along the central part. The sand that is removed from the beach is mainly transported along the beach and does not result in large-scale deposition on the sea-floor further offshore. Apart from these small changes the coastline has remained approximately at its present position during the past 125 years so that the coast is near to a dynamic equilibrium, without large-scale mud accumulation.

Off the Thames Estuary and the East Anglian

coast a muddy admixture is present in a similar way in the form of mud laminae in sandy deposits or is trapped between the gravel present along the western side of the deep water channel, with mud contents reaching $> 10\%$ but generally below 5%. Here also soft mud was found locally during calm weather where normally the bottom is sandy. Small mud deposits are present off Yarmouth (McCave, 1979) and Lowestoft (Joyce, 1979). The mud is probably supplied from eroding cliffs further to the north, moved southward by the nearshore southward residual current and concentrated in the estuary of the river Yare. There are no estimates of the amount of mud deposited annually off the Thames Estuary and East Anglia but the dimensions of the thin mud-containing layers indicate that it must be very small in relation to the total amount of suspended matter present in the southern North Sea.

Estuaries and Wadden

Mud accumulation in estuaries is very variable. In the Scheldt and the Ems estuaries the river mud hardly reaches the sea. In the Scheldt there is a net inflow of suspended material from the sea, the river mud not being found west of Terneuzen, which is about half way along the Western Scheldt (Terwindt, 1967, 1977). The mud from the Ems is not found beyond the Dollart area (Favejee, 1960). Mud from the Rhine and the Meuse is deposited in a number of artificial lakes such as the IJssel Lake and the Haringvliet Lake, and after the enclosure of the main outlets the remainder flows out through the Water-weg west of Rotterdam. Outflow of suspended material is also evident at the Elbe, the Weser and the Thames. McCave (1973) came to a figure of 1.8×10^6 ton of mud that would be deposited yearly in the estuaries. Harbours are dredged so that there is no net accumulation of mud supplied either from a river or from the sea. Dredged material, however, is partly dropped on land and in this way is withdrawn from circulation, but it is not always clear, as in the Rotterdam harbours, how much of this mud has come directly from the river. McCave (1973) tentatively estimated that *c.* 2×10^6 tons of mud that would have been deposited in the North Sea is withdrawn yearly from circulation by dumping on land.

The sediment distribution in the Wadden Sea is given by De Glopper (1967) for the Dutch part, by the Geologische Uebersichtskarte 1:200,000 for the part between the Ems and the Elbe, by Ostendorff (1943) and Iwersen (1943) for Schleswig-Holstein and by Smidt (1951) for the Danish Wadden Sea. The sediments of the Wash have been described by Evans (1965); a sediment map of the Wash is given by Wingfield *et al.* (1978). The muddy deposits in the Wadden Sea occur along the inner margins and on the tidal watersheds. Most of the mud comes from the North Sea (Favejee, 1951; Postma, 1961; Salomons *et al.*, 1975) and not from local erosion. The rate of mud accumulation in the Wadden Sea was estimated by McCave (1973) to be *c.* 0.8×10^6 ton yr^{-1}, which was based on data of Hansen (1951), Verwey (1952) and Postma (1961). Along the coasts of Friesland and Groningen no mud is being deposited at the moment: the land reclamation works, aimed at the entrapment of mud between groynes and cribs, are being continued only in a very limited way but are not completely stopped because stopping would result in erosion of the coast (Dijkema, 1975). Sedimentation is therefore concentrated on the tidal watersheds. In general the sedimentation of *c.* 10^6 ton yr^{-1} is sufficient to keep pace with the rate of subsidence. For the entire Wadden Sea, however, this value is too low because large amounts of mud are deposited in the Dollart and probably also in other embayments like the Leybucht, the Jade and the embayments along the coasts of Schleswig-Holstein and Denmark. This supposition is based on data for the Dollart: De Smet & Wiggers (1960) arrive at an average deposition of 10^6 ton yr^{-1} in the Dollart during the past 450 yr and Reenders & Van der Meulen (1972) arrive at *c.* 0.6×10^6 ton yr^{-1} during the past 18 yr. For the other embayments no data are available. Reineck (1960) gives a sedimentation rate of 29 cm/100 yr for the Jade, over a surface of 184 km². Assuming that 50% of this material is smaller than 60 μm, the rate of deposition is in the order of 0.5×10^6 ton yr^{-1}. It follows that deposition in the Wadden Sea is at least more than 2.5×10^6 ton yr^{-1} and, assuming that in the other embayments some mud is also being deposited, it will be of the order of 3×10^6 ton yr^{-1}. For the Wash Kestner (1975) calculated the volume changes based on contour areas from surveys made in 1828 and 1917–18. The increase in net sediment volume was 27×10^6 m³, with sedimentation above low water and erosion chiefly below low water. Assuming a water content of the sediment of 30% and a specific gravity of 2.6 this volume is equivalent to 23.4×10^6 tons, or a deposition rate of 0.26×10^6 ton yr^{-1}. It is not clear to what extent only suspended material has been deposited, but this value can be regarded as a maximum average.

German Bight, 'Elbe-Rinne' and Outer Silver Pit

South-east of Helgoland a deposit is located with > 50 % of material of < 63 μm (Reineck, Gutman & Hertweck, 1967; Reineck *et al.*, 1968; Müller, Reineck & Staesche, 1968). It is surrounded by an area of less fine sediment (10–15 % < 63 μm) which continues in a north-western direction in the so-called 'Elbe-Rinne' (Hertweck & Reineck, 1969). The latter is a Pleistocene valley which has been filled up with (in the top 3·5 m) fine-grained sediment, with a maximum thickness of *c.* 15 m (unpublished data NIOZ). The mud area south-east of Helgoland has a maximum thickness of 20 m (Reineck, 1969). The latter deposit wedges out towards the west, south of Helgoland, in water depths of *c.* 40 m. Here Saale glacial deposits are exposed on the seafloor or are covered by a thin layer of muddy sand (10–15 % < 63 μm). The relation between the infill in the 'Elbe-Rinne' and the mud deposit south-east of Helgoland is not clear. For the mud deposit south-east of Helgoland Reineck (1963) calculated, on the basis of pieces of slag present in the sediment, a sedimentation rate of 20 cm/100 yr. Reineck *et al.* (1967) and Gadow (1969) arrived at a rate of 30–50 cm/100 yr, based on the presence of slag as well as layers of gypsum, eroded from Helgoland. On the basis of the thickness of the deposit Reineck (1969) calculated a sedimentation rate of 50 cm/100 yr, assuming that sedimentation began around 4000 yr ago, when the area was flooded during the Holocene transgression. McCave (1970) arrived at a mud-sedimentation rate of 15·5 cm/100 yr, with 4·5 cm/100 yr of sandy material being deposited during storms. A ^{210}Pb dating by Goldberg (personal communication) gave 70 cm/100 yr, and a ^{210}Pb dating by Dominik *et al.* (1978) made on a different core 77 cm/100 yr (average) with a period of very low sedimentation (1·8 mm yr^{-1}) between 1930 and 1954, and periods of higher sedimentation in 1963–4 (18·2 mm yr^{-1}), 1954–63 (> 6·7 mm yr^{-1}) and 1915–30 (16 mm yr^{-1}). Sedimentation has been discontinuous and is strongly related to the frequency of storm floods. Förstner & Reineck (1974) analysed the trace element content of the upper 4 m of the mud (in the same core as used by Dominik *et al.*, 1978): it starts to increase towards the top at *c.* 120 cm deep in the bottom. A sedimentation rate of 70 cm/100 yr locates the onset of trace-metal contamination in the first half of the nineteenth century, which relates it to human activity, whereas a rate of 50 cm or less would place it around 1750 or earlier, which is too early for industrial contamination. It follows that there is good agreement between the ^{210}Pb rates and trace-element data. Moreover, the layered and laminated structure of the core analysed by Dominik *et al.* (1978) and Förstner & Reineck (1974) excluded a large-scale effect of reworking by bottom fauna or water movements, so that a sedimentation rate of 70–77 cm/100 yr can be considered reliable. This gives a deposition rate of 3×10^6 ton yr^{-1}, taking into account the mud content and the extension of the deposit. Dominik *et al.* (1978) give for the most recent period (1963–74) a sedimentation rate of *c.* 182 cm/100 yr, which gives a recent deposition rate of $7·5 \times 10^6$ ton yr^{-1}.

On the sedimentation of mud in the 'Elbe-Rinne' only limited data are available: if deposition started here also around 8000 yr ago, as in the nearby Oyster Grounds, and still continues, the rate of sedimentation is up to 19 cm/100 yr in the north-western part where the infill is thickest, and near to zero south of Helgoland, resulting in an average deposition rate of *c.* 10 cm/100 yr. Conflicting data, however, have been obtained on sedimentation rates. A ^{210}Pb date in the north-western part gave 26 cm/100 yr (Goldberg, personal communication); a date based on ^{14}C in organic matter in a core from the same area gave 2 cm/100 yr and a similar rate was found with ^{14}C in carbonate (Erlenkeuser, personal communication). Trace-metal concentrations, determined by Salomons (personal communication) indicated a higher trace-element content in the upper 10–15 cm of the sediment.

These results, coupled with (a) absence of layering and (b) evidence for reworking by bottom fauna, indicate that these results are strongly influenced by reworking of the sediment and that recent sedimentation rates are probably low. Also the distribution of suspended matter and the residual current pattern in the southern North Sea indicate that only small amounts of suspended material reach this area.

Mud-containing sediments in the Outer Silver Pit area were extensively discussed by Veenstra (1965), Zagwijn & Veenstra (1966) and Eisma (1975). These sediments have been deposited in an old (Pleistocene) valley (the Outer Silver Pit) and in two closed basins (probably former tunnel valleys) situated south-west of Dogger Bank: Botney Cut and Markham's Hole. In similar closed basins situated further west (Well Hole, Coal Pit, Sole Pit, Silver Pit) only a thin sandy infill is found, probably because in that area the currents are stronger so that the fine-grained material is largely kept in suspension. Also the large waves from the NW–NE may have an effect: the area of mud deposition is more sheltered towards that direction by Dogger Bank. The sediment infill is of variable thickness, containing 30–60% of material < 80 μm in the upper few metres, with a maximum thickness of 52 m. Assuming that deposition started at the time the area was flooded *c.* 10,000–9500 yr ago and is still continuing, the average sedimentation rate is of the order of 4–10 cm yr^{-1} with a maximum of 52 cm/100 yr. Pollen records however, give 11–45 cm/100 yr for the top of the infill (Zagwijn & Veenstra, 1966). Based on these data the deposition of mud is 0·2–4·1 × 10^6 ton yr^{-1}; using only the pollen data, which reflect the more recent conditions, gives a deposition of 1–4 × 10^6 ton yr^{-1}.

Kattegat, Skagerrak, Norwegian Channel and fjords

In the Skagerrak fine-grained deposits are present in the deeper parts at water depths of 100 m or more, especially on the northern side where the finest deposits occur (Van Weering, 1981). The thickness of these deposits is very variable and may be up to 120 m (Van Weering *et al.*, 1973). In the Norwegian Channel this infill becomes much thinner, except for a small area off Egersund, and north of Bergen is locally absent where Pleistocene (glacial) deposits are exposed on the seafloor. Lange (1956) estimated on the basis of microfossils that the recent deposition rate in the Skagerrak is 14 cm/100 yr, but Combaz *et al.* (1974) found in other samples 0·4–2·0 cm/100 yr for the last 2500 yr on the basis of pollen analysis. Van Weering (1981) found Holocene rates between 0 and 12·5 cm/100 yr. In the Kattegat Rohde (1973) found a thick mud deposit situated east of Cape Skagen. From measurements of flow and suspended-matter content he calculated that *c.* 8 million tons of mud are supplied annually by the Jutland current along northern Denmark and deposited in the Kattegat. A general estimate of the amount of suspended material deposited in the Skagerrak and the Norwegian Channel can be obtained from the decrease in the suspended-matter content in the northward-flowing surface water between the inner Skagerrak and the Atlantic Ocean. This material is partly diluted by advection of water from the northern North Sea and partly sinks down into the Atlantic water that flows inward to the Skagerrak along the bottom of the Norwegian Channel. This circulation results in a transport back

into the Skagerrak where the fine-grained deposits are concentrated. The fact that the northward (outward) decrease in suspended-matter content in the surface water is matched by an inward increase in suspended-matter content in the bottom water indicates that this mechanism is operating. On the basis of the outward loss and the inward gain it can be estimated (using three sets of data collected in 1977–9) that 4–7 × 10^6 ton is deposited annually in the Skagerrak and the southern part of the Norwegian Channel (Eisma & Kalf, unpublished data). The data of Rohde (1973), however, indicate that near-bottom transport can be of great importance in the Skagerrak. This implies that actually much more fine-grained material is picked up and transported near to the bottom than is known at present, and also that much more may be deposited in areas like the Skagerrak than is estimated from the suspended matter content in the overlying water column.

In the Norwegian and Swedish fjords bottom deposits are fine-grained or sandy because of slides and turbidity currents (Holtedahl, 1975). A sedimentation rate of 20 cm/100 yr in Gullmarfjord is mentioned by Lange (1956). Probably all material supplied by rivers that enter the fjords is deposited within the fjords; suspended-matter concentrations are high inside the fjords and drop off towards the outer parts (Aarthun, 1961). Moreover, inside the fjords there is probably deposition of fine material flowing in from the sea with the bottom water (Eisma & Kalf, 1978). The data of Price & Skei (1975) indicate that during the summer, when runoff is at its maximum and the density stratification in the fjord is strongest, fine particles may be moved out

Table 2. Outflow + deposition.

	ton yr^{-1} (dry weight) (× 10^6)
Outflow to North Atlantic Ocean	11·4
	+ < 3
	——14·4
Deposition	
Estuaries (McCave, 1973)	1·8
Wadden Sea + Wash	≃ 3
German Bight	3–7·5
Elbe Rinne	?
Outer Silver Pit	1–4
Kattegat	8
Skagerrak	4–7
	——21–31·5
Dumped on land (McCave, 1973)	2
	Total: 37·5–48

into the North Sea with the surface water. McCave (1973), on the basis of a rough calculation, estimated that the entrapment of mud in the fjords is of the same order as supply by runoff ($1\cdot3 \times 10^6$ ton yr^{-1}).

On the basis of the data given above the total amount deposited annually in the North Sea is $23-33\cdot5 \times 10^6$ ton (Table 2). This includes an average of 2×20^6 ton which is dumped annually on land. The large range is primarily due to the uncertainties concerning sedimentation rates.

CONCLUSIONS

From Tables 1 and 2 it can be seen that the total estimated supply comes near to the sum of estimated outflow into the Atlantic and estimated deposition. Adding some supply through the coastal waters of the Dover–Calais Straits and some additional supply from the seafloor (indicated with question marks in Table 1) brings the estimated supply well within the range of estimated outflow + deposition. But although this result looks rather good, it should not obscure the uncertainties in the data that have been used; the figures used above should only be regarded as broad estimates. More precise data, however, cannot be obtained at present, chiefly because of uncertainties about the amounts supplied by seafloor erosion and about sedimentation rates. Deposition rates based on one single method of determination appear very unreliable, as is clear from the data for the area south-east of Helgoland; only agreement between different sets of data gives reliable results. A third uncertainty involves transport of fine-grained material near the bottom, which was left out of the discussion. In the southern North Sea, at least in the shallower parts, this is not of special importance except in harbours and river mouths, but it may be of great importance in areas like the Skagerrak.

REFERENCES

AARTHUN, K.E. (1961) Submarine daylight in a glacier-fed Norwegian fjord. The history of the Hardangerfjord. *Sarsia*, **1**, 7–20.

BORLEY, J.O. (1923). The marine deposits of the Southern North Sea. *Fishery Invest., Lond.* **11**, 4–6.

BROGMUS, W. (1952) Eine revision des wasserhaushaltes der Ostsee. *Kieler Meeresforsch.* **9**, 15–42.

BRUNS, E. (1958) Ozeanologie, I. *VEB Dtsch. Verl. Wiss.* Berlin, 420 pp.

COMBAZ, A., BELLET, J., POULAIN, D., CARATINI, Cl. & TISSOT, C. (1974) Etude microscopique de la matière organique de sédiments quaternaires de Mer de Norvège. *Orgon 1, CNEXO*, pp. 139–150.

CREUTZBERG, F. & POSTMA, H. (1979). An experimental approach to the distribution of mud in the Southern North Sea. *Neth. J. Sea Res.* **13**, 99–116.

DIETRICH, G. (1950) Die anomale Jahresschwankung des Wärmeinhalts im Englischen Kanal, ihre Ursachen und Auswirkungen. *Dt. hydrogr. Z.* **3**, 184–201.

DIETRICH, G. (1955) Ergebnisse synoptisch ozeanographischen Arbeiten in der Nordsee. *TagBer. Geographentag Hamburg*, pp. 376–383.

DIJKEMA, K.S. (1975) Vegetatie en beheer van de kwelders en landaanwinningswerken aan de Waddenzeekust van Noord-Groningen. *Meded. no. 2.* Werkgroep Waddengebied. 49 pp.

DOMINIK, J., FÖRSTNER, U., MANGINI, A. & REINECK, H.E. (1978) Pb-210 and Cs-137 chronology of heavy metal pollution in a sediment core from the German Bight (North Sea). *Senckenberg. Mar.* **10**, 213–228.

EISMA, D. (1966) The distribution of benthic marine molluscs off the main Dutch coast. *Neth. J. Sea Res.* **3**, 107–163.

EISMA, D. (1975) Holocene sedimentation in the Outer Silver Pit area (Southern North Sea). *Comm. Mar. Sci.* **1**, 407–426.

EISMA, D. (1976) Deeltjesgrootte van dood gesuspendeerd materiaal in de Zuidelijke Bocht van de Noordzee. *Intern Rapport Nederlands Instituut voor Onderzoek der Zee*, p. 19.

EISMA, D. & KALF, J. (1978) Suspended matter between Norway and Shetland and in the Sognefjord. *Intern Rapport Nederlands Instituut voor Onderzoek der Zee*, p. 13.

EISMA, D. & KALF, J. (1979) Distribution and particle size of suspended matter in the Southern Bight of the North Sea and the eastern Channel. *Neth. J. Sea Res.* **13**, 298–324.

ERLENKEUSER, H. (1978) The use of radiocarbon in estuarine research. In: *Biogeochemistry of Estuarine Sediments. Proc. UNESCO/SCOR Workshop Melreux*, pp. 140–153.

EVANS, G. (1965) Intertidal flat sediments and their environments of deposition in the Wash. *Q. Jl geol. Soc. Lond.* **121**, 209–245.

FAVEJEE, J.Ch.L. (1951) The origin of the "Wadden" mud. *Meded. Landb. Hoogesch. Wageningen*, **51**, 113–141.

FAVEJEE, J.Ch.L. (1960) On the origin of the mud deposits in the Ems-estuary. In: *Das Ems-estuarium* (Ed. by J.H. Voorthuysen). *Verh. Kned. geol.-mijnb. Genoot.* **XIX**, 147–151.

FIGGE, K. (1974) Sediment distribution mapping in the German Bight (North Sea). *Mém. Inst. Géol. Bassin d'Aquitaine*, **7**, 253–257.

FÖRSTNER, U. & REINECK, H.E. (1974) Die Anreicherung von Spurenelementen in den rezenten Sedimenten eines Profilkerns aus der Deutschen Bucht. *Senckenberg. Mar.* **6**, 175–184.

GADOW, S. (1969) Gips als Leitmineral für das Liefergebiet Helgoland und für den Transport bei Sturmfluten. *Natur Mus., Frankf.* **99**, 537–540.

DE GLOPPER, R.J. (1967) Over de bodemgesteldheid van het Waddengebied. *Rapp. Meded. Rijksd. IJsselmeerpolders*, **43**, 67.

GOSSÉ, J.G. (1977) A preliminary investigation into the possibility of erosion in the area of the Flemish banks. *RijkswatSt. Dir. Waterh. Waterbew. Fys. Afd. Nota FA 7702*, 20 pp.

GULLENTOPS, F., MOENS, M., RINGELE, A. & SENGIER, R. (1977) Geologisce kenmerken van de suspensie en de sedimenten. *Projet Mer, Rapport Final (Sedimentologie)*, **4**, 1–121.

HANSEN, K. (1951) Preliminary report on the sediments of the Danish Wadden Sea. *Meddr dansk geol. Foren.* **12**, 1–26.

HERTWECK, G. & REINECK, H.E. (1969) Sedimentologie der Meeresbodensenke NW von Helgoland (Nordsee). *Senckenberg. Mar.* **50**, 147–152.

HINRICH, H. (1974) Schwebstoffgehalt, Gebietsniederschlag, Abfluss und Schwebstofffracht der Ems bei Reiner und Verzen in den Jahren 1965–1971. *Dt. gewässerk. Mitt.* **18**, 85–95.

HINRICH, H. (1975) Die Schwebstoffbelastung der Weser und der Vergleich mit Elbe und Ems. *Dt. gewässerk. Mitt.* **19**, 113–120.

HOLTEDAHL, H. (1975) The geology of the Hardangerfjord, West Norway. *Norg. geol. Unders.* **323**, 87 pp.

IWERSEN, J. (1943) Zur bodenkundlichen Kartierung des nordfriesischen Wattergebietes. *Westküste*, pp. 47–71.

JACOBS, M.B. & EWING, M. (1961) Suspended particulate matter: concentration in the major oceans. *Science, N.Y.* **163**, 380–383.

JANSEN, J.H.F. (1976) Late Pleistocene and Holocene history of the Northern North Sea, based on acoustic reflection records. *Neth. J. Sea Res.* **10**, 1–43.

JANSEN, J.H.F., DOPPERT, J.W.C., HOOGENDOORN-TOERING, K., DE JONG, K. & SPAINK, G. (1979) Late Pleistocene and Holocene deposits in the Witch and Fladen Ground Area, Northern North Sea. *Neth. J. Sea Res.* **13**, 1–39.

JANSEN, J.H.F., VAN WEERING, T.C.E. & EISMA, D. (1979) Late Quaternary sedimentation in the North Sea. In: *The Quaternary History of the North Sea* (Ed. by E. Oele, R.T.E. Schüttenhelm and A.J. Wiggers), pp. 175–187. Uppsala.

JARKE, J. (1956) Der Boden der südlichen Nordsee. *Dt. hydrogr. Z.* **9**, 1–9.

JOHNSON, T.C. & ELKINS, S.R. (1979) Holocene deposits from the northern North Sea: evidence for dynamic control of their mineral and chemical composition. *Geologie Mijnb.* **58**, 353–366.

JOYCE, J. (1979) On the behaviour of dumped dredge spoil. *Bull. Mar. Pollut. N. S.* **10**, 158–160.

KESTNER, F.J.T. (1975) The loose-boundary regime of the Wash. *Geogr. J.* **141**, 388–414.

LANGE, W. (1956) Grundproben aus Skagerrak und Kattegat, mikrofaunisch und sedimentpetrographisch untersucht. *Meyniana*, **5**, 51–86.

LEE, A.J. & FOLKARD, A.R. (1969) Factors affecting turbidity in the Southern North Sea. *J. Cons. perm. int. Explor. Mer.* **32**, 291–302.

LÜDERS, K. (1939) Sediments of the North Sea. In: *Recent Marine Sediments* (Ed. by P.D. Trask). *Publs Am. Ass. petrol. Geol.* 322–342.

LUDWIG, G. & FIGGE, K. (1979) Schwozmineral rozhommen und Sandverteilung in der Deutschen Bucht. *Geol. Jb.* D32, 23–68.

McCAVE, I.N. (1970) Deposition of finegrained suspended sediment from tidal currents. *J. geophys. Res.* **75**, 4151–4159.

McCAVE, I.N. (1973) Mud in the North Sea. In: *North Sea Science* (Ed. by E.D. Goldberg), pp. 75–100. M.I.T. Press.

McCAVE, I.N. (1979) Coastal accumulation of fine sediments: interpretative problems related to pollution. *ICES Workshop Sed. Poll. Interchange paper no. 9.*

MORRA, R.H.J., OUDSHOORN, H.M., SVAŠEK, J.N. & DE VOS, F.J. (1961) De zandbeweging in het getijgebied van zuidwest-Nederland. *Rapport Deltacommissie, Deel 5, Bijdrage IV.6.*

MÜLLER, G., REINECK, H.E. & STAESCHE, W. (1968) Mineralogisch-sedimentpetrografische Untersuchungen an Sedimenten der Deutschen Bucht (südöstliche Nordsee). *Senckenberg. Leth.* **49**, 347–365.

OELE, E. (1969) The Quaternary geology of the Dutch part of the North Sea, north of the Frisian Isles. *Geologie Mijnb.* **48**, 467–480.

OSTENDORFF, E. (1943) Die Grund- und Bodenverhältnisse der Watten zwischen Sylt und Eiderstedt. *Westküste*, pp. 1–6.

OTTO, L. (1976) Problems in the application of reservoir theory to the North Sea. *ICES, Hydr. Comm. CM 1976/C*, **18**, 17 pp.

PETIT, D. (1977) *Etudes sur la pollution de l'environnement par le plomb en Belgique—les isotopes stables du plomb en tant qu'indicateurs de son origine.* Thèse, Université Libre de Bruxelles.

POSTMA, H. (1961) Transport and accumulation of suspended matter in the Dutch Waddensea. *Neth. J. Sea Res.* **1**, 148–190.

POSTMA, H. (1973) Transport and budget of organic matter in the North Sea. In: *North Sea Science*, pp. 326–333. M.I.T. Press.

POSTMA, H. (1978) The nutrient contents of North Sea water: changes in recent years particularly in the Southern Bight. *Rapp. P.-v. Réun. Cons. perm. int. Explor. Mer*, **172**, 350–357.

POSTMA, H. & KALLE, K. (1955) Die Entstehung von Trübungszonen im Unterlauf der Flüsse, speziell im Hinblick auf die Verhältnisse in der Unterelbe. *Dt. hydrogr. Z.* **8**, 137–144.

PRANDLE, D. (1978) Monthly-mean residual flows through the Dover Strait 1949–1972. *J. mar. biol. Ass. U.K.* **58**, 965–973.

PRATJE, O. (1949) Die Bodenbedeckung der nordeuropäischen Meere. *Handb. Seefisch. Nordeur.* **1** (3).

PRATJE, O. (1952) Erfahrungen bei der Gewinnung von Grundproben vom fahrenden Schiff. *Dt. hydrogr. Z.* **5**, 28–31.

PRICE, N.B. & SKEI, J.M. (1975) Areal and seasonal variations in the chemistry of suspended particulate matter in a deep water fjord. *Est. Coastal mar. Sci.* **3**, 349–369.

RAMSTER, J.W. (1965) Studies with the Woodhead sea-bed drifter in the Southern North Sea. *Lab. Leafl. Fish. Lab. Lowestoft*, **6**, 1–4.

REENDERS, R. & VAN DER MEULEN, D.H. (1972) De ontwikkeling van de Dollard over de periode 1952–1969/'70. *RijkswatSt Dir. Groningen, Nota nr. 72.1.*

REINECK, H.E. (1960) Ueber Zeitlücken in Rezenten Flachsee-sedimenten. *Geol. Rdsch.* **49**, 149–161.

REINECK, H.E. (1963) Sedimentgefüge im Bereich der südlichen Nordsee. *Abh. senckenb. naturforsch. Ges. 503.*

REINECK, H.E. (1969) Zwei Sparkerprofile südöstlich Helgoland. *Natur. Mus., Frankf.* **99,** 9–14.

REINECK, H.E., DÖRJES, J., GADOW, S. & HERTWECK, G. (1968) Sedimentologie, Faunenzonierung und Faciesabfolge vor der Ostküste der inneren Deutschen Bucht. *Senckenberg. Leth.* **49,** 261–309.

REINECK, H.E., GUTMAN, W.F. & HERTWECK, G. (1967) Das Schlickgebiet südlich Helgoland als Beispiel rezenter Schelfablagerungen. *Senckenberg. Leth.* **48,** 219–275.

ROHDE, J. (1973) Sediment transport and accumulation at the Skagerrak–Kattegat border. *Report no. 8.* Oceanografiska Institutionen, Göteborgs Universitet.

SALOMONS, W., HOFMAN, P., BOELENS, R. & MOOK, W.G. (1975) The oxygen isotopic composition of the fraction less than 2 microns (clay fraction) in recent sediments from Western Europe. *Mar. Geol.* **18,** M23–M28.

SCHÜTTENHELM, R.T.E. (1980) The superficial geology of the Dutch Sector of the North Sea. *Mar. Geol.* **34,** M27–M37.

SHELDON, R.W. (1968) Sedimentation in the estuary of the river Crouch, Essex, England. *Limnol. Oceanogr.* **13,** 72–83.

DE SMET, L.A.H. & WIGGERS, A.J. (1960) Einige Bemerkungen über die Herkunft und die Sedimentationsgeschwindigkeit der Dollartablagerungen. *Verh. K. ned. geol.-mijnb. Genoot.* **XIX,** 129–133.

SMIDT, E.L.B. (1951) Animal production in the Danish Waddensea. *Meddr. Kommn Danm. Fisk.-og Havunders,* **XI,** 151.

VAN STRAATEN, L.M.J.U. (1961) Directional effects of winds, waves and currents along the Dutch North Sea coast. *Geologie Mijnb.* **40,** 333–346, 363–391.

VAN STRAATEN, L.M.J.U. (1965) Coastal barrier deposits in South and North Holland. *Meded. geol. Sticht.* **17,** 41–76.

VAN STRAATEN, L.M.J.U. & VAN KUENEN, Ph. H. (1957) Accumulation of fine-grained sediments in the Dutch Wadden Sea. *Geologie Mijnb.* **19,** 329–354.

TERWINDT, J.H.J. (1967) Mud transport in the Dutch Delta area and along the adjacent coastline. *Neth. J. Sea Res.* **3,** 505–531.

TERWINDT, J.H.J. (1977) Mud in the Dutch Delta Area. *Geologie Mijnb.* **56,** 203–210.

VEENSTRA, H.J. (1965) Geology of the Dogger Bank area, North Sea. *Mar. Geol.* **3,** 245–262.

VEENSTRA, H.J. (1971) Sediments of the southern North Sea. In: *The Geology of the East Atlantic Continental Margin* (Ed. by F.M. Delany). *Inst. Geol. Sci. Rep.* 70/15, pp. 9–23.

VERWEY, J. (1952) On the ecology and distribution of cockle and mussel in the Dutch Wadden Sea, their role in sedimentation and the source of their food supply. *Archs néerl. Zool.* **10,** 171–239.

VAN WEERING, TJ.C.E. (1981) Recent sediments and sediment transport in the northern North Sea: surface sediments of the Skagerrak. In: *Holocene Marine Sedimentation in the North Sea Basin* (Ed. by S.-D. Nio et al.). *Spec. Publs int. Ass. Sediment.* **5,** 335–359. Blackwell Scientific Publications, Oxford. 524 pp.

VAN WEERING, TJ.C.E., JANSEN, J.H.F. & EISMA, D. (1973) Acoustic reflection profiles of the Norwegian Channel between Oslo and Bergen. *Neth. J. Sea Res.* **6,** 241–263.

WINGFIELD, R.T.R., EVANS, C.D.R., DEEGAN, S.E. & FLOYD, R. (1978) Geological and geophysical survey of the Wash. *Inst. geol. Sci. London Rep.* 78/18, 32 pp.

ZAGWIJN, W.H. & VEENSTRA, H.J. (1966) A pollen-analytical study of cores from the Outer Silver Pit, North Sea. *Mar. Geol.* **4,** 539–551.

Spec. Publs int. Ass. Sediment. (1981) **5**, 429–450

Pathways of mud and particulate trace metals from rivers to the southern North Sea

W. SALOMONS *and* W. D. EYSINK*

Delft Hydraulics Laboratory, Haren Branch, c/o Institute for Soil Fertility,
P.O. Box 30003, 9750 RA Haren (Gr), the Netherlands and
* *Delft Hydraulics Laboratory, 'De Voorst', P.O. Box 152, 8300 AD Emmeloord, the Netherlands*

INTRODUCTION

One of the sources of trace metals in the North Sea are the rivers. The major rivers from Belgium, the Netherlands and Germany which influence the southern North Sea are shown in Fig. 1. Estuaries are 'normal' pathways by which the particulate and dissolved trace metals enter the marine environment. However, civil engineering activities have created two additional pathways which affect the transport of metals into the North Sea.

(a) The closing of river mouths and coastal lagoons has created large freshwater basins in which the river water, before it enters the marine environment, is subject to a variety of hydrodynamic and geochemical processes which affect the amount of trace metals entering the marine environment.

(b) The construction of deep harbours has also influenced the sedimentation of the suspended matter and its associated heavy metals. Part of the dredged material from the harbours is dumped on land and is thus permanently removed from the aquatic environment, while another part is transported to dumping sites in the North Sea.

To determine the input of trace metals into the marine environment, the geochemical and hydrodynamical processes associated with each of these three pathways should be known. Intensive research carried out over the past five years in the estuaries, the harbour areas and the freshwater basins has made it possible to construct a first tentative balance of the amount of trace metals entering the North Sea by way of these three pathways.

0141–3600/81/1205–0429 $02.00

THE RIVERS

General

The particulate metals in the rivers are carried by the suspended mud in the water, and so the transport of the particulate trace metals of a river is determined by its mud transport and the metal concentrations in the suspended mud.

In general the (suspended) mud content of the water (mud is defined as matter finer than 50 μm) shows some relation with the discharge of the river: low discharge corresponds to low mud concentrations, and high discharge to high suspended-matter concentrations. However, many individual factors are involved in the rate of erosion of the catchment area, the mud content of the water and hence the sediment transport by the river (Hinrich, 1974). This results in a considerable scatter in the data showing the relationship between the suspended mud concentrations and the river discharge. An example of this is given in Fig. 2 for the river Meuse, which shows the effects of the seasons (distinct difference between the winter months and the rest of the year), the influence of a high water wave (extremely high suspended mud concentrations during the front of a wave increasing discharge) and presumably the effects of agricultural activities (e.g. accelerated erosion due to harvesting and ploughing) during September, October and November. The relatively high mud contents during an increasing discharge have also been shown by Kreugel & Terwindt (1969) for discharges exceeding 800 m³ sec⁻¹ in the River Meuse.

The trace-metal concentrations in the suspended

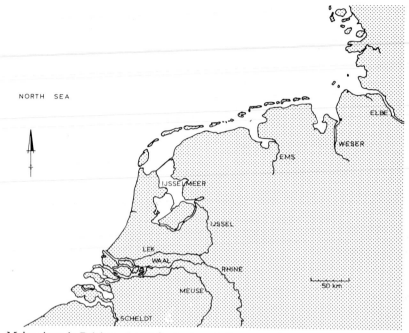

Fig. 1. Major rivers in Belgium, the Netherlands and Germany which influence the southern North Sea.

matter depend on the discharge of the river (Hell-mann, 1970; Schleichert, 1975). With an increase in the discharge, the suspended mud concentrations increase but the metal concentrations in the suspended matter decrease. This relationship sometimes makes it possible to detect the influence of tributaries on the composition of suspended mud in the main river. An example has been given by Schleichert (1975) for the influence of the rivers Neckar and Main on the river Rhine.

In spring, most of the water in the Rhine is melt-water, so that as a consequence the suspended matter also mainly originates from its upper course. In the autumn period, precipitations in the upper course accumulate as snow, and the contribution of distributaries like the (polluted) rivers Main and Neckar increases. Since the cadmium concentrations in the river Neckar (and also other trace metals, with the exception of chromium) are much higher compared with the river Rhine, two different relationships are found between the discharge and particulate metal concentrations: one for the autumn period (Neckar influence) and one for the remainder of the year (Fig. 3). Also in the lower course of the Rhine, in the Nieuwe Merwede, particulate metal concentrations depend on the discharge (Fig. 3).

The decrease in particulate metal concentrations

at high discharge is due to the contribution of contaminated eroded soil particles into the river system by the surface runoff. Also, during the high discharge the residence time of the sediment in the rivers is reduced because of the increased flow velocity, and consequently the particles have less time to pick up trace metals from polluted sources. Moreover, dissolved metal concentrations are much lower because of dilution by relatively uncontaminated water, so far less metal will be adsorbed. The above-mentioned phenomena make it difficult to determine the proper amount of mud transport and hence of metal transport. During high water waves large amounts of mud and metals may be transported. To determine these effects as well as the right transport quantities, frequent sampling of the river water is required. However, most data are based on weekly, bi-weekly or even monthly samplings, especially with regard to the trace metals, and often only individual measurements are available.

The river Scheldt

The river Scheldt, a relatively small river draining part of western Belgium, discharges into the Western Scheldt estuary (Fig. 1). Depending on the rainfall, the discharge rate may show considerable fluctua-

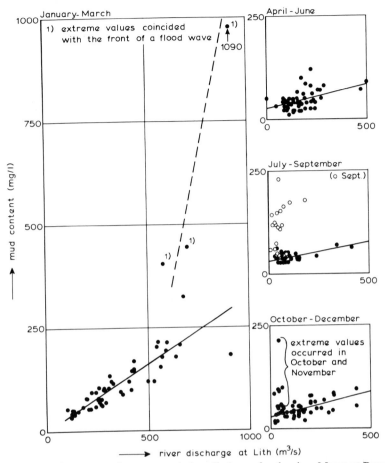

Fig. 2. Relationship between mud content and river discharge for the river Meuse at Ravenstein in 1971.

tions with time while the monthly mean discharge changes with the season. With the discharge of the water the concentration and the transport of suspended mud also vary with time. The annual mean discharges of water and mud amount to 3.2×10^9 m^3 yr^{-1} and about 1×10^6 tons yr^{-1} respectively (Table 1). The latter figure is a rough estimate based on various data from the literature (McCave, 1973; Wollast *et al.*, 1973; Wollast, 1976; Terwindt, 1977; Eisma, 1978), ranging from 0·1 to 1·41 × 10^6 tons yr^{-1}. Only a negligible amount of this mud supply seems to reach the North Sea as most of it is deposited in the estuary of the Western Scheldt. Metal concentrations in the suspended mud sampled in 1978 at Hoboken and those in sediments collected downstream from Antwerp are presented in Table 2.

The rivers Rhine and Meuse

The rivers Rhine and Meuse influence the same areas in the Netherlands (Fig. 4). The river Meuse and the lower course of the river Rhine, as well as the entire delta area, have been subject to extensive interference and alteration by man, so that the many civil works affect the regime of the rivers Rhine and Meuse and the distribution of the water over the lower courses. The enclosure of the Zuiderzee (1932) and the Haringvliet (1979), especially, have had a great influence on the hydrology of the rivers Rhine and Meuse and the estuaries in the south-west of the Netherlands.

The distribution of the water of the Rhine is controlled by weirs in the lower Rhine and Lek (completed in 1970) and sluices in the closure dam of the Volkerak (completed in 1976) and the Haring-

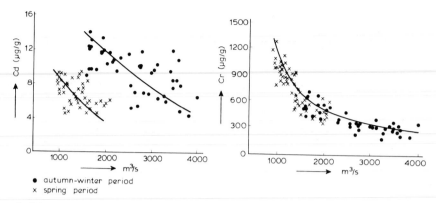

A Rhine at Koblenz

- autumn-winter period
x spring period

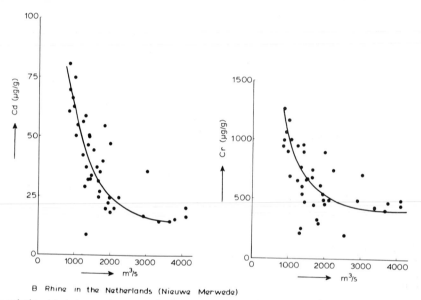

B Rhine in the Netherlands (Nieuwe Merwede)

Fig. 3. The relationship between the water discharge and the metal concentrations in the suspended matter. (A) River Rhine at Koblenz (Schleichert, 1975). (B) River Rhine in the Netherlands: locality 3 of Fig. 4.

vliet (completed in 1970). The distribution is planned to comply with the so-called 'normal discharge scheme (NLP '70)' as given in Fig. 5. In fact, this makes the Rotterdam Waterway the main estuary of the rivers Rhine and Meuse at present. At high discharge most of the Rhine water enters the North Sea by way of the Haringvliet. Data on the discharge rates of water and mud of the rivers Rhine and Meuse are presented in Table 1.

The heavy metal pollution of the river Rhine is not a recent phenomenon, because early in the twentieth century the river Rhine could be regarded as heavily polluted with trace metals. Cadmium concentrations of $4.4\ \mu g\ g^{-1}$ are already 15 times the

baseline level ($0.3\ \mu g\ g^{-1}$), whereas the zinc concentrations are more than $1000\ \mu g\ g^{-1}$. Between 1920 and 1958 all trace-metal concentrations increased (Fig. 6), while between 1958 and 1979 this rise continued for cadmium only. Striking are the decreases for arsenic (after 1958) and for mercury (after 1973). In fact arsenic is close to the baseline level at present. Zinc and lead concentrations are also slightly decreasing.

The anthropogenic contribution to the metal concentrations in sediments from the river Rhine and in some of the other rivers studied exceeds background values by one order of magnitude (Salomons & de Groot, 1978). It can even be estimated that less than

Table 1. Water discharge, mud transport and some tentative data for particulate metal transport by the rivers Rhine, Meuse, Ems, Scheldt, Weser and Elbe. Me_t = total metal transport

River	Water discharge 10^9 m^3 yr^{-1}	Mud transport 10^6 tons yr^{-1}	Particulate metal transport* tons yr^{-1}						
			Cu	Ni	Zn	Pb	Cd	Cr	Me_t
Scheldt	3·2	1†	270	130	1430	400	55·6	390	2675
Meuse	10·3	0·7‡	144	70	2821	466	57	186	3743
Rhine	69·4	3·4§	884	391	4845	1547	120	1870	9657
Ems	3·2	0·07§	6	3	41	6	0·2	8	63
Weser	11	0·37§							
Elbe	22·5	0·8§	288	68	2000	208	14	296	2874

* Based on the most recent data from Table 2 and the annual mud transport. For the rivers Ems and Elbe no data on the suspended matter were available, therefore, the deposited sediment concentrations were used.

† Data on the annual silt discharge of the Scheldt range from 0·1 to 1·42 × 10^6 tons yr^{-1} (McCave, 1973; Wollast *et al.*, 1973; Wollast, 1976; Eisma, 1978). Based on Wollast (1976) and Terwindt (1977) it is taken at approximately 1 × 10^6 tons yr^{-1}, although this figure seems rather high for this small river.

‡ Based on Terwindt (1977).

§ Hinrich (1971, 1974), Hellmann *et al.* (1977).

Table 2. Metal concentrations in bottom sediments and in suspended matter from a number of rivers

	Cu	Ni	Zn	Pb	Cd	Cr
Scheldt						
Bottom sediments 1974	165	66	1080	230	26·4	380
Bottom sediments 1978/79	180	61	1015	270	35·4	290
Suspended matter 1978	270	130	1430	400	55·6	390
Meuse						
Bottom sediments 1958	160	43	1520	380	28·5	215
Bottom sediments 1970	165	48	1430	325	24·2	325
Bottom sediments 1975	185	67	3050	475	41·5	330
Bottom sediments 1977	170	64	2630	440	40·4	255
Suspended matter 1977/78	205	100	4030	665	81·7	265
Rhine						
Bottom sediments 1922	68	36	1050	275	4·4	110
Bottom sediments 1958	295	54	2420	535	14	640
Bottom sediments 1970	325	62	1855	445	27	790
Bottom sediments 1977	285	76	1665	390	37·4	825
Suspended matter 1977/78	260	115	1425	455	35·4	550
Ems						
Bottom sediments 1964	95	38	504	66	2·6	115
Bottom sediments 1971	80	42	590	82	3·0	110
Elbe						
Bottom sediments	360	85	2500	260	16·9	370

The data on the bottom sediments have been corrected for differences in grain-size composition; the values refer to the calculated concentrations at 50 % < 16 μm (Salomons & Mook, 1977). Bottom sediments from the Scheldt were taken between the river Rupel and Antwerp; the suspended matter data refer to samples taken at Hoboken. Data on the suspended matter from the rivers Rhine and Meuse are from Rijkswaterstaat (1977–8). The bottom sediments from the rivers Rhine and Meuse were sampled in the Biesbosch area, those from the river Ems at the locality of Diele. The data for the river Elbe have been taken from Brummer & Lichtfuss, 1977. All data in μg g^{-1}.

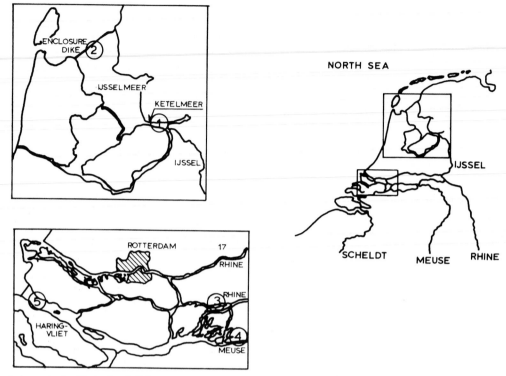

Fig. 4. Areas in the Netherlands influenced by the rivers Rhine and Meuse: IJsselmeer, Haringvliet and Rotterdam harbour.

1% of the cadmium found in sediments from the river Rhine originates from natural sources.

The zinc concentrations in the sediments of the river Meuse are higher than those in the Rhine; they show a large increase between 1958 and 1979. The chromium levels in sediments from the Meuse are lower compared with the Rhine; cadmium levels, however, are slightly higher. In Table 1, a tentative estimate is given of the total amount of particulate metals transported by the various rivers entering the southern North Sea, showing that the rivers Rhine and Meuse taken together account for 70% of the total metal transport.

The rivers Ems, Weser and Elbe

The rivers Ems and Weser are relatively small rivers which drain the northern part of western Germany, with the Elbe, draining water and suspended mud from areas up to Czechoslovakia, being more important. Data on mud and water discharges of these rivers are obtained from Hinrich (1971, 1974) and Hellmann *et al.* (1977) and presented in

Table 1. Metal concentrations in the sediments from the river Ems are given in Table 2. Zinc and lead concentrations increased slightly between 1964 and 1975. Metal concentrations in the river Ems are low compared with other rivers influencing the southern North Sea. Metal concentrations in the river Elbe have been studied in detail by Lichtfuss & Brümmer (1977), and some results are presented in Table 2. The total amount of particulate metals transported by the river Elbe is about equal to that of the river Scheldt.

PROCESSES AFFECTING THE TRANSPORT OF SEDIMENT AND METALS BY RIVERS TO THE NORTH SEA

Estuaries

In the estuaries the fresh river water and the salt sea-water meet and are mixed by the strong tidal motion and the drift currents generated by the wind. The mixing ratio can often be determined from the salinity data. The hydrodynamics of an estuary, in

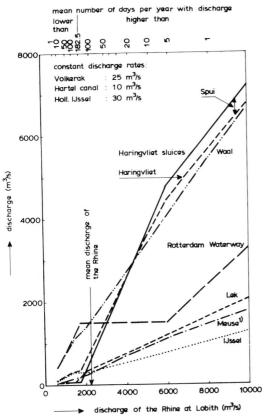

Fig. 5. Distribution of the discharges of the Rhine and Meuse according to the discharge scheme of the weirs in the lower Rhine and the Haringvliet sluices (NLP 70).

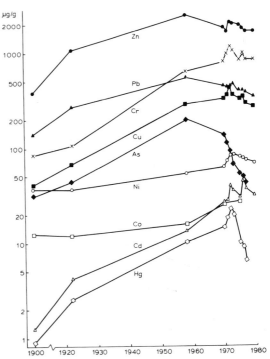

Fig. 6. History of heavy metal pollution of the river Rhine. Metal concentrations in the deposited sediments. All data corrected for grain-size differences.

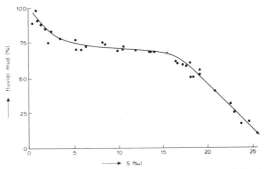

Fig. 7. Relationship between the percentage of fluvial matter in suspended matter from the Scheldt estuary and salinity.

general, are very complex because of the effects of such independent variables as tide, river discharge and wind and the density differences between water masses in the estuary caused by differences in salinity and temperature. In this system also marine and fluvial suspended matter meet and are mixed by the hydrodynamic forces. The mixing ratio of marine and fluvial suspended mud (and also of the particulate metals) cannot simply be obtained from the mud concentration in the river and the sea respectively, because settling (differential) processes occur. These processes can be caused by different phenomena which do not necessarily affect the fluvial and marine mud transports in the same way.

Knowledge of the mixing ratios of marine to fluvial material in the suspended matter, deposited sediments and in dredged material is essential for an understanding of the transport processes as well as

for making a balance of fluvial mud and particulate metals in the estuarine environment. Studies using radioactive or activable tracers provide answers to the movement of the sediments over only relatively short time periods. More useful for long-term transport studies are natural tracers. Several natural tracers (e.g. natural differences in composition between two or more sediment sources) are available in principle for tracing argillaceous sediments:

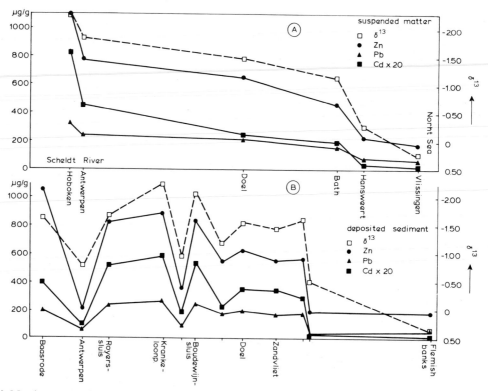

Fig. 8. (A) Metal concentrations and isotopic composition of the carbonates in suspended matter from the Scheldt estuary. (B) Metal concentrations and isotopic composition of the carbonates in deposited sediments from the Scheldt estuary.

differences in mineralogy; differences in chemical composition; and differences in the isotopic composition. The first two differences, in most cases, refer to the differences in the overall composition between two sediment sources; the third refers to differences in the isotopic composition of individual components (e.g. clays) making up the sediment. A difference in composition between two sediment sources may be used as a natural tracer provided that the tracer shows conservative behaviour, i.e. the amount of tracer per unit weight of sediment in a specific source (river or sea) is not subject to variations in time, either during transport or after deposition (Salomons *et al.*, 1978). Useful tracers for the estuaries along the southern North Sea are the isotopic composition of the carbonates and clay minerals and, in a few cases, the multi-element method (as determined with neutron activation analysis) (Salomons, 1975; Salomons *et al.*, 1975; Salomons & Mook, 1977).

The ratio of marine to fluvial mud in the suspended matter is not a simple function of the salinity. The relationship between salinity and the percentage of fluvial mud in the suspended matter for the Scheldt estuary shows three distinct ranges (Fig. 7). With low salinities the percentage of fluvial mud decreases relatively quickly to about 70%, and stays more or less constant between 5 and 15‰. The amount of fluvial mud decreases again with salinities of 15‰ and higher. This second decrease coincides with the outflow of the Scheldt in the Western Scheldt.

In general the fluvial mud content in the bottom sediments does not compare with that in the suspended matter (Fig. 8). The isotopic composition of the carbonate fraction (representative of the amount of fluvial mud) in the suspended matter decreases in the seaward direction; however, the isotopic composition of the carbonates in the bottom sediments shows an erratic behaviour. High values, indicating the presence of marine mud, are found at two localities in the estuary. The pattern of the heavy-metal concentrations in the bottom sediments is identical to that of the isotopic composition of the carbonates; e.g. low metal concentrations (influence of marine mud) correlate with high values for the isotopic composition of the carbonates.

Table 3. The calculated and observed trace-metal concentrations at Leerort (Ems estuary) and in the Haringvliet (Rhine–Meuse estuary)

	Concentration (μg g^{-1})			
	Leerort		Haringvliet	
Metal	Calculated	Measured	Calculated	Measured
Cadmium	0·7	0·8	5·0	5·0
Chromium	86	97	195	224
Copper	17·1	22·6	82	99
Nickel	28·0	33·3	25·8	30·8
Lead	49·0	58·6	177	213
Zinc	196	220	675	870

The mixing of relatively uncontaminated marine sediments with contaminated fluvial sediments explains to a large extent the decrease in metal concentrations in bottom sediments as observed in the Rhine–Meuse, Ems, Elbe and Scheldt estuaries (Salomons & Mook, 1977; Müller and Förstner, 1975) (see also Figs 8 and 11). However, superimposed on this physical process of mixing, chemical processes such as adsorption, precipitation and mobilization may occur. To determine whether these processes affect the composition of the particulate metals, it is essential to know the mixing ratio of marine to fluvial sediments. If that mixing ratio and

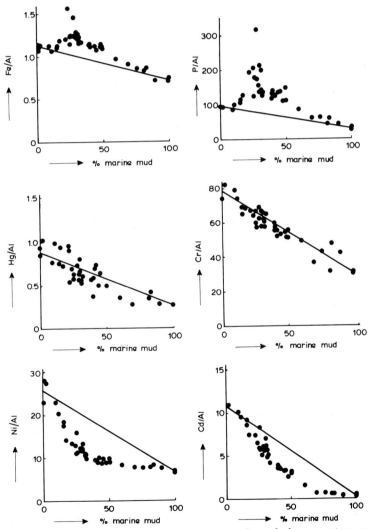

Fig. 9. The relationship between trace element to aluminium ratio and the percentage marine mud in the suspended matter in the Scheldt estuary. The straight line is the theoretical mixing curve connecting the fluvial and marine end-members.

the relevant metal concentrations are known, it is possible to calculate the trace-metal concentrations in the estuarine deposits for trace metals behaving conservatively. If so, the computed concentrations should be equal to those actually measured. Values that exceed those actually observed would point to mobilization processes; lower values, on the other hand, would indicate precipitation or adsorption processes.

Computed and measured metal concentrations for estuarine deposits from the Rhine–Meuse and Ems estuaries are presented in Table 3 (Salomons & Mook, 1977). The computed and measured values are very similar, showing that the mixing of marine and fluvial sediments is the main cause for the observed decrease in metal concentrations. However, the computed values tend to be slightly lower, which can be explained by the possible occurrence of precipitation–adsorption processes. Evidence for adsorption–precipitation processes has also been given by Duinker & Nolting (1977) for the Rhine–Meuse estuary. The dissolved trace metals show a drastic decrease in their concentrations over the range 0–5‰ salinity.

More detailed investigations have been carried out in the Scheldt estuary. Not only the deposited sediments, but also the suspended matter were studied. To determine the mixing ratio of marine to fluvial sediments, the stable isotopic composition of the carbonates was used. Results on the bottom sediments, presented in Fig. 8, have already been discussed. The behaviour of 24 trace elements during estuarine mixing was studied by analysing the suspended matter and determining the ratio of marine to fluvial sediments, and some of the results are presented in Fig. 9, with the straight line giving the theoretical mixing curve for the suspended matter. Positive deviations from the mixing curve (addition processes) are found for Fe and P (shown in Fig. 9), but also for V, As, Eu, Tb and Mn. Conservative mixing was observed for Yb, Ta, Rb, Th, Cs, Hg and Cr. For the elements Co, Pb, Zn, Cu, Sb, Cd and Ni negative deviations from the mixing curve were found, indicating the transfer of these elements from the particulate phase to the estuarine waters.

These results, together with those for the rivers Rhine, Ems and Elbe, show that the mixing of marine and fluvial sediments is the dominant process which affects the particulate trace-metal concentrations. However superimposed on the mixing process, addition and/or removal processes, depending on the metal and on environmental conditions, are taking place.

The differences in behaviour of trace metals in the Scheldt and in the Rhine–Meuse estuary show the need for more research into their estuarine behaviour. The influence of fluvial sediments on the composition of the sediments deposited close to the mouths of the estuaries appears to be limited. In the Ems estuary at Leerort, situated above the mean salinity limit at high water slack, already 90% of the sediments are derived from the marine environment. In the Western Scheldt, the influence of River Scheldt sediments can be detected only down to Hansweert, whereas in the Rotterdam Harbour area, at least downstream of the Oude Maas, most of the deposited sediments originate from the North Sea.

Harbour areas

Important harbour areas for sea-going vessels along the European coast of the southern North Sea are the harbours of Antwerp on the Scheldt, Rotterdam on the Rotterdam Waterway, IJmuiden (Amsterdam), Emden on the Ems–Dollard estuary, Wilhelmshaven on the Jade estuary, Bremerhaven on the Weser estuary and Hamburg on the Elbe estuary. In these areas artificial navigation channels and harbour basins have been dredged, thereby disturbing natural conditions. These new situations can only be maintained by considerable maintenance dredging, as the channels and harbour basins act as sediment traps. The amount of sediment removed by dredging is considerable as can be seen from the total amount removed each year from the harbours given in Table 4. This $75-80 \times 10^6$ tons is already twice the total mud budget of the North Sea. Part of this spoil is permanently removed from the system (applied as landfill), but most of it is dumped at selected dumping sites in the North Sea and so it is not lost from the aquatic system.

The artificial deepening of the estuaries in the navigation channels and in front of jetties at the bank locally reduces the natural current velocities. This decreases the capacities both of the suspended sediment transport and the sediment transport along the bottom, so siltation of the dredged areas occurs. Theoretically, the following formula can be derived, which can be used to estimate the silting rate of suspended sediment in those areas:

$$\Delta \Upsilon_s = wc_1 A \left\{ 1 - \left(1 + \frac{18}{C_2} \log \frac{h_1}{h_2} \right) \frac{u_2}{u_1} \right\}$$

Table 4. Mean annual maintenance dredging of harbours along the coast of the southern North Sea in the period around 1975

Harbour	Mean annual maintenance dredging (approximate values in 10^6 m³ yr^{-1})
Dutch harbours at the Western Scheldt estuary	2·5 (Data Rijkswaterstaat)
Antwerp	10 (Wollast, 1976)
Rotterdam	21 (van Oostrum, 1978)
Scheveningen	0·2 (Data Rijkswaterstaat)
IJmuiden	2·5 (Data Rijkswaterstaat)
Emden/Delfzijl	15 (Data Rijkswaterstaat)
Wilhelmshaven	?
Bremen/Bremerhaven	10 (Niebuhr *et al.*, 1961; D'Angremond *et al.*, 1977)
Hamburg/Cuxhaven	11 (D'Agremond *et al.*, 1977)
London	0·7 (D'Angremond *et al.*, 1977)
Hull	4·8 (D'Angremond *et al.*, 1977)

where

$\Lambda\Upsilon_s$ = silting rate of suspended sediment in a dredged area;

w = settling velocity of sediment grains in stagnant water;

c_1 = sediment concentration (weight) of the water upstream of the deepened area;

A = horizontal area of the deepened part of the estuary;

C_2 = roughness coefficient of Chézy of the deepened area;

h_1/h_2 = ratio of water depths upstream and at the deepened area respectively;

u_2/u_1 = ratio of flow velocities upstream and at the deepened area respectively, which can be estimated by:

$u_2/u_1 = h_1/h_2$ (flow perpendicular to a dredged channel)

or

$u_2/u_1 = F_1/F_2$ (flow parallel to a dredged channel) where F_1/F_2 is the ratio of the cross-sections of the estuary at the dredged region before and after the dredging of the channel.

This shows the dependence of the silting rate on the dimensions of the dredged area (A, h_1/h_2 and/or F_1/F_2), the sediment characteristics (w), and the natural sediment transport (c_1).

The siltation of the harbour basins along the estuaries, however, is governed by a different mechanism. In general, the flow velocities in these basins are far lower than those outside and so are the capacities of sediment transport. This means that a larger part of the suspended sediment that penetrates into a basin will settle there. The silting rate of a harbour basin depends upon the volume of water which is exchanged with each tide and the sediment concentration of the water which is penetrating. The water exchange is determined by: (a) the tidal volume of the harbour basin; (b) the horizontal exchange in the harbour entrance by the flow in the estuary; and (c) the vertical exchange, caused by density differences between the water in the basin and in the estuary. The first quantity is simply given by the area of the harbour times the tidal range. The second quantity is determined by a more complex mechanism.

A passing flow in front of the harbour entrance generates an eddy at that point. Silty water from the passing flow is exchanged with relatively clean water from the eddy by diffusion, with the mud penetrating in the slowly flowing eddy slowly settling, etc. This horizontal exchange depends on the flow velocity in the estuary and on the size and shape of the harbour entrance. Both the exchange by tidal volume and the horizontal exchange mainly affect the harbour at and near the entrance. The latter particularly is an effective exchange mechanism that affects the whole harbour basin. The density differences generating this type of exchange flow are caused by the salinity of the water in front of the basin fluctuating with the tide. Around high water slack the water in front of the basin will be more saline than the water in the basin and consequently the denser estuary water will penetrate along the bottom into the basin, replacing the same amount of water which is flowing out at the surface. Around low water slack the opposite occurs. In both cases harbour water, of which the suspended sediment has partly settled in the basin, is replaced by water with a higher mud content from the estuary, with the density-induced exchange velocities being proportional to the square root of the density differences and the water depth. The exchanged volume per tide can be expressed as:

$$V_e = F * hb * 0.45\sqrt{\left(\frac{\Delta\rho}{\rho}gh\right)} * T,$$

where

V_e = density-induced exchanged volume per tide;

f = empirical coefficient a.o. depending on the harbour layout;

hb = cross-sectional area of the harbour entrance;

$0.45 \sqrt{\dfrac{\Delta\rho}{\rho}} \, gh$ = initial velocity of lock exchange flow (Yih, 1965; Abraham & Eysink, 1971);

$\Delta\rho$ = characterizing density difference, e.g. maximum density difference during a tidal period;

ρ = density of water;

T = tidal period.

So the mechanism of vertical exchange is most effective in deep harbours with a wide entrance in the area of a salt-water wedge which is moving forth and back with the tide.

In addition, however, there are two other mechanisms that enhance the siltation of harbour basins in the area of the salt-water wedge. First, in the transition region where fresh water mixes with saline water, mud particles start to flocculate. This increases the effective fall velocities of the silt particles and so stimulates the settling of mud in relatively calm regions like the harbour basin. Secondly, the density difference between fresh water and sea-water distorts the flow-velocity profiles and causes the water to circulate in the transition zone with a residual inflow along the bottom and a residual outflow at the surface. The salt penetrating in this way along the bottom is balanced by entrainment of water from the lower layer into the upper layer, where it is discharged to the sea. Mud particles also follow this path, although they may sink back into the lower layer after they have been entrained into the upper layer. So, in the transition zone between fresh and salt water a circulation of suspended sediment is possible, which in general causes it to accumulate, as often is found in increased sediment concentrations of the water in the transition zone (Postma, 1960, 1967b; Wollast *et al.*, 1973; Peters & Sterling, 1976). This increases the sediment contents of the exchanged water volumes and so the siltation of the harbour basins. The sediment settled in the artificial channels and basins is regularly removed by maintenance dredging, the spoil being partly brought to land and partly returned to the aquatic system by dumping it at sea or at particular dumping sites in the estuary. During the dumping process a large part of the fine sediment is washed out and entrained by

the local currents. The rest of the spoil sinks to the bottom where, in general, it is gradually eroded and redistributed. In fact, the natural transport of sediment is temporarily interrupted by settling in the man-made settling basins and sediment is, at least partly, brought back again into the aquatic system for further transportation by nature by the dumping of maintenance dredging material in the estuary or at sea.

Harbour of Rotterdam: case study

The history of the harbour of Rotterdam, situated along today's main branches of the Rhine delta, gives a good impression of the present importance of the above-described processes in the transport of mud to the southern North Sea. The Rotterdam Waterway is only relatively young; it is a man-made branch of the Rhine delta which was dug in 1868. During the next three decades it took a great deal of effort to achieve a stable access channel which was satisfactory for the harbour of Rotterdam. This situation with a stable profile of the Rotterdam Waterway lasted from about 1897 to 1909 but then gradually changed due to harbour extensions (such as the first and second Petrol harbours and the Maas-, Rijn-, Waal- and Eem-harbour basins) and improvements of the Rotterdam Waterway until a second equilibrium (including the effect of maintenance dredging) was reached in the period from about 1950 to 1958. Since then, considerable new harbour extensions (such as the Botlek and Europoort basins) and improvements of the fairway to Rotterdam have led to further changes in the estuary. During these developments the storage area increased with the increasing harbour area, and the hydraulic resistance against the penetration of the tide and the salt sea-water decreased with the increasing water depth in the Rotterdam Waterway. This resulted in a considerable increase of the tidal range in Rotterdam, and further inland in a drastic increase of the tidal volumes, the salt penetration and, which is particularly important for the subject under consideration, the maintenance dredging (Table 4). Since the closure of the Haringvliet in November 1970, the mean tidal volume of the Rotterdam Waterway has decreased from about 210×10^6 m^3 (mean ebb volume plus mean flood volume at Hook of Holland) to about 170×10^6 m^3, which also has had an unfavourable effect with regard to the maintenance dredging (Haring, 1978). The amount of spoil from the basins increased from 0.3–0.4×10^6

Table 5. Some characteristic figures of the Rotterdam Waterway in relation to the historical development of the Rotterdam harbour

Year	1900	1923	1955	1962	1974
Area of harbour basins (10^4 m²)	60	570	850	1600	2100
Cross-sectional area of KMR 1030 at Hook of Holland (m² below LW)	3200	3625	5100		7900
Sum of mean ebb and flood volumes at KMR 1030 (10^6 m³)	112	138	173	196	170*
Mean tidal range at Rotterdam (m)	1·45	1·50	1·70	1·75	1·72
Mean annual maintenance dredging of the basins (10^6 m³)†		0·3–0·4	2–2·5	4–4·5	7·5–8

* On 2 November 1970 the Haringvliet was closed off from the sea, which reduced the mean tidal volume of the Rotterdam Waterway at the Hook of Holland from approximately 200×10^6 m³ to about 170×10^6 m³.

† Based on Terwindt (1971) and van de Ridder (1978).

m³ in 1923 to $7 \cdot 7$–8×10^6 m³ in 1974; a 20- to 25-fold increase in about 50 years of harbour development.

At present the total annual maintenance dredging amounts to about 21×10^6 m³, of which approximately 70% is mud. Expressed in tons of dry matter per year, this corresponds with about 10×10^6 tons yr^{-1}, which is about 5 times the amount supplied by the river. More detailed information is shown in Table 5, which gives the data of maintenance dredging as provided by the Public Works Department of Rotterdam and by the Rijkswaterstaat.

The siltation of the fairway and harbour basins does not occur gradually but shows a seasonal effect with, in general, the highest silting rates being during the winter (van de Ridder, 1978; Ottevanger, 1978; Van Bochove & Nederlof, 1979) (see Fig. 10). The seasonal effects in river discharges and meteorological conditions (storm periods) play a role in this phenomenon, and there are indications that particularly stormy weather has a great influence on the supply of sediment that settles in the harbour area (Terwindt, 1977). It is obvious that the harbour basins and fairways act as important sinks for marine sediments. According to the results of stable isotope studies the contribution of marine sediments in the spoil decreases in a landward direction. In the Europoort area the amount of marine mud is estimated to be about 90%, in the Botlek area about 60% and in the most easterly harbours about 25%.

The spoil of the maintenance dredging is partly dumped on land, but the greater part is dumped at sea at a site (Loswal Noord) 5 km offshore northwest of the mouth of the Rotterdam Waterway. Since 1970 a mean annual discharge of about $14 \cdot 5 \times 10^6$ m³ of spoil has been released there, representing about 7×10^6 tons of dry sediment per year. At least half to three-quarters of this spoil is mud (Terwindt, 1971,

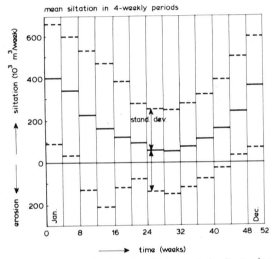

Fig. 10. Seasonal effect in siltation of the Rotterdam harbours (after van Bochove & Nederlof, 1979).

1978; Van Oostrum, 1978), which has meant a mean annual mud discharge here of 4–5×10^6 tons (dry matter). In the past few years this amount of spoil dumped at sea has been increased to about $17 \cdot 5 \times 10^6$ m³ yr^{-1}, containing 5–$6 \cdot 5 \times 10^6$ tons of dry mud. A large proportion of this muddy part of the spoil is resuspended during the dumping process as well as by wave and current action on the bottom (Terwindt, 1978). In fact, it is an important source of (reactivated) mud in the North Sea.

The spoil from the westerly harbours, which is less contaminated because of low admixtures of fluvial sediments (Fig. 11) is dumped at sea (Table 6), while the most severely contaminated spoil from the easterly harbour areas with high admixtures of fluvial sediment (Fig. 11) is dumped on land. So the landfill

Table 6. Dredging data of the Rotterdam harbour area. Based on data supplied by Rotterdam (period 1970–5, see also Stuurgroep Berging Baggerspecie (1978) and Rijkswaterstaat (period 1974–8)

	Maintenance dredging	
		Dumped at sea
Location	Quantity $(10^6 \text{ m}^3 \text{ yr}^{-1})$	Quantity $(10^6 \text{ m}^3 \text{ yr}^{-1})$
Fairways		
Euro channel	1·1	
Maas channel	1·5	1·5
Maas entrance	4·5	4·5
Caland and Beer channels (entrance)	4·5	4·5
Rotterdam Waterway KMR 1000–1033	2·8	2·8
Adjacent river area	0·9	
Sub-total	15·3	13·3
Harbours		
Europoort	1·2	1·2
Botlek	3·9	1·8
Other harbours	2·4	
Total quantity	22·8	16·3

areas act as important sinks for fluvial sediments and their associated pollutants. However, not only 'fluvial' metals end up in the landfill areas, but also relatively large amounts of 'marine metals' are put on land as well (Fig. 12).

Using the approximate mixing ratios of marine to fluvial sediments in the Rotterdam harbour area, the amount of dredged material and its metal concentrations, the amount of metal of fluvial origin which ends up in the landfill areas and which is dumped at Loswal Noord has been calculated. The results for 1977 are shown in Table 7. The balance shown in Table 7, however, is subject to a number of uncertainties. First, the survey of the Rotterdam Harbour was conducted in September 1977. The preceding period was one of decreased siltations; moreover, the major period of siltation is in winter (Fig. 10). And as the supply of mud is mainly from the sea, the percentage of marine mud in the dredged material may therefore be higher and thus, by using the data of September, the amount of fluvial mud transported to the North Sea (Loswal Noord) may be overestimated.

A second uncertainty concerns the ratio of marine to fluvial sediments in the major siltation area (Europoort/Maasmond). The natural tracers are rather insensitive at low admixtures of either marine to fluvial, or fluvial to marine sediments. The best

estimate of 10% fluvial sediments in the dredged material from the Europoort/Maasmond may be out by 5%, and this may result in a large uncertainty about the metal balance. Additional data on the dredged material as well as further research into natural tracers are therefore required to produce a reasonably clear and accurate picture.

Freshwater basins

Two freshwater basins have been created by man within the delta area of the rivers Rhine and Meuse, which affect the amounts of mud and of particulate and dissolved metals entering the North Sea. The first basin, the IJsselmeer, has existed since 1932 when the former Zuiderzee was closed off from the Wadden Sea, while the second is the Haringvliet basin which was created in November 1970 (Fig. 4).

After the creation of the IJsselmeer the tidal motion has disappeared completely and the basin gradually became fresh by flushing with Rhine water supplied by the river IJssel and discharging through the sluices in the Enclosure dyke. Basically the water in this basin was practically stagnant and had a long residence time. Then the area of the basin gradually diminished as a consequence of the reclamation of polders, causing a decrease in the residence time of the water. At present, the mean residence time is still considerable: about 6 months. Primarily most of the suspended matter from the river IJssel settles close to the river mouth (the Keteldiep area); consequently, the deposits there have very high concentrations of trace metals (Table 8). Part of the suspended matter is directly transported into the lake itself, the bottom sediments of which are easily stirred up during stormy periods and transported by wind-induced drift currents. This mechanism of diffusive sediment transport is possible due to the restricted depth and the great surface area of the basin, allowing the wind energy to create sufficient turbulence in the water to stir up the bottom sediments. Fig. 13 shows the relationship between the wave height and the suspended-matter concentration. The amount of mud brought into suspension during storms is several times the yearly supply by the river IJssel. This and the long residence time of the water in the lake indicates that the sediment transports in the lake will be strongly affected by diffusion processes. Due to this the bottom sediments, with low metal concentrations, are mixed intensively with the recent contaminated suspended matter. As a consequence

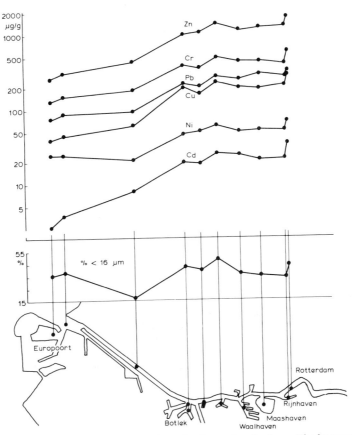

Fig. 11. Mean metal concentrations in dredged material from the Rotterdam harbour. The lower curve represents the mean grain-size composition (percentage < 16 μm) of the samples analysed. This survey was conducted in September 1977, and the data are based on about 150 sediment samples.

hardly any gradients in the metal concentrations in the bottom sediments are found (intensive reworking during storm) and, due to the admixture of relatively uncontaminated bottom sediments, the metal concentrations in the suspended matter at the Enclosure dyke are lower compared with the Ketelmeer (Table 9).

Apart from physical mixing processes, three more processes affect metal concentrations in the IJsselmeer, all directly or indirectly related to the high nutrient load of the river IJssel. The high phosphorus input into the IJsselmeer and the relatively long residence time (6 months) cause massive algal blooms. As a consequence, large amounts of organic matter are produced, part of which accumulates in the bottom deposits. Also the large amount of phosphorus taken up by algae becomes part of the sediment. The phosphorus balance of the lake,

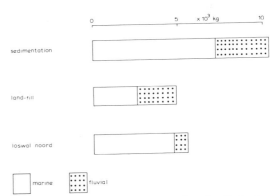

Fig. 12. Tentative metal balance for the Rotterdam harbour. (Total amounts of trace metals (Cu + Cr + Pb + Zn + Ni + Cd) are given.)

Table 7. Tentative balance of trace metals of fluvial origin in dredged material from the Rotterdam harbours (1977) (in tons (1000 kg))

	Zn	Cu	Cr	Pb	Cd	Ni	Me$_t$
Landfill							
Rijn-, Maas-Waalhaven	462	73	156	105	8	20	823
Eem-, 2nd Petroleum-, 1 Petroleum harbour, Botlek	807	150	318	175	16·3	40	1507
Total amount fluvial metals to landfill	1269	223	474	280	24·3	60	2330
North Sea							
Europoort	229	33	103	62	3·2	16	446
Botlek	233	45	92	54	4·4	12	440
Total amount of fluvial metals to the North Sea	462	78	195	116	7·6	28	887

Table 8. Metal concentrations in sediments from the IJsselmeer. All concentrations in μg g^{-1} and corrected for grain size differences by using the calculated data at 50 % < 16 μm (Salomons & de Groot, 1978)

	Zn	Cu	Ni	Pb	Cd	Cr
Ketelmeer 1977	1960	250	67	320	34	570
IJsselmeer 1977	430	40	30	73	2·8	94
IJsselmeer 1933	133	19	39	39	0·4	88

therefore, can be used to estimate the amount of organic matter which accumulates in the bottom deposits (Postma, 1967a). In 1977 and 1978 the accumulation of organic matter was estimated at about 160,000 tons yr^{-1}, while Postma (1967a) had calculated for the period 1932–65 a mean accumulation of 120,000 tons yr^{-1}. Although metals are accumulated by algae, the metal concentrations in the algae are lower compared with those in the inorganic suspended matter (Salomons & Mook, 1980). Therefore the addition of newly formed organic matter to the suspended matter and to the bottom sediments causes a decrease in metal concentrations (dilution effect).

The uptake of carbon dioxide by algae and carbon dioxide exchange between the surface water and the atmosphere are responsible for the pH increase in the lake from 7·3 (river value) to more than 9 (in summer) in the northern part of the lake. At these high pH values, calcium carbonate precipitates. The pH of the lake in winter reaches values of only 8, and no precipitation of calcium carbonate takes place. In fact the dissolved calcium concentrations in the lake show distinct seasonal cycles. The production of calcium carbonate in the lake was estimated at 400,000 tons yr^{-1} for 1977 and 1978 (Salomons & Mook, 1980). Because metal concentrations in calcium carbonate are low, its addition to the suspended

matter causes an apparent decrease in metal concentrations (dilution effect).

The third process affecting metal concentrations is also related to the pH increase, which causes an adsorption of some dissolved trace metals. The adsorption process is important for cadmium, chromium and zinc, because the adsorption of these three metals is strongly dependent on the pH over the range observed in the lake. On the other hand, the adsorption of copper is not strongly dependent on the pH between 7 and 9 and therefore adsorption plays a minor role in the removal of copper from solution (Fig. 14). A comparison between dissolved and particulate metal concentrations in the mouth of the river IJssel (Ketelmeer) and the Enclosure dyke shows the extent of these various (physical and chemical) processes. The IJsselmeer not only acts as a sink for particulate metals but also for dissolved metals (Tables 9 and 10).

The lake not only acts as a sink for sediment transported by the river IJssel into the lake but, due to the biogeochemical processes, an internal sediment production takes place which is twice as high as the input of sediment by the river. A rough mud balance can be obtained by using the mud supply by the IJssel and the mud discharge of the sluices at Den Oever and Kornwerderzand (Table 11).

The freshwater basin Hollands Diep–Haringvliet

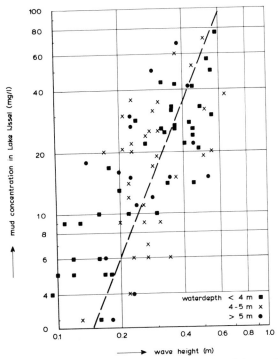

Fig. 13. Relationship between waves (wind) and the mud content of the water of the IJsselmeer.

was created by the closure of the Volkerak in 1969 and the Haringvliet in 1970. Since then the tidal motion, now penetrating mainly via the Rotterdam Waterway, Oude Maas, Spui and Dordtse Kil, has been strongly reduced. Basically, three phases can be distinguished in this new situation (Fig. 5).

(1) During Rhine discharges lower than 1730 m³ sec⁻¹ the discharge through the Harvingvliet sluices is negligible. The waters of the rivers Meuse and Rhine are discharged to the sea through the Dordtse Kil and Spui. The tidal range in the basin is only 0·1–0·2 m. The flow velocities in this situation are very weak, and wind-induced currents may dominate.

(2) With Rhine discharges increasing from 1730 m³ sec⁻¹ up to 6000 m³ sec⁻¹ the gate opening of the Haringvliet sluices gradually increases until its maximum. The discharges vary with the tide at sea, resulting in a tidal range in the basin which slowly increases from about 0·2 m at low Rhine discharges to several decimetres up to a maximum of roughly 0·5 m with increasing flow velocities.

(3) During Rhine discharges exceeding 6000 m³ sec⁻¹ the gates of the Haringvliet sluices are fully opened at low tide at sea. The tidal range and the motion of water then approaches the former situation. These conditions, however, occur on average only 6 days per year.

Thus the tidal range and the flow velocities in the basin are generally low. The basin is much deeper than the IJsselmeer which strongly reduces the effect of wind on the mud transports in this freshwater basin. The effect of the discharge of the Rhine is in this respect more important. It appears that in the new situation, after the closure of the Haringvliet in November 1970, mud is deposited in the Nieuwe Merwede and Amer near the Biesbosch during low Rhine discharges. During high Rhine discharges part of this may be eroded again and, with other mud, may be partly redeposited further downstream, in particular in the Hollands Diep (Haring, 1978; Ferguson *et al.*, 1976; Terwindt, 1977). An approximate mud budget for the Hollands Diep–Haringvliet basin is presented in Table 12 (see also Fig. 15), showing a siltation of about 1·8 × 10⁶ tons of mud per year.

Processes affecting trace metals in the Haringvliet are the same as in the IJsselmeer. However, the differences in the hydrodynamic conditions cause some notable variations in their extent. The residence time of the water is much shorter (weeks) compared

Table 9. Metal concentrations (μg g⁻¹) in the suspended matter from the freshwater basins. The numbers in parentheses refer to the localities shown in Fig. 4. Data are based on 3–4 weekly surveys conducted in the freshwater basins in 1977 and 1978

	Zn	Cu	Ni	Pb	Cd	Cr
IJsselmeer						
Ketelmeer (1)	2160	283	94	428	47·7	644
Enclosure dike (2)	751	62	56	158	5·6	184
Haringvliet						
Nieuw Merwede (3)	1743	294	88	489	44·5	765
Amer (4)	3152	204	88	537	40·3	315
Haringvlietdam (5)	1869	172	77	470	21·1	518

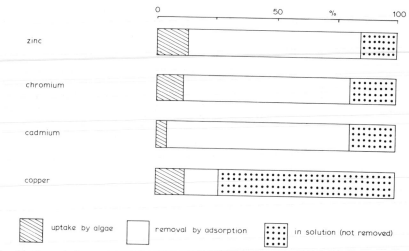

Fig. 14. Fate of dissolved trace metals in the IJsselmeer (based on Salomons & Mook, 1980).

Table 10. Dissolved metal concentrations ($\mu g\,l^{-1}$) in the freshwater basins. The numbers in parentheses refer to the localities shown in Fig. 4. Data are based on 3–4 weekly surveys conducted in the freshwater basins in 1977 and 1978

	Zn	Cu	Ni	Pb	Cd	Cr
IJsselmeer						
Ketelmeer (1)	40	5·5	6·0	1·5	0·8	2·3
Enclosure dike (2)	4·3	2·7	5·3	1·6	0·1	0·3
Haringvliet						
Nieuw Merwede (3)	70	4·6	6·1	1·3	1·6	2·7
Amer (4)	41	2·9	14	3·1	0·2	0·6
Haringvlietdam (5)	25	2·8	5·0	3·4	0·6	1·1

Table 11. Approximate silt budget of the IJsselmeer (exclusive of Markermeer) and the water discharge of the river IJssel and through the sluices in the Enclosure dyke

	Sediment ($\times 10^3$ ton yr^{-1})	Water* ($\times 10^6$ m³)
IJssel†	330	9600
Organic matter‡	160§	
Carbonates	440	
Sluices Enclosure dike¶		−12430**
Den Oever	−180	
Kornwerderzand	−105	

* Water discharge data based on Rijkswaterstaat and Rijksdienst IJsselmeerpolders (1976). Mean values over the period 1969–74.
† Based on mean river discharges according to plan S 300 and mean silt contents at Kampen (period 1972–8; 11 to 14 samples per quarter) (Rijkswaterstaat, 1972–8).
‡ Approximately 120,000 tons yr^{-1} according to Postma (1976b) during the period 1932–65.
¶ Based on discharge data and silt content of the period 1973–7; three water samples per quarter (Rijkswaterstaat, 1972–8).
** Apart from the river IJssel, polders and some small streams contribute water to the lake.

with the IJsselmeer. No massive algal blooms occur and the pH reaches values of about 8 only in the western part of the basin, thus causing a smaller removal of dissolved trace metals (Table 10). As the depth of the Haringvliet is greater than that of the shallow IJsselmeer, the resuspension of (relatively unpolluted) bottom sediment, therefore, is less frequent, causing only small admixtures of bottom sediments to the suspended matter. Particulate metal concentrations, therefore, stay high in the Haringvliet, which acts mainly as a sink for suspended matter and its associated pollutants but less so for the dissolved trace metals.

TENTATIVE MUD AND METAL BALANCE FOR THE RIVERS RHINE AND MEUSE

The data for the rivers Rhine and Meuse, which account for a large part of the metals transported by the rivers in Belgium, the Netherlands and Germany

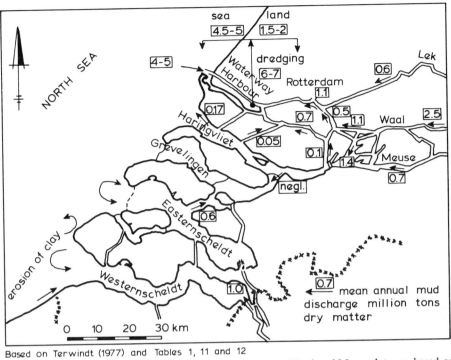

Fig. 15. Mean annual discharge of mud in the Dutch delta area. (Lek, Waal and Meuse data are based on Terwindt, 1977, all other data in Tables 6 and 12).

Table 12. Approximate silt budget of the Hollands Diep–Haringvliet basin

	Major mean silt discharges and productions ($\times 10^3$ tons yr^{-1})
Nieuwe Merwede	1400*
Meuse	700*
Dordtse Kil ⎫	
Volkerak sluices ⎬	−320†
Spui ⎪	
Haringvliet sluices ⎭	
Siltation	1780 10³ tons yr⁻¹

* Terwindt (1967, 1977).

† Based on quarterly mean discharge and silt content data of the Rijkswaterstaat during the period 1973–7 (generally 3–7 water samples per location per quarter). These data did not allow a reliable estimate of the separate mean annual discharges as the silt discharges of the Dordtse Kil, the Spui and of the Haringvliet sluices distinctly depend on the discharges of the Rhine and during the period considered the discharges of the Meuse and the Rhine did not exceed the average discharges. However, the sum of these silt discharges, also taking into account the NLP 70 scheme (see Fig. 3), seems to be rather constant over a rather wide range of discharge conditions at an average of about 320,000 tons yr⁻¹. For average discharge conditions, this total silt discharge can

be subdivided as follows: Dordtse Kil: 100,000 tons yr⁻¹, Volkerak sluices: 5000 tons yr⁻¹, Spui: 40,000 tons yr⁻¹, Haringvliet sluices: 175,000 tons yr⁻¹.

in the direction of the North Sea (Table 1), can be used to show the relative importance of the three pathways for the input of trace metals and mud into the southern North Sea.

Data on the amount of trace metals transported by the rivers Rhine and Meuse have been obtained from Rijkswaterstaat (1977–8). The weekly measurements of both dissolved and particulate trace metals were simply averaged, because insufficient data are available to estimate the influence of high water waves on metal transport. The amounts, in tons yr⁻¹, are presented in Table 13. The calculations on the input and accumulation of trace metals in the freshwater basins are based on data presented in Tables 9 and 10. The water discharge is estimated from the NLP '70 scheme (Fig. 5) and the mud discharge (particulate metal transport) are based on Tables 11 and 12 (see also Fig. 15). To facilitate the reading of the data, they are given as a percentage of the total metal transport by the rivers Rhine and Meuse.

Table 13. Preliminary metal balance for the Netherlands

	Cu	Ni	Zn	Pb	Cd	Cr	Me_t
Input (tons yr^{-1})							
Rhine	1138	835	8380	1548	173	2383	14,457
Meuse	77	51	1671	159	22	80	2060
Total	1215	886	10,051	1707	195	2463	16,517
Accumulation (in percentage of input by Rhine and Meuse)							
Haringvliet basin	43	27	49	53	51	46	48
IJsselmeer	7	1	8	5	9	7	7
Landfill	18	7	13	16	13	19	14
Total	68	35	70	74	73	72	69
Output							
Haringvliet basin	7	20	8	4	8	5	8
IJsselmeer	4	10	3	4	2	2	3
Loswal Noord	6	3	5	7	4	8	5
Total	17	33	16	15	14	15	16
Not accounted for	15	32	14	11	13	13	15
(Any other direct transport by the estuary to the North Sea)							

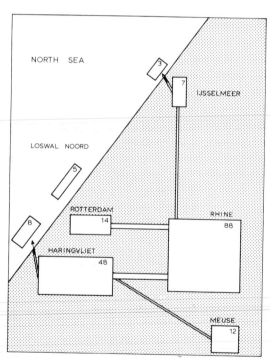

Fig. 16. Metal balance for the rivers Rhine and Meuse. Transport of metals (Cu + Cd + Cr + Pb + Ni + Zn) by the rivers Rhine and Meuse is 100 %. The major sinks and the transport to the southern North Sea are shown. Insufficient data were available on the direct transport of trace metals through the Rhine estuary to the North Sea; it is probably the major part of the 15 % not accounted for.

Concentrations of trace metals in dredged material from the Rotterdam harbour were obtained during a survey in September 1977. Data on the amount of dredged material used for landfill and transported to the North Sea have been obtained from the municipality of Rotterdam and from the Rijkswaterstaat. These data, together with the approximate ratios of marine to fluvial sediments in dredged material (see 'Harbour of Rotterdam; case study') were used to calculate the amount of trace metals of fluvial origin which were transported to the landfill areas or to the North Sea. Insufficient data were available on the direct transport of trace metals through the Rhine estuary to the North Sea, so this part of the metal balance could not be directly calculated; it is probably the major part of 'not accounted for' in Table 13.

The trace metals (with the exception of nickel) behave more or less similarly. Therefore, it is possible to make a schematic diagram showing the major sinks of trace metals from the rivers Rhine and Meuse and the output to the North Sea by using the total metal transport quantities (Fig. 16). Of the total amount of trace metals transported by the rivers Rhine and Meuse, 48 % accumulates in the Haringvliet. For the individual metals (with the exception of nickel), this amount varies between 43 and 53 %. More than two-thirds of the trace metals entering the Netherlands through the rivers Rhine and Meuse accumulate in the freshwater basins or end up in the landfill areas (dredged material), with less than one-third entering the North Sea. The freshwater basins account for 11 % of the input into the North Sea and the dredging activities for 5 %. Not accounted for, which may be partly the direct transport through the estuary to the North Sea, is about 15 %.

CONCLUSIONS

The two additional pathways (dredging and freshwater basins) determine to a large extent the amount of trace metals entering the North Sea through the rivers Rhine and Meuse. Together they account for about 70 % of the metal load transported by the rivers in Belgium, the Netherlands and Germany to the North Sea.

The freshwater basins act as sinks for particulate and for dissolved trace metals. Relatively small amounts of suspended matter escape in the freshwater basins and enter the North Sea. Due to the mixing of bottom sediments (relatively uncontaminated) with fluvial (highly contaminated) suspended matter in the freshwater basins, the metal concentrations in the suspended matter entering the North Sea are lower compared with the river. This process is, due to its shallowness, important in the IJsselmeer but less so in the Haringvliet. The pH increase in the freshwater basins, due to algal growth and CO_2 exchange with the atmosphere, is responsible for the adsorption of trace metals on the suspended matter. Also the uptake of metals by algae causes a removal of dissolved trace metals from solution. The long residence time of the water in the IJsselmeer and the massive algal blooms in this lake are responsible for the fact that the removal of dissolved metals in the IJsselmeer is greater when compared with the Haringvliet.

The creation of artificial deep navigation channels and the construction of harbour basins cause an increased settling of suspended matter. In the case of the Rotterdam harbour the supply of mud originates mainly from the North Sea. But as part of the dredged material is used for landfill, both fluvial and marine mud and their associated trace metals are removed from the aquatic environment. The major part of the dredged material is transported to the North Sea, and the dumping sites act as important point sources for sediments in the North Sea. The total quantity of sediments dredged along the southern North Sea coast is twice the mud budget of the North Sea.

The result of the two additional pathways with regard to pollutants entering the North Sea has been that two-thirds of the trace-metal load of the rivers Rhine and Meuse accumulate in the Netherlands. This situation may be regarded as beneficial for the North Sea, but poses problems for the management of the inland waters and the landfill areas in the Netherlands.

ACKNOWLEDGMENTS

Part of this work has been made possible by several investigations carried out on behalf of the Rijkswaterstaat and the municipality of Rotterdam. In addition these two organizations provided us with unpublished data. The investigations carried out in the IJsselmeer were partly supported by a grant from the EEC (contract No. 199-77-1 ENV N).

REFERENCES

ABRAHAM, G. & EYSINK, W.D. (1971) Magnitude of interfacial shear in exchange flow. *J. Hydrol. Res.* **9**, 125–151.

ANGREMOND, K.d'., BRAKEL, J., HOEKSTRA, A.J., KLEINBLOESEM, W.C., NEDERLOF, L. & DE NEKKER, J. (1977) *Assessment of certain European dredging practices and dredging material containment and reclamation methods.* Adriaan Volker, Rotterdam. Contract study for W.E.S., Vicksburg, U.S.A.

BOCHOVE, G. VAN & NEDERLOF, L. (1979) Vaargedrag van diepstekende schepen in slibrijke gebieden. *Ingenieur*, **91**, 525–530.

DUINKER, J.C. & NOLTING, R.F. (1977) Dissolved and particulate trace metals in the Rhine estuary and the Southern Bight. *Bull. Mar. Pollut.* **8**, 65–69.

EISMA, D. (1978) Sedimentatie van gesuspendeerd materiaal in de Noordzee. In: 'Texel 1978', *Waterbeweging en menging in het zuidelijk gedeelte van de Noordzee. Verslagen van tweedaagse werkbijeenkomst op het NIOZ te Texel* (Ed. by Zimmerman), pp. 141–162. 22–23 February 1978.

FERGUSON, H.A., TERWINDT, J.H.J., DEKKER, J., BAKKER, W.T. & VAN DER TUIN, H. (1976) *Slib in de Deltawateren.* Rijkswaterstaat.

HARING, J. (1978) De geschiedenis van de ontwikkeling van de waterbeweging en het profiel van de rivieren in het noordelijk Deltabekken over de perioden 1870–1970–1976. Rijkswaterstaat, *Dir. W. en W. dist. ZW, nota nr.* 44.011.02.

HELLMAN, H. (1970) Die Adsorption von Schwermetallen an den Schwebstoffen des Rheins. Eine Untersuchung zur Entgiftung des Rheinwassers (ein Nachtrag). *Dt. gewässerk. Mitt.* **14**, 42–47.

HELLMAN, H., HINRICH, H., KNÖPP, H., KOLB, S., KOTHÉ, P., MÜLLER, D., MUNDSCHENK, H., SCHWILLE, F. & TIPPNER, M. (1977) Schwebstoffe und Schlammablagerungen in Bundeswasserstrassen. *Jber. Anst. Bund. Gewässerkunde*, 1–28.

HINRICH, H. (1971) Schwebstoffgehalt und Schwebstoff-fracht der Haupt- und einigen Nebenflüsse in der Bundesrepublik Deutschland. *Dt. gewässerk. Mitt.* **15**, 113–129.

HINRICH, H. (1974) Schwebstoffgehalt, Gebietsnieder-schlag, Abfluss und Schwebstofffracht der Ems bei Rheine und Versen in den Jahren 1965 bis 1971. *Dt. gewässerk. Mitt.* **18**, 85–95.

KREUGEL, A.L. & TERWINDT, J.H.J. (1969) Onderzoek slibafvoer in de Maas bij Lith en Alphen. *RijkswatSt. W.A. K.234.*

LICHTFUSS, R. & BRÜMMER, G. (1977) Schwermetallbelas-tung von Elbe Sedimenten. *Naturwiss.* **64**, 122–125.

McCAVE, J.N. (1973) Mud in the North Sea. In: *North Sea Science* (Ed. by Goldberg), pp. 75–100.

MÜLLER, G. & FÖRSTNER, U. (1975) Heavy metals in sediments of the Rhine and Elbe estuaries: mobiliza-tion and mixing effect. *Environ. Geol.* **1**, 33–39.

NIEBUHR, W., JANSSEN, T., KRAUSE, M., MÜLLER, F., SCHREIER, M. & WALTHER, F. (1961) Orientation and layout of the approaches to the German North Sea ports and improvement of the navigable channels. *20th Int. Navigation Congress, Baltimore, USA.* **S-II-2**, 5–37.

OOSTRUM, W.H.A. VAN (1978) MKO-project. *Ingenieur*, **90**, 776–779.

OTTEVANGER, G. (1978) Het onderhoudsbaggerwerk in de Rotterdamse regio onder verantwoordelijkheid van de Rijksoverheid. *Ingenieur*, **90**, 785–793.

PETERS, J.J. & STERLING, A. (1976) Hydrodynamique et transport de sédiments de l'estuaire de l'Escaut. In: *Projet Mer, Rapport final*, **10**, 1–71. (Ed. by Nihoul, J.C.J. and Wollast, R.)

POSTMA, H. (1960) Einige Bemerkungen über den Sinkstofftransport im Ems-Dollart Estuarium. *Verh. K. ned. geol.-mijnb. Genoot.* **19**, 103–110.

POSTMA, H. (1967a) Observations on the hydrochemistry of inland waters in the Netherlands. In: *Chemical Environment in the Aquatic Habitat* (Ed. by H.L. Golterman and R.S. Clymo), pp. 30–38.

POSTMA, H. (1967b) Sediment transport and sedimenta-tion in the estuarine environment. In: *Estuaries* (Ed. by G.H. Lauff), *Publs Am. Ass. Advmt Sci.* **83**, 158–179.

RIDDER, K.H. VAN DE (1978) Rotterdam en het havenslib. *Ingenieur*, **90**, 781–785.

RIJKSWATERSTAAT (1972–1978) *Kwaliteitsonderzoek in de Rijkswateren.* RIZA Kwartaalverslagen.

RIJKSWATERSTAAT (1974) Slibafvoer Lek, Waal en Maas. Deltadienst W.A., Projectgroep Slibtransportmech-anisme. *Werkgroep 1.1, Nota W 74–055.*

RIJKSWATERSTAAT & RIJKSDIENST IJSSELMEERPOLDERS (1976) Waterstaatkundige werken en waterkwaliteit in het IJsselmeergebied. *Band 1: hoofdnota en samenvatting, band 2: kwaliteit IJsselmeerwater gedurende de periode, 1970–1975.*

SALOMONS, W. (1975) Chemical and isotopic composition of carbonates in recent sediments and soils from western Europe. *J. sedim. Petrol.* **45**, 440–449.

SALOMONS, W., BOELENS, R., HOFFMAN, P. & MOOK W.G. (1975) The oxygen isotopic composition of the fraction less than 2 microns (clay fraction) in recent sediments from Western Europe. *Mar. Geol.* **18**, M23–M28.

SALOMONS, W., DE BRUIN, M., DUIN, R.P.W. & MOOK, W.G. (1978) Mixing of marine and fluvial sediments in estuaries. *Abstr. 16th Coastal Engineering Conference*, Hamburg, F.R.G.

SALOMONS, W. & DE GROOT, A.J. (1978) Pollution history of trace metal in sediments as affected by the river Rhine. In: *Environmental Biogeochemistry and Geo-microbiology* (Ed. by W.E. Krumbein), **I**, 149–164.

SALOMONS, W. & MOOK, W.G. (1977) Trace metal con-centrations in estuarine sediments: mobilization, mix-ing or precipitation. *Neth. J. Sea Res.* **11**, 199–209.

SALOMONS, W. & MOOK, W.G. (1980) Biogeochemical processes affecting trace metal concentrations in lake sediments (IJsselmeer, the Netherlands). *Sci. Tot. Env.* **16**, 217–229.

SCHLEICHERT, U. (1975) Schwermetallgehalte der Schweb-stoff des Rheins bei Koblenz im Jahresablauf. *Dt. gewäss. Mitt.* **19**, 150–157.

STUURGROEP BERGING BAGGERSPECIE (1978) *Voorlopig Milieu Effect Rapport Berging Baggerspecie.* Den Haag, July.

TERWINDT, J.H.J. (1967) Mud transport in the Dutch delta area and along the adjacent coast line. *Neth. J. Sea Res.* **3**, 505–531.

TERWINDT, J.H.J. (1971) Nader onderzoek naar de slibbewegingen in het kustwater voor het Deltagebied en de Schone kust in relatie tot het storten van slib op Loswal Noord. *RijkswatSt. Deltadienst W.A. W-71.030.*

TERWINDT, J.H.J. (1977) Mud in the Dutch delta area. *Geologie Mijnb.* **56**, 203–210.

TERWINDT, J.H.J. (1978) Dispersion of dredged material in the North Sea. *WES Symp. Environmental Impact of Dredging and Dredged Materials.* Vicksburg.

WOLLAST, R. (1976) Transport et accumulation de polluants dans l'estuaire de l'Escaut. In: *Projet Mer, Rapport Final* (Ed. by J.C.J. Nihoul and R. Wollast), **10**, 191–218.

WOLLAST, R. et al. (1973) *Origine et mécanisme de l'envasement de l'estuaire de l'Escaut.* Université Bruxelles. Rapport de synthèse.

YIH, CHIA-SHUN (1965) *Dynamics of Nonhomogeneous Fluids*, pp. 134–138. Macmillan, New York.

Spec. Publs int. Ass. Sediment. (1981) **5**, 451–459

Partition of Fe, Mn, Al, K, Mg, Cu and Zn between particulate organic matter and minerals, and its dependence on total concentrations of suspended matter

JAN C. DUINKER

Netherlands Institute for Sea Research, Texel, the Netherlands

ABSTRACT

Particulate suspended matter samples of the Southern North Sea were analysed for Fe, Mn, Al, K, Mg, Cu, Zn and organic C. It was found that the properties of seston obtained at low seston concentrations are different from properties obtained at high seston concentrations. Evidence is given for the existence of a seston fraction, consisting of small, low-density particles. This fraction resists settling, thus being in suspension continuously. The fraction has higher contents of Mg, Cu, Zn and organic C, and lower contents of Fe, Al and K and higher elemental/Al ratios (for all elements studied here) than a fraction consisting of larger, more dense particles, mainly derived from bottom sediments by resuspension. The varying contribution of the latter fraction results in practically constant elemental contents and elemental/Al ratios at seston concentrations above $10 \ mg \ l^{-1}$, elemental/Al ratios being practically identical to those reported for fine-grained sediments. Metals are partitioned over organic matter and mineral components. The higher X/Al ratios at low seston concentrations suggest that specific concentrations in the continuously suspended fraction are determined primarily by organic matter and mineral components such as hydrous oxides of Fe and Mn rather than Al-silicates. It remains unsolved whether some metals (Cu, Zn) are associated with organic matter or with minerals, that co-vary with organic matter. It is suggested that detrital organic matter rather than living phytoplankton should be considered as a potentially significant site for trace metals in the Southern Bight.

INTRODUCTION

Metals associated with particles in sea-water are distributed between organic matter and various mineral components such as hydrous oxides of Fe and Mn and clay minerals. It is important to distinguish these forms as they are subjected to different transport mechanisms and biogeochemical processes in the marine environment. The study is complicated as a result of the common occurrence of aggregates of various mineral components and organic matter.

It has been suggested that elemental partition and its dependence on particle size in bottom sediments can be studied by application of chemical procedures, extracting elements from specific binding sites (Patchineelam & Förstner, 1977; Tessier, Campbell

0141–3600/81/1205–0451 $02.00

& Bisson, 1979). The possibilities for application of these techniques to the analysis of suspended matter may be limited because of appreciable modifications in the natural state of aggregation of suspended matter during sampling and separating techniques such as settling, filtration and centrifugation. Alternatively, elemental X/Al and X/organic C ratios in bulk suspended matter may assist in studying elemental partition and thus in understanding transport mechanisms and biogeochemical processes of suspended matter and its associated elements.

This approach has been applied to study the degree of association of some metals (Fe, Mn, Al, K, Mg, Cu, Zn) with inorganic and organic particulate constituents in the southern North Sea. Additionally, a study has been made of its dependence on total suspended matter concentrations.

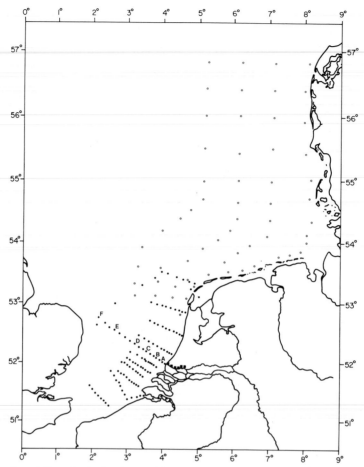

Fig. 1. Sampling stations in the Southern Bight. ○, April–May; ●, October 1975; A–F, April 1979.

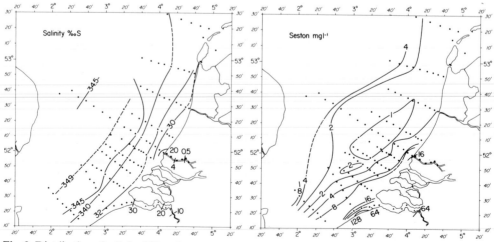

Fig. 2. Distribution of salinity (‰) and seston concentration in the surface layer (mg l⁻¹) in October 1975.

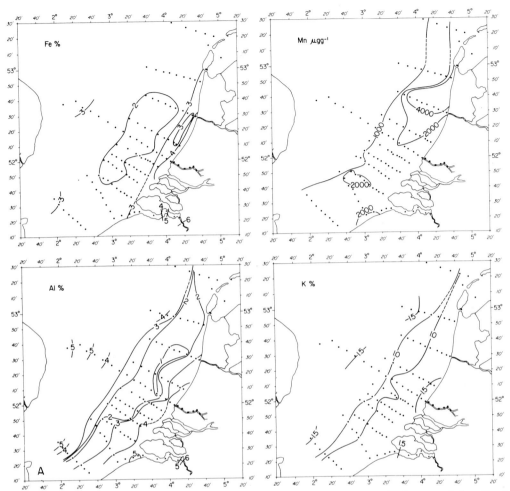

Fig. 3. Distribution of contents of Fe, Mn, Al, K (A), Mg, Cu, Zn and organic C (B) in suspended matter (weight per weight, μg g^{-1} and %) in October 1975.

SAMPLING AND METHODS

Teflon-coated Go-flo (General Oceanics) bottles were used to obtain 1 or 5 litre samples (depending on seston concentration) at stations off the Dutch coast in October 1975 and in the German Bight and the adjacent region off the Danish coast in April–May 1974 (Fig. 1). Samples for metal analysis were filtered in an all-teflon unit over pre-cleaned Millipore filters (0·45 μm). These were dried at 70 °C and analysed by flame (Fe, Mn, Al, K, Mg, Zn) and flameless (Cu) atomic absorption spectrometry (Duinker & Nolting, 1978). Particulate organic C, obtained on Whatman GF/C filters, was determined by a wet oxidation method (Menzel & Vaccaro,

1964) using a total carbon system (Oceanography International Corp.). Samples were obtained for phytoplankton analysis, at stations A–F in April 1979. Seston at stations E and F was composed of mineral components mainly. Seston at stations A–D was essentially composed of phytoplankton species, mainly μ-flagellates (0·5–7·0 × 10^6 l^{-1}) with small contributions of diatoms (2–4 × 10^3 l^{-1}). We are obliged to Dr W. W. C. Gieskes for the analysis.

RESULTS AND DISCUSSION

The distribution of salinity and seston concentration for the stations in transects off the Dutch coast

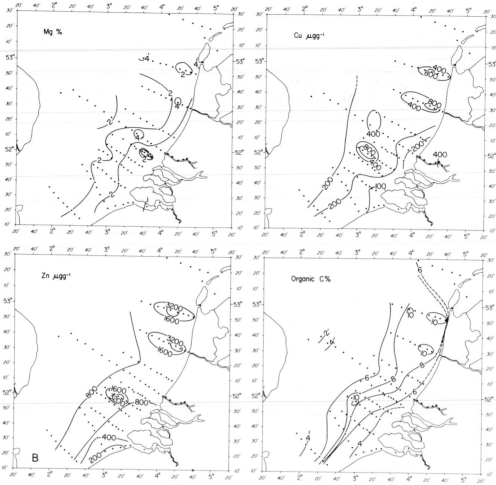

Fig. 3. For legend see p. 453.

is given in Fig. 2. Seston concentrations are lowest ($0.5–1.0$ mg l^{-1}) in an area east of the axis of maximum salinity. The distribution of metal concentrations per water volume (μg l^{-1}) is, at least qualitatively, similar to the distribution of total suspended matter, values decreasing in transects from the coast toward the central part of the Bight. However, the rate of change is distinctly different for the various elements. This can be observed from plots of the distribution of elemental contents in terms of seston dry weight (μg g^{-1} or %): those of Fe, Al and K pass through a minimum and those of Mg, Cu, Zn and organic C through a maximum (Fig. 3). Minima and maxima do not coincide with the axis of maximum salinity but rather with the minima in seston concentration and maxima in organic C content.

Elemental contents of Fe, Al and K are greater at higher than at lower seston concentrations while those of Mg, Cu, Zn and organic C are greater at lower seston concentrations. A critical change occurs at about 5 mg l^{-1} (Fig. 4). The behaviour of Mn is different: its more complicated distribution pattern and the less regular relations of its content in seston with both salinity (Fig. 3) and seston concentrations (Fig. 4) result from the relatively slow oxidation of Mn(II) into particulate Mn(IV) in the lower estuaries of the Rhine and Scheldt rivers and the adjacent coastal area. Strongly increased concentrations of dissolved Mn(II) in the upper parts of the estuaries result from processes that cause a reduction of particulate manganese species in response to gradients in chemical and physicochemical parameters (Wol-

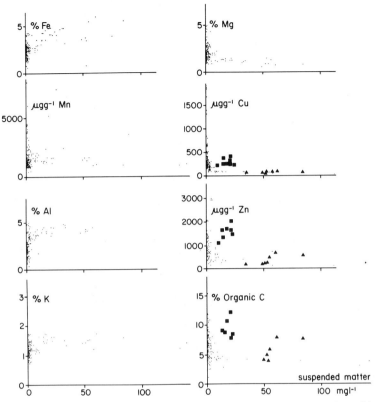

Fig. 4. Elemental contents of Fe, Mn, Al, K, Mg, Cu, Zn and organic C (in µg g⁻¹ or %) in relation to total suspended matter concentration (mg l⁻¹) in October 1975. ■, Rhine, ▲, Scheldt (as in Fig. 1).

last, Billen & Duinker, 1979; Duinker, Wollast & Billen, 1979).

Aluminium is closely related to total suspended matter (Fig. 5); it may be considered as almost entirely associated with Al-silicates, mainly clay minerals. The elemental content of an element X in particulate matter is altered when varying amounts of components with low content of X (e.g. quartz) are mixed with fine-grained Al-minerals with higher contents of X. However, X/Al ratios are not altered. Elemental/Al ratios have been used to explain sources and transport of suspended matter (Spencer & Sachs, 1970; Price & Calvert, 1973). The present data show that X/Al ratios are practically constant at seston concentrations above 10–20 mg l⁻¹. This applies to the plots of all elements considered here. In addition, the X/Al ratios increase strongly with decreasing seston concentration below 5 mg l⁻¹. This is obvious (from the data of Fig. 4) for Mg, Cu, Zn and organic C: these relations are not given here.

It also applies to Fe, Mn and K: these relations are represented in Fig. 6.

Different seston fractions at different seston concentrations

The data indicate that the properties of seston obtained at low and high seston concentrations are different. We suggest the existence of a seston fraction consisting of small, low-density particles. It may resist settling; it is described here as continuously suspended. This fraction dominates at low seston concentrations. Total suspended matter also contains varying contributions of larger, more dense particles belonging to either bottom sediment or suspension, depending on conditions that allow settling or re-suspension. This fraction dominates at higher seston concentrations. The fractions not only have different settling properties; they also differ in their content of organic matter and various minerals and thus in elemental contents. Thus relatively high contents of

456 J. C. Duinker

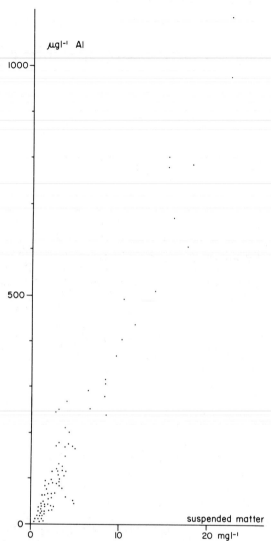

Fig. 5. Concentration of particulate Al (mg l⁻¹), in relation to the concentration of suspended matter (mg l⁻¹) in the samples obtained in October 1975. No river data included.

seston concentrations contains a relatively high fraction of bottom-derived material. Evidence for the significance of this mechanism was also obtained in de Varde Å (Denmark) by measurements of mineral and chemical composition of suspended matter in vertical profiles during a tidal cycle (Duinker *et al.*, 1981). Additional evidence for higher contents of, e.g. Cu and Zn in particulate suspended matter at low seston concentrations was obtained from the samples obtained in the region off the Danish and German coasts in April–May 1974 (Figs 7 and 8).

Variations of elemental contents with total seston concentration as a result of varying contributions of larger, denser particles may remain undetected when concentrations are expressed on a volume (μg l⁻¹) rather than on a sediment dry weight basis (μg l⁻¹). The mechanism suggested above may also prove to be a useful tool in explaining data reported by others, e.g. for estuaries (Sholkovitz, 1979), coastal regions (Sundby & Loring, 1978) and open ocean regions (Wallace, Hoffman & Duce, 1979).

Elemental partition

The metals studied here are partitioned over organic matter and minerals. The partition of a number of metals over various components (Al-silicates and others) has also been suggested for deep water of the Gulf of Maine (Spencer & Sachs, 1970), Loch Etive (Price & Calvert, 1973) and Saguenay fjord (Sundby & Loring, 1978). The strong increase in the X/Al ratios at low seston concentrations suggests that the elemental contents in the continuously suspended fraction are determined primarily by mineral components such as Fe and Mn hydrous oxides rather than by Al-silicates.

Contents of particulate Mg, Cu and Zn co-vary with organic carbon content; a negative correlation is observed for Al (Fig. 9), Fe and K. No such conclusion can be drawn for Mn. Therefore Mg, Cu and Zn may be associated with organic matter or, alternatively, with mineral fragments such as oxide coatings, occurring in association with organic matter. Humic substances might be involved (Pillai *et al.*, 1971; Neihof & Loeb, 1972; Senesi *et al.*, 1977). Although chemical extraction techniques have been suggested to study elemental partition in bottom sediments and its dependence on particle size in great detail (Patchineelam & Förstner, 1977; Tessier, Campbell & Bisson, 1979), it remains a problem how similar techniques could give reliable results for suspended matter, taking into account the modifica-

Fe, Al and K occur in material present at high seston concentrations and high contents of Mg, Cu, Zn and organic C in material present at low seston concentrations (Fig. 4).

The X/Al ratios above 10–20 mg l⁻¹ for the major elements (Fe/Al 0·5, Mn/Al 0·02, K/Al 0·3, Mg/Al 0·2) are very similar to the values reported for fine-grained sediments (0·5, 0·01, 0·29 and 0·17 respectively: Goldschmidt, 1954; Krauskopf, 1965). This supports the idea that suspended matter at high

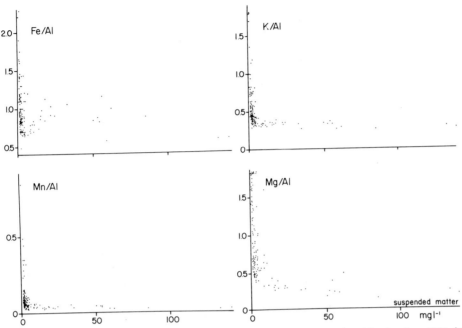

Fig. 6. Elemental/Al ratios for Fe, Mn, K and Mg in suspended matter obtained in October 1975. No river data included.

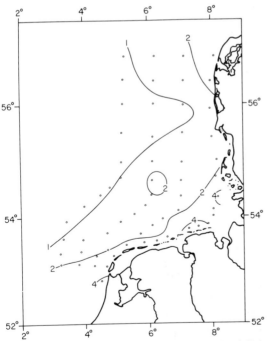

Fig. 7. Distribution of seston concentration (mg l⁻¹) in the surface layer in April–May 1974.

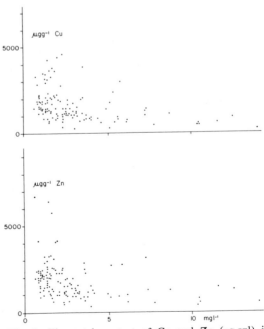

Fig. 8. Elemental content of Cu and Zn (μg g⁻¹) in relation to total suspended matter concentration (mg l⁻¹) in April–May 1974.

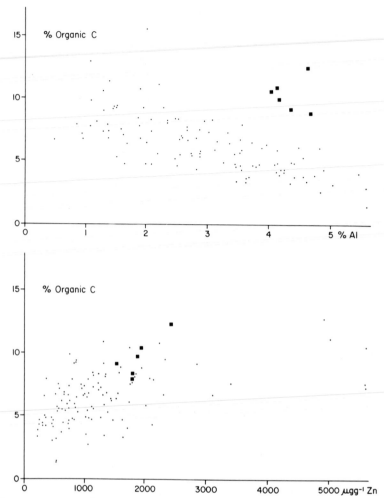

Fig. 9. Relation between contents of Zn and organic C and of Al and organic C in samples obtained in October 1975, Scheldt values included. ■, Rhine.

Table 1. Range of elemental contents in pure samples of dinoflagellates and diatoms at stations A–D in April 1979, at seston concentrations in the 0·3–1·6 mg l^{-1} range

Fe	0·3–0·4 %		Mg	0·6–2·3 %
Mn	15–114 µg g^{-1}		Cu	21–100 µg g^{-1}
K	0·2–1·2 %		Zn	85–184 µg g^{-1}

tions in the properties of the particles expected to occur during separation of water and particles by either settling, filtration or centrifugation.

We have tried to determine the role of living phytoplankton as a possible site for trace elements. Practically pure samples of dinoflagellates and diatoms were obtained at some stations at low seston concentrations off the Dutch coast in April 1979.

The contents of Cu, Zn, Fe and Mn were found to be far below the levels found in total suspended matter at low concentrations (Table 1). The relatively unimportant role of living phytoplankton in the Southern Bight was also reported in earlier work (Duinker & Nolting, 1976). It is supported by the data of Martin & Knauer (1973). Increased particulate trace-metal concentrations below 100 m in the

Antarctic region were related to adsorption by dead cells (Harris & Fabris, 1979). Wallace *et al.* (1977) concluded from the relative variations of particulate trace metals with respect to particulate Al and organic C in north-west Atlantic water that organic matter is a probable regulator of particulate trace-metal abundance in open-ocean surface water. Wollast (1979) reported a strong correlation between trace metals and organic matter in bottom sediments of the North Sea. These results, in combination with the strong correlation between the contents of several elements and organic C in seston (Fig. 9), suggest that detrital organic matter rather than living phytoplankton should be considered as a potentially significant site for the metals in our data.

However, mineral components such as Fe and Mn hydrous oxides, co-varying with organic matter, possibly as aggregates, may also be important sites for trace metals, in particular at low seston concentrations. We have not been able to distinguish between these possibilities and we feel that this aspect deserves more attention than it has been given in most reports in the literature, stating the association of trace metals with organic matter in cases where only a positive correlation with organic C has been found. Application of advanced optical methods (scanning electron microscopy) and significant improvements in chemical leaching techniques are important steps required in future work.

REFERENCES

DUINKER, J.C. & NOLTING, R.F. (1976) Distribution model for particulate trace metals in the Rhine estuary, Southern Bight and Dutch Wadden Sea. *Neth. J. Sea Res.* **10**, 71–102.

DUINKER, J.C. & NOLTING, R.F. (1978) Mixing, removal and mobilization of trace metals in the Rhine estuary. *Neth. J. Sea Res.* **12**, 205–223.

DUINKER, J.C., WOLLAST, R. & BILLEN, G. (1979) Manganese in the Rhine and Scheldt estuaries. Part 2. Geochemical cycling. *Est. coastal Mar. Sci.* **9**, 727–738.

DUINKER, J.C., HILLEBRAND, M.TH.J., NOLTING, R.F., WELLERSHAUS, S. & KINGO JACOBSEN, N. (1981) The river Varde Å: processes affecting the behaviour of metals and organochlorines during estuarine mixing. *Neth. J. Sea Res.* **14**, 237–267.

GOLDSCHIMDT, V.M. (1954) In: *Geochemistry* (Ed. by E. Muir). Oxford University Press. 730 pp.

HARRIS, J.E. & FABRIS, G.J. (1979) Concentrations of suspended matter and particulate cadmium, copper, lead and zinc in the Indian sector of the Antarctic ocean. *Mar. Chem.* **8**, 163–179.

KRAUSKOPF, K.B. (1965) *Introduction to Geochemistry*. McGraw-Hill, New York. 721 pp.

MARTIN, J.H. & KNAUER, G.A. (1973) The elemental composition of plankton. *Geochim. cosmochim. Acta*, **37**, 1639–1653.

MENZEL, D.W. & VACCARO, R.F. (1964) The measurement of dissolved organic and particulate carbon in seawater. *Limnol. Oceanogr.* **9**, 138–142.

NEIHOF, R.H. & LOEB, G.I. (1972) The surface charge of particulate matter in seawater. *Limnol. Oceanogr.* **17**, 7–16.

PATCHINEELAM, S.R. & FÖRSTNER, U. (1977) Bindungsformen von Schwermetallen in marinen Sedimenten. *Senckenberg Mar.* **9**, 75–104.

PILLAI, T.N.V., DESAI, M.V.M., MATHEW, E., GANAPATHY, S. & GANGULY, A.K. (1971) Organic materials in the marine environment and the associated metallic elements. *Curr. Sci.* **40**, 75–81.

PRICE, N.B. & CALVERT, S.E. (1973) A study of the geochemistry of suspended particulate matter in coastal waters. *Mar. Chem.* **1**, 169–189.

SENESI, N., GRIFFITH, S.M., SCHNITZER, M. & TOWNSEND, M.G. (1977) Binding of Fe^{3+} by humic materials. *Geochim. cosmochim. Acta*, **41**, 969–976.

SHOLKOVITZ, E.R. (1979) Chemical and physical processes controlling the chemical composition of suspended material in the river Tay estuary. *Est. coastal mar. Sci.* **8**, 523–545.

SPENCER, D.W. & SACHS, P.L. (1970) Some aspects of the distribution, chemistry, and mineralogy of suspended matter in the Gulf of Maine. *Mar. Geol.* **9**, 117–136.

SUNDBY, B. & LORING, D.H. (1978) Geochemistry of suspended particulate matter in the Saguenay fjord. *Can. J. Earth Sci.* **15**, 1002–1011.

TESSIER, A., CAMPBELL, P.G.C. & BISSON, M. (1979) Sequential extraction procedure for the speciation of particulate trace metals. *Analyt. Chem.* **51**, 844–851.

WALLACE, G.T., HOFFMAN, G.L. & DUCE, R.A. (1977) The influence of organic matter and atmospheric deposition on the particulate trace metal concentration of Northwest Atlantic surface seawater. *Mar. Chem.* **5**, 143–170.

WOLLAST, R. (1979) Heavy metals in the suspended matter and in the recent sediments of the Southern Bight. *Int. Meet. Holocene Marine Sedimentation in the North Sea Basin. IAS abstracts*, p. 68. Texel, September 17–23.

WOLLAST, R., BILLEN, G. & DUINKER, J.C. (1979) Manganese in the Rhine and Scheldt estuaries. Part 1. Physico-chemical aspects. *Est. coastal Mar. Sci.* **9**, 161–169.

Spec. Publs int. Ass. Sediment. (1981) **5**, 461–468

The entrapment of pollutants in Norwegian fjord sediments
—a beneficial situation for the North Sea

JENS SKEI

Norwegian Institute for Water Research, P.O. Box 333, Blindern, Oslo 3, Norway

ABSTRACT

Fjords in Norway are frequently used as recipients of industrial and domestic wastes. Due to the estuarine circulation, multi-layered flow and long periods of stagnancy in the bottom water, effluents containing metals and organic micro-pollutants are to a large extent removed from the water and incorporated in the sediments. Some fjords show symptoms of eutrophication and the increased primary production enhances the settling of pollutants tied up with organic debris. Additionally, Norwegian fjords are often topographically confined, which also restricts the transport of pollutants to the North Sea and the Skagerrak.

Studies of the chemical composition of sediment cores in a number of polluted fjords in Norway suggest that metals and non-degradable organic compounds are fixed in the sediments close to the source. This is illustrated by steep horizontal gradients of pollutants (Hg, Zn, Pb, Cd, HCB, PAH) in the surface sediments.

Preliminary results indicates that discharge of pollutants to fjords causes local problems with elevated levels in biota, waters and sediments. The waste discharge often takes place below the seaward-flowing surface layer, i.e. in the compensating current, transporting the pollutants towards the head of fjords. The result is an efficient trapping of pollutants in the fjord sediments and the amount of waste being transported out of the fjords and into the open ocean is apparently small. This is not to say that the sediments act as a permanent sink for pollutants, but any pollutants reintroduced into the water are not likely to impose any threat to the environmental quality of the North Sea.

INTRODUCTION

Prior to 1970 the disposal of pollutant substances to coastal waters was acknowledged as an acceptable and unavoidable way of getting rid of non-utilizable material. It was previously accepted on terms of a fundamental lack of knowledge about the behaviour of pollutants in marine waters. It had been considered unavoidable because no better alternative had been found. In recent years there has been a change in attitude towards marine disposal and there is now much more concern about the dispersal and sedimentation of pollutants in coastal waters and the interaction with the open ocean. However, the problem of pollutant disposal is apprehended very

differently. Do we want to sacrifice a small area for the benefit of a larger and more influential area? In other words: do we want to concentrate the pollutants in a small marine enclosure or is dilution a solution to the problem? The answer to this dilemma is partly given by nature itself. Any pollutant, dissolved or particulate, organic or inorganic is to a large extent quickly removed from the water mass and incorporated in the bottom sediments. This scavenging of pollutants is efficient in inshore waters where inorganic detritus and plankton are abundant and where sulphide regimes exist in the near-bottom waters. This implies that the most influential processes with respect to dispersal of pollutant substances take place in the immediate vicinity of the

0141–3600/81/1205–0461 $02.00

462   J. Skei

source of pollution. If we want to study the behaviour of pollutants in the marine environment we should look for a well-defined point source and start our investigation in the adjacent area.

SEDIMENTATION IN NORWEGIAN FJORDS

Fjords in Norway are frequently used as recipients of industrial and domestic wastes. They represent relatively enclosed coastal waters which the pollutants have to pass through before eventually being transported to the open ocean. Most fjords have a relatively high content of inorganic particulate matter due to river input. The silts and clays are often transported in the seaward-flowing surface water situated above the halocline. The extent of density stratification shows seasonal variations, and this has an influence on the vertical distribution of inorganic silicates. This may be demonstrated by the distribution of suspended particulate aluminium (Fig. 1) which almost exclusively is present in aluminium silicates (Joyner, 1964). However, it should be emphasized that normally discharge of industrial waste takes place at some depth in fjords and below the density boundary. As a consequence the dynamic of pollutant dispersal may be different from that of natural sediments. In a typical fjord with estuarine circulation, there is an opposite-directed current to the surface flow located below this boundary. If the waste is discharged in this water mass, the pollutants tend to be transported towards the head of the fjord. This implies a pile-up and a local accumulation of pollutants. Water underlying the surface layer normally has a longer residence time in the fjord, and this creates a situation where sedimentation is accelerated due to flocculation and/or zooplankton grazing. Consequently, steep concentration gradients of pollutants appear in the bottom sediments away from the source in cases where this is located near the head of the fjord. The fixation of, for instance, metals in the sediments may be enhanced by eutrophication in the water and massive algal bloom. The algae scavenge the metals from the water and eventually die and sink to the bottom. Due to a high rate of sedimentation of organic material anoxic conditions develop rapidly in the sediments. This may lead to formation of metal sulphides with low solubility and the sediments may be considered as a sink for pollutants. Alternatively, formation of metal polysulphide and

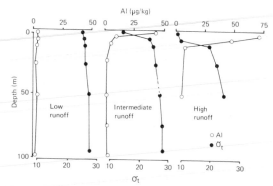

Fig. 1. The relationship between particulate Al and the density of the water (δ_t) at low, intermediate and high runoff.

bisulphide complexes may enhance the solubility of trace metals over that predicted by simple sulphides (Gardner, 1974). Organic chelators may also influence the mobility of metals in anoxic sediments (Hallberg, Bubela & Fergusson, 1980).

There are very few fjords in Norway with permanently anoxic bottom waters. Intermittently anoxic fjords, however, are numerous. This implies that the surface sediments are periodically exposed to oxygenated water. When anoxic bottom water is replaced by oxic water the chemical and microbiological environment at the sediment–water interface changes drastically. This changeover in redox conditions may enhance the mobilization of sedimented pollutants (Khalid, Patrick & Cambrell, 1978), but by no means would the rate of mobilization be so high that it could impose any environmental threat to the coastal area outside the fjords. Fluxes of metals from sediments to water are thought to be relatively small, except for manganese (cf. Elderfield & Hepworth, 1975).

AREAS OF STUDY

Three fjords in southern Norway with direct connection to the North Sea and the Skagerrak were selected for the present study. These fjords receive various types of effluents causing horizontal and vertical gradients of pollutants in the bottom sediments. Three groups of pollutants are considered: (1) heavy metals, (2) hexachlorobenzene (HCB), (3) polycyclic aromatic hydrocarbons (PAH).

The location of the fjords is shown on Fig. 2. To the east Iddefjord is situated on the border between Sweden and Norway. A Norwegian pulp and paper

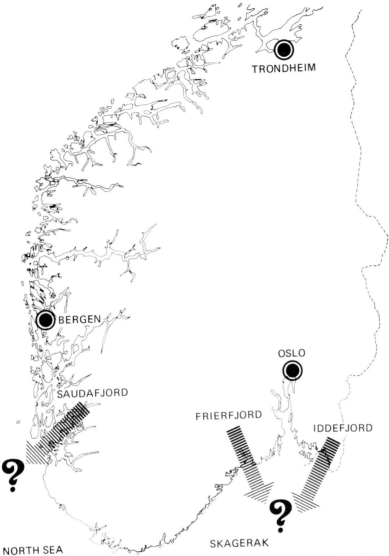

Fig. 2. Map of southern Norway. Arrows indicate possible pathways of pollutants from Iddefjord, Frierfjord and Saudafjord.

factory is polluting the water with organic matter and metals from SO_2-production. Frierfjord (Fig. 2) is well known as one of the most polluted fjords in Norway (Skei, 1978). There is a concentration of industry surrounding this fjord including a chlor-alkali plant which discharges Hg and HCB, a pulp and paper factory which prior to 1970 discharged Hg and a ferro-manganese smelter which in addition

to metals also discharges PAH. Finally, Saudafjord (Fig. 2) receives effluents consisting of heavy metals, particularly Cd, and PAH from a ferro-alloy smelter (Bjørseth, Knutzen & Skei, 1979).

The question we may now ask is whether the effluents released into these fjords mainly remain within the fjords or are they transported into the Skagerrak and the North Sea.

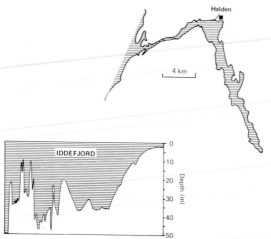

Fig. 3. Map of Iddefjord with depth profile from the head (right) to the mouth (left) of the fjord. Source of pollution indicated (■, pulp and paper).

METHODS

Sediment samples have been collected using a gravity corer (Niemistö, 1974) and sectioned at 2 cm intervals. Samples for metal analysis were dried at 80 °C, homogenized in an agate mortar and digested using concentrated HNO_3. The analyses were performed by atomic absorption (flame/flameless).

Analysis of chlorinated hydrocarbons was done by extracting into a cyclohexane/isopropanol phase, treatment with H_2SO_4 and determination using a Perkin-Elmer 3920 gas chromatograph with an electron capture detector (ECD) (Ofstad *et al.*, 1978).

Polycyclic aromatic hydrocarbons (PAH) were determined using glass capillary gas chromatography according to Björseth *et al.* (1979). The analysis of chlorinated as well as polycyclic aromatic hydrocarbons was performed at the Central Institute for Industrial Research in Oslo.

RESULTS AND DISCUSSION

Iddefjord

The pulp and paper factory in Iddefjord is located near the river mouth at Halden and the fjord is connected to the sea or Skagerrak towards the southwest (Fig. 3). Iddefjord has a shallow sill (9 m) and a maximum depth of 45 m. Frequently the bottom water turns anoxic and sometimes the H_2S-water

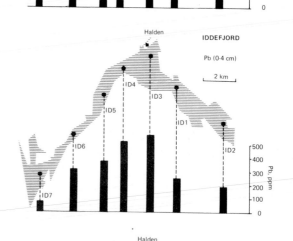

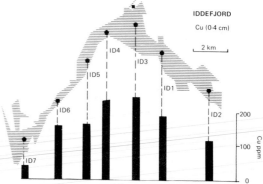

Figs 4, 5 and 6. Concentrations of Zn, Pb and Cu in surface sediments (0–4 cm) of Iddefjord.

reaches the very surface. This is due to a combination of little water exchange because of a shallow sill and an extensive load of degradable organic matter from the pulp and paper factory. Inside the sill the bottom sediments are black and anoxic. The distribution of Zn, Pb and Cu in the surface sediments shown in Figs 4, 5 and 6 is more or less identical,

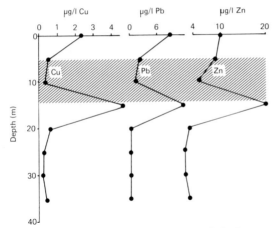

Fig. 7. Vertical distribution of Cu, Pb and Zn in the water of Iddefjord. Shaded area indicates vertically displaced, low oxygenated water.

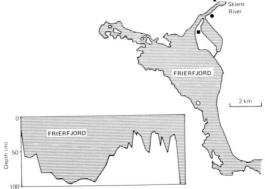

Fig. 8. Map of Frierfjord with depth profile from the head (left) to the mouth (right) of the fjord. Sources of pollution indicated (▲, pulp and paper; ●, ferro-manganese smelter; ■, chlor-alkali).

with an extensive accumulation in the sediments within 5 km from the source. Apparently, most of the metals are retained inside the sill and the sediments outside the sill show approximately normal concentrations. The efficient trapping in the sediments may be enhanced by the anoxic conditions. But what happens if oxic conditions are re-established in the bottom water following deep-water renewal?

The vertical distribution of Zn, Pb and Cu in the water of Iddefjord following bottom water exchange and vertical displacement of old water is illustrated in Fig. 7. Immediately below the old water a maximum concentration of the metals occurs while the old water itself is characterized by low metal content. This may suggest some mobilization of metals when the conditions change from anoxic to oxic in the bottom water (cf. Khalid *et al.*, 1978). The re-suspension of bottom sediments which takes place during the water exchange may contribute to this mobilization. Nevertheless, it may be concluded that the pollution of Iddefjord is mainly a local problem, and small amounts of metals are transported into the Skagerrak. This is also confirmed by analysis of mussels outside the Iddefjord, showing no significant metal contamination (Phillips, 1977).

Frierfjord

The situation in Frierfjord (Fig. 8) is complicated due to the diversity of pollutants. This fjord is also intermittently anoxic, with problems of eutrophication. The algal bloom presumably contributes to the

removal of pollutants from the water mass to the bottom sediments.

The discharge of Hg from a local chlor-alkali plant to a semi-enclosed bay connected to Frierfjord and the river through channels, has caused an immense accumulation of Hg in the bottom sediments of the bay (Fig. 9; Skei, 1978). Here concentrations of more than 300 ppm Hg were measured while the concentration in sediments of Frierfjord and the Skiens river never exceeded 15 ppm. This again is an example of spontaneous sedimentation of a pollutant. Naturally this creates environmental problems in the source area, but it seems to pose no direct danger to the coastal water outside (the Skagerrak and the North Sea). A minimum of 10 tons of Hg resides in the bottom sediments of this bay (Skei, 1978) and presumably it causes less harm here than evenly distributed in the coastal water. The concentrations of dissolved Hg in the water of this semi-enclosed bay are 0·5–1·0 μg l^{-1}. Measurements of Hg in the coastal water outside Frierfjord have shown normal concentrations of 0·05 μg l^{-1}, while within Frierfjord a two- to three-fold increase has been observed. This again supports the statement that fjord pollution is mainly a local problem.

Measurements of dissolved metals immediately above the sediment in Frierfjord when the bottom water contained H$_2$S and after a deep water renewal when oxygen was present, showed an increase in the concentration of Zn from 5 to 27 μg l^{-1}. No such increase was observed for Hg, Cu and Pb. This suggests that the release of Zn from contaminated sediments is substantial when the redox condition changes from anoxic to oxic (cf. Holmes, Slade &

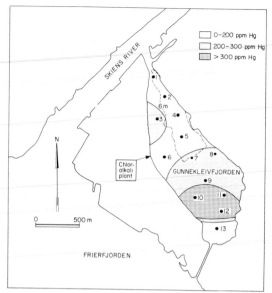

Fig. 9. Location of sediment stations and distribution of Hg in the surface sediments of Gunnekleivfjorden (after Skei, 1978).

McLerran, 1974). This, however, is nevertheless only a local problem as the flux of metals from the sediments is not sufficiently large to have any large-scale effect.

Measurement of chlorinated hydrocarbons in Frierfjord has been carried out regularly since 1974 due to discharge of such compounds from a magnesium plant (Ofstad et al., 1978). Measurements in the water and the sediments indicate steep gradients of HCB away from the source (Fig. 10). This suggests that HCB is quickly removed from the water and incorporated in the sediments. The concentrations of HCB in the surface sediments decrease from 11 ppm near the point of discharge to 0·006 ppm some 15 km away. The vertical distribution of HCB in a sediment core from Frierfjord is shown on Fig. 11. Apparently the sediments of Frierfjord contain non-detectable amounts of HCB below 8 cm depth, which presumably is prior to 1952 when the magnesium plant was established.

Just as with metals, the sediments also appear to entrap chlorinated hydrocarbons. This minimizes the widespread effect of these pollutants.

Saudafjord

The third area selected is Saudafjord, on the southwest coast of Norway (Fig. 12). This fjord is

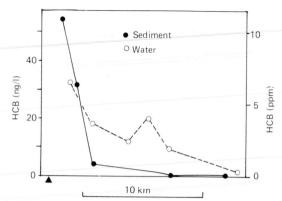

Fig. 10. Concentrations of HCB in water and sediments of Frierfjord away from the source (▲).

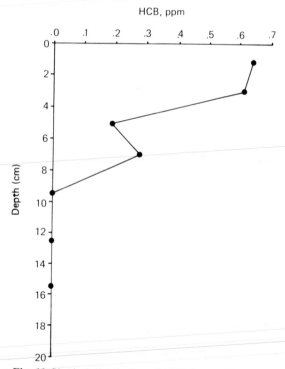

Fig. 11. Vertical distribution of HCB in a sediment core from Frierfjord.

considerably different from the two other areas considered. It has a very deep sill (~200 m) and the bottom water (~400 m) is well oxygenated. As a consequence the surface sediments are oxic. The water flow in the river entering at the head of the fjord is regulated and sediment transport is low. Hence there are few inorganic particles suspended in the fjord water. The fjord is not eutrophicated, and

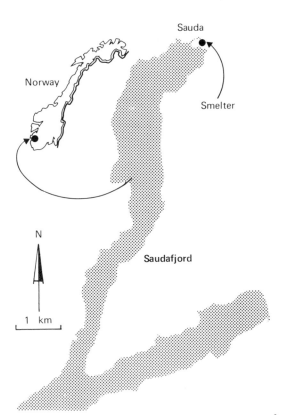

Fig. 12. Map of Saudafjord, on the south-west coast of Norway.

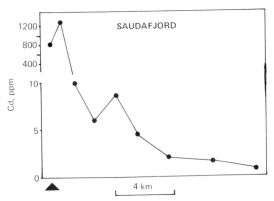

Fig. 13. Lateral distribution of Cd in surface sediments of Saudafjord away from the source (▲).

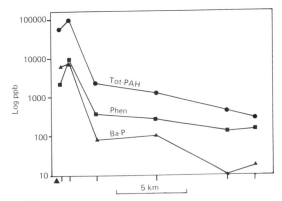

Fig. 14. Lateral distribution of total PAH, phenanthrene and benzo(a)pyrene in surface sediments of Saudafjord away from the source (▲) (after Björseth *et al.*, 1979).

the sediments receive little organic matter. The overall rate of sedimentation calculated from ^{210}Pb analysis is approximately 1·4 mm yr^{-1} in the middle of Saudafjord over the past 50 years. This is less than sedimentation rates measured in Iddefjord (~ 3 mm yr^{-1}) and Frierfjord (~ 2 mm yr^{-1}) (Skei, unpublished).

Saudafjord receives waste water from a ferro-alloy smelter, principally containing PAH, but also metals like Cd, Pb and Zn. Some metals may also be contributed by a closed Zn/Pb mine. The bottom sediments at the head of the fjord and close to the source of pollution contain high concentrations of Cd, with a maximum of more than 1200 ppm (Fig. 13). However, within a couple of kilometres from the source the sediments contain 10 ppm and some 15 km away less than 1 ppm. This demonstrates clearly a rapid incorporation of Cd into the bottom sediments. As the scavenging of Cd on to clays or organic matter is supposed to be inefficient due to the scarcity of these scavengers in the water of Sauda-

fjord, it is most likely that Cd is introduced in the fjord tightly bound in particles. However, re-mobilization of Cd in the bottom sediments may take place in view of the mobility of Cd (Salomons & Eysink, 1981). This is to be investigated in the near future.

The other pollutant which is abundant in the environment of Saudafjord is PAH, due to a daily discharge of 3–4 tons (in 1977). The interest for PAH is due to the acknowledgement of the carcinogenic effect of certain PAH-components. There is a considerable lack of knowledge about the behaviour of these components in the marine environment. However, several investigations have shown that PAH tends to be enriched in bottom sediments (Windsor & Hites, 1979). The horizontal distribution of total PAH, phenanthrene and Ba-P is shown on Fig. 14 (Bjørseth *et al.*, 1979). The concentrations of total PAH near the source of pollution were 100

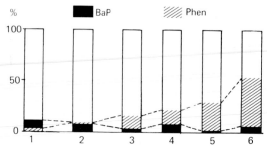

Fig. 15. The relative abundance of benzo(a)pyrene and phenanthrene in surface sediments of Saudafjord with increasing distance from the source (stations 1 → 6) (after Björseth *et al.*, 1979).

ppm, decreasing to 1 ppm some 10 km away. The relative abundance of PAH components in the sediments varies with increasing distance from the source of pollution.

Figure 15 shows an increase of the percentage of phenanthrene seaward, while the abundance of Ba-P in the surface sediments decreases or stays more or less constant at about 10%. This lateral change in the PAH composition in the fjord sediments suggests a selective settling of components or dissolution or biodegradation of certain compounds relative to others in the sediments (Björseth *et al.*, 1979). Preferential settling could be due to differences of water solubility of the various PAH components in the sediments. Alternatively, sediments deposited far away from the smelter are relatively more influenced by naturally occurring PAH which could be of a different composition from that discharged by the smelter. The similarity in the distribution of total PAH and Cd suggests an identical source.

CONCLUSION

The overall conclusion from the study of three fjords in southern Norway must be that these fjords tend to entrap the pollutants relatively efficiently. Any recycling of pollutants may occur internally, but the net input of pollutants from these fjords to the North Sea and the Skagerrak is thought to be negligible, compared with other sources like atmospheric fallout (see Cambray, Jefferies & Topping, 1979). It should be emphasized, however, that a need exists for a better assessment of pollutant input to fjords as well as annual fluxes of pollutants to bottom sediments. The latter is now being looked

at, using ^{210}Pb. By comparing the annual flux to the fjord sediments with pollutant discharges, the input of pollutants to the North Sea and the Skagerrak may be tentatively quantified.

REFERENCES

BJØRSETH, A., KNUTZEN, J. & SKEI, J. (1979) Determination of polycyclic aromatic hydrocarbons in sediments and mussels from Saudafjord, W. Norway, by glass capillary gas chromatography. *Sci. Total Environ.* **13**, 77–86.

CAMBRAY, R.S., JEFFERIES, D.F. & TOPPING, G. (1979) The atmospheric input of trace elements to the North Sea. *Comm. Mar. Sci.* **5**, 175–194.

ELDERFIELD, H. & HEPWORTH, A. (1975) Diagenesis, metals and pollution in estuaries. *Bull. Mar. Pollut.* **6**, 85–87.

GARDNER, L.B. (1974) Organic versus inorganic trace metal complexes in sulfide marine waters—some speculative calculations based on available stability constants. *Geochim. Cosmochim. Acta*, **38**, 1297–1302.

HALLBERG, R., BUBELA, B. & FERGUSON, J. (1980) Simulation of metal-chelating in two reducing sedimentary systems. *Geomicrobiology*, **2**, 1–15.

HOLMES, C.W., SLADE, E.A. & McLERRAN, C.J. (1974) Migration and redistribution of zinc and cadmium in marine estuarine system. *Environ. Sci. Tech.* **8**, 255–259.

JOYNER, T. (1964) The determination and distribution of particulate aluminium and iron in the coastal waters of the Pacific North-West. *J. Mar. Res.* **22**, 259–268.

KHALID, R.A., PATRICK, W.H. & GAMBRELL, R.P. (1978) Effect of dissolved oxygen on chemical transformation of heavy metals, phosphorus and nitrogen in an estuarine sediment. *Est. Coastal. Mar. Sci.* **6**, 21–35.

NIEMISTÖ, L. (1974) A gravity corer for studies of soft sediments. *Hav-forsknings-Inst. Skr.*, Helsinki, **238**, 33–38.

OFSTAD, E.B., LUNDE, G., MARTINSEN, K. & RYGG, B. (1978) Chlorinated aromatic hydrocarbons in fish from an area polluted by industrial effluents. *Sci. Total Environ.* **10**, 219–230.

PHILLIPS, D.J.H. (1977) The concentration of zinc, lead, iron, copper, cadmium and manganese in mussels, *Mytilus edulis* (L.), from Singlefjorden, Western Sweden. In: *The National Swedish Environment Protection Report, P.M. 843*, pp. 51–81.

SALOMONS, W. & EYSINK, W.D. (1981) Pathways of mud and particulate trace metals from rivers to the southern North Sea. In: *Holocene Marine Sedimentation in the North Sea Basin* (Ed. by S.-D. Nio *et al.*). *Spec. Publs int. Ass. Sediment.* **5**, 429–450. Blackwell Scientific Publications, Oxford. 524 pp.

SKEI, J. (1978) Serious mercury contamination of sediments in a Norwegian semi-enclosed bay. *Bull. Mar. Pollut.* **9**, 191–193.

WINDSOR, J.G. & HITES, R.A. (1979) Polycyclic aromatic hydrocarbons in Gulf of Maine sediments and Nova Scotia soils. *Geochim. Cosmochim. Acta*, **43**, 27–33.

Spec. Publs int. Ass. Sediment. (1981) **5**, 469–495

Benthonic foraminiferal distributions in a southern Norwegian fjord system: a re-evaluation of Oslo Fjord data

J. THIEDE, G. QVALE, O. SKARBOE* *and* J. E. STRAND†

Department of Geology, University of Oslo, P.O. Box 1047, *Blindern, Oslo* 3, *Norway*

ABSTRACT

Benthonic foraminifers are some of the most frequent and ubiquitous biogenic components of the marine late Quaternary sediments of the Oslo Fjord, which opens through the Skagerrak into the North Sea and whose hydrography is dominated by water masses entering the fjord from the Skagerrak. The main foraminiferal assemblages are limited to specific water masses with non-horizontal upper and lower boundaries. Since surface and bottom water mass hydrographies are quite independent of each other, it can be expected that shallow and deep water faunas responded in a different manner to the late Quaternary climatic fluctuations. It seems doubtful how useful benthonic foraminifers are for biostratigraphical zonations in an area which has been affected by important eustatic sea-level fluctuations and isostatic uplift of more than 200 m during the course of the last deglaciation. Historical studies of benthonic foraminifers can therefore be expected to permit reconstruction of the palaeohydrography of this fjord system rather than to be useful as biostratigraphic markers to subdivide the Holocene.

Using multivariate statistical techniques, the modern benthonic foraminiferal faunas of the Oslo Fjord can be subdivided into six different assemblages which explain 90 % of the variability of the census data. *Bolivina* cf. *B. robusta* is dominating the most important assemblage (27 % var.) in the deepest part of the Oslo Fjord basins, but this assemblage is essentially not reaching the inner basins. Its upper limits are found at different levels along the eastern and western flank of the outer fjord. The second most important assemblage (25 % var.) is dominated by *Bulimina marginata* and *Nonion labradoricum*; it is found in a narrow intermediate depth range in the outer and middle fjord; its upper limit climbs to shallower depth in the inner basins of the fjord where, however, this assemblage also occupies water depths far below its range of occurrence in the outer and middle basin. *Eggerella scabra* is the most abundant species of a shallow water assemblage (20 % var.) of the outer and middle fjord basins, but this assemblage does not succeed to invade the inner fjord at similar depths. The remaining three assemblages are quantitatively relatively unimportant. *Cassidulina laevigata*, *Nonion barleeanum* and *Hyalinea balthica* characterize another outer fjord assemblage (8 % var.) at intermediate water depths. *Ammotium cassis* and *Ammoscalaria runiana* are important members of an assemblage (4 % var.) along the east side of the outer fjord basin, while an assemblage (5 % var.) with *Nonion depressulus asterotuberculatus*, *Ammonia batavus* and *Elphidium incertum* is restricted to narrow and small inlets on the east and west side of the outer Oslo Fjord.

INTRODUCTION

Benthonic foraminifers have been used for many years as biostratigraphic tools to date late Quaternary shallow and deep water marine sediments along

* Now at Continental Shelf Institute, I.K.U., Trondheim, Norway.
† Now at Norsk Hydro, Sandvika, Norway.

0141-3600/81/1205-0469 $02.00

the continental margins, in and around the shelf seas of the North Atlantic Ocean. They are attractive because they occur frequently and often well preserved, because their occurrence seems to change in the stratigraphic record and their distribution seemed to be related mainly to water depth. Marine benthonic foraminifers live on the seafloor or in the uppermost few millimetres/centimetres of the sediment, their distributions today are therefore believed to be controlled by the hydrography of the bottom-water masses rather than by water depth, which is a static variable. The main theme of this paper will therefore be a demonstration of such distribution patterns and their relation to water masses.

The Oslo Fjord, its general setting and hydrography

As an example we are using benthonic foraminiferal distributions from the Oslo Fjord (Fig. 1) which represents an approximately 100 km long marine basin in the northern prolongation of the Skagerrak between the North and Baltic Seas. The water depths of the Oslo Fjord and Skagerrak exceed those of the North Sea by several hundred metres. Both basins are filled by water masses whose dominant portion flows from the southern Norwegian Sea through the North Sea into the deep Skagerrak and Oslo Fjord. The Oslo Fjord is a particularly attractive small marine basin for studying the interaction between hydrography and the underlying sediment cover because of its proximity to the North Sea, its wide water depth range, its hydrography and its apparently homogeneous fine-grained modern sediment cover.

The geologic Oslo Graben structure (Størmer, 1935; Ramberg, 1976) which includes the Oslo Fjord, and its Quaternary evolution (Sørensen, 1979) have led to the subdivision of the fjord into sub-basins > 150 m (Bunnefjorden) and > 300 m (outer Oslo Fjord) deep (Munthe-Kaas, 1967). The basins are separated from each other by thresholds, the shallowest reaching into < 20 m water depth (Drøbak sill) which prevent an effective and rapid exchange or renewal of the bottom-water masses of the inner basin. The Oslo Fjord can therefore be considered as an example of a silled fjord regime, whose surface-water masses are affected by strong seasonal climatic changes, but whose bottom waters are only intermittently renewed (Gade, 1970). We have tried to illustrate some important hydrographic features by plotting sea-water temperatures and salinities of two seasons (Figs 2 and 3). They represent averages of measurements obtained throughout the past 40 years because geologic samples of surface sediments in a region with accumulation rates of 50–150 cm 10^{-3} yr as known from the Oslo Fjord (Richards, 1976) can be assumed to correspond roughly to this time span. The raw hydrographic data have been collected from a variety of sources but their quality has been screened and they are presently stored at the Norwegian Oceanographic Data Center (NOD, Bergen, Norway). The stations of the hydrographic measurements have been arranged according to their latitude along a depth profile following the axis of the main Oslo Fjord basin. Where available, the values of each two-minute-of-latitude interval have been averaged, then plotted into the centre of this interval and used for contouring.

The general hydrographic characteristics of the Oslo Fjord are well known (NIVA, 1970; Gade, 1970), and are also reflected by the average values used here. The deep water below 80–90 m in the outer Oslo Fjord and below 40–50 m in its inner basin remain relatively uniform throughout the year with temperatures of 7–8 °C and salinities of 33–34‰. Deep-water exchange, which (Gade, 1970) is disturbing the uniform hydrography of the inner Oslo Fjord basin-bottom waters, and which replenishes their oxygen concentrations, is usually confined to the late winter months (NIVA, 1970). Due to different sill depths the source of this bottom-water renewal has to be sought in different parts of the water column, resulting in slightly different hydrographies of the individual isolated deep basins.

The shallow part of the Oslo Fjord water column is strongly affected by seasonal changes because temperatures at the water surface reach 20 °C during the summer. During winter time the inner basin surface waters cool to 2 °C or less, whereas the outer Oslo Fjord remains a bit warmer. Because of the moderate influx of fresh water the brackish layer at the surface is relatively thin and it is separated by a steep gradient from the deeper water masses with normal marine salinities which originate from the Skagerrak trough (Figs 2 and 3).

Benthonic foraminiferal distributions in a fjord system

Benthonic foraminiferal shells are common sediment components in many fjords (Aarseth *et al.*, 1975, Herman, O'Neil & Drahe, 1972, Feyling-Hansen, 1964, and many others). With few exceptions these studies emphasized stratigraphic distributions of these fossils in one or few cores. This seems a dangerous thing to do because fjords are filled by intensively stratified water masses, the stratification being controlled by the climatic situation and by the morphologic shape of the fjord. Fjords are also in general situated in regions of the world which had been glaciated during the Quaternary at least once if not several times. As a result their water depth must have fluctuated drastically as a response to isostatic

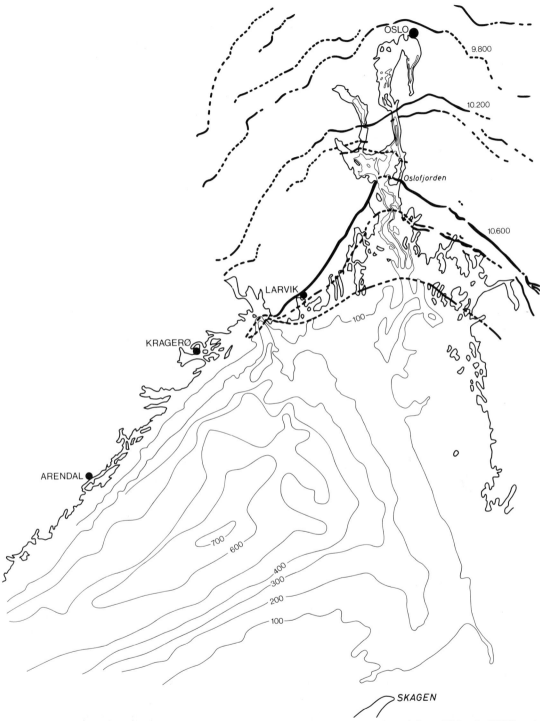

Fig. 1. Location of the Oslo Fjord. The bathymetry of the Oslo Fjord has been adapted from Richards (1976) and Munthe-Kaas (1967), the distribution of the main ice-margin deposits from Sørensen (1979). Water depths in metres. Ages in yr BP.

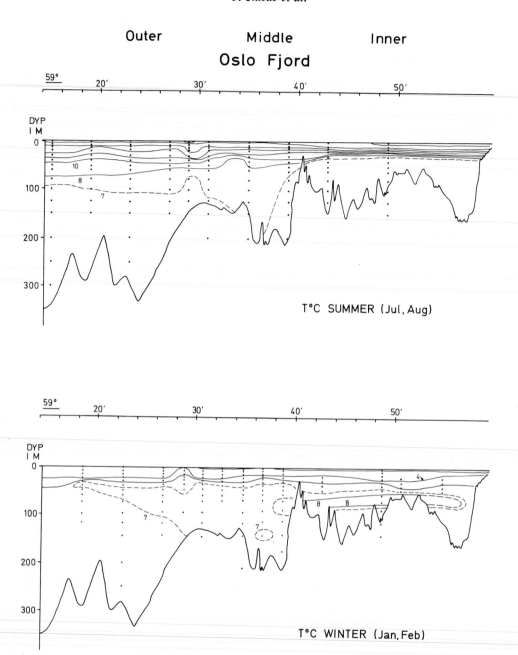

Fig. 2. Distribution of summer (July and August) and winter (January and February) temperatures (in °C) of the Oslo Fjord waters. Compiled from various sources. The solid line traces the morphology along the axis of the Oslo Fjord.

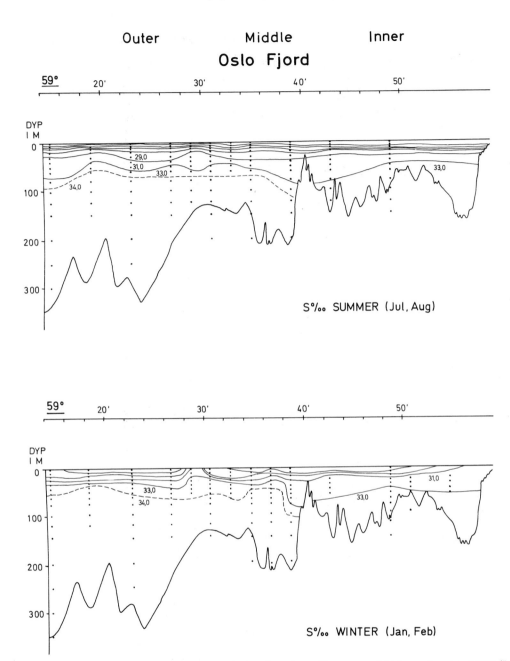

Fig. 3. Salinities (‰) of Oslo Fjord water masses during the winter (January and February) and summer (July and August) seasons. Compiled from various sources.

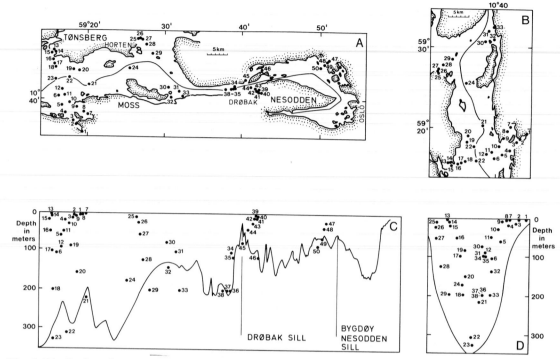

Fig. 4. Distribution of sample localities. (A) Map of the outer, middle and inner Oslo Fjord with sample localities. (B) Map of the middle and outer Oslo Fjord with sample localities, (A) and (B) after Risdal (1964). (C) North–south depth profile along the axis of the Oslo Fjord with samples locations of (A) projected into the profile plane. (D) East–west depth profile across the outer and middle Oslo Fjord with all sample localities of (B) projected into the profile plane.

and eustatic glacial and postglacial sea-level changes, especially during the late Quaternary. It is therefore difficult to see how compositional changes of benthonic foraminiferal assemblages, whose distribution was believed to be controlled by water depth, could be used without a detailed knowledge of the depth dependence of all important species. However, this information has been lacking hitherto for most fjords, the Oslo Fjord being one of the exceptions because of the well-documented study of benthonic foraminiferal distributions in its surface sediments by Risdal (1964).

The data matrix of benthonic foraminiferal distributions which had been published by Risdal (1964) originally comprised 112 species found at 50 stations in the inner, middle and outer Oslo Fjord (Fig. 4). He distinguished five depth-related assemblages (Table 1) which were believed to account for the distributions of the bulk of the species, but he also mentioned that his depth zonation could not be applied to the entire region and that the hydrography

of the water masses these animals lived in and under might have had an important function in controlling their distributions.

METHODS

Data and plots

The sediment samples of Risdal's (1964) analysis represent the uppermost 1–2 cm of small gravity cores. The sediments were very fine-grained and mostly clayey, and changes in the distributions of the benthonic foraminiferal faunas are therefore probably not related to changes of the textural properties of the sediments. Many, but not all of the samples, have been preserved in alcohol and stained with rose bengal which allowed the distinction of living specimens from empty shells. Risdal (1964) has marked all stations where he stained the samples and indicated which species had been found living, but he failed to quantify their proportions. The distribution

Table 1. Depth distribution of benthonic foraminifers in the Oslo Fjord (after Risdal, 1964)

Depth zone (no.)	Range (m)	Dominant species	Concentration (%)	Other important species
1	3–4	*Ammotium cassis*	34–90	*Ammoscalaria runiana* *Milliammina fusca* *Ammonia batavus* *Nonion depressulus asterotuberculatus*
2	4–6 (a) (b)	*Nonion depressulus astro-tuberculatus* and *Ammonia batavus* *Elphidium excavatum* *Ammoscalaria runiana*		
3	10–30	*Eggerella scabra*	25–100	*Reophax scorpiurus* *Reophax subfusiformis* *Ammoscalaria pseudospiralis* *Ammoscalaria runiana* *Ammotium cassis*
4	30–100	No dominant species		*Bulimina marginata* (max. 75 %) *Verneulina media* *Hyalinea balthica* *Cassidulina laevigata* *Nonion labradoricum* *Nonion barleeanum* *Virgulina fusiformis* *Uvigerina peregrina* *Adercotryma glomeratum*
5	100–330	*Bolivina* cf. *robusta*	39–74	*Globulimina turgida* *Verneuilina media* *Biliminia marginata* *Hyalinea balthica* *Cassidulina laevigata* *Nonion barleeanum* *Nonion labradoricum*

of living foraminifers could therefore only be analysed for presence and absence. Otherwise Risdal (1963, 1964) has used the methodology of Höglund's (1947) study of foraminifers from the Gullmar Fjord (western Sweden) and eastern Skagerrak; because of Höglund's washing technique it is not precisely known in which grain-size interval the investigated foraminiferal shell assemblages have to be sought. However, since all of Risdal's (1964) samples have been treated the same way, we can still assume that his counting results are based on comparable foraminiferal shell assemblages and that he produced a homogeneous data matrix.

The sample locations from where the foraminiferal data have been obtained have been plotted in a way different from Risdal (1964), who dealt with the foraminiferal distributions station by station (Fig. 4A, B). We wanted to compare these data with the Oslo Fjord hydrography and therefore chose to project all sample stations into depth profiles (a) from south to north along the axis of the entire main basin (Fig. 4C) and (b) from west to east across the axis of the outer and middle Oslo Fjord (Fig. 4D). We have excluded the data from the inner Oslo Fjord from the latter plots because the foraminiferal distributions inside the Drøbak sill were controlled by the different hydrography of the deep-water masses of the inner basin. The two plots allowed us to understand the foraminiferal distributions in a three-dimensional space. These projections are somewhat disadvantageous if interfaces or gradients follow levels oblique but not perpendicular to the plotting plain. The sampling locations which fall below the fjord bottom (cf. Fig. 4A, C) come from basins deeper than the fjord along its axis.

It was found that both depth profiles were reasonably well documented although additional sample points from the inner Oslo Fjord would have been desirable. Data of the same quantitative properties as the ones presented by Risdal (1964) were not available although Risdal (1963) had studied a considerable number of sediment cores from that part of

Table 2. Benthonic foraminiferal species and their maximum percentages of Risdal's (1964) raw data. (a) Species which have been deleted from further analysis (with maximum percentages and absolute abundances). (b) Species retained with maximum percentages (cf. Table 3 for absolute abundances)

A			B	
Species	%	No.	Species	%
Alveolophragmium subglobosum (A. subglo)	0·6	1	*Adercotryma glomeratum (A. glomer)*	15·0
Ammodiscus catinus (A. catinw)	1·0	4	*Alveolophragmium kosterensis (A. koster)*	5·4
Ammolagena clavata (A. clavat)	0·3	1	*Alveolophragmium nitidum (A. nitidu)*	3·0
Amphicoryna scalaris (A. scalar)	0·3	1	*Ammobaculites agglutinans (A. agglut)*	18·8
Bolivina albatrossa (B. albatr)	0·1	1	*Ammodiscus gullmarensis (A. gullma)*	1·5
Bolivina pseudoplicata (B. pseupl)	0·0	0	*Ammonia batavus (A. batavu)*	33·3
Bolivina pseudopunctata (B. pseupu)	0·4	1	*Ammoscalaria pseudospiralis (A. pseudo)*	12·6
Buliminella elegantissima (B. elegan)	0·5	6	*Ammoscalaria runiana (A. runian)*	30·2
Ceratobulimina arctica (C. arctic)	0·6	2	*Ammotium cassis (A. cassis)*	89·9
Cibicides pseudoungerianus (C. pseudo)	1·0	2	*Angulogerina angulosa (A. angulo)*	7·5
Crithionina goësi (C. goësi)	0·0	0	*Astrononion gallowayi (A. gallow)*	3·3
Dentalina drammeniesis (D. dramme)	0·1	1	*Bolivina cf. robusta (B. cf. rob)*	74·0
Elphidium margaritaceum (E. margar)	0·3	4	*Bulimina marginata (B. margin)*	73·4
Epistominella exigua (E. exigua)	0·5	1	*Cassidulina crassa (C. crassa)*	1·6
Fissurina apiculata (F. apicul)	0·0	0	*Cassidulina laevigata (C. laevig)*	23·0
Fissurina fasciata (F. fascia)	0·2	2	*Cibicides lobatulus (C. lobatu)*	15·0
Fissurina pseudoglobosa (F. pseudo)	0·7	3	*Eggerella scabra (E. scabra)*	100·0
Fissurina serrata (F. serrat)	0·1	1	*Elphidium excavatum (E. excava)*	45·3
Glomospira charoides (G. charoi)	0·2	1	*Elphidium incertum clavatum (E. I. clav)*	6·5
Hyperammina fragilis (H. fragil)	0·3	4	*Elphidium incertum incertum (E. I. ince)*	26·6
Haplophragmoides membranaceum (H. membra)	0·3	4	*Elphidium subarcticum (E. subarc)*	8·2
Hauerinella inconstans (H. incons)	0·3	4	*Elphidium* sp. *(E. sp.)*	2·2
Lagena distoma (L. distom)	0·6	1	*Globobulimina turgida (G. turgid)*	18·4
Lagena filicosta (L. filico)	0·5	1	*Guttulina lactea (G. lactea)*	1·4
Lagena gracillima (L. gracil)	0·3	1	*Haplophragmoides bradyi (H. Bradyi)*	3·6
Lagena hispida (L. hispid)	0·2	1	*Hyalinea balthica (H. balthi)*	24·7
Lagena nebulosa (L. nebulo)	0·5	1	*Hyperammina laevigata (H. laevig)*	2·3
Lagena setigera (L. setige)	0·5	1	*Liebusella (L. goësi)*	3·0
Lagena striata (L. striat)	0·2	1	*Miliammina fusca (M. fusca)*	16·0
Lagena substriata (L. substr)	0·4	1	*Nonion barleeanum (N. barlee)*	13·6
Lenticulina cf. angulata (L. cf. ang)	0·6	1	*Nonion depressulus asterotuberculatus (N. depres)*	87·0
Miliolinella subtrotunda (M. subrot)	0·6	1	*Nonion labradoricum (N. labrad)*	27·5
Nonionella iridea (N. iridea)	0·1	1	*Nonionella turgida (N. turgid)*	2·1
Nummuloculina irregularis (N. irregu)	0·5	1	*Oolina hexagona (O. hexago)*	1·3
Oolina melo (O. melo)	0·1	1	*Psammosphaera bowmanni (P. bowman)*	3·8
Parafissurina lateralis (P. latera)	0·2	2	*Pullenia bulloides (P. bulloi)*	4·2
Pullenia subcarinata (P. subcar)	0·4	4	*Pullenia osloensis (P. osloen)*	3·4
Pyrgo comata (P. comata)	0·1	1	*Pullenia* spp. *(P. spp.)*	3·4
Pyrgo williamsoni (P. willia)	0·8	9	*Quinqueloculina seminulum (Q. seminu)*	2·0
Quinqueloculina agglutinata (Q. agglut)	0·1	1	*Recurvoides trochamminiforme (R. trocha)*	3·8
Quinqueloculina stalkeri (Q. stalke)	0·4	2	*Reophax rostrata (R. rostra)*	1·4
Rectoglandulina rotundata (R. rotund)	0·3	1	*Reophax scorpiurus (R. scorpi)*	12·6
Rectoglandulina torrida (R. torrid)	0·3	1	*Reophax scotti (R. scotti)*	1·4
Reophax guttifera (R. guttif)	0·5	2	*Reophax subfusiformis (R. subfus)*	18·2
Saccammina sphaerica (S. sphaer)	0·5	2	*Reophax* spp. *(R. spp.)*	5·5
Textularia skagerakensis (T. skager)	0·6	1	*Rhabdammina discreta (R. discre)*	2·0
Trochammina adaperta (T. adaper)	0·8	6	*Rosalina columbiensis (R. columb)*	3·4
Trochammina cf. quadriloba (T. cf. qua)	0·6	2	*Spiroplectammina biformis (S. biform)*	7·6
Virgulina concava (V. concav)	0·9	2	*Textularia earlandi (T. earlan)*	2·3
Virgulina skagerakensis (V. skager)	0·3	4	*Textularia truncata (T. trunca)*	1·2
Virgulina schreibersiana (V. schrei)	0·6	1	*Tritaxis conica (T. conica)*	4·2
Tritaxis fusca (T. fusca)	0·2	1	*Uvigerina peregrina (U. peregr)*	10·0
Trochammina quadriloba (T. quadri)	0·2	1	*Verneuilina media (V. media)*	28·5
Reophax dentaliniformis (R. dental)	0·2	2	*Virgulina fusiformis (V. fusifo)*	17·9
Reophax gracilis (R. gracil)	0·3	2	*Glomospira glomerata (G. glomer)*	2·6
Unidentified *(unidenti)*	0·0	0	*Haplophragmoides fragile (H. fragil)*	3·2
			Rosalina williamsoni (R. willia)	1·3

the fjord. The extensive collection of quantitative distributions of benthonic foraminifers presented by Christiansen (1958) lack information about the calcareous species which are so important in the fossil record, but it is entirely devoted to the arenaceous species.

Statistical treatment of the data

Absolute abundances for all foraminiferal species observed were presented by Risdal (1964), station by station, for 50 localities. We have adopted the raw counts as they are and we will not discuss further the taxonomy or the spelling of Risdal's systematic categories. We have then removed those species which occurred with 1% or less of the total fora- miniferal fauna at any station because they were expected to be statistically insignificant. They, the remaining 58 species, and their maximum percent- ages are listed in Table 2. The percentages of the revised list of species (not row-normalised) which had been found with frequencies exceeding 1% of the total fauna and the total number of foraminiferal shells counted in each sample have been listed in Table 3. The general statistics of the revised raw data can be found in Table 4.

The revised raw data have then been reduced to a few variables by factor analysis using the program CABFAC (Imbrie & Kipp, 1971; Klovan & Imbrie, 1971). The varimax factor matrix indicates the im- portance of each factor (here equivalent to benthonic foraminiferal species assemblage) at each sample location, the varimax factor score matrix the impor- tance of each category (here benthonic foraminiferal species) in each factor. The varimax factor score matrix is listed in Table 5. Species distributions using a percentage of total benthonic foraminiferal faunas and the values of the varimax factor matrix have been contoured to establish the biogeographic distribu- tions of the most important species and of the foraminiferal assemblages.

SPECIES DISTRIBUTIONS: EMPTY SHELLS AND LIVING SPECIMENS

Only those species or subspecies which appeared to represent dominant elements of Risdal's (1964) depth zonation (Table 1), or of the quantitatively defined species assemblages (Table 5), will be dis- cussed. The only species not illustrated though thought to be important in zone 2b of Risdal (1964)

is *E. incertum excavatum*. We found that it occurred frequently at only one location near the coast on the western side of the outer Oslo Fjord. In addition we have discussed the distributions of a few species which were either very important components of the Oslo Fjord faunas or which composed elements of stratigraphic importance as observed in studies of the marine Quaternary sediment cover in and around the Oslo Fjord (Fig. 5).

Adercotryma glomeratum

This species occurs (Fig. 5) most abundantly between 20 and 100 m water depths on the outer and middle fjord basin, and of between depth of 50 and 100 m in the inner fjord basin. *A. glomeratum* has been found in shallower waters on the western than on the eastern side of the fjord. The distribution pattern probably reflects its response to salinity (a greater influx of fresh water on the eastern side and less mixing of low-salinity surface-layer and deeper more saline bottom-waters in the inner basin). Living specimens have not been found in depths greater than 160 m in any part of the Oslo Fjord.

A. glomeratum occurs in the Gullmar Fjord and Skagerrak (Høglund, 1947) and is common in Arctic waters (Cushman, 1948; Jarke, 1960). Leslie (1965) reported the species from Hudson Bay from water depths ranging from 50 to 230 m, with its maximum occurrence between 90 and 205 m. He found it in waters ranging from -1.04 to $1.05\,°C$ and varying in salinity from 30·4 to 33‰. Bandy & Echols (1964) observed *A. glomeratum* both in the Antarctic and the Gulf of Mexico. In Antarctic waters it occurs most abundantly below 1400 m water depth at about 0 °C. In the Gulf of Mexico they found the species most frequently above 1400 m depth in waters of about 5 °C. *A. glomeratum* is found living on both hard and soft bottoms (Christiansen, 1958; Leslie, 1965). It seems to tolerate a wide range in temperatures (eurytherm) but its distribution appears to be restricted mostly by the salinity of the water masses (stenohaline).

Ammonia batavus

A. batavus occurs in considerable numbers only in shallow waters (less than 50 m) on the eastern side of the Oslo Fjord. It is most numerous in the two small estuaries (Kurefjorden and Krokstadfjorden). North of the Drøbak sill it has been identified in only two samples. Jarke (1961) reported *A. batavus* from the

Table 3. Percentages of revised species list of benthonic foraminifers from the Oslo Fjord, and total number of foraminiferal shells counted in each sample (after Risdal, 1964)

Sample	A. glomer	A. koster	A. nitidu	A. agglut	A. gullma	A. batavu	A. pseudo	A. runian	A. cassis	A. angulo	A. gallow	B. cf. rob	B. margin
1	0·0	0·0	0·0	0·0	0·0	11·9	0·0	0·0	0·0	0·0	0·0	0·0	0·0
2	0·0	0·0	0·0	0·0	0·0	33·3	0·0	0·0	0·0	0·0	0·0	0·0	0·0
3	1·0	0·0	0·0	0·0	0·0	0·0	0·0	0·3	0·0	0·0	0·0	0·0	1·3
4	0·8	0·0	0·0	0·0	0·0	1·6	1·6	0·0	0·0	0·0	0·0	0·0	2·0
5	3·9	0·0	0·0	0·0	0·0	0·2	0·7	0·2	0·0	0·0	0·2	0·2	16·4
6	2·5	0·0	0·2	0·0	0·2	0·0	0·0	0·0	0·0	0·5	3·3	12·5	15·0
7	0·0	0·0	0·0	5·4	0·0	11·7	0·0	21·1	34·1	0·0	0·0	0·0	0·0
8	0·0	0·0	0·0	18·8	0·0	14·7	0·0	4·9	0·0	0·0	0·0	0·8	0·0
9	0·0	0·0	0·0	0·0	0·0	0·5	0·0	0·7	0·1	0·0	0·0	0·1	0·0
10	1·1	0·0	0·0	0·0	0·0	4·4	9·8	0·0	0·0	0·0	0·5	0·5	21·1
11	0·6	0·3	0·0	0·0	0·3	1·2	0·0	0·0	0·0	3·4	1·5	0·9	27·2
12	5·3	0·0	0·0	0·0	0·0	0·0	0·0	0·0	0·0	0·4	0·9	13·4	23·1
13	0·0	0·0	0·0	0·0	0·0	0·0	0·0	30·2	0·0	0·0	0·0	0·0	0·0
14	0·0	0·0	0·0	0·0	0·0	0·0	0·0	9·6	0·0	0·0	0·0	0·0	0·0
15	0·0	0·0	0·0	0·0	0·0	3·5	0·0	0·7	0·0	0·0	0·0	0·0	1·4
16	3·1	0·0	0·0	0·0	0·0	0·0	0·0	0·0	0·0	0·2	0·0	1·0	39·5
17	4·5	0·2	1·4	0·0	0·0	0·0	0·0	0·0	0·0	0·2	0·0	8·5	16·0
18	0·9	0·0	1·1	0·0	0·0	0·0	0·0	0·2	0·0	0·2	0·0	63·4	7·3
19	7·4	0·0	0·0	0·0	0·0	0·0	0·0	0·0	0·0	0·3	0·6	1·6	17·9
20	1·3	0·0	0·0	0·0	0·0	0·0	0·0	0·0	0·0	0·1	0·0	58·5	5·2
21	0·2	0·0	0·8	0·0	0·0	0·0	0·0	0·0	0·0	0·0	0·2	74·0	1·9
22	9·3	1·8	3·0	0·0	0·0	0·0	0·0	0·0	0·0	0·0	0·0	39·3	1·5
23	1·0	0·0	1·0	0·0	0·0	0·0	0·0	0·0	0·0	0·0	0·0	67·2	4·5
24	0·0	0·0	0·7	0·0	0·0	0·0	0·0	0·0	0·0	0·0	0·0	67·5	3·6
25	0·2	0·0	0·0	0·0	0·0	0·5	12·6	0·2	1·0	0·0	0·0	0·0	0·0
26	8·4	0·4	0·0	0·0	0·4	0·4	2·1	0·0	0·0	0·0	0·0	0·0	22·6
27	5·7	0·0	0·3	0·0	0·0	0·0	0·0	0·0	0·0	0·0	0·0	0·7	27·4
28	2·6	0·0	0·3	0·0	0·0	0·0	0·0	0·0	0·0	0·6	0·3	47·0	13·2
29	0·5	0·0	2·1	0·0	0·0	0·0	0·0	0·0	0·0	0·0	0·0	69·7	4·3
30	2·7	0·0	0·0	0·0	0·0	0·0	0·0	0·0	0·0	0·0	0·0	15·1	29·6
31	4·0	1·0	1·0	0·0	0·0	0·0	0·0	0·0	0·0	0·0	0·0	19·8	17·8
32	0·2	0·2	0·2	0·0	0·0	0·0	0·0	0·0	0·0	0·0	0·2	60·9	8·5
33	0·5	0·5	1·9	0·0	0·2	0·0	0·0	0·0	0·0	0·2	0·0	69·8	4·1
34	0·0	5·4	0·5	0·0	0·0	0·0	0·0	0·0	0·0	1·6	0·5	2·7	4·9
35	1·3	5·4	0·3	0·0	0·0	0·0	0·0	0·0	0·0	7·5	2·7	4·7	2·0
36	0·5	0·0	1·1	0·0	0·0	0·0	0·0	0·0	0·0	0·5	0·5	54·9	6·3
37	0·0	1·0	0·4	0·0	0·0	0·1	0·0	0·0	0·0	1·4	0·1	52·5	4·0
38	0·7	0·3	0·7	0·0	0·0	0·0	0·0	0·0	0·0	0·3	0·1	53·0`	4·3
39	0·0	0·0	0·0	0·0	0·0	0·0	1·5	0·0	89·9	0·0	0·0	0·0	0·0
40	0·0	0·0	0·0	0·0	0·0	0·0	0·0	10·0	23·3	0·0	0·0	0·0	0·0
41	0·0	0·0	0·0	0·0	0·0	0·0	0·0	0·0	0·0	0·0	0·0	0·0	0·0
42	2·0	0·0	0·0	0·0	0·0	2·0	0·4	0·0	5·6	0·0	0·0	0·0	0·0
43	6·4	0·1	0·0	0·0	0·3	0·4	0·6	0·0	0·4	0·0	0·1	0·3	56·5
44	4·3	0·1	0·0	0·0	0·0	0·0	0·0	0·0	0·0	1·8	0·1	2·4	27·9
45	2·8	0·4	0·0	0·0	0·4	0·0	0·0	0·0	0·0	0·0	0·0	1·2	38·0
46	15·0	0·9	0·0	0·0	0·0	0·0	0·0	0·0	0·0	0·0	0·0	4·0	46·3
47	1·0	0·0	0·0	0·0	1·5	0·5	0·0	0·0	0·0	0·0	0·0	0·0	67·6
48	1·6	1·6	0·0	0·0	0·0	0·0	0·0	0·0	0·0	0·0	0·0	0·0	73·4
49	8·9	1·3	0·0	0·0	1·3	0·0	0·0	0·0	0·0	0·0	0·0	0·0	24·7
50	13·7	1·0	0·0	0·0	1·0	0·0	0·0	0·0	0·0	0·0	0·0	0·0	47·7

Table 3 (cont.)

Sample	C. crassa	C. laevig	C. lobatu	E. scabra	E. excava	E. I. clav	E. I. ince	E. subarc	E. sp.	G. turgid	G. lactea	H. bradyi	H. balthi	H. laevig
1	0·0	0·0	0·0	0·0	0·9	0·0	0·0	0·0	0·0	0·0	0·0	0·0	0·0	0·0
2	0·0	0·0	0·0	0·0	0·0	0·0	26·6	0·0	2·2	0·0	0·0	0·0	0·0	0·0
3	0·0	0·0	0·0	75·7	0·0	0·0	0·2	0·7	1·5	0·0	0·0	0·0	0·2	0·0
4	0·0	0·0	0·0	65·9	0·0	0·4	0·6	0·4	0·0	0·0	0·0	2·6	1·8	0·0
5	0·0	2·1	0·0	1·0	0·0	5·2	0·0	0·2	0·0	2·7	0·2	3·6	4·1	0·0
6	0·0	18·6	0·0	0·0	0·2	1·6	0·0	0·0	0·0	0·0	0·0	0·0	0·0	0·0
7	0·0	0·0	0·0	2·7	0·0	1·9	0·0	0·0	0·0	0·0	0·0	0·0	0·0	0·0
8	0·0	0·0	0·0	27·0	0·0	1·6	1·6	0·0	0·0	0·0	0·0	0·0	0·0	0·0
9	0·0	0·0	0·0	97·8	0·0	0·0	0·0	0·1	0·0	0·0	0·0	0·0	5·0	0·0
10	0·0	5·3	0·2	25·3	0·0	0·2	0·0	0·5	0·0	1·2	0·0	0·3	17·8	0·0
11	0·9	11·6	0·0	0·3	0·0	0·6	0·3	0·6	0·0	3·4	0·0	0·4	11·2	0·0
12	0·0	9·0	0·0	0·0	0·0	0·7	0·2	0·0	0·0	0·0	0·0	0·0	0·0	0·0
13	0·0	0·0	0·0	5·8	45·3	3·3	2·5	0·0	0·0	0·0	0·0	0·0	0·0	0·0
14	0·0	0·0	0·0	79·2	0·0	3·8	0·3	0·0	0·0	0·0	1·4	0·0	0·0	0·0
15	0·0	0·0	0·0	71·8	0·0	0·0	0·0	4·9	0·0	0·0	0·0	0·2	11·6	0·0
16	0·2	3·1	0·0	2·0	0·0	0·0	0·0	0·5	0·0	0·5	0·0	1·4	10·2	0·2
17	0·0	7·7	0·0	0·0	0·0	0·0	0·0	0·0	0·0	2·7	0·0	0·0	6·2	0·2
18	0·0	2·9	0·0	0·0	0·0	0·2	0·0	0·0	0·0	1·9	0·0	0·6	10·0	0·0
19	0·0	3·5	0·0	0·0	0·0	1·6	0·0	0·0	0·0	7·2	0·0	0·0	4·9	0·0
20	0·0	10·2	0·0	0·0	0·0	0·0	0·0	0·0	0·0	2·8	0·0	1·4	1·7	0·0
21	0·0	5·6	0·2	0·0	0·0	0·2	0·0	0·0	0·0	1·2	0·0	0·6	2·4	0·0
22	0·0	5·1	0·0	0·0	0·0	2·4	0·0	0·3	0·0	2·5	0·0	0·4	5·6	1·0
23	0·0	1·7	0·0	0·0	0·0	0·0	0·0	0·0	0·0	5·4	0·0	0·7	1·5	0·0
24	0·0	3·1	0·0	0·0	0·0	0·7	0·0	0·0	0·0	0·0	0·0	0·0	0·2	0·0
25	0·0	0·0	0·0	83·4	0·0	0·0	0·0	0·0	0·0	0·0	0·0	0·4	6·3	0·0
26	0·0	0·8	0 0	21·8	0·0	0·0	0·0	0·0	0·0	0·0	0·0	0·3	14·2	0·3
27	0·0	1·1	0·0	4·2	0·0	0·0	0·0	0·0	0·0	0·0	0·0	0·0	3·6	0·0
28	0·0	5·6	0·0	0·0	0·0	0·0	0·6	0·0	0·0	1·8	0·0	0·2	3·7	0·8
29	0·0	1·6	0·0	0·0	0·0	0·0	0·0	0·0	0·0	2·7	0·0	1·3	14·4	0·0
30	0·0	6·8	0·0	0·0	0·0	0·6	0·0	0·0	0·0	1·5	0·0	0·0	11·7	0·5
31	0·0	9·6	0·0	0·0	0·0	0·0	0·0	0·0	0·0	9·2	0·0	0·2	3·9	0·0
32	0·0	1·4	0·2	0·0	0·0	0·4	0·2	0·0	0·0	1·3	0·0	0·0	3·0	0·5
33	0·0	4·1	0·2	0·0	0·0	0·0	0·0	0·0	0·0	15·3	0·0	2·7	7·1	0·0
34	1·6	23·0	0·5	0·0	0·0	1·6	0·0	0·0	0·0	3·4	0·0	0·0	6·8	0·0
35	1·3	19·1	15·0	0·0	0·0	0·0	0·0	0·5	0·0	18·4	0·0	0·0	0·0	2·3
36	0·0	2·8	0·0	0·0	0·0	0·0	0·0	0·1	0·0	13·2	0·0	0·7	3·4	0·0
37	0·8	1·4	1·4	0·0	0·1	0·4	0·1	0·1	0·0	18·0	0·0	0·0	2·8	0·3
38	0·0	4·3	0·0	0·0	0·0	0·3	0·0	0·0	0·0	0·0	0·0	0·0	0·0	0·0
39	0·0	0·0	0·0	1·5	0·0	0·0	0·0	0·0	0·0	0·5	0·0	0·0	0·0	0·0
40	0·0	0·5	0·0	61·6	0·0	0·0	0·0	0·0	0·0	0·0	0·0	0·0	0·0	0·0
41	0·0	0·0	0·0	100·0	0·0	0·0	0·0	0·0	0·0	0·0	0·0	0·0	0·0	0·0
42	0·0	0·0	0·0	82·9	0·0	0·0	0·0	0·6	0·0	0·0	0·0	0·0	5·3	0·0
43	0·0	8·3	0·0	8·5	0·0	0·6	0·0	1·3	0·0	0·5	0·1	0·0	24·7	0·0
44	0·5	20·9	0·9	0·0	0·0	0·5	0·0	2·0	0·0	2·8	0·0	0·0	3·6	0·0
45	0·0	5·2	0·4	0·0	0·0	0·4	0·0	1·3	0·0	0·0	0·0	0·0	2·7	0·0
46	0·0	14·0	0·0	0·0	0·0	0·9	0·0	0·7	0·0	0·0	0·0	0·0	0·0	0·0
47	0·0	0·0	0·0	3·3	0·0	0·0	0·0	1·0	0·0	0·5	0·0	0·0	0·2	0·0
48	0·0	1·3	0·0	0·0	0·0	0·2	0·0	8·2	0·0	4·1	0·0	0·0	1·0	0·0
49	0·0	4·4	0·0	0·0	0·0	6·5	0·0	3·1	0·0	1·5	0·0	0·0	3·9	0·0
50	0·0	7·9	0·0	0·0	0·0	2·3	0·0							

J. Thiede et al.

Table 3 (cont.)

Sample	R. scorpi	R. scotti	R. subfus	R. spp	R. discre	R. columb	S. biform	T. earlan	T. trunca	T. conica	U. peregr	V. media	V. fusifo
1	0·0	0·0	0·0	0·0	0·0	0·0	0·0	0·0	0·0	0·0	0·0	0·0	
2	0·0	0·0	0·0	0·0	0·0	0·0	0·0	0·0	0·0	0·0	0·0	0·0	0·0
3	0·0	0·0	0·0	5·5	0·0	0·0	7·6	1·0	0·0	0·0	0·0	0·0	0·0
4	0·0	0·2	18·2	0·0	0·0	0·0	0·8	1·8	0·0	0·0	0·0	0·0	2·3
5	0·0	1·0	5·2	0·0	0·0	0·0	0·5	2·3	0·0	0·0	0·5	11·1	10·5
6	0·0	0·0	0·2	0·0	0·0	0·0	0·0	0·0	0·2	0·0	1·3	10·2	0·5
7	0·0	0·0	0·3	0·0	0·0	0·0	0·0	0·0	0·0	0·0	0·0	0·0	0·0
8	0·0	0·0	0·8	0·0	0·0	0·0	0·0	0·0	0·0	0·0	0·0	0·0	0·0
9	0·0	0·0	0·0	0·0	0·0	0·0	0·0	0·0	0·0	0·0	0·0	0·0	0·0
10	0·0	0·0	10·7	0·0	0·0	0·0	0·0	0·0	0·0	0·0	0·0	0·0	0·0
11	1·5	0·0	0·9	0·0	0·0	0·0	0·3	0·0	0·0	1·2	0·0	3·5	0·0
12	0·0	0·0	0·0	0·0	0·0	0·0	0·0	0·0	0·0	0·0	6·4	1·8	0·0
13	0·0	0·0	0·8	0·0	0·0	0·0	0·0	0·0	0·0	0·0	5·6	10·0	0·0
14	1·1	0·0	0·3	0·0	0·0	0·0	0·3	0·0	0·0	0·0	0·0	0·0	0·0
15	12·6	1·4	0·0	0·0	0·0	0·0	2·1	0·0	0·0	0·0	0·0	0·0	0·0
16	3·8	0·0	0·0	0·0	0·0	0·0	0·0	0·0	0·0	0·0	4·3	15·5	0·0
17	0·0	0·0	0·0	0·0	0·0	0·0	0·0	0·0	0·0	0·0	6·0	21·7	0·0
18	0·0	0·0	0·0	0·0	0·6	0·0	0·0	0·0	0·0	0·0	0·9	0·2	0·0
19	2·6	0·0	3·2	0·0	0·0	0·0	0·0	0·0	0·0	0·0	0·2	3·6	0·0
20	0·0	0·0	0·1	0·0	0·0	0·0	0·0	0·0	0·0	0·0	10·0	14·6	0·0
21	0·0	0·0	0·2	0·0	0·2	0·0	0·0	0·0	0·0	0·0	0·4	1·8	0·0
22	0·0	0·0	0·0	0·0	0·9	0·0	0·0	0·9	0·0	0·5	0·0	0·8	0·0
23	0·0	0·0	1·9	0·0	0·4	0·0	0·0	0·0	0·0	4·2	0·0	14·7	0·0
24	0·0	0·0	0·5	0·0	0·0	0·0	0·0	0·0	0·0	1·7	0·0	2·3	0·0
25	0·2	0·0	0·7	0·0	0·0	0·0	0·0	0·0	0·0	0·5	0·0	6·7	0·0
26	7·5	0·0	4·6	0·0	0·0	0·0	0·0	0·0	0·0	0·0	0·0	0·0	0·4
27	0·3	0·0	0·3	0·0	0·0	0·0	0·0	0·0	0·0	0·0	0·0	18·9	0·0
28	0·0	0·0	0·0	0·0	0·0	0·0	0·0	0·0	0·0	0·0	6·9	28·5	0·0
29	0·0	0·0	0·8	0·0	0·0	0·0	0·0	0·0	0·0	0·0	1·6	17·5	0·0
30	0·0	0·0	0·0	0·0	0·0	0·0	0·0	0·6	0·0	1·6	0·0	6·2	0·0
31	0·0	0·0	0·5	0·0	0·0	0·5	0·0	0·0	0·0	1·0	2·0	19·3	4·8
32	0·0	0·0	0·2	0·0	0·0	0·0	0·0	0·7	0·0	0·2	0·4	3·6	0·0
33	0·0	0·0	0·2	0·0	0·2	0·0	0·0	0·2	0·0	1·6	0·2	3·0	0·2
34	0·0	0·0	0·5	0·0	0·5	0·0	0·0	0·0	0·0	3·2	1·0	4·9	0·0
35	0·0	0·0	0·0	0·0	2·0	3·4	0·0	0·0	0·0	2·0	1·3	0·6	1·6
36	0·0	0·0	0·0	0·0	1·1	0·5	0·0	0·0	0·0	0·0	2·3	2·3	0·0
37	0·0	0·0	0·0	0·0	0·1	0·4	0·0	0·0	0·0	0·1	0·8	2·8	0·7
38	0·0	0·0	0·7	0·0	0·0	0·3	0·0	0·0	0·0	0·3	0·0	5·0	0·0
39	0·0	0·0	2·1	0·0	0·0	0·0	0·0	0·0	0·0	0·0	0·0	0·0	0·0
40	0·0	0·0	0·5	0·0	0·0	0·0	0·5	0·0	0·0	0·0	0·0	0·0	0·0
41	0·0	0·0	0·0	0·0	0·0	0·0	0·0	0·0	0·0	0·0	0·0	0·0	0·0
42	0·0	0·0	4·4	0·0	0·0	0·0	0·0	0·0	0·0	0·0	0·0	0·0	0·0
43	0·0	0·0	0·4	0·0	0·0	0·0	0·0	0·0	0·0	0·0	0·0	0·0	0·0
44	0·0	0·0	0·5	0·0	0·0	0·1	0·0	0·1	0·0	0·0	0·3	0·3	1·6
45	0·0	0·0	0·0	0·0	0·0	2·4	0·0	0·0	0·0	0·0	0·0	0·0	13·6
46	0·0	0·0	0·0	0·0	0·0	0·4	0·0	0·0	0·0	0·0	0·0	0·9	5·9
47	0·0	1·0	0·2	0·0	0·0	0·0	5·8	0·7	0·0	0·0	0·0	0·2	16·0
48	0·0	0·0	1·3	0·0	0·0	0·0	0·8	0·0	0·0	0·0	0·0	0·0	13·6
49	0·0	0·0	2·4	0·0	0·0	0·0	0·0	0·3	0·0	0·0	0·0	0·3	17·9
50	0·0	0·0	1·5	0·0	0·0	0·2	0·0	0·0	0·0	0·0	0·0	0·2	5·8

Table 3 (cont.)

Sample	*L. goesi*	*M. fusca*	*N. barlee*	*N. depres*	*N. labrad*	*N. turgid*	*O. hexago*	*P. bowman*	*P. bulloi*	*P. osloen*	*P.* spp.	*Q. seminu*	*R. trocha*	*R. rostra*
1	0·0	0·0	0·0	87·0	0·0	0·0	0·0	0·0	0·0	0·0	0·0	0·0	0·0	0·0
2	0·0	6·6	0·0	31·1	0·0	0·0	0·0	0·0	0·0	0·0	0·0	0·0	0·2	0·0
3	0·0	0·5	0·0	0·0	0·3	0·0	0·0	0·1	0·0	0·0	0·0	0·0	0·0	0·0
4	0·0	0·0	0·0	0·0	0·8	0·6	0·0	0·4	0·0	0·0	0·0	0·0	0·0	0·0
5	1·5	0·0	0·0	0·0	27·5	2·1	0·0	0·2	0·0	0·5	0·0	0·0	0·2	0·0
6	1·3	0·0	13·6	0·0	1·1	0·5	0·0	0·0	1·1	1·1	0·0	0·2	0·0	0·0
7	0·0	16·0	0·0	6·2	0·0	0·0	0·0	0·0	0·0	0·0	0·0	0·0	0·0	0·0
8	0·0	0·0	0·0	29·5	0·0	0·0	0·0	0·0	0·0	0·0	0·0	0·0	0·0	0·0
9	0·0	0·1	0·0	0·0	0·0	0·0	0·0	0·0	0·0	0·0	0·0	2·0	0·0	1·4
10	0·2	0·2	0·2	0·0	4·7	0·5	0·0	0·0	0·0	0·0	0·0	0·6	0·0	0·0
11	0·6	0·0	8·9	0·0	3·7	1·5	0·0	0·0	0·0	0·0	0·0	0·0	0·0	0·0
12	0·7	0·0	11·7	0·0	2·1	1·2	0·0	0·0	0·0	0·0	0·0	0·0	0·0	0·0
13	0·0	0·8	0·0	10·9	0·0	0·0	0·0	0·0	0·0	0·0	0·0	0·0	0·0	0·0
14	0·0	0·3	0·0	0·7	0·0	0·0	0·0	3·8	0·0	0·0	0·0	0·0	0·0	0·0
15	0·0	0·0	0·0	0·0	0·0	0·0	0·0	0·0	0·0	0·0	0·0	0·0	0·0	0·0
16	1·5	0·0	7·7	0·0	2·8	1·2	0·2	0·0	0·0	0·0	0·0	0·2	1·1	0·5
17	1·7	0·0	13·4	0·0	2·8	0·2	0·0	0·0	0·0	0·0	0·0	0·0	0·6	0·0
18	0·2	0·0	3·4	0·0	0·2	0·2	0·0	0·0	2·0	0·0	0·0	1·6	2·9	0·3
19	1·6	0·0	4·5	0·0	4·8	0·9	0·0	0·0	0·0	0·0	0·0	0·1	0·0	0·0
20	0·6	0·0	5·2	0·0	0·9	0·0	0·0	0·0	0·9	1·3	0·0	0·0	0·0	0·0
21	1·1	0·0	5·6	0·0	0·2	0·0	0·0	0·0	0·8	0·0	0·0	0·0	2·1	0·0
22	0·9	0·0	0·6	0·0	0·0	0·0	0·0	0·0	4·2	0·9	0·0	0·0	2·1	0·0
23	0·6	0·0	0·8	0·0	0·4	0·0	0·0	0·0	2·5	0·0	0·0	0·0	0·2	0·0
24	0·7	0·0	3·8	0·0	0·2	0·0	0·0	0·0	1·0	0·2	0·0	0·0	0·0	0·0
25	0·0	0·0	0·5	0·0	0·0	0·0	0·0	0·0	0·0	0·0	0·0	0·0	2·5	0·0
26	2·1	0·0	0·0	0·0	0·0	0·0	0·0	0·0	0·0	0·0	0·0	0·0	3·8	0·0
27	3·0	0·0	0·3	0·0	1·1	0·0	0·0	0·0	0·0	0·0	0·0	0·9	0·6	0·0
28	0·6	0·0	3·6	0·0	0·3	0·0	0·0	0·0	0·0	0·0	0·0	0·0	0·2	0·0
29	1·3	0·0	0·2	0·0	0·2	0·0	0·0	0·0	3·2	0·0	0·0	1·3	0·0	0·0
30	0·0	0·0	4·1	0·0	0·6	0·6	0·0	0·0	0·0	3·4	0·0	0·5	0·5	0·0
31	3·0	0·0	3·0	0·0	0·5	0·0	0·0	0·0	0·0	0·0	0·0	0·2	0·0	0·0
32	0·2	0·0	3·9	0·0	0·7	0·2	0·0	0·0	0·2	0·0	0·0	0·2	0·8	0·0
33	0·2	0·0	0·2	0·0	1·3	0·0	0·0	0·0	2·2	1·0	0·0	0·0	0·0	0·0
34	0·5	1·0	12·6	0·0	0·0	0·0	0·0	0·0	0·0	0·0	0·0	1·3	2·7	0·0
35	0·0	0·0	10·9	0·0	0·0	0·0	1·3	0·0	1·1	0·0	0·0	0·0	0·0	0·0
36	0·0	0·0	3·4	0·0	0·0	0·0	0·0	0·0	0·0	0·0	3·4	0·0	0·1	0·0
37	0·8	0·0	5·4	0·0	0·0	0·1	0·1	0·0	0·7	0·0	0·0	0·0	0·0	0·0
38	0·3	0·0	2·1	0·0	2·8	0·0	0·0	0·0	0·0	0·0	0·0	0·0	0·0	0·0
39	0·0	1·5	0·0	2·6	0·0	0·0	0·0	0·0	0·0	0·0	0·0	0·0	0·0	0·0
40	0·0	2·7	0·0	0·0	0·0	0·0	0·0	0·0	0·0	0·0	0·0	0·0	0·0	0·0
41	0·0	0·0	0·0	0·0	0·0	0·0	0·0	0·0	0·0	0·0	0·0	0·0	0·0	0·0
42	0·0	0·0	0·0	1·6	0·0	0·0	0·0	0·0	0·0	0·0	0·0	0·0	0·1	0·0
43	0·1	0·1	0·0	0·0	9·4	0·1	0·0	0·0	0·0	0·0	0·0	0·3	0·5	0·0
44	0·1	0·0	2·2	0·0	5·8	0·0	0·1	0·0	0·0	0·0	0·0	0·0	0·0	0·0
45	0·4	0·0	0·0	0·0	25·6	0·0	0·4	0·0	0·0	0·0	0·0	0·9	0·0	0·0
46	0·0	0·0	0·0	0·0	3·1	0·0	0·4	0·0	0·0	0·0	0·0	0·0	0·0	0·0
47	0·0	0·0	0·0	0·0	0·5	0·0	0·0	0·0	0·0	0·0	0·0	0·0	0·0	0·0
48	0·2	0·0	0·0	0·0	3·5	0·0	0·0	0·0	0·0	0·0	0·0	0·0	0·0	0·0
49	2·4	0·0	0·0	0·0	14·8	0·0	0·0	0·0	0·0	0·0	0·0	0·0	0·0	0·0
50	0·5	0·0	0·0	0·0	7·9	0·0	0·0	0·0	0·0	0·0	0·0	0·0	0·0	0·0

J. *Thiede et al.*

Table 3 (cont.)

Sample	G. glomer	H. fragil	R. willia	Other sp.	Total CN
1	0·0	0·0	0·0	0·0	302
2	0·0	0·0	0·0	0·0	45
3	0·0	0·0	0·0	0·7	1136
4	0·0	0·0	0·0	0·4	1972
5	0·0	0·0	0·0	0·6	756
6	0·0	0·0	0·0	1·6	720
7	0·0	0·0	0·0	0·0	255
8	0·0	0·0	0·0	0·0	122
9	0·0	0·0	0·0	0·0	1533
10	0·0	0·0	0·0	0·8	1005
11	0·0	0·0	0·0	2·7	1293
12	0·0	0·0	0·0	0·0	410
13	0·0	0·0	0·0	0·0	119
14	0·0	0·0	0·0	0·0	260
15	0·0	0·0	0·0	0·0	142
16	0·0	0·0	0·0	0·7	387
17	0·0	0·0	0·0	0·5	350
18	0·0	0·0	0·0	1·6	435
19	2·6	3·2	0·0	0·6	307
20	0·0	0·0	0·0	0·7	882
21	0·0	0·0	0·0	0·2	351
22	0·0	0·0	0·0	3·6	333
23	0·0	0·0	0·0	0·8	1855
24	0·0	0·0	0·0	1·8	385
25	0·0	0·0	0·0	0·0	1140
26	0·0	0·0	0·0	0·0	238
27	0·0	0·0	0·0	0·3	259
28	0·0	0·0	0·0	0·3	302
29	0·0	0·0	0·0	0·4	1110
30	0·0	0·0	0·0	5·4	145
31	0·0	0·0	0·0	1·5	196
32	0·0	0·0	0·0	2·1	410
33	0·0	0·0	0·0	1·6	362
34	0·0	0·0	0·0	4·0	182
35	0·0	0·0	1·3	2·4	293
36	0·0	0·0	0·0	0·5	173
37	0·0	0·0	0·0	1·9	693
38	0·0	0·0	0·0	1·6	554
39	0·0	0·0	0·0	0·5	378
40	0·0	0·0	0·0	0·0	180
41	0·0	0·0	0·0	0·0	500
42	0·0	0·0	0·0	0·8	741
43	0·0	0·0	0·0	0·0	3790
44	0·0	0·0	0·3	0·0	4802
45	0·0	0·0	0·0	0·4	250
46	0·0	0·0	0·0	2·5	220
47	0·0	0·0	0·0	0·4	392
48	0·0	0·0	0·0	0·0	730
49	0·0	0·0	0·0	0·7	581
50	0·0	0·0	0·0	0·6	754

Table 4. General statistics of revised raw data

Variable	Mean	Standard deviation	Minimum value	Maximum value	Range
A. glomer	2·518	3·5377	0·0	15·0	15·0
A. koster	0·438	1·1146	0·0	5·4	5·4
A. nitidu	0·340	0·6389	0·0	3·0	3·0
A. agglut	0·484	2·7512	0·0	18·8	18·8
A. gullma	0·112	0·3160	0·0	1·5	1·5
A. batavu	1·738	5·5202	0·0	33·3	33·3
A. pseudo	0·586	2·2491	0·0	12·6	12·6
A. runian	1·562	5·4447	0·0	30·2	30·2
A. cassis	3·088	13·8088	0·0	89·9	89·9
A. angulo	0·384	1·1915	0·0	7·5	7·5
A. gallow	0·234	0·6365	0·0	3·3	3·3
B. cf. rob	17·362	26·2547	0·0	74·0	74·0
B. margin	14·720	18·5615	0·0	73·4	73·4
C. crassa	0·106	0·3328	0·0	1·6	1·6
C. laevig	4·672	5·8791	0·0	23·0	23·0
C. lobatu	0·380	2·1243	0·0	15·0	15·0
E. scabra	16·434	31·0293	0·0	100·0	100·0
E. excava	0·930	6·4042	0·0	45·3	45·3
E. I. clav	0·782	1·3773	0·0	6·5	6·5
E. I. ince	0·664	3·7671	0·0	26·6	26·6
E. subarc	0·540	1·4100	0·0	8·2	8·2
E. sp.	0·074	0·3730	0·0	2·2	2·2
G. turgid	2·524	4·5595	0·0	18·4	18·4
G. lactea	0·034	0·1996	0·0	1·4	1·4
H. bradyi	0·360	0·7643	0·0	3·6	3·6
H. balthi	4·334	5·3808	0·0	24·7	24·7
H. laevig	0·122	0·3765	0·0	2·3	2·3
L. goesi	0·578	0·7934	0·0	3·0	3·0
M. fusca	0·596	2·4469	0·0	16·0	16·0
N. barlee	2·636	3·9901	0·0	13·6	13·6
N. depres	3·392	13·5513	0·0	87·0	87·0
N. labrad	2·612	5·6963	0·0	27·5	27·5
N. turgid	0·198	0·4461	0·0	2·1	2·1
O. hexago	0·050	0·1992	0·0	1·3	1·3
P. bowman	0·090	0·5392	0·0	3·8	3·8
P. bulloi	0·398	0·9086	0·0	4·2	4·2
P. osloen	0·168	0·5560	0·0	3·4	3·4
P. spp.	0·068	0·4808	0·0	3·4	3·4
Q. seminu	0·206	0·4587	0·0	2·0	2·0
R. trocha	0·424	0·8975	0·0	3·8	3·8
R. rostra	0·044	0·2120	0·0	1·4	1·4
R. scorpi	0·592	2·1299	0·0	12·6	12·6
R. scotti	0·072	0·2763	0·0	1·4	1·4
R. subfus	1·302	3·0709	0·0	18·2	18·2
R. spp.	0·110	0·7778	0·0	5·5	5·5
R. discre	0·120	0·3546	0·0	2·0	2·0
R. columb	0·170	0·5877	0·0	3·4	3·4
S. biform	0·368	1·3621	0·0	7·6	7·6
T. earlan	0·172	0·4594	0·0	2·3	2·3
T. trunca	0·028	0·1715	0·0	1·2	1·2
T. conica	0·356	0·8572	0·0	4·2	4·2
U. peregr	1·036	2·2060	0·0	10·0	10·0
V. media	4·734	7·0255	0·0	28·5	28·5
V. fusifo	1·968	4·4602	0·0	17·9	17·9
G. glomer	0·052	0·3677	0·0	2·6	2·6
H. fragil	0·064	0·4525	0·0	3·2	3·2
R. willia	0·032	0·1878	0·0	1·3	1·3
Other spp.	0·904	1·1683	0·0	5·4	5·4

J. Thiede et al.

Table 5. Varimax factor score matrix

Variable	1	2	3	4	5	6
A. glomer	0·011	0·140	0·005	−0·003	0·039	−0·006
A. koster	−0·003	−0·011	−0·001	0·000	0·091	0·004
A. nitidu	0·017	−0·003	0·000	−0·000	0·014	−0·001
A. agglut	0·002	−0·002	0·011	0·133	−0·001	0·039
A. gullma	−0·000	0·010	−0·000	−0·000	−0·008	0·001
A. batavu	0·003	0·009	−0·000	0·409	−0·002	0·067
A. pseudo	−0·002	0·015	0·038	−0·003	−0·004	−0·002
A. runian	0·003	−0·001	0·022	0·120	0·002	0·288
A. cassis	0·003	0·003	0·016	−0·059	0·007	0·932
A. angulo	−0·005	−0·015	−0·002	0·001	0·103	0·004
A. gallow	−0·002	−0·004	−0·000	0·001	0·056	0·001
B. cf. rob	0·992	−0·025	0·002	−0·009	−0·065	−0·002
B. margin	0·017	0·927	−0·005	−0·003	−0·066	−0·003
C. crassa	−0·002	−0·004	−0·000	0·000	0·030	0·001
C. laevig	0·016	0·050	−0·011	0·004	0·587	0·020
C. lobatu	−0·009	−0·033	−0·000	0·001	0·144	0·005
E. scabra	−0·002	0·001	0·995	0·029	0·001	−0·020
E. excava	0·001	−0·000	−0·003	0·110	0·001	0·070
E. i. clav	0·004	0·036	0·002	0·020	−0·003	0·026
E. i. ince	0·003	0·001	−0·014	0·210	−0·000	−0·019
E. subarc	−0·000	0·041	0·005	−0·001	−0·036	0·002
E. sp.	0·000	−0·000	0·001	0·016	−0·000	−0·002
G. turgid	0·096	−0·013	−0·004	−0·001	0·148	0·013
G. lactea	−0·000	−0·000	0·002	−0·000	0·001	−0·001
H. bradyi	0·003	0·002	−0·001	0·000	0·050	0·001
H. balthi	0·015	0·126	−0·002	0·004	0·396	−0·009
H. laevig	0·007	0·000	0·000	−0·000	−0·000	−0·000
L. goësi	0·007	0·021	0·001	−0·000	0·031	−0·003
M. fusca	0·000	−0·000	−0·005	0·065	0·009	0·179
N. barlee	0·022	−0·015	−0·010	0·003	0·435	0·004
N. depres	0·008	−0·001	−0·032	0·857	−0·006	−0·032
N. labrad	−0·005	0·223	−0·010	0·001	−0·128	0·012
N. turgid	−0·001	0·011	0·000	0·000	0·011	−0·001
O. hexago	−0·001	−0·001	−0·000	0·000	0·011	0·001
P. bowman	0·000	−0·000	0·006	−0·001	−0·000	−0·001
P. bulloi	0·024	−0·003	0·000	−0·000	−0·000	−0·001
P. osloen	0·004	0·003	−0·001	0·000	−0·000	−0·000
P. spp.	0·005	−0·001	0·000	−0·000	0·014	0·001
Q. seminu	−0·000	0·006	0·002	0·001	0·022	−0·001
R. trocha	0·003	0·003	0·003	−0·001	0·050	−0·004
R. rostra	−0·001	0·002	0·002	0·000	0·003	−0·001
R. scorpi	−0·006	0·017	0·029	−0·003	0·021	−0·012
R. scotti	−0·000	0·004	0·002	−0·000	−0·004	−0·000
R. subfus	−0·001	0·044	0·057	0·002	−0·005	0·002
R. spp.	0·000	−0·001	0·007	−0·001	−0·000	−0·002
R. discre	0·003	−0·006	−0·000	−0·000	0·022	0·001
R. columb	−0·001	−0·001	−0·001	0·000	0·027	0·002
S. biform	0·000	0·011	0·015	−0·003	−0·014	−0·002
T. earlan	0·003	0·007	0·004	−0·001	−0·007	−0·001
T. trunca	−0·001	0·001	−0·000	0·000	0·005	0·000
T. conica	0·012	−0·012	−0·000	−0·000	0·046	0·001
U. peregr	−0·008	0·030	−0·005	0·002	0·145	−0·007
V. media	0·066	0·111	0·023	−0·003	0·391	−0·043
V. fusifo	−0·000	0·175	−0·008	−0·002	−0·159	0·013
G. glomer	−0·002	0·003	−0·000	−0·001	0·009	0·000
H. fragil	−0·002	0·003	−0·001	−0·001	0·011	0·000
R. willia	−0·001	−0·003	−0·000	0·000	0·011	0·000
Other sp.	0·018	0·009	0·001	0·005	0·070	−0·000

English Channel. It also occurs in the shallowest parts of the North Sea and Skagerrak. According to van Voorthuysen (1960) *A. batavus* has a temperature tolerance from −1·86 to 24 °C, but it seems to be most common in temperate shallow waters.

Ammoscalaria runiana

Living specimens of this species have been found in shallow water (3–20 m) of the Oslo Fjord (Fig. 5), but empty shells occasionally occur down to 200 m. Christiansen (1958) also found *A. runiana* restricted to shallow waters down to a maximum depth of 42 m in the Drøbak Sound. In the Gullmar Fjord the lower limit of its occurrence is at 33 m (Høglund, 1947) whereas it has not yet been found in the Skagerrak. Apart from the Oslo Fjord and the Gullmar Fjord *A. runiana* occurs in shallow waters around the British Isles (Heron-Allen & Earland, 1916, 1930).

Ammotium cassis

In the Oslo Fjord this species has been found living at very shallow stations (Fig. 5), down to 25 m, but it is most numerous above a water depth of 10 m. At station 39 (4 m depth) it constitutes 90 % of the total foraminiferal assemblage.

A. cassis has been reported from a wide variety of environments. It has been found in lagoonal waters in southern Brazil (Closs, 1962; Closs & Madeira, 1965; Closs & Medeiros, 1968) and in eastern Canada (Bartlett, 1966). Lutze (1965) described it from the Baltic. Occasional specimens have been found in samples from the Barents Sea (Lorange, 1977), but these probably did not represent part of the living assemblage.

Bolivina cf. robusta

Living specimens of *B.* cf. *B. robusta* occur mainly in the outer Oslo Fjord basin, where it exceeds 50 % of the total foraminiferal assemblage in the deepest part. Inside the Drøbak sill it was only found in small numbers. Its distribution seems related to well-oxygenated waters with salinities of about 33–35‰. Accordingly Høglund (1947) found only a few specimens of *B.* cf. *B. robusta* in the Gullmar Fjord, but reported it as a common species in the Skagerrak. Further westwards along the coast of southern Norway the abundance of *B.* cf. *B. robusta* decreases, and it only occurs occasionally on the shelf off

western Norway. Investigations of Lange (1956) and Jørgensen *et al.* (1981) show that *B.* cf. *B. robusta* immigrated into Skagerrak during Holocene time.

Bulimina marginata

Today this species is living at depths of between 20 and 300 m in the Oslo Fjord (Fig. 5). Between 20 and 100 m more than 20 % of the total foraminiferal assemblage consists of *B. marginata*. Its abundance increases towards the inner fjord basin. *B. marginata* is also particularly common in the Gullmar Fjord (Høglund, 1947) below 20 m water depth, but it is most numerous at 40–50 m. *B. marginata* also occurs in the Skagerrak and the North Sea (Lange, 1956; Jarke, 1961). Occasional specimens have also been recognized in material from the Barents Sea (Lorange, 1977). According to Murray (1971) *B. marginata* is a shelf species, but it has not been found in the Baltic Sea (Lutze, 1965), presumably because its salinity is too low for this species to live there. Gevirtz *et al.* (1971), in their study of foraminifers from the shelf off Long Island, found that *B. marginata* preferred salinities of 33–35‰ and temperatures about 10–14 °C.

Cassidulina laevigata

This species is common at all stations in the Oslo Fjord, except in the upper 20 m, where it has not been observed. The maximum occurrence of *C. laevigata* is found in an interval of 100–150 m water depth outside the Drøbak sill, whereas it occurs in waters deeper than 40–50 m north of this sill.

This is a widely distributed species with a continuous occurrence from a depth of 20 m in the Oslo Fjord to Skagerrak and the Norwegian Channel down to at least 1000 m at the continental slope off Norway (Kihle, 1971; Ofstad, 1977; Moyes *et al.*, 1974). These areas cover a wide range with respect both to temperatures and salinities.

Eggerella scabra

In all samples between 10 and 20 m *E. scabra* accounts for more than 50 % of the fauna. Five metres above and below this belt the percentage decreases to about 20–25 % of the total fauna, but it is also observed down to 60 m water depth (Fig. 5). Outside this 5–25 m zone it is only a minor constituent. Inside the Drøbak sill *E. scabra* seems to

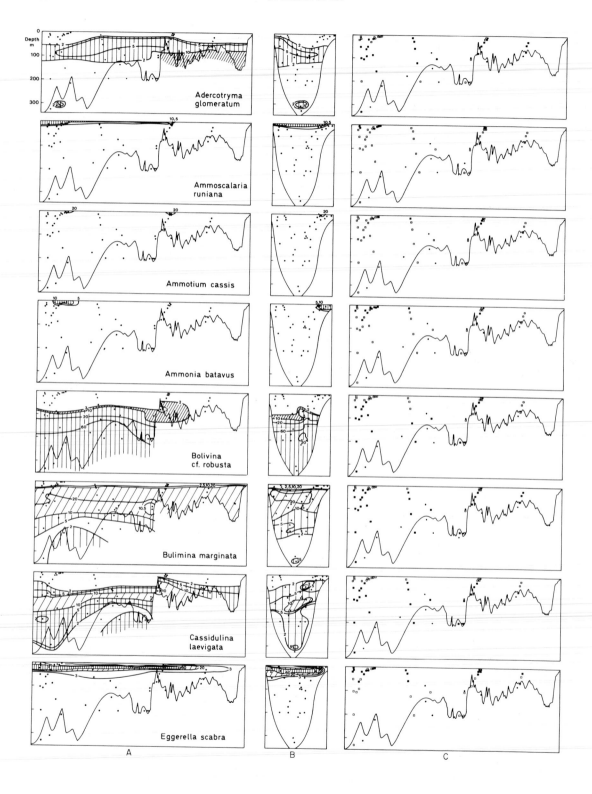

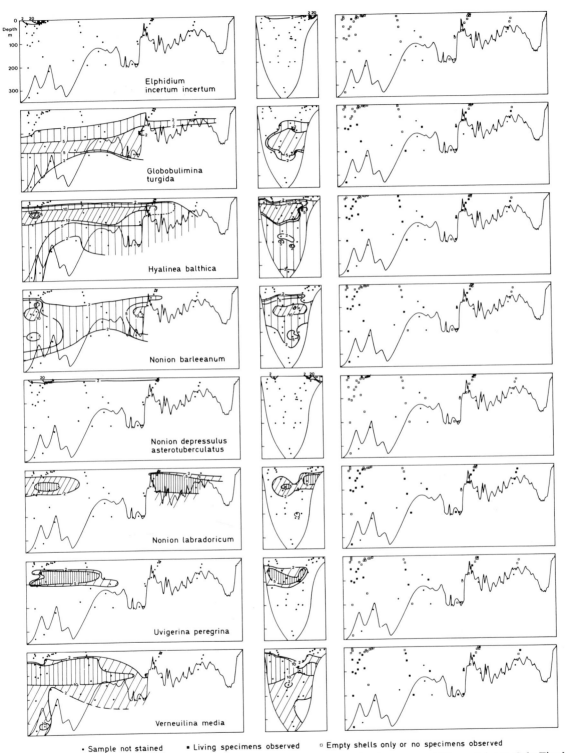

• Sample not stained   ■ Living specimens observed   □ Empty shells only or no specimens observed

Fig. 5. Percentage distribution of selected benthonic foraminiferal species in outer, middle and inner Oslo Fjord surface sediments. (A) North–south profile. (B) West–east profile across middle and outer Oslo Fjord. (C) Distribution of stations where living specimens have been observed. Symbols apply to (C) only.

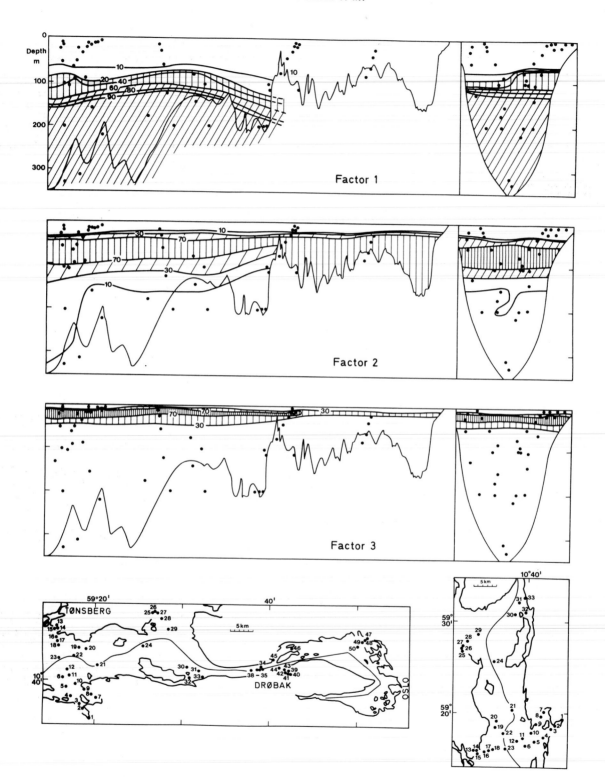

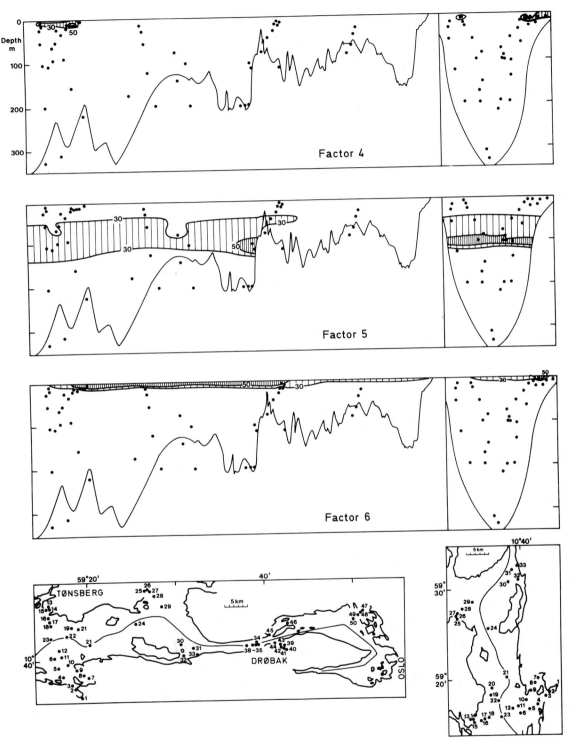

Fig. 6. Distribution of major benthonic foraminiferal assemblages in the Oslo Fjord.

dominate in a more shallow facies, namely 5–10 m water depth.

Høglund (1947) has found a similar distributional pattern in the Gullmar Fjord area. Lutze (1965) described *E. scabra* as a dominant form below 18–20 m in the south-western Baltic and showed how the species decreases in abundance in an easterly direction. He then postulated that *E. scabra* needs more than 20‰ salinity, at least during parts of the year, to survive. Feyling-Hanssen (1964) used this species as an important indicator of his zone G in late Holocene deposits in the Oslo Fjord.

Elphidium incertum incertum

Risdal (1964) referred to two publications when identifying this species:

> 1858 *Polystomella umbilicatula* var. *incerta* Williamson,
> 1954 *Elphidium incertum incertum* (Williamson) Feyling-Hanssen.

However, Feyling-Hanseen (1972) argued that this determination comprises two different species. The former, in addition to other differences, has a granulate instead of a radiate wall structure.

The species which Risdal (1964) described as *E. i. incertum* (Williamson, 1858) most probably corresponds to the form which Feyling-Hanssen (1972) renamed as *E. excavatum* (Terquem) forma *alba*. Wilkinson (1979) believed that the correct name of this morphologically variable species is *E. clavatum* Cushman. He then described *E. clavatum album* as a species confined to boreal environments.

Very often forma *alba* occurs together with forma *clavata*, and numerous transition forms bridge differences between the two (Feyling-Hanssen, 1972). But in some cases (Feyling-Hanssen, 1972) particularly in Weichselian deposits, they could be separated. Thus forma *clavata* dominates in glacial or late glacial deposits, whereas forma *alba* occurs mainly in deposits of more temperate environments.

In the Oslo Fjord this species is rare except in three stations, all between 5 and 6 m water depth in the south-eastern area. It never reaches the area inside the Drøbak sill. Parker (1952) found that this species constituted a large percentage of the North American faunas found in Long Island Sound, Gardiners Bay and Buzzards Bays, with temperatures varying from 1° to 21 °C and salinities from 28 to 30‰. It can henceforth be concluded that this species can tolerate brackish environments of large temperature fluctuations.

Globobulimina turgida

G. turgida, like a number of other species, has different distributions at each side of the Drøbak sill. In the outer and middle basins of the Oslo Fjord we find the highest relative abundances (Fig. 5) in the 100–200 m interval, but it is also common in the deepest sample (330 m). The belt of maximum occurrence has a gentle dip to the south, and is not found along the western side of the Oslo Fjord. Inside the Drøbak sill *G. turgida* seems to live in a narrow zone at about 60–80 m, with the highest abundance in sample no. 49 (62 m, 4·1 %).

G. turgida has been reported from the western Atlantic continental shelf deeper than 90 m. In the Skagerrak it occurs frequently in depths of 200–300 m. It is abundant on the basin floor of the Gullmar Fjord (Høglund, 1947). The species is also reported in cores and surface samples at 3–4 km depth in the North Atlantic (Phleger, Parker & Peirson, 1953). Murray (1973, p. 249) gives the following ecological data for the genus *Globobulimina*: slightly hyposaline (32‰) to normal marine, < 10 °C and 20–2000 m depth range.

Hyalinea balthica

The species has a wide distribution in samples deeper than 20 m north and south of the Drøbak sill. In the interval 40–100 m we find its maximum abundance (< 10 %), as opposed to the distribution of *Globobulimina turgida*, is mainly situated at the western side of the Oslo Fjord.

According to Nørvang (1945) its main Recent distribution is under Boreal–Lusitanian conditions in the Atlantic and the Pacific, in water depths from 40 to 4500 m. *H. balthica* appeared in the Mediterranean at the beginning of the Quaternary period where it is believed to be a Quaternary index fossil together with *Cassidulina laevigata carinata*.

Nonion barleeanum

N. barleeanum has its greatest occurrence in the Oslo Fjord between 50 and 300 m water depth (Fig. 5). Except for one sample, the species is missing inside the Drøbak sill. In a cross-section from the outer and middle Oslo Fjord *N. barleeanum* seems to have its maximum occurrence around 100 m depth

(Fig. 5). The living specimens follow the same pattern. Lorange (1977) found *Nonion barleeanum* in the Barents Sea between 80 and 310 m in waters with salinities around 35‰ and temperatures between 1·5° and 1·0 °C. Leslie (1965) observed this species in the central part of Hudson Bay living most abundantly at 100 and 210 m depth in waters of salinities around 33‰.

Nonion depressulus asterotuberculatus

In modern taxonomy (Feyling-Hanssen *et al.*, 1971) this species has been classified as *Elphidium albiumbilicatum*. In the Oslo Fjord it occurs in the upper 10 m with its highest frequencies from 4 to 6 m (Fig. 5). The living specimens have the same distribution as the empty shells. Lutze (1965) found the species mainly in shallow and brackish waters of the Baltic at depths of less than 15 m. Rottgardt (1952) found it (recorded as *E. asklundi*) frequently in low-salinity environments of an estuary in eastern Schleswig and in the Kiel Canal, where salinities are as low as 0·35‰.

Nonion labradoricum

N. labradoricum has been found in two regions of the examined area. The species comprises 5 % of the fauna inside the Drøbak sill and in the outer part of the Oslo Fjord at depths between 25 and 100 m. However, both empty shells and living specimens of this species have been found in waters as deep as 330 m. In a section across the outer and middle part of the fjord, *N. labradoricum* occurs most frequently on the eastern side. The distribution of the living specimens follows a similar pattern.

N. labradoricum is typical for the polar–subpolar environments of the North Atlantic (Nørvang, 1945), but it does not live at shallow water depths (Knudsen, 1977). In the Altafjord–Tromsøflaket region off northern Norway, Strand (1979) found it only inside the fjord sill and concluded that the species did not prefer the North Atlantic water (> 35‰) which is situated outside the sill. Bandy (1953) found a similar distribution pattern outside San Francisco, because this species disappeared where the salinity reached 35‰.

Uvigerina peregrina

U. peregrina is most frequent in the outer Oslo Fjord in the 50–110 m depth interval. Except for two samples, it does not occur inside the Drøbak sill (Fig. 5). Nørvang (1945) mentioned that *U. peregrina* lives in the Lusitanian and Boreal parts of the Atlantic between 30 and 4350 m water depths. Høglund (1947) recorded it between 28 and 72 m in the Gullmar Fjord and down to 700 m in the Skagerrak. Boltovskoy & Streeter (1976) registered it outside California between 40 and 265 m where salinities range from 33 to 34·5‰. *U. peregrina* is also widely distributed in the deep basins of the Mediterranean (Cita & Zocchi, 1978) with its relatively warm (12–14 °C) and saline (38–39‰) water masses.

Verneuilina media

In the Oslo Fjord this species is mainly found outside the sill at Drøbak, but it has also occasionally been observed inside the sill (Fig. 5). Its highest abundances have been registered in the 25–125 m depth interval; both empty shells and living forms are found as deep as 330 m, but it has not been found living in depths shallower than 25 m. Haman (1971) observed it between 20 and 40 m where the salinity was between 33·5 and 34·0‰ in Tremadoc Bay, North Wales. Høglund (1947) registered it mainly between 150 and 250 m in the Gullmar Fjord, but never shallower than 22 m, while Christiansen (1958) found it living between 10 and 211 m in the area close to the Drøbak Sound.

BENTHONIC FORAMINIFERAL ASSEMBLAGES IN THE OSLO FJORD

After the large matrix of compositional data of benthonic foraminiferal faunas in the Oslo Fjord has been reduced by Q-mode factor analysis to a few variables (= factors), we are able to discuss the distribution of species assemblages whose composition (Table 5) has been defined by means of objective statistical techniques (Klovan & Imbrie, 1971). Of the model of 10 factors which explained 96 % of the cumulative variance of these data with communalities of 0·81 or better (> 0·9 in 90 % of the samples) we are only presenting the six most important factors because they had a regionally coherent distribution (Fig. 6). The last four factors together accounted for only 7 % of the variance. They were either restricted to single locations or they did not reveal a regionally coherent distributional pattern. They have therefore been omitted.

Factor 1

The most important benthonic foraminiferal assemblage (= factor) found at the network of stations described in this study is restricted to the deeper parts of the outer and middle Oslo Fjord (Fig. 6). It accounts for 27% of the variance of the revised raw data, and its dominant species is *Bolivina* cf. *B. robusta*. High abundances of this assemblage are confined to the deep-water masses (in general below 100 m) which are not affected by the strong seasonal temperature and salinity changes (Figs 2 and 3) of the surface-water masses. This assemblage is therefore unable to cross over the Drøbak sill into the inner Oslo Fjord, and water masses of the same hydrographic characteristics as those inhabited by this assemblage are not believed to flow over into the inner Oslo Fjord. The interface limiting the upper occurrence of this assemblage revealed considerable morphology because it reaches into shallower water depths along the eastern than along the western flank of the outer and middle Oslo Fjord.

Factor 2

The second most important assemblage accounts for 25% of the raw data variance. Its dominant species are *Bulimina marginata* and, to a lesser degree, *Nonion labradoricum* (Table 5), two calcareous benthonic foraminiferal species widely known from shallow areas around the north-east Atlantic (Feyling-Hanssen, 1964). This assemblage occurs in the outer and in the inner Oslo Fjord (Fig. 6). Its sharp upper limit is situated at approximately 40 m water depth in the outer Oslo Fjord, but this limit gradually shallows towards the inner Oslo Fjord where it is found at approximately only 20 m water depth. This assemblage is therefore able to cross over the Drøbak sill together with the water masses overflowing from the middle Oslo Fjord into the Vestfjord (Gade, 1970). As these water masses fill the entire deep part of the inner Oslo Fjord, this assemblage frequently has been found at all inner Oslo Fjord stations below the 20 m level. The depth distribution outside the Drøbak sill, however, is considerably different from the inner Oslo Fjord because the importance of this assemblage decreases below 100 m water depth in the outer Oslo Fjord, but it gradually shallows to approximately 60 m in the middle Oslo Fjord close to the Drøbak sill. The upper boundary of this assemblage coin-cides with the lower boundary of the surface water layer which warms beyond 10–12 °C during the summer (Fig. 2).

Factor 3

Eggerella scabra is the dominant species of an assemblage which accounts for 20% of the raw data variance and which is restricted to shallow-water masses of the outer and middle Oslo Fjord (Fig. 6). It is rare in the shallowest samples, but it has its main occurrence at 10–40 m water depth in the outermost Oslo Fjord, whereas it is restricted to a depth interval of 10–25 m in the Drøbak region. Although this assemblage is able to cross the Drøbak sill into the inner Oslo Fjord, its highest abundance is restricted to stations very close to the Drøbak region, which we are at present unable to explain because of the scarce sample coverage in the inner Oslo Fjord.

Factor 4

Factor 4 describes an assemblage accounting for 8% of the variance of the raw data. It is dominated by shallow-water species such as *Nonion depressulus asterotuberculatus*, *Ammonia batavus* and *Elphidium i. incertum*, and it is confined in its distribution to the restricted environments of small inlets along the eastern and western side of the outer Oslo Fjord (for example Krokstadfjorden and Kurefjorden).

Factor 5

An intermediate depth interval (approximately 50–150 m) in the outer and middle Oslo Fjord is also inhabited by an assemblage (accounting for 8% variance) which is composed of *Cassidulina laevigata*, *Nonion barleeanum*, *Verneuilina media* and *Hyalinea balthica*. Although found throughout the outer and middle Oslo Fjord, this assemblage is most important in a region where the hydrography is distributed due to the presence of the Drøbak sill.

Factor 6

The last assemblage, which is dominated by *Ammotium cassis* and *Ammoscalaria runiana*, accounted for 5% of the variance. Both arenaceous species are typical of very shallow locations near the coast. It is interesting to note that this assemblage is apparently restricted to the brackish water masses along the eastern side of the outer and middle Oslo Fjord.

CONCLUSIONS

It is evident from the discussion of species and assemblage distributions of benthonic foraminifers in the Oslo Fjord that the validity of biostratigraphic results obtained from marine Quaternary sediments is highly speculative if they are based on this fossil group alone. Neither the single species nor the assemblages of the Recent Oslo Fjord benthonic foraminifers inhabit the same depth interval throughout the entire area studied here, whereas their depth distributions often follow undulating interfaces due to slight shifts of the hydrography. In essence they follow water bodies of the same hydrographic characteristics rather than water depth as claimed by Risdal (1964). Eustatic and isostatic sea-level changes throughout the Holocene have affected the inner Oslo Fjord region at rates different from the outer and middle Oslo Fjord (Brøgger, 1901). In the course of this evolution not only water depth of deposition and habitat of the benthonic foraminifers has changed, but possibly more important, the sills controlling the bottom-water exchange between the various sub-basins of this fjord regime have become much shallower. Consequently the palaeogeography of the Oslo Fjord and its hydrography have changed in a way which cannot be reconstructed quantitatively at the present time, but almost certainly they have changed the foraminiferal habitats drastically. The usefulness of any biostratigraphic boundary based on benthonic foraminifers, even if they have been deposited in a supposedly known water depth, will therefore have to be controlled by other stratigraphic techniques.

The data discussed here have a number of qualities which are worth further discussion. It is known that slumping and redeposition are typical for fjords (Holtedahl, 1975) because they commonly have very steep slopes and because their sediments are deposited rapidly (Richards, 1976). The coincidence of distributions of living specimens with the empty shells of the same species of benthonic foraminifers in the Oslo Fjord (Fig. 5) demonstrated convincingly that the surface cover of sediments in the Oslo Fjord is only slightly affected by these processes at present. We can therefore expect fossil benthonic foraminifers in fjord sediments to represent excellent tools to reconstruct ancient palaeoenvironments of their habitats. Attempts to achieve such reconstructions

based on quantitative hydrographic data have been carried out in the deep Atlantic Ocean (Lohmann, 1978). Most of the studies from the North Atlantic (Streeter, 1973; Schnitker, 1974; Belanger & Streeter, 1980) and from the surrounding shallow-water regions, including our own, had to settle for a qualitative comparison between the hydrography of the bottom-water masses in the studied area with the benthonic foraminiferal distributions.

In the case of the Oslo Fjord there is a clear relationship between foraminiferal distributions (Figs 5 and 6) and the stratification of the water masses (Figs 2 and 3). The water column can be subdivided into an upper (upper 40–60 m) brackish layer under the influence of seasonal temperature fluctuations where eurytherm and euryhaline species find a favourable habitat, and the lower homogeneous deep-water masses where hydrographic characteristics change at a slower pace. It was particularly important to study the impact of the presence of the Drøbak sill upon the source of the water masses which enter the deep inner Oslo Fjord and the consequences this had for the composition of its benthonic foraminiferal faunas.

We hope that the relationship of benthonic foraminiferal surface-sediment shell assemblages to the hydrography of the bottom-water masses will hold for the major part of the Holocene. This relationship can then be used in the future in attempts to reconstruct the ancient palaeoenvironments of the Holocene Oslo Fjord which is preserved in great detail in the fine-grained, homogeneous fjord sediments with their extremely high accumulation rates (Richards, 1976).

ACKNOWLEDGMENTS

We are grateful to the Norwegian Oceanographic Data Center under R. Leinebø (NOD, Bergen, Norway) for supplying us with the hydrographic data from the Oslo Fjord. An early version of this paper has been reviewed by J. Nagy (Oslo) and B. M. Funnell (Norwich). This study has been supported by Norsk Teknisk-naturvitenskapelig Forskningråd (NTNF), Project no. 1810.8110. It is a contribution to OSKAP (Oslofjord-Skagerrak-Project) at the Department of Geology, University of Oslo.

REFERENCES

AARSETH, I. *et al.* (1975) Late Quaternary sediments from Korsfjorden, western Norway. *Sarsia*, **58**, 43–66.

BANDY, O.L. (1953) Ecology and paleoecology of some California foraminifera. Pt. 1. The frequency distribution of recent foraminifera off California. *J. Paleont.* **27**, 161–182.

BANDY, O.L. & ECHOLS, R.J. (1964) Antarctic foraminiferal zonation. Biol. Antarct. Seas. *Antarct. Res. Ser.* **1**, 73–91.

BARTLETT, G.A. (1966) Foraminifera distribution on Tradacie Bay, Prince Edward Island. *Geol. Surv. Pap. Can.* **66-20**, 54 pp.

BELANGER, P.E. & STREETER, S.S. (1980) Distribution and ecology of benthic foraminifera in the Norwegian-Greenland Sea. *Mar. Micropaleont.* **5**, 401–428.

BOLTOVSKØY, E. & WRIGHT, R. (1976) *Recent Foraminifera.* Junk, The Hague. 515 pp.

BRØGGER, W.C. (1901) Om de senglaciale and postglaciale nivåforandringer i Kristianiafeltet. *Norg. geol. Unders.* **31**, 731 pp.

CHRISTIANSEN, B. (1958) The foraminifer fauna in the Drøbak sound in the Oslo Fjord (Norway). *Nytt Mag. Zool.* **6**, 5–91.

CITA, M.B. & ZOCCHI, M. (1978) Distribution pattern of benthic foraminifera on the floor of the Mediterranean Sea. *Oceanol. Acta*, **1**, 445–462.

CLOSS, D. (1962) Foraminiferos e Tecamebas da Lagoa dos Patos. *Escol. Geol. Univ. Rio Grande do Sul*, **11**, 130 pp.

CLOSS, D. & MADEIRA, M. (1965) Seasonal distributions of brackish foraminifera in the Patos Lagoon, southern Brazil. *Escol. Geol. Univ. Rio Grande do Sul*, **15**, 41 pp.

CLOSS, D. & MEDEIROS, V.M.F. (1968) New observations on the ecological subdivision of the Patos Lagoon in southern Brazil. *Inst. Cienc. Natur. Univ. Rio Grande do Sul*, **24**, 7–33.

CUSHMAN, J.A. (1930) The foraminifera of the Atlantic Ocean. *Bull. U.S. Natn. Mus.* **104**, 1–79.

CUSHMAN, J.A. (1948) Arctic foraminifera. *Spec. Publs Cushman Lab.* **12**, 37 pp.

FEYLING-HANSSEN, R.W. (1954) Late-Pleistocene foraminifera from the Oslofjord area, southeast Norway. *Norsk geol. Tidsskr.* **33**, 109–152.

FEYLING-HANSSEN, R.W. (1964) Foraminifera in the late Quaternary deposits from the Oslofjord area. *Norg. geol. Unders.* **225**, 383 pp.

FEYLING-HANSSEN, R.W. (1972) The foraminifer *Elphidium excavatum* (Terquem) and its variant forms. *Micropaleontology* **18**, 337–354.

FEYLING-HANSSEN, R., JØRGENSEN, J.A., KNUDSEN, K.L. & LYKKE ANDERSEN, A.-L. (1971) Late Quaternary Foraminifera from Vendsyssel, Denmark and Sandnes, Norway. *Bull. geol. Denm.* **21**, 67–317.

GADE, H.G. (1970) Hydrographic investigations in the Oslofjord, a study of water circulation and exchange processes. *Univ. Bergen, Geophys. Inst. Div. A (Phys. Oceanogr.) Rep.* 24 (three volumes).

GEVIRTZ, J.L. *et al.* (1971) Paraecology of benthonic foraminifera and associated microorganisms of the continental shelf off Long Island, New York. *J. Paleont.* **45**, 153–177.

HAMAN, D. (1971) Foraminiferal assemblages in Tremadoc Bay, North Wales, U.K. *J. foramin. Res.* **1**, 126–143.

HERMAN, Y., O'NEIL, J.R. & DRAKE, C.L. (1972) Micropaleontology and paleotemperatures of postglacial SW Greenland fjord cores. *Acta geol. Univ. Ouluensis*, **1**, 357–407.

HERON-ALLEN, E. & EARLAND, A. (1916) The foraminifera of the West of Scotland. *Trans. Linn. Soc. Zool.* **11**, 197–299.

HERON-ALLEN, E. & EARLAND, A. (1930) The foraminifera of the Plymouth district. *Jl R. microsc. Soc.* **50**, 46–84.

HØGLUND, H. (1947) Foraminifera of the Gullmar Fjord and the Skagerrak. *Zool. Bidr. Upps.* **26**, 328 pp.

HOLTEDAHL, H. (1975) The geology of the Hardangerfjord, West Norway. *Norg. geol. Unders.* **323**, 87 pp.

IMBRIE, J. & KIPP, N. (1971) A new micropaleontological method for quantitative paleoclimatology: application to a late Pleistocene Caribbean core. In: *Late Cenozoic Glacial Ages* (Ed. by K. K. Turekian), pp. 71–181. Yale University Press.

JARKE, J. (1960) Beitrag zur Kenntnis der Foraminiferenfauna der mittleren und westlichen Barents-See. *Int. Revue ges. Hydrobiol. Hydrogr.* **45**, 581–654.

JARKE, J. (1961) Die Beziehungen zwischen hydrographischen Verhältnissen, Faziesentwicklung und Foraminiferenverbreitung in der heutigen Nordsee als Vorbild für die Verhältnisse während der Miocän-Zeit. *Meyniana*, **10**, 21–36.

JØRGENSEN, P., ERLENKEUSER, H., LANGE, H., NAGY, J., RUMOHR, J. & WERNER, F. (1981) Sedimentological and stratigraphical studies of two cores from the Skagerrak. In: *Holocene Marine Sedimentation in the North Sea Basin* (Ed. by S.-D. Nio *et al.*). *Spec. Publs int. Ass. Sediment.* **5**, 397–414. Blackwell Scientific Publications, Oxford. 524 pp.

KIHLE, R. (1971) Foraminifera in five sediment cores in a profile across the Norwegian Channel south of Mandal. *Norsk geol. Tidsskr.* **51**, 261–286.

KLOVAN, J.E. & IMBRIE, J. (1971) An algorithm and FORTRAN-IV program for large scale Q-mode factor analysis and calculation of factor scores. *J. Int. Ass. Math. Geol.* **3**, 61–77.

KNUDSEN, K.L. (1977) Foraminiferal faunas of the Quaternary Hostrup Clay from northern Jutland, Denmark. *Boreas*, **6**, 229–245.

LANGE, W. (1956) Grundproben aus Skagerrak und Kattegat, mikrofaunistisch und sedimentpetrographisch untersucht. *Meyniana*, **5**, 51–86.

LESLIE, R.J. (1965) Ecology and paleoecology of Hudson Bay foraminifera. *Rep. Bedford Inst. Oceanogr.* **65-6**, 192 pp.

LOHMANN, G.P. (1978) Abyssal benthonic foraminifera as hydrographic indicators in the western South Atlantic Ocean. *J. foramin. Res.* **8**, 6–34.

LORANGE, K. (1977) *En mikropaleontologisk stratigrafisk undersøkelse av kvartaere sedimenter i nordvestre del av Barentshavet.* Unpublished Thesis. University of Oslo. 237 pp.

LUTZE, G.F. (1965) Zur Foraminiferen-Fauna der Ostsee. *Meyniana*, **15**, 75–142.

MOYES, J., GAYET, J., PUJOL, C. & PUJOS-LAMY, A. (1974) Etude stratigraphique et sedimentologique. *Orgon*, **1** (Mer de Norvége), 81–137.

MUNTHE-KAAS, H. (1967) Fjordens topografi. *Norsk Inst. Vannforsk. (NIVA) Oslofjord prosj. Delrapp.* **15**, 10 pp.

MURRAY, J.W. (1971) *An Atlas of British Recent Foraminiferids.* Heinemann, London. 244 pp.

MURRAY, J.W. (1973) *Distribution and Ecology of Living Benthic Foraminiferids.* Heinemann, London. 274 pp.

NIVA (Norwegian Institute for Water Research) (1970) The Oslofjord and its pollutional problems. In: *Eutrophication in Large Lakes and Impoundments* (Ed. by C. P. Milway), pp. 445–525. Uppsala Symposium, May 1968, OCDE, Paris.

NØRVANG, A. (1945) The zoology of Iceland, *Foraminifera,* **2**, 79 pp.

OFSTAD, K. (1977) *Kvartære foraminiferer og stratigrafi i sedimentkjerner fra Norskerenna utenfor Vestlandet.* Unpublished Thesis, University of Oslo. 152 pp.

PARKER, F.L. (1952) Foraminiferal distribution in the Long Island Sound-Buzzard Bay area. *Bull. Harv. Coll. Mus. Comp. Zool.* **106**, 425–473.

PHLEGER, F.B., PARKER, F.L. & PEIRSON, J.F. (1953) North Atlantic foraminifera. *Rep. Swed. deep Sea Exped.* **7**, 149 pp.

RAMBERG, I.B. (1976) Gravity interpretation of the Oslo Graben and associated igneous rocks. *Norg. geol. Unders.* **325**, 194 pp.

RICHARDS, A.F. (1976) Marine geotechnics of the Oslofjorden region. In: *Laurits Bjerrum Memorial Volume* (Ed. by N. Janbu, F. Jørstad and B. Kjaernsli), pp. 41–63. Norwegian Geotechnical Institute, Oslo.

RISDAL, D. (1963) Foraminiferfaunaen i en del sedimentkjerner fra indre Oslofjord. *Norg. geol. Unders.* **224**, 90 pp.

RISDAL, D. (1964) Foraminiferfaunaens relasjon til dybdeforholdene i Oslofjorden, med en diskusjon av de senkvartære foraminifersoner. *Norg. geol. Unders.* **226**, 142 pp.

ROTTGARDT, D. (1952) Mikropaläontologisch wichtige Bestandteile recenter brackischer Sedimente an den Küsten Schleswig-Holsteins. *Meyniana,* **1**, 169–228.

SCHNITKER, D. (1974) West Atlantic abyssal circulation during the past 125,000 years. *Nature,* **248**, 385–387.

SØRENSEN, R. (1979) Late Weichselian deglaciation in the Oslofjord area, South Norway. *Boreas,* **8**, 241–246.

STØRMER, L. (1935) Contribution to the geology of the southern part of the Oslofjord. *Norsk geol. Tidsskr.* **15**, 43–113.

STRAND, J.E. (1979) *Paleoklimatisk og stratigrafisk undersøkelse av senkvartære marine sedimenter fra Altafjorden og Tromsøflaket, Nord-Norge ved hjelp av bentoniske foraminiferer.* Unpublished Thesis, University of Oslo. 107 pp.

STREETER, S.S. (1973) Bottom water and benthonic foraminifera in the North Atlantic–glacial-interglacial contrasts. *Quat. Res.* **3**, 131–141.

TERQUEM, O. (1876) Essai sur le classement des animaux qui vivent sur la plage et dans les environs de Dunkerque. *Mém. Soc. dunkerq. Encour. Sci.* **19**, 405–457.

VAN VOORTHUYSEN, J.H. (1960) Die Foraminiferen des Dollart-Ems Estuarium. *Verh. K. ned. Geol.-mijnb. Genoot.* **19**, 237–269.

WILKINSON, I.P. (1979) The taxonomy, morphology and distribution of the Quaternary and Recent foraminifer *Elphidium clavatum* Cushman. *J. Paleont.* **53**, 628–641.

WILLIAMSON, W.C. (1858) *On the Recent Foraminifera of Great Britain.* Royal Society, London. 197 pp.

Spec. Publs int. Ass. Sediment. (1981) **5**, 497–503

The Holocene marine transgression and its molluscan fauna in the Skagerrak–Limfjord region, Denmark

KAJ STRAND PETERSEN

Geological Survey of Denmark,
31 Thoravej, DK-2400 Copenhagen NV, Denmark

ABSTRACT

Late Weichselian erosional furrows in the Chalk and Pleistocene deposits of northern Jutland are demonstrated by data from wells and seismic profiles. The Holocene marine sediments in the furrows have a thickness of 30 m in cored sections. C-14 dates on various core levels indicate an early Holocene transgression around 9000–8000 BP corresponding to a depth of 25 m below local present-day sea-level. At 7000 BP the transgression reached a maximum of 3 m above present-day sea-level. The partly infilled furrows had a water depth of 25 m and connected the Limfjord system with the North Sea. With a mean sedimentation rate of 6 m per 1000 years the furrows shallowed to a water depth of 5 m between 7000–4000 BP. Beach ridges subsequently developed during the Subatlantic and disconnected the North Sea and Limfjord systems. Judging from the molluscan fauna the main part of the marine sequence deposited between 7000–5000 BP is characterized by the inshore 'fjord' group furnishing *Abra alba* and *Corbula gibba*.

INTRODUCTION

Studies of the marine sequence found in an investigation well at Vust in northern Jutland were inspired partly by the unfinished work of V. Nordmann on the fauna in the *Tapes* sea from this area, and partly by the investigations connected with the publication of the geological map of the Central Limfjord Region (Gry, 1979).

The locality of the investigation well is the beach ridge complex south of the village of Vust, east of the Lund Fjord in Hanherred (Fig. 1). The beach ridges at Vust, which are amongst the most marked in Hanherred, were referred to the Bronze Age by V. Nordmann, who investigated the site at the beginning of this century (Nordmann, 1905, p. 90).

Well data compiled for the Pre-Quaternary surface map of the Central Limfjord Region (Gry, 1979) showed that chalk constitutes an important part of the former islands which formed an archipelago during the time of the *Tapes* sea. Furthermore, it

became clear that deep furrows occurred in the Chalk between the 'islands' as depicted in Fig. 1. From some of the borings mollusc-rich material indicated a Holocene age for the sediment deposited in the furrows.

The furrows are interpreted as valleys of erosion formed during the late Weichselian to early Holocene, a period with a lower base-level of erosion than today. Consequently the Holocene marine transgression is to be found at an earlier date within the furrows than in the raised marine deposits exposed elsewhere in northern Denmark (see Krog, 1979).

Therefore, when it was demonstrated that a deep furrow system existed below the young beach ridge complex at Vust, it was decided to sink a borehole at the site in 1976 (DGU no. 23,250). This was expected to yield a maximum average of the Holocene marine sequence.

In order to obtain material sufficient for quantitive molluscan fauna analyses and C-14 datings from the cored sections, drilling equipment with a diameter

0141-3600/81/1205-0497 $02.00

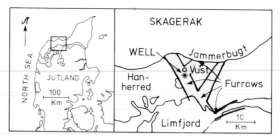

Fig. 1. Location map, illustrating the locality of the study area in relation to Denmark and in relation to the North Sea. The directions of the furrows are shown.

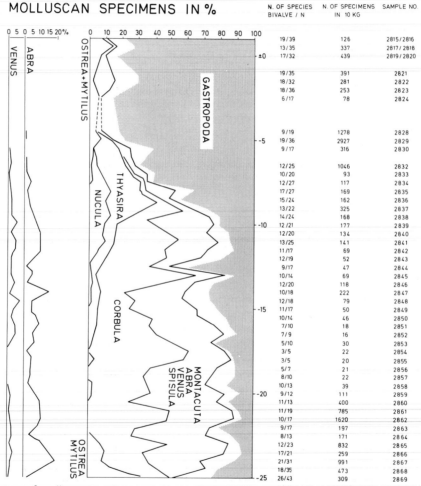

Fig. 2. Percentage of molluscan specimens emphasizing the bivalve record. The number of species found in each sample and the number of specimens counted in each sample are adapted to 10 kg sediment.

Table 1. Radiocarbon dates from the Vust borehole 57° 06′ 02″ N. lat. 9° 04′ 19″ E. long. The calibrated dates are after Damon, Long & Wellik (1973). The six oldest dates are after the Scandinavian varve chronology (Tauber, 1970)

Sample elevation m	Dated material	Field sample no.	Lab. no.	C-14 years BP	Calibrated dates BP
+1·22 to +1·70	*Tapes aureus*	2815 2816	K-3221	3130 ± 70	3390
+0·89	Stub of *Tilia* sp.	2817 2818	K-2747	3250 ± 70	3545
−1·50 to −2·00	*Lucinoma borealis*	2823	K-2754	4050 ± 65	4620
−4·50 to −5·00	*L. borealis*	2829	K-2871	4210 ± 85	4870
−13·50 to −14·50	*Cardium, Thracia, Natica, Nassa*	2848 2849	K-2872	5460 ± 95	6330
−20·50 to −21·00	*Corbula gibba*	2861	K-2873	6550 ± 110	7400
−23·00 to −23·50	*C. gibba*	2866	K-2874	6810 ± 110	7610
−24·50 to −25·00	*Cyprina islandica*	2869	K-3282	7380 ± 110	7980
−24·50 to −25·00	*Ostrea edulis*	2869	K-2755	7580 ± 120	8090
−24·50 to −25·00	*Mytilus edulis*	2869	K-3280	7660 ± 115	8170
−24·50 to −25·00	*Cirripedia*	2869	K-3281	7860 ± 115	8260
−25·50 to −26·00	Peat	2871	K-2875	9830 ± 115	9830

of 31 cm has been used. From each sampling interval of 0·5 m all material was kept yielding 0·04 m³. Besides a bulk sample weighing about 10 kg the rest was sieved on a 0·5 mm screen during the operation in the field. A total of 55 samples was taken in the Holocene sequence.

From the bulk sample one part has been kept for later investigations on microfossils and another part has been analysed for grain-size distribution by H. Bahnson at the laboratory of the Geological Survey of Denmark. The median grain size is presented on Fig. 4, depicting the coarser material in what constitutes the beach ridges at Vust and at the bottom, sample 2869, the coarser material from the transgression overlying the gyttja and peat found between − 26·0 and − 25·0 m.

The gamma-log which was made for the borehole by K. Klitten at the Technical University of Denmark and shown in Fig. 3 reveals the same changes as seen from the grain-size analyses, i.e. the coarser material in the uppermost part of the transgression layers—giving 200–300 counts min⁻¹, while the main part of the cored section has 400–500 count min⁻¹ except for the gyttja and peat in the interval between − 25·0 and − 26·0 m and the interval − 18·0 to − 14·0 m with clayey silt where 600–700 counts min⁻¹ are met with.

On account of the molluscan analyses of the samples from the cored sections 111 species have been found. However, as in Fig. 2, it is seen how the changes in environment might be characterized by rather few bivalves when figured in number of specimens. The faunal diversity is shown in the column: *N* of species—bivalve/*N*.

The number of specimens in 10 kg shown in Fig. 2 reflects the weight of the bulk sample, which forms the base for the quantitative analyses given in Fig. 4.

The datings from the different parts of the cored section are based mainly on shell material listed in Table 1. These have been analysed by H. Tauber at the Copenhagen C-14 Laboratory. The C-14 datings on marine shells are calculated in such a way that synchronogenic terrestrial and marine material get the same C-14 age, i.e. the reservoir-effect for marine samples has been taken into account in the C-14 age. The calibrated dates have been used in Fig. 3 in order to present the sedimentation rate.

The basal peat sample 2871, 9830 ± 115 BP (conv. C-14) together with the overlying gyttja have been investigated pollen analytically by H. Krog at the Survey.

The use of the terms Preboreal, Boreal, Atlantic, Subboreal and Subatlantic in the text refers to the chronozones defined by Mangerud *et al.* (1974).

RESULTS

On the basis of former C-14 datings from the Central Limfjord Region (Petersen, 1976, table 2, shown in the upper part of Fig. 3) combined with the C-14 dates from the well (shown in the lower part of Fig. 3), the rise of the sea-level can be

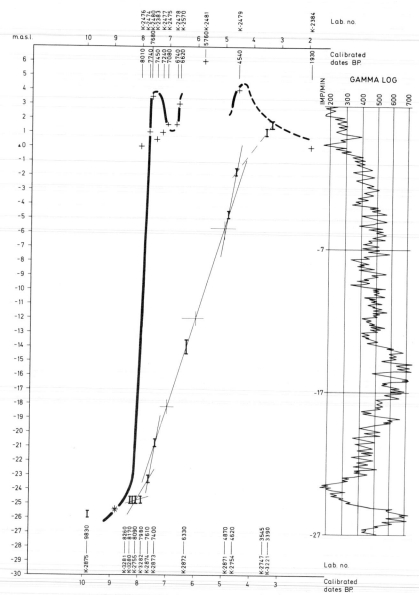

Fig. 3. Radiocarbon dates from the central part of the Limfjord area at the top of the figure with crosses. The succession of C-14 dates from the borehole at Vust (57° 06′ 02″ N. lat. 9° 04′ 19″ E. long.) from the depth of −26 m to +1·5 m mainly based on marine shells, indicating the sedimentation rate (calibrated C-14 dates) in the furrow. The asterisk marks the transgression, dated by pollen analyses, and the start of the curve for the relative rise of sea-level. To the right a gamma-log from a depth of −27 to +3 m is shown, illustrating the sedimentological variations by counts min⁻¹ (imp min⁻¹).

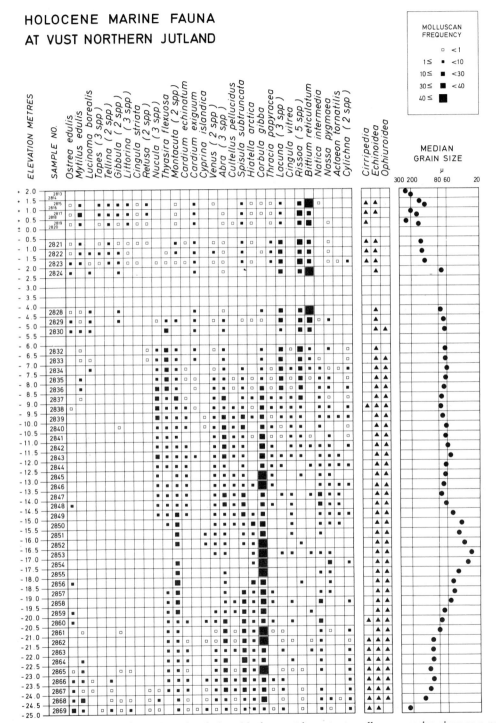

Fig. 4. The Holocene marine fauna, Vust borehole, with the most important molluscan species given as a percentage of frequency and grouped according to the occurrence in the littoral and sublittoral zone. The occurrence of *Cirripedia*, *Echinoidea* and *Ophiuroidea* are shown. Median grain size of each sample is given to the right.

estimated. The lowermost sample, from −26 m, is a peat *in situ* dated to 9830±115 BP (conv. C-14). The superimposed fresh- to brackish water gyttja was found by pollen analysis to be early Boreal, referred to pollen zone V (cf. Jessen, 1935). The pollen spectrum was dominated by *Corylus*, *Pinus* and *Betula*. *Alnus* and trees of the oak mixed forest were lacking. This dating indicates the onset of a transgression at a level of −25·0 to −25·5 m below present-day sea-level. The sample above, from −24·5 to −25·0 m, consists of shelly gravel and represents a fully marine environment as seen from the fauna (Fig. 4, sample No. 2869). From this level four fauna elements have been dated, *Cirripedia*, *Mytilus edulis*, *Ostrea edulis* and *Cyprina islandica*, giving C-14 years from 7860±115 to 7380±110 BP. From a faunistic point of view increasing water depth is indicated through these 500 years. In the following, main part of the succession a deeper water fauna is established. This is seen from the presence of the molluscan species, *Nucula* to *Cylichna* (Fig. 4), dominated by species of *Montacuta*, *Venus*, *Abra*, *Spisula* and especially by *Corbula gibba* (Fig. 2). Furthermore, *Echinoidea* and *Ophiuroidea* are found without interruption throughout this sequence.

Three main stages can be demonstrated:

(1) A lower littoral deposit (Fig. 4; samples 2869–2866) with such species as *Mytilus edulis*, *Ostrea edulis* and *Littorina littorea*, from the early Atlantic.

(2) A deeper water fauna with *Abra alba* and *Corbula gibba* (Fig. 4, samples 2865–2835). From these 15 m of the succession gastropods are found which do not occur in the exposed marine deposits of Denmark, but are closely related to the recent North Sea fauna at a water depth of 20–30 m. The deposition of this second stage occurred during middle- to late Atlantic.

(3) In the upper part of the cored section (Fig. 4, samples 2834–2815) sublittoral to littoral species occur. This corresponds to early Subboreal to late Subboreal time. The species found within this part of the sequence belong to the fauna known from the *Tapes* Sea in the northern part of Denmark.

DISCUSSION

It is seen from Fig. 4 that species found in the upper part of the cored sections—the sublittoral to littoral deposits—already occurred in the transgression layers in the lowermost part of the section. This indicates that no marked climate-induced changes in the fauna took place during the 5000 years, from early Atlantic to late Subboreal. Changes in the assemblage can be explained as a shift in facies.

On the basis of sedimentological data obtained from gamma-logs and grain-size analyses in connection with the C-14 datings, it is shown on Fig. 3 that from 7000–5000 BP the mean sedimentation rate for mainly sandy sediments was 6 m per 1000 years.

In the interval 7000–6000 BP (Fig. 3) the sediment is more fine-grained compared to what is found for the sediment deposited during the following 1000 years, with a similar sedimentation rate per 1000 years. The fauna is very poor in species as seen in the interval −18·0 to −14·0 m on Fig. 4. The changes are also reflected in the disappearance of *Venus* within this interval—samples 2858–2853 on Fig. 2.

It is tempting to regard the finely grained sediment at the level from −20 to −14 m as representing deposition from a north-easterly drift of sediment from the south along the coast of Jutland captured in the bay (Jammerbugt). The area under consideration was part of this bay at that time. From the Jammerbugt sediment passed further around the northern tip of Jutland and gave rise to the so-called Skagen formation in the Kattegat (Flodén, 1973).

Considering the hypothetical early Atlantic shorelines as depicted by Jelgersma (1979, Fig. V-16) it should be stressed that the above-mentioned furrows were fully functional as marine straits connecting the Limfjord system with the North Sea during the middle Atlantic. This was a consequence of the high positive eustatic rise of the sea-level which occurred in the early Atlantic. The southern part of the Skagerrak—the Jammerbugt—was thus part of the North Sea basin during the early Atlantic. Huge quantities of fine-grained sediment were deposited before the present-day topography of the sea bottom came into existence within the Skagerrak-Limfjord region.

ACKNOWLEDGMENTS

I would like to thank Dr H. Tauber (Copenhagen C-14 Laboratory), Dr K. Klitten (The Institute for Applied Geology) and Dr H. Krog (Geological Survey of Denmark) for discussions and contributions. To Dr Ella Hoch (Geological Museum) I express my appreciation for critically reading the manuscript.

REFERENCES

DAMON, P.E., LONG, A. & WALLIK, E.I. (1973) Dendro-chronologic calibration of the carbon-14 time scale. *Proc. 8th Int. Conf. Radiocarbon Dating*, pp. A28–A43. New Zealand.

FLODÉN, T. (1973). Notes on the bedrock of the eastern Skagerrak with remarks on the Pleistocene deposits. Acta Univ. Stockholmiensis. *Stockholm Contr. Geol.* **24**, 79–102.

GRY, H. (1979). Beskrivelse til Geologiske Kort over Danmark. Kortbladet Løgstør. *Danm. geol. Unders.* **26**, 58 pp.

JELGERSMA, S. (1979) Sea-level changes in the North Sea basin. In: *The Quaternary History of the North Sea* (Ed. by E. Oele, R. T. E. Schüttenhelm and A. J. Wiggers). *Acta Univ. Ups. Symp. Univ. Ups. Annum Quingentesimum Celebrantis*, **2**, 233–248.

JESSEN, K. (1935) Archaeological dating in the history of North Jutland's vegetation. *Acta Arch.* **5**, 185–214.

KROG, H. (1979) Late Pleistocene and Holocene shore-lines in Western Denmark. In: *The Quaternary History of the North Sea* (Ed. by E. Oele, R. T. E. Schüttenhelm and A. J. Wiggers). *Acta Univ. Ups. Symp. Univ. Ups. Annum Quingentesimum Celebrantis*, **2**, 75–83.

MANGERUD, J., ANDERSEN, S. T., BERGLUND, B. E. & DONNER, J. J. (1974) Quaternary stratigraphy of Norden, a proposal for terminology and classification. *Boreas*, **3**, 109–128.

NORDMANN, V. (1905) Danmarks Pattedyr i Fortiden. *Danm. geol. Unders.* **5**, 133 pp.

PETERSEN, K. STRAND (1976) Om Limfjordens post-glaciale marine udvikling og niveauforhold, belyst ved molluskfaunaen og C-14 dateringer. *Danm. geol. Unders.* 75–103.

TAUBER, H. (1970) The Scandinavian varve chronology and C-14 dating. In: *Proc. XIIth Nobel Symposium, Radiocarbon Variations and Absolute Chronology* (Ed. by J. U. Olsson), pp. 173–196. Stockholm.

Spec. Publs int. Ass. Sediment. (1981) **5**, 505–515

Rheological characteristics of Rappahannock Estuary muds, Southeastern Virginia, U.S.A.

RICHARD W. FAAS

Department of Geology, Lafayette College, Easton, Pennsylvania 18042, U.S.A.

ABSTRACT

Samples of naturally occurring fluid mud were taken from the upper, middle, and lower reaches of the Rappahannock Estuary, a relatively straight, funnel-shaped, partially mixed coastal plain estuary, possessing a well-developed 'turbidity maximum'. Samples were analysed to determine their viscosity characteristics using a conventional rotational viscometer. Analyses were made aboard ship immediately after retrieval and were followed by laboratory analyses to determine the effects of settling times on viscosity behaviour. Fluid mud samples exhibited various forms of non-Newtonian behaviour, varying from pseudoplastic to dilatant, depending upon shear rates and stresses, and position within the estuary relative to salt content.

Viscosity of sediments in the upper freshwater portion of the estuary is less than that of sediments from the lower, more saline portion. Pseudoplastic behaviour (shear rate thinning) is exhibited by all samples with dilatant behaviour (shear rate thickening) occurring through intermediate shear rates in a 20-mile segment of the middle estuary. Viscosity variations, including thixotropic effects, are believed to influence sediment transport throughout the estuary and maintain the 'turbidity maximum' in the middle estuary.

INTRODUCTION

Studies of the mechanics of transportation and deposition of fine-grained sediments have generally been concerned with the magnitude of shear stresses needed to initiate motion of particles and consequent erosion of the sediment bed. The problem has been studied in flumes (Partheniades, 1965; Postma, 1967); in nature (Neuman, Gebelin & Scoffin, 1970); and most recently, in flumes in nature (Young & Southard, 1978). In nearly all instances most investigators agree that the important parameters governing erosion include water content, interstitial water salinity, mineralogy, grain-size distribution, organic matter composition, and other factors.

This paper examines the role of viscosity of bottom muds as it relates to erosion and resuspension of fine-grained sediments in an estuarine environment. It is not concerned with variation in values of the individual parameters indicated above, but only considers them as they affect the main parameter, i.e. mud viscosity. Viscosity is considered to be a

significant variable governing sedimentary processes. The factors determining viscosity may be determinable themselves, yet it is their interaction which specifies viscosity and, consequently the susceptibility of the sediment to erosion and resuspension.

Rheological properties of sediments

Newton (1687) described viscosity as 'the resistance which arises from the lack of slipperiness originating in a fluid—all things being equal—is proportional to the velocity by which the parts of the fluid are being separated from each other'. A more workable description defines the coefficient of viscosity as the ratio of the rate of applied shearing stress to the rate of shear for ideal bodies (Van Wazer *et al.* 1963). On a stress-strain rate diagram (Fig. 1a), the slope of the line represents the viscosity of the liquid and remains constant for a Newtonian fluid.

Sedimentary (clay–water) systems exhibit various forms of non-Newtonian behaviour. The viscosity of such systems is not a constant but is a function of

0141-3600/81/1205-0505 $02.00
© 1981 International Association of Sedimentologists

SED 5

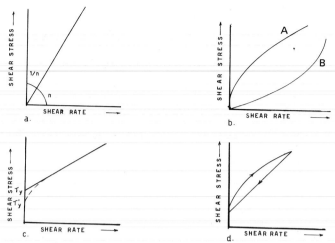

Fig. 1. Rheological behaviour patterns. (a) Newtonian flow. (b) Pseudoplastic and dilatant flow—non-Newtonian. (c) Bingham plastic flow with yield stress. (d) Thixotropic behaviour.

shear rate or shear stress and also of time. Figure 1(b) indicates the two most common types of non-Newtonian behaviour. Curve A represents pseudoplastic flow in which the viscosity of the material decreases as the shearing rate increases. Curve B illustrates dilatant flow in which the viscosity of the material increases as the shear rate increases. In some systems no flow occurs until a shearing stress reaches a certain value, τ_y, (Fig. 1c). The value of shearing stress at which the system begins to yield to the stress is called the *yield stress*. Commonly curves of this type follow the dashed line of Fig. 1(c), and the behaviour is called 'Bingham plastic flow' (Van Olphen, 1963). Extrapolation of the straight line portion of the curve to the x-axis gives a yield stress of τ_y, called the Bingham yield stress, which is slightly higher than the measured yield stress at τ_y'.

A commonly observed behaviour of sediment-water dispersal systems is *thixotropy*. Such a system exhibits a reversible reduction in yield stress and viscosity, with a distinct time dependence on application of shear strain. Resumption of the original properties occurs some time after the disturbing stresses have ceased (Bauer & Collins, 1967). A thixotropic system shows a hysteresis 'loop' when analysed through accelerating and decelerating shear rates, due to the lack of complete recovery of strength (Fig. 1d).

The form of behaviour exhibited by the system can be best observed on a flow curve, with log shear rate (rpm) on the horizontal axis and log shear stress (dial reading) on the vertical axis. This plot expands the lower values of each parameter and enables a

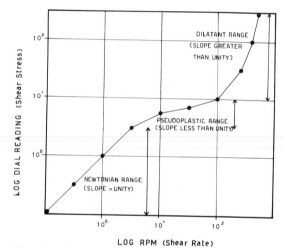

Fig. 2. Flow curve showing Newtonian and non-Newtonian flow (pseudoplastic and dilatant).

visual determination to be made of flow behaviour between successive increases in shear rate. Figure 2 is an idealized flow curve showing Newtonian, pseudoplastic, and dilatant flow patterns. During Newtonian flow, shear stress increases in direct proportion with shear rate; pseudoplastic behaviour results when shear stress increases less rapidly than shear rate; dilatant behaviour results when shear stress increases more rapidly than shear rate.

Previous work

Investigation of the viscosity of clay-water suspensions is discussed by Van Olphen (1963), particularly with respect to the properties of drilling muds. Other applications of viscosity to clay-water systems deal with paper coatings (Dalal & Kline, 1974) and ceramic glazes (Pinnow & Merwin, 1952). However, viscosity studies of naturally occurring sedimentary systems are quite limited. Krone (1962, 1963) determined viscosity of muds from San Francisco Bay during a study of particle flocculation. Several studies by Dutch workers (Delft Hydraulics Laboratory, 1962; NEDECO, 1965, 1968) showed the viscosity of freshwater muds to be lower than that of salt-water muds, regardless of the percentage of solids. Migniot (1968), using muds from various marine, estuarine, and fluvial environments substantiated the relationship between viscosity, salinity, and solids concentration and developed an equation relating the yield value (rigidity) to the fourth or fifth power of the solids concentration. Wells (1978) measured viscosities from 2·0 to >2030 cPs in muds from the central Surinam coast having bulk densities from 1·025 to 1·184 g cm^{-3} at a salinity of 36‰ (J. T. Wells, 1978, personal communication). With the exception of a few publications dealing with coarser-grained sediments (e.g. Myers, 1977), the above represents nearly all known published reports of viscosity determinations of naturally occurring fine-grained sediments.

Objectives and study area

The purpose of this investigation was to determine the apparent viscosity of the surficial muds of the Rappahannock Estuary under near natural conditions. Figure 3 shows the location of the estuary and sampling localities. The Rappahannock is a narrow (maximum width, 2·4 km), relatively straight, funnel-shaped, partially mixed estuary. Depths range from 6 m in the upper estuary to 23 m in a silled basin in the lower estuary. Tidal range is less than 80 cm with a mean tide of 48 cm. Current velocity, measured 30 cm above the bottom, ranges from near zero at slack water to 65 cm sec^{-1} at maximum current. Tidal asymmetry is characteristic and time-velocity curves show relatively low velocity occurring for approximately 2 h during slack before ebb and 1 h during slack before flood (Nichols & Poor, 1967). Sediments in the estuary are largely silt- and clay-sized, with sand-sized material comprizing less than 3%. Median diameters of suspended sediment range between 2 and 16 μm. The clay mineral assemblage consists of a mixture of illite, kaolinite, and chlorite (Nelson, 1960). Distinct biogeochemical differences exist within the bottom sediments throughout the estuary (Nelson, 1972). A pronounced 'turbidity maximum' in the middle estuary (Nichols & Poor,

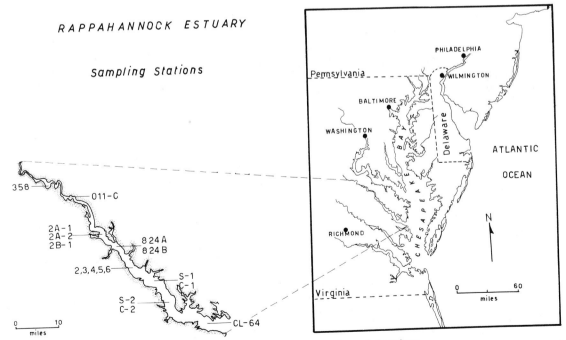

Fig. 3. The Rappahannock Estuary and sample locations.

1967) results in deposits of fluid mud, 10–20 cm thick, nearly covering the entire bottom. Fluid mud is a dense, static suspension, ranging in density between 1.005 g cm^{-3} and 1.30 g cm^{-3}, corresponding to concentrations of 10–480 g l^{-1}. It is found in significantly greater quantities in the Severn and Rhine estuaries (Kirby & Parker, 1977); the Gironde (Allen Castaing & Klingebiel, 1974), the Thames (Odd & Owen, 1972), and other similar estuarine systems. These static suspensions develop during neap-tides from dense, layered, mobile suspensions (Kirby & Parker, 1977).

Sampling methods and instrumentation

Sampling of the fluid muds was done through the use of a specially designed large volume hand corer. Other samples were obtained through standard box coring techniques. Samples were analysed in their natural water and in the concentrations at which they were sampled. No pre-treatment of the samples was involved, and all were analysed as soon after collection as possible. In some cases, they were analysed directly in the surfaces of box cores, or were transferred from box cores to polyethelyene containers for immediate analysis. Several samples were taken by Scuba divers who filled large-mouth polyethylene containers with the mobile bottom mud. These were brought to the laboratory and analysed directly in the container within a few hours after collection. All samples (with the exception of 824A, 824B, 2A-1, 2A-2, 2B-1) were thoroughly shaken to effect complete dispersal before analysis. Sub-samples for size analysis, water content/bulk density, Atterberg Limits, and salinity were taken

after initial viscosity measurements had been made. Size analyses were made with a Coulter Counter, following established procedures (Table 1).

Viscosity measurements were made with the Brookfield RVT eight-speed Synchro-Lectric viscometer (manufactured by the Brookfield Engineering Laboratories, Stoughton, Massachusetts). A synchronous motor drives a cylinder, spindle, or disc element which is immersed in the suspension, at a constant speed of rotation. The force required to overcome the resistance offered by the sample to this motion is derived from a beryllium copper spring which drives the immersed element. A dial at the top of the viscometer records the extent to which the spring is wound in surmounting the viscous resistance. Consequently, the dial reading is proportional to the viscosity of the sample (Sherman, 1970).

ANALYTICAL RESULTS

Viscosity 'notch' development

Inspection of the flow curves and rheograms indicates sequential development of a 'viscosity notch' (a decrease in apparent viscosity, followed by a sharp increase) at low rates of shear. This feature (Fig. 4) records a distinct change in the rheological behaviour of the fluid mud suspension throughout the range of shear rate. From the moment of the initiation of shear in the mud, apparent viscosity decreases as shear rate increases. This is *pseudoplastic* behaviour and is related to the rapid breakdown of a loose, flocculent particle fabric under low shear rates. At greater shear rates, apparent viscosity increases, giving rise to *dilatant* behaviour. In this case, individual clay

Table 1. Physical properties of Rappahannock Estuary sediments

Sample no.	Size distribution			Atterberg limits			Unified* soil classification
	Median (μm)	Silt (%)	Clay (%)	W_L	W_P	I_P	
358	13·4	89	11	—	—	—	—
011-C	8·8	76	24	133·5	55·3	78·2	MH
2A-1	7·2	69	31	—	—	—	—
2 to 6	6·6	82	18	—	—	—	—
C-1	7·0	73·5	26·5	166·7	52·6	114·0	CH
S-1	7·1	70·5	29·5	178·7	53·3	125·4	CH
C-2	6·9	76	24	181·8	54·9	126·9	CH
S-2	6·6	80·5	19·5	—	—	—	—
CL-64	6·5	87	13	152·0	57·3	94·7	OH

* Unified Soil Classification (Wagner, 1957) provides the following description, based upon the liquid limit (W_L) and plasticity index (I_P): CH—inorganic clays of high plasticity; fat clays. OH—organic clays of medium to high plasticity. MH—inorganic silts, micaceous or diatomaceous fine sandy or silty soils, elastic silts.

particles have reoriented themselves into parallel alignment with closer packing. Layers of particles begin to glide over adjacent layers forming laminae which increase the effective mass, causing a momentum resistance to shearing stresses. This subject has been discussed in detail by Metzner & Whitlock (1958) and given the substitute term, 'shear-rate thickening', by Bauer & Collins (1967).

At high shear rates the dilatant structure disintegrates and fabric breakdown becomes complete. Water moves into spaces between the particles, which then behave hydrodynamically as single entities. The apparent viscosity approaches that of the water medium, dependent upon temperature and density. The 'viscosity notch' first occurs in hand core samples 2A-1, 2A-2, and 2B-1 from the upper part of the middle estuary (Fig. 3). It is minimal and hardly noticeable in 2B-1; however, 2A-1 and 2A-2 show a significant 'notch' occurring between 2·5–5·0 rpm.

The 'notch' is significantly developed in hand core samples 824A and 824B (Fig. 3) taken 5 miles down river from cores 2A-1, 2A-2, and 2B-1, in the middle estuary. The 'notch' occurs between 2·5 and 5·0 rpm in 824A and between 2·5 and 10 rpm in 824B.

Maximum development of the viscosity 'notch' occurs in box core samples 2–6, 5 miles down river from 824A and 824B (Fig. 3). Sample 5 shows a typical profile of this series (Fig. 4). The 'notch' begins at 5 rpm and extends to 10 rpm; other samples from this site showed the 'notch' restricted between either 2·5–10 rpm, or between 5 and 20 rpm.

Five miles farther downstream, samples were taken from the Rappahannock Channel (C-1) and adjacent shoal area (S-1) by divers (Fig. 3). A small, but noticeable 'notch' occurred between 5·0 and 10 rpm in sample S-1 after it had been allowed to settle for 2 h. No indication of a 'notch' was shown for several analyses performed on sample C-1, although

the viscosity decrease appears less rapid than with previous samples.

Another 5 miles downstream, samples C-2 (channel) and S02 (adjacent shoal) were taken (Fig. 3). A slight, but noticeable viscosity 'notch' occurred in sample C-2, between 2·5 and 5·0 rpm. However, no 'notch' was observed in sample S-1.

A final sample (CL-64) was taken 10 miles farther downstream from a deep trough (20 m) within the lower estuary (Fig. 3). Rheograms of this sample show no 'viscosity notch'.

Two sets of samples taken from the upper estuary (Sample 358—Portobago Bay, and 011-C—approximately 13 miles downstream but still in the upper estuary) (Fig. 3) showed no indication of 'viscosity notch' development.

Development of the 'viscosity notch' follows an orderly progression and appears related to the salinity distribution throughout the estuary. Figure 5 shows the shape of the rheograms of sediments from throughout the estuary. The interstitial water salinity for each sample is plotted in the upper half of the diagram. Samples from the upper estuary (at 358 and 011-C) are entirely pseudoplastic and viscosity increases between Mile 60–65 and Mile 40–45. The salinity of these samples is 0·421‰ and 0·493‰, respectively. 'Viscosity notch' development occurs between Mile 40–20, coinciding with a steep increase in salinity. Dilatant behaviour extends through the maximum range of shear rate (5–20 rpm) between Mile 30–25, at a salinity of 8·301‰. Dilatant behaviour decreases between Mile 25–20 (salinity is 12·466‰) and becomes entirely pseudoplastic between Mile 20–5 as salinity increases in excess of 14‰. Unfortunately, samples were not taken between salinities of 0·5–5·0‰, so it is not possible to determine the minimum salinity at which 'notch' development occurred.

Thixotropy

Nearly all the natural samples analysed show thixotropic behaviour to some extent. The degree of thixotropy appears dependent upon settling time and possibly interstitial water salinity, although other factors, e.g. nature and composition of organic matter may not be ruled out.

Comparison of thixotropic behaviour of sediments from the upper, middle, and lower reaches of the estuary was made. Samples were initially shaken to effect maximum dispersal, then allowed to settle for up to 74 h. Measurements were made at different

HAND CORE 5
Salinity - 8.3 ppt

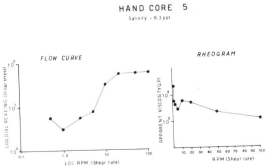

Fig. 4. Flow diagram and rheogram of sample 5—shows well-developed 'viscosity notch'.

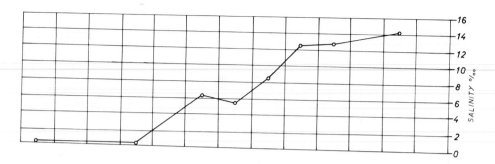

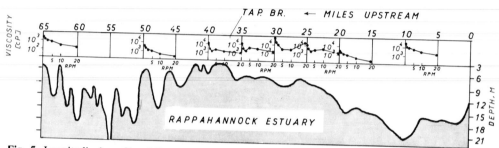

Fig. 5. Longitudinal profile of Rappahannock Estuary showing salinity distribution and viscosity profiles.

THIXOTROPIC BEHAVIOR

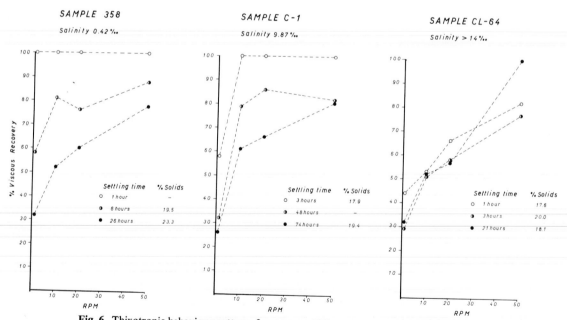

Fig. 6. Thixotropic behaviour patterns for upper, middle, and lower estuary sediments.

spindle speeds (to 100 rpm) on an accelerating and decelerating cycle. Readings were taken every 2 min; consequently, the total analysis took 32 min. Apparent viscosity values of the sheared sediment from the decelerating cycle were compared with values of the unsheared sediment from the accelerating cycle to determine the percentage of original viscosity recovered. The degree of thixotropy is shown as the percent of recovery of apparent viscosity at 1, 10, 20, and 50 rpm (Fig. 6).

Sample 358 (upper estuary) and C-1 (middle estuary) show thixotropy only after 3 h of settling. Maximum thixotropy occurs at lowest rates of shear (rpm). Sample CL-64, taken from the lower estuary, shows quite different thixotropic behaviour. Recovery of original viscosity appears less dependent on time than on rate of shear. Consequently, this analysis suggests that sediment from the lower estuary is more thixotropic than sediment from the middle and upper estuary, and would be easily resuspended by low currents occurring after slack water. On the other hand, middle and upper estuary sediments exhibit nearly complete recovery of original viscosity during a short (1 h) slack water settling interval and would require stronger currents before resuspension would occur.

DISCUSSION

Rheological behaviour patterns shown by the sediments throughout the Rappahannock Estuary may be, in part, responsible for the existence and maintenance of the 'turbidity maximum' in the intermediate reaches of the estuary. Specifically, repeated occurrences of 'shear-rate thickening', as shown by the 'viscosity notch' throughout intermediate salinity reaches appears significant.

Figure 7 records suspended sediment concentrations during an ebb tide in the 'turbidity maximum' in the Rappahannock River, taken over a 6-h period on June 21 1978. A Partech transmissometer, calibrated to determine suspensions to a maximum concentration of 1000 ppm, was mounted 60 cm above the river bed on a bottom-setting tripod device. Of particular interest is the double peak in suspended sediment concentration, the first occurring between 1030 and 1130 h; the second larger peak occurring between 1230 and 1330 h.

Nichols & Poor (1967) presented graphs of current velocity and suspended sediment concentration in the area of the 'turbidity maximum' in the Rappa-

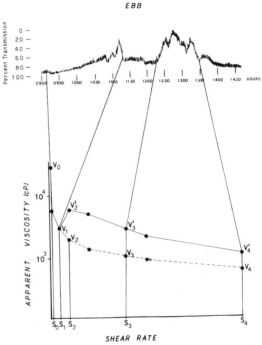

Fig. 7. Transmissometer record and rheogram analysis—hand core 5.

hannock (their fig. 5). These graphs show tidal asymmetry, with approximately 2 h of slack water occurring before ebb and 1 h of slack water occurring before flood. Corresponding suspended sediment concentrations also show two peaks occurring on both flood and ebb, the smaller occurring shortly after slack water on both tides, followed by maximum peaks just after maximum velocity. Their data, collected in 1965, closely resemble Fig. 7, indicating relatively constant and predictable behaviour of the suspended sediments in the 'turbidity maximum'.

Analysis of the rheogram of hand core 5 immediately below the transmissometer record (Fig. 7) provides a physical explanation for the two peaks of suspended sediment concentration in an accelerating shear field. The rheogram is interpreted as expressing the sediment response to accelerating shear rates developed during the ebb tide.

Pseudoplastic behaviour begins at apparent viscosity V_0, at shear rate S_0, and continues to V_1 and shear rate S_1, resulting in the first peak which terminates at 1115 h. Dilatant behaviour (shear rate thickening) occurs through the shear rate interval S_1 to S_2. Apparent viscosity at S_2 should normally be the value of V_2, but has increased to the value of V_2'. Pseudoplastic behaviour begins again at V_2' and continues to the end of the run at S_4 with a corresponding apparent viscosity of V_4'. The lowered concentration of suspended sediment existing between 1115 and 1215 h results from dilatant behaviour which increased the apparent viscosity through the shear rate interval S_2 to S_3. The value of apparent viscosity (V_1) existing at low shear rate S_1 occurs again (V_3') at the higher rate existing at S_3. Resuspension again occurs through the shear rate interval S_3 to S_4, corresponding to the second peak of suspended sediments which begins at 1215 and extends to 1330 h. Had behaviour been completely pseudoplastic, the apparent viscosity would have continued to decrease to its value of V_3 with continuous resuspension. Since V_3 is lower than the final apparent viscosity of V_4' and occurs at the lower shear rate S_3, it is likely that all available sediment would have been resuspended earlier in the tidal cycle, perhaps by 1230 h. It is believed that the changes in rheological behaviour of the bottom sediments in this region are responsible for the double peaks in suspended sediment concentration (Fig. 7) and, because of resuspension through a longer period of time, are also responsible for the existence of the 'turbidity maximum' in these reaches.

Sediment transport patterns throughout the estuary may result from the viscosity characteristics of the sediments. Nichols (1973) has shown that, in general, a greater amount of sediment is contributed to the 'turbidity maximum' from the river than from the estuary. Only at lowest flow rates (23 m³ s⁻¹) is more sediment supplied to the 'turbidity maximum' from the estuary than from the river. Much of this material is retained temporarily within the 'turbidity maximum' as the estuary changes from an escape mode to a trapping mode, depending upon flow conditions. Resuspension of low viscosity upper estuary sediments occurs at lower shear stresses and shear rates than in similar sediments in the lower estuary. Lower estuary sediments, although initially more viscous, are resuspended due to thixotropy and are transported back up-estuary in the lower layer. Dilatant behaviour of middle estuary sediments serves to increase the period of resuspension and adds to the high concentration of suspended sediments in the 'turbidity maximum'. In addition, tidal

asymmetry effects associated with the shorter slack water interval before flood will encourage further resuspension of the more thixotropic lower estuary sediments due to less time for thixotropic recovery (Fig. 6). Nichols & Poor (1967) indicate a net up-estuary transport due to tidal asymmetry. Thixotropic differences within the various reaches of the estuary may be responsible for this net up-estuary transport.

Postma (1967) suggested that packing of sediment particles is a function of salinity. Particles exist wholly dispersed in fresh water and are surrounded by thick water layers. These particles settle to the bottom, creating a substrate with large pore spaces and low interparticle attraction. In brackish water, particles become partially flocculated and show closest packing with increased interparticle attraction. Marine muds are wholly flocculated and have large void spaces. Edzwald & O'Melia (1975) show that destabilization of clay minerals (increasing particle aggregation) occurs with increasing electrolyte, with destabilization occurring more rapidly in the order: illite < kaolinite < montmorillonite. Numerous other studies have shown the increase in aggregation (flocculation) with increasing electrolyte concentrations and the VODL theory (Verwey & Overbeek, 1948) has been generally accepted to explain this phenomena.

The rheological behaviour of the Rappahannock Estuary sediments (Fig. 8) can be explained in accordance with the observations of Postma (1967) and Edzwald & O'Melia (1975). The low viscosity observed in the upper fresh water portion of the estuary (sample 358) results from the more stabilized, water-rich, loosely packed sediment. Pseudoplastic behaviour in such sediments results from the development of repulsive forces between similarly charged double layers. In the middle estuary (samples 2–6), increased electrolyte content causes a reduction in thickness of the double layers and partial destabilization of the clays. Packing increases and, during shearing, parallel orientation of individual clay particles form oriented masses which, because of inertial effects, cause a temporary 'thickening' of the clay/water mixture. At greater shear rates, total destruction of the thickened fabric occurs and leads again to pseudoplastic behaviour. Sediments in the lower estuary (CL-64) show maximum destabilization and interparticle attraction. High viscosity results from this attraction, but dilatant behaviour does not occur as the aggregates do not decompose into individual clay particles. Instead, viscosity decrease is totally pseudoplastic as the individual aggregates

RHEOLOGICAL BEHAVIOR OF LOW, INTERMEDIATE, AND HIGH SALINITY MUDS

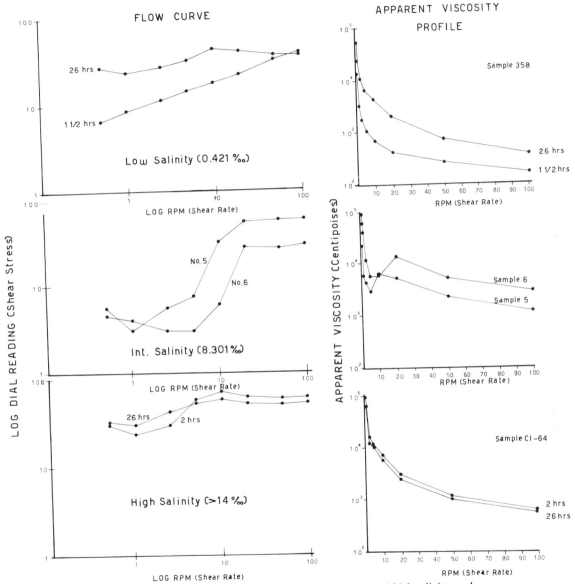

Fig. 8. Rheological behaviour of low, intermediate and high salinity muds.

become detached from the aggregate framework during increased shear rates.

CONCLUSIONS

Young & Southard (1978) state that in view of differences between laboratory and field measurements of threshold erosion velocities for cohesive sediments, these differences must be related to more subtle and: 'as yet unmeasured physical or biochemical parameters'. The work presented here has attempted to evaluate the effect of little known viscosity behaviour patterns of cohesive sediments in the erosion and resuspension mechanics in the Rappahannock Estuary with the following conclusions:

(1) Significant differences in apparent viscosity exist between each of the estuarine segments (Fig. 8). Apparent viscosity increases down-estuary with increasing salinity in the bottom sediments. In the middle estuary pseudoplastic behaviour at low shear rates changes abruptly to dilatant behaviour at intermediate shear rates, followed by pseudoplastic flow at higher shear rates. This pattern is expressed only in sediments having a salinity between 4 and 12‰ (Fig. 5), and results from the destruction of a partially flocculated, partially aggregated type of open domain fabric. During dilatant flow, parallel alignment and 'sandwiching' of clay aggregates increases the density of a unit volume of sheared sediment and results in an increase in apparent viscosity, effectively resisting applied shears. At highest shears, the total accumulated shear strength of the mass is exceeded and erosion begins. Records of suspended sediment 60 cm above the bottom of the estuary show two maxima: one occurring several hours after slack before ebb; the other occurring at maximum ebb velocity. The rheological behaviour described above is believed responsible for the two suspended sediment maxima.

(2) The general effect of dilatant behaviour in the sediments within the area of the 'turbidity maximum' is to increase the amount of time sediment remains in suspension within a tidal cycle. Sediment that would normally have been resuspended through pseudoplastic behaviour during low shear rates becomes resistant (shear-rate thickened) at greater shear rates. This has the effect of increasing the apparent viscosity of the sediment and delaying its resuspension until appropriately higher shear rates are reached. Thus, a dilatant interval of 1 h will result in a period of resuspension spanning 3 h, rather than 2 h, if behaviour has been entirely psuedoplastic (Fig. 7).

(3) The thixotropic behaviour of lower estuary sediments may be responsible for the net up-estuary transport as noted by Nichols & Poor (1967). The 1-h slack water period before flood is insufficient to allow the newly settled sediments to recover their original strength (Fig. 6) and they become easily resuspended at low shear rates and are carried back into the middle estuary. This corresponds with the up-estuary distribution of *Elphidium incertum* tests as noted by Nichols & Ellison (1967).

(4) Although no data are presented in this paper, similar rheological effects have been observed to occur in the sediments of Schelde Estuary. Pseudoplastic flow behaviour is common, however, dilatant flow at low shear rates occurs in samples possessing from 4 to 14‰ salinity after 19 h of settling. This differs from the Rappahannock sediments where dilatancy appears independent of settling time. Yield stress development occurs quite slowly in Schelde sediments, becoming maximum after 100 h of settling. In the Rappahannock, yield stress becomes maximum after only 40 h of settling. Additional work is in progress to specify the differences and similarities in rheological properties of cohesive sediments from different estuaries.

The above conclusions concerning rheological changes of cohesive sediments are based upon conditions of laminar flow, as required by the fundamental concept of viscosity. Thus, the behaviour patterns so described relate specifically to cohesive sediments within the viscous sublayer (McCave, 1970). Periodic disruption of this sub-layer may occur due to the rheological response of the sediments under varying rates of shear, causing entrainment of the sediment into the turbulent region where it may be transported in the manner suggested by Gordon (1975).

It seems clear that fine-grained cohesive sediments behave quite differently in a stress field than non-cohesive sediments. The geochemical environment in which they are found and the variability of the associated stress field evoke dynamic changes in the forms of rheological behaviour patterns exhibited by the sediments which, in turn, determine when and where erosion and resuspension will occur.

ACKNOWLEDGMENTS

The results reported here were obtained during a study entitled, 'Estuarine Fluid Mud: its Behavior and Accumulation', Grant No. DAA629-76-6-008 by the U.S. Army Research Office to the Virginia Institute of Marine Science, Gloucester Point, Virginia. Maynard Nichols provided sediment data and current and suspended sediment records, in addition to encouragement and support facilities. Galen Thompson and Craig Lukin aided in obtaining samples and in various other aspects of the field programme. Their contributions are gratefully acknowledged. The viscometer was provided by the Department of Chemical Engineering, Lafayette College.

REFERENCES

ALLEN, G. P., CASTAING, P. & KLINGEBIEL, A. (1974) Suspended sediment transport and deposition in the Gironde Estuary and adjacent shelf. *Mem. Inst. Geol. Bassin Aquitaine*, **7**, 27–36.

BAUER, W. H. & COLLINS, E. A. (1967) Thixotropy and dilatancy. In: *Rheology: Theory and Applications* (Ed. by F. R. Eirich), pp. 423–459. Academic Press, New York.

DALAL, C. S. & KLINE, J. E. (1974) The relationship between degree of clay dispersion and the optical and pore properties of starch-clay coatings. *J. Tech. Ass. Pulp Paper Ind.* **57**, 91–95.

DEMERARA COASTAL INVESTIGATION (1962) *Report on siltation of Demerara Bar Channel and coastal erosion in British Guiana.* Delft Hydraulics Laboratory, Delft, the Netherlands.

EDZWALD, J. K. & O'MELIA, C. R. (1975) Clay distribution in Recent estuarine sediments. *Clays Clay Miner.* **23**, 39–44.

GORDON, C. M. (1975) Sediment entrainment and suspension in a turbulent tidal flow. *Mar. Geol.* **18**, M57–M64.

KIRBY, R. & PARKER, W. (1977) The physical characteristics and environmental significance of fine-sediment suspensions in estuaries. In: *Estuaries, Geophysics and the Environment*, pp. 110–120. Studies in Geophysics. National Academy of Sciences.

KRONE, R. B. (1962) A field study of flocculation as a factor in estuarial shoaling processes. *Committee on Tidal Hydraulics, U.S. Army Corps Engrs, Tech. Bull. No. 19.*

KRONE, R. B. (1963) A study of rheologic properties of estuarial sediments. *Committee on Tidal Hydraulics, U.S. Army Corps Engrs, Tech. Bull. No. 7.*

MCCAVE, I. N. (1970) Deposition of fine-grained suspended sediment from tidal currents. *J. geophys. Res.* **75**, 4151–4159.

METZNER, A. B. & WHITLOCK, M. (1958) Flow behavior of concentrated (dilatant) suspensions. *Trans. Soc. Rheol.* **II**, 239–254.

MIGNIOT, C. (1968) A study of the physical properties of various forms of fine sediment and their behavior under hydrodynamic action. *Houille blanche*, **7**, 591–620.

MYERS, A. C. (1977) Sediment processing in a marine subtidal sandy bottom community: I. Physical aspects. *J. mar. Res.* **35**, 609–632.

NEDECO (1965) *A Study on the Siltation of the Bangkok Channel, vol II.* Netherlands Engineering Consultants, The Hague. 474 pp.

NEDECO (1968) *Surinam Transportation Study.* Netherlands Engineering Consultants, The Hague. 293 pp.

NELSON, B. W. (1960) Clay mineralogy of the bottom sediments, Rappahannock River, Virginia. In: *Proc. 7th Nat. Conf. Clays and Clay Minerals*, pp. 135–147.

NELSON, B. W. (1972) Biogeochemical variables in bottom sediments of the Rappahannock River Estuary. In: *Environmental Framework of Coastal Plain Estuaries* (Ed. by B. W. Nelson). *Mem. geol. Soc. Am.* **133**, 417–451.

NEUMAN, A. C., GEBELIN, C. D. & SCOFFIN, T. P. (1970) The composition, structure, and erodability of subtidal mats, Abaco, Bahamas. *J. sedim. Petrol.* **40**, 274–297.

NEWTON, I. S. (1687) *Philosophiae Naturalis Principia Mathematica*, Book 2, Section IX.

NICHOLS, M. M. (1973) Development of the turbidity maximum in a coastal plain estuary. *AROD Final Report.* Contracts G1164 and G47, 47 pp.

NICHOLS, M. M. & ELLISON, R. L. (1967) Sedimentary patterns of microfauna in a coastal plain estuary. In: *Estuaries* (Ed. by G. H. Lauff) *Am. Ass. Adv. Sci. Publ.* **83**, 283–288.

NICHOLS, M. M. & POOR, G. (1967) Sediment transport in a coastal plain estuary. *J. Watways Harb. Div. Proc. Am. Soc. civ. Engrs.* **9**, 83–95.

ODD, N. V. M. & OWEN, M. W. (1972) A two-layer model of mud transport in the Thames Estuary. *Proc. Inst. Civ. Eng., Suppl.* **IX**, paper 75175, 175–205.

PARTHENIADES, E. (1965) Erosion and deposition of cohesive soils. *J. Hydraul. Div. Am. Soc. civ. Engrs* **91**, 105–139.

PINNOW, H. R. & MERWIN, B. W. (1952) Control methods for the consistency of one-fire glazes. *Bull. Am. Ceram. Soc.* **31**, 332–333.

POSTMA, H. (1967) Sediment transport and sedimentation in the estuarine environment. In: *Estuaries* (Ed. by G. H. Lauff). *Am. Ass. Adv. Sci. Publ.* **83**, 158–179.

SHERMAN, P. (1970) *Industrial Rheology.* Academic Press, London. 432 pp.

VAN OLPHEN, H. (1963) *An Introduction to Clay Colloid Chemistry.* Interscience Publishers, New York. 301 pp.

VAN WAZER, J. R., LYONS, J. W., KIM, K. Y. & COLWELL, R. E. (1963) *Viscosity and Flow Measurements. A Laboratory Handbook of Rheology.* Interscience Publishers, New York. 406 pp.

VERWEY, E. J. W. & OVERBEEK, J. TH. G. (1948) *Theory of the Stability of Lyophobic Colloids.* Elsevier, New York.

WAGNER, A. A. (1957) The uses of the Unified Soil Classification System by the Bureau of Reclamation. *Proc. 4th Int. Conf. Soil Mech. Found. Eng. London*, **1**, 125.

WELLS, J. T. (1978) Shallow-water waves and fluid-mud dynamics, coast of Surinam, South America. *Coastal Studies Institute Tech. Rpt No. 257.* Louisiana State University, Baton Rouge. 56 pp.

YOUNG, R. N. & SOUTHARD, J. B. (1978) Erosion of fine-grained marine sediments: sea floor and laboratory experiments. *Bull. geol. Soc. Am.* **89**, 663–672.